BITANGENTIAL DIRECT AND INVERSE PROBLEMS FOR SYSTEMS OF INTEGRAL AND DIFFERENTIAL EQUATIONS

This largely self-contained treatment surveys, unites and extends some 20 years of research on direct and inverse problems for canonical systems of integral and differential equations and related systems. Five basic inverse problems are studied in which the main part of the given data is either a monodromy matrix; an input scattering matrix; an input impedance matrix; a matrix-valued spectral function; or an asymptotic scattering matrix. The corresponding direct problems are also treated.

The book incorporates introductions to the theory of matrix-valued entire functions, reproducing kernel Hilbert spaces of vector-valued entire functions (with special attention to two important spaces introduced by L. de Branges), the theory of J-inner matrix-valued functions and their application to bitangential interpolation and extension problems, which can be used independently for courses and seminars in analysis or for self-study. A number of examples are presented to illustrate the theory.

Encyclopedia of Mathematics and Its Applications

This series is devoted to significant topics or themes that have wide application in mathematics or mathematical science and for which a detailed development of the abstract theory is less important than a thorough and concrete exploration of the implications and applications.

Books in the **Encyclopedia of Mathematics and Its Applications** cover their subjects comprehensively. Less important results may be summarized as exercises at the ends of chapters. For technicalities, readers can be referred to the bibliography, which is expected to be comprehensive. As a result, volumes are encyclopedic references or manageable guides to major subjects.

All the titles listed below can be obtained from good booksellers or from Cambridge University Press.
For a complete series listing visit www.cambridge.org/mathematics.

93 G. Gierz *et al. Continuous Lattices and Domains*
94 S. R. Finch *Mathematical Constants*
95 Y. Jabri *The Mountain Pass Theorem*
96 G. Gasper and M. Rahman *Basic Hypergeometric Series, 2nd edn*
97 M. C. Pedicchio and W. Tholen (eds.) *Categorical Foundations*
98 M. E. H. Ismail *Classical and Quantum Orthogonal Polynomials in One Variable*
99 T. Mora *Solving Polynomial Equation Systems II*
100 E. Olivieri and M. Eulália Vares *Large Deviations and Metastability*
101 A. Kushner, V. Lychagin and V. Rubtsov *Contact Geometry and Nonlinear Differential Equations*
102 L. W. Beineke and R. J. Wilson (eds.) with P. J. Cameron *Topics in Algebraic Graph Theory*
103 O. J. Staffans *Well-Posed Linear Systems*
104 J. M. Lewis, S. Lakshmivarahan and S. K. Dhall *Dynamic Data Assimilation*
105 M. Lothaire *Applied Combinatorics on Words*
106 A. Markoe *Analytic Tomography*
107 P. A. Martin *Multiple Scattering*
108 R. A. Brualdi *Combinatorial Matrix Classes*
109 J. M. Borwein and J. D. Vanderwerff *Convex Functions*
110 M.-J. Lai and L. L. Schumaker *Spline Functions on Triangulations*
111 R. T. Curtis *Symmetric Generation of Groups*
112 H. Salzmann *et al. The Classical Fields*
113 S. Peszat and J. Zabczyk *Stochastic Partial Differential Equations with Lévy Noise*
114 J. Beck *Combinatorial Games*
115 L. Barreira and Y. Pesin *Nonuniform Hyperbolicity*
116 D. Z. Arov and H. Dym *J-Contractive Matrix Valued Functions and Related Topics*
117 R. Glowinski, J.-L. Lions and J. He *Exact and Approximate Controllability for Distributed Parameter Systems*
118 A. A. Borovkov and K. A. Borovkov *Asymptotic Analysis of Random Walks*
119 M. Deza and M. Dutour Sikirić *Geometry of Chemical Graphs*
120 T. Nishiura *Absolute Measurable Spaces*
121 M. Prest *Purity, Spectra and Localisation*
122 S. Khrushchev *Orthogonal Polynomials and Continued Fractions*
123 H. Nagamochi and T. Ibaraki *Algorithmic Aspects of Graph Connectivity*
124 F. W. King *Hilbert Transforms I*
125 F. W. King *Hilbert Transforms II*
126 O. Calin and D.-C. Chang *Sub-Riemannian Geometry*
127 M. Grabisch *et al. Aggregation Functions*
128 L. W. Beineke and R. J. Wilson (eds.) with J. L. Gross and T. W. Tucker *Topics in Topological Graph Theory*
129 J. Berstel, D. Perrin and C. Reutenauer *Codes and Automata*
130 T. G. Faticoni *Modules over Endomorphism Rings*
131 H. Morimoto *Stochastic Control and Mathematical Modeling*
132 G. Schmidt *Relational Mathematics*
133 P. Kornerup and D. W. Matula *Finite Precision Number Systems and Arithmetic*
134 Y. Crama and P. L. Hammer (eds.) *Boolean Models and Methods in Mathematics, Computer Science, and Engineering*
135 V. Berthé and M. Rigo (eds.) *Combinatorics, Automata and Number Theory*
136 A. Kristály, V. D. Rădulescu and C. Varga *Variational Principles in Mathematical Physics, Geometry, and Economics*
137 J. Berstel and C. Reutenauer *Noncommutative Rational Series with Applications*
138 B. Courcelle *Graph Structure and Monadic Second-Order Logic*
139 M. Fiedler *Matrices and Graphs in Geometry*
140 N. Vakil *Real Analysis through Modern Infinitesimals*
141 R. B. Paris *Hadamard Expansions and Hyperasymptotic Evaluation*
142 Y. Crama and P. L. Hammer *Boolean Functions*
143 A. Arapostathis, V. S. Borkar and M. K. Ghosh *Ergodic Control of Diffusion Processes*
144 N. Caspard, B. Leclerc and B. Monjardet *Finite Ordered Sets*
145 D. Z. Arov and H. Dym *Bitangential Direct and Inverse Problems for Systems of Integral and Differential Equations*
146 G. Dassios *Ellipsoidal Harmonics*

ENCYCLOPEDIA OF MATHEMATICS AND ITS APPLICATIONS

Bitangential Direct and Inverse Problems for Systems of Integral and Differential Equations

DAMIR Z. AROV
South-Ukrainian National Pedagogical University, Odessa

HARRY DYM
Weizmann Institute of Science, Rehovot, Israel

CAMBRIDGE UNIVERSITY PRESS
Cambridge, New York, Melbourne, Madrid, Cape Town,
Singapore, São Paulo, Delhi, Mexico City

Cambridge University Press
The Edinburgh Building, Cambridge CB2 8RU, UK

Published in the United States of America by Cambridge University Press, New York

www.cambridge.org
Information on this title: www.cambridge.org/9781107018877

First published 2012
Reprinted 2012

A catalogue record for this publication is available from the British Library

Library of Congress Cataloguing in Publication Data
Arov, Damir Z.
Bitangential direct and inverse problems for systems of integral and differential equations / Damir Z. Arov, Harry Dym.
p. cm. – (Encyclopedia of mathematics and its applications ; 145)
Includes bibliographical references.
ISBN 978-1-107-01887-7 (hardback)
1. Inverse problems (Differential equations) 2. Integral equations.
I. Dym, Harry, 1934– II. Title.
QA378.5.A76 2012
515′.45 – dc23 2012000304

ISBN 978-1-107-01887-7 Hardback

Dedicated to our wives Natasha and Irene,
for their continued support and encouragement, and
for being ideal companions on the path of life.

CONTENTS

PREFACE

This book is devoted to direct and inverse problems for canonical integral and differential systems. Five basic problems are considered: those in which an essential part of the data is either (1) a monodromy matrix; or (2) an input scattering matrix; or (3) an input impedance matrix; or (4) a spectral function; or (5) an asymptotic scattering matrix.

There is a rich literature on direct and inverse problems for canonical integral and differential systems and for first- and second-order differential equations that can be reduced to such systems. However, the intersection between most of this work and this book is relatively small. The approach used here combines and extends ideas that originate in the fundamental work of M.G. Krein, V.P. Potapov and L. de Branges:

M.G. Krein studied direct and inverse problems for Dirac systems (and differential equations that may be reduced to Dirac systems) by identifying the matrizant of the system with a family of resolvent matrices for assorted classes of extension problems that are continuous analogs of the classical Schur and Carathéodory extension problems.

In this monograph we present bitangential generalizations of the Krein method that is based on identifying the matrizants of canonical systems of equations as resolvent matrices of an ordered family of bitangential generalized interpolation/extension problems that were studied earlier by the authors and are also reviewed in reasonable detail in the text.

The exposition rests heavily on the theory of J-inner mvf's (matrix-valued functions) that was developed and applied to a number of problems in analysis (including the inverse monodromy problem for canonical differential systems) by V.P. Potapov in his study of J-contractive mvf's. The parts of this theory that are needed here are taken mainly from our earlier monograph *J-Contractive Matrix Valued Functions and Related Topics*. Nevertheless, in order to keep this monograph reasonably self-contained, material that is needed for the exposition is repeated as needed, though usually without proof, and often in simpler form if that is adequate for the application at hand.

Extensive use is made of reproducing kernel Hilbert spaces of vector-valued entire functions of the kind introduced by L. de Branges. This theory is developed further and used to provide alternate characterizations of certain classes of entire J-inner mvf's and some useful results on the isometric inclusion of some nested families of reproducing kernel Hilbert spaces.

We have tried to give a reasonable sample of the literature that we felt was most relevant to the topics developed in the monograph. But for every item listed, there are tens if not hundreds of articles that are somewhat connected. To keep the length of the list of references reasonable, we have not referenced articles that deal with canonical systems on the full line, or non-Hermitian Hamiltonians and Hamiltonians with negative eigenvalues.

The authors gratefully acknowledge and thank the administration of South-Ukrainian National Pedagogical University for authorizing extended leaves of absence to enable the first author to visit the second and finally, and most importantly, the Minerva Foundation, the Israel Science Foundation, the Arthur and Rochelle Belfer Institute of Mathematics and Computer Science, and the Visiting Professorship program at the Weizmann Institute for the financial support that made these visits possible and enabled the authors to work together under ideal conditions.

1

Introduction

This book focuses on direct and inverse problems for canonical differential systems of the form

$$\frac{d}{dx}u(x,\lambda) = i\lambda u(x,\lambda)H(x)J \quad \text{a.e. on } [0,d), \tag{1.1}$$

where $\lambda \in \mathbb{C}$, $J \in \mathbb{C}^{m\times m}$ is a **signature matrix**, i.e., $J^* = J$ and $J^*J = I_m$, and $H(x)$ is an $m \times m$ mvf (matrix-valued function) that is called the **Hamiltonian** of the system and is assumed to satisfy the conditions

$$H \in L_{1,\text{loc}}^{m\times m}([0,d)) \quad \text{and} \quad H(x) \geq 0 \quad \text{a.e. on } [0,d). \tag{1.2}$$

The solution $u(x,\lambda)$ of (1.1) under condition (1.2) is a locally absolutely continuous $k \times m$ mvf that is uniquely defined by specifying an initial condition $u(0,\lambda)$; it is also the unique locally absolutely continuous solution of the integral equation

$$u(x,\lambda) = u(0,\lambda) + i\lambda \int_0^x u(s,\lambda)H(s)ds J, \quad 0 \leq x < d. \tag{1.3}$$

The signature matrix J that is usually considered in (1.1) and (1.3) is either $\pm I_m$ or $\pm j_{pq}$, $\pm J_p$ and $\pm \mathcal{J}_p$, where

$$j_{pq} = \begin{bmatrix} I_p & 0 \\ 0 & -I_q \end{bmatrix}, \quad J_p = \begin{bmatrix} 0 & -I_p \\ -I_p & 0 \end{bmatrix}, \quad \mathcal{J}_p = \begin{bmatrix} 0 & -iI_p \\ iI_p & 0 \end{bmatrix} \tag{1.4}$$

and $p + q = m$ in the first displayed matrix and $2p = m$ in the other two. Every signature matrix $J \neq \pm I_m$ is unitarily similar to j_{pq}, where

$$p = \text{rank}(I_m + J) \quad \text{and} \quad q = \text{rank}(I_m - J) \quad \text{with } p + q = m. \tag{1.5}$$

Thus, $\pm J_p$ and $\pm \mathcal{J}_p$ are unitarily similar to $j_p = j_{pp}$:

$$J_p = \mathfrak{V}^* j_p \mathfrak{V}, \quad \text{where} \quad \mathfrak{V} = \frac{1}{\sqrt{2}}\begin{bmatrix} -I_p & I_p \\ I_p & I_p \end{bmatrix} \tag{1.6}$$

and

$$\mathcal{J}_p = \mathfrak{V}_1^* J_p \mathfrak{V}_1, \quad \text{where} \quad \mathfrak{V}_1 = \begin{bmatrix} -iI_p & 0 \\ 0 & I_p \end{bmatrix}. \tag{1.7}$$

Consequently a canonical system (1.1) with arbitrary signature matrix $J \neq \pm I_m$ may (and will) be reduced to a corresponding differential system with j_{pq} in place of J, or with J_p or $\mathcal{J}_p$ if $q = p$. The choice of J depends upon the problem under consideration. Thus, for example, $J = j_{pq}$ is appropriate for direct and inverse scattering problems, whereas the choice $J = J_p$ (or $J = \mathcal{J}_p$) is appropriate for direct and inverse spectral and input impedance (alias Weyl–Titchmarsh function) problems.

A number of different second-order differential equations and systems of differential equations of first-order may be reduced to the canonical differential system (1.1): the Feller–Krein string equation, the Dirac–Krein differential system, the Schrödinger equation with a matrix-valued potential of the form $q(x) = v'(x) \pm v(x)^2$ and Sturm–Liouville equations with appropriate restrictions on the coefficients.

The canonical system (1.1) arises (at least formally) by applying the Fourier transform

$$u(x, \lambda) = \widehat{y}(x, \lambda) = \int_0^\infty e^{i\lambda t} y(x, t) dt$$

to the solution $y(x, t)$ of the Cauchy problem

$$\begin{aligned} &\frac{\partial y}{\partial x}(x, t) = -\frac{\partial y}{\partial t}(x, t) H(x) J, \quad 0 \leq x \leq d, \ 0 \leq t < \infty, \\ &y(x, 0) = 0. \end{aligned} \tag{1.8}$$

The $m \times m$ matrix-valued solution $U_x(\lambda)$ of the system (1.1) that satisfies the initial condition $U_0(\lambda) = I_m$ is called the **matrizant** of the system. It may be interpreted as the **transfer function** from the input data $y(0, t)$ to the output $y(x, t)$ on the interval $[0, x]$ of the system in which the evolution of the data is described by equation (1.8), since

$$\widehat{y}(x, \lambda) = \widehat{y}(0, \lambda) U_x(\lambda) \quad \text{for } 0 \leq x < d. \tag{1.9}$$

1.1 The matrizant as a chain of entire J-inner mvf's

The matrizant $U_x(\lambda)$ is an entire mvf in the variable λ for each $x \in [0, d)$ such that

$$\frac{d}{dx}\left\{U_x(\lambda) J U_x(\omega)^*\right\} = i(\lambda - \overline{\omega}) U_x(\lambda) H(x) U_x(\omega)^* \quad \text{a.e. on the interval } (0, d),$$

which in turn implies that

$$
\begin{aligned}
U_{x_2}(\lambda)JU_{x_2}(\omega)^* - U_{x_1}(\lambda)JU_{x_1}(\omega)^* \\
= i(\lambda - \overline{\omega}) \int_{x_1}^{x_2} U_x(\lambda)H(x)U_x(\omega)^* dx.
\end{aligned}
\tag{1.10}
$$

Thus, as $U_0(\lambda) = I_m$, the kernel

$$
K_\omega^U(\lambda) = \begin{cases} \dfrac{J - U(\lambda)JU(\omega)^*}{-2\pi i(\lambda - \overline{\omega})} & if\ \lambda \neq \overline{\omega} \\ \\ \dfrac{1}{2\pi i}\left(\dfrac{\partial U}{\partial \lambda}\right)(\overline{\omega})JU(\omega)^* & if\ \lambda = \overline{\omega} \end{cases}
\tag{1.11}
$$

with $U(\lambda) = U_x(\lambda)$ is positive on $\mathbb{C} \times \mathbb{C}$ in the sense that

$$
\sum_{i,j=1}^{n} v_i^* K_{\omega_j}^U(\omega_i) v_j \geq 0
\tag{1.12}
$$

for every choice of points $\omega_1, \ldots, \omega_n$ in $\mathbb{C}$ and vectors $v_1, \ldots, v_n$ in $\mathbb{C}^m$, since

$$
\sum_{i,j=1}^{n} v_i^* K_{\omega_j}^U(\omega_i) v_j = \frac{1}{2\pi} \int_0^x \left(\sum_{i=1}^{n} v_i^* U_s(\omega_i)\right) H(s) \left(\sum_{j=1}^{n} v_j^* U_s(\omega_j)\right)^* ds,
$$

by formula (1.10). Therefore, by the matrix version of a theorem of Aronszajn in [Arn50], there is an RKHS (**reproducing kernel Hilbert space**) $\mathcal{H}(U)$ with RK (**reproducing kernel**) $K_\omega^U(\lambda)$ defined by formula (1.11); see Section 4.6.

Moreover, $U_x(\lambda)$ belongs to the class $\mathcal{E} \cap \mathcal{U}(J)$ of entire mvfs that are J-inner with respect to the open upper half plane $\mathbb{C}_+$:

$$
U_x(\lambda)^* J U_x(\lambda) \leq J \quad \text{for } \lambda \in \mathbb{C}_+
\tag{1.13}
$$

and

$$
U_x(\lambda)^* J U_x(\lambda) = J \quad \text{for } \lambda \in \mathbb{R}
\tag{1.14}
$$

for every $x \in [0, d)$, and, as follows easily from (1.10) with $x_2 = x$ and $x_1 = 0$ (since $U_0(\lambda) \equiv I_m$),

$$
U_x^{\#}(\lambda) J U_x(\lambda) = J \quad \text{for every point } \lambda \in \mathbb{C},
\tag{1.15}
$$

in which

$$
f^{\#}(\lambda) = f(\overline{\lambda})^*
$$

for any mvf f that is defined at $\overline{\lambda}$.

The matrizant $U_x(\lambda)$, $0 \leq x < d$, is **nondecreasing** with respect to x in the sense that

$$
U_{x_1}^{-1} U_{x_2} \in \mathcal{E} \cap \mathcal{U}(J) \quad \text{when } 0 \leq x_1 \leq x_2 < d.
\tag{1.16}
$$

It is also locally absolutely continuous in $[0, d)$ with respect to x and normalized by the conditions

$$U_x(0) = I_m \quad \text{for} \quad 0 \leq x < d \quad \text{and} \quad U_0(\lambda) = I_m.$$

The symbol $\mathcal{U}(J)$ will denote the class of **J-inner** mvf's with respect to $\mathbb{C}_+$. This is the class of $m \times m$ mvf's that are: (1) meromorphic in $\mathbb{C}_+$; (2) meet the constraint (1.13) at points of holomorphy in $\mathbb{C}_+$; (3) meet the constraint (1.14) a.e. on $\mathbb{R}$ for the nontangential limits; and (4) are extended into the open lower half plane $\mathbb{C}_-$ by (1.15) with U in place of U_x; this class will be considered in more detail in Section 3.8.

Let

$\mathfrak{h}_f$ denote the set of points in $\mathbb{C}$ at which the mvf f is holomorphic

and let

$$\mathcal{U}^\circ(J) = \{U \in \mathcal{U}(J) : 0 \in \mathfrak{h}_U \quad \text{and} \quad U(0) = I_m\}. \tag{1.17}$$

Thus, the matrizant $U_x \in \mathcal{E} \cap \mathcal{U}^\circ(J)$ for every $x \in [0, d)$.

1.2 Monodromy matrices of regular systems

A system (1.1) with Hamiltonian H subject to (1.2) is said to be a **regular canonical differential system** if

$$d < \infty \quad \text{and} \quad H \in L_1^{m\times m}([0, d]). \tag{1.18}$$

In this case the solutions $u(x, \lambda)$ of the system (1.1) are considered on the closed interval $[0, d]$ and $u(x, \lambda)$ is absolutely continuous with respect to x on $[0, d]$. Thus, the value $U_d(\lambda)$ of the matrizant $U_x(\lambda)$ at the right-hand end point of the interval is well defined; it is called the **monodromy matrix** of the canonical differential system (1.1).

The monodromy matrix $U_d(\lambda)$ belongs to the class $\mathcal{E} \cap \mathcal{U}^\circ(J)$. It is a significant frequency characteristic of a regular system in which the evolution of the data $y(x, t)$ on the interval $0 \leq x \leq d$ is described by (1.8). The property (1.14) means that the system is **lossless**. In particular, equation (1.8) with $J = J_p$ describes the evolution $y(x, t) = [v(x, t),\ i(x, t)]$ of voltages $v(x, t) = [v_1(x, t), \ldots, v_p(x, t)]$ and currents $i(x, t) = [i_1(x, t), \ldots, i_p(x, t)]$ of a lossless ideal linear circuit that is realized by p-lines with distributed parameters

$$H(x) = \begin{bmatrix} C(x) & G(x) \\ G(x)^* & L(x) \end{bmatrix} \quad \text{for } 0 \leq x \leq d,$$

where $C(x)$, $G(x)$ and $L(x)$ are the distributed capacitance, gyratance and inductance, respectively. Thus, the **inverse monodromy problem**, which is to recover the Hamiltonian $H(x)$, given a monodromy matrix $U \in \mathcal{E} \cap \mathcal{U}^\circ(J)$, is a significant inverse problem for lossless p-line circuits with distributed parameters (for $J = J_p$).

A fundamental theorem of V.P. Potapov [Po60] guarantees that the inverse monodromy problem has at least one solution $H(x)$ that satisfies the normalization conditions

$$H \in L_1^{m\times m}([0,d]), \quad H(x) \geq 0 \quad \text{and} \quad \operatorname{trace} H(x) = 1 \text{ a.e. on } [0,d]. \tag{1.19}$$

A Hamiltonian $H(x)$ on an interval $[0, d]$ is said to be a **normalized Hamiltonian** if it meets the three conditions in (1.19). However, there may be more than one such solution $H(x)$.

If $J = I_m$, then $\mathcal{U}(J)$ coincides with

$$\mathcal{S}_{\text{in}}^{m\times m} = \{\text{the class of } m \times m \text{ inner functions with respect to } \mathbb{C}_+\}$$

and the inverse monodromy problem for a given monodromy matrix $U \in \mathcal{E} \cap \mathcal{S}_{\text{in}}^{m\times m}$ with $U(0) = I_m$ has a unique solution $H(x)$ subject to the conditions in (1.19) if and only if $U(\lambda)$ and $\det U(\lambda)$ have the same exponential type. This criterion, which is due to Brodskii and Kisilevskii (see, e.g., [Bro72]), will be discussed in Chapter 8.

There are also uniqueness theorems for the inverse monodromy problem when $J \neq \pm I_m$, under extra conditions on the given monodromy matrix and the unknown Hamiltonian $H(x)$: If $J = \mathcal{J}_p$ and the monodromy matrix is **symplectic** (i.e., $U(\lambda)^\tau \mathcal{J}_p U(\lambda) = \mathcal{J}_p$, where $U(\lambda)^\tau$ denotes the transpose of $U(\lambda)$) and $p = 1$, then, by a fundamental theorem of L. de Branges [Br68a], there exists exactly one real solution $H(x) = \overline{H(x)} \geq 0$ a.e. on $[0, d]$ of the inverse monodromy problem that meets the normalization condition (1.19). This theorem and a number of its implications (including a uniqueness theorem for the inverse monodromy problem for the Feller–Krein string equation) will be discussed in Chapter 2 and again, in more detail, in Chapter 8.

1.3 Canonical integral systems

Every canonical differential system (1.1) with Hamiltonian $H(x)$ that satisfies (1.2) is equivalent to the **canonical integral system**

$$u(x,\lambda) = u(x,0) + i\lambda \int_0^x u(s,\lambda)dM(s)J, \quad 0 \leq x < d, \tag{1.20}$$

with

$$M(x) = \int_0^x H(s)ds.$$

We shall consider the system (1.20) under less restrictive conditions on $M(x)$, i.e., assuming only that

$$M(x) \quad \text{is a continuous nondecreasing} \\ m \times m \text{ mvf on } [0, d) \text{ with } M(0) = 0. \tag{1.21}$$

The continuous solution $U_x(\lambda) = U(x, \lambda)$ of the system (1.20) with $U_0(\lambda) = I_m$ is called the **matrizant** of the system. In view of (1.21), the matrizant satisfies an analog of (1.10) with $dM(x)$ in place of $H(x)dx$. Thus, the matrizant is a family of entire J-inner mvf's with respect to λ that is continuous and nondecreasing with respect to x on the interval $[0, d)$. Moreover, U_x, $0 \le x < d$, is a normalized family, since $U_x \in \mathcal{U}^\circ(J)$ for $0 \le x < d$ and $U_0(\lambda) = I_m$.

If $d < \infty$ and $M(x)$ is bounded on $[0, d)$, then $M(x)$ extends continuously to $[0, d]$ and the continuous solution $U_x(\lambda)$ of the system (1.20) is considered on $[0, d]$. The system (1.20) is a said to be a **regular canonical integral system** if

$$d < \infty \quad \text{and} \quad M(x) \text{ is a continuous nondecreasing} \\ m \times m \text{ mvf on } [0, d] \text{ with } M(0) = 0. \tag{1.22}$$

The matrizant $U_x(\lambda)$ of a regular integral system is a normalized nondecreasing continuous family of entire J-inner mvf's on the closed interval $[0, d]$; the mvf $U_d(\lambda)$ is called the **monodromy matrix** of the system.

1.4 Singular, right regular and right strongly regular matrizants

The main results on direct and inverse problems for canonical integral systems that are presented in this monograph are obtained for the class of systems with matrizants that satisfy the extra condition

$$U_x \in \mathcal{U}_{rR}(J) \quad \text{for } 0 \le x < d, \tag{1.23}$$

in which $\mathcal{U}_{rR}(J)$ denotes the class of **right regular** J-inner mvf's $U(\lambda)$ that may be defined in terms of the RKHS $\mathcal{H}(U)$ as

$$\mathcal{U}_{rR}(J) = \{U \in \mathcal{U}(J) : \mathcal{H}(U) \cap L_2^m(\mathbb{R}) \quad \text{is dense in } \mathcal{H}(U)\}. \tag{1.24}$$

Existence and uniqueness will be obtained for the solutions of a number of inverse problems in the class of systems with matrizants that meet the condition (1.23).

Formulas for U_x and $M(x)$ will be obtained under the more restrictive condition

$$U_x \in \mathcal{U}_{rsR}(J) \quad \text{for } 0 \le x < d, \tag{1.25}$$

where $\mathcal{U}_{rsR}(J)$ denotes the class of **right strongly regular** J-inner mvf's that may be defined in terms of the RKHS $\mathcal{H}(U)$ as

$$\mathcal{U}_{rsR}(J) = \{U \in \mathcal{U}(J) : \mathcal{H}(U) \subset L_2^m(\mathbb{R})\}. \tag{1.26}$$

If $U \in \mathcal{E} \cap \mathcal{U}(J)$, then $\mathcal{H}(U)$ is a Hilbert space of entire $m \times 1$ vvf's (vector-valued functions) f. The condition in (1.26) means that the restriction of $f \in \mathcal{H}(U)$ to $\mathbb{R}$ belongs to $L_2^m(\mathbb{R})$.

The class

$$\mathcal{U}_{AR}(J) = \{U \in \mathcal{U}(J) : \text{ every left divisor of } U \text{ belongs to the class } \mathcal{U}_{rR}(J)\}$$

that sits between $\mathcal{U}_{rR}(J)$ and $\mathcal{U}_{rsR}(J)$:

$$\mathcal{U}_{rsR}(J) \subset \mathcal{U}_{AR}(J) \subset U_{rR}(J),$$

will also play a significant role; it is discussed in detail in Section 4.9.

In Chapter 12 it will be shown that the matrizants of Dirac–Krein systems with locally summable potentials belong to the class $\mathcal{U}_{rsR}(J)$. On the other hand, the matrizant $Y_x(\lambda)$, $0 \le x < d$, of a Schrödinger equation

$$-y''(x,\lambda) + y(x,\lambda)q(x) = \lambda y(x,\lambda), \quad 0 \le x < d,$$

with potential

$$q \in L_{1,\mathrm{loc}}^{p\times p}([0,d)) \quad \text{and} \quad q(x) = q(x)^* \text{ a.e. on } [0,d)$$

belongs to the class

$$\mathcal{U}_S(J) = \{U \in \mathcal{U}(J) : \mathcal{H}(U) \cap L_2^m(\mathbb{R}) = \{0\}\} \tag{1.27}$$

of **singular** J-inner mvf's. Nevertheless, if the Riccati equation

$$q(x) = v'(x) + v(x)^2 \quad \text{or} \quad q(x) = v'(x) - v(x)^2$$

admits a locally summable solution $v(x)$ on $[0,d)$, then the Schrödinger equation may be reduced to the Dirac system

$$\frac{d}{dx}u(x,\lambda) = i\lambda u(x,\lambda)j_p + u(x,\lambda)\begin{bmatrix} 0 & v(x) \\ v(x)^* & 0 \end{bmatrix}, \quad 0 \le x < d, \tag{1.28}$$

and the matrizant of this system satisfies the condition (1.25); see Chapter 12 for the details.

In Chapter 5 it will be shown that any continuous normalized nondecreasing family of entire J-inner mvf's $U_x(\lambda)$, $0 \le x < d$, that satisfies the condition (1.23) is the matrizant of exactly one canonical integral system (1.20) with $M(x)$ satisfying (1.21). Moreover,

$$M(x) = 2\pi K_0^{U_x}(0) = \lim_{\lambda\to 0} \frac{U_x(\lambda) - I_m}{i\lambda} J = -i\frac{\partial U_x}{\partial\lambda}(0)J \tag{1.29}$$

for every system (1.20) with a mass function $M(x)$ that satisfies (1.21); see Theorem 5.8.

In view of the characterization (1.26) and the fact that

$$\mathcal{H}(U_x) \subseteq \mathcal{H}(U_d) \quad \text{for every } x \in [0,d],$$

it follows that

$$U_d \in \mathcal{U}_{rsR}(J) \Longrightarrow U_x \in \mathcal{U}_{rsR}(J) \quad \text{for every } x \in [0, d].$$

Moreover, if $U_x(\lambda)$, $0 \le x \le d$, is a nondecreasing continuous normalized family of entire J inner mvf's and $U_d \in \mathcal{U}_{rsR}(J)$, then $U_x(\lambda)$, $0 \le x \le d$, is the matrizant of exactly one canonical integral system on $[0, d]$. This system is regular; U_d is its monodromy matrix and formula (1.29) holds on the closed interval $[0, d]$.

Since

$$L_\infty^{m\times m}(\mathbb{R}) \cap \mathcal{U}(J) \subset \mathcal{U}_{rsR}(J),$$

a monodromy matrix U belongs to the class

$$U \in \mathcal{E} \cap \mathcal{U}_{rsR}^{\circ}(J) \tag{1.30}$$

if it is bounded on $\mathbb{R}$. Thus,

$$\mathcal{S}_{\text{in}}^{m\times m} = \mathcal{U}_{rsR}(I_m).$$

The condition (1.30) on a monodromy matrix does not guarantee the uniqueness of a normalized solution $H(x)$ for the inverse monodromy problem even when $J = I_m$ (in view of the Brodskii–Kisilevskii criterion).

In this book, we shall focus attention on direct and inverse problems for canonical integral and differential systems with $J \neq \pm I_m$ with matrizants that meet the condition (1.23). In this case it may be assumed that $J = j_{pq}$ or $J = J_p$, if $q = p$.

The matrizant of a canonical integral system (1.20) will be denoted by $W_x(\lambda)$ if $J = j_{pq}$ and by $A_x(\lambda)$ if $J = J_p$. The matrizant $W_x(\lambda)$ (resp., $A_x(\lambda)$) will play an important role in the study of direct and inverse input scattering (resp., input impedance and spectral) problems.

1.5 Input scattering matrices

If $W \in \mathcal{U}(j_{pq})$, then the **linear fractional transformation**

$$T_W[\varepsilon] = (w_{11}\varepsilon + w_{12})(w_{21}\varepsilon + w_{22})^{-1} \tag{1.31}$$

based on the four-block decomposition

$$W = \begin{bmatrix} w_{11} & w_{12} \\ w_{21} & w_{22} \end{bmatrix} \tag{1.32}$$

with blocks w_{11} of size $p \times p$ and w_{22} of size $q \times q$ maps the **Schur class**

$$\mathcal{S}^{p\times q} = \{\text{holomorphic contractive } p \times q \text{ mvf's } \varepsilon \text{ in } \mathbb{C}_+\} \tag{1.33}$$

into itself, i.e.,

$$T_W[\mathcal{S}^{p\times q}] \stackrel{\text{def}}{=} \{T_W[\varepsilon] : \varepsilon \in \mathcal{S}^{p\times q}\} \subseteq \mathcal{S}^{p\times q}. \tag{1.34}$$

Since the matrizant W_x, $0 \le x < d$, is a nondecreasing family of j_{pq}-inner mvf's, this inclusion implies that $T_{W_{x_2}}[\mathcal{S}^{p\times q}] \subseteq T_{W_{x_1}}[\mathcal{S}^{p\times q}]$ if $0 \le x_1 \le x_2 < d$ and that the set

$$\mathcal{S}^d_{\text{scat}}(dM) \stackrel{\text{def}}{=} \bigcap_{0\le x<d} T_{W_x}[\mathcal{S}^{p\times q}] \tag{1.35}$$

is a nonempty subset of $\mathcal{S}^{p\times q}$. The mvf's s that belong to $\mathcal{S}^d_{\text{scat}}(dM)$ will be called **input scattering matrices** of the system (1.20) for $J = j_{pq}$. A physical interpretation of this notion is presented in Section 6.1 for regular systems, in which case

$$\mathcal{S}^d_{\text{scat}}(dM) = T_{W_d}[\mathcal{S}^{p\times q}].$$

There is another characterization of the class $\mathcal{U}_{rsR}(J)$ in addition to (1.26) when $J = j_{pq}$:

$$W \in \mathcal{U}_{rsR}(j_{pq}) \Longleftrightarrow T_W[\mathcal{S}^{p\times q}] \cap \mathring{\mathcal{S}}^{p\times q} \ne \emptyset, \tag{1.36}$$

where

$$\mathring{\mathcal{S}}^{p\times q} = \{s \in \mathcal{S}^{p\times q} : \|s\|_\infty < 1\}. \tag{1.37}$$

Thus, the matrizant W_x, $0 \le x < d$, of every solution of the inverse input scattering problem with $s \in \mathring{\mathcal{S}}^{p\times q}$ will be a family of strongly right regular j_{pq}-inner mvf's.

1.6 Chains of associated pairs of the first kind

The set of canonical integral systems (1.20) with $J = j_{pq}$ and a given input scattering matrix $s \in \mathcal{S}^{p\times q}_{sz}$ that is defined in (1.56) (and the set of regular canonical systems with a given monodromy matrix $W \in \mathcal{E} \cap \mathcal{U}^\circ_{rR}(j_{pq})$) with matrizant

$$W_x \in \mathcal{U}^\circ_{rR}(j_{pq}) \quad \text{for } 0 \le x < d \tag{1.38}$$

will be parametrized in terms of a **chain** $\{b_1^x, b_2^x\}$, $0 \le x < d$, **of pairs** of entire inner mvf's $b_1^x \in \mathcal{S}^{p\times p}_{\text{in}}$, $b_2^x \in \mathcal{S}^{q\times q}_{\text{in}}$ that is

nondecreasing in the sense that

$$(b_1^{x_1})^{-1} b_1^{x_2} \in \mathcal{E} \cap \mathcal{S}^{p\times p}_{\text{in}} \quad \text{and} \quad b_2^{x_2}(b_2^{x_1})^{-1} \in \mathcal{E} \cap \mathcal{S}^{q\times q}_{\text{in}}, \quad 0 \le x_1 \le x_2 < d;$$

continuous in the sense that $b_1^x(\lambda)$ and $b_2^x(\lambda)$ are continuous mvf's of x on $[0, d)$ for each fixed $\lambda \in \mathbb{C}$; and

normalized by the condition

$$b_1^x(0) = I_p, \quad b_2^x(0) = I_q \quad \text{for } 0 \le x < d, \quad b_1^0(\lambda) \equiv I_p \quad \text{and} \quad b_2^0(\lambda) \equiv I_q.$$

This parametrization rests on the association of a pair of inner mvf's $\{b_1, b_2\}$ to each $W \in \mathcal{U}(j_{pq})$ that is obtained from the inner–outer factorization and outer–inner factorizations of the mvf's $(w_{11}^{\#})^{-1}$ and w_{22}^{-1}, which belong to $\mathcal{S}^{p\times p}$ and $\mathcal{S}^{q\times q}$, respectively:

$$(w_{11}^{\#})^{-1} = b_1\varphi_1 \quad \text{and} \quad w_{22}^{-1} = \varphi_2 b_2, \tag{1.39}$$

where

$$b_1 \in \mathcal{S}_{\text{in}}^{p\times p}, \ b_2 \in \mathcal{S}_{\text{in}}^{q\times q}, \ \varphi_1 \in \mathcal{S}_{\text{out}}^{p\times p}, \ \varphi_2 \in \mathcal{S}_{\text{out}}^{q\times q}$$

and

$$\mathcal{S}_{\text{out}}^{p\times p} = \{\text{the set of outer mvf's in } \mathcal{S}^{p\times p}\}.$$

Such a pair $\{b_1, b_2\}$ will be called an **associated pair** of W and the set of such pairs will be denoted $ap(W)$. In view of the implication

$$W \in \mathcal{E} \cap \mathcal{U}(j_{pq}) \quad \text{and} \quad \{b_1, b_2\} \in ap(W) \Longrightarrow$$
$$b_1 \in \mathcal{E} \cap \mathcal{S}_{\text{in}}^{p\times p} \text{ and } b_2 \in \mathcal{E} \cap \mathcal{S}_{\text{in}}^{q\times q}, \tag{1.40}$$

the pair $\{b_1, b_2\}$ may be uniquely specified by the normalization conditions $b_1(0) = I_p$ and $b_2(0) = I_q$ when $W \in \mathcal{E} \cap \mathcal{U}(j_{pq})$.

The converse of the implication (1.40) is valid when W belongs to the class $\mathcal{U}_{rR}(j_{pq})$ of right regular j_{pq}-inner mvf's, but not in general.

Since $W_x(\lambda), 0 \leq x < d$, is a nondecreasing family, there is a unique normalized nondecreasing chain of pairs of entire inner mvf's $\{b_1^x, b_2^x\}, 0 \leq x < d$, corresponding to the matrizant $W_x(\lambda), 0 \leq x < d$, of the system (1.20) with $J = j_{pq}$ and $M(x)$ subject to the restrictions in (1.21) such that

$$\{b_1^x, b_2^x\} \in ap(W_x) \quad \text{for } 0 \leq x < d. \tag{1.41}$$

If also

$$W_x \in \mathcal{U}_{rR}(j_{pq}) \quad \text{for } 0 \leq x < d, \tag{1.42}$$

then, as will be shown in Theorem 5.13, the chain of pairs is continuous too. Consequently, if (1.42) is in force, then there exists exactly one continuous normalized nondecreasing chain of pairs $\{b_1^x, b_2^x\}$ of entire inner mvf's such that (1.41) holds.

Associated pairs are also defined for J-inner mvf's U with

$$J = V_1^* j_{pq} V_1 \quad \text{for some } m \times m \text{ unitary matrix } V_1 \tag{1.43}$$

upon noting that if

$$W(\lambda) = V_1 U(\lambda) V_1^*, \tag{1.44}$$

then

$$U \in \mathcal{U}(J) \Longleftrightarrow W \in \mathcal{U}(j_{pq}),$$

$$U \in \mathcal{U}_{rR}(J) \Longleftrightarrow W \in \mathcal{U}_{rR}(j_{pq}),$$

$$U \in \mathcal{U}_{rsR}(J) \Longleftrightarrow W \in \mathcal{U}_{rsR}(j_{pq});$$

and $\{b_1, b_2\}$ is called an **associated pair of the first kind** for U if $\{b_1, b_2\} \in ap(V_1 U V_1^*)$; the set of these pairs is denoted $ap_I(U)$. If U is entire, then b_1 and b_2 are both entire and may be uniquely specified by imposing the normalization conditions $b_1(0) = I_p$ and $b_2(0) = I_q$. The normalized associated pair depends upon the choice of the unitary matrix V_1 in a nonessential way: $\{b_1, b_2\}$ transforms to $\{u^* b_1 u, v^* b_2 v\}$ for some pair of constant unitary matrices.

Thus, to each canonical integral system (1.20) with mass function $M(x)$ subject to (1.21), $J \neq \pm I_m$ and matrizant $U_x(\lambda)$, there is exactly one normalized nondecreasing chain of pairs of entire inner mvf's

$$\{b_1^x, b_2^x\} \in ap_I(U_x) \quad \text{for } 0 \leq x < d.$$

If (1.23) is in force, then this chain is continuous in x on $[0, d)$ for each fixed $\lambda \in \mathbb{C}$.

The set of input scattering matrices for a canonical system with J as in (1.43) can also be defined in terms of the input scattering matrices for $W_x = V_1 U_x V_1^*$; see Section 6.1.

1.7 The bitangential direct input scattering problem

The study of the connections between the mass function $M(x)$ of a given canonical integral system (1.20) with $J = j_{pq}$, the chain $\{b_1^x, b_2^x\}$, $0 \leq x < d$, of normalized pairs of entire inner mvf's associated with the matrizant $W_x(\lambda)$ of this system and the set $\mathcal{S}_{\text{scat}}^d(dM)$ will be referred to as the **bitangential direct input scattering problem**. Chapter 6 is devoted to a discussion of this problem.

The Weyl–Titchmarsh classification of systems (1.20) with $J = j_{pq}$ is obtained by consideration of the limit ball

$$\mathcal{B}_d(\omega) = \{s(\omega) : s \in \mathcal{S}_{\text{scat}}^d(dM)\} \quad \text{for } \omega \in \mathbb{C}_+ \tag{1.45}$$

with left and right semiradii $\mathfrak{r}_\ell^d(\omega)$ and $\mathfrak{r}_r^d(\omega)$, respectively; formulas for the semiradii are presented in Theorem 6.4. A fundamental theorem of S.A. Orlov [Or76] guarantees that the numbers

$$\mathfrak{n}_\ell^d = \operatorname{rank} \mathfrak{r}_\ell^d(\omega) \quad \text{and} \quad \mathfrak{n}_r^d = \operatorname{rank} \mathfrak{r}_r^d(\omega) \tag{1.46}$$

are independent of the choice of $\omega \in \mathbb{C}_+$. If

$$\mathcal{S}_{\text{scat}}^d(dM) \cap \mathring{\mathcal{S}}^{p\times q} \neq \emptyset, \tag{1.47}$$

then

$$\mathfrak{n}_\ell^d = \lim_{x\uparrow d} \operatorname{rank}\{b_1^x(\omega)b_1^x(\omega)^*\} \quad \text{and} \quad \mathfrak{n}_r^d = \lim_{x\uparrow d} \operatorname{rank}\{b_2^x(\omega)^* b_1^x(\omega)\}. \tag{1.48}$$

Two extreme cases are of special interest:

(1) **The limit point case:** $\mathfrak{n}_\ell^d = 0$ or $\mathfrak{n}_r^d = 0$. A system in the limit point case has only one input scattering matrix, i.e.,

$$\mathcal{S}_{\text{scat}}^d(dM) = \{s^\circ\}.$$

(2) **The full rank case:** $\mathfrak{n}_\ell^d = p$ and $\mathfrak{n}_r^d = q$.

A regular system is always in the full rank case. Moreover, if $d < \infty$ and (1.47) is in force, then the converse is also true, i.e., $\mathfrak{n}_\ell^d = p$ and $\mathfrak{n}_r^d = q \Longrightarrow$ the system is regular.

A characterization in the spirit of Weyl of the unique input scattering matrix s° in the limit point case when $\mathfrak{n}_\ell^d = 0$ is given in Theorem 6.24.

1.8 Bitangential inverse monodromy and inverse scattering problems

Chapter 8 is devoted to the **bitangential inverse monodromy problem** for regular integral systems. The data for this problem when $J = j_{pq}$ is

(1) a **monodromy matrix** $W \in \mathcal{E} \cap \mathcal{U}^\circ(j_{pq})$; and
(2) a normalized nondecreasing continuous chain $\{b_1^x, b_2^x\}$, $0 \le x \le d$, of pairs of entire inner mvf's with $\{b_1^d, b_2^d\} \in ap(W)$.

The problem is to find a regular canonical integral system (1.20) with $J = j_{pq}$ and $M(x)$ subject to (1.21) such that the matrizant $W_x(\lambda)$ satisfies (1.41) and $W_d(\lambda) = W(\lambda)$ (i.e., W is the monodromy matrix for the system (1.20)).

In Chapter 10 we shall discuss a **bitangential inverse input scattering problem** in which the given data is

(1) an **input scattering matrix** $s \in \mathcal{S}^{p\times q}$; and
(2) a normalized nondecreasing continuous chain $\{b_1^x, b_2^x\}$, $0 \le x < d$, of pairs of entire inner mvf's b_1^x and b_2^x of sizes $p \times p$ and $q \times q$, respectively.

The problem is to find a canonical integral system (1.20) with $J = j_{pq}$ and $M(x)$ subject to (1.21) such that the matrizant $W_x(\lambda)$ satisfies (1.41) and $s \in \bigcap_{0\le x<d} T_{W_x}[\mathcal{S}^{p\times q}]$ (i.e., s is an input scattering matrix for the system (1.20)).

The extra condition (1.42) on the matrizant guarantees that each of these two inverse problems has at most one solution. If the given mvf W in the first problem (resp., s in the second problem) belongs to the class $\mathcal{E} \cap \mathcal{U}_{AR}^\circ(j_{pq})$ (resp., $\mathring{\mathcal{S}}^{p\times q}$) then this problem has exactly one solution. Formulas for the matrizant $W_x(\lambda)$

of this solution (and hence for $M(x)$, in view of (1.29)) will be obtained in Chapters 8 and 10 when $W_x \in \mathcal{U}_{rsR}^{\circ}(j_{pq})$ for $0 \leq x \leq d$ (resp., $0 \leq x < d$).

The bitangential inverse monodromy problem with given data $W \in \mathcal{E} \cap \mathcal{U}_{AR}^{\circ}(j_{pq})$ and $\{b_1^x, b_2^x\}$, $0 \leq x \leq d$, may be reduced to a bitangential inverse input scattering problem with $s^{\circ} \in T_W[\mathcal{S}^{p\times q}]$ and the same chain of pairs. The latter problem will be solved by identifying the matrizant W_x, $0 \leq x \leq d$, with specified resolvent matrices of a family of approriately chosen interpolation problems in the Schur class; see Sections 1.9 and 1.10.

1.9 The generalized Schur interpolation problem

The **generalized Schur interpolation problem** GSIP$(b_1, b_2; s^{\circ})$ is to describe the set

$$\mathcal{S}(b_1, b_2; s^{\circ}) = \{s \in \mathcal{S}^{p\times q} : b_1^{-1}(s - s^{\circ})b_2^{-1} \in H_{\infty}^{p\times q}\}, \tag{1.49}$$

given $b_1 \in \mathcal{S}_{\text{in}}^{p\times p}$, $b_2 \in \mathcal{S}_{\text{in}}^{q\times q}$ and $s^{\circ} \in \mathcal{S}^{p\times q}$.

The GSIP$(b_1, b_2; s^{\circ})$ is said to be **completely indeterminate** if for every nonzero vector $\eta \in \mathbb{C}^q$

$$s(\lambda)\eta \not\equiv s^{\circ}(\lambda)\eta \quad \text{for at least one mvf } s \in \mathcal{S}(b_1, b_2; s^{\circ}); \tag{1.50}$$

it is said to be **strictly completely indeterminate** if

$$\mathcal{S}(b_1, b_2; s^{\circ}) \cap \mathring{\mathcal{S}}^{p\times q} \neq \emptyset. \tag{1.51}$$

A mvf $W \in \mathcal{U}(j_{pq})$ for which

$$\mathcal{S}(b_1, b_2; s^{\circ}) = T_W[\mathcal{S}^{p\times q}]. \tag{1.52}$$

holds is called a **resolvent matrix** for the corresponding interpolation problem. Resolvent matrices $W \in \mathcal{U}(j_{pq})$ for a completely indeterminate (resp., strictly completely indeterminate) interpolation problems always exist and automatically belong to the class $\mathcal{U}_{rR}(j_{pq})$ (resp., $\mathcal{U}_{rsR}(j_{pq})$). Conversely, if (1.52) holds for some $W \in \mathcal{U}(j_{pq})$, then the GSIP$(b_1, b_2; s^{\circ})$ is completely indeterminate and hence $W \in \mathcal{U}_{rR}(j_{pq})$. Moreover, the interpolation problem is strictly completely indeterminate if and only if $W \in \mathcal{U}_{rsR}(j_{pq})$. Thus, the class $\mathcal{U}_{rR}(j_{pq})$ (resp., $\mathcal{U}_{rsR}(j_{pq})$) is identified with the class of resolvent matrices of completely indeterminate (resp., strictly completely indeterminate) generalized Schur interpolation problems.

A GSIP$(b_1, b_2; s^{\circ})$ may have many resolvent matrices. However, if the GSIP$(b_1, b_2; s^{\circ})$ is completely indeterminate, then there exists an essentially unique mvf $W \in \mathcal{U}(j_{pq})$ such that

$$\mathcal{S}(b_1, b_2; s^{\circ}) = T_W[\mathcal{S}^{p\times q}] \quad \text{and} \quad \{b_1, b_2\} \in ap(W). \tag{1.53}$$

Here, essentially unique means up to a right constant j_{pq}-unitary multiplier. If b_1 and b_2 are entire, then there is exactly one resolvent matrix $W \in \mathcal{E} \cap \mathcal{U}^{\circ}(j_{pq})$

for this completely indeterminate problem for which (1.53) holds. Conversely, if $W \in \mathcal{U}_{rR}(j_{pq})$ and if

$$\{b_1, b_2\} \in ap(W) \quad \text{and} \quad s^\circ \in T_W[\mathcal{S}^{p\times q}], \tag{1.54}$$

then (1.52) holds. If (1.52) holds for a mvf $W \in \mathcal{U}(j_{pq})$, then $W \in \mathcal{U}_{rsR}(j_{pq})$ if and only if (1.51) is also in force.

An algorithm for computing the normalized resolvent matrix $W \in \mathcal{E} \cap \mathcal{U}^\circ_{rsR}(j_{pq})$ from the given data in the strictly completely indeterminate case is given in Theorem 4.75.

If $s^\circ \in \mathring{\mathcal{S}}^{p\times q}$ it is clear that the GSIP$(b_1, b_2; s^\circ)$ is strictly completely indeterminate. However, this is far from necessary: If

$$s^\circ(\lambda) = \int_0^1 e^{i\lambda t} dt$$

and both $b_1(\lambda)$ and $b_2(\lambda)$ are entire inner functions, then, in view of Theorem 9.47, the corresponding GSIP$(b_1, b_2; s^\circ)$ is strictly completely indeterminate even though s° is not strictly contractive.

Characterizations of the class of strictly completely indeterminate problems and the less restrictive class of completely indeterminate problems may be formulated in terms of the operators based on the data that are defined in (4.24). Another characterization of the latter class is presented in Theorem 4.16, where it is shown that a GSIP$(b_1, b_2; s^\circ)$ is completely indeterminate if and only if

$$\mathcal{S}(b_1, b_2; s^\circ) \cap \mathcal{S}^{p\times q}_{sz} \neq \emptyset, \tag{1.55}$$

where

$$\mathcal{S}^{p\times q}_{sz} = \left\{ s \in \mathcal{S}^{p\times q} : \int_{-\infty}^{\infty} \frac{|\ln \det(I_q - s^*(\mu)s(\mu))|}{1+\mu^2} d\mu < \infty \right\} \tag{1.56}$$

is the class of mvf's in $\mathcal{S}^{p\times q}$ that meet the indicated Szegö condition. Thus, if a GSIP$(b_1, b_2; s^\circ)$ is strictly completely indeterminate, then it is completely indeterminate, but not conversely.

1.10 Identifying matrizants as resolvent matrices when $J = j_{pq}$

The preceding discussion implies that if $W_x(\lambda)$, $0 \leq x < d$, is the matrizant of a solution to the inverse input scattering problem with a given input scattering matrix $s^\circ \in \mathcal{S}^{p\times q}_{sz}$ and W_x satisfies the condition (1.42), then there exists a unique normalized nondecreasing continuous chain of pairs $\{b_1^x, b_2^x\}$, $0 \leq x < d$, of entire inner mvf's such that (1.41) holds. Moreover, W_x is the unique resolvent matrix for the GSIP$(b_1^x, b_2^x; s^\circ)$ that satisfies the conditions

$$W_x \in \mathcal{U}^\circ_{rR}(j_{pq}) \quad \text{and} \quad \{b_1^x, b_2^x\} \in ap(W_x) \quad \text{for } 0 \leq x < d. \tag{1.57}$$

Conversely, if $s^\circ \in \mathcal{S}_{sz}^{p\times q}$ and $\{b_1^x, b_2^x\}$ is any normalized nondecreasing continuous chain of pairs of entire inner mvf's of sizes $p \times p$ and $q \times q$, respectively, then the GSIP$(b_1^x, b_2^x; s^\circ)$ is completely indeterminate and there exists exactly one resolvent matrix W_x of this problem such that (1.57) holds. Moreover, the inclusions

$$\mathcal{S}(b_1^{x_2}, b_2^{x_2}; s^\circ) \subseteq \mathcal{S}(b_1^{x_1}, b_2^{x_1}; s^\circ) \quad \text{for } 0 \leq x_1 \leq x_2 < d \tag{1.58}$$

(and hence the monotonicity of the chain W_x, $0 \leq x < d$) follow from the monotonicity of the chain $\{b_1^x, b_2^x\}$.

This family of resolvent matrices is a normalized nondecreasing chain of mvf's in the class $\mathcal{E} \cap \mathcal{U}_{rR}^\circ(j_{pq})$. If $W_x(\lambda)$ is also continuous on $[0, d)$ for each fixed $\lambda \in \mathbb{C}$, then W_x, $0 \leq x < d$, is the matrizant of a canonical system of the form (1.20) with $J = j_{pq}$ and a mass function $M(x)$ that satisfies the constraints (1.21); see Theorem 5.8. In fact there is only one such system with an input scattering matrix s° and a matrizant that satisfies the conditions (1.41) and (1.42).

The extra condition of continuity will hold if $W_x \in \mathcal{U}_{AR}(j_{pq})$ for every $x \in [0, d)$. The identification of the matrizant with the resolvent matrices of a family of GSIP's is exploited to solve the bitangential inverse input scattering problem in Chapter 10.

If $W \in \mathcal{E} \cap \mathcal{U}_{rR}(j_{pq})$ is given instead of $s^\circ \in \mathcal{S}_{sz}^{p\times q}$, then the bitangential inverse monodromy problem may be reformulated as a bitangential inverse input scattering problem with $s^\circ \in T_W[\mathcal{S}^{p\times q}]$ and the same chain $\{b_1^x, b_2^x\}$, $0 \leq x \leq d$. This is discussed in Chapter 8.

1.11 Input impedance matrices and spectral functions

In the study of bitangential direct and inverse input impedance and spectral problems, a central role is played by mvf's c in the **Carathéodory class**

$$\mathcal{C}^{p\times p} = \{p \times p \text{ mvf's } c \text{ that are holomorphic with } \Re c \geq 0 \text{ in } \mathbb{C}_+\} \tag{1.59}$$

and mvf's $A \in \mathcal{U}(J_p)$. This stems from the fact that if $A \in \mathcal{U}(J_p)$, then the linear fractional transformation

$$T_A[\tau] = (a_{11}\tau + a_{12})(a_{21}\tau + a_{22})^{-1}$$

based on the four-block decomposition

$$A = \begin{bmatrix} a_{11} & a_{12} \\ a_{21} & a_{22} \end{bmatrix} \quad (\text{with } p \times p \text{ blocks } a_{ij})$$

maps $\tau \in \mathcal{C}^{p\times p}$ into $\mathcal{C}^{p\times p}$, whenever it is well defined, i.e., when

$$\det(a_{21}(\lambda)\tau(\lambda) + a_{22}(\lambda)) \not\equiv 0 \text{ in } \mathbb{C}_+.$$

It turns out to be more useful to consider the linear fractional transformation

$$T_B[\varepsilon] = (b_{11}\varepsilon + b_{12})(b_{21}\varepsilon + b_{22})^{-1} \quad \text{based on} \quad B = \begin{bmatrix} b_{11} & b_{12} \\ b_{21} & b_{22} \end{bmatrix}, \tag{1.60}$$

where $B = A\mathfrak{V}$ and

$$\varepsilon \in \mathcal{S}^{p\times p} \cap \mathcal{D}(T_B) \stackrel{\text{def}}{=} \{\varepsilon \in \mathcal{S}^{p\times p} : \det(b_{21}\varepsilon + b_{22}) \not\equiv 0 \text{ in } \mathbb{C}_+\}.$$

Then

$$\mathcal{C}(A) \stackrel{\text{def}}{=} \{T_B[\varepsilon] : \varepsilon \in \mathcal{S}^{p\times p} \cap \mathcal{D}(T_B)\} \subseteq \mathcal{C}^{p\times p}. \tag{1.61}$$

Moreover, the monotonicity of the matrizant $A_x(\lambda)$, $0 \le x < d$, of a canonical integral system (1.20) with $J = J_p$ and $M(x)$ subject to (1.21) implies that

$$\mathcal{C}(A_{x_2}) \subseteq \mathcal{C}(A_{x_1}) \quad \text{for } 0 \le x_1 \le x_2 < d. \tag{1.62}$$

A mvf c that belongs to the set

$$\mathcal{C}^d_{\text{imp}}(dM) \stackrel{\text{def}}{=} \bigcap_{0\le x<d} \mathcal{C}(A_x) \tag{1.63}$$

will be called an **input impedance matrix** for the corresponding canonical integral system (or in more traditional terminology, a **Weyl–Titchmarsh function**). Conditions which insure that $\mathcal{C}^d_{\text{imp}}(dM) \neq \emptyset$ will be presented in Sections 7.1 and 7.2.

The **Cayley transform**

$$T_{\mathfrak{V}}[s] = (I_p - s)(I_p + s)^{-1}$$

based on the mvf $\mathfrak{V}$ defined in (1.6) maps the set

$$\mathcal{S}^{p\times p} \cap \mathcal{D}(T_{\mathfrak{V}}) \quad \text{onto} \quad \mathcal{C}^{p\times p}.$$

Moreover,

$$\mathcal{C}^d_{\text{imp}}(dM) = \left\{ T_{\mathfrak{V}}[s] : s \in \left(\bigcap_{0\le t<d} T_{W_t}[\mathcal{S}^{p\times p}] \right) \bigcap \mathcal{D}(T_{\mathfrak{V}}) \right\}$$

and hence,

$$\mathcal{C}^d_{\text{imp}}(dM) \neq \emptyset \Longleftrightarrow \left(\bigcap_{0\le t<d} T_{W_t}[\mathcal{S}^{p\times p}] \right) \bigcap \mathcal{D}(T_{\mathfrak{V}}) \neq \emptyset.$$

Thus, $\mathcal{C}^d_{\text{imp}}(dM) \neq \emptyset$ if and only if there exists at least one input scattering matrix s such that $\det(I_p + s(\lambda)) \not\equiv 0$. If the system is regular with monodromy matrix A_d, then

$$\mathcal{C}^d_{\text{imp}}(dM) = \mathcal{C}(A_d) \neq \emptyset. \tag{1.64}$$

The $p \times p$ nondecreasing mvf $d\sigma(\mu)$ in the **Riesz–Herglotz–Nevanlinna representation**

$$c(\lambda) = i\alpha - i\lambda\beta + \frac{1}{\pi i}\int_{-\infty}^{\infty}\left\{\frac{1}{\mu-\lambda} - \frac{\mu}{1+\mu^2}\right\}d\sigma(\mu) \quad \text{for } \lambda \in \mathbb{C}_+, \tag{1.65}$$

of a mvf $c \in \mathcal{C}^{p\times p}$ will be called the **spectral function** of c; it is subject to the constraint

$$\int_{-\infty}^{\infty}(1+\mu^2)^{-1}\text{trace}\, d\sigma(\mu) < \infty. \tag{1.66}$$

Let

$$(\mathcal{C}^d_{\text{imp}}(dM))_{\text{sp}} \stackrel{\text{def}}{=} \{\text{the spectral functions of mvf's } c \in \mathcal{C}^d_{\text{imp}}(dM)\}.$$

Under appropriate conditions, the set $(\mathcal{C}^d_{\text{imp}}(dM))_{\text{sp}}$ coincides with the set $\Sigma^d_{\text{psf}}(dM)$ of **pseudospectral functions** $\sigma(\mu)$ of the system (1.20), subject to (1.21), with $J = J_p$, which is the set of nondecreasing $p \times p$ mvf's on $\mathbb{R}$ for which the **Parseval–Plancherel** equality

$$\int_{-\infty}^{\infty}(\mathcal{F}_2 f)(\mu)^* d\sigma(\mu)(\mathcal{F}_2 f)(\mu) = \int_0^d f(x)^* dM(x) f(x) \tag{1.67}$$

holds for every $f \in L_2^m(dM) \ominus \ker \mathcal{F}_2$, where $\mathcal{F}_2$ is the **generalized Fourier transform**

$$(\mathcal{F}_2 f)(\lambda) = \frac{1}{\sqrt{\pi}}\int_0^d [a_{21}(x,\lambda) \quad a_{22}(x,\lambda)]dM(x)f(x) \tag{1.68}$$

based on the second block row of $A_x(\lambda)$. If $\ker \mathcal{F}_2 = \{0\}$, then $\Sigma^d_{\text{psf}}(dM)$ concides with the set $\Sigma^d_{\text{sf}}(dM)$ of spectral functions of the system.

1.12 de Branges spaces

A $p \times 2p$ entire mvf

$$\mathfrak{E}(\lambda) = \begin{bmatrix} E_-(\lambda) & E_+(\lambda) \end{bmatrix}$$

with blocks $E_\pm(\lambda)$ of sizes $p \times p$ will be called an entire **de Branges matrix** if

$$\det E_+(\lambda) \not\equiv 0 \quad \text{and} \quad E_+^{-1}E_- \in \mathcal{S}_{\text{in}}^{p\times p}. \tag{1.69}$$

If $A \in \mathcal{E} \cap \mathcal{U}(J_p)$, then

$$\begin{aligned}\mathfrak{E}(\lambda) &= \begin{bmatrix} E_-(\lambda) & E_+(\lambda) \end{bmatrix} = \sqrt{2}[0 \quad I_p]A(\lambda)\mathfrak{V} \\ &= [a_{22}(\lambda) - a_{21}(\lambda) \quad a_{22}(\lambda) + a_{21}(\lambda)]\end{aligned} \tag{1.70}$$

is an entire de Branges matrix. Moreover, the blocks $E_-(\lambda)$ and $E_+(\lambda)$ have the following extra properties:

$$\rho_i^{-1}E_+^{-1} \in H_2^{p\times p} \quad \text{and} \quad \rho_i^{-1}(E_-^\#)^{-1} \in H_2^{p\times p}, \tag{1.71}$$

where

$$\rho_\omega(\lambda) = -2\pi i(\lambda - \overline{\omega}) \tag{1.72}$$

and $H_2^{p\times p}$ denotes the class of $p \times p$ mvf's with entries in the Hardy class H_2 over $\mathbb{C}_+$ that is specified in the list of spaces considered near the beginning of Chapter 3. Thus,

$$\det E_\pm(\lambda) \neq 0 \text{ in } \overline{\mathbb{C}_\pm}. \tag{1.73}$$

In view of (1.69), the kernel

$$K_\omega^{\mathfrak{E}}(\lambda) = \frac{E_+(\lambda)E_+(\omega)^* - E_-(\lambda)E_-(\omega)^*}{\rho_\omega(\lambda)} \quad \text{for } \lambda \neq \overline{\omega}$$

(extended continuously for $\lambda = \overline{\omega}$) is positive on $\mathbb{C} \times \mathbb{C}$. Therefore, by a matrix version of a theorem of Aronszajn there is a RKHS $\mathcal{B}(\mathfrak{E})$ with $K_\omega^{\mathfrak{E}}(\lambda)$ as its RK. This RKHS will be called a **de Branges space**. The vectors in $\mathcal{B}(\mathfrak{E})$ are entire $p \times 1$ vvf's. The set $(\mathcal{B}(\mathfrak{E}))_{\mathrm{sf}}$ of **spectral functions** of $\mathcal{B}(\mathfrak{E})$ is the set of $p \times p$ nondecreasing mvf's $\sigma(\mu)$ on $\mathbb{R}$ with $\sigma(0) = 0$ and $\sigma(\mu) = (\sigma(\mu+) + \sigma(\mu-))/2$ such that

$$\int_{-\infty}^{\infty} f(\mu)^* d\sigma(\mu) f(\mu) = \int_{-\infty}^{\infty} f(\mu)^* E_+(\mu)^{-*} E_+(\mu)^{-1} f(\mu) d\mu \tag{1.74}$$

for $f \in \mathcal{B}(\mathfrak{E})$.

If $A \in \mathcal{E} \cap \mathcal{U}(J_p)$, $c^\circ = T_A[I_p]$,

$$\lim_{\nu\uparrow\infty} \nu^{-1} c^\circ(i\nu) = 0 \tag{1.75}$$

and $\mathfrak{E} = \sqrt{2}[0 \quad I_p]A\mathfrak{V}$, then the set $(\mathcal{H}(A))_{\mathrm{sf}}$ of spectral functions of the RKHS $\mathcal{H}(A)$ coincides with $(\mathcal{B}(\mathfrak{E}))_{\mathrm{sf}}$. Moreover, if $\mathfrak{E}$ and A are related by (1.70), then the mapping

$$U_2 : f \in \mathcal{H}(A) \longrightarrow \sqrt{2}[0 \quad I_p]f$$

is a coisometry from $\mathcal{H}(A)$ into $\mathcal{B}(\mathfrak{E})$, which is unitary if and only if (1.75) holds. If $A \in \mathcal{U}_{AR}(J_p)$, then (1.75) will hold.

If the blocks of an entire de Branges matrix $\mathfrak{E} = [E_- \quad E_+]$ satisfy the conditions in (1.71) and $E_-(0) = E_+(0) = I_p$, then there exists exactly one mvf $A \in \mathcal{E} \cap \mathcal{U}^\circ(J_p)$ such that (1.70) and (1.75) hold.

If $A_d(\lambda)$ is the monodromy matrix of a regular canonical system (1.20) with $J = J_p$ and $M(x)$ subject to (1.21) and $\mathfrak{E}_d$ is the de Branges matrix defined in terms of A_d by formula (1.70), then

$$\Sigma_{\rm psf}^d(dM) = (\mathcal{B}(\mathfrak{E}))_{\rm sf} \neq \emptyset$$

and

$$(\mathcal{H}(A))_{\rm sf} = (\mathcal{B}(\mathfrak{E}))_{\rm sf} \Longleftrightarrow \text{(1.75) is in force;}$$

see Lemma 7.20.

1.13 Bitangential direct and inverse input impedance and spectral problems

The bitangential direct and inverse input impedance and spectral problems that are considered in Chapters 7 and 11 are related to interpolation problems in the Carathéodory class in much the same way that input scattering problems are related to interpolation problems in the Schur class. The main differences are:

1. The system (1.20) is considered with $J = J_p$ instead of $J = j_{pq}$ and the matrizant is denoted $A_t(\lambda)$ instead of $W_t(\lambda)$.
2. The set $\{b_3, b_4\} \in ap(B)$ for $B = A\mathfrak{V}$ and $A \in \mathcal{U}(J_p)$ will be considered (instead of $\{b_1, b_2\} \in ap(W)$ for the mvf $W = \mathfrak{V}A\mathfrak{V}$ in the class $\mathcal{U}(j_p)$). This new set of associated pairs is defined by the factorizations

$$(b_{21}^{\#})^{-1} = b_3\varphi_3 \quad \text{and} \quad b_{22}^{-1} = \varphi_4 b_4, \tag{1.76}$$

where b_{ij} are $p \times p$ blocks of the mvf $B(\lambda) = A(\lambda)\mathfrak{V}$, b_3 and b_4 are $p \times p$ inner mvf's and φ_3 and φ_4 are outer mvf's in the **Smirnov class**

$$\mathcal{N}_+^{p\times p} = \{g^{-1}h : g \in \mathcal{S}_{\rm out}^{1\times 1} \quad \text{and} \quad h \in \mathcal{S}^{p\times p}\}. \tag{1.77}$$

The notation $\{b_3, b_4\} \in ap_{II}(A)$ and $\{b_1, b_2\} \in ap_I(A)$ will be used to distinguish between these two sets of pairs; and they will be called **associated pairs of the second kind** for $A \in \mathcal{U}(J_p)$ and **associated pairs of the first kind** for $A \in \mathcal{U}(J_p)$, respectively.
3. The generalized Carathéodory interpolation problem GCIP$(b_3, b_4; c^\circ)$ with given mvf's $b_3 \in \mathcal{S}_{\rm in}^{p\times p}$, $b_4 \in \mathcal{S}_{\rm in}^{p\times p}$ and $c^\circ \in \mathcal{C}^{p\times p}$, which is to describe the set

$$\mathcal{C}(b_3, b_4; c^\circ) = \{c \in \mathcal{C}^{p\times p} : b_3^{-1}(c - c^\circ)b_4^{-1} \in \mathcal{N}_+^{p\times p}\}, \tag{1.78}$$

is considered instead of the GSIP.
4. The subclass

$$\mathcal{C}_{sz}^{p\times p} = \left\{c \in \mathcal{C}^{p\times p} : \int_{-\infty}^{\infty} \frac{|\ln\det\{c(\mu) + c(\mu)^*\}|}{1+\mu^2} d\mu < \infty\right\} \tag{1.79}$$

is considered instead of the class $\mathcal{S}_{sz}^{p\times q}$.

A mvf $A \in \mathcal{U}(J_p)$ for which

$$\mathcal{C}(b_3, b_4; c^\circ) = \mathcal{C}(A) \tag{1.80}$$

is called a **resolvent** matrix of the GCIP$(b_3, b_4; c^\circ)$.

The GCIP$(b_3, b_4; c^\circ)$ is said to be **completely indeterminate** if for every nonzero vector $\eta \in \mathbb{C}^p$

$$c(\lambda)\eta \not\equiv c^\circ(\lambda)\eta \quad \text{for at least one mvf } c \in \mathcal{C}(b_3, b_4; c^\circ). \tag{1.81}$$

If the GCIP$(b_3, b_4; c^\circ)$ is completely indeterminate, then there exists an essentially unique mvf $A \in \mathcal{U}(J_p)$ such that

$$\mathcal{C}(b_3, b_4; c^\circ) = \mathcal{C}(A) \quad \text{and} \quad \{b_3, b_4\} \in ap_{II}(A). \tag{1.82}$$

If $c^\circ \in \mathcal{C}_{sz}^{p\times p}$, then the GCIP$(b_3, b_4; c^\circ)$ is completely indeterminate. Conversely, if the GCIP$(b_3, b_4; c^\circ)$ is completely indeterminate, then $c = T_A[I_p]$ belongs to $\mathcal{C}_{sz}^{p\times p}$. Moreover, if $A \in \mathcal{U}(J_p)$, then:

(1) A is a resolvent matrix for a completely indeterminate GCIP if and only if $A \in \mathcal{U}_{rR}(J_p)$;
(2) if (1.82) holds, then

$$b_3 \in \mathcal{E} \cap \mathcal{S}_{\text{in}}^{p\times p} \text{ and } b_4 \in \mathcal{E} \cap \mathcal{S}_{\text{in}}^{p\times p} \Longleftrightarrow A \in \mathcal{E} \cap \mathcal{U}(J). \tag{1.83}$$

Thus, to each completely indeterminate GCIP$(b_3, b_4; s^\circ)$ based on a pair of entire inner mvf's b_3 and b_4, there exists exactly one mvf $A \in \mathcal{E} \cap \mathcal{U}^\circ(J_p)$ such that (1.82) holds. Moreover, this resolvent matrix automatically belongs to the class $\mathcal{U}_{rR}(J_p)$. Formulas for this resolvent matrix in terms of the given data are presented in Chapter 11 when the interpolation problem is **strictly completely indeterminate**, i.e., when

$$\mathcal{C}(b_3, b_4; c^\circ) \cap \mathring{\mathcal{C}}^{p\times p} \neq \emptyset, \tag{1.84}$$

where

$$\mathring{\mathcal{C}}^{p\times p} = \{c \in \mathcal{C}^{p\times p} : c \in H_\infty^{p\times p} \quad \text{and } (\Re c)^{-1} \in L_\infty^{p\times p}\}. \tag{1.85}$$

If b_3 and b_4 are entire and (1.82) is in force, then (1.85) will hold if and only if $A \in \mathcal{U}_{rsR}(J_p)$, since

$$\mathcal{U}_{rsR}(J_p) = \{A \in \mathcal{U}(J_p) : \mathcal{C}(A) \cap \mathring{\mathcal{C}}^{p\times p} \neq \emptyset\}. \tag{1.86}$$

The constraint $c^\circ \in \mathring{\mathcal{C}}^{p\times p}$ is a sufficient condition for (1.84) to hold. However, it is not a necessary condition for (1.84) to hold. Thus, for example, (1.84) will hold if b_3 and b_4 are entire and $c^\circ \in \mathcal{C}^{p\times p}$ admits a representation of the form

$$c^\circ(\lambda) = \gamma + \int_0^\infty e^{i\lambda t} h(t)dt \quad \text{for } \lambda \in \overline{\mathbb{C}_+}, \text{ with } h \in L_1^{p\times p}(\mathbb{R}_+) \text{ and } \Re\gamma > 0;$$

see, e.g., Theorem 11.10.

If $A_x(\lambda)$, $0 \leq x < d$, is the matrizant of the canonical integral system (1.20) with $J = J_p$ and mass function $M(x)$ that meets the constraints (1.21), then there exists exactly one normalized nondecreasing chain $\{b_3^x, b_4^x\}$, $0 \leq x < d$, of pairs of entire inner $p \times p$ mvf's such that

$$\{b_3^x, b_4^x\} \in ap_{II}(A_x) \quad \text{for } 0 \leq x < d. \tag{1.87}$$

Moreover, if

$$A_x \in \mathcal{U}_{rR}(J_p) \quad \text{for } 0 \leq x < d, \tag{1.88}$$

then, as will be shown in Theorem 5.13, this chain is also continuous. The **bitangential direct input impedance** problem is to clarify the connections between the properties of the chains A_x and $\{b_3^x, b_4^x\}$ for $0 \leq x < d$ and the set $\mathcal{C}_{\text{imp}}^d(dM)$. A Weyl–Titchmarsh like classification of the systems (1.20) with $J = J_p$ and mass function $M(x)$ subject to the constraints (1.21) is obtained by considering the ranks of the left and right semiradii of the limit ball that is defined in Section 7.2. The Weyl characterization of the unique input impedance for the case when the left semiradius is zero is given. Bitangential direct input impedances and spectral functions are studied in Chapter 7. The identification of the matrizant $A_x(\lambda)$ as the normalized resolvent matrix of the completely indeterminate GCIP$(b_3^x, b_4^x; c^\circ)$ that satisfies the condition (1.87) is exploited to solve the bitangential inverse input impedance problem in Chapter 11.

1.14 Krein extension problems and Dirac systems

Let $\mathcal{G}_\infty^{p\times p}$ denote the class of continuous $p \times p$ mvf's on $\mathbb{R}$ for which $g(-t) = g(t)^*$ and the kernel

$$k(t, s) = g(t - s) - g(t) - g(-s) + g(0) \tag{1.89}$$

is positive. A mvf $g \in \mathcal{G}_\infty^{p\times p}$ will be called a **helical** mvf on $\mathbb{R}$. The formula

$$c(\lambda) = \lambda^2 \int_0^\infty e^{i\lambda t} g(t) dt \quad \text{for } \lambda \in \mathbb{C}_+ \tag{1.90}$$

defines a one-to-one correspondence between mvf's $c \in \mathcal{C}^{p\times p}$ and

$$g \in \mathcal{G}_\infty^{p\times p}(0) \stackrel{\text{def}}{=} \{g \in \mathcal{G}_\infty^{p\times p} : g(0) \leq 0\}. \tag{1.91}$$

The problem of extending a helical mvf from a finite interval $[-a, a]$ to all of $\mathbb{R}$ (due to M.G. Krein) is equivalent to a GCIP$(b_3, b_4; c^\circ)$ with $b_3 = e^{i\lambda a_1} I_p$, $b_4 = e^{i\lambda a_2} I_p$, $a_1 \geq 0$, $a_2 \geq 0$ and $a_1 + a_2 = 2a$. In the bitangential generalization of this problem that is considered in Chapter 9, b_3 and b_4 are taken to be arbitrary entire inner mvf's of exponential type a_1 and a_2, respectively.

A bitangential generalization of another extension problem that originates with M.G. Krein and corresponds to restricting c° to be of the form

$$c(\lambda) = I_p + 2\int_0^\infty e^{i\lambda t} h(t)dt \quad \text{for some } h \in L_1^{p\times p}(\mathbb{R}_+) \tag{1.92}$$

is also considered in Chapter 9. If $h(t)$ is extended from

$$\mathbb{R}_+ = [0, \infty) \quad \text{to} \quad \mathbb{R}_- = (-\infty, 0]$$

by setting $h(-t) = h(t)^*$ a.e., then a mvf c of the form (1.92) will be in the Carathéodory class if and only if

$$\int_0^a \varphi(t)^* \left\{ \varphi(t) + \int_0^a h(t-s)\varphi(s)ds \right\} dt \geq 0 \tag{1.93}$$

for every $\varphi \in L_2^p([0, a])$ and every $a > 0$.

Krein called the mvf h in the integral representation (1.92) the **accelerant** of c and considered the problem of extending a given mvf h° from the class

$$\mathcal{A}_a^{p\times p}(I_p) = \{h \in L_1^{p\times p}((-a, a)) : \text{(1.93) holds for every } \varphi \in L_2^p((0, a))\} \tag{1.94}$$

with $a < \infty$ to mvf's in the class $\mathcal{A}_\infty^{p\times p}(I_p)$ and obtained a description of the set

$$\mathcal{A}(h^\circ; a) = \{h \in \mathcal{A}_\infty^{p\times p} : h(t) = h^\circ(t) \quad \text{a.e. on } [-a, a]\}$$

when h° belongs to the class

$$\mathring{\mathcal{A}}_a^{p\times p} = \left\{ h \in L_1^{p\times p}((-a, a)) : \text{(1.93) holds with lower bound} \right. \\ \left. \varepsilon \int_0^a \varphi(t)^*\varphi(t)dt \quad \text{and } \varepsilon > 0 \text{ in place of } 0 \right\}. \tag{1.95}$$

Krein exploited the fact that the input impedance of a Dirac system with summable potential admits an integral representation of the form (1.92) to reformulate the inverse spectral problem for Dirac systems and second-order differential systems that can be reduced to Dirac systems as extension problems from mvf's in $\mathcal{A}_a^{p\times p}$ to mvf's in $\mathcal{A}_\infty^{p\times p}$. Krein viewed these extension problems as continuous analogs of the classical Carathéodory extension problem, in which the first n coefficients in the power series expansion of a holomorphic function with nonnegative real part in the open unit disk $\mathbb{D}$ are specified.

1.15 Direct and inverse problems for Dirac–Krein systems

Chapter 12 focuses primarily on direct and inverse problems for DK (Dirac–Krein)-systems. If $J = j_{pq}$, then these are systems of the form

$$u'(x,\lambda) = i\lambda u(x,\lambda)\begin{bmatrix}\alpha I_p & 0\\ 0 & \beta I_q\end{bmatrix} j_{pq} + u(x,\lambda)\begin{bmatrix}0 & \mathfrak{v}(x)\\ \mathfrak{v}(x)^* & 0\end{bmatrix}, \quad 0 \le x < d, \tag{1.96}$$

where

$$\alpha \ge 0, \quad \beta \ge 0, \kappa \overset{\text{def}}{=} \alpha + \beta > 0 \quad \text{and } \mathfrak{v} \in L_{1,\text{loc}}([0,d)). \tag{1.97}$$

The study of these problems is connected with the Wiener classes

$$\mathcal{W}^{k\times\ell}(\gamma) = \left\{\gamma + \int_{-\infty}^{\infty} e^{i\mu t} h(\mu)d\mu : h \in L_1^{k\times\ell}(\mathbb{R})\right\}, \tag{1.98}$$

where $\gamma \in \mathbb{C}^{k\times\ell}$ and the subclasses $\mathcal{W}_\pm^{k\times\ell}(\gamma)$ of $\mathcal{W}^{k\times\ell}(\gamma)$ in which the support of h is restricted to $\mathbb{R}_\pm$. The matrizant $W_x(\lambda)$ of (1.96) admits two representations:

$$W_x(\lambda) = W_x^0(\lambda)\mathfrak{A}_r^x(\lambda) \quad \text{and} \quad W_x(\lambda) = \mathfrak{A}_\ell^x(\lambda)W_x^0(\lambda), \quad 0 \le x < d, \tag{1.99}$$

where

$$W_x^0(\lambda) = \begin{bmatrix} e_{\alpha x} I_p & 0\\ 0 & e_{-\beta x} I_q\end{bmatrix}, \quad 0 \le x < d, \tag{1.100}$$

is the matrizant of the system (1.96) with $\mathfrak{v}(x) \equiv 0$ and $\mathfrak{A}_r^x$ and $\mathfrak{A}_\ell^x$ belong to the classes $\mathcal{W}_r(j_{pq})$ and $\mathcal{W}_\ell(j_{pq})$ of j_{pq}-unitary mvf's on $\mathbb{R}$ that belong to $\mathcal{W}^{m\times m}(I_m)$ with block decompositions

$$\mathfrak{A}_r = \begin{bmatrix}\mathfrak{a}_- & \mathfrak{b}_-\\ \mathfrak{b}_+ & \mathfrak{a}_+\end{bmatrix} \quad \text{and} \quad \mathfrak{A}_\ell = \begin{bmatrix}\mathfrak{d}_- & \mathfrak{c}_+\\ \mathfrak{c}_- & \mathfrak{d}_+\end{bmatrix} \tag{1.101}$$

in which

$$\mathfrak{a}_-^{\pm 1}, \mathfrak{d}_-^{\pm 1} \in \mathcal{W}_-^{p\times p}(I_p), \quad \mathfrak{a}_+^{\pm 1}, \mathfrak{d}_+^{\pm 1} \in \mathcal{W}_+^{q\times q}(I_q), \quad \mathfrak{b}_- \in \mathcal{W}_-^{p\times q}(0),$$
$$\mathfrak{b}_+ \in \mathcal{W}_+^{q\times p}(0), \quad \mathfrak{c}_+ \in \mathcal{W}_+^{p\times q}(0) \quad \text{and} \quad \mathfrak{c}_- \in \mathcal{W}_-^{q\times p}(0). \tag{1.102}$$

In view of (1.99) and the special properties of $\mathfrak{A}_r^x$, it turns out that

$$W_x \in \mathcal{E} \cap \mathcal{U}_{rsR}(j_{pq}) \quad \text{and} \quad \{e_{\alpha x}I_p, e_{\beta x}I_q\} \in ap(W_x) \quad \text{for } 0 \le x < d. \tag{1.103}$$

Moreover, the mvf

$$\widetilde{W}_x(\lambda) = W_x(\lambda)W_x(0)^{-1}, \quad 0 \le x < d, \tag{1.104}$$

is the matrizant of the canonical differential system (1.1) with $J = j_{pq}$ and Hamiltonian

$$H(x) = W_x(0) \begin{bmatrix} \alpha I_p & 0 \\ 0 & \beta I_q \end{bmatrix} W_x(0)^* \quad \text{for } 0 \le x < d. \tag{1.105}$$

Because of the special properties of this connection, which will be explained in detail in Chapter 12, the study of direct and inverse problems for the DK-system (1.96) is equivalent to the study of these problems for canonical differential systems with Hamiltonians of the form (1.105). Moreover, as (1.103) holds for the matrizant $\widetilde{W}_x(\lambda)$, $0 \le x < d$, of the corresponding canonical differential system, the chain of associated pairs is inherited from the structure of the DK-system and need not be specified seperately. Thus, the main results for canonical differential systems with matrizants in the class $\mathcal{U}_{rsR}(j_{pq})$ are applicable to these canonical systems and hence also to the associated DK-systems.

If $d = \infty$ and $\mathfrak{v} \in L_1^{p\times q}(\mathbb{R}_+)$, then $\mathfrak{A}_\ell^x$ tends to a limit

$$\mathfrak{A}_\ell^\infty = \begin{bmatrix} \mathfrak{d}_-^\infty & \mathfrak{c}_+^\infty \\ \mathfrak{c}_-^\infty & \mathfrak{d}_+^\infty \end{bmatrix} \in \mathcal{W}_\ell(j_{pq})$$

in the Wiener norm as $x \uparrow \infty$. Moreover, the DK-system (1.96) is in the limit point case, i.e., there is only one input scattering matrix: $s_{\text{in}} = \mathfrak{c}_+^\infty(\mathfrak{d}_+^\infty)^{-1}$; and $s_{\text{in}} \in \mathcal{W}_+^{p\times q}(0) \cap \mathring{\mathcal{S}}^{p\times q}$.

Asymptotic scattering matrices s_ε^∞ for DK-systems are defined by considering the solutions

$$u(x, \lambda) = \begin{bmatrix} u_1(x, \lambda) & u_2(x, \lambda) \end{bmatrix}$$

of (1.96) with components of sizes $q \times p$ and $q \times q$ that are subject to the boundary condition

$$u_1(0, \lambda) = u_2(0, \lambda)\varepsilon(\lambda) \quad \text{for } \varepsilon \in \mathcal{S}^{q\times p} \tag{1.106}$$

and the asymptotic conditions

$$\|u_2(x, \cdot) - e_{-\beta x} I_q\|_\infty \to 0 \quad \text{as } x \uparrow \infty. \tag{1.107}$$

Then there exists exactly one mvf s_ε^∞ in the set

$$\mathbb{B}^{q\times p} = \{f \in L_\infty^{q\times p}(\mathbb{R}) : \|f\|_\infty \le 1\}$$

such that

$$\|u_1(x, \cdot) - e_{\alpha x} s_\varepsilon^\infty\|_\infty \to 0 \quad \text{as } x \uparrow \infty. \tag{1.108}$$

Furthermore, the set $\{s_\varepsilon^{q\times p} : \varepsilon \in \mathcal{S}^{q\times p}\}$ coincides with the set of solutions of the following version of the Nehari problem NP($\mathfrak{g}$) that is formulated in terms of a given mvf $\mathfrak{g} \in L_1^{q\times p}(\mathbb{R}_+)$:

Describe the set

$$\mathcal{N}(\mathfrak{g}) = \left\{ f \in \mathbb{B}^{q\times p} : f(\mu) - \int_0^\infty e^{-i\mu t}\mathfrak{g}(t)dt \text{ belongs to } H_\infty^{p\times q} \right\}.$$

It turns out that

$$\mathcal{N}(\mathfrak{g}) = \{(\varepsilon \mathfrak{c}_+^\infty + \mathfrak{d}_+^\infty)^{-1}(\varepsilon \mathfrak{d}_-^\infty + \mathfrak{c}_-^\infty) : \varepsilon \in \mathcal{S}^{q\times p}\}.$$

This connection between $\mathcal{N}(\mathfrak{g})$ and the set of asymptotic scattering matrices yields an algorithm for computing $\mathfrak{A}_\ell^\infty$, ε and s_{in}.

There are analogous connections of direct and inverse impedance and spectral problems when $q = p$. In particular, there is exactly one input impedance $c_{\text{in}} = T_{\mathfrak{V}}[s_{\text{in}}]$ such that

$$\sigma'(\mu) = \Re c_{\text{in}}(\mu) = (\mathfrak{c}_+^\infty(\mu) + \mathfrak{d}_+^\infty(\mu))^{-*}(\mathfrak{c}_+^\infty(\mu) + \mathfrak{d}_+^\infty(\mu))^{-1} \quad \text{on } \mathbb{R},$$

is also the unique spectral function of the corresponding DK-system. Moreover, $(\mathfrak{c}_+^\infty + \mathfrak{d}_+^\infty)^{\pm 1} \in \mathcal{W}_+^{p\times p}(I_p)$, $c_{\text{in}} \in \mathcal{W}_+^{p\times p}(I_p) \cap \mathring{\mathcal{C}}^{p\times p}$, $\sigma' \in \mathcal{W}^{p\times p}(I_p)$; and for each $x \in [0, \infty)$ $A_x(\lambda) = \mathfrak{V}W_x(\lambda)\mathfrak{V}$ may be identified with an appropriately chosen resolvent matrix of the problem GCIP$(e_{\kappa x}I_p, I_p; c_{\text{in}})$, or, equivalently of the extension problem of the accelerant h of c_{in} from $[-\kappa x, \kappa x]$ to $\mathbb{R}$. In fact, $h \in \mathring{\mathcal{A}}_\infty^{p\times p}$ and its restriction to any interval $[-\kappa x, \kappa x]$ has exactly one resolvent matrix $A_x(\lambda)$ (subject to some natural constraints). This resolvent matrix is the matrizant of a DK-system.

1.16 Supplementary notes

The connection between inverse problems for differential systems of first- and second-order that may be reduced to Dirac systems and continuous analogs of Schur, Carathéodory and Nehari extension problems was first discovered by M.G. Krein. He exploited this connection to find solutions of inverse problems. In [Kr51a] he wrote

> Just as to any Jacobi matrix J there corresponds a power moment problem the solutions of which (mass distribution functions) uniquely determine J, to any second order differential operator L with a boundary condition at one end point, there corresponds a "generalized" moment problem such that every distribution function of this problem completely determines the boundary conditions and the operator (if it is given in a certain "canonical" form). For operators of "sufficiently regular type" the "generalized moment problem is the extension problem for Hermitian positive functions" that was developed by the author in [Kr40], [Kr44a], [Kr44b] and [Kr49].

Krein presented his method and some of its applications in a number of short *Doklady* notes, mostly without proof [Kr51a], [Kr52a], [Kr54], [Kr55] and [Kr56]. Additional details were provided in [Kr52b] and in subsequent publications with

H. Langer [KrL85] and F.E. Melik-Adamyan [KrMA68], [KrMA70], [KrMA84] and [KrMA86].

Proofs of Krein's main results on the inverse spectral problem for the Feller–Krein string equation seem to have been first presented in [DMc76].

In the class of problems that Krein considered, the chain of associated pairs is

$$\{e^{i\lambda x\alpha}I_p, e^{i\lambda x\beta}I_q\}, \quad 0 \le x < d,$$

where the parameters, $\alpha \ge 0$, $\beta \ge 0$ with $\alpha + \beta > 0$, are uniquely defined by the DK-system to which the given system is reduced.

In the general setting of canonical integral and differential systems that are studied in this book, the associated pairs of a matrizant may be any normalized nondecreasing continuous chain of pairs of entire inner mvf's of sizes $p \times p$ and $q \times q$. In these inverse problems, the given data includes such a chain, in addition to the usually specified frequency characteristic; the survey articles [ArD04a], [ArD05b] and [ArD08a] provide additional perspective.

2

Canonical systems and related differential equations

This chapter focuses on the connections between canonical integral and differential systems and the properties of their matrizants and monodromy matrices. The necessary properties of entire J-inner mvf's will be presented. A number of differential systems of first order and differential equations of second order that can be reduced to Dirac systems will be considered: The Dirac–Krein differential system with potential, the Feller–Krein string equation and the Schrödinger equation with Hermitian potential $q(x)$ such that the Riccati equation $q(x) = v'(x) + v(x)^2$ (or $q(x) = v'(x) - v(x)^2$) has a locally summable Hermitian solution $v(x)$.

2.1 Canonical integral systems

Recall that a **canonical integral system** is an integral system of the form

$$y(t, \lambda) = y(0, \lambda) + i\lambda \int_0^t y(s, \lambda) dM(s) J, \qquad 0 \leq t < d, \tag{2.1}$$

where J is an $m \times m$ signature matrix, the **mass function** $M(t)$ is subject to the constraints (1.21) and the solution $y_t(\lambda) = y(t, \lambda)$ is a $k \times m$ mvf that is continuous in t on the interval $[0, d)$ for each fixed $\lambda \in \mathbb{C}$. The usual choices of J are $\pm I_m$, the signature matrices $\pm j_{pq}$, $\pm J_p$, $\pm \mathcal{J}_p$ that are defined in (1.4) and $\pm j_p \stackrel{\text{def}}{=} \pm j_{pp}$.

Lemma 2.1 *If $\lambda \in \mathbb{C}$, $y^\circ \in \mathbb{C}^{k\times m}$ and the mass function $M(s)$ in (2.1) satisfies the conditions in (1.21), then there exists exactly one continuous solution $y(t, \lambda)$ of the canonical integral system (2.1) with this mass function on $[0, d)$ such that $y(0, \lambda) = y^\circ$. This mvf $y(t, \lambda)$ is entire in the variable λ for each fixed $t \in [0, d)$ and is subject to the bound*

$$\|y(t, \lambda)\| \leq \|y^\circ\| \exp\{|\lambda| \operatorname{trace} M(t)\} \quad \textit{for } 0 \leq t < d. \tag{2.2}$$

The solution $y(t,\lambda)$ may be obtained by the method of successive approximation:

$$y(t,\lambda) = \lim_{n\uparrow\infty}(t,\lambda) \quad \text{for } 0 \le t < d. \tag{2.3}$$

where

$$y_n(t,\lambda) = y^\circ + i\lambda \int_0^t y_{n-1}(s,\lambda)dM(s)J \quad \text{for } 0 \le t < d \text{ and } n \ge 1 \tag{2.4}$$

and

$$y_0(t,\lambda) \equiv y^\circ \quad \text{for } 0 \le t < d. \tag{2.5}$$

Proof The proof is broken into three parts:

1. **Existence:** Let

$$u_n(t,\lambda) = y_n(t,\lambda) - y_{n-1}(t,\lambda) \quad \text{for } 0 \le t < d \text{ and } n \ge 1,$$

where the mvf's $y_n(t,\lambda)$ are defined by formulas (2.4) and (2.5) and let

$$\psi(t) = \operatorname{trace} M(t) \quad \text{for } 0 \le t < d.$$

Then

$$u_n(t,\lambda) = i\lambda \int_0^t u_{n-1}(s,\lambda)dM(s)J \quad \text{for } 0 \le t < d \text{ and } n > 1,$$

$$u_1(t,\lambda) = i\lambda y^\circ M(t)J \quad \text{for } 0 \le t < d,$$

and $\psi(t)$ is a continuous nondecreasing scalar-valued function on $[0,d)$. Our first main objective is to establish the bound

$$\|u_k(t,\lambda)\| \le \frac{|\lambda|^k \psi(t)^k}{k!}\|y^\circ\| \quad \text{for } 0 \le t < d \text{ and } k \ge 1. \tag{2.6}$$

The proof is by induction. The case $k = 1$ is self-evident. Suppose next that the bound is valid for $k = n-1$ and let $0 = t_0 < t_1 < \cdots < t_\ell < t < d$. Then the inequalities

$$\begin{aligned}
&\left\|\sum_{j=1}^{\ell} u_{n-1}(t_{j-1},\lambda)[M(t_j) - M(t_{j-1})]J\right\| \\
&\quad \le \sum_{j=1}^{\ell} \|u_{n-1}(t_{j-1})\|\,\|M(t_j) - M(t_{j-1})\|\,\|J\| \\
&\quad \le \sum_{j=1}^{\ell} \frac{|\lambda|^{n-1}\psi(t_{j-1})^{n-1}[\psi(t_j) - \psi(t_{j-1})]\|y^\circ\|}{(n-1)!} \\
&\quad \le \frac{|\lambda|^{n-1}}{(n-1)!}\|y^\circ\| \int_0^t \psi(s)^{n-1} d\psi(s) \\
&\quad = \frac{|\lambda|^{n-1}\psi(t)^n}{n!}\|y^\circ\|
\end{aligned}$$

imply that

$$\left\| \int_0^t u_{n-1}(s,\lambda)dM(s)J \right\| \le \frac{|\lambda|^{n-1}\psi(t)^n}{n!}\|y^\circ\| \quad \text{for } 0 \le t < d,$$

which, together with the recursion for $u_n(t,\lambda)$, justifies the bound (2.6) for $k = n$ also, and hence by induction for every integer $k \ge 1$.

Moreover, since the mvf's $u_n(t,\lambda)$ are continuous in t on $[0,d)$ and polynomials in λ and, by the preceding estimates,

$$\sum_{n=1}^{\infty} u_n(t,\lambda)$$

converges uniformly for $t \in [0, d_1]$ and $|\lambda| \le R$ when $d_1 < d$ and $R < \infty$, it follows that

$$y_n(t,\lambda) = y^\circ + \sum_{k=1}^{n} u_k(t,\lambda) = y^\circ + \int_0^t y_{n-1}(s,\lambda)dM(s)J$$

converges uniformly to a solution $y(t,\lambda)$ of the integral equation (2.1) that is continuous in the variable t on the interval $[0,d)$ for each fixed $\lambda \in \mathbb{C}$ and entire in the variable λ for each fixed $t \in [0,d)$. The bound

$$\begin{aligned} \|y_n(t,\lambda)\| &\le \|y^\circ\| + \sum_{k=1}^{n} \|y^\circ\| \frac{|\lambda|^k}{k!}\psi(t)^k \\ &\le \|y^\circ\| \exp\{|\lambda|\psi(t)\} \end{aligned}$$

is used to justify bringing the limit inside the integral and also serves to justify the stated inequality (2.2).

2. Uniqueness: Let $\widetilde{y}(t,\lambda)$ be any other continuous solution of the equation (2.1) on the interval $[0,d)$ with initial condition $\widetilde{y}(0,\lambda) = y^\circ$. Then the mvf

$$z(t,\lambda) = y(t,\lambda) - \widetilde{y}(t,\lambda) \quad \text{for } 0 \le t < d,$$

is a continuous solution of the equation

$$z(t,\lambda) = i\lambda \int_0^t z(s,\lambda)dM(s)J \quad \text{for } 0 \le t < d.$$

Fix $t < d$, choose λ so that $|\lambda|\,\psi(t) < 1$ and let

$$\delta_t = \max\{\|z(s,\lambda)\| : s \in [0,t]\}.$$

Then, since

$$\|z(s,\lambda)\| \le |\lambda|\delta_s\psi(s) \le |\lambda|\delta_t\psi(t),$$

it follows that

$$(1 - |\lambda|\psi(t))\delta_t \le 0.$$

Therefore, $\delta_t = 0$. But this implies that $z_t(\lambda) = 0$ in a neighborhood of zero, and hence, as z_t is entire, $z_t(\lambda) = 0$ in the whole complex plane. Thus, the solution $y(t, \lambda)$ that was obtained by successive approximation in the earlier part of the proof is in fact the only solution of the canonical integral system (6.1) that is continuous in the variable t on the interval $[0, d)$ for each fixed $\lambda \in \mathbb{C}$.

3. The bound (2.2): The bound (2.2) was established for the solution that was obtained by successive approximation in Part 1. However, in view of Part 2, the integral equation (2.1) has only one continuous solutions and hence it is subject to this bound. □

2.2 Connections with canonical differential systems

If the mass function $M(t)$, $0 \le t < d$, in the canonical integral system (2.1) is locally absolutely continuous on the interval $[0, d)$, and $M'(t) = H(t)$ a.e. on $[0, d)$, then the system (2.1) can be written as

$$y(t, \lambda) = y(0, \lambda) + i\lambda \int_0^t y(s, \lambda)H(s)dsJ \quad \text{for } 0 \le t < d. \tag{2.7}$$

Thus, in this case the unique continuous solution $y(t, \lambda)$ of system (2.1) is locally absolutely continuous in t on the interval $[0, d)$ for each fixed $\lambda \in \mathbb{C}$ and it is a solution of the canonical differential system

$$y'(t, \lambda) = i\lambda y(t, \lambda)H(t)J, \quad \text{a.e. on } 0 \le t < d, \tag{2.8}$$

where the Hamiltonian $H(t)$ satisfies the conditions

$$H(t) \ge 0 \quad \text{a.e. on } [0, d), \quad \text{and} \quad H \in L_{1,\text{loc}}^{m\times m}([0, d)). \tag{2.9}$$

Conversely, if $y(t, \lambda)$, $0 \le t < d$, is a locally absolutely continuous solution of a canonical differential system of the form (2.8) in which the mvf $H(t)$ satisfies the conditions (2.9), then $y(t, \lambda)$ is also a continuous solution of the canonical integral system (2.1) with locally absolutely continuous mass function $M(t) = \int_0^t H(s)ds$.

A canonical integral system (2.1) may always be reduced to a corresponding canonical differential system even when the mass function $M(t)$, $0 \le t < d$, is not locally absolutely continuous:

Lemma 2.2 *Let $M(t)$ be a continuous nondecreasing $m \times m$ mvf on the interval $[0, d]$ with $M(0) = 0$ and let*

$$\psi(t) = \text{trace}\, M(t). \tag{2.10}$$

Then $\psi(t)$ is a continuous nondecreasing function on $[0, d]$ with $\psi(0) = 0$ and $M(t)$ is absolutely continuous with respect to $\psi(t)$ on the interval $0 \le t \le d$.

Proof The claim follows from the bound

$$\|M(t_2) - M(t_1)\| \leq \text{trace}\{M(t_2) - M(t_1)\} = \psi(t_2) - \psi(t_1), \tag{2.11}$$

which is valid for $0 \leq t_1 \leq t_2 < d$. □

If $\psi(t)$ is a strictly increasing continuous function on $[0, d)$ with $\psi(0) = 0$ and

$$\ell = \lim_{t \uparrow d} \psi(t),$$

then there exists a continuous strictly increasing function $\varphi(x)$ on $[0, \ell)$ such that

$$\varphi(\psi(t)) = t \text{ for } 0 \leq t < d \text{ and } \psi(\varphi(x)) = x \text{ for } 0 \leq x < \ell.$$

Let $y(t, \lambda)$, $0 \leq t < d$, be a continuous solution of the system (2.1) and

$$u(x, \lambda) = y(\varphi(x), \lambda), \qquad 0 \leq x < \ell.$$

Then $u(x, \lambda)$ is a continuous solution of the canonical integral system with mass function

$$\widetilde{M}(x) = M(\varphi(x)) \tag{2.12}$$

on the interval $[0, \ell)$:

$$u(x, \lambda) = u(0, \lambda) + i\lambda \int_0^x u(s, \lambda) d\widetilde{M}(s) J, \quad 0 \leq x < \ell. \tag{2.13}$$

Moreover, since (2.11) holds, the mvf $\widetilde{M}(x)$ is locally absolutely continuous on the interval $[0, \ell)$ and $\widetilde{M}(0) = M(\varphi(0)) = M(0) = 0$. Consequently,

$$\widetilde{M}(x) = \int_0^x H(s) ds, \ \ 0 \leq x < \ell, \ \ \text{where } H(x) = \widetilde{M}'(x) \text{ a.e.on } [0, \ell),$$

and hence $u(x, \lambda)$ is a locally absolutely continuous solution of the canonical differential system

$$u'(x, \lambda) = i\lambda u(x, \lambda) H(x) J \quad \text{a.e. on } [0, \ell), \tag{2.14}$$

where the $m \times m$ mvf $H(x)$, the Hamiltonian of the system, satisfies the conditions

$$H \in L_{1,\text{loc}}^{m \times m}([0, \ell)) \quad \text{and} \quad H(x) \geq 0 \quad \text{a.e. on } [0, \ell). \tag{2.15}$$

Moreover, since

$$\text{trace}\, \widetilde{M}(x) = \text{trace}\, M(\varphi(x)) = \text{trace}\, M(\varphi(\psi(t))) = \text{trace}\, M(t) = x, \tag{2.16}$$

it follows that

$$\text{trace}\, H(x) = 1 \quad \text{a.e. on the interval } [0, \ell).$$

Conversely, if $u(x, \lambda)$ is a locally absolutely continuous solution of the system (2.14), then it is a solution of the integral system (2.13) with mass function $\widetilde{M}(x)$

defined in (2.12). Consequently, the function

$$y(t,\lambda) = u(\psi(t),\lambda), \qquad 0 \le t < d$$

is the continuous solution of the original canonical integral system (2.1) with mass function $M(t) = \widetilde{M}(\psi(t)),\ 0 \le t < d$. Moreover,

$$y(0,\lambda) = u(0,\lambda),$$

since $\psi(0) = 0$.

If trace $M(t)$ is not a strictly increasing function on $[0, d]$, then it turns out to be more convenient (though less traditional) to consider the strictly increasing function

$$x(t) = \text{trace}\, M(t) + t \quad \text{for } 0 \le t < \ell. \tag{2.17}$$

Then $x(t)$ is a strictly increasing continuous function on $[0, d)$ and the inverse function $t = \vartheta(x)$ is a strictly increasing continuous function on $[0, \ell)$, where

$$\ell = d + \lim_{t \uparrow d} \text{trace}\, M(t) \le \infty.$$

The mvf $M(t)$ is absolutely continuous with respect to $x(t)$, i.e.,

$$\widetilde{M}(x) \stackrel{\text{def}}{=} M(\vartheta(x)) = \int_0^x \widetilde{H}(s)ds$$

and

$$\text{trace}\,\{\widetilde{M}(x_2) - \widetilde{M}(x_1)\} = x_2 - x_1 - \{\vartheta(x_2) - \vartheta(x_1)\} \le x_2 - x_1$$

for $0 \le x_1 \le x_2 < \ell$. Consequently, $\widetilde{M}(x)$ is locally absolutely continuous on $[0, \ell)$ and

$$\widetilde{M}'(x) = \widetilde{H}(x) \quad \text{with trace}\, \widetilde{H}(x) \le 1 \ \text{ a.e. on } [0, \ell).$$

Thus, the vvf

$$u(x,\lambda) = y(\vartheta(x),\lambda),\ \ 0 \le x < \ell,$$

is a locally absolutely continuous solution of the canonical differential system

$$u'(x,\lambda) = i\lambda u(x,\lambda)\widetilde{H}(x)J \quad \text{for } 0 \le x < \ell$$

with

$$u(0,\lambda) = y(0,\lambda).$$

Conversely, if $u(x,\lambda)$ is the solution of this system, then $y(t,\lambda) = u(x(t),\lambda)$, $0 \le t < d$, is the continuous solution of the integral system (2.7).

2.3 The matrizant and its properties

The continuous solution $U_t(\lambda) = U(t,\lambda)$, $0 \le t < d$, of the integral equation

$$U(t,\lambda) = I_m + i\lambda \int_0^t U(s,\lambda)dM(s)J, \quad 0 \le t < d \tag{2.18}$$

is called the **matrizant** (or fundamental solution) of the system (2.1). The existence and uniqueness of the matrizant and the fact that it is an entire $m \times m$ mvf in the variable λ is established in Lemma 2.1. Moreover, in view of the uniqueness, every continuous solution $y(t,\lambda)$ of the system (2.1) can be expressed as

$$y(t,\lambda) = y(0,\lambda)U(t,\lambda), \quad 0 \le t < d. \tag{2.19}$$

A number of important properties of the matrizant may be obtained from the following fact:

Lemma 2.3 *The matrizant $U_t(\lambda)$, $0 \le t < d$, of the system (2.1) satisfies the identity*

$$J - U_t(\lambda)JU_t(\omega)^* = -i(\lambda - \overline{\omega}) \int_0^t U_s(\lambda)dM(s)U_s(\omega)^*\,, \quad 0 \le t < d, \tag{2.20}$$

for every choice of $\lambda, \omega \in \mathbb{C}$.

Proof In Section 2.2, it was shown that there is no loss of generality in assuming that $M(t)$ is locally absolutely continuous on $[0,d)$, because this can always be achieved by introducing an appropriate change of the independent variable. Granting this, we may replace $dM(s)$ by $H(s)ds$ and then (2.20) follows from (1.10). □

Remark 2.4 *Formula (2.20) serves to identify the kernel*

$$K_\omega^{U_t}(\lambda) = \begin{cases} \dfrac{J - U_t(\lambda)JU_t(\omega)^*}{-2\pi i(\lambda - \overline{\omega})} & \text{for } \lambda \ne \overline{\omega} \text{ and } 0 \le t < d \\[2ex] \dfrac{1}{2\pi i}\left(\dfrac{\partial U_t}{\partial \lambda}\right)(\overline{\omega})JU_t(\omega)^* & \text{for } \lambda = \overline{\omega} \text{ and } 0 \le t < d \end{cases} \tag{2.21}$$

based on the matrizant as a positive kernel in the sense of (1.12); since

$$K_\omega^{U_t}(\lambda) = \frac{1}{2\pi}\int_0^t U_s(\lambda)dM(s)U_s(\omega)^* \quad \text{for } 0 \le t < d.$$

Theorem 2.5 *Let $U_t(\lambda) = U(t,\lambda)$, $0 \le t < d$, be the matrizant of the canonical integral system (2.1) with a continuous nondecreasing mass function $M(t)$ with $M(0) = 0$. Then the family of mvf's $\{U_t\}$, $0 \le t < d$, enjoys the following properties:*

(1) $U_t \in \mathcal{E} \cap \mathcal{U}^\circ(J)$ *for each fixed* $t \in [0, d)$ *and* $U_0(\lambda) = I_m$.
(2) *The family* $\{U_t\}$, $0 \le t < d$, *is nondecreasing in the sense that*

$$U_{t_1}^{-1} U_{t_2} \in \mathcal{E} \cap \mathcal{U}(J) \;\; if \;\; 0 \le t_1 \le t_2 < d. \qquad (2.22)$$

(3) *The mvf* $U_t(\lambda)$ *is continuous in* t *on the interval* $[0, d)$ *for each fixed* $\lambda \in \mathbb{C}$.
(4) $\|U_t(\lambda)\| \le \exp\{|\lambda| \operatorname{trace} M(t)\}$ *for* $0 \le t < d$.
(5) $\det U_t(\lambda) = \exp\{i\lambda \operatorname{trace} M(t)J\}$.

Proof Assertions (3), (4) and the fact that $U_t(\lambda)$ is an entire mvf in the variable λ are immediate from the definition of the matrizant and Lemma 2.1; (2) and the rest of (1) follow easily from Lemma 2.3: Formula (2.20) implies that $U_t \in \mathcal{U}(J)$ for $t \in [0, d)$ and that

$$U_{t_1}(\lambda)JU_{t_1}(\lambda)^* - U_{t_2}(\lambda)JU_{t_2}(\lambda) = 2(\Im\lambda) \int_{t_1}^{t_2} U_s(\lambda)dM(s)U_s(\lambda)^*. \qquad (2.23)$$

Therefore,

$$U_{t_1}(\lambda)JU_{t_1}(\lambda)^* - U_{t_2}(\lambda)JU_{t_2}(\lambda)^* \ge 0 \text{ if } 0 \le t_1 \le t_2 < d \text{ and } \lambda \in \mathbb{C}_+$$

and hence, as $U_t(\lambda)JU_t^\#(\lambda) = J$, $U_t(\lambda)$ is invertible for every point $\lambda \in \mathbb{C}$,

$$U_t(\lambda)^{-1} = JU_t^\#(\lambda)J$$

is entire, and (2.22) holds too.

To verify (5), let $\psi_t(\lambda) = \det U_t(\lambda)$ and suppose first that $M(t)$ is locally absolutely continuous on $[0, d)$ with $M'(t) = H(t)$ a.e. on $[0, d)$. Then, since $\det U_t(\lambda)$ is linear in each row of $U_t(\lambda)$ separately, it is readily checked that

$$\psi_t'(\lambda) = \operatorname{trace}\left\{U_t'(\lambda)U_t(\lambda)^{-1}\right\}\psi_t(\lambda) = \operatorname{trace}\left\{U_t(\lambda)^{-1}U_t'(\lambda)\right\}\psi_t(\lambda)$$

and hence that

$$\frac{\psi_t'(\lambda)}{\psi_t(\lambda)} = \operatorname{trace}\{i\lambda H(t)J\}.$$

Consequently,

$$\int_0^t \frac{d}{ds} \ln \psi_s(\lambda)ds = \int_0^t \operatorname{trace}\{i\lambda H(s)J\}ds = \operatorname{trace}\left\{\int_0^t i\lambda H(s)Jds\right\},$$

which leads easily to the advertised formula in the setting under consideration.

□

A family of mvf's $\{U_t\}$, $0 \le t < d$, for which properties (1)–(3) of Theorem 2.5 hold is called a **normalized nondecreasing continuous chain of entire J-inner (or inner, if $J = I_m$) mvf's**. This chain is called a **strictly increasing chain** if the ratio $U_{t_1}^{-1}U_{t_2}$ is not a constant mvf when $0 \le t_1 < t_2 < d$.

Remark 2.6 *A family of mvf's $U_t \in \mathcal{E} \cap \mathcal{U}(J)$, $0 \le t < d$, may also be nondecreasing in the sense that*

$$U_{t_2}U_{t_1}^{-1} \in \mathcal{E} \cap \mathcal{U}(J) \quad when \quad 0 \le t_1 \le t_2 < d. \tag{2.24}$$

To distinguish between these two senses, a family of mvf's $U_t \in \mathcal{E} \cap \mathcal{U}(J)$, $0 \le t < d$, that is nondecreasing in the sense of (2.22) (resp., (2.24)) will be referred to as a nondecreasing $\curvearrowright$ chain (resp., nondecreasing $\curvearrowleft$ chain).

Theorem 2.7 *Let $U_t(\lambda)$, $0 \le t < d$, be the matrizant of the canonical integral system (2.1). Then the mass function $M(t)$, $0 \le t < d$, of the system may be recovered from the matrizant by the formula*

$$M(t) = 2\pi K_0^{U_t}(0) = -i\left(\frac{\partial U_t}{\partial \lambda}\right)(0)J, \quad 0 \le t < d, \tag{2.25}$$

or, equivalently by the formula

$$U_t(\lambda) = I_m + i\lambda M(t)J + O(\lambda^2) \text{ as } \lambda \to 0 \textit{ for each fixed } t \in [0, d). \tag{2.26}$$

Proof Formula (2.1) implies that

$$\begin{aligned} U_t(\lambda) &= I_m + i\lambda \int_0^t \left[I_m + i\lambda \int_0^s U_{s'}(\lambda) dM(s')J \right] dM(s)J \\ &= I_m + i\lambda \int_0^t dM(s)J - \lambda^2 \int_0^t \int_0^s U_{s'}(\lambda) dM(s')J dM(s)J, \end{aligned}$$

which yields (2.26) and, subsequently (2.25). □

Conversely, the matrizant of a canonical integral system (2.1) with mass function $M(t)$ can be expressed as a multiplicative integral of that mass function; see Theorem 2.9 below.

The matrizant $U_x(\lambda) = U(x, \lambda)$ of the canonical differential system (2.14) is the locally absolutely continuous solution of the Cauchy problem

$$\begin{aligned} \frac{d}{dx} U(x, \lambda) &= i\lambda U(x, \lambda) H(x) J \quad \text{a.e. on } [0, \ell) \\ U(0, \lambda) &= I_m. \end{aligned}$$

Since $U_x(\lambda)$ is also the matrizant of the integral system (2.1) with $M(x) = \int_0^x H(s)ds$, it also has the properties listed in Theorem 2.5 and is even a locally absolutely continuous mvf on $[0, \ell)$ for each choice of $\lambda \in \mathbb{C}$.

The Hamiltonian $H(x)$ of the canonical differential system (2.14) is uniquely defined by the matrizant, for example by

$$H(x) = -U_x(i)^{-1} \left(\frac{\partial U_x}{\partial x}\right)(i)J \quad \text{a.e., on the interval} \quad [0, \ell).$$

If the differential system (2.14) is obtained from the integral system (2.1) by replacing the variable $t = \varphi(x)$, $0 \le x \le \ell$, as in Section 2.2, then the matrizants $U_x(\lambda)$ and $Y_t(\lambda)$ of these systems are connected by the formula

$$U(x,\lambda) = Y(\varphi(x),\lambda), \quad 0 \le x < \ell; \quad Y(t,\lambda) = U(\psi(t),\lambda), \quad 0 \le t < d. \tag{2.27}$$

If the mass function $M(t)$ is locally absolutely continuous on the interval $[0,d)$, then we may choose $\psi(t) = t$, $0 \le t < d$, and then $x = t$, $\ell = d$, $M(x) = M(t)$ and $Y(t,\lambda) = U(x,\lambda)$, $0 \le t < d$.

Remark 2.8 *If $U_t(\lambda)$, $0 \le t < d$, is the matrizant of the integral system (2.1) with a mass function $M(x)$ that is subject to (1.21), then the bounds*

$$\begin{aligned}\|M(t_2) - M(t_1)\| &= \left\| \frac{1}{2\pi} \int_0^{2\pi} \frac{U_{t_2}(Re^{i\theta}) - U_{t_1}(Re^{i\theta})}{Re^{i\theta}} d\theta J \right\| \\ &\le \frac{1}{2\pi R} \int_0^{2\pi} \|U_{t_2}(Re^{i\theta}) - U_{t_1}(Re^{i\theta})\| d\theta\end{aligned}$$

and

$$\begin{aligned}\|U_{t_2}(\lambda) - U_{t_1}(\lambda)\| &= \left\| i\lambda \int_{t_1}^{t_2} U_s(\lambda) dM(s) J \right\| \\ &\le |\lambda|\, e^{|\lambda| \operatorname{trace} M(t_2)} \|M(t_2) - M(t_1)\|\end{aligned}$$

imply that $M(t)$ is locally absolutely continuous on $[0,d)$ if and only if $U_t(\lambda)$ is locally absolutely continuous on $[0,d)$ for λ restricted to any compact subset of $\mathbb{C}$.

2.4 Regular case: Monodromy matrix

Recall that an integral system (2.1) is called a **regular canonical integral system** if (1.22) is in force. Analogously, a differential system (2.8) with Hamiltonian subject to (2.9) is called a **regular canonical differential system** if

$$d < \infty \quad \text{and} \quad H \in L_1^{m\times m}([0,d]). \tag{2.28}$$

The continuous solutions $y(t,\lambda)$ of regular canonical systems may be considered on the closed interval $[0,d]$ too. In particular, in this setting, the matrizant $U_t(\lambda)$, $0 \le t \le d$, is a normalized continuous nondecreasing $\curvearrowright$ chain of entire J-inner mvf's on the closed interval $[0,d]$ and the properties listed in Theorem 2.5 now hold for $0 \le t \le d$. The mvf

$$U(\lambda) = U_d(\lambda)$$

is called the **monodromy matrix** of the system (2.1)/(2.8) on the closed interval $[0,d]$. The monodromy matrix $U(\lambda)$ of a canonical integral/differential system is an entire J-inner mvf with $U(0) = I_m$, i.e.,

$$U \in \mathcal{E} \cap \mathcal{U}^\circ(J). \tag{2.29}$$

Conversely, Potapov's theorem (Theorem 2.11, below) guarantees that every mvf $U \in \mathcal{E} \cap \mathcal{U}^\circ(J)$ is the monodromy matrix of a regular canonical integral/differential system.

The Hamiltonian $H(x) = dM(\varphi(x))/dx$ of the canonical differential system (2.14) that is obtained from a regular canonical integral system (2.1) as in Section 2.2 with $\varphi(0) = 0$ and $\varphi(d) = \ell$ is positive semi-definite a.e. and summable on the interval $[0, \ell]$. Moreover, the matrizant $U_x(\lambda)$ of such a regular system (2.14) is a normalized nondecreasing $\curvearrowright$ absolutely continuous chain of entire J-inner mvf's on $[0, \ell]$, i.e., it has the following properties:

(1) $U_x \in \mathcal{E} \cap \mathcal{U}(J)$ for each $x \in [0, \ell]$.
(2) $U_{x_1}^{-1} U_{x_2} \in \mathcal{E} \bigcap \mathcal{U}(J)$ for $0 \le x_1 \le x_2 \le \ell$.
(3) $U_0(\lambda) = I_m$ for each $\lambda \in \mathbb{C}$ and $U_x(0) = I_m$ for each $x \in [0, \ell]$.
(4) $U_x(\lambda)$ is an absolutely continuous mvf in the variable x on the interval $[0, \ell]$ for each $\lambda \in \mathbb{C}$.

The value

$$U(\lambda) = U_\ell(\lambda)$$

of the matrizant $U_x(\lambda)$ of a regular differential system at the right end point ℓ is called the **monodromy matrix** of the system (2.14). If the system (2.14) corresponds to a canonical integral system (2.1) as above, then the monodromy matrices of these two systems coincide, because the relations (2.27) hold for $x = \ell$ and $t = d$ too. Thus, there is a two-sided connection between the canonical integral and differential systems (2.1) and (2.14) and consequently results that are obtained for one of them may be transferred to the other.

If $H(x)$ is locally summable, but not summable on the interval $[0, \ell]$ or $\ell = \infty$, (i.e., in the singular case), then the properties (1)–(3) hold for the matrizant $U_x(\lambda)$ on the interval $[0, \ell)$ and in (4) the property of absolute continuity on $[0, \ell]$ is replaced by the property of local absolute continuity on the interval $[0, \ell)$.

2.5 Multiplicative integral formulas for matrizants and monodromy matrices; Potapov's theorems

The matrizant (fundamental solution) of the canonical differential system (2.14) when $m = 1$ is given by the formula

$$U(x, \lambda) = \exp\left\{ i\lambda \int_0^x H(s)ds J \right\} \quad \text{for } 0 \le x < \ell,$$

where $J = \pm 1$. Similarly, the matrizant of the canonical integral system (2.1) when $m = 1$ is given by the formula

$$Y(t, \lambda) = \exp\left\{ i\lambda \int_0^t dM(s)J \right\} \quad \text{for } 0 \le t < d.$$

Analogous formulas for the matrizant exist in the case $m > 1$, but with multiplicative integrals in place of additive integrals:

$$U(x, \lambda) = \int_0^{\overset{x}{\curvearrowright}} \exp\{i\lambda H(s)Jds\} \quad \text{for } 0 \le x < \ell, \tag{2.30}$$

and

$$Y(t, \lambda) = \int_0^{\overset{t}{\curvearrowright}} \exp\{i\lambda d(M(s)J)\} \quad \text{for } 0 \le t < d, \tag{2.31}$$

where the multiplicative integral in formula (2.30) may be defined as the integral in (2.31) with $t = x$, $d = \ell$ and $M(t) = \int_0^t H(s)ds$ for $t \in [0, d)$.

If $H(x)$ is locally integrable in the Riemann sense on the interval $[0, \ell)$ (and not just in the Lebesgue sense), then the multiplicative integral in (2.30) can also be defined directly.

The integral in (2.31) fits into the general setting of multiplicative integrals

$$\mathfrak{I}_a^b(\varphi, \sigma) = \int_a^{\overset{b}{\curvearrowright}} e^{\varphi(t)d\sigma(t)},$$

in which $\varphi(t)$ is a scalar continuous function on $[a, b]$ and $\sigma(t)$ is an $m \times m$ mvf of bounded variation on $[a, b]$. It is defined as the limit of the products

$$\prod_{j=1}^{\overset{n}{\curvearrowright}} e^{\varphi(\xi_j)[\sigma(t_j)-\sigma(t_{j-1})]} = e^{\varphi(\xi_1)[\sigma(t_1)-\sigma(t_0)]} \cdots e^{\varphi(\xi_n)[\sigma(t_n)-\sigma(t_{n-1})]}$$

where $a = t_0 < t_1 < \ldots < t_n = b$ and $\xi_j \in [t_{j-1}, t_j]$ for $j = 1, \ldots n$ are arbitrary, when $\max\{t_j - t_{j-1}\} \to 0$. The verification of the existence of this limit may be adapted from the discussion in section 1.7 of [DF79]. Moreover, if

$$c = \max_{t \in [a,b]} \{|\varphi(t)|\} \text{ and } L = \operatorname{Var}\{\sigma(t) : a \le t \le b\},$$

then

$$\|\mathfrak{I}_a^b\| \le e^{cL}.$$

If, in particular, $\sigma(t) = M(t)J$, where $M(t)$ is a nondecreasing $m \times m$ mvf on $[a, b]$, then

$$\|\sigma(t_j) - \sigma(t_{j-1})\| = \|M(t_j) - M(t_{j-1})\| \le \operatorname{trace}\{M(t_j) - M(t_{j-1})\}$$

and hence

$$L \le \text{trace}\,\{M(b) - M(a)\}.$$

If $\mathfrak{I}_a^b(\varphi, \sigma)$ exists, then

$$\mathfrak{I}_a^b(\varphi, \sigma) = \mathfrak{I}_a^c(\varphi, \sigma)\mathfrak{I}_c^b(\varphi, \sigma) \quad \text{for } a < c < b. \tag{2.32}$$

Theorem 2.9 *The matrizant $U_t(\lambda)$, $0 \le t < d$, of a canonical integral system (2.1) with a continuous nondecreasing $m \times m$ mass function $M(t)$ on the interval $[0, d)$ with $M(0) = 0$ may be expressed in terms of $M(t)$ as a multiplicative integral:*

$$U_t(\lambda) = \int_0^{\overset{t}{\curvearrowright}} e^{i\lambda dM(s)J} \quad \textit{for } 0 \le t < d. \tag{2.33}$$

If $d < \infty$ and $M(t)$ is continuous on the closed interval $[0, d]$, then formula (2.33) is valid for $t = d$ also, i.e., for the monodromy matrix $U_d(\lambda)$ of the system (2.1).

The matrizant $U_t(\lambda)$, $0 \le t < d$, of a canonical differential system (2.8) with Hamiltonian $H(t)$, $0 \le t < d$, that satisfies the conditions (2.9) may be expressed in terms of $H(t)$ by formula (2.33) with

$$M(t) = \int_0^t H(s)ds \quad \textit{for } 0 \le t < d. \tag{2.34}$$

If $d < \infty$ and $H \in L_1^{m\times m}([0, d])$, then formulas (2.33) and (2.34) are valid for $t = d$ too, i.e., for the monodromy matrix $U_d(\lambda)$ of the system (2.8).

Proof This theorem is a special case of a general result due to Volterra that expresses the continuous solution $y(t)$, $0 \le t < d$, of the linear integral equation

$$y(t) = I_m + \int_0^t y(s)dG(s), \quad 0 \le t < d, \tag{2.35}$$

in which $G(t)$ is a given continuous $m \times m$ mvf on the interval $[0, d)$ that is of bounded variation on each closed subinterval of $[0, d)$ as the multiplicative integral

$$y(t) = \int_0^{\overset{t}{\curvearrowright}} e^{dG(s)} \quad \text{for } 0 \le t < d. \tag{2.36}$$

If

$$G(t) = \int_0^t g(s)ds \quad \text{for } 0 \le t < d \quad \text{with } g \in L_{1,\text{loc}}^{m\times m}([0, d)), \tag{2.37}$$

then the mvf $y(t)$ defined by formula (2.35) is the locally absolutely continuous solution of the Cauchy problem

$$\begin{aligned} &y'(t) = y(t)g(t) \quad \text{a.e. on the interval } [0, d) \\ &y(t) = 0. \end{aligned} \tag{2.38}$$

These facts are justified in [DF79]. □

We next present a simple instructive example that helps to explain why the matrizant of the canonical differential system (2.14) can be expressed as the multiplicative integral (2.30).

Let $H(x)$ be the mvf on the interval $[0, \ell]$ defined by

$$H(x) = H_j, \quad \text{if } x \in [x_{j-1}, x_j), \text{ for } 1 \le j \le n, \tag{2.39}$$

where

$$0 = x_0 < x_1 < \cdots < x_n = \ell, \; H_j \in \mathbb{C}^{m\times m}, \; H_j \ge 0, \text{ and } 1 \le j \le n.$$

Then the matrizant $U_x(\lambda)$ of the differential system (2.14) is given by the formulas

$$U_x(\lambda) = \begin{cases} \exp\{i\lambda H_1(x-x_0)\}, \text{ if } x_0 \le x \le x_1, \\ \exp\{i\lambda H_1(x_1-x_0)\}\exp\{i\lambda H_2(x-x_1)\} \text{ if } x_1 \le x \le x_2, \\ \exp\{i\lambda H_1(x_1-x_0)\}\exp\{i\lambda H_2(x_2-x_1)\}\exp\{i\lambda H_3(x-x_2)\}, \\ \qquad \text{if } x_2 \le x \le x_3, \\ \qquad \vdots \\ \overset{\curvearrowright}{\prod}_{j=1}^{n-1} \exp\{i\lambda H_j(x_j - x_{j-1}\}\exp\{i\lambda H_n(x-x_{n-1})\} \\ \qquad \text{if } x_{n-1} \le x \le x_n. \end{cases}$$

All these formulas for $U_x(\lambda)$ on the indicated subintervals may be rewritten uniformly as a multiplicative integral (2.30) for the mvf $H(x)$, defined in (2.39).

Theorem 2.9 is developed further in the following two equivalent theorems of V.P. Potapov, that are presented as an application of his theory of multiplicative representations of meromorphic J-contractive $m \times m$ mvf's in $\mathbb{C}_+$ in his fundamental study [Po60].

Theorem 2.10 (V.P. Potapov) *If $U \in \mathcal{E} \cap \mathcal{U}^\circ(J)$, then it admits a multiplicative integral representation*

$$U(\lambda) = \int_0^{\overset{\curvearrowright}{\ell}} e^{i\lambda dM(t)J}, \tag{2.40}$$

where $M(t)$ is a continuous nondecreasing $m \times m$ mvf on a closed interval $[0, \ell]$ *with $M(0) = 0$. Moreover, $M(t)$ may be chosen so that*

$$M(t) = \int_0^t H(s)ds \quad \textit{for } 0 \le t \le \ell, \textit{ where } H \in L_1^{m\times m}([0, \ell]), \quad H(t) \ge 0 \quad \textit{a.e. on } [0, \ell] \tag{2.41}$$

and may be normalized by the condition

$$\operatorname{trace} M'(t) = \operatorname{trace} H(t) = 1 \quad \textit{a.e. on } [0, \ell]. \tag{2.42}$$

If (2.42) is in force, then

$$\ell = \ell_U = -i\,\mathrm{trace}\left(\frac{dU}{d\lambda}\right)(0)J. \tag{2.43}$$

Proof See [Po60]. □

Theorem 2.11 (V.P. Potapov) *If $U \in \mathcal{E} \cap \mathcal{U}^\circ(J)$, then $U(\lambda)$ is the monodromy matrix of a canonical integral system (2.1) with a mass function $M(t)$ that is subject to (1.21). Moreover, there exists a system (2.1) on a closed interval $[0, d]$ with monodromy matrix $U(\lambda)$ such that the mass function $M(t)$, $0 \le t < d$, is absolutely continuous on the interval $[0, d]$, i.e.,*

$$M(t) = \int_0^t H(s)ds, \quad 0 \le t < d, \quad H(t) = M'(t) \text{ a.e. on } [0, d], \tag{2.44}$$

where

$$H(t) \ge 0 \quad \text{a.e. on } [0, d] \quad \text{and} \quad H \in L_1^{m \times m}([0, d]) \tag{2.45}$$

and

$$\mathrm{trace}\, H(t) = 1 \quad \text{a.e. on } [0, d]. \tag{2.46}$$

The mvf $U(\lambda)$ is the monodromy matrix of the canonical differential equation (1.1) with this Hamiltonian. Under condition (2.46), $d = \ell_U$, where ℓ_U is given by formula (2.43).

Proof See [Po60]. □

The equivalence of Theorems 2.10 and 2.11 follows from Theorem 2.9.

Given an mvf $U \in \mathcal{E} \cap \mathcal{U}^\circ(J)$, the **inverse monodromy problem** for canonical integral (resp., differential) systems is to describe the set of mass functions $M(t)$ (resp., Hamiltonians $H(t)$) that meet the conditions (1.21) (resp., (2.45)), $0 \le t \le d$, such that $U(\lambda)$ is the monodromy matrix of the corresponding canonical integral (resp., differential) system. Potapov's theorem (2.11) guarantees that both of these inverse monodromy problems have at least one solution. However, even under the extra normalization condition (2.46), there may be more than one solution. Nevertheless, in view of formula (2.25), $M(d)$ is uniquely defined by the monodromy matrix $U(\lambda)$ by the formula

$$M(d) = -iU'(0)J = 2\pi K_0^U(0). \tag{2.47}$$

Consequently, for canonical differential systems on $[0, d]$ with unknown Hamiltonian $H(t)$, $M(d) = \int_0^d H(s)ds$ is uniquely defined by the monodromy matrix $U(\lambda)$ and thus, if $\mathrm{trace}\, H(t) = 1$ a.e., on $[0, d]$, then d is also uniquely defined by the monodromy matrix: $d = \ell_U$, where ℓ_U is given by (2.43).

Another proof of the existence of solutions to the inverse monodromy problem that does not depend upon Potapov's theorem will be presented in Chapter 8 for monodromy matrices $U \in \mathcal{E} \cap \mathcal{U}_{AR}^{\circ}(J)$. A parametrization of the set of solutions and criteria for the uniqueness of solutions will also be presented there.

Remark 2.12 *Theorem 2.10 gives a parametrization of mvf's $U \in \mathcal{E} \cap \mathcal{U}^{\circ}(J)$ in terms of mass functions $M(t)$, $0 \le t \le d$, or Hamiltonians $H(t)$, $0 \le t \le \ell$.*

In some future applications the normalization det $H(t) = 1$ (or more generally det $H(t) = \rho^2$, a positive constant) will be imposed instead of (2.46). The next remark explains how to achieve this.

Remark 2.13 *If the Hamiltonian in (1.1) meets the constraints*

$$H \in L_1^{m\times m}([0,d]) \quad \textit{and} \quad H(x) > 0 \textit{ a.e. on } [0,d],$$

then

$$\varphi(x) = \int_0^x (\det H(s))^{1/m} ds$$

is an absolutely continuous strictly increasing function on $[0,d]$. Therefore, there exists a strictly increasing absolutely continuous function ψ on the interval $[0, \varphi(d)]$ such that $\psi(\varphi(x)) = x$. Moreover,

$$y(t,\lambda) = u(\psi(t),\lambda) \quad \textit{for } 0 \le t \le \varphi(d)$$

is the solution of the canonical differential system

$$y'(t,\lambda) = i\lambda y(t,\lambda)\psi'(t)H(\psi(t))J$$

with a Hamiltonian $\psi'(t)H(\psi(t))$ that is subject to the constraint

$$\begin{aligned}\det \psi'(t)H(\psi(t)) &= \psi'(t)^m \det H(\psi(t)) = \psi'(t)^m \varphi'(t)^m \\ &= \left\{\frac{d}{dt}\varphi(\psi(t))\right\}^m = 1 \quad \textit{a.e. on } [0,\varphi(d)].\end{aligned}$$

Furthermore, the monodromy matrix of the first system $U(d,\lambda)$ coincides with the monodromy matrix $Y(\varphi(d),\lambda)$ of the second system.

Remark 2.14 *If $A \in \mathbb{C}^{m\times m}$ and $A \ge 0$, then the mvf $U(\lambda) = e^{i\lambda AJ}$ belongs to the class $\mathcal{E} \cap \mathcal{U}^{\circ}(J)$, since it is the monodromy matrix of the canonical differential system with Hamiltonian $H(t) = A$ on the interval $[0,1]$. Since $\mathcal{E} \cap \mathcal{U}^{\circ}(J)$ is a multiplicative semigroup, the finite products*

$$\overset{\curvearrowright}{\prod_{k=1}^{n}} e^{i\lambda A_k J} \quad \textit{with } A_k \in \mathbb{C}^{m\times m} \textit{ and } A_k \ge 0 \textit{ for } k = 1,\ldots,n \tag{2.48}$$

also belong to the class $\mathcal{E} \cap \mathcal{U}^{\circ}(J)$.

Theorem 2.15 *An $m \times m$ mvf $U(\lambda)$ belongs to the class $\mathcal{E} \cap \mathcal{U}^\circ(J)$ if and only if there exists a sequence of products $\Pi_n(\lambda)$, $n = 1, 2, \ldots$, of the form (2.48) such that*

$$U(\lambda) = \lim_{n\uparrow\infty} \Pi_n(\lambda) \tag{2.49}$$

uniformly on compact subsets of $\mathbb{C}$.

Proof If $U \in \mathcal{E} \cap \mathcal{U}^\circ(J)$, then, by Theorem 2.10, $U(\lambda)$ admits a multiplicative integral representation of the form (2.40) with some continuous nondecreasing $m \times m$ mass function $M(t)$. Thus, (2.49) holds with

$$\Pi_n(\lambda) = \prod_{j=1}^{\frown n} e^{i\lambda(M(t_j)-M(t_{j-1}))J}, \quad \text{where } t_j = j\ell/n \text{ for } j = 0, 1, \ldots, n.$$

Moreover, the convergence is uniform on each compact subset of $\mathbb{C}$, since

$$\|\Pi_n(\lambda)\| \le e^{R\mathrm{trace}\,\{M(\ell)-M(0)\}} \quad \text{for } |\lambda| \le R.$$

The converse is self-evident, since $\Pi_n \in \mathcal{E} \cap \mathcal{U}^\circ(J)$. □

Corollary 2.16 *If $U \in \mathcal{E} \cap \mathcal{U}^\circ(J)$, then*

$$\|U(\lambda)\| \le \exp\{|\lambda| 2\pi \,\mathrm{trace}\, K_0^U(0)\}. \tag{2.50}$$

and

$$\det U(\lambda) = \exp\{i\lambda 2\pi \,\mathrm{trace}\,[K_0^U(0)J]\}. \tag{2.51}$$

Proof In view of Theorem 2.15, it suffices to establish the asserted statements for $U(\lambda) = \exp\{i\lambda AJ\}$ for positive semidefinite matrices $A \in \mathbb{C}^{m\times m}$. But then clearly

$$\|U(\lambda)\| \le \sum_{k=0}^{\infty} \frac{\|i\lambda AJ\|^k}{k!} \le \sum_{k=0}^{\infty} \frac{|\lambda|^k \|A\|^k}{k!} \le \exp\{|\lambda|\|A\|\}$$

and, since $\det e^B = \exp\{\mathrm{trace}\, B\}$,

$$\det U(\lambda) = \exp\{\mathrm{trace}\,[i\lambda AJ]\}.$$

The proof is now easily completed upon noting that

$$A = -iU'(0)J = 2\pi K_0^U(0) \quad \text{and} \quad \|A\| \le \mathrm{trace}\, A.$$ □

2.6 The Feller–Krein string equation

Let $m(s)$ denote the mass of a string on the interval $[0, s]$, $0 \le s < d$, i.e., $m(s)$ is an arbitrary left continuous nondecreasing scalar function on the interval $[0, d)$.

The transverse harmonic oscillation of the string with frequency $\lambda \in \mathbb{R}$ may be written in complex form as

$$a(s, t, \lambda) = a(s, \lambda) \exp(i\lambda t), \quad \lambda \in \mathbb{R}$$

so that $\Re a(s, t, \lambda)$, $\lambda \in \mathbb{R}$, has a physical interpretation as the displacement along the axis ordinate of the string with abscissa s at the instant t under a small oscillation of the string with frequency λ. The complex amplitude $a(s, \lambda)$ of this oscillation is the continuous solution of the integral equation

$$a(s, \lambda) = a(0, \lambda) + a'_-(0, \lambda)s - \lambda^2 \int_0^s (s-u)a(u, \lambda)dm(u), \ 0 \le s < d,$$

i.e., $a(s, \lambda)$ is a solution of the Feller–Krein string equation

$$-\frac{d^2}{dm\,ds} a(s, \lambda) = \lambda^2 a(s, \lambda). \tag{2.52}$$

The left derivative $a'_-(s, \lambda)$ of the function $a(s, \lambda)$ is equal to

$$a'_-(s, \lambda) = a'_-(0, \lambda) - \lambda^2 \int_0^{s-} a(u, \lambda)\, dm(u), \ 0 < s < d, \tag{2.53}$$

whereas the right derivative $a'_+(s, \lambda)$ of $a(s, \lambda)$ is equal to

$$a'_+(s, \lambda) = a'_-(0, \lambda) - \lambda^2 \int_0^{s} a(u, \lambda)\, dm(u), \ 0 \le s < d. \tag{2.54}$$

Therefore,

$$a'_+(s, \lambda) - a'_-(s, \lambda) = -\lambda^2\, m[s],$$

where $m[s] = m(s) - m(s-)$, the jump in m at s, if any. Moreover, since

$$\begin{aligned}
-\lambda^2 \int_0^s (s-u)a(u, \lambda)\, dm(u) &= -\lambda^2 \int_0^s \int_u^s a(u, \lambda) dv dm(u) \\
&= -\lambda^2 \int_0^s \int_0^v a(u, \lambda)\, dm(u) dv \\
&= \int_0^s (a'_+(v, \lambda) - a'_+(0, \lambda)) dv
\end{aligned}$$

it is readily seen that

$$a(s, \lambda) = a(0, \lambda) + \int_0^s a'_+(v, \lambda) dv, \ 0 \le s \le v. \tag{2.55}$$

Let

$$y_2(s, \lambda) = a(s, \lambda) \quad \text{and} \quad y_1(s, \lambda) = -(i\lambda)^{-1} a'_+(s, \lambda), \ 0 \le s < d.$$

Then (2.55) and (2.53) may be rewritten as the system of equations

$$\begin{cases} y_1(s,\lambda) = y_1(0,\lambda) - i\lambda \int_0^s y_2(u,\lambda)dm(u) \\ y_2(s,\lambda) = y_2(0,\lambda) - i\lambda \int_0^s y_1(u,\lambda)du \\ \qquad \text{on the interval} \quad 0 \le s < d, \end{cases} \tag{2.56}$$

or equivalently, upon setting

$$y(s,\lambda) = [y_1(s,\lambda) \quad y_2(s,\lambda)] \quad \text{and} \quad M(s) = \begin{bmatrix} s & 0 \\ 0 & m(s) \end{bmatrix}$$

for $0 \le s < d$, as

$$y(s,\lambda) = y(0,\lambda) + i\lambda \int_0^s y(u,\lambda)dM(u)J_1, \quad 0 \le s < d. \tag{2.57}$$

This is an equation of the form (2.1) with $J = J_1 = \begin{bmatrix} 0 & -1 \\ -1 & 0 \end{bmatrix}$, but with a nondecreasing 2×2 matrix-valued mass function $M(s)$ that is only left continuous on the interval $[0, d)$. Neverthenless, since $y_2(s,\lambda)$ is continuous on $[0, d)$, the Riemann–Stieltjes integral in the first equation in the system (2.56) is well defined and $y_1(s,\lambda)$ will be left continuous on this interval for every choice of $y_1(0,\lambda) \in \mathbb{C}$ and $y_2(0,\lambda) \in \mathbb{C}$. Therefore, the Riemann integral in the second equation in the system (2.56) is well defined too.

The canonical integral system (2.57) may be reduced to the corresponding differential system (2.14), just as above, by introducing the new variable $x = \psi(s)$, where

$$\psi(s) = \operatorname{trace} M(s) = m(s) + s, \quad 0 \le s < d,$$

and considering the inverse function $s = \varphi(x), \ 0 \le x < \ell$, where

$$\ell = \lim_{s \uparrow d} \operatorname{trace} M(s).$$

The vvf $u(x,\lambda) = y(\varphi(x),\lambda)$ satisfies the canonical differential system

$$u'(x,\lambda) = i\lambda u(x.\lambda)H(x)J_1 \quad \text{a.e. on } [0,\ell) \tag{2.58}$$

with

$$H \in L_{1,\mathrm{loc}}^{2\times 2}([0,\ell)), \quad H(x) = \overline{H(x)} = \begin{bmatrix} h_1(x) & 0 \\ 0 & h_2(x) \end{bmatrix} \ge 0 \quad \text{a.e. in } [0,\ell), \tag{2.59}$$

and

$$\operatorname{trace} H(x) = h_1(x) + h_2(x) = 1 \quad \text{a.e. in the interval } [0,\ell). \tag{2.60}$$

The mass distribution function $m(s)$, $0 \leq s < d$, of the string is defined by a Hamiltonian $H(x)$ with properties (2.59) and (2.60): if

$$s = \int_0^x h_1(u)du,$$

then

$$m(s) = \int_0^x h_2(u)du \quad \text{for } 0 \leq x < \ell, \quad \text{and} \quad d = \int_0^\ell h_1(x)dx.$$

Thus, $d < \infty$ if and only if $h_1 \in L_1([0, \ell))$ and it has a finite mass if and only if $h_2 \in L_1([0, \ell))$. In view of the normalization condition (2.60), both of these conditions are met if and only if $\ell < \infty$, i.e., if and only if the string is a **regular string**; if $\ell = \infty$, the string is called a **singular string**.

Let $a_1(s, \lambda)$ and $a_2(s, \lambda)$ be the unique solutions of the Feller–Krein string equation that satisfy the initial conditions

$$a_1(0, \lambda) = 1, \quad (a_1)'_-(0) = 0, \quad a_2(0, \lambda) = 0, \quad (a_2)'_-(0, \lambda) = 1 \tag{2.61}$$

and let $A_x(\lambda)$ denote the matrizant of the canonical differential system (2.58) with H as in (2.59) and (2.60). Then

$$\begin{bmatrix} (a_2)'_+(\varphi(x), \lambda) & -i\lambda a_2(\varphi(x), \lambda) \\ (-i\lambda)^{-1}(a_1)'_+(\varphi(x), \lambda) & a_1(\varphi(x), \lambda) \end{bmatrix} = \begin{bmatrix} a_{11}(x, \lambda) & a_{12}(x, \lambda) \\ a_{21}(x, \lambda) & a_{22}(x, \lambda) \end{bmatrix}. \tag{2.62}$$

Following Krein [Kr52a], a string that is subject to the initial condition $a'_-(0, \lambda) = 0$ (resp., $a(0, \lambda) = 0$) will be called an S_1-string (resp., S_2-string).

The generalized Fourier transform for an S_1-string

$$(\mathcal{F}_{st}f)(\lambda) = \int_{0-}^d a_1(s, \sqrt{\lambda})f(s)dm(s) \quad \text{for } f \in L_2(dm; [0, d]) \tag{2.63}$$

is based on the solution $a_1(s, \lambda)$ of the Feller–Krein string equation. A nondecreasing function $\tau(\mu)$ on $\mathbb{R}$ is said to be a spectral function of an S_1-string if the Parseval–Plancherel formula

$$\int_{-\infty}^\infty |(\mathcal{F}_{st}f)(\mu)|^2 d\tau(\mu) = \int_0^d |f(s)|^2 dm(s) \tag{2.64}$$

holds for every $f \in L_2(dm; [0, d))$. If the string is a singular string, then there is exactly one spectral function $\tau(\mu)$ with

$$\tau(\mu) = 0 \quad \text{for } \mu < 0 \text{ and} \quad \tau(\mu) = \frac{\tau(\mu+) + \tau(\mu-)}{2} \quad \text{for } \mu > 0. \tag{2.65}$$

Moreover, it turns out that

$$\lim_{s\uparrow\infty} \frac{a_2(s, i\nu)}{a_1(s, i\nu)} = \lim_{s\uparrow\infty} \frac{(a_2)'_+(s, i\nu)}{(a_1)'_+(s, i\nu)} = G(-\nu^2) \quad \text{for } \nu > 0, \tag{2.66}$$

$$G(\lambda) = \beta + \int_{0-}^\infty \frac{d\tau(\mu)}{\mu - \lambda} \quad \text{for } \lambda \in \mathbb{C} \setminus \{[0, \infty)\} \text{ and some } \beta \geq 0, \tag{2.67}$$

and

$$\int_0^\infty \frac{d\tau(\mu)}{1+\mu} < \infty. \tag{2.68}$$

Moreover, $\beta = 0$ if and only if $m(0+) = m(0) = 0$.

Krein called $\tau(\mu)$ the **principal spectral function** of the S_1-string. The function

$$c(\lambda) = -i\lambda G(\lambda^2) \quad \text{for } \lambda \in \mathbb{C}_+ \tag{2.69}$$

belongs to the Carathéodory class $\mathcal{C}$ and meets the reality condition $\overline{c(-\overline{\lambda})} = c(\lambda)$. Formulas (2.62) and (2.67) imply that

$$c(i\nu) = \lim_{x\uparrow\infty} \frac{a_{11}(x, i\nu)}{a_{21}(x, i\nu)} = \lim_{x\uparrow\infty} \frac{a_{12}(x, i\nu)}{a_{22}(x, i\nu)}. \tag{2.70}$$

It may be shown that the mvf c defined in (2.70) is the only mvf in the set $\mathcal{C}_{\text{imp}}(H)$ that is defined for the canonical system (2.58) with H subject to (2.59) and (2.60) that is real. Moreover, its spectral function

$$\sigma(\mu) = \begin{cases} (\pi/2)\tau(\mu^2) & \text{if } \mu \geq 0 \\ -(\pi/2)\tau(\mu^2) & \text{if } \mu < 0 \end{cases} \tag{2.71}$$

is the only symmetrical spectral function for this system; see [KaKr74b] and Theorem 8.79.

Lemma 2.17 *The matrizant $Y_s(\lambda) = Y(s, \lambda)$, $0 \leq s < d$, of the canonical integral system (2.56) that corresponds to the Feller–Krein string equation is a normalized nondecreasing family of entire J_1-inner mvf's that is real and symplectic, i.e.,*

$$\overline{Y_s(-\overline{\lambda})} = Y_s(\lambda) \quad \text{and} \quad Y_s(\lambda)^\tau \mathcal{J}_1 Y_s(\lambda) = \mathcal{J}_1 \quad \text{for } 0 \leq s < d.$$

If the string is a regular string, then these properties hold on the closed interval $[0, d]$ and so are inherited by the monodromy matrix $Y_d(\lambda)$.

Proof This will follow from general results on canonical integral and differential systems with real and symplectic matrizants that will be discussed in Section 8.6. □

2.7 Differential systems with potential

Particular attention will be paid to differential systems of the form

$$y'(x, \lambda) = i\lambda y(x, \lambda) NJ + y(x, \lambda)\, \mathcal{V}(x), \quad 0 \leq x < d, \tag{2.72}$$

where J is an $m \times m$ signature matrix, $N \geq 0$ is an $m \times m$ constant matrix that commutes with J,

$$\mathcal{V}(x) \in L_{1,\text{loc}}^{m\times m}([0, d)) \quad \text{and} \quad \mathcal{V}(x)J + J\mathcal{V}(x)^* = 0 \quad \text{a.e. on } [0, d). \tag{2.73}$$

Lemma 2.18 *If the $m \times m$ mvf $\mathcal{V}(x)$ meets the properties in (2.73), then the Cauchy problem*

$$\begin{aligned} Y'(x) &= Y(x)\mathcal{V}(x) \quad \text{for } 0 \le x < d \\ Y(0) &= I_m \end{aligned} \tag{2.74}$$

has exactly one locally absolutely continuous solution on the interval $[0, d)$ and this solution is J-unitary at each point of $[0, d)$.

Proof The first assertion may be verified by rewriting (2.74) as the integral equation

$$Y(x) = I_m + \int_0^x Y(s)\mathcal{V}(s)ds \quad \text{for } 0 \le x < d,$$

and modifying the proof of Lemma 2.1. The asserted J-unitarity of the solution follows from the fact that $Y(0)JY(0)^* = J$ and

$$\frac{d}{dx}\left\{Y(x)JY(x)^*\right\} = 0 \quad \text{a.e. on } [0, d).$$

□

Lemma 2.19 *Let $Y_x(\lambda) = Y(x, \lambda)$ denote the matrizant of the differential system*

$$y'(x, \lambda) = i\lambda y(x, \lambda)H_0(x)J + y(x, \lambda)\mathcal{V}(x), \quad 0 \le x < d, \tag{2.75}$$

in which the potential $\mathcal{V}(x)$ satisfies the conditions in (2.73),

$$H_0 \in L_{1,\mathrm{loc}}^{m\times m}([0, d)) \quad \text{and} \quad H_0(x) \ge 0 \quad \text{a.e. on } [0, d). \tag{2.76}$$

Then $Y(x) = Y(x, 0)$ is a solution of the Cauchy problem (2.74) and

$$U(x, \lambda) = Y(x, \lambda)Y(x)^{-1}, \quad \text{for } 0 \le x < d \tag{2.77}$$

is the matrizant of the canonical differential system (2.8) with Hamiltonian

$$H(x) = Y(x)H_0(x)Y(x)^* \quad \text{a.e. on the interval } [0, d). \tag{2.78}$$

Conversely, if $U_x(\lambda)$ is the matrizant of the canonical differential system (2.8) in which the Hamiltonian is of the form (2.78), where $Y(x)$ is a locally absolutely continuous mvf with J-unitary values on $[0, d)$, then

$$Y(x, \lambda) = U(x, \lambda)Y(x)$$

is the matrizant of the system (2.75) with potential

$$\mathcal{V}(x) = Y(x)^{-1}Y'(x)$$

and this potential meets the conditions in (2.73).

Proof A straightforward calculation yields the formula

$$U'(x, \lambda) = i\lambda U(x, \lambda)H(x)J \quad \text{with} \quad H(x) = Y(x)H_0(x)JY(x)^{-1}J.$$

Then (2.78) follows from the fact that $Y(x)$ is J-unitary on $[0, d)$. Moreover, in view of (2.78), $H(x)$ satisfies the conditions in (2.9), since $H_0(x)$ satisfies the conditions in (2.76) and $Y(x)$ is locally absolutely continuous on $[0, d)$. □

Lemma 2.19 will be of particular interest when $H_0(x)$ is a constant $m \times m$ matrix that commutes with J, i.e., when

$$H(x) = Y(x)NY(x)^* \quad \text{for } 0 \leq x < d, \tag{2.79}$$

where $N \in \mathbb{C}^{m\times m}$, $N \geq 0$, $NJ = JN$ and $Y(x)$ is a locally absolutely continuous J-unitary valued mvf on $[0, d)$.

2.8 Dirac–Krein systems

For each signature matrix $J \in \mathbb{C}^{m\times m}$, let

$$P_+ = \frac{1}{2}(I_m + J) \quad \text{and} \quad P_- = \frac{1}{2}(I_m + J). \tag{2.80}$$

A differential system of the form (2.72) with potential $\mathcal{V}(x)$ that is subject to the constraints (2.73) will be called a

Dirac system if $N = \alpha I_m$ for some $\alpha > 0$;
Krein system if $N = \alpha P_+$ for some $\alpha > 0$ or $N = \beta P_-$ for some $\beta > 0$;
Dirac–Krein system if

$$N = N_{\alpha,\beta} \stackrel{\text{def}}{=} \alpha P_+ + \beta P_- \quad \text{for some } \alpha \geq 0 \text{ and } \beta \geq 0$$

with $\kappa \stackrel{\text{def}}{=} \alpha + \beta > 0$.

Dirac–Krein systems will be called **DK-systems** for short.

Lemma 2.20 *If $Y_{\alpha,\beta}(x, \lambda)$ denotes the matrizant of the DK-system (2.72) with $N = N_{\alpha,\beta}$ and $-\alpha \leq \gamma \leq \beta$, then $e^{i\gamma\lambda x}Y_{\alpha,\beta}(x, \lambda)$ is the matrizant of the DK-system with $N = N_{\alpha+\gamma,\beta-\gamma}$.*

Proof Let $Y_{\alpha,\beta}(x, \lambda)$ denote the matrizant of the DK-system (2.72) with $N = N_{\alpha,\beta}$. Then since

$$\gamma I_m + N_{\alpha,\beta}J = \gamma P_+ + \gamma P_- + \alpha P_+ - \beta P_- = N_{\alpha+\gamma,\beta-\gamma}J,$$

it is readily checked that $e^{i\gamma\lambda x}Y_{\alpha,\beta}(x, \lambda)$ is the matrizant of the Dirac–Krein system with $N = N_{\alpha+\gamma,\beta-\gamma}$. □

Remark 2.21 *The formulas*

$$Y_{\alpha,\beta}(x, \lambda) = e^{i\lambda\alpha x}Y_{0,\kappa}(x, \lambda) = e^{i\lambda(\alpha-\beta)x/2}Y_{\kappa/2,\kappa/2}(x, \lambda)$$

$$= e^{-i\lambda\beta x}Y_{\kappa,0}(x, \lambda)$$

permit one to reduce a DK-system with $N = N_{\alpha,\beta}$ *to a Dirac system with* $N = (\kappa/2)I_m$ *or to a Krein system with* $N = N_{\kappa,0}$ *or* $N = N_{0,\kappa}$.

Lemma 2.22 *If* $\mathcal{V}(x)$ *satisfies the conditions in (2.73), then the Cauchy problem*

$$\begin{aligned} G'(x) &= iG(x)\frac{\mathcal{V}(x) - \mathcal{V}(x)^*}{2i}, \quad 0 \le x < d, \\ G(0) &= I_m \end{aligned} \tag{2.81}$$

has exactly one locally absolutely continuous solution $G(x)$ *on the interval* $[0, d)$. *Moreover, this solution is unitary and* J*-unitary on* $[0, d)$, *i.e.,*

$$G(x)^*G(x) = I_m \quad \text{and} \quad G(x)^*JG(x) = J \quad \text{for } 0 \le x < d. \tag{2.82}$$

Proof The proof that the Cauchy problem has exactly one locally absolutely continuous solution is much the same as the verification of the analogous statement in Lemma 2.18. The rest follows by checking that $\{G(x)G(x)^*\}' = 0$ and $\{G(x)JG(x)^*\}' = 0$ a.e. on $[0, d)$. □

Lemma 2.23 *Let* $y(x, \lambda)$ *be a solution of the differential system*

$$y'(x, \lambda) = i\lambda y(x, \lambda)N_{\alpha,\beta}J + y(x, \lambda)\mathcal{V}(x), \quad 0 \le x < d, \tag{2.83}$$

with a potential $\mathcal{V}(x)$ *that satisfies the conditions in (2.73) and let* $G(x)$ *be the unique locally absolutely continuous solution of the Cauchy problem (2.81) corresponding to this potential. Then the mvf*

$$\widetilde{y}(x, \lambda) = y(x, \lambda)G(x)^{-1} \quad \text{for } 0 \le x < d \tag{2.84}$$

is a solution of the differential system

$$\widetilde{y}'(x, \lambda) = i\lambda\widetilde{y}N_{\alpha,\beta}J + \widetilde{y}(x, \lambda)\mathcal{V}_1(x), 0 \le x < d, \tag{2.85}$$

where

$$\mathcal{V}_1(x) = G(x)\frac{\mathcal{V}(x) + \mathcal{V}(x)^*}{2}G(x)^* \quad \text{a.e. on } [0, d). \tag{2.86}$$

Moreover, $\mathcal{V}_1(x)$ *meets the conditions*

$$\begin{aligned} \mathcal{V}_1(x) \in L_{1,\mathrm{loc}}^{m\times m}([0, d)), \quad \mathcal{V}_1(x) = \mathcal{V}_1(x)^* \quad \text{and} \\ \mathcal{V}_1(x)J + J\mathcal{V}_1(x)^* = 0 \quad \text{a.e. on } [0, d). \end{aligned} \tag{2.87}$$

Proof The verification of (2.85) is a straightforward calculation, with the help of the observation that

$$GP_\pm G^* = P_\pm = G^*P_\pm G,$$

which is an easy consequence of (2.82). Finally, the third assertion in (2.87) follows from the fact that

$$\mathcal{V}_1J + J\mathcal{V}_1^* = 0 \Longleftrightarrow G^*(\mathcal{V}_1J + J\mathcal{V}_1^*)G = 0 \Longleftrightarrow (V + V^*)J + J(V + V^*) = 0.$$

□

It is readily checked that the dependence of the four-block decomposition of a potential $\mathcal{V}_1(x)$ that meets the constraints (2.87) on the signature matrix J is as follows:

J	j_{pq}	J_p	$\mathcal{J}_p$
$\mathcal{V}_1(x)$	$\begin{bmatrix} 0 & v(x) \\ v(x)^* & 0 \end{bmatrix}$	$\begin{bmatrix} v_1(x) & iv_2(x) \\ -iv_2(x) & -v_1(x) \end{bmatrix}$	$\begin{bmatrix} v_1(x) & v_2(x) \\ v_2(x) & -v_1(x) \end{bmatrix}$

where the exhibited mvf's $v(x)$, $v_1(x)$ and $v_2(x)$ are summable on every closed subinterval of $[0, d)$ and $v_k(x)^* = v_k(x)$ for $k = 1, 2$.

2.9 The Schrödinger equation

In this section we shall establish a connection between the fundamental solution $\Psi_x(\lambda)$ of the matrix Schrödinger equation

$$- \psi''(x, \lambda) + \psi(x, \lambda)q(x) = \lambda\psi(x, \lambda), \quad 0 \leq x < \ell, \tag{2.88}$$

with a $p \times p$ matrix-valued potential $q = q^* \in L_{1,\mathrm{loc}}^{p\times p}([0, \ell))$ such that at least one of the Riccati equations

$$q(x) = v(x)^2 + v'(x) \quad \text{or} \quad q(x) = v(x)^2 + v'(x) \tag{2.89}$$

has a locally absolutely continuous Hermitian solution $v(x)$ on $[0, \ell)$, and the matrizant $Y_x(\lambda)$ of the Dirac system

$$y' = i\lambda y J_p + y\mathcal{V} \quad \text{with } \mathcal{V}(x) = \begin{bmatrix} -v(x) & 0 \\ 0 & v(x) \end{bmatrix} = \mathcal{V}(x)^*. \tag{2.90}$$

To begin with, it is readily checked that the matrizant $Y_x(\lambda)$ of the Dirac system (2.90) is a solution of the Schrödinger equation

$$-Y_x''(\lambda) + Y_x(\lambda)(\mathcal{V}(x)^2 + \mathcal{V}'(x)) = \lambda^2 Y_x(\lambda)$$

with potential

$$\mathcal{V}^2(x) + \mathcal{V}'(x) = \begin{bmatrix} v^2(x) - v'(x) & 0 \\ 0 & v^2(x) + v'(x) \end{bmatrix}$$

such that $Y_0(\lambda) = I_m$. Moreover, in view of (2.90),

$$Y_x'(\lambda)\,|_{x=0} = i\lambda J_p + \begin{bmatrix} -v(0) & 0 \\ 0 & v(0) \end{bmatrix} = \begin{bmatrix} -v(0) & -i\lambda I_p \\ -i\lambda I_p & v(0) \end{bmatrix}.$$

Next, upon writing $Y_x(\lambda)$ in block form as

$$Y_x(\lambda) = \begin{bmatrix} y_{11}(x,\lambda) & y_{12}(x,\lambda) \\ y_{21}(x,\lambda) & y_{22}(x,\lambda) \end{bmatrix} \quad \text{(with } p \times p \text{ blocks)},$$

it is easily seen that the first block column of $Y_x(\lambda)$ is a solution of the matrix Schrödinger equation

$$-\begin{bmatrix} y''_{11}(x,\lambda) \\ y''_{21}(x,\lambda) \end{bmatrix} + \begin{bmatrix} y_{11}(x,\lambda) \\ y_{21}(x,\lambda) \end{bmatrix} (v(x)^2 - v'(x)) = \lambda^2 \begin{bmatrix} y_{11}(x,\lambda) \\ y_{21}(x,\lambda) \end{bmatrix} \tag{2.91}$$

that satisfies the initial conditions

$$\begin{bmatrix} y_{11}(0,\lambda) \\ y_{21}(0,\lambda) \end{bmatrix} = \begin{bmatrix} I_p \\ 0 \end{bmatrix} \quad \text{and} \quad \begin{bmatrix} y'_{11}(0,\lambda) \\ y'_{21}(0,\lambda) \end{bmatrix} = \begin{bmatrix} -v(0) \\ -i\lambda I_p \end{bmatrix}, \tag{2.92}$$

whereas the second block column of $Y_x(\lambda)$ is a solution of the matrix Schrödinger equation

$$-\begin{bmatrix} y''_{12}(x,\lambda) \\ y''_{22}(x,\lambda) \end{bmatrix} + \begin{bmatrix} y_{12}(x,\lambda) \\ y_{22}(x,\lambda) \end{bmatrix} (v(x)^2 + v'(x)) = \lambda^2 \begin{bmatrix} y_{12}(x,\lambda) \\ y_{22}(x,\lambda) \end{bmatrix} \tag{2.93}$$

that satisfies the initial conditions

$$\begin{bmatrix} y_{12}(0,\lambda) \\ y_{22}(0,\lambda) \end{bmatrix} = \begin{bmatrix} 0 \\ I_p \end{bmatrix} \quad \text{and} \quad \begin{bmatrix} y'_{12}(0,\lambda) \\ y'_{22}(0,\lambda) \end{bmatrix} = \begin{bmatrix} -i\lambda I_p \\ v(0) \end{bmatrix}. \tag{2.94}$$

Thus, $y_{11}(x,\lambda)$ and $y_{21}(x,\lambda)$ can be expressed in terms of the $p \times p$ solutions $\psi_j(x,\lambda)$, $j = 1, 2$, of (2.88) that meet the initial conditions $\psi_1(0,\lambda) = I_p$, $\psi'_1(0,\lambda) = 0$, $\psi_2(0,\lambda) = 0$ and $\psi'_2(0,\lambda) = I_p$ with potential $q(x) = v(x)^2 - v'(x)$:

$$\begin{bmatrix} y_{11}(x,\lambda) \\ y_{21}(x,\lambda) \end{bmatrix} = \begin{bmatrix} I_p & -v(0) \\ 0 & -i\lambda I_p \end{bmatrix} \begin{bmatrix} \psi_1(x,\lambda^2) \\ \psi_2(x,\lambda^2) \end{bmatrix} \tag{2.95}$$

and, since the matrizant is a solution of (2.90),

$$\begin{bmatrix} y'_{11}(x,\lambda) \\ y'_{21}(x,\lambda) \end{bmatrix} = -i\lambda \begin{bmatrix} y_{12}(x,\lambda) \\ y_{22}(x,\lambda) \end{bmatrix} - \begin{bmatrix} y_{11}(x,\lambda) \\ y_{21}(x,\lambda) \end{bmatrix} v(x), \tag{2.96}$$

which yields a formula for the second block column of $Y_x(\lambda)$ in terms of the first and hence, upon combining and rearranging these formulas,

$$Y_x(\lambda) = L_\lambda \begin{bmatrix} I_p & -v(0) \\ 0 & -iI_p \end{bmatrix} \Psi_x(\lambda^2) \begin{bmatrix} I_p & iv(x) \\ 0 & iI_p \end{bmatrix} L_\lambda^{-1}, \quad 0 \le x < d, \tag{2.97}$$

where

$$L_\lambda = \begin{bmatrix} I_p & 0 \\ 0 & \lambda I_p \end{bmatrix} \quad \text{and} \quad \Psi_x(\lambda) = \begin{bmatrix} \psi_1(x,\lambda) & \psi'_1(x,\lambda) \\ \psi_2(x,\lambda) & \psi'_2(x,\lambda) \end{bmatrix}. \tag{2.98}$$

A similar connection exists between $Y_x(\lambda)$ and the fundamental solution of (2.88) with potential $q(x) = v(x)^2 + v'(x)$.

Formulas (2.97) and (2.98) imply that for each $x \in [0, d)$ there exists a constant c_x such that

$$\|Y_x(\lambda)\| \leq c_x|\lambda|\|\Psi_x(\lambda^2)\| \quad \text{and} \quad \|\Psi_x(\lambda^2)\| \leq c_x|\lambda|\|Y_x(\lambda)\|$$

for $|\lambda| \geq 1$. Thus, as $\{e_xI_p, e_xI_p\} \in ap_{II}(Y_x)$ for every $x \in [0, d)$, as will be shown below in Theorem 12.6, it follows that

$$\limsup_{r\uparrow\infty} \frac{\ln \max\{\|\Psi_x(\lambda)\| : |\lambda| \leq r\}}{r^{1/2}} = \limsup_{\nu\uparrow\infty} \frac{\ln \|Y_x(i\nu)\|}{\nu} = x.$$

Therefore, since Ψ_x is a $-\mathcal{J}_p$-inner mvf of minimal exponential type, it belongs to the class $\mathcal{U}_S(-\mathcal{J}_p)$; see Theorem 3.56. Nevertheless, since $Y_x \in \mathcal{U}_{rsR}(J_p)$ and $\{e_xI_p, e_xI_p\} \in ap(Y_x)$ for every $x \in [0, d)$, the connection (2.97) can be exploited to establish existence, uniqueness and algorithms for solving associated inverse problems on the basis of corresponding results for canonical differential systems with right strongly regular matrizants; see [ArD04b], [ArD05c] and Chapter 12.

2.10 Supplementary notes

Additional information on multiplicative integrals and applications may be found in [Po60], [DF79], [GJ90] and [Zol03]. M.S. Livsic exploited the fact that the monodromy matrix of a canonical differential system belongs to the class $\mathcal{E} \cap \mathcal{U}^\circ(J)$ of entire J-inner mvfs U with $U(0) = I_m$ in his development of the theory of characteristic functions for Volterra operators and in his construction of triangular models for Volterra operators with imaginary parts of finite rank; see Section 8.4. This class also arose in M.G. Krein's studies of entire symmetric operators with defect indices (m, m); see, e.g., [GoGo97] and the references cited therein.

Potapov's original proof of Theorem 2.10 in [Po60] was based on approximating a mvf $U \in \mathcal{E} \cap \mathcal{U}^\circ(J)$ by rational J-inner mvf's (expressed as finite products of what are now known as elementary Blaschke–Potapov products). In [Po60] he presented two applications of this theorem: In the first, he used the multiplicative representation of $U \in \mathcal{E} \cap \mathcal{U}^\circ(J)$ that he obtained in Theorem 2.10 to identify every mvf U in this class as the monodromy matrix of a canonical system (see Theorem 2.11 for the precise statement); in the second he used the triangular model of Volterra operators constructed by M.S. Livsic to show that every mvf $U \in \mathcal{E} \cap \mathcal{U}^\circ(J)$ can be realized as the characteristic mvf of a Volterra operator.

Later, M.S. Brodskii developed triangular integral representations of Volterra operators and used these representations to obtain Livsic's triangular model and Theorems 2.10 and 2.11 by purely operator theoretical methods; see sections 25 and 26 of [Bro72].

The operator $-\frac{d^2}{dmdx}$ was studied independently by W. Feller in [Fe55], [Fe57], [Fe59a] and [Fe59b] in his work on diffusion processes and by M.G. Krein as an application of his theory of the resolvent matrices of entire symmetric operators to the development of the spectral theory of strings. The main results on the direct problem may be found in [KaKr74b]. Krein conjectured that every nondecreasing function $\tau(\mu)$ on $\mathbb{R}_+$ that meets the constraint $\int_0^\infty (1+\mu)^{-1} d\tau(\mu) < \infty$ is the spectral function of exactly one singular S_1-string in [Kr53], but did not supply a proof. This conjecture was verified in the monograph [DMc76] by H. Dym and H.P. McKean with the help of a fundamental theorem of de Branges [Br68a] that served to guarantee the uniqueness of real Hamiltonians H subject to the constraints (2.59) and (2.60) in 2×2 canonical differential systems (2.58) with $J = \mathcal{J}_1$; see Theorem 8.3 for the full statement of de Branges' theorem. Historical remarks may be found in [KaKr74b], appendix 3 of [GoGo97], and from the other side of what was then the iron curtain [Dy97].

In relativistic quantum theory, the differential system

$$i\mathcal{J}_p u'(x,\lambda) + P(x)u(x,\lambda) = \lambda u(x,\lambda) \quad \text{with}$$
$$u(x,\lambda) = \begin{bmatrix} u_1(x,\lambda) \\ u_2(x,\lambda) \end{bmatrix}, \quad 0 \le x < d, \tag{2.99}$$

where

$$P \in L_{1,\mathrm{loc}}^{m\times m} \quad \text{and} \quad \overline{P(x)} = P(x)^\tau = P(x) \quad \text{a.e. on } [0,d) \tag{2.100}$$

is called a p-dimensional stationary Dirac system. This system is equivalent to the system (2.72) for $y(x,\lambda) = u(x,\lambda)^\tau$, with $J = -\mathcal{J}_p$, $m = 2p$, $N = I_m$, and $\mathcal{V}(x) = -iP(x)\mathcal{J}_p$; and the properties (2.100) are equivalent to the constraints (2.73) together with the extra constraint

$$\overline{\mathcal{V}(x)} = \mathcal{V}(x) \quad \text{a.e. on } [0,d). \tag{2.101}$$

Lemma 2.23 is adapted from [Ad68].

The class of Schrödinger equations with potentials of the form (2.89) that are expressed in terms of solutions of Riccati equations was studied by M.G. Krein [Kr56], where he presented a method for solving the inverse spectral problem for such equations. He also studied inverse problems for differential equations with potential and the Sturm–Liouville problem by reducing them to Dirac–Krein systems. The latter class of systems will be discussed in Chapter 12. Matrix Schrödinger equations are also studied at length in the treatises of Z.S. Agranovich and V.M. Marchenko [AM63] and [Ma11]; direct and inverse spectral problems for scalar Schrödinger equations are probably discussed in hundreds, if not thousands, of papers. The monographs [LS75], [LS91] and [Le87] of B.M. Levitan and I.S. Sargsjan are a good place to start. Additional references will be provided in later chapters.

In [Si99] B. Simon introduced a method for solving the inverse spectral problem for the Schrödinger equation that is very close to the Krein method that was discussed in the Supplementary Notes to the preceding chapter. It is based on the asymptotic representation

$$m(-\lambda^2) = -\lambda - \int_0^a e^{-2s\lambda} A(s) ds + O(e^{-2a\lambda})$$

of the function $m(\lambda) = \psi'(0, \lambda)/\psi(0, \lambda)$ that is defined in terms of the essentially unique solution $\psi \in L_1(\mathbb{R}_+)$ of (2.88). By theorem 7.1 in [Si99], the potential $q(x)$ is uniquely defined on the interval $[0, a]$ by the values of the function $A(x)$ on $[0, a]$. Thus, the inverse problem for q is solved by solving an extension problem for $A(x)$. On pp. 1031–1032 in [Si99] Simon explains his method by comparing it with known methods of recovering a Jacobi matrix from its spectral function with the aid of orthogonal polynomials. For additional perspective, see [GS00], [Re02] and [Re03],

M.G. Krein's explanation of his method in terms of Jacobi matrices in [Kr51a] was already noted in Section 1.16. Krein viewed the blocks of the matrizant of (2.72) with $J = j_p$ and $N = P_+$ as continuous analogs of orthogonal polynomials on the circle. He used the solutions of these systems in his study of inverse problems for Dirac systems and differential equations of second order that can be reduced to Dirac systems; continuous analogs of orthogonal polynomials are discussed in [Kr55].

3

Matrix-valued functions in the Nevanlinna class

In this chapter a number of results that will be needed to study bitangential direct and inverse problems in future chapters will be reviewed briefly. A more detailed and leisurely development may be found in chapters 3 and 4 of [ArD08b].

Recall that the symbols $\mathbb{C}$, $\mathbb{C}_+$ and $\mathbb{C}_-$ denote the complex plane, the open upper half plane and open lower half plane, respectively; $\mathbb{R}$ denotes the real line; $\mathbb{R}_+ = [0, \infty)$, $\mathbb{R}_- = (-\infty, 0]$; $\mathbb{T} = \{\lambda \in \mathbb{C} : |\lambda| = 1\}$; $\mathbb{C}^{p\times q}$ denotes the set of $p \times q$ matrices with complex entries and $\mathbb{C}^p = \mathbb{C}^{p\times 1}$; if $A \in \mathbb{C}^{p\times q}$, then A^τ denotes its transpose and $\|A\|$ its maximum singular value. The closure of a set Ω in a topological space is denoted $\overline{\Omega}$.

Basic classes of functions

A measurable $p \times q$ mvf $f(\mu)$ on $\mathbb{R}$ is said to belong to:

$L_r^{p\times q}$ for $1 \leq r < \infty$ if

$$\|f\|_r^r \stackrel{\text{def}}{=} \int_{-\infty}^{\infty} \text{trace}\{f(\mu)^* f(\mu)\}^{r/2} d\mu \quad \text{is finite.}$$

$\widetilde{L}_r^{p\times q}$ for $1 \leq r < \infty$ if

$$\int_{-\infty}^{\infty} (1+\mu^2)^{-1} \text{trace}\{f(\mu)^* f(\mu)\}^{r/2} d\mu \quad \text{is finite.}$$

$L_\infty^{p\times q}$ if

$$\|f\|_\infty \stackrel{\text{def}}{=} \text{ess sup}\{\|f(\mu\| : \mu \in \mathbb{R}\} \quad \text{is finite.}$$

$\mathbb{B}^{p\times q}$ if $f \in L_\infty^{p\times q}$ and $\|f\|_\infty \leq 1$.

$\mathcal{W}^{p\times q}(\gamma)$ (the **Wiener class**) for a fixed $\gamma \in \mathbb{C}^{p\times q}$, if it admits a representation of the form

$$f(\mu) = \gamma + \int_{-\infty}^{\infty} e^{i\mu t} h(t) dt, \ \text{for } \mu \in \mathbb{R},$$

where $h \in L_1^{p\times q}$;
$\mathcal{W}^{p\times q}$ if it belongs to $\mathcal{W}^{p\times q}(\gamma)$ for some $\gamma \in \mathbb{C}^{p\times q}$;

A $p \times q$ mvf $f(\lambda)$ is said to belong to:

$\mathcal{W}_+^{p\times q}(\gamma)$ for a fixed $\gamma \in \mathbb{C}^{p\times q}$, if it admits a representation of the form

$$f(\lambda) = \gamma + \int_0^\infty e^{i\lambda t} h(t)dt, \ \text{ for } \lambda \in \mathbb{R} \cup \mathbb{C}_+,$$

where $h \in L_1^{p\times q}(\mathbb{R}_+)$;
$\mathcal{W}_+^{p\times q}$ if it belongs to $\mathcal{W}_+^{p\times q}(\gamma)$ for some $\gamma \in \mathbb{C}^{p\times q}$;
$\mathcal{W}_-^{p\times q}(\gamma)$ for a fixed $\gamma \in \mathbb{C}^{p\times q}$, if it admits a representation of the form

$$f(\lambda) = \gamma + \int_{-\infty}^0 e^{i\lambda t} h(t)dt, \ \text{ for } \lambda \in \mathbb{R} \cup \mathbb{C}_-,$$

where $h \in L_1^{p\times q}(\mathbb{R}_-)$;
$\mathcal{W}_-^{p\times q}$ if it belongs to $\mathcal{W}_-^{p\times q}(\gamma)$ for some $\gamma \in \mathbb{C}^{p\times q}$;
$\mathcal{S}^{p\times q}$ (the **Schur class**) if it is holomorphic in $\mathbb{C}_+$ and if $f(\lambda)^* f(\lambda) \le I_q$ for every point $\lambda \in \mathbb{C}_+$;
$H_\infty^{p\times q}$ if it is holomorphic in $\mathbb{C}_+$ and if

$$\|f\|_\infty = \sup\{\|f(\lambda)\| : \lambda \in \mathbb{C}_+\} < \infty;$$

$H_r^{p\times q}$ (the **Hardy class**), for $1 \le r < \infty$, if it is holomorphic in $\mathbb{C}_+$ and if

$$\|f\|_r^r = \sup_{\nu>0} \int_{-\infty}^\infty \operatorname{trace}\{f(\mu + i\nu)^* f(\mu + i\nu)\}^{r/2} d\mu \ < \ \infty;$$

$\mathcal{C}^{p\times p}$ (the **Carathéodory class**) if $q = p$ and it is holomorphic in $\mathbb{C}_+$ and

$$(\Re f)(\lambda) \ = \ \frac{f(\lambda) + f(\lambda)^*}{2} \ \ge \ 0$$

for every point $\lambda \in \mathbb{C}_+$;
$\mathcal{N}_+^{p\times q}$ (the **Smirnov class**) and $\mathcal{N}_{\text{out}}^{p\times q}$ the **subclass of outer mvf's in** $\mathcal{N}_+^{p\times q}$ will be defined in (3.2) and (3.3), respectively;
$\mathcal{N}^{p\times q}$ (the **Nevanlinna class** of mvf's with bounded Nevanlinna characteristic) if it can be expressed in the form $f = h^{-1}g$, where $g \in H_\infty^{p\times q}$ and $h \in H_\infty$ $(= H_\infty^{1\times 1})$;
$\mathcal{E}^{p\times q}$ for the class of entire $p \times q$ mvf's.

For each class of $p \times q$ mvf's $\mathcal{X}^{p\times q}$ we shall use the symbols

$$\mathcal{X} \ \text{ instead of } \mathcal{X}^{1\times 1} \quad \text{and} \quad \mathcal{X}^p \ \text{ instead of } \mathcal{X}^{p\times 1}, \tag{3.1}$$

$\mathcal{X}_{\text{const}}^{p\times q}$ for the set of mvf's in $\mathcal{X}^{p\times q}$ that are constant

and

$$\mathcal{E} \cap \mathcal{X}^{p\times q} \quad \text{for the class of entire mvf's in } \mathcal{X}^{p\times q}.$$

The notation listed below will be used extensively.

$\langle g, h\rangle_{st} = \int_{-\infty}^{\infty} \text{trace}\, h(\mu)^* g(\mu) d\mu$ for the **standard inner product** in $L_2^{p\times q}$;
Π_+ to denote the orthogonal projection from the Hilbert space $L_2^{p\times q}$ onto the closed subspace $H_2^{p\times q}$;
$\Pi_- = I - \Pi_+$ for the complementary projection;
$\Pi_{\mathcal{L}}$ denotes the orthogonal projection onto a closed subspace $\mathcal{L}$ of a Hilbert space;
$f^\#(\lambda) = f(\overline{\lambda})^*, \quad f^\sim(\lambda) = f(-\overline{\lambda})^*$;
$\rho_\omega(\lambda) = -2\pi i(\lambda - \overline{\omega})$;
$\bigvee_{\alpha\in A}\{\mathcal{L}_\alpha\}$ for the closed linear span of subsets $\mathcal{L}_\alpha$ in a Hilbert space $\mathcal{X}$;
$e_t = e_t(\lambda) = \exp(it\lambda)$;

$$\ln^+ |a| = \begin{cases} \ln|a| & \text{if} |a| \geq 1 \\ 0 & \text{if} \quad |a| < 1. \end{cases}$$

3.1 Preliminaries on the Nevanlinna class $\mathcal{N}^{p\times q}$

The class $\mathcal{N}^{p\times q}$ is closed under addition, and, when meaningful, multiplication and inversion. Moreover, even though $\mathcal{N}^{p\times q}$ is listed last, it is the largest class in the classes of meromorphic $p \times q$ mvf's in $\mathbb{C}_+$ that are listed above.

The well-known theorem of Fatou on the existence of boundary values for functions in H_∞ guarantees that every mvf $f \in \mathcal{N}^{p\times q}$ has nontangential boundary values $f(\mu)$ a.e. on $\mathbb{R}$. In particular,

$$f \in \mathcal{N}^{p\times q} \Longrightarrow f(\mu) = \lim_{\nu\downarrow 0} f(\mu + i\nu) \quad \text{a.e. on } \mathbb{R}.$$

Moreover, this $f(\mu)$ may be extended to a measurable $p \times q$ mvf on the full axis $\mathbb{R}$ and f is uniquely defined by its boundary values on a subset of $\mathbb{R}$ of positive Lebesgue measure; see e.g., theorem 3.4 in [ArD08b] and the references cited there. In view of this, a mvf $f \in \mathcal{N}^{p\times q}$ will often be identified with its boundary value.

A mvf $f \in \mathcal{S}^{p\times q}$ is said to belong to the class

(1) $\mathcal{S}_{in}^{p\times q}$ of **inner** mvf's if $f(\mu)^* f(\mu) = I_q$ a.e. on $\mathbb{R}$;
(2) $\mathcal{S}_{*in}^{p\times q}$ of **$*$-inner** mvf's if $f(\mu) f(\mu)^* = I_p$ a.e. on $\mathbb{R}$;
(3) $\mathcal{S}_{out}^{p\times q}$ of **outer** mvf's if $\{fh : h \in H_2^q\}$ is dense in H_2^p;
(4) $\mathcal{S}_{*out}^{p\times q}$ of **$*$-outer** mvf's if $f^\sim$ is outer.

It is also readily checked that

$$\mathcal{S}_{in}^{p\times q} \neq \emptyset \Longleftrightarrow p \geq q, \quad \mathcal{S}_{out}^{p\times q} \neq \emptyset \Longleftrightarrow p \leq q$$

and

$$f \in \mathcal{S}_{\rm in}^{p\times q} \Longleftrightarrow f^{\sim} \in \mathcal{S}_{\rm *in}^{q\times p}.$$

However, if, as in most of this monograph, attention is restricted to square inner and outer mvf's in $\mathcal{S}^{p\times p}$, then it is not necessary to consider the classes $\mathcal{S}_{\rm *in}^{p\times p}$ and $\mathcal{S}_{\rm *out}^{p\times p}$ separately, since

$$\mathcal{S}_{\rm *in}^{p\times p} = \mathcal{S}_{\rm in}^{p\times p} \quad \text{and} \quad \mathcal{S}_{\rm *out}^{p\times p} = \mathcal{S}_{\rm out}^{p\times p}.$$

In keeping with the convention (3.1),

$$\mathcal{S}_{\rm in} = \mathcal{S}_{\rm in}^{1\times 1} \quad \text{and} \quad \mathcal{S}_{\rm out} = \mathcal{S}_{\rm out}^{1\times 1}.$$

The **Smirnov class** $\mathcal{N}_+^{p\times q}$ and the subclass $\mathcal{N}_{\rm out}^{p\times q}$ of outer $p\times q$ mvf's in $\mathcal{N}_+^{p\times q}$ may be defined by the formulas

$$\mathcal{N}_+^{p\times q} = \{h^{-1}g:\ g \in \mathcal{S}^{p\times q} \text{ and } h \in \mathcal{S}_{\rm out}\} \tag{3.2}$$

and

$$\mathcal{N}_{\rm out}^{p\times q} = \{h^{-1}g:\ g \in \mathcal{S}_{\rm out}^{p\times q} \text{ and } h \in \mathcal{S}_{\rm out}\}. \tag{3.3}$$

The Smirnov maximum principle

We turn next to the **Smirnov maximum principle**. In the formulation, $f \in \mathcal{N}_+^{p\times q}$ is identified with its boundary values.

Theorem 3.1 *If $f \in \mathcal{N}_+^{p\times q}$, then*

$$\sup_{\nu>0} \int_{-\infty}^{\infty} (\text{trace}\{f(\mu+i\nu)^* f(\mu+i\nu)\})^{r/2} d\mu = \int_{-\infty}^{\infty} (\text{trace}\{f(\mu)^* f(\mu)\})^{r/2} d\mu$$

for $1 \le r < \infty$,

$$\sup_{\lambda\in\mathbb{C}_+} \text{trace}\{f(\lambda)^* f(\lambda)\} = \text{ess sup}_{\mu\in\mathbb{R}} \text{trace}\{f(\mu)^* f(\mu)\},$$

$$\sup_{\nu>0} \int_{-\infty}^{\infty} \|f(\mu+i\nu)\|^r d\mu = \int_{-\infty}^{\infty} \|f(\mu)\|^r d\mu \quad \textit{for } 1 \le r < \infty$$

and

$$\sup_{\lambda\in\mathbb{C}_+} \|f(\lambda)\| = \text{ess sup}_{\mu\in\mathbb{R}}\{\|f(\mu)\|\},$$

where in these equalities both sides can be infinite. In particular,

$$\mathcal{N}_+^{p\times q} \cap L_r^{p\times q}(\mathbb{R}) = H_r^{p\times q} \tag{3.4}$$

for $1 \le r \le \infty$.

Proof See theorem A on p. 88 of [RR85]. □

In view of the inclusion $H_r^{p\times q} \subset \mathcal{N}^{p\times q}$ for $1 \le r \le \infty$, every $f \in H_r^{p\times q}$, $1 \le r \le \infty$, has nontangential boundary values. Moreover, the norm in $H_r^{p\times q}$ can be computed in terms of boundary values only, and, by Theorem 3.1, the corresponding spaces $H_r^{p\times q}$ can be identified as closed subspaces of the Lebesgue spaces $L_r^{p\times q}$ on the line, for $1 \le r \le \infty$. In particular, $H_2^{p\times q}$ is isometrically included in the Hilbert space $L_2^{p\times q}$ with inner product

$$\langle f, g\rangle_{\mathrm{st}} = \int_{-\infty}^{\infty} \mathrm{trace}\, \{g(\mu)^* f(\mu)\} d\mu. \tag{3.5}$$

Analogous classes will be considered for the open lower half plane $\mathbb{C}_-$. In particular, a $p \times q$ mvf f is said to belong to

the Nevanlinna class with respect to $\mathbb{C}_-$ if $f^\# \in \mathcal{N}^{q\times p}$;
the Smirnov class $\mathcal{N}_-^{p\times q}$ if $f^\# \in \mathcal{N}_+^{q\times p}$.

Every mvf f in the Nevanlinna class with respect to $\mathbb{C}_-$ also has nontangential boundary values. Thus,

$$f(\mu) = \lim_{\nu\downarrow 0} f(\mu - i\nu) \quad \text{a.e. on } \mathbb{R}.$$

Moreover, the orthogonal complement $(H_2^{p\times q})^\perp$ of $H_2^{p\times q}$ in $L_2^{p\times q}$ will be identified as the boundary values of the set of $p \times q$ mvf's f that are holomorphic in $\mathbb{C}_-$ and meet the constraint

$$\|f\|_2^2 = \sup_{\nu>0} \int_{-\infty}^{\infty} \mathrm{trace}\{f(\mu - i\nu)^* f(\mu - i\nu)\} d\mu \; < \; \infty.$$

Thus, $f \in (H_2^{p\times q})^\perp \iff f^\# \in H_2^{q\times p}$. Therefore,

$$L_2^{p\times q} = H_2^{p\times q} \oplus (H_2^{p\times q})^\perp \quad \text{and} \quad L_2^p = H_2^p \oplus (H_2^p)^\perp \quad \text{if} \quad q = 1.$$

The **Cauchy formula**

$$f(\omega) = \frac{1}{2\pi i} \int_{-\infty}^{\infty} \frac{f(\mu)}{\mu - \omega} d\mu \quad \text{for } \omega \in \mathbb{C}_+ \tag{3.6}$$

and the **Poisson formula**

$$f(\omega) = \frac{\Im \omega}{\pi} \int_{-\infty}^{\infty} \frac{f(\mu)}{|\mu - \omega|^2} d\mu \quad \text{for } \omega \in \mathbb{C}_+ \tag{3.7}$$

are valid for every $f \in H_r^{p\times q}$ for $1 \le r < \infty$. Formula (3.7) is also valid for $f \in H_\infty^{p\times q}$.

This follows from theorems 11.2 and 11.8 in [Du70], since it suffices to verify the asserted formulas for each entry in the mvf f.

Inner–outer factorization

Let $f(\lambda)$ be a mvf that is meromorphic in some open nonempty subset Ω of $\mathbb{C}$ (not necessarily connected). Then $\mathfrak{h}_f$ denotes the set of points $\omega \in \Omega$ at which f is holomorphic,

$$\mathfrak{h}_f^+ = \mathfrak{h}_f \cap \mathbb{C}_+, \quad \mathfrak{h}_f^- = \mathfrak{h}_f \cap \mathbb{C}_- \quad \text{and} \quad \mathfrak{h}_f^0 = \mathfrak{h}_f \cap \mathbb{R}.$$

The **rank** of a meromorphic $p \times q$ mvf $f(\lambda)$ in $\mathbb{C}_+$ is defined by the formula

$$\operatorname{rank} f = \max\{\operatorname{rank} f(\lambda) : \lambda \in \mathfrak{h}_f^+\}.$$

The following implications will be useful:

$$f \in \mathcal{N}_{\text{out}}^{p\times q} \Longrightarrow \operatorname{rank} f = p \quad \text{and} \quad f \in \mathcal{S}_{\text{in}}^{p\times q} \Longrightarrow \operatorname{rank} f = q. \tag{3.8}$$

Theorem 3.2 *Every mvf $f \in \mathcal{N}_+^{p\times q}$ that is not identically equal to zero admits an inner–outer factorization of the form*

$$f(\lambda) = b_L(\lambda)\varphi_L(\lambda), \quad \text{where } b_L \in \mathcal{S}_{\text{in}}^{p\times r} \text{ and } \varphi_L \in \mathcal{N}_{\text{out}}^{r\times q} \tag{3.9}$$

and a $*$-outer–$*$-inner factorization of the form

$$f(\lambda) = \varphi_R(\lambda)b_R(\lambda), \quad \text{where } \varphi_R \in \mathcal{N}_{*\text{out}}^{p\times r} \text{ and } b_R \in \mathcal{S}_{*\text{out}}^{r\times q}. \tag{3.10}$$

In both of these factorizations,

$$r = \operatorname{rank} f. \tag{3.11}$$

The factors in each of these factorizations are defined uniquely up to replacement of

$$b_L \text{ and } \varphi_L \quad \text{by} \quad b_L u \text{ and } u^*\varphi_L$$

and

$$\varphi_R \text{ and } b_R \quad \text{by} \quad \varphi_R v \text{ and } v^* b_R,$$

where u and v are constant unitary $r \times r$ matrices. Moreover

$$f \in H_t^{p\times q} \Longleftrightarrow \varphi_L \in H_t^{r\times q} \Longleftrightarrow \varphi_R \in H_t^{p\times r}, \quad 1 \le t \le \infty, \tag{3.12}$$

$$\rho_i^{-1} f \in H_t^{p\times q} \Longleftrightarrow \rho_i^{-1}\varphi_L \in H_t^{r\times q} \Longleftrightarrow \rho_i^{-1}\varphi_R \in H_t^{p\times r}, \quad 1 \le t \le \infty, \tag{3.13}$$

and

$$f \in \mathcal{S}^{p\times q} \Longleftrightarrow \varphi_L \in \mathcal{S}_{\text{out}}^{r\times q} \Longleftrightarrow \varphi_R \in \mathcal{S}_{*\text{out}}^{p\times r}, \quad 1 \le t \le \infty, \tag{3.14}$$

Proof See theorem 3.71 in [ArD08b] for the first part. The nontrivial directions in the last three sets of equivalences follow from Theorem 3.1, the Smirnov maximum principle. □

The Beurling–Lax theorem

A number of results connected with the Beurling–Lax theorem are now presented without proof; proofs may be found on pp. 108–111 of [ArD08b] and the references cited there.

Theorem 3.3 (Beurling–Lax) *Let $\mathcal{L}$ be a proper closed nonzero subspace of H_2^p such that*

$$e_t f \in \mathcal{L} \ \ \text{for every } f \in \mathcal{L} \ \ \text{and every } \ t \geq 0.$$

Then there exists a positive integer $r \leq p$ and an inner mvf $b \in \mathcal{S}_{in}^{p\times r}$ such that

$$\mathcal{L} = bH_2^r. \tag{3.15}$$

Moreover, this mvf $b(\lambda)$ is uniquely defined by $\mathcal{L}$ up to a unitary constant right multiplier.

Remark 3.4 *For many of the applications in this monograph, the following reformulation of Theorem 3.3 is useful:*

Let $\mathcal{M}$ be a proper closed nonzero subspace of H_2^p such that

$$\Pi_+ e_{-t} f \in \mathcal{M} \ \ \text{for every } f \in \mathcal{M} \ \ \text{and every } \ t \geq 0.$$

Then there exists a positive integer $r \leq p$ and an inner mvf $b \in \mathcal{S}_{in}^{p\times r}$ such that

$$\mathcal{M} = \mathcal{H}(b) \stackrel{\text{def}}{=} H_2^p \ominus bH_2^r. \tag{3.16}$$

Moreover, this mvf $b(\lambda)$ is uniquely defined by $\mathcal{M}$ up to a unitary constant right multiplier.

The connection between this formulation and Theorem 3.3 rests on the observation that if $f \in \mathcal{M}$ and $g \in \mathcal{M}^\perp$, the orthogonal complement of $\mathcal{M}$ in H_2^p, then

$$\langle e_t g, f\rangle_{\text{st}} = \langle g, \Pi_+ e_{-t} f\rangle_{\text{st}} \quad \text{for every } t \geq 0.$$

Thus, a closed subspace $\mathcal{M}$ of H_2^p meets the conditions formulated just above if and only if $\mathcal{M}^\perp$ meets the conditions formulated in Theorem 3.3.

Lemma 3.5 *Let $b_\alpha(\lambda) = (\lambda - \alpha)/(\lambda - \overline{\alpha})$ for $\alpha \in \mathbb{C}_+$ and let $\mathcal{L}$ be a proper closed subspace of H_2^p. Then the following assertions are equivalent:*

1. *$e_t\mathcal{L} \subseteq \mathcal{L}$ for every $t \geq 0$.*
2. *$b_\alpha\mathcal{L} \subseteq \mathcal{L}$ for at least one point $\alpha \in \mathbb{C}_+$.*
3. *$b_\alpha\mathcal{L} \subseteq \mathcal{L}$ for every point $\alpha \in \mathbb{C}_+$.*

The **generalized backwards shift** operator R_α is defined for vvf's and mvf's by the rule

$$(R_\alpha f)(\lambda) = \begin{cases} \dfrac{f(\lambda) - f(\alpha)}{\lambda - \alpha} & \text{if } \lambda \neq \alpha \\ f'(\alpha) & \text{if } \lambda = \alpha \end{cases} \tag{3.17}$$

for every $\lambda, \alpha \in \mathfrak{h}_f$. In order to keep the typography simple, the space in which R_α acts will not be indicated in the notation.

Lemma 3.6 *If $\mathcal{L}$ is a proper closed subspace of H_2^p, then the following assertions are equivalent:*

(1) *$\Pi_+ e_{-t} f \in \mathcal{L}$ for every $f \in \mathcal{L}$ and every $t \geq 0$.*
(2) *$R_\alpha \mathcal{L} \subseteq \mathcal{L}$ for at least one point $\alpha \in \mathbb{C}_+$.*
(3) *$R_\alpha \mathcal{L} \subseteq \mathcal{L}$ for every point $\alpha \in \mathbb{C}_+$.*
(4) *There exists a positive integer $r \leq p$ and an essentially unique mvf $b \in \mathcal{S}_{\text{in}}^{p\times r}$ such that $\mathcal{L} = H_2^p \ominus bH_2^r$.*

Lemma 3.7 *If $bH_2^r \supseteq b_1 H_2^p$, where $b \in \mathcal{S}_{\text{in}}^{p\times r}$ and $b_1 \in \mathcal{S}_{\text{in}}^{p\times p}$, then $r = p$ and $b^{-1}b_1 \in \mathcal{S}_{\text{in}}^{p\times p}$.*

Corollary 3.8 *If $bH_2^r \supseteq \beta H_2^p$, where $b \in \mathcal{S}_{\text{in}}^{p\times r}$ and $\beta \in \mathcal{S}_{\text{in}}$, then $r = p$ and $\beta b^{-1} \in \mathcal{S}_{\text{in}}^{p\times p}$.*

Theorem 3.9 *If $b_\alpha \in \mathcal{S}_{\text{in}}^{p\times p}$ for $\alpha \in \mathcal{A}$, then:*

I *There exists an essentially unique mvf $b \in \mathcal{S}_{\text{in}}^{p\times p}$ such that*
(1) *$b^{-1}b_\alpha \in \mathcal{S}_{\text{in}}^{p\times p}$ for every $\alpha \in \mathcal{A}$.*
(2) *If $\widetilde{b} \in \mathcal{S}_{\text{in}}^{p\times p}$ and $\widetilde{b}^{-1}b_\alpha \in \mathcal{S}_{\text{in}}^{p\times p}$ for every $\alpha \in \mathcal{A}$, then $\widetilde{b}^{-1}b \in \mathcal{S}_{\text{in}}^{p\times p}$.*
Moreover, if b_α is entire for some $\alpha \in \mathcal{A}$, then b is entire.

II *If*

$$b_\alpha^{-1}b^\circ \in \mathcal{S}_{\text{in}}^{p\times p} \quad \text{for some } b^\circ \in \mathcal{S}_{\text{in}}^{p\times p} \text{ and every } \alpha \in \mathcal{A}, \tag{3.18}$$

then there exists an essentially unique mvf $b \in \mathcal{S}_{\text{in}}^{p\times p}$ such that
(1) *$b_\alpha^{-1}b \in \mathcal{S}_{\text{in}}^{p\times p}$ for every $\alpha \in \mathcal{A}$.*
(2) *$b^{-1}b^\circ \in \mathcal{S}_{\text{in}}^{p\times p}$ for every $b^\circ \in \mathcal{S}_{\text{in}}^{p\times p}$ for which (3.18) holds.*
Moreover,

$$bH_2^p = \bigvee_{\alpha\in\mathcal{A}} b_\alpha H_2^p \quad \text{in setting I and} \quad bH_2^p = \bigcap_{\alpha\in\mathcal{A}} b_\alpha H_2^p \quad \text{in setting II.}$$

Furthermore, if b_α is entire for every $\alpha \in \mathcal{A}$, then b is entire.

A mvf b that satisfies the two conditions in I (resp., II) is called a **greatest common left divisor** (resp., **least common right multiple**) of the family $\{b_\alpha : \alpha \in \mathcal{A}\}$.

Corollary 3.10 *If b_1 and b_2 belong to $\mathcal{S}_{in}^{p\times p}$, then there exists an essentially unique $b \in \mathcal{S}_{in}^{p\times p}$ such that*

$$bH_2^p = b_1 H_2^p \bigvee b_2 H_2^p.$$

Moreover, if $b \in \mathcal{S}_{in}^{p\times p}$, then

$$\mathcal{H}(b) = \mathcal{H}(b_1) \cap \mathcal{H}(b_2) \Longleftrightarrow bH_2^p = b_1 H_2^p \bigvee b_2 H_2^p$$

and

$$b_1 \in \mathcal{E} \cap \mathcal{S}_{in}^{p\times p} \quad or \quad b_2 \in \mathcal{E} \cap \mathcal{S}_{in}^{p\times p} \Longrightarrow b \in \mathcal{E} \cap \mathcal{S}_{in}^{p\times p}.$$

The next two theorems summarize a number of well-known results for the convenience of the reader; proofs may be found on pp. 116–121 of [ArD08b] and the references cited there.

Theorem 3.11 *If $s \in \mathcal{S}^{p\times p}$, then:*

(1) $\det s \in \mathcal{S}$.
(2) $s \in \mathcal{S}_{out}^{p\times p} \iff \det s \in \mathcal{S}_{out}$.
(3) $s \in \mathcal{S}_{in}^{p\times p} \iff \det s \in \mathcal{S}_{in}$.
(4) $\det\{I_p - s(\lambda)\} \not\equiv 0$ *in* $\mathbb{C}_+ \Longrightarrow$ *that the mvf* $I_p - s$ *is outer in* $H_\infty^{p\times p}$.

If $s \in \mathcal{S}^{p\times q}$ and $s(\omega)^* s(\omega) < I_q$ at a point $\omega \in \mathbb{C}_+$, then $s(\lambda)^* s(\lambda) < I_q$ at every point $\lambda \in \mathbb{C}_+$.

Theorem 3.12 *If $f \in \mathcal{N}^{p\times p}$ and* $\det f(\lambda) \not\equiv 0$ *in $\mathbb{C}_+$, then*

(1) $f \in \mathcal{N}_{out}^{p\times p}$ *if and only if both f and f^{-1} belong to $\mathcal{N}_+^{p\times p}$.*
(2) $f \in \mathcal{N}_{out}^{p\times p}$ *if and only if $f^{-1} \in \mathcal{N}_{out}^{p\times p}$.*

Moreover, if $f \in \mathcal{N}_+^{p\times p}$. then

$$f \in \mathcal{N}_{out}^{p\times p} \Longleftrightarrow \det f \in \mathcal{N}_{out}. \tag{3.19}$$

Lemma 3.13 *If $b \in \mathcal{S}_{in}^{p\times p}$ and $d =$* $\det b$*, then:*

(1) $b^{-1}d \in \mathcal{S}_{in}^{p\times p}$.
(2) *If* $\det b(\lambda)$ *is constant in $\mathbb{C}_+$, then $b(\lambda) \equiv$ constant.*

Proof Let $f = db^{-1}$. Then $f \in H_\infty^{p\times p}$ and, as follows with the help of (3) of Theorem 3.11, $f(\mu)^* f(\mu) = I_p$ a.e. on $\mathbb{R}$. Therefore, $f \in \mathcal{S}_{in}^{p\times p}$ by the maximum principle in $H_\infty^{p\times p}$. This proves (1); (2) follows from (1). □

Lemma 3.14 *If $b \in \mathcal{S}_{in}^{p\times q}$, then $R_\alpha b\xi \in \mathcal{H}(b)$ for every $\alpha \in \mathbb{C}_+$ and every $\xi \in \mathbb{C}^q$. If $b \in \mathcal{S}_{in}^{p\times p}$ and $d =$* $\det b$*, then $\mathcal{H}(b) \subseteq \mathcal{H}(dI_p)$.*

Proof If $\alpha \in \mathbb{C}_+$ and $\xi \in \mathbb{C}^q$, then $R_\alpha b\xi \in H_2^p$ and, for every $h \in H_2^q$,

$$\langle R_\alpha b\xi, bh\rangle_{st} = \left\langle \frac{\xi}{\mu - \alpha}, h\right\rangle_{st} - \left\langle \frac{b(\alpha)\xi}{\mu - \alpha}, bh\right\rangle_{st} = 0 - 0 = 0,$$

since $\xi/(\lambda - \alpha)$ belongs to $(H_2^q)^\perp$ and $b(\alpha)\xi/(\lambda - \alpha)$ belongs to $(H_2^p)^\perp$. The last assertion follows from Lemma 3.13 and Corollary 3.8. □

3.2 Linear fractional transformations and Redheffer transformations

The linear fractional transformations T_U based on an $m \times m$ mvf

$$U(\lambda) = \begin{bmatrix} u_{11}(\lambda) & u_{12}(\lambda) \\ u_{21}(\lambda) & u_{22}(\lambda) \end{bmatrix} \tag{3.20}$$

that is meromorphic in $\mathbb{C}_+$ with blocks $u_{11}(\lambda)$ of size $p \times p$ and $u_{22}(\lambda)$ of size $q \times q$, respectively, is defined by the formula

$$T_U[x] = \{u_{11}(\lambda)x(\lambda) + u_{12}(\lambda)\}\{u_{21}(\lambda)x(\lambda) + u_{22}(\lambda)\}^{-1} \tag{3.21}$$

for $p \times q$ mvf's $x(\lambda)$ that belong to the set

$$\begin{aligned} \mathcal{D}(T_U) &= \{x(\lambda) : x \text{ is } p \times q \text{ meromorphic mvf in } \mathbb{C}_+ \\ &\qquad \text{with } \det[u_{21}(\lambda)x(\lambda) + u_{22}(\lambda)] \not\equiv 0 \text{ in } \mathbb{C}_+\}, \end{aligned} \tag{3.22}$$

which is the domain of definition of T_U. The notation

$$T_U[X] = \{T_U[x] : x \in X\} \quad \text{for subsets } X \text{ of } \mathcal{D}(T_U)$$

will be useful.

If U_1 and U_2 are $m \times m$ meromorphic mvf's in $\mathbb{C}_+$, $x \in \mathcal{D}(T_{U_1})$ and $T_{U_1}[x] \in \mathcal{D}(T_{U_2})$, then

$$x \in \mathcal{D}(T_{U_2U_1}) \quad \text{and} \quad T_{U_2}[T_{U_1}[x]] = T_{U_2U_1}[x]. \tag{3.23}$$

Moreover, if $U_2(\lambda)U_1(\lambda) = I_m$, then T_{U_1} maps $\mathcal{D}(T_{U_1})$ bijectively onto $\mathcal{D}(T_{U_2})$ and $T_{U_1}^{-1} = T_{U_2}$, i.e.,

$$\begin{aligned} &T_{U_1}[\mathcal{D}(T_{U_1})] = \mathcal{D}(T_{U_2}), \quad T_{U_2}[\mathcal{D}(T_{U_2})] = \mathcal{D}(T_{U_1}), \\ &\qquad T_{U_2}T_{U_1}|_{\mathcal{D}(T_{U_1})} = I|_{\mathcal{D}(T_{U_1})} \quad \text{and} \quad T_{U_1}T_{U_2}|_{\mathcal{D}(T_{U_2})} = I|_{\mathcal{D}(T_{U_2})}. \end{aligned} \tag{3.24}$$

Remark 3.15 *In some applications it is convenient to introduce a second linear fractional transformation: the* **left linear fractional transformation**

$$T_U^\ell[y] = \{y(\lambda)u_{12}(\lambda) + u_{22}(\lambda)\}^{-1}\{y(\lambda)u_{11}(\lambda) + u_{21}(\lambda)\}, \tag{3.25}$$

which acts in the set of $q \times p$ mvf's $y(\lambda)$ that are meromorphic in $\mathbb{C}_+$ and belong to the set

$$\mathcal{D}(T_U^\ell) = \{y(\lambda) : \det\{y(\lambda)u_{12}(\lambda) + u_{22}(\lambda)\} \not\equiv 0 \text{ in } \mathbb{C}_+\},$$

the domain of definition of T_U^ℓ. The transformation T_U in (3.21) is sometimes referred to as a right linear fractional transformation and denoted T_U^r.

The notation

$$s_{12} = T_U^r[0_{p\times q}] = u_{12}u_{22}^{-1} \quad \text{and} \quad \chi = T_U^\ell[0_{q\times p}] = u_{22}^{-1}u_{21} \tag{3.26}$$

will be used for $m \times m$ mvf's U that are meromorphic in $\mathbb{C}_+$ with $\det u_{22}(\lambda) \not\equiv 0$ in $\mathbb{C}_+$.

Lemma 3.16 *If $U(\lambda)$ is a meromorphic $m \times m$ mvf in $\mathbb{C}_+$ with block decomposition (3.20) such that* $\det u_{22}(\lambda) \not\equiv 0$ *in $\mathbb{C}_+$, then:*

(1) $\mathcal{S}^{p\times q} \subseteq \mathcal{D}(T_U^r) \iff \chi \in \mathcal{S}^{q\times p}$ *and $\chi(\lambda)^*\chi(\lambda) < I_p$ for each point $\lambda \in \mathbb{C}_+$.*
(2) $\mathcal{S}^{q\times p} \subseteq \mathcal{D}(T_U^\ell) \iff s_{12} \in \mathcal{S}^{p\times q}$ *and $s_{12}(\lambda)^*s_{12}(\lambda) < I_q$ for each point $\lambda \in \mathbb{C}_+$.*

Proof This is lemma 4.64 in [ArD08b]. □

In the future, we shall usually drop the superscript r and write T_U instead of T_U^r.

Lemma 3.17 *Let $U(\lambda)$ be an $m \times m$ mvf that is meromorphic in $\mathbb{C}_+$ with block decomposition (3.20) and assume that $\mathcal{S}^{p\times q} \subseteq \mathcal{D}(T_U)$ (so that (1) of Lemma 3.16 is applicable). Then the mvf*

$$\Sigma = PG(U) = (P_- + P_+U)(P_+ + P_-U)^{-1} \quad \textit{with } P_\pm = (I_m \pm j_{pq})/2$$

is meromorphic in $\mathbb{C}_+$ and the blocks σ_{ij} of the four block decomposition of Σ are given by the formulas

$$\sigma_{11} = u_{11} - u_{12}u_{22}^{-1}u_{21}, \ \sigma_{12} = u_{12}u_{22}^{-1}, \ \sigma_{21} = -u_{22}^{-1}u_{21}, \ \sigma_{22} = u_{22}^{-1} \tag{3.27}$$

and the **Redheffer transform**

$$R_\Sigma[\varepsilon] = \sigma_{12} + \sigma_{11}\varepsilon(I_q - \sigma_{21}\varepsilon)^{-1}\sigma_{22} \tag{3.28}$$

is a $p \times q$ mvf that is meromorphic in $\mathbb{C}_+$ and holomorphic in $\mathfrak{h}_\Sigma^+$ for every choice of $\varepsilon \in \mathcal{S}^{p\times q}$. Moreover,

$$T_U[\varepsilon] = R_\Sigma[\varepsilon] \quad \textit{for every } \varepsilon \in \mathcal{S}^{p\times q}. \tag{3.29}$$

Proof Under the given assumptions, $\sigma_{21} \in \mathcal{S}^{q\times p}$ and

$$\sigma_{21}(\lambda)^*\sigma_{21}(\lambda) < I_p \quad \text{for } \lambda \in \mathbb{C}_+.$$

Thus, $\det\{I_q - s_{21}(\lambda)\varepsilon(\lambda)\} \neq 0$ for every point $\lambda \in \mathbb{C}_+$ and every $\varepsilon \in \mathcal{S}^{p\times q}$. Therefore, the $p \times q$ mvf $R_\Sigma[\varepsilon]$ is meromorphic in $\mathbb{C}_+$ and holomorphic in $\mathfrak{h}_\Sigma^+$ for every $\varepsilon \in \mathcal{S}^{p\times q}$. Formula (3.29) is a straightforward calculation. □

Lemma 3.18 *Let $U(\lambda)$ be an $m \times m$ mvf that is meromorphic in $\mathbb{C}_+$ with block decomposition (3.20) and assume that $\mathcal{S}^{p\times q} \subseteq \mathcal{D}(T_U)$ (so that (1) of Lemma 3.16 is applicable and the formulas (3.27)–(3.29) are in force). Then for each point $\omega \in \mathfrak{h}_U^+ \cap \mathfrak{h}_{u_{22}^{-1}}$ there exists exactly one matrix $W \in \mathcal{U}_{\rm const}(j_{pq})$ such that the blocks $\widetilde{u}_{ij}$ of the new mvf $\widetilde{U}(\lambda) = U(\lambda)W$ meet the following normalization conditions at the point ω:*

$$\widetilde{u}_{11}(\omega) \geq 0, \quad \widetilde{u}_{22}(\omega) > 0 \quad \text{and} \quad \widetilde{u}_{21}(\omega) = 0. \tag{3.30}$$

Moreover,

$$T_{\widetilde{U}}[\varepsilon] = R_{\widetilde{\Sigma}}[\varepsilon], \tag{3.31}$$

where $\widetilde{\Sigma} = PG[\widetilde{U}]$ with $P_\pm = (I_m \pm j_{pq})/2$ and the blocks $\widetilde{\sigma}_{ij}$ of $\widetilde{\Sigma}$ meet the normalization conditions

$$\widetilde{\sigma}_{11}(\omega) \geq 0, \quad \widetilde{\sigma}_{22}(\omega) > 0 \quad \text{and} \quad \widetilde{\sigma}_{21}(\omega) = 0. \tag{3.32}$$

Furthermore,

$$\det U(\omega) \neq 0 \Longleftrightarrow \widetilde{u}_{11}(\omega) > 0.$$

Proof The matrix W may be defined by formula (3.127), with

$$k = -\chi(\omega)^* = -u_{21}(\omega)^* u_{22}(\omega)^{-*}$$

and unitary matrices u and v such that

$$\widetilde{u}_{11}(\omega) = \{u_{11}(\omega) + u_{12}(\omega)k^*\}(I_p - kk^*)^{-1/2}u \geq 0$$

and

$$\widetilde{u}_{22}(\omega) = \{u_{21}(\omega)k + u_{22}(\omega)\}(I_q - k^*k)^{-1/2}v > 0. \qquad \square$$

Matrix balls

A **matrix ball** $\mathcal{B}$ in $\mathbb{C}^{p\times q}$ with **center** $\gamma_c \in \mathbb{C}^{p\times q}$ is a set of the form

$$\{\gamma_c + R_\ell Q R_r : Q \in \mathcal{S}_{\rm const}^{p\times q}, \ R_\ell \in \mathbb{C}^{p\times p}, R_r \in \mathbb{C}^{q\times q}, R_\ell \geq 0 \text{ and } R_r \geq 0\};$$

R_ℓ is called the **left semiradius** and R_r the **right semiradius**. More information on matrix balls may be found, e.g., in section 2.4 of [ArD08b]. In particular, it is known that the center γ_c of a matrix ball is uniquely defined by the ball and that if the ball contains more than one point, then the left and right semiradii $R_\ell \geq 0$ and $R_r \geq 0$ of $\mathcal{B}$ are defined by the ball up to the transformation

$$R_\ell \longrightarrow \rho R_\ell \quad \text{and} \quad R_r \longrightarrow \rho^{-1} R_r \quad \text{for some } \rho > 0.$$

Lemma 3.19 *Let $U(\lambda)$ be an $m \times m$ mvf that is meromorphic in $\mathbb{C}_+$ with block decomposition (3.20) and assume that $\mathcal{S}^{p\times q} \subseteq \mathcal{D}(T_U)$ (so that (1) of Lemma 3.16 is applicable and formulas (3.27)–(3.28) are in force). Fix a point $\omega \in \mathfrak{h}_U^+ \cap \mathfrak{h}_{u_{22}^{-1}}$ and let $\widetilde{\Sigma} = [\widetilde{\sigma}_{ij}]$ be defined as in Lemma 3.18. Then the set*

$$\mathcal{B}_U(\omega) \stackrel{\text{def}}{=} \{T_{U(\omega)}[\varepsilon(\omega)] : \varepsilon \in \mathcal{S}^{p\times q}\} \tag{3.33}$$

is a matrix ball with center $\widetilde{\sigma}_{12}(\omega)$ and left and right semiradii $R_\ell(\omega) = \widetilde{\sigma}_{11}(\omega)$ and $R_r(\omega) = \widetilde{\sigma}_{22}(\omega)$, respectively:

$$\mathcal{B}_U(\omega) \stackrel{\text{def}}{=} \{\widetilde{\sigma}_{12}(\omega) + \widetilde{\sigma}_{11}(\omega)Q\widetilde{\sigma}_{22}(\omega) : Q \in \mathcal{S}_{\text{const}}^{p\times q}\}. \tag{3.34}$$

In this ball $R_r(\omega) > 0$ and

$$R_\ell(\omega) > 0 \Longleftrightarrow \det U(\omega) \neq 0.$$

Proof This follows easily from Lemmas 3.17 and 3.18 and the fact that

$$T_U[\mathcal{S}^{p\times q}] = T_{\widetilde{U}}[\mathcal{S}^{p\times q}] = R_{\widetilde{\Sigma}}[\mathcal{S}^{p\times q}].$$

□

Remark 3.20 *The center $\gamma_c(\omega)$ and the semiradii $R_\ell(\omega)$ and $R_r(\omega)$ of the ball $\mathcal{B}_U(\omega)$ may be expressed directly in terms of the blocks of $U(\omega)$ by the formulas*

$$\begin{aligned} R_\ell(\omega)^2 &= \widetilde{\sigma}_{11}(\omega)^2 = \widetilde{u}_{11}(\omega)^2 \\ &= \{u_{11}(\omega) + u_{12}(\omega)k^*\}(I_p - kk^*)^{-1}\{u_{11}(\omega) + u_{12}(\omega)k^*\}^*, \end{aligned} \tag{3.35}$$

$$\begin{aligned} R_r(\omega)^2 &= \widetilde{\sigma}_{22}(\omega)^2 = \widetilde{u}_{22}(\omega)^{-2} \\ &= (u_{22}(\omega)u_{22}(\omega)^* - u_{21}(\omega)u_{21}(\omega)^*)^{-1} \end{aligned} \tag{3.36}$$

and

$$\begin{aligned} \gamma_c(\omega) &= \widetilde{\sigma}_{12}(\omega) = T_{\widetilde{U}(\omega)}[0] = T_{U(\omega)}[-\chi(\omega)^*] \\ &= (u_{12}(\omega)u_{22}(\omega)^* - u_{11}(\omega)u_{21}(\omega)^*)R_r(\omega)^2, \end{aligned} \tag{3.37}$$

where

$$k = -\chi(\omega)^* \quad \textit{and} \quad \chi(\omega) = u_{22}(\omega)^{-1}u_{21}(\omega).$$

The linear fractional transformation

$$T_{\mathfrak{V}}[x] = (I_p - x)(I_p + x)^{-1} \tag{3.38}$$

based on the matrix $\mathfrak{V}$ defined by formula (1.6) is called the **Cayley transform**. It is readily checked that

$$\mathcal{C}^{p\times p} \subset \mathcal{D}(T_{\mathfrak{V}}), \quad \mathcal{C}^{p\times p} = T_{\mathfrak{V}}[\mathcal{S}^{p\times p} \cap \mathcal{D}(T_{\mathfrak{V}})] \tag{3.39}$$

and

$$\mathcal{S}^{p\times p}\cap\mathcal{D}(T_{\mathfrak{V}})=\{s\in\mathcal{S}^{p\times p}:\ \det\{I_p+s(\lambda)\}\neq 0 \text{ in } \mathbb{C}_+\}=T_{\mathfrak{V}}[\mathcal{C}^{p\times p}]. \tag{3.40}$$

Thus,

$$c\in\mathcal{C}^{p\times p}\Longleftrightarrow c=T_{\mathfrak{V}}[s]\quad\text{for some } s\in\mathcal{S}^{p\times p}\cap\mathcal{D}(T_{\mathfrak{V}}). \tag{3.41}$$

The inclusion

$$\mathcal{C}^{p\times p}\subset\mathcal{N}^{p\times p},$$

which follows from (3.41), serves to guarantee that every mvf $c\in\mathcal{C}^{p\times p}$ has non-tangential boundary limits $c(\mu)$ at almost all points $\mu\in\mathbb{R}$. In particular,

$$c(\mu)=\lim_{\varepsilon\downarrow 0}c(\mu+i\varepsilon)\quad\text{for almost all points } \mu\in\mathbb{R}.$$

3.3 The Riesz–Herglotz–Nevanlinna representation

It is well known that a $p\times p$ mvf $c(\lambda)$ belongs to the Carathéodory class $\mathcal{C}^{p\times p}$ if and only if it admits an integral representation via the **Riesz–Herglotz–Nevanlinna** formula

$$c(\lambda)=i\alpha-i\lambda\beta+\frac{1}{\pi i}\int_{-\infty}^{\infty}\left\{\frac{1}{\mu-\lambda}-\frac{\mu}{1+\mu^2}\right\}d\sigma(\mu)\quad\text{for } \lambda\in\mathbb{C}_+, \tag{3.42}$$

where $\alpha=\alpha^*\in\mathbb{C}^{p\times p}$, $\beta\in\mathbb{C}^{p\times p}$, $\beta\geq 0$ and $\sigma(\mu)$ is a nondecreasing $p\times p$ mvf on $\mathbb{R}$ such that

$$\int_{-\infty}^{\infty}\frac{d(\text{trace}\,\sigma(\mu))}{1+\mu^2}<\infty. \tag{3.43}$$

The parameters α and β are uniquely defined by $c(\lambda)$ via the formulas

$$\alpha=\Im c(i)\quad\text{and}\quad\beta=\lim_{\nu\uparrow\infty}\nu^{-1}\Re c(i\nu). \tag{3.44}$$

A mvf $\sigma(\mu)$ in (3.42) will be called a **spectral function of** $c(\lambda)$; it always may (and will be) chosen so that

$$\sigma(0)=0\quad\text{and}\quad\sigma(\mu)=\frac{\sigma(\mu+)+\sigma(\mu-)}{2}. \tag{3.45}$$

Under these normalization conditions, $\sigma(\mu)$ is uniquely defined by $c(\lambda)$.

The particular formula

$$\frac{i^{t-1}}{\sin\pi t}\left(\frac{i}{\lambda}\right)^t=-i\frac{\cos(\pi t/2)}{\sin\pi t}+\frac{1}{\pi i}\int_0^{\infty}\left\{\frac{1}{\mu-\lambda}-\frac{\mu}{1+\mu^2}\right\}\frac{1}{\mu^t}d\mu \tag{3.46}$$

$$\text{for } t\in(-1,0)\cup(0,1) \text{ and } \lambda\in\mathbb{C}\setminus\mathbb{R}_+$$

and variations thereof play a useful role in a number of applications.

In view of the normalization (3.45), the **Stieltjes inversion formula**

$$\sigma(\mu_2) - \sigma(\mu_1) = \lim_{\nu \downarrow 0} \int_{\mu_1}^{\mu_2} \Re(c(\mu + i\nu))d\mu \tag{3.47}$$

is valid at every pair of points $\mu_1, \mu_2 \in \mathbb{R}$, and not just at points of continuity of σ; see, e.g., [KaKr74a].

The spectral function $\sigma(\mu)$ can be decomposed into the sum

$$\sigma(\mu) = \sigma_s(\mu) + \sigma_a(\mu), \tag{3.48}$$

of two nondecreasing $p \times p$ mvf's $\sigma_s(\mu)$ and $\sigma_a(\mu)$, where $\sigma_s(\mu)$ is the **singular component** of $\sigma(\mu)$, i.e., $\sigma_s'(\mu) = 0$ for almost all points $\mu \in \mathbb{R}$, and $\sigma_a(\mu)$ is the **locally absolutely continuous** part of $\sigma(\mu)$ normalized by the condition $\sigma_a(0) = 0$, i.e.,

$$\sigma_a(\mu) = \int_0^{\mu} f(b)db, \tag{3.49}$$

where

$$f(\mu) \geq 0 \quad \text{a.e. on } \mathbb{R} \quad \text{and} \quad \int_{-\infty}^{\infty} \frac{\text{trace } f(\mu)}{1+\mu^2} d\mu < \infty. \tag{3.50}$$

The convergence of the last integral follows from (3.43). In fact, condition (3.43) is equivalent to the two conditions

$$\int_{-\infty}^{\infty} \frac{d(\text{trace } \sigma_s(\mu))}{1+\mu^2} < \infty \quad \text{and} \quad \int_{-\infty}^{\infty} \frac{\text{trace } f(\mu)}{1+\mu^2} d\mu < \infty.$$

Moreover,

$$f(\mu) = \sigma'(\mu) = \Re c(\mu) \quad \text{a.e. on } \mathbb{R} \tag{3.51}$$

and hence, in view of formula (3.50),

$$\int_{-\infty}^{\infty} \frac{\text{trace}\{c(\mu) + c(\mu)^*\}}{1+\mu^2} d\mu < \infty. \tag{3.52}$$

If $\sigma(\mu)$ is locally absolutely continuous, i.e., if $\sigma(\mu) = \sigma_a(\mu)$, then the $p \times p$ mvf $f(\mu) = \sigma'(\mu)$ is called the **spectral density** of $c(\lambda)$.

Formula (3.42) implies that

$$\Re c(i) = \beta + \frac{1}{\pi} \int_{-\infty}^{\infty} \frac{d\sigma(\mu)}{1+\mu^2},$$

which leads easily to the following conclusions:

Lemma 3.21 *Let $c \in \mathcal{C}^{p\times p}$. Then in formula (3.42)*

$$\beta = 0 \iff \Re c(i) = \frac{1}{\pi} \int_{-\infty}^{\infty} \frac{d\sigma(\mu)}{1+\mu^2}, \tag{3.53}$$

whereas $\beta = 0$ *and* $\sigma(\mu)$ *is locally absolutely continuous if and only if*

$$\Re c(i) = \frac{1}{\pi}\int_{-\infty}^{\infty} \frac{\Re c(\mu)}{1+\mu^2} d\mu. \tag{3.54}$$

In view of the integral representation formula (3.42) and the decomposition (3.48), every mvf $c \in \mathcal{C}^{p\times p}$ has an additive decomposition

$$c(\lambda) = c_s(\lambda) + c_a(\lambda), \quad \text{with} \quad c_s \in \mathcal{C}_{\text{sing}}^{p\times p} \quad \text{and} \quad c_a \in \mathcal{C}_a^{p\times p}, \tag{3.55}$$

where

$$\mathcal{C}_a^{p\times p} = \{c \in \mathcal{C}^{p\times p} : \beta = 0 \quad \text{and} \quad \sigma_s(\mu) = 0\} \tag{3.56}$$

and

$$\mathcal{C}_{\text{sing}}^{p\times p} = \{c \in \mathcal{C}^{p\times p} : \sigma_a(\mu) = 0\} = \{c \in \mathcal{C}^{m\times m} : \Re c(\mu) = 0 \quad \text{a.e. on } \mathbb{R}\}. \tag{3.57}$$

This decomposition is unique up to an additive purely imaginary constant $p \times p$ matrix. Thus, in terms of the notation introduced in (3.42), (3.48) and (3.51), we may set

$$c_s(\lambda) = i\alpha - i\beta\lambda + \frac{1}{\pi i}\int_{-\infty}^{\infty} \left\{\frac{1}{\mu-\lambda} - \frac{\mu}{1+\mu^2}\right\} d\sigma_s(\mu) \tag{3.58}$$

and

$$c_a(\lambda) = \frac{1}{\pi i}\int_{-\infty}^{\infty} \left\{\frac{1}{\mu-\lambda} - \frac{\mu}{1+\mu^2}\right\} f(\mu) d\mu, \tag{3.59}$$

where $f(\mu) = \Re c(\mu)$ a.e. on $\mathbb{R}$. This decomposition corresponds to the normalization $c_a(i) \geq 0$, and is uniquely determined by this normalization. There are other normalizations that may be imposed on c_a that insure uniqueness of the decomposition (3.55).

Some proper subclasses of the Carathéodory class $\mathcal{C}^{p\times p}$

(a) $\mathcal{C}^{p\times p} \cap H_\infty^{p\times p}$.

If $c \in \mathcal{C}^{p\times p} \cap H_\infty^{p\times p}$, then, by a well-known theorem of the brothers Riesz (see, e.g., p. 74 [RR94]), $\beta = 0$ in formula (3.42), the spectral function $\sigma(\mu)$ of $c(\lambda)$ is locally absolutely continuous and (3.58) reduces to $c_s(\lambda) = i\alpha$, i.e., $\mathcal{C}^{p\times p} \cap H_\infty^{p\times p} \subset \mathcal{C}_a^{p\times p}$. Therefore,

$$c(\lambda) = i\alpha + \frac{1}{\pi i}\int_{-\infty}^{\infty} \left\{\frac{1}{\mu-\lambda} - \frac{\mu}{1+\mu^2}\right\} f(\mu) d\mu, \tag{3.60}$$

where $\alpha^* = \alpha \in \mathbb{C}^{p\times p}$,

$$f(\mu) = (\Re c)(\mu) \geq 0 \quad \text{a.e. on } \mathbb{R} \quad \text{and} \quad f \in L_\infty^{p\times p}(\mathbb{R}). \tag{3.61}$$

It is clear that if condition (3.61) is in force, then the function $c(\lambda)$ defined by formula (3.60) belongs to $\mathcal{C}_a^{p\times p}$ and that

$$\Re c(\lambda) \leq \|f\|_\infty I_p \quad \text{for } \lambda \in \mathbb{C}_+.$$

(b) $\mathring{\mathcal{C}}^{p\times p} = \{c \in \mathcal{C}^{p\times p} \cap H_\infty^{p\times p} : (\Re c)(\mu) \geq \delta_c I_p > 0 \quad \text{a.e. on } \mathbb{R}\}$, where $\delta_c > 0$ depends upon c.

The class $\mathring{\mathcal{C}}^{p\times p}$ is simply related to the class $\mathring{\mathcal{S}}^{p\times p}$ that was defined in (1.37):

$$\mathring{\mathcal{C}}^{p\times p} = T_{\mathfrak{V}}[\mathring{\mathcal{S}}^{p\times p}]. \tag{3.62}$$

If $c \in \mathring{\mathcal{C}}^{p\times p}$, then the mvf $f(\mu)$ in the integral representation (3.60) is subject to the bounds

$$\delta_1 I_p \leq f(\mu) \leq \delta_2 I_p \quad \text{a.e. on } \mathbb{R},$$

where $0 < \delta_1 \leq \delta_2$.

The classes $\mathring{\mathcal{S}}^{p\times q}$ and $\mathring{\mathcal{C}}^{p\times p}$ play a significant role in the study of the inverse problems considered in this monograph, in view of the characterizations (1.36) and (1.86) of the classes $\mathcal{U}_{rsR}(j_{pq})$ and $\mathcal{U}_{rR}(j_{pq})$, respectively.

(c) $\mathcal{C}_{sz}^{p\times p} = \{c \in \mathcal{C}^{p\times p} : \ln\{\det(c + c^*)\} \in \widetilde{L}_1\}$.

The condition refers to the nontangential boundary values of c. The inclusion $\mathring{\mathcal{C}}^{p\times p} \subset \mathcal{C}_{sz}^{p\times p}$ is proper.

(d) $\mathcal{C}_0^{p\times p} = \{c \in \mathcal{C}^{p\times p} : \sup\{\|\nu c(i\nu)\| : \nu > 0\} < \infty\}$.

It is known that a mvf $c(\lambda)$ belongs to $\mathcal{C}_0^{p\times p}$ if and only if it admits a representation of the form

$$c(\lambda) = \frac{1}{\pi i}\int_{-\infty}^{\infty} \frac{1}{\mu - \lambda} d\sigma(\mu), \tag{3.63}$$

where σ is a bounded nondecreasing $p \times p$ mvf on $\mathbb{R}$, or, equivalently, if and only if $c \in \mathcal{C}^{p\times p}$ and

$$\sup_{\nu>0}\{\nu \operatorname{trace} \Re c(i\nu)\} < \infty \quad \text{and} \quad \lim_{\nu\uparrow\infty} c(i\nu) = 0. \tag{3.64}$$

The mvf σ in formula (3.63) coincides with the spectral function of c; in the representation (3.42) for this mvf c,

$$\beta = 0 \quad \text{and} \quad \alpha = -\frac{1}{\pi}\int_{-\infty}^{\infty} \frac{\mu}{1+\mu^2} d\sigma(\mu).$$

(e) $\mathcal{C}^{p\times p} \cap \mathit{Real} \cap \mathit{Symm} = \{c \in \mathcal{C}^{p\times p} : \overline{c(-\overline{\lambda})} = c(\lambda) \quad \text{and} \quad c(\lambda)^\tau = c(\lambda)\}$.

The integral representation formula (3.42) for $c(\lambda)$ implies that

$$\overline{c(-\overline{\lambda})} = -i\overline{\alpha} - i\overline{\beta}\lambda - \frac{1}{\pi i}\int_{-\infty}^{\infty} \left\{\frac{1}{\mu + \lambda} - \frac{\mu}{1+\mu^2}\right\} d\sigma(\mu) \quad \text{for } \lambda \in \mathbb{C}_+.$$

Thus, with the help of the integral representation formula for $c(\lambda)^\tau$, it is readily checked that a mvf $c \in \mathcal{C}^{p\times p}$ belongs to the subclass of real symmetric mvf's in

$\mathcal{C}^{p\times p}$ if and only if the parameters in the integral representation (3.42) for $c(\lambda)$ are subject to the following restrictions:

$$\begin{aligned} &\alpha = 0, \quad \beta = \overline{\beta} = \beta^\tau \geq 0, \quad \sigma(\mu) = \overline{\sigma(\mu)} \quad \text{and} \\ &\sigma(\mu) = -\sigma(-\mu) \text{ for } \mu < 0, \end{aligned} \tag{3.65}$$

where σ is assumed to be normalized as in (3.45). Therefore,

$$c(\lambda) = \frac{c(\lambda) + \overline{c(-\overline{\lambda})}}{2} = -i\beta\lambda + \frac{\lambda}{i\pi}\int_{-\infty}^{\infty} \frac{1}{\mu^2 - \lambda^2} d\sigma(\mu)$$

and hence, upon setting $\tau(\mu) = (2/\pi)\sigma(\sqrt{\mu})$ for $\mu > 0$, formula (3.42) can be rewritten as

$$c(\lambda) = -i\lambda\left\{\beta + \int_0^\infty \frac{d\tau(\mu)}{\mu - \lambda^2}\right\} \quad \text{for } \lambda \in \mathbb{C}_+, \tag{3.66}$$

where $\beta \in \mathbb{R}^{p\times p}$ is positive semidefinite and τ is a nondecreasing mvf on $\mathbb{R}_+$ that is subject to the constraint

$$\int_0^\infty \frac{d\,\mathrm{trace}\,\tau(\mu)}{1 + \mu} < \infty \tag{3.67}$$

and the normalization conditions

$$\tau(0) = 0 \quad \text{and} \quad \tau(\mu) = \frac{\tau(\mu-) + \tau(\mu+)}{2} \quad \text{for } \mu > 0. \tag{3.68}$$

The mvf

$$G(\lambda) = \beta + \frac{1}{\pi}\int_{0-}^\infty \frac{d\tau(\mu)}{\mu - \lambda} \quad \text{for } \lambda \in \mathbb{C} \setminus \{[0, \infty)\} \tag{3.69}$$

has the following special properties:

$$-iG \in \mathcal{C}^{p\times p}, \quad \overline{G(\overline{\lambda})} = G(\lambda), \quad G(\lambda)^\tau = G(\lambda) \quad \text{for } \lambda \in \mathbb{C} \setminus \mathbb{R}_- \tag{3.70}$$

and

$$c(\lambda) = -i\lambda\, G(\lambda^2) \quad \text{belongs to the class } \mathcal{C}^{p\times p} \cap \mathit{Real} \cap \mathit{Symm}. \tag{3.71}$$

The conditions (3.70) define the **Stieltjes class** of mvf's $G(\lambda)$. Formula (3.69) with a positive semidefinite matrix $\beta \in \mathbb{R}^{p\times p}$ and a nondecreasing $p \times p$ mvf τ on $\mathbb{R}_+$ that is subject to the constraint (3.67) provides a complete parametrization of this class. Moreover, formula (3.71) establishes a one-to-one correspondence between mvf's $G(\lambda)$ in the Stieltjes class and real symmetric mvf's $c \in \mathcal{C}^{p\times p}$; see [KaKr74a] for more information on these classes in the scalar case.

Lemma 3.22 *The following assertions hold:*

(1) $\mathcal{C}^{p\times p} = T_{\mathfrak{V}}[\mathcal{S}^{p\times p} \cap \mathcal{D}(T_{\mathfrak{V}})]$.
(2) $\mathcal{C}^{p\times p} \subset \mathcal{N}_+^{p\times p}$.

(3) *If* $c \in \mathcal{C}^{p\times p}$, *then* $c \in \mathcal{N}_{\text{out}}^{p\times p}$ *if* $\Re c(\omega) > 0$ *for at least one (and hence every) point* $\omega \in \mathbb{C}_+$.
(4) $\mathring{\mathcal{C}}^{p\times p} = T_{\mathfrak{V}}[\mathring{\mathcal{S}}^{p\times p}]$ *and* $\mathring{\mathcal{C}}^{p\times p} \subset \mathcal{N}_{\text{out}}^{p\times p}$.
(5) *If* $c \in \mathcal{C}^{p\times p}$, *then:* $c^{-1} \in \mathcal{N}_+^{p\times p} \iff \det c(\lambda) \not\equiv 0$ *in* $\mathbb{C}_+$ $\iff c \in \mathcal{N}_{\text{out}}^{p\times p} \iff c^{-1} \in \mathcal{C}^{p\times p}$.
(6) $\mathcal{C}_{sz}^{p\times p} = T_{\mathfrak{V}}[\mathcal{S}_{sz}^{p\times p}]$.

Proof The first five assertions follow from lemmas 3.57 and 3.58 in [ArD08b]; whereas (6) follows from the identity

$$\Re c = (I_p + s^*)^{-1}(I_p - s^*s)(I_p + s)^{-1} \quad \text{a.e. on } \mathbb{R}$$

and the fact that $\ln \det(I_p + s) \in \widetilde{L}_1$ since $\det(I_p + s) \in \mathcal{N}$ for $s = T_{\mathfrak{V}}[c]$ and $c \in \mathcal{C}^{p\times p}$. □

3.4 The class $\mathcal{E} \cap \mathcal{N}^{p\times q}$ of entire mvf's in $\mathcal{N}^{p\times q}$

A $p \times q$ mvf $f(\lambda) = [f_{jk}(\lambda)]$ is **entire** if each of its entries $f_{jk}(\lambda)$ is an entire function. The class of entire $p \times q$ mvf's $f(\lambda)$ will be denoted $\mathcal{E}^{p\times q}$. If f also belongs to some other class $\mathcal{X}^{p\times q}$, then we shall simply write $f \in \mathcal{E} \cap \mathcal{X}^{p\times q}$.

An entire $p \times q$ mvf is said to be of ***exponential type*** if there is a constant $\tau \geq 0$ such that

$$\|f(\lambda)\| \leq \gamma \exp\{\tau|\lambda|\}, \quad \text{for all points } \lambda \in \mathbb{C} \tag{3.72}$$

for some $\gamma > 0$. In this case, the exact type $\tau(f)$ of f is defined by the formula

$$\tau(f) = \inf\{\tau : (3.72) \text{ holds}\}. \tag{3.73}$$

Equivalently, an entire $p \times q$ mvf f, $f \not\equiv 0$, is said to be of exponential type $\tau(f)$, if

$$\tau(f) = \limsup_{r\to\infty} \frac{\ln \| M(r)\|}{r} < \infty, \tag{3.74}$$

where

$$M(r) = \max\{\|f(\lambda)\| : |\lambda| = r\}.$$

The simple bounds in the next lemma will be useful.

Lemma 3.23 *If* $V \in \mathbb{C}^{p\times q}$, *then*

$$\|V\|^2 \leq \text{trace}(V^*V) \leq q\|V\|^2 \tag{3.75}$$

and

$$\det(V^*V) \leq \|V\|^{2q}. \tag{3.76}$$

If $p = q$ and V is invertible with $\|V^{-1}\| \le 1$, then

$$\|V\| \le |\det V| \le \|V\|^p. \tag{3.77}$$

Proof Let $\mu_1^2 \ge \cdots \ge \mu_q^2$ denote the eigenvalues of the positive semidefinite matrix V^*V. Then (3.75) and (3.76) are immediate from the observation that

$$\operatorname{trace}(V^*V) = \mu_1^2 + \cdots + \mu_q^2, \quad \|V\| = \mu_1$$

and

$$\det(V^*V) = \mu_1^2 \cdots \mu_q^2.$$

Finally, (3.77) is easily obtained from the same set of formulas when $q = p$ and V is invertible with $\|V^{-1}\| \le 1$, because then $\mu_j^2 \ge 1$ for $j = 1, \ldots, p$. □

The inequalities (3.75) applied to the matrix $f(\lambda)$ yield the auxiliary formula

$$\tau(f) = \limsup_{r\to\infty} \frac{\ln \max\{\operatorname{trace}(f(\lambda)^* f(\lambda)) : |\lambda| = r\}}{2r}. \tag{3.78}$$

Moreover, a mvf $f \in \mathcal{E}^{p\times q}$ is of exponential type if and only if all the entries $f_{ij}(\lambda)$ of the mvf f are entire functions of exponential type. Furthermore, if $f \in \mathcal{E}^{p\times q}$ is of exponential type, then

$$\tau(f) = \max\{\tau(f_{ij}) : f_{ij} \not\equiv 0, 1 \le i \le p, 1 \le j \le q\}. \tag{3.79}$$

To verify inequality (3.79), observe first that the inequality

$$|f_{ij}(\lambda)|^2 \le \operatorname{trace}\{f(\lambda)^* f(\lambda)\}$$

implies that $\tau(f_{ij}) \le \tau(f)$, if $f_{ij} \not\equiv 0$. On the other hand, if

$$\tau = \max\{\tau(f_{ij}) :\ f_{ij} \not\equiv 0, 1 \le i \le p, 1 \le j \le q\},$$

then there exists a number $\gamma > 0$ such that

$$\operatorname{trace} f(\lambda)^* f(\lambda) = \sum_{i=1}^{p} \sum_{j=1}^{q} |f_{ij}(\lambda)|^2 \le \gamma \exp\{2(\tau + \varepsilon)|\lambda|\}$$

for every $\varepsilon > 0$. Consequently, $\tau(f) \le \tau$. Also, it is clear that if

$$M_\pm(r) = \max\{|f(\lambda)| :\ |\lambda| \le r \quad \text{and} \quad \lambda \in \overline{\mathbb{C}_\pm}\},$$

then

$$\tau(f) = \max\{\tau_+(f),\ \tau_-(f)\}, \tag{3.80}$$

where

$$\tau_\pm(f) = \limsup_{r\to\infty} \frac{\ln M_\pm(r)}{r};$$

and that

$$\tau_\pm(f) = \max\{\tau_\pm(f_{ij}) : f_{ij} \not\equiv 0,\ 1 \le i \le p,\ 1 \le j \le q\}.$$

Analogs of formula (3.78) are valid for $\tau_\pm(f)$. Moreover, if a mvf $f \in \mathcal{E}^{p\times q}$ has exponential type $\tau(f)$, then the types $\tau_+(f)$ and $\tau_-(f)$ of the mvf f in the closed upper and lower half planes $\overline{\mathbb{C}_+}$ and $\overline{\mathbb{C}_-}$ are not less than the exponential types τ_f^+ and τ_f^- of the mvf f on the upper and lower imaginary half axis respectively, i.e.,

$$\tau_f^\pm \stackrel{\text{def}}{=} \limsup_{\nu\uparrow\infty} \frac{\ln \| f(\pm i\nu) \|}{\nu} \le \tau_\pm(f). \tag{3.81}$$

The notation

$$\delta(f) = \tau(\det f) \quad \text{and} \quad \delta_f^\pm = \tau^\pm(\det f) \tag{3.82}$$

will be used for $f \in \mathcal{E}^{p\times p}$ with $\tau(f) < \infty$. The latter notation will also be used for functions in the Nevanlinna class in the appropriate half plane: Let

$$\delta_f^+ = \limsup_{\nu\uparrow\infty} \frac{\ln |\det f(i\nu)|}{\nu} \quad \text{and} \quad \delta_f^- = \limsup_{\nu\uparrow\infty} \frac{\ln |\det f^\#(i\nu)|}{\nu} \tag{3.83}$$

for $f \in \mathcal{N}^{p\times p}$ with $\det f(\lambda) \not\equiv 0$ and for $f^\# \in \mathcal{N}^{p\times p}$ with $\det f(\lambda) \not\equiv 0$, respectively.

The notation τ_f and type(f) (resp., δ_f and type$(\det f)$) will also be used in place of $\tau(f)$ (resp., $\delta(f)$).

Theorem 3.24 (M.G. Krein) *Let $f \in \mathcal{E}^{p\times q}$. Then $f \in \mathcal{N}^{p\times q}$ if and only $\tau_+(f) < \infty$ and f satisfies the Cartwright condition*

$$\int_{-\infty}^{\infty} \frac{\ln^+ \|f(\mu)\|}{1+\mu^2} d\mu < \infty. \tag{3.84}$$

Moreover, if $f \in \mathcal{E} \cap \mathcal{N}^{p\times q}$, then $\tau_+(f) = \tau_f^+$.

Proof See [Kr47], [Kr51b] and section 6.11 of [RR85]. □

Lemma 3.25 *If $f \in \mathcal{E} \cap \mathcal{N}^{p\times q}$ and $f(\lambda) \not\equiv 0$, then*

$$\tau_f^+ = \limsup_{\nu\uparrow\infty} \frac{\ln \operatorname{trace}\{f(i\nu)^* f(i\nu)\}}{2\nu}. \tag{3.85}$$

Proof This is immediate from (3.75). □

Theorem 3.26 *If $f(\lambda) = [f_{jk}(\lambda)]$ is a $p \times q$ mvf of class $\mathcal{N}^{p\times q}$ such that f is holomorphic in $\mathbb{C}_+$ and $f(\lambda) \not\equiv 0$, then*

$$\tau_f^+ = \max\{\tau_{f_{jk}}^+ : j = 1, \ldots, p,\ k = 1, \ldots, q \quad \textit{and} \quad f_{jk}(\lambda) \not\equiv 0\}.$$

Proof The proof is the same as that of (3.79). □

Theorem 3.27 *Let $f \in \mathcal{E}^{p\times p}$ be invertible in $\mathbb{C}_+$ and let $f^{-1} \in \mathcal{N}_+^{p\times p}$. Then:*

(1) $f \in \mathcal{E} \cap \mathcal{N}^{p\times p}$.
(2) $0 \le \tau_f^+ < \infty$.
(3) $\delta_f^+ = \limsup_{\nu\uparrow\infty} \dfrac{\ln|\det f(i\nu)|}{\nu} = \lim_{\nu\uparrow\infty} \dfrac{\ln|\det f(i\nu)|}{\nu}$ *exists as a limit and* $\tau_f^+ \le \delta_f^+ \le p\tau_f^+$.
(4) $e^{-i\delta_f^+\lambda} \det f(\lambda)^{-1} \in \mathcal{N}_{\rm out}$.
(5) $\delta_f^+ = p\tau_f^+$ *if and only if* $e^{i\tau_f^+\lambda} f(\lambda) \in \mathcal{N}_{\rm out}^{p\times p}$.

Proof See theorem 3.94 in [ArD08b]. □

Corollary 3.28 *Let $f \in \mathcal{E}^{p\times p}$ be invertible in $\mathbb{C}_+$ with $f^{-1} \in H_\infty^{p\times p}$. Then the following are equivalent:*

(1) $\lim\limits_{\nu\uparrow\infty} \dfrac{\ln \|f(i\nu)\|}{\nu} = 0$.
(2) $\lim\limits_{\nu\uparrow\infty} \dfrac{\ln[\text{trace}\{f(i\nu)^* f(i\nu)\}]}{\nu} = 0$.
(3) $\lim_{\nu\uparrow\infty} \dfrac{\ln|\det f(i\nu)|}{\nu} = 0$.
(4) $f \in \mathcal{N}_{\rm out}^{p\times p}$.

Proof This formulation takes advantage of the fact that for a sequence of non-negative numbers $\{x_k\}$,

$$\limsup_{k\to\infty} x_k = 0 \iff \lim_{k\to\infty} x_k = 0.$$

The rest is immediate from the preceding two theorems and the inequalities in Lemma 3.23. □

3.5 The class $\Pi^{p\times q}$ of mvf's in $\mathcal{N}^{p\times q}$ with pseudocontinuations

A $p \times q$ mvf f_- defined in $\mathbb{C}_-$ is said to be a **pseudocontinuation** of a mvf $f \in \mathcal{N}^{p\times q}$, if

(1) $f_-^\# \in \mathcal{N}^{q\times p}$, i.e., if f_- is a meromorphic $p \times q$ mvf in $\mathbb{C}_-$ with bounded Nevanlinna characteristic in $\mathbb{C}_-$ and
(2) $\lim\limits_{\nu\downarrow 0} f_-(\mu - i\nu) = \lim\limits_{\nu\downarrow 0} f(\mu + i\nu)\ (= f(\mu))$ a.e. on $\mathbb{R}$.

The subclass of all mvf's $f \in \mathcal{N}^{p\times q}$ that admit pseudocontinuations f_- into $\mathbb{C}_-$ will be denoted $\Pi^{p\times q}$. Since a pseudocontinuation f_- is uniquely defined by its boundary values on a subset of $\mathbb{R}$ of positive Lebesgue measure, each $f \in \Pi^{p\times q}$ admits only one pseudocontinuation f_-.

Although

- $$\Pi^{p\times q} \subset \mathcal{N}^{p\times q},$$

by definition, and $f \in \mathcal{N}^{p\times q}$ is defined only on $\mathbb{C}_+$, we will consider mvf's $f \in \Pi^{p\times q}$ in the full complex plane $\mathbb{C}$ via the formulas

$$f(\lambda) = f_-(\lambda) \quad \text{for } \lambda \in \mathfrak{h}_{f_-} \cap \mathbb{C}_-$$

and

$$f(\mu) = \lim_{\nu \to 0} f(\mu + i\nu) \quad \text{a.e. on } \mathbb{R}.$$

The symbol $\mathfrak{h}_f$ will be used to denote the domain of holomorphy of this extended mvf $f(\lambda)$ in the full complex plane and

$$\mathfrak{h}_f^+ = \mathfrak{h}_f \cap \mathbb{C}_+, \quad \mathfrak{h}_f^- = \mathfrak{h}_f \cap \mathbb{C}_- \quad \text{and} \quad \mathfrak{h}_f^0 = \mathfrak{h}_f \cap \mathbb{R}.$$

Thus, $\mathfrak{h}_f^\pm$ is all of $\mathbb{C}_\pm$ except for the poles of f in this set; but $\mathfrak{h}_f^0$ may be empty. We shall also write

$$\Pi^p = \Pi^{p\times 1}, \quad \Pi = \Pi^1 \quad \text{and} \quad \Pi^{p\times q} \cap \mathcal{X}^{p\times q} = \Pi \cap \mathcal{X}^{p\times q}$$

for short.

It is clear that a $p \times q$ mvf f belongs to the class $\Pi^{p\times q}$ if and only if all the entries in the mvf f belong to the class Π. Moreover, $\Pi^{p\times q}$ is a linear space and

$$f \in \Pi^{p\times r}, \ g \in \Pi^{r\times q} \Longrightarrow fg \in \Pi^{p\times q};$$

$$f \in \Pi^{p\times p} \Longrightarrow \det f \in \Pi \text{ and trace } f \in \Pi;$$

$$f \in \Pi^{p\times p} \text{ and } \det f \not\equiv 0 \Longrightarrow f^{-1} \in \Pi^{p\times p};$$

$$f \in \Pi^{p\times q} \Longrightarrow f^\# \in \Pi^{q\times p}, \ f^\sim \in \Pi^{q\times p} \text{ and } f(-\lambda) \in \Pi^{p\times q}.$$

The next result follows from the characterization of scalar functions $f \in \Pi \cap H_2$ that is due to Douglas, Shapiro and Shields [DSS70].

Theorem 3.29 *Let $f \in H_2^{p\times q}$. Then $f \in \Pi \cap H_2^{p\times q}$ if and only if $b^{-1}f \in (H_2^{p\times q})^\perp$ for some $b \in \mathcal{S}_{in}$.*

Proof See corollary 3.107 in [ArD08b]. □

If $f \in \Pi^{p\times q}$ and if the restriction of f to $\mathbb{C}_+$ has a holomorphic extension to $\mathbb{C}_-$, then this extension coincides with the pseudocontinuation f_- of f, as follows from the uniqueness of the holomorphic extension. In particular, the pseudocontinuation f_- of an entire mvf $f \in \Pi^{p\times q}$ is the restriction of f to $\mathbb{C}_-$. Thus, if $f \in \mathcal{E}^{p\times q}$, then

$$\begin{aligned} f \in \Pi^{p\times q} &\Longleftrightarrow f \text{ has bounded Nevanlinna characteristic} \\ &\qquad \text{in both half planes } \mathbb{C}_+ \text{ and } \mathbb{C}_- \\ &\Longleftrightarrow f \in \mathcal{N}^{p\times q} \text{ and } f^\# \in \mathcal{N}^{q\times p}. \end{aligned}$$

Theorem 3.30 (M.G. Krein) *Let $f \in \mathcal{E}^{p\times q}$. Then $f \in \Pi^{p\times q}$ if and only if f is an entire mvf of exponential type and satisfies the Cartwright condition*

$$\int_{-\infty}^{\infty} \frac{\ln^+ \| f(\mu) \|}{1+\mu^2} d\mu < \infty. \tag{3.86}$$

Moreover, if $f \in \mathcal{E} \cap \Pi^{p\times q}$, then

(1) $\tau_+(f) = \tau_f^+, \quad \tau_-(f) = \tau_f^-$.
(2) $\tau_f^- + \tau_f^+ \geq 0$.
(3) $\tau(f) = \max \{\tau_f^+, \tau_f^-\}$.

Proof Assertions (1) and (3) follow from Theorem 3.24. To verify (2), suppose to the contrary that $\tau_f^+ + \tau_f^- < 0$ and let $g = e_\delta f$, where $\tau_f^+ < \delta < -\tau_f^-$. Then $\tau_g^+ = -\delta + \tau_f^+ < 0$ and $\tau_g^- = \delta + \tau_f^- < 0$, which is impossible, since in this case $g(\lambda) \equiv 0$. Therefore, (2) is also valid. □

3.6 Fourier transforms and Paley–Wiener theorems

The **Fourier transform and inverse transform**

$$\widehat{f}(\mu) = \int_{-\infty}^{\infty} e^{i\mu t} f(t)dt \quad \text{and} \quad f^\vee(t) = \frac{1}{2\pi} \int_{-\infty}^{\infty} e^{-i\mu t} f(\mu)d\mu \tag{3.87}$$

will be considered mainly for $f \in L_1^{p\times q}$ and $f \in L_2^{p\times q}$.

If $f \in L_2^{p\times q}$, then the integral is understood as the limit of the integrals $\int_{-A}^{A}$ in $L_2^{p\times q}$ as $A \uparrow \infty$. Moreover, the mapping

$$f \to (2\pi)^{-1/2}\widehat{f} \quad \text{is a unitary operator in } L_2^{p\times q},$$

i.e., it is onto, the **Plancherel formula** holds:

$$\langle \widehat{f}, \widehat{g} \rangle_{\text{st}} = 2\pi \langle f, g \rangle_{\text{st}} \quad \text{for } f, g \in L_2^{p\times q},$$

and

$$f(t) = (\widehat{f})^\vee(t) \quad \text{a.e. on } \mathbb{R}. \tag{3.88}$$

If $f \in L_2^{p\times q}$, then $\mu\widehat{f}(\mu)$ belongs to $L_2^{p\times q}$ if and only if

$$f \quad \text{is locally absolutely continuous on } \mathbb{R} \quad \text{and} \quad f' \in L_2^{p\times q}. \tag{3.89}$$

Moreover, if these conditions hold, then

$$\mu\widehat{f}(\mu) = i\widehat{f'}(\mu). \tag{3.90}$$

If $f \in L_1^{p\times r}$ and $g \in L_1^{r\times q}$, then $\widehat{f} \in \mathcal{W}^{p\times r}(0)$, $\widehat{g} \in \mathcal{W}^{r\times q}(0)$, $\widehat{f}\,\widehat{g} \in \mathcal{W}^{p\times q}(0)$ and

$$(\widehat{f}\,\widehat{g})^\vee(t) = \int_{-\infty}^{\infty} f(t-u)g(u)du \quad \text{a.e. on } \mathbb{R}. \tag{3.91}$$

Formula (3.91) is also valid if $f \in L_1^{p\times r}$, $g \in L_s^{r\times q}$ and $1 < s < \infty$.

Theorem 3.31 (N. Wiener and R. Paley) *If $\widehat{f} \in \mathcal{W}^{p\times p}(0)$ and $\gamma \in \mathbb{C}^{p\times p}$, then there exists a matrix $\delta \in \mathbb{C}^{p\times p}$ and a mvf $\widehat{g} \in \mathcal{W}^{p\times p}(0)$ such that*

$$(\gamma + \widehat{f}(\mu))(\delta + \widehat{g}(\mu)) = I_p \tag{3.92}$$

if and only if

$$\det(\gamma + \widehat{f}(\mu)) \neq 0 \quad \textit{for every point } \mu \in \mathbb{R} \textit{ and } \gamma \textit{ is invertible.} \tag{3.93}$$

If $\widehat{f} \in \mathcal{W}_+^{p\times p}(0)$ (resp., $\mathcal{W}_-^{p\times p}(0)$) and $\gamma \in \mathbb{C}^{p\times p}$, then there exists a matrix $\delta \in \mathbb{C}^{p\times p}$ and a mvf $\widehat{g} \in \mathcal{W}_+^{p\times p}(0)$ (resp., $\mathcal{W}_-^{p\times p}(0)$) such that (3.92) holds for all points $\mu \in \mathbb{R}$ if and only if

$$\det(\gamma + \widehat{f}(\lambda)) \neq 0 \quad \textit{for every point } \lambda \in \overline{\mathbb{C}_+} \textit{ (resp., } \overline{\mathbb{C}_-}\textit{) and } \gamma \textit{ is invertible.} \tag{3.94}$$

Proof The stated assertions for mvf's are easily deduced from the scalar versions, the first of which is due to N. Wiener; the second (and third) to Paley and Wiener; see [PaW34] for proofs and, for another approach, [GRS64]. □

If (3.92) holds, then (by the Riemann–Lebesgue lemma) $\gamma\delta = I_p$.

Theorem 3.32 (Paley–Wiener) *Let f be a $p \times q$ mvf that is holomorphic in $\mathbb{C}_+$. Then $f \in H_2^{p\times q}$ if and only if*

$$f(\lambda) = \int_0^\infty e^{i\lambda x} f^\vee(x)dx \quad \textit{for } \lambda \in \mathbb{C}_+$$

and some $f^\vee \in L_2^{p\times q}(\mathbb{R}_+)$. Moreover, if $f \in H_2^{p\times q}$, then its boundary values $f(\mu)$ admit the one-sided Fourier representation

$$f(\mu) = \int_0^\infty e^{i\mu x} f^\vee(x)dx \quad \textit{a.e. on } \mathbb{R}.$$

Proof This follows easily from the scalar Paley–Wiener theorem; see, e.g., pp. 158–160 of [DMc72]. □

Theorem 3.33 (Paley–Wiener) *A $p \times q$ mvf f admits a representation of the form*

$$f(\lambda) = \int_{-\alpha}^{\beta} e^{i\lambda x} f^\vee(x)dx \quad \textit{for } \lambda \in \mathbb{C}$$

and some $f^\vee \in L_2^{p\times q}([-\alpha, \beta])$ with $0 \leq \alpha,\ \beta < \infty$ if and only if $f(\lambda)$ is an entire $p \times q$ mvf of exponential type with $\tau_+(f) \leq \alpha$ and $\tau_-(f) \leq \beta$ and $f \in L_2^{p\times q}$.

Proof This follows easily from the scalar Paley–Wiener theorem; see, e.g., pp. 162–164 of [DMc72]. □

3.7 Entire inner mvf's

A scalar entire inner function $f(\lambda)$ is automatically of the form

$$f(\lambda) = f(0)e^{i\lambda d}, \quad \text{where} \quad |f(0)| = 1 \quad \text{and} \quad d \geq 0.$$

The set of entire inner $p \times p$ mvf's is much richer. It includes mvf's of the form

$$f(\lambda) = f(0)e^{i\lambda D},$$

where

$$f(0) \in \mathbb{C}^{p\times p}, \quad D \in \mathbb{C}^{p\times p}, \quad f(0)^* f(0) = I_p \quad \text{and} \quad D \geq 0,$$

as well as products of mvf's of this form. A complete description of the class $\mathcal{E} \cap \mathcal{S}_{in}^{p\times p}$ is given by Potapov's theorem (Theorem 2.10) with $J = I_p$, and by Theorem 2.15 (the proof of which is based on Theorem 2.10 and is also due to Potapov).

Lemma 3.34 *If $s \in \mathcal{S}^{p\times p}$ and $d(\lambda) = \det s(\lambda)$, then*

$$s \in \mathcal{E} \cap \mathcal{S}_{\text{in}}^{p\times p} \Longleftrightarrow d \in \mathcal{E} \cap \mathcal{S}_{\text{in}}$$

$$\Longleftrightarrow d(\lambda) = \gamma e_\beta(\lambda) \text{ for some } \gamma \in \mathbb{T} \text{ and } \beta \geq 0.$$

Proof Theorem 3.11 guarantees that $s \in \mathcal{S}_{\text{in}}^{p\times p}$ if and only if $d \in \mathcal{S}_{\text{in}}$. Thus, the implication $s \in \mathcal{E} \cap \mathcal{S}_{\text{in}}^{p\times p} \Longrightarrow d \in \mathcal{E} \cap \mathcal{S}_{\text{in}}$ is clear. Conversely, if $d \in \mathcal{E} \cap \mathcal{S}_{\text{in}}$, then $d = \gamma e_\delta$ for some choice of $\gamma \in \mathbb{T}$ and $\delta > 0$, $s \in \mathcal{S}_{\text{in}}^{p\times p}$ and

$$s(\lambda)^{-1} = e^{-i\delta\lambda} h(\lambda)$$

for some choice of $h \in H_\infty^{p\times p}$. Therefore,

$$e^{-i\delta\mu} s(\mu) = h(\mu)^*$$

for almost all points $\mu \in \mathbb{R}$, which in turn implies that

$$\frac{s(\lambda)}{\lambda - \omega}\xi \in e^{i\delta\lambda}(H_2^p)^\perp$$

for $\omega \in \mathbb{C}_+$ and $\xi \in \mathbb{C}^p$. But this in turn implies that

$$\frac{s(\lambda) - s(\omega)}{\lambda - \omega}\xi \in H_2^p \ominus e_\delta H_2^p,$$

and hence that s is entire, in view of Theorem 3.33. □

Similar considerations lead easily to the following supplementary result.

Lemma 3.35 *If $b \in \mathcal{E} \cap \mathcal{S}_{\text{in}}^{p\times p}$ and $b = b_1 b_2$, where $b_1, b_2 \in \mathcal{S}_{\text{in}}^{p\times p}$, then $b_1, b_2 \in \mathcal{E}^{p\times p}$.*

Proof By Lemma 3.34,

$$b \in \mathcal{E} \cap \mathcal{S}_{\text{in}}^{p\times p} \Longrightarrow \det b(\lambda) = \gamma e^{i\lambda\delta},$$

where $|\gamma| = 1$ and $\delta \geq 0$. Therefore, det $b_1(\lambda)$ is of the same form and hence $b_1 \in \mathcal{E} \cap \mathcal{S}_{\text{in}}^{p\times p}$. The same argument applies to b_2. □

The preceding analysis also leads easily to the following set of conclusions, which will be useful in the sequel.

Theorem 3.36 *If* $b \in \mathcal{E} \cap \mathcal{S}_{\text{in}}^{p\times p}$, *then:*

(1) $\det b(\lambda) = e^{i\lambda\alpha} \times \det b(0)$ *for some* $\alpha \geq 0$.
(2) *The limits in (3.81) and (3.83) satisfy the inequalities*

$$0 \leq \tau_b^- \leq \delta_b^- \leq p\tau_b^- < \infty. \tag{3.95}$$

(3) $\delta_b = \delta_b^- = \lim_{\nu\uparrow\infty} \dfrac{\ln|\det b(-i\nu)|}{\nu} = \alpha.$
(4) $p\tau_b^- \geq \alpha$, *with equality if and only if*

$$b(\lambda) = e^{i\lambda\alpha/p} b(0).$$

(5) $b(\lambda)$ *is an entire mvf of exponential type* τ_b^-.

Proof Item (1) is established in Lemma 3.34. Moreover, in view of (1), $f(\lambda) = b(\lambda)^{-1} = b^{\#}(\lambda)$ is entire and satisfies the hypotheses of Theorem 3.27. The latter serves to establish (2); (3) is immediate from (1) and then (4) is immediate from (5) of Theorem 3.27.

Finally, the proof of (5) rests mainly on the observation that

$$\tau_f^+ = \limsup_{\nu\uparrow\infty} \frac{\ln\|f(i\nu)\|}{\nu} = \limsup_{\nu\uparrow\infty} \frac{\ln\|b(-i\nu)\|}{\nu}$$

bounds the growth of $b(\lambda)$ on the negative imaginary axis. Theorem 3.24 applied to entire mvf's that are of the Nevanlinna class in the lower half plane $\mathbb{C}_-$ implies that $\tau_-(b) = \tau_f^+$. Therefore, $\tau(b) = \max\{\tau_-(b), \tau_+(b)\} = \tau_-(b)$, since $\|b(\lambda)\| \leq 1$ in $\mathbb{C}_+$. □

In subsequent developments, the structure of entire inner mvf's will be of central importance. The following three examples illustrate some of the possibilities:

Example 3.37 *If* $b(\lambda) = e^{i\lambda a} I_p$ *with* $a \geq 0$, *then*

$$p\tau_b = \delta_b = pa.$$

Example 3.38 *If* $b(\lambda) = e^{i\lambda a} \oplus I_{p-1} = \text{diag}\{e^{i\lambda a}, 1, \ldots, 1\}$ *with* $a \geq 0$, *then*

$$\tau_b = \delta_b = a.$$

Example 3.39 *If* $b(\lambda) = \mathrm{diag}\{e^{i\lambda a_1}, e^{i\lambda a_2}, \ldots, e^{i\lambda a_p}\}$ *with* $a_1 \geq a_2 \geq \cdots \geq a_p \geq 0$, *then*

$$\tau_b = a_1 \quad \text{and} \quad \delta_b = a_1 + \cdots + a_p.$$

For future use, we record the following observation which is relevant to Example 3.38.

Lemma 3.40 *If* $b \in \mathcal{E} \cap \mathcal{S}_{\text{in}}^{p\times p}$ *admits a factorization* $b(\lambda) = b_1(\lambda)b_2(\lambda)$ *with* $b_i \in \mathcal{S}_{\text{in}}^{p\times p}$ *for* $i = 1, 2$ *and if* $\delta_b = \tau_b$, *then:*

(1) $b_i \in \mathcal{E} \cap \mathcal{S}_{\text{in}}^{p\times p}$ *and*
(2) $\delta_{b_i} = \tau_{b_i}$

for $i = 1, 2$.

Proof Since (1) is available from Lemma 3.35, we can invoke Theorem 3.36 to help obtain the chain of inequalities

$$\delta_b = \delta_{b_1} + \delta_{b_2} \geq \tau_{b_1} + \tau_{b_2} \geq \tau_b.$$

But now as the upper and lower bounds are presumed to be equal, equality must prevail throughout. This leads easily to (2), since $\delta_{b_i} \geq \tau_{b_i}$. □

Lemma 3.41 *If* $b \in \mathcal{E} \cap \mathcal{S}_{\text{in}}^{p\times p}$ *and* $\tau_b = a$, *then:*

(1) $\mathcal{H}(b) \subseteq \mathcal{H}(e_a I_p)$, *or, equivalently,* $e_a b^{-1} \in \mathcal{S}_{\text{in}}^{p\times p}$.
(2) *Every vvf* $f \in \mathcal{H}(b)$ *is an entire vvf of exponential type* $\tau_f \leq a$.
(3) $a = \max\{\tau_f : f \in \mathcal{H}(b)\}$.
(4) $R_\alpha b\xi \in \mathcal{H}(e_a I_p)$ *for every* $\alpha \in \mathbb{C}$ *and* $\xi \in \mathbb{C}^p$.

Proof Assertion (2) follows from the corresponding assertion for mvf's $U \in \mathcal{U}(J)$ in Theorem 4.43, below. Then, since $\mathcal{H}(b) \subset H_2^p$, (1) follows from (2) and the Paley–Wiener theorem. Assertion (3) follows from (2), the equality (3.79) and the fact that $R_0 b\xi \in \mathcal{H}(b)$ for every $\xi \in \mathbb{C}^p$. Assertion (4) then follows from (1) and Lemma 3.14 when $\alpha \in \mathbb{C}_+$. However, since the entries of $b(\lambda)$ are entire functions of exponential type no larger than a that are bounded on $\mathbb{R}$, the Paley–Wiener theorem guarantees the existence of a mvf $h \in L_2^{p\times p}$ such that

$$b(\lambda) = b(0) + i\lambda \int_0^a e^{i\lambda s} h(s)ds \quad \text{for every } \lambda \in \mathbb{C}$$

and this in turn leads to the formula

$$(R_\alpha b)(\lambda) = i \int_0^a e^{i\lambda t} \left\{ h(t) - i\alpha e^{-i\alpha t} \int_t^a e^{i\alpha s} h(s)ds \right\} dt,$$

which serves to justify (4). □

3.8 J contractive, J-inner and entire J-inner mvf's

The **Potapov class** $\mathcal{P}(J)$ is the class of $m \times m$ mvf's $U(\lambda)$ that are meromorphic in $\mathbb{C}_+$ and are J-contractive in $\mathfrak{h}_U^+$, i.e.,

$$U(\lambda)^* J U(\lambda) \leq J \quad \text{for all points } \lambda \in \mathfrak{h}_U^+. \tag{3.96}$$

In [Po60], V.P. Potapov obtained a multiplicative representation of the mvf's in the multiplicative semigroup

$$\mathcal{P}^\circ(J) = \{U \in \mathcal{P}(J) : \det U(\lambda) \not\equiv 0\} \tag{3.97}$$

that generalizes the well-known Blaschke–Riesz–Herglotz factorization for scalar-valued functions s in the Schur class.

It is easily seen that

$$\mathcal{P}(I_m) = \mathcal{S}^{m\times m}, \quad \mathcal{P}^\circ(I_m) = \{s \in \mathcal{S}^{m\times m} : \det s(\lambda) \not\equiv 0\}$$

and

$$\mathcal{P}(-I_m) = \{U : U^{-1} \in \mathcal{S}^{m\times m}\} = \mathcal{P}^\circ(-I_m).$$

The Potapov–Ginzburg transform of $\mathcal{P}(J)$ into $\mathcal{S}^{m\times m}$ and $\mathcal{U}(J)$ into $\mathcal{S}_{in}^{m\times m}$

If $U \in \mathcal{P}(J)$ and $J \neq \pm I_m$, then the **PG (Potapov–Ginzburg) transform** $S = PG(U)$ is given by the formula

$$\begin{aligned} S = PG(U) &= (P_- + P_+ U)(P_+ + P_- U)^{-1} \\ &= (P_+ - U P_-)^{-1}(U P_+ - P_-), \end{aligned} \tag{3.98}$$

where

$$P_\pm = \frac{1}{2}(I_m \pm J) \tag{3.99}$$

are complementary orthogonal projections and $P_+ - P_- = J$. It may be checked that if $U \in \mathcal{P}(J)$, then

$$\det\{(P_+ + P_- U(\lambda))(P_+ - U(\lambda) P_-)\} \neq 0 \quad \text{for } \lambda \in \mathfrak{h}_U^+, \tag{3.100}$$

$$I_m - S(\lambda)^* S(\lambda) = (P_+ + U(\lambda)^* P_-)^{-1}\{J - U(\lambda)^* J U(\lambda)\}(P_+ + P_- U(\lambda))^{-1} \quad \text{for } \lambda \in \mathfrak{h}_U^+ \tag{3.101}$$

and

$$I_m - S(\lambda) S(\lambda)^* = (P_+ - U(\lambda) P_-)^{-1}\{J - U(\lambda) J U(\lambda)^*\}(P_+ - P_- U(\lambda)^*)^{-1} \quad \text{for } \lambda \in \mathfrak{h}_U^+; \tag{3.102}$$

see, e.g., lemma 2.3 in [ArD08b]. Thus, $S(\lambda)$ is a holomorphic contractive mvf on $\mathfrak{h}_U^+$ and hence has a unique holomorphic extension to a mvf in $\mathcal{S}^{m\times m}$. Moreover,

$$\det\{(P_+ + P_-S(\lambda))(P_+ - S(\lambda)P_-)\} \neq 0 \quad \text{for } \lambda \in \mathbb{C}_+, \tag{3.103}$$

and

$$U = PG(S) = (P_- + P_+S)(P_+ + P_-S)^{-1} = (P_+ - SP_-)^{-1}(SP_+ - P_-), \tag{3.104}$$

Formula (3.104) implies the inclusion

$$\mathcal{P}(J) \subset \mathcal{N}^{m\times m}. \tag{3.105}$$

Thus, as mvf's $f \in \mathcal{N}^{m\times m}$ have nontangential boundary values a.e. on $\mathbb{R}$, the same holds true for $U \in \mathcal{P}(J)$. Consequently,

$$U(\mu) = \lim_{\nu\downarrow 0} U(\mu + i\nu)$$

exists a.e. on $\mathbb{R}$ and

$$U(\mu)^*JU(\mu) \leq J \quad \text{a.e. on } \mathbb{R} \tag{3.106}$$

for a mvf $U \in \mathcal{P}(J)$. This implies that $U(\mu) \neq 0$ a.e. on $\mathbb{R}$ if $J \neq I_m$.

A mvf $U \in \mathcal{P}(J)$ belongs to the class $\mathcal{U}(J)$ of J-inner mvf's if its nontangential boundary values are J-unitary a.e. on $\mathbb{R}$, i.e., if

$$U(\mu)^*JU(\mu) = J \quad \text{for almost all points } \mu \in \mathbb{R}. \tag{3.107}$$

If $U \in \mathcal{U}(J)$, then its domain of definition extends into $\mathbb{C}_-$ by the **symmetry principle**

$$U(\lambda) = JU^{\#}(\lambda)^{-1}J \quad \text{for } \overline{\lambda} \in \mathfrak{h}_U^+ \text{ such that } U(\overline{\lambda}) \text{ is invertible}, \tag{3.108}$$

and into $\mathbb{R}$ by nontangential limits. Consequently $U(\lambda)$ is holomorphic in

$$\mathfrak{h}_U = \mathfrak{h}_U^+ \cup \mathfrak{h}_U^- \cup \mathfrak{h}_U^0,$$

where $\mathfrak{h}_U^{\pm}$ and $\mathfrak{h}_U^0$ are the domains of holomorphy of $U(\lambda)$ in $\mathbb{C}_{\pm}$ and $\mathbb{R}$, respectively, and

$$\mathfrak{h}_U^- \supseteq \{\lambda \in \mathbb{C}_- : \overline{\lambda} \in \mathbb{C}_+ \quad \text{and} \quad \det U(\overline{\lambda}) \neq 0\}.$$

Moreover, the restriction $U_- = U|_{\mathfrak{h}_U^-}$ of the mvf U to $\mathfrak{h}_U^-$ is a meromorphic pseudocontinuation of the restriction $U_+ = U|_{\mathfrak{h}_U^+}$ of the mvf U to $\mathfrak{h}_U^+$, i.e., U_- is a meromorphic mvf in $\mathbb{C}_-$ with bounded Nevanlinna characteristic in $\mathbb{C}_-$ and

$$U(\mu) = \lim_{\nu\downarrow 0} U(\mu + i\nu) = \lim_{\nu\downarrow 0} U(\mu - i\nu) \quad \text{a.e. on } \mathbb{R}. \tag{3.109}$$

Thus,

$$\mathcal{U}(J) \subset \Pi^{m\times m} \tag{3.110}$$

and the PG transform maps $\mathcal{U}(J)$ onto

$$\{S \in \mathcal{S}_{\text{in}}^{m\times m} : \det(P_+ + P_-S) \not\equiv 0 \text{ in } \mathbb{C}_+\}.$$

The class $\mathcal{U}(J)$ is a **multiplicative semigroup**; a mvf $U_1 \in \mathcal{U}(J)$ will be called a **left divisor** (resp., **right divisor**) of U if $U_1^{-1}U \in \mathcal{U}(J)$ (resp., $UU_1^{-1} \in \mathcal{U}(J)$).

If $U \in \mathcal{E} \cap \mathcal{U}(J)$, then, in view of (3.110), $U \in \mathcal{E} \cap \Pi^{m\times m}$ and, by Theorem 3.30, it has finite exponential type τ_U and

$$\tau_U = \max\{\tau_U^+, \tau_U^-\}. \tag{3.111}$$

If $U \in \mathcal{U}(J)$, then

$$|\det U(\mu)| = 1 \quad \text{a.e. on } \mathbb{R}$$

and hence

$$\mathcal{U}(J) \subseteq \mathcal{P}^\circ(J) \subseteq \mathcal{P}(J),$$

where the last inclusion is proper if $J \neq -I_m$. Thus, Potapov's results on the multiplicative structure of mvf's in the class $\mathcal{P}^\circ(J)$ are automatically applicable to mvf's in the class $\mathcal{U}(J)$. Moreover,

$$U \in \mathcal{U}(J) \Longrightarrow \det U \in \mathcal{N} \stackrel{\text{def}}{=} \mathcal{N}^{1\times 1} \quad \text{and} \quad |\det U(\mu)| = 1 \text{ a.e. on } \mathbb{R}$$

and hence

$$U \in \mathcal{U}(J) \Longrightarrow \det U(\lambda) = \frac{b(\lambda)}{d(\lambda)}, \quad \text{where } b,\, d \in \mathcal{S}_{in} \stackrel{\text{def}}{=} \mathcal{S}_{in}^{1\times 1}.$$

Thus,

$$U \in \mathcal{E} \cap \mathcal{U}(J) \Longrightarrow \det U(\lambda) = \gamma e^{i\alpha\lambda}, \quad \text{where } \alpha \in \mathbb{R},\, \gamma \in \mathbb{C} \text{ and } |\gamma| = 1. \tag{3.112}$$

In view of the equivalences

$$S^*S \leq I_m \Longleftrightarrow SS^* \leq I_m$$

and

$$\det(P_+ - P_-S) \neq 0 \Longleftrightarrow \det(P_+ + P_-S^*) \neq 0$$

for any $S \in \mathbb{C}^{m\times m}$, the PG transform implies that

$$U^*JU \leq J \Longleftrightarrow UJU^* \leq J \quad \text{for every } U \in \mathbb{C}^{m\times m}. \tag{3.113}$$

Moreover, since $J^2 = I_m$,

$$U^*JU = J \Longleftrightarrow UJU^* = J \quad \text{for every } U \in \mathbb{C}^{m\times m}.$$

Thus, in terms of the notation

$$f^\sim(\lambda) = f(-\overline{\lambda})^*,$$

it follows that

$$U \in \mathcal{P}(J) \Longleftrightarrow U^{\sim} \in \mathcal{P}(J) \quad \text{and} \quad U \in \mathcal{U}(J) \Longleftrightarrow U^{\sim} \in \mathcal{U}(J).$$

Formulas (3.101) and (3.102) and the preceding discussion yield the following conclusions:

Theorem 3.42 *The PG transform (3.98) maps the class*

$$\mathcal{P}(J) \quad \textit{onto} \quad \{S \in \mathcal{S}^{m\times m} : \det\{P_+ + P_-S(\lambda)\} \not\equiv 0\}$$

and

$$\mathcal{U}(J) \quad \textit{onto} \quad \{S \in \mathcal{S}_{\text{in}}^{m\times m} : \det\{P_+ + P_-S(\lambda)\} \not\equiv 0\}.$$

If $J = j_{pq}$ with $p \geq 1$ and $q \geq 1$, then the orthogonal projectors defined in (3.99) can be written explicitly as

$$P_+ = \begin{bmatrix} I_p & 0 \\ 0 & 0_{q\times q} \end{bmatrix} \quad \text{and} \quad P_- = \begin{bmatrix} 0_{p\times p} & 0 \\ 0 & I_q \end{bmatrix}.$$

Correspondingly, $W \in \mathcal{P}(j_{pq})$ and $S = PG(W)$ may be written in block form as

$$W = \begin{bmatrix} w_{11} & w_{12} \\ w_{21} & w_{22} \end{bmatrix} \quad \text{and} \quad S = \begin{bmatrix} s_{11} & s_{12} \\ s_{21} & s_{22} \end{bmatrix} \tag{3.114}$$

with blocks w_{11} and s_{11} of size $p \times p$ and w_{22} and s_{22} of size $q \times q$.

Remark 3.43 *If* $W \in \mathcal{E} \cap \mathcal{U}(j_{pq})$, *then the identities*

$$W(\lambda) = j_{pq}\{W^{\#}(\lambda)\}^{-1} j_{pq} \quad \textit{and} \quad w_{11} = \{s_{11}^{\#}\}^{-1}$$

and the inequalities (3.121) and (3.123) imply that

$$\tau_W^+ = \tau_{w_{22}}^+ \quad \textit{and} \quad \tau_W^- = \tau_{w_{11}}^-. \tag{3.115}$$

Formulas (3.98)–(3.104) lead to the following conclusions:

Lemma 3.44 *Let* $W \in \mathcal{P}(j_{pq})$, *let* $S = PG(W)$ *be written in the standard block form (3.114) and fix* $\lambda \in \mathfrak{h}_W^+$. *Then*

(1) $w_{22}(\lambda)$ *and* $s_{22}(\lambda)$ *are invertible.*

(2) $$S = \begin{bmatrix} w_{11} & w_{12} \\ 0 & I_q \end{bmatrix} \begin{bmatrix} I_p & 0 \\ w_{21} & w_{22} \end{bmatrix}^{-1} = \begin{bmatrix} I_p & -w_{12} \\ 0 & -w_{22} \end{bmatrix}^{-1} \begin{bmatrix} w_{11} & 0 \\ w_{21} & -I_q \end{bmatrix}, \textit{ i.e.,}$$

$$\begin{aligned} s_{11} &= w_{11} - w_{12}w_{22}^{-1}w_{21}, \; s_{12} = w_{12}w_{22}^{-1}, \\ s_{21} &= -w_{22}^{-1}w_{21} \textit{ and } s_{22} = w_{22}^{-1}. \end{aligned} \tag{3.116}$$

(3) $W = \begin{bmatrix} s_{11} & s_{12} \\ 0 & I_q \end{bmatrix} \begin{bmatrix} I_p & 0 \\ s_{21} & s_{22} \end{bmatrix}^{-1} = \begin{bmatrix} I_p & -s_{12} \\ 0 & -s_{22} \end{bmatrix}^{-1} \begin{bmatrix} s_{11} & 0 \\ s_{21} & -I_q \end{bmatrix}$,
i.e.,

$$\begin{aligned} w_{11} &= s_{11} - s_{12}s_{22}^{-1}s_{21}, \; w_{12} = s_{12}s_{22}^{-1}, \\ w_{21} &= -s_{22}^{-1}s_{21} \text{ and } w_{22} = s_{22}^{-1}. \end{aligned} \tag{3.117}$$

(4) $s_{11} \in \mathcal{S}^{p\times p}$, $s_{12} \in \mathcal{S}^{p\times q}$, $s_{21} \in \mathcal{S}^{q\times p}$, $s_{22} \in \mathcal{S}^{q\times q}$ *and*

$$s_{12}(\lambda)^* s_{12}(\lambda) < I_q \quad \text{and} \quad s_{21}(\lambda)^* s_{21}(\lambda) < I_p. \tag{3.118}$$

(5) $\det W(\lambda) = \dfrac{\det s_{11}(\lambda)}{\det s_{22}(\lambda)}$ *and* $\det S = \dfrac{\det w_{11}(\lambda)}{\det w_{22}(\lambda)}$.

(6) $W(\lambda)$ *is invertible* $\Longleftrightarrow$ $s_{11}(\lambda)$ *is invertible, i.e.,*

$$\det W(\lambda) \neq 0 \Longleftrightarrow \det s_{11}(\lambda) \neq 0.$$

(7) *The formulas for W and S can also be expressed as*

$$W = \begin{bmatrix} I_p & s_{12} \\ 0 & I_q \end{bmatrix} \begin{bmatrix} s_{11} & 0 \\ 0 & w_{22} \end{bmatrix} \begin{bmatrix} I_p & 0 \\ -s_{21} & I_q \end{bmatrix} \tag{3.119}$$

and

$$S = \begin{bmatrix} I_p & w_{12} \\ 0 & I_q \end{bmatrix} \begin{bmatrix} w_{11} & 0 \\ 0 & s_{22} \end{bmatrix} \begin{bmatrix} I_p & 0 \\ -w_{21} & I_q \end{bmatrix}. \tag{3.120}$$

(8) $w_{22}(\lambda) = s_{22}^{-1}(\lambda)$ *and*

$$\|w_{22}(\lambda)\| \leq \|W(\lambda)\| \leq 3\|w_{22}(\lambda)\|. \tag{3.121}$$

(9) *If* $W(\lambda)$ *is invertible (or, equivalently,* $s_{11}(\lambda)$ *is invertible), then*

$$W(\lambda)^{-1} = \begin{bmatrix} I_p & 0 \\ s_{21}(\lambda) & I_q \end{bmatrix} \begin{bmatrix} s_{11}(\lambda)^{-1} & 0 \\ 0 & s_{22}(\lambda) \end{bmatrix} \begin{bmatrix} I_p & -s_{12}(\lambda) \\ 0 & I_q \end{bmatrix} \tag{3.122}$$

and

$$\|s_{11}(\lambda)^{-1}\| \leq \|W(\lambda)^{-1}\| \leq 3\|s_{11}(\lambda)^{-1}\|. \tag{3.123}$$

(10) *If* $W \in \mathcal{U}(j_{pq})$ *and* $\det W(\overline{\lambda}) \neq 0$, *then*

$$w_{11}^{\#}(\lambda)s_{11}(\lambda) = I_p. \tag{3.124}$$

Proof Items (1), (2), (3) and the inclusions in (4) are immediate from formulas (3.100), (3.103), (3.98) and (3.104). Moreover, since $S^*S \leq I_m$ and $SS^* \leq I_m$, it is readily seen that

$$s_{12}^* s_{12} + s_{22}^* s_{22} \leq I_q \quad \text{and} \quad s_{21}s_{21}^* + s_{22}s_{22}^* \leq I_q.$$

The inequalities in (3.118) then follow from the fact that s_{22} is invertible at points $\lambda \in \mathfrak{h}_W^+$.

Item (6) is immediate from (5), which, in turn, follows from (3) and (2). Item (7) is obtained from the Schur complement formula

$$\begin{bmatrix} w_{11} & w_{12} \\ w_{21} & w_{22} \end{bmatrix} = \begin{bmatrix} I_p & w_{12}w_{22}^{-1} \\ 0 & I_q \end{bmatrix} \begin{bmatrix} w_{11} - w_{12}w_{22}^{-1}w_{21} & 0 \\ 0 & w_{22} \end{bmatrix} \begin{bmatrix} I_p & 0 \\ w_{22}^{-1}w_{21} & I_q \end{bmatrix} \tag{3.125}$$

and the formulas in (3.116). Similar arguments serve to justify formula (3.120). The upper bound in (8) follows from (3.119), since the matrices s_{ij} are contractive. Finally, (9) follows from (7), and (10) follows from (3.108) with $J = j_{pq}$ and (3.122). □

The preceding analysis implies that the PG transform maps

$$\mathcal{P}(j_{pq}) \quad \text{bijectively onto } \{S \in \mathcal{S}^{m\times m} : \det s_{22}(\lambda) \not\equiv 0\}$$

and

$$\mathcal{U}(j_{pq}) \quad \text{bijectively onto } \{S \in \mathcal{S}_{in}^{m\times m} : \det s_{22}(\lambda) \not\equiv 0\}.$$

Let

$$\mathcal{U}_{\text{const}}(J) = \{U \in \mathbb{C}^{m\times m} : U^*JU = J\}.$$

Lemma 3.45 *If $W \in \mathcal{U}_{\text{const}}(j_{pq})$, then there exists a unique choice of parameters*

$$k \in \mathbb{C}^{p\times q},\ u \in \mathbb{C}^{p\times p},\ v \in \mathbb{C}^{q\times q} \quad \text{with} \quad k^*k < I_q, \\ u^*u = I_p \text{ and } v^*v = I_q \tag{3.126}$$

such that

$$W = \begin{bmatrix} (I_p - kk^*)^{-1/2} & k(I_q - k^*k)^{-1/2} \\ k^*(I_p - kk^*)^{-1/2} & (I_q - k^*k)^{-1/2} \end{bmatrix} \begin{bmatrix} u & 0 \\ 0 & v \end{bmatrix}. \tag{3.127}$$

Conversely, if k, u, v is any set of three matrices that meet the conditions in (3.126), then the matrix W defined by formula (3.127) is j_{pq}-unitary.

Proof Let $W \in \mathcal{U}_{\text{const}}(j_{pq})$, let w_{ij} and s_{ij} be the blocks in the four-block decomposition of W and $S = PG(W)$, respectively, and let $k = s_{12}$. Then, by Lemma 3.44, $k^*k < I_q$,

$$s_{11}s_{11}^* = I_p - kk^*, \quad s_{22}^*s_{22} = I_q - k^*k \quad \text{and} \quad s_{21}s_{11}^* = -s_{22}k^*,$$

since $S^*S = I_m$, where $m = p + q$. The first two equalities imply that

$$s_{11} = (I_p - kk^*)^{1/2}u \quad \text{and} \quad s_{22} = v^*(I_q - k^*k)^{1/2}, \tag{3.128}$$

where $u^*u = uu^* = I_p$, $v^*v = vv^* = I_q$ and u and v are uniquely defined by s_{11} and s_{22}, respectively. Moreover, these formulas lead easily to the asserted parametrization of W with parameters that are uniquely defined by W and satisfy the constraints in (3.126), since

$$w_{11} = s_{11}^{-*}, \ w_{22} = s_{22}^{-1}, \ w_{12} = ks_{22}^{-1} \text{ and } w_{21} = -s_{22}^{-1}s_{21} = k^*s_{11}^{-*}. \tag{3.129}$$

The converse is easily checked by direct calculation. □

Lemma 3.46 *If $W \in \mathcal{U}_{\text{const}}(j_{pq})$, then $u = I_p$ and $v = I_q$ in the parametrization formula (3.127) if and only if $W > 0$.*

Proof If $W \in \mathcal{U}_{\text{const}}(j_{pq})$ and $W > 0$, then the parametrization formula (3.127) implies that

$$w_{11} = (I_p - kk^*)^{-1/2}u > 0 \quad \text{and} \quad w_{22} = (I_q - k^*k)^{-1/2}v > 0.$$

Therefore, by the uniqueness of polar decompositions, $u = I_p$ and $v = I_q$.

Conversely, if $W \in \mathcal{U}_{\text{const}}(j_{pq})$ and $u = I_p$ and $v = I_q$, then, by Schur complements, formula (3.127) can be written as

$$W = \begin{bmatrix} I_p & k \\ 0 & I_q \end{bmatrix} \begin{bmatrix} (I_p - kk^*)^{1/2} & 0 \\ 0 & (I_q - k^*k)^{-1/2} \end{bmatrix} \begin{bmatrix} I_p & 0 \\ k^* & I_q \end{bmatrix},$$

since $w_{11} - w_{12}w_{22}^{-1}w_{21} = (I_p - kk^*)^{1/2}$. But this clearly displays W as a positive definite matrix. □

Remark 3.47 *In view of Lemma 3.46, the product in (3.127) is the polar decomposition of the constant j_{pq}-unitary matrix W, in which the first term on the right is a positive definite j_{pq}-unitary matrix and the second term is both unitary and j_{pq}-unitary.*

Connections with the classes $\mathcal{C}^{m\times m}$ and $\mathcal{C}_{\text{sing}}^{m\times m}$

Recall that $\mathcal{C}_{\text{sing}}^{m\times m}$ denotes the subclass of mvf's $c(\lambda)$ in the Carathéodory class $\mathcal{C}^{m\times m}$ with singular spectral functions $\sigma(\mu)$ (in the integral representation (1.65)), i.e.,

$$\mathcal{C}_{\text{sing}}^{m\times m} = \{c \in \mathcal{C}^{m\times m} : \Re c(\mu) = 0 \quad \text{a.e. on } \mathbb{R}\}. \tag{3.130}$$

It is easily checked that the **Cayley transform**

$$S = T_{\mathfrak{V}}(C) = (I_m - C)(I_m + C)^{-1} \tag{3.131}$$

maps $\mathcal{C}^{m\times m}$ bijectively onto

$$\{S \in \mathcal{S}^{m\times m} : \det\{I_m + S(\lambda)\} \not\equiv 0\} \tag{3.132}$$

and that $C(\lambda)$ may be recovered from $S(\lambda)$ by the same formula:

$$C = T_{\mathfrak{V}}(S) = (I_m - S)(I_m + S)^{-1}. \tag{3.133}$$

Moreover, $T_{\mathfrak{V}}$ maps $\mathcal{C}_{\text{sing}}^{m\times m}$ onto

$$\{S \in \mathcal{S}_{\text{in}}^{m\times m} : \det\{I_m + S(\lambda)\} \not\equiv 0\}. \tag{3.134}$$

Analogous connections exist between the classes $\mathcal{P}(J)$ and $\mathcal{C}^{m\times m}$ and between the classes $\mathcal{U}(J)$ and $\mathcal{C}_{\rm sing}^{m\times m}$:

Lemma 3.48 *The transform*

$$C = J(I_m - U)(I_m + U)^{-1} \tag{3.135}$$

maps

$$\{U \in \mathcal{E} \cap \mathcal{U}^\circ(J) : \det(I_m + U(\lambda)) \not\equiv 0\}$$

injectively into

$$\{C \in \mathcal{C}_{\rm sing}^{m\times m} : C \text{ is meromorphic in } \mathbb{C},\ \mathfrak{h}_C \supset \mathbb{C}_+ \cup \mathbb{C}_- \cup \{0\} \text{ and } C(0) = 0\}. \tag{3.136}$$

Proof See Lemma 6.3 and the proof of theorem 5.22 in [ArD08b]. □

3.9 Associated pairs of the first kind

Associated pairs of the first kind will be defined initially for mvf's $W \in \mathcal{U}(j_{pq})$ in terms of inner–outer and outer–inner factorizations of the blocks s_{11} and s_{22} in the PG transform of W; see (3.138) below. Subsequently. associated pairs of the first kind will be defined for mvf's $U \in \mathcal{U}(J)$ when J is unitarily equivalent to j_{pq}.

Let $W \in \mathcal{U}(j_{pq})$ and let $S = PG(W)$ be the Potapov–Ginzburg transform of W. Then, by Lemma 3.44, the blocks on the diagonal in the standard four-block decomposition of S are such that

$$s_{11} \in \mathcal{S}^{p\times p}, \quad \det s_{11}(\lambda) \not\equiv 0, \quad s_{22} \in \mathcal{S}^{q\times q} \quad \text{and} \quad \det s_{22}(\lambda) \not\equiv 0. \tag{3.137}$$

Therefore, both of these blocks admit inner–outer and outer–inner factorizations. In particular,

$$s_{11} = b_1\varphi_1 \quad \text{where} \quad b_1 \in \mathcal{S}_{\rm in}^{p\times p} \quad \text{and} \quad \varphi_1 \in \mathcal{S}_{\rm out}^{p\times p} \tag{3.138}$$

and

$$s_{22} = \varphi_2 b_2 \quad \text{where} \quad b_2 \in \mathcal{S}_{\rm in}^{q\times q} \quad \text{and} \quad \varphi_2 \in \mathcal{S}_{\rm out}^{q\times q}. \tag{3.139}$$

A pair $\{b_1, b_2\}$ of mvf's $b_1 \in \mathcal{S}_{\rm in}^{p\times p}$ and $b_2 \in \mathcal{S}_{\rm in}^{q\times q}$ that is obtained from the factorizations (3.138) and (3.139) is called an **associated pair** of W and the set of such pairs is denoted $ap(W)$. If $\{b_1, b_2\} \in ap(W)$, then

$$ap(W) = \{b_1u, vb_2 : u \in \mathbb{C}^{p\times p} \quad \text{and} \quad v \in \mathbb{C}^{q\times q} \quad \text{are unitary matrices}\}.$$

If

$$J = V^* j_{pq} V, \quad V^*V = VV^* = I_m \quad \text{and} \quad W(\lambda) = VU(\lambda)V^*, \tag{3.140}$$

then

$$U \in \mathcal{U}(J) \Longleftrightarrow W \in \mathcal{U}(j_{pq}). \tag{3.141}$$

If $U \in \mathcal{U}(J)$ and $W(\lambda)$ is defined by (3.140), then an associated pair $\{b_1, b_2\} \in ap(W)$ is called an **associated pair of the first kind** for U. The set of such pairs is denoted $ap_I(U)$, i.e.,

$$ap_I(U) = ap(VUV^*).$$

This definition depends upon the choice of the unitary matrix V in the formula $J = V^* j_{pq} V$, in a nonessential way:

If V_1 is also a unitary matrix such that $J = V_1^* j_{pq} V_1$, then

$$(V_1 V^*) j_{pq} = j_{pq}(V_1 V^*)$$

and hence

$$V_1 V^* = \begin{bmatrix} u & 0 \\ 0 & v \end{bmatrix}, \quad \text{i.e.,} \quad V_1 = \begin{bmatrix} u & 0 \\ 0 & v \end{bmatrix} V,$$

where $u \in \mathbb{C}^{p\times p}$ and $v \in \mathbb{C}^{q\times q}$ are both unitary matrices. Thus, if $W_1(\lambda) = V_1 U(\lambda) V_1^*$, then

$$\begin{aligned} W_1(\lambda) &= V_1 V^*(VU(\lambda)V^*)VV_1^* = (V_1V^*)W(\lambda)(V_1V^*)^* \\ &= \begin{bmatrix} u & 0 \\ 0 & v \end{bmatrix} W(\lambda) \begin{bmatrix} u^* & 0 \\ 0 & v^* \end{bmatrix}. \end{aligned}$$

Consequently,

$$\{b_1, b_2\} \in ap(W) \Longleftrightarrow \{ub_1u^*, v^*b_2v\} \in ap(W_1). \tag{3.142}$$

Thus, the set $ap_I(U)$ depends upon the choice of the unitary matrix V such that $J = V^* j_{pq} V^*$ only up to the transformation

$$b_1(\lambda) \longrightarrow ub_1(\lambda)u^* \quad \text{and} \quad b_2(\lambda) \longrightarrow v^*b_2(\lambda)v$$

for some pair of unitary matrices $u \in \mathbb{C}^{p\times p}$ and $v \in \mathbb{C}^{q\times q}$, which depend upon the choice of the unitary matrices V and V_1.

Lemma 3.49 *If $U \in \mathcal{U}(J)$, J is unitarily equivalent to j_{pq} and $\{b_1, b_2\} \in ap_I(U)$, then*

$$\det U = \gamma \frac{\det b_1}{\det b_2} \quad \textit{for some } \gamma \in \mathbb{T}. \tag{3.143}$$

Proof We may assume that $J = j_{pq}$, and $\{b_1, b_2\} \in ap(W)$, where $W \in \mathcal{U}(j_{pq})$. Then, in view of formulas (3.138) and (3.139),

$$\varphi_1^* \varphi_1 = s_{11}^* s_{11} = I_p - s_{21}^* s_{21}$$

and

$$\varphi_2\varphi_2^* = s_{22}s_{22}^* = I_q - s_{21}s_{21}^*.$$

Therefore,

$$|\det \varphi_2(\mu)| = |\det \varphi_1(\mu)| \quad \text{a.e. on } \mathbb{R} \tag{3.144}$$

and hence, as $\det \varphi_1$ and $\det \varphi_2$ are both outer functions,

$$\det \varphi_1(\lambda) = \gamma \det \varphi_2(\lambda) \quad \text{for some } \gamma \in \mathbb{T}. \tag{3.145}$$

Formula (3.143) now follows easily from (5) of Lemma 3.44 and formula (3.145). □

Lemma 3.50 *Let $U = U_1U_2$, where $U_j \in \mathcal{U}(J)$ for $j = 1, 2$ and $J = V^* j_{pq} V$ for some unitary matrix $V \in \mathbb{C}^{m\times m}$ and suppose $\{b_1, b_2\} \in ap_I(U)$ and $\{b_1^{(1)}, b_2^{(1)}\} \in ap_I(U_1)$. Then $(b_1^{(1)})^{-1}b_1 \in \mathcal{S}_{\text{in}}^{p\times p}$ and $b_2(b_2^{(1)})^{-1} \in \mathcal{S}_{\text{in}}^{q\times q}$.*

Proof This follows from lemma 4.28 in [ArD08b], which treats the case $J = j_{pq}$. □

3.10 Singular and right (and left) regular J-inner mvf's

Recall that a mvf $U \in \mathcal{U}(J)$ is said to belong to the class $\mathcal{U}_S(J)$ of **singular J-inner mvf's** if it belongs to $\mathcal{N}_{\text{out}}^{m\times m}$, i.e.,

$$\mathcal{U}_S(J) = \mathcal{U}(J) \cap \mathcal{N}_{\text{out}}^{m\times m}. \tag{3.146}$$

Clearly,

$$U \in \mathcal{U}_S(J) \Longleftrightarrow U^\sim \in \mathcal{U}_S(J).$$

If $J = \pm I_m$, then $\mathcal{U}_S(J)$ is just the class of unitary matrices in $\mathbb{C}^{m\times m}$.

If $J \neq \pm I_m$, then every elementary Blaschke–Potapov factor

$$U(\lambda) = I_m + \frac{\varepsilon}{\pi i(\lambda - \omega)} \quad \text{with } \omega \in \mathbb{R},\ \varepsilon \in \mathbb{C}^{m\times m},\ \varepsilon J \geq 0,\ \varepsilon^2 = 0, \tag{3.147}$$

with a pole in $\mathbb{R}$ belongs to the class $\mathcal{U}_S(J)$, since $(\lambda - \omega)$ is an outer function in the Smirnov class $\mathcal{N}_+$ and hence both $U(\lambda)$ and

$$U(\lambda)^{-1} = I_m - \frac{\varepsilon}{\pi i(\lambda - \omega)}$$

belong to the Smirnov class $\mathcal{N}_+^{m\times m}$. Similarly every elementary Blaschke–Potapov factor

$$U(\lambda) = I_m + \lambda\varepsilon \quad \text{with } \varepsilon \in \mathbb{C}^{m\times m},\ \varepsilon J \geq 0,\ \varepsilon^2 = 0, \tag{3.148}$$

with a pole at ∞ belongs to $\mathcal{U}_S(J)$ as do finite products and convergent infinite products of such factors. The latter assertion is a consequence of the following result:

Theorem 3.51 *Let $U = U_1U_2$, where $U_k \in \mathcal{U}(J)$ for $k = 1, 2$. Then $U \in \mathcal{U}_S(J)$ if and only if $U_1 \in \mathcal{U}_S(J)$ and $U_2 \in \mathcal{U}_S(J)$. Moreover, convergent left or right products of mvf's in $\mathcal{U}_S(J)$ belong to $\mathcal{U}_S(J)$.*

Proof See theorems 4.33 and 4.44 in [ArD08b]. □

A mvf $U \in \mathcal{U}(J)$ is said to belong to the class $\mathcal{U}_{rR}(J)$ of **right regular** J-inner mvf's if

$$U = U_1U_2 \quad \text{with } U_1 \in \mathcal{U}(J) \text{ and } U_2 \in \mathcal{U}_S(J) \Longrightarrow U_2 \in \mathcal{U}_{\text{const}}(J).$$

Analogously, a mvf $U \in \mathcal{U}(J)$ is said to belong to the class $\mathcal{U}_{\ell R}(J)$ of **left regular** J-inner mvf's if

$$U = U_2U_1 \quad \text{with } U_1 \in \mathcal{U}(J) \text{ and } U_2 \in \mathcal{U}_S(J) \Longrightarrow U_2 \in \mathcal{U}_{\text{const}}(J).$$

It is useful to note that

$$U \in \mathcal{U}_{rR}(J) \Longleftrightarrow U^{\sim} \in \mathcal{U}_{\ell R}(J). \tag{3.149}$$

If $J = \pm I_m$, then $\mathcal{U}_{rR}(J) = \mathcal{U}_{\ell R}(J) = \mathcal{U}(J)$. Thus, $\mathcal{U}_{rR}(J)$ and $\mathcal{U}_{\ell R}(J)$ are only proper subclasses of $\mathcal{U}(J)$ if $J \neq \pm I_m$.

Lemma 3.52 *If $U \in \mathcal{U}(J)$, $P_\pm = (I_m \pm J)/2$, $J \neq \pm I_m$ and $S = PG(U)$, then*

$$U \in \mathcal{U}_S(J) \Longleftrightarrow (P_+ + P_-S)/2 \in \mathcal{S}_{\text{out}}^{m\times m} \quad \textit{and} \quad (P_- + P_+S)/2 \in \mathcal{S}_{\text{out}}^{m\times m}.$$

Proof See lemma 4.40 in [ArD08b]. □

Lemma 3.53 *If $U \in \mathcal{U}(J)$, $J \neq \pm I_m$ and $\{b_1, b_2\} \in ap_I(U)$, then*

(1) $U \in \mathcal{N}_+^{m\times m} \Longleftrightarrow b_2(\lambda)$ *is a constant unitary matrix.*
(2) $U^{-1} \in \mathcal{N}_+^{m\times m} \Longleftrightarrow b_1(\lambda)$ *is a constant unitary matrix.*
(3) $U \in \mathcal{U}_S(J) \Longleftrightarrow b_1(\lambda)$ *and* $b_2(\lambda)$ *are both constant unitary matrices.*

Proof This follows from lemmas 4.33 and 4.39 in [ArD08b]. □

Lemma 3.54 *Let $U = U_1U_2$, where $U_j \in \mathcal{U}(J)$ for $j = 1, 2$ and $J \neq \pm I_m$. Then*

(1) $U \in \mathcal{N}_+^{m\times m} \Longleftrightarrow U_1 \in \mathcal{N}_+^{m\times m}$ *and* $U_2 \in \mathcal{N}_+^{m\times m}$.
(2) $U^\# \in \mathcal{N}_+^{m\times m} \Longleftrightarrow U_1^\# \in \mathcal{N}_+^{m\times m}$ *and* $U_2^\# \in \mathcal{N}_+^{m\times m}$.
(3) $U_2 \in \mathcal{U}_S(J) \Longleftrightarrow ap_I(U_1) = ap_I(U)$.

Proof It suffices to consider the case $J = j_{pq}$. But then the blocks of the PG transforms $S = PG(W)$, $S_1 = PG(W_1)$ and $S_2 = PG(W_2)$ are connected by the formulas

$$s_{11} = s_{11}^{(1)}(I_p - s_{12}^{(2)}s_{21}^{(1)})^{-1}s_{11}^{(2)} \tag{3.150}$$

and

$$s_{22} = s_{22}^{(2)}(I_q - s_{21}^{(1)}s_{12}^{(2)})^{-1}s_{22}^{(1)}. \tag{3.151}$$

Since the middle factors on the right in these two formulas are both outer mvf's, thanks to (4) of Theorem 3.11, it follows that

$$s_{11} \in \mathcal{S}_{\text{out}}^{p\times p} \iff s_{11}^{(1)} \in \mathcal{S}_{\text{out}}^{p\times p} \quad \text{and} \quad s_{11}^{(2)} \in \mathcal{S}_{\text{out}}^{p\times p},$$

$$s_{22} \in \mathcal{S}_{\text{out}}^{q\times q} \iff s_{22}^{(1)} \in \mathcal{S}_{\text{out}}^{q\times q} \quad \text{and} \quad s_{22}^{(2)} \in \mathcal{S}_{\text{out}}^{q\times q}$$

and

$$\begin{aligned} ap(W) = ap(W_1) &\iff s_{22}^{(2)} \in \mathcal{S}_{\text{out}}^{q\times q} \quad \text{and} \quad s_{11}^{(2)} \in \mathcal{S}_{\text{out}}^{p\times p} \\ &\iff W_2 \in \mathcal{U}_S(j_{pq}). \end{aligned}$$

□

Theorem 3.55 *Let $U \in \mathcal{U}(J)$, $J \neq \pm I_m$, and let $\{b_1, b_2\} \in ap_I(U)$. Then*

$$\mathfrak{h}_U^+ = \mathfrak{h}_{b_2^\#}^+, \quad \mathfrak{h}_U^- = \mathfrak{h}_{b_1}^- \quad \textit{and} \quad \mathfrak{h}_U \subseteq \mathfrak{h}_{b_2^\#} \cap \mathfrak{h}_{b_1}. \tag{3.152}$$

Moreover:

(1) *If U is entire, then b_1 and b_2 are entire.*
(2) *If $U \in \mathcal{U}_{rR}(J)$, then $\mathfrak{h}_U = \mathfrak{h}_{b_2^\#} \cap \mathfrak{h}_{b_1}$ and hence, U is entire if and only if b_1 and b_2 are entire.*

Proof This is theorem 4.54 in [ArD08b]. □

Theorem 3.56 *Let $U \in \mathcal{E} \cap \mathcal{U}(J)$, where $J = V^* j_{pq} V$ for some unitary $m \times m$ matrix V and let $\{b_1, b_2\} \in ap_I(U)$. Then:*

(1) *$b_1 \in \mathcal{E} \cap \mathcal{S}_{\text{in}}^{p\times p}$ and $b_2 \in \mathcal{E} \cap \mathcal{S}_{in}^{q\times q}$.*
(2) *$\tau_U^+ = \tau_{b_2^{-1}}^+$ and $\tau_U^- = \tau_{b_1}^-$.*
(3) *$U \in \mathcal{U}_S(J) \iff \tau_U = 0$.*
(4) *$U \in \mathcal{U}_S(J) \iff \tau_{b_1} = 0$ and $\tau_{b_2} = 0$, i.e., if and only if $b_1(\lambda) = b_1(0)$ and $b_2(\lambda) = b_2(0)$.*

Proof The first assertion is contained in Theorem 3.55; the second follows from theorem 4.58 in [ArD08b]. Assertion (4) is covered by Lemma 3.54; (3) then follows from (2) and (4). □

Remark 3.57 *The converse of assertion (1) in Theorem 3.55 is not true. The elementary Blaschke–Potapov factor in (3.147) belongs to the class $\mathcal{U}_S(J)$. Consequently, if $\{b_1, b_2\} \in ap_I(U)$, then $b_1(\lambda)$ and $b_2(\lambda)$ are both unitary constant* *matrices and therefore entire but U is not entire. The matrices*

$$J = J_p, \quad \text{and} \quad \varepsilon = \begin{bmatrix} 0 & -\delta \\ 0 & 0 \end{bmatrix} \quad \text{with} \quad \delta \in \mathbb{C}^{p\times p} \quad \text{and} \quad \delta > 0$$

meet the stated conditions.

Theorem 3.56 implies that if $J \neq \pm I_m$, then an entire J-inner mvf $U(\lambda)$ is a singular J-inner mvf if and only if it is of minimal exponential type.

The resolvent matrices of completely indeterminate moment problems and bitangential generalizations of such problems are products of elementary Blaschke–Potapov factors with poles at infinity and hence they belong to the class $\mathcal{E} \cap \mathcal{U}_S(J)$. The matrizants of the matrix Schrödinger equation (2.88) with a locally summable potential $q(x)$ on $[0, d)$ also belong to this class, since (as was shown in Section 2.9) they are entire mvf's of minimal exponential type.

Characterizations of the mvf's U in the classes $\mathcal{U}_S(J)$ and $\mathcal{U}_{rR}(J)$ in terms of the corresponding RKHS's $\mathcal{H}(U)$ as in (1.27) and (1.24) will be discussed later in Chapter 4.

3.11 Linear fractional transformations of $\mathcal{S}^{p\times q}$ into itself

The proofs of the next two lemmas may be found on pp. 210–213 of [ArD08b] and the references cited therein.

Lemma 3.58 *If $W \in \mathcal{P}(j_{pq})$, then:*

(1) $\mathcal{S}^{p\times q} \subseteq \mathcal{D}(T_W)$ *and* $T_W[\mathcal{S}^{p\times q}] \subseteq \mathcal{S}^{p\times q}$.
(2) $\mathcal{S}^{q\times p} \subseteq \mathcal{D}(T_W^\ell)$ *and* $T_W^\ell[\mathcal{S}^{q\times p}] \subseteq \mathcal{S}^{q\times p}$.

If $W \in \mathcal{U}(j_{pq})$, then also:

(3) *If* $p \geq q$, *then* $T_W[\mathcal{S}_{\text{in}}^{p\times q}] \subseteq \mathcal{S}_{\text{in}}^{p\times q}$ *and* $T_W^\ell[\mathcal{S}_{*\text{in}}^{q\times p}] \subseteq \mathcal{S}_{*\text{in}}^{q\times p}$.
(4) *If* $p \leq q$, *then* $T_W[\mathcal{S}_{*\text{in}}^{p\times q}] \subseteq \mathcal{S}_{*\text{in}}^{p\times q}$ *and* $T_W^\ell[\mathcal{S}_{\text{in}}^{q\times p}] \subseteq \mathcal{S}_{\text{in}}^{q\times p}$.
(5) *If* $W = W_1W_2$, *where* $W_1, W_2 \in \mathcal{P}(j_{pq})$, *then* $W_1, W_2 \in \mathcal{U}(j_{pq})$ *and* $T_W[\mathcal{S}^{p\times q}] \subseteq T_{W_1}[\mathcal{S}^{p\times q}]$.

Lemma 3.59 *If $W_n \in \mathcal{U}(j_{pq})$, $n \geq 1$, is a nondecreasing $\curvearrowright$ sequence such that*

$$W(\lambda) = \lim_{n\uparrow\infty} W_n(\lambda) \quad \text{for} \quad \lambda \in \bigcap_{n\geq 1} \mathfrak{h}_{W_n}^+ \quad \text{and} \quad W \in \mathcal{P}^\circ(j_{pq}),$$

then

$$T_{W_{n+1}}[\mathcal{S}^{p\times q}] \subseteq T_{W_n}[\mathcal{S}^{p\times q}],\ T_W[\mathcal{S}^{p\times q}] = \bigcap_{n\geq 1} T_{W_n}[\mathcal{S}^{p\times q}]$$

and $W \in \mathcal{U}(j_{pq})$. Moreover, if $\{b_1^{(n)}, b_2^{(n)}\} \in ap(W_n)$ for $n \geq 1$ and this sequence is normalized by the conditions

$$b_1^{(n)}(\omega) > 0, \quad b_2^{(n)}(\omega) > 0 \quad \text{at a point } \omega \in \mathbb{C}_+,$$

then the limits

$$b_1(\lambda) = \lim_{n\uparrow\infty} b_1^{(n)}(\lambda) \quad \text{and} \quad b_2(\lambda) = \lim_{n\uparrow\infty} b_2^{(n)}(\lambda)$$

exist at each point $\lambda \in \mathbb{C}_+$, $b_1 \in \mathcal{S}_{\text{in}}^{p\times p}$, $b_2 \in \mathcal{S}_{\text{in}}^{q\times q}$ and $\{b_1, b_2\} \in ap(W)$.

If $W_n \in \mathcal{U}^\circ(j_{pq})$ for every integer $n \geq 1$, then the same conclusion holds for $\omega = 0$, i.e., for the normalization $b_1^{(n)}(0) = I_p$ and $b_2^{(n)}(0) = I_q$.

Remark 3.60 *If $W \in \mathcal{P}(j_{pq})$, $\varepsilon_n \in \mathcal{S}^{p\times q}$ for $n \geq 1$ and*

$$s(\lambda) = \lim_{n\uparrow\infty} T_W[\varepsilon_n],$$

then there exists a subsequence $\{\varepsilon_{n_k}\}$, $k = 1, 2, \ldots$ of $\{\varepsilon_n\}$ that converges to a mvf $\varepsilon_0 \in \mathcal{S}^{p\times q}$ at each point of $\mathbb{C}_+$. Therefore,

$$s(\lambda) = \lim_{k\uparrow\infty} T_{W(\lambda)}[\varepsilon_{n_k}(\lambda)] = (T_W[\varepsilon_0])(\lambda);$$

i.e., the set $T_W[\mathcal{S}^{p\times q}]$ is a closed subspace of $\mathcal{S}^{p\times q}$ with respect to pointwise convergence.

Proofs of the next several results may be found on pp. 230–233 of [ArD08b].

Lemma 3.61 *If $U \in \mathcal{E} \cap \mathcal{U}(J)$, $J \neq \pm I_m$, and $\{b_1, b_2\} \in ap_I(U)$, then $\rho U \in \mathcal{P}(J)$ for some scalar function ρ if and only if*

$$\rho(\lambda) = e^{i\gamma} e_\beta(\lambda), \quad \text{where } \overline{\gamma} = \gamma, \ \beta = \beta_2 - \beta_1, \tag{3.153}$$

and β_1 and β_2 are nonnegative numbers such that

$$e_{\beta_1}^{-1} b_1 \in \mathcal{S}_{\text{in}}^{p\times p} \quad \text{and} \quad e_{\beta_2}^{-1} b_2 \in \mathcal{S}_{\text{in}}^{q\times q}. \tag{3.154}$$

Moreover, for such a choice of ρ, $\rho U \in \mathcal{E} \cap \mathcal{U}(J)$.

Theorem 3.62 *Let W be a nondegenerate meromorphic $m \times m$ mvf in $\mathbb{C}_+$ and let*

$$\mathcal{S}^{p\times q} \subseteq \mathcal{D}(T_W) \quad \text{and} \quad T_W[\mathcal{S}^{p\times q}] = \mathcal{S}^{p\times q}. \tag{3.155}$$

Then there exists a scalar meromorphic function $\rho(\lambda)$ in $\mathbb{C}_+$ such that $\rho W \in \mathcal{U}_{\text{const}}(j_{pq})$. If $W \in \mathcal{P}^\circ(j_{pq})$ and (3.155) holds, then $W \in \mathcal{U}_{\text{const}}(j_{pq})$.

Lemma 3.63 *Let W and W_1 both belong to $\mathcal{U}(j_{pq})$. Then*

$$T_W[\mathcal{S}^{p\times q}] = T_{W_1}[\mathcal{S}^{p\times q}] \quad \text{and} \quad ap(W) = ap(W_1)$$

if and only if

$$W_1(\lambda) = W(\lambda)U \quad \text{and} \quad U \in \mathcal{U}_{\text{const}}(j_{pq}).$$

Theorem 3.64 *Let W and W_1 both belong to $\mathcal{U}(j_{pq})$, let $\{b_1, b_2\} \in ap(W)$ and $\{b_1^{(1)}, b_2^{(1)}\} \in ap(W_1)$. Then the conditions*

$$(b_1^{(1)})^{-1}b_1 \in \mathcal{S}_{\text{in}}^{p\times p}, \quad b_2(b_2^{(1)})^{-1} \in \mathcal{S}_{\text{in}}^{q\times q} \quad \text{and} \quad T_W[\mathcal{S}^{p\times q}] \subseteq T_{W_1}[\mathcal{S}^{p\times q}]$$

hold if and only if $W_1^{-1}W \in \mathcal{U}(j_{pq})$. Moreover, if $W_1^{-1}W \in \mathcal{U}(j_{pq})$, then

$$W_1^{-1}W \in \mathcal{U}_S(j_{pq}) \Longleftrightarrow ap(W_1) = ap(W). \tag{3.156}$$

Lemma 3.65 *Let $W \in \mathcal{E} \cap \mathcal{U}(j_{pq})$, $W_1 \in \mathcal{P}^\circ(j_{pq})$, $\{b_1, b_2\} \in ap(W)$ and suppose that*

$$T_W[\mathcal{S}^{p\times q}] \subseteq T_{W_1}[\mathcal{S}^{p\times q}]. \tag{3.157}$$

Then:

(1) *$W_1 \in \mathcal{E} \cap \mathcal{U}(j_{pq})$ and*

$$W = \frac{e_{\beta_1}}{e_{\beta_2}} W_1 W_2, \tag{3.158}$$

where $W_2 \in \mathcal{E} \cap \mathcal{U}(j_{pq})$, $\beta_1 \geq 0$, $\beta_2 \geq 0$,

$$e_{\beta_1}^{-1} b_1 \in \mathcal{E} \cap \mathcal{S}_{\text{in}}^{p\times p} \quad \text{and} \quad e_{\beta_2^{-1}} b_2 \in \mathcal{E} \cap \mathcal{S}_{\text{in}}^{q\times q}. \tag{3.159}$$

(2) *If equality prevails in (3.157), then $W_2 \in \mathcal{U}_{\text{const}}(j_{pq})$.*

3.12 Linear fractional transformations in $\mathcal{C}^{p\times p}$ and from $\mathcal{S}^{p\times p}$ into $\mathcal{C}^{p\times p}$

Let $m = 2p$ and let

$$A(\lambda) = \begin{bmatrix} a_{11}(\lambda) & a_{12}(\lambda) \\ a_{21}(\lambda) & a_{22}(\lambda) \end{bmatrix}$$

be a meromorphic $m \times m$ mvf in $\mathbb{C}_+$ with blocks $a_{ij}(\lambda)$ of size $p \times p$, and let

$$B(\lambda) = A(\lambda)\mathfrak{V} = \begin{bmatrix} b_{11}(\lambda) & b_{12}(\lambda) \\ b_{21}(\lambda) & b_{22}(\lambda) \end{bmatrix} \tag{3.160}$$

and

$$W(\lambda) = \mathfrak{V}A(\lambda)\mathfrak{V} = \begin{bmatrix} w_{11}(\lambda) & w_{12}(\lambda) \\ w_{21}(\lambda) & w_{22}(\lambda) \end{bmatrix}, \tag{3.161}$$

where $\mathfrak{V}$ is defined in formula (1.6).

If $A \in \mathcal{U}(J_p)$, then it is convenient to consider linear fractional transformations based on the mvf's

$$W(\lambda) = \mathfrak{V}A(\lambda)\mathfrak{V}, \quad B(\lambda) = A(\lambda)\mathfrak{V} \tag{3.162}$$

and $A(\lambda)$. The mvf's $B(\lambda)$ belong to the class $\mathcal{U}(j_p, J_p)$ of $m \times m$ mvf's that are meromorphic in $\mathbb{C}_+$ and (j_p, J_p)-contractive in $\mathfrak{h}_B^+$, i.e.,

$$B(\lambda)^* J_p B(\lambda) \leq j_p \quad \text{for every } \lambda \in \mathfrak{h}_B^+ \tag{3.163}$$

and are also (j_p, J_p)-unitary a.e. on $\mathbb{R}$:

$$B(\mu)^* J_p B(\mu) = j_p \quad \text{a.e. on } \mathbb{R}. \tag{3.164}$$

Moreover,

$$A \in \mathcal{U}(J_p) \Longleftrightarrow B \in \mathcal{U}(j_p, J_p) \Longleftrightarrow W \in \mathcal{U}(j_p) \tag{3.165}$$

and

$$A \in \mathcal{E} \cap \mathcal{U}(J_p) \Longleftrightarrow B \in \mathcal{E} \cap \mathcal{U}(j_p, J_p) \Longleftrightarrow W \in \mathcal{E} \cap \mathcal{U}(j_p). \tag{3.166}$$

If $A \in \mathcal{P}(J_p)$ and B and W are defined by (3.162), then, in view of Lemma 3.58 and formulas (3.39) and (3.40),

$$\begin{aligned}\mathcal{S}^{p\times p} \cap \mathcal{D}(T_B) &= \{\varepsilon \in \mathcal{S}^{p\times p} : \det(b_{21}\varepsilon + b_{22}) \not\equiv 0 \text{ in } \mathbb{C}_+\} \\ &= \{\varepsilon \in \mathcal{S}^{p\times p} : T_W[\varepsilon] \in \mathcal{D}(T_{\mathfrak{V}})\}\end{aligned}$$

and

$$T_B[\mathcal{S}^{p\times p} \cap \mathcal{D}(T_B)] = T_{\mathfrak{V}}[T_W[\mathcal{S}^{p\times p} \cap \mathcal{D}(T_B)]] \subseteq \mathcal{C}^{p\times p}.$$

Recall that

$$\mathcal{C}(A) = T_B[\mathcal{S}^{p\times p} \cap \mathcal{D}(T_B)]. \tag{3.167}$$

Then it is easy to check that

$$T_A[\mathcal{C}^{p\times p} \cap \mathcal{D}(T_A)] \subseteq \mathcal{C}(A) \subseteq \mathcal{C}^{p\times p} \subset \mathcal{D}(T_{\mathfrak{V}}), \tag{3.168}$$

$$T_{\mathfrak{V}}[\mathcal{C}(A)] \subseteq T_W[\mathcal{S}^{p\times p}], \quad T_{\mathfrak{V}}[\mathcal{S}_{\text{in}}^{p\times p} \cap \mathcal{D}(T_{\mathfrak{V}})] \subseteq \mathcal{C}_{\text{sing}}^{p\times p}, \tag{3.169}$$

$$\mathring{\mathcal{C}}^{p\times p} \subset \mathcal{D}(T_A), \tag{3.170}$$

and, if $\tau \in \mathring{\mathcal{C}}^{p\times p}$ and $c = T_A[\tau]$, then $(\Re c)(\lambda) > 0$ for every point $\lambda \in \mathbb{C}_+$. Moreover, it is readily seen that if $A \in \mathcal{U}(J_p)$, then also

$$T_A[\mathcal{C}_{\text{sing}}^{p\times p} \cap \mathcal{D}(T_A)] \subseteq T_B[\mathcal{S}_{\text{in}}^{p\times p} \cap \mathcal{D}(T_B)] \subseteq \mathcal{C}_{\text{sing}}^{p\times p} \subset \mathcal{D}(T_{\mathfrak{V}}) \tag{3.171}$$

and

$$T_{\mathfrak{V}B}[\mathcal{S}_{\text{in}}^{p\times p} \cap \mathcal{D}(T_B)] \subseteq T_W[\mathcal{S}_{\text{in}}^{p\times p}]. \tag{3.172}$$

In view of (3.162) and (3.113), condition (3.163) is equivalent to the condition

$$B(\lambda)j_pB(\lambda)^* \leq J_p \quad \text{for every} \quad \lambda \in \mathfrak{h}_B^+, \tag{3.173}$$

whereas condition (3.164) is equivalent to the condition

$$B(\mu)j_pB(\mu)^* = J_p \quad \text{for almost all points } \mu \in \mathbb{R}. \tag{3.174}$$

In particular, (3.173) implies that:

$$r(\lambda) \stackrel{\text{def}}{=} b_{22}(\lambda)b_{22}(\lambda)^* - b_{21}(\lambda)b_{21}(\lambda)^* \geq 0 \quad \text{for } \lambda \in \mathfrak{h}_B^+, \tag{3.175}$$

whereas (3.163) implies that

$$b_{12}(\lambda)^*b_{22}(\lambda) + b_{22}(\lambda)^*b_{12}(\lambda) \geq I_p \quad \text{for } \lambda \in \mathfrak{h}_B^+, \tag{3.176}$$

and (3.174) implies that

$$b_{22}(\mu)b_{22}(\mu)^* - b_{21}(\mu)b_{21}(\mu)^* = 0 \quad \text{a.e. on } \mathbb{R}. \tag{3.177}$$

Lemma 3.66 *If $B \in \mathcal{U}(j_p, J_p)$, then:*

(1) $\det b_{22}(\lambda) \neq 0$ *for every point* $\lambda \in \mathfrak{h}_B^+$.
(2) *The mvf* $c_0 = b_{12}b_{22}^{-1} = T_B^r[0_{p\times p}]$ *belongs to* $\mathcal{C}^{p\times p}$.
(3) *The mvf* $\chi = b_{22}^{-1}b_{21} = T_B^\ell[0_{p\times p}]$ *belongs to* $\mathcal{S}_{\text{in}}^{p\times p}$.
(4) *The mvf* $\rho_i^{-1}(b_{22})^{-1}$ *belongs to* $H_2^{p\times p}$.
(5) $\rho_i^{-1}(b_{21}^\#)^{-1} \in H_2^{p\times p}$.
(6) $b_{11} - b_{12}b_{22}^{-1}b_{21} = -(b_{21}^\#)^{-1}$.

Proof This follows from lemma 4.35 in [ArD08b]. □

Lemma 3.67 *Let $A \in \mathcal{U}(J_p)$, let $B = A\mathfrak{V}$ and let $\chi = b_{22}^{-1}b_{21}$. Then the following conditions are equivalent:*

(1) $I_p - \chi(\omega)\chi(\omega)^* > 0$ *for at least one point* $\omega \in \mathbb{C}_+$.
(2) $I_p - \chi(\lambda)\chi(\lambda)^* > 0$ *for every point* $\lambda \in \mathbb{C}_+$.
(3) $b_{22}(\omega)b_{22}(\omega)^* - b_{21}(\omega)b_{21}(\omega)^* > 0$ *for at least one point* $\omega \in \mathfrak{h}_B^+$.
(4) $b_{22}(\lambda)b_{22}(\lambda)^* - b_{21}(\lambda)b_{21}(\lambda)^* > 0$ *for every point* $\lambda \in \mathfrak{h}_B^+$.
(5) $\mathcal{S}^{p\times p} \subseteq \mathcal{D}(T_B)$.

If $A \in \mathcal{E} \cap \mathcal{U}(J)$, then conditions (1)–(5) are equivalent to the condition

$$ia_{21}'(0) > 0. \tag{3.178}$$

Proof This follows from lemma 4.70 in [ArD08b]. □

Lemma 3.68 *Let $A \in \mathcal{U}(J_p)$, $B = A\mathfrak{V}$ and $W = \mathfrak{V}A\mathfrak{V}$. Then*

$$T_{\mathfrak{V}}[T_A[\mathcal{C}^{p\times p} \cap \mathcal{D}(T_A)]] \subseteq T_W[\mathcal{S}^{p\times q}]. \tag{3.179}$$

If $s \in T_W[\mathcal{S}^{p\times q}]$, there is a sequence of mvf's $s_n \in T_{\mathfrak{V}A}[\mathcal{C}^{p\times p} \cap \mathcal{D}(T_A)]$ such that

$$s(\lambda) = \lim_{n\uparrow\infty} s_n(\lambda).$$

Moreover, if $s \in T_W[\mathcal{S}_{\text{in}}^{p\times q}]$, then the mvf's s_n may be chosen from $T_{\mathfrak{V}A}[\mathcal{C}_{\text{sing}}^{p\times p} \cap \mathcal{D}(T_A)]$.

Proof This follows lemma 4.71 in [ArD08b]. □

Lemma 3.69 *Let $A \in \mathcal{P}(J_p)$, $A_1 \in \mathcal{P}^\circ(J_p)$ and $A_1^{-1}A \in \mathcal{P}(J_p)$. Then*

$$\mathcal{C}(A) \subseteq \mathcal{C}(A_1). \tag{3.180}$$

Proof This is lemma 4.72 in [ArD08b]. □

Theorem 3.70 *Let $A \in \mathcal{E} \cap \mathcal{U}(J_p)$, $B(\lambda) = A(\lambda)\mathfrak{V}$,*

$$\mathfrak{E}(\lambda) = [E_-(\lambda) \quad E_+(\lambda)] = \sqrt{2}\begin{bmatrix}0 & I_p\end{bmatrix} B(\lambda),$$

$\chi = E_+^{-1}E_-$, $\omega \in \mathbb{C}_+$ and $I_p - \chi(\omega)\chi(\omega)^* > 0$. Then the set

$$\mathcal{B}(\omega) = \{c(\omega) : c \in \mathcal{C}(A)\}. \tag{3.181}$$

is a matrix ball:

$$\mathcal{B}(\omega) = \{\gamma_c(\omega) + R_\ell(\omega) v R_r(\omega) : v \in \mathcal{S}_{\text{const}}^{p\times p}\}. \tag{3.182}$$

with center $\gamma_c(\omega)$, left semiradius $R_\ell(\omega) > 0$ and right semiradius $R_r(\omega) > 0$ that are given by the formulas

$$R_\ell(\omega)^2 = -r(\bar{\omega})^{-1} = 2E_-^\#(\omega)^{-1}\{I_p - \chi(\omega)^*\chi(\omega)\}^{-1}E_-^\#(\omega)^{-*} \tag{3.183}$$

$$R_r(\omega)^2 = r(\omega)^{-1} = 2E_+(\omega)^{-*}\{I_p - \chi(\omega)\chi(\omega)^*\}^{-1}E_+(\omega)^{-1} \tag{3.184}$$

$$\gamma_c(\omega) = T_{B(\omega)}[-\chi(\omega)^*]. \tag{3.185}$$

Proof This follows from Lemma 3.19, formulas (3.32), (3.36), (3.37) and (3.175) and Lemma 3.66. □

Remark 3.71 *The particular left and right semiradii that are defined by formulas (3.184) and (3.183) can also be expressed in terms of the RK $K_\omega^{\mathfrak{E}}(\lambda)$ of the RKHS $\mathcal{B}(\mathfrak{E})$ for $\omega \in \mathbb{C}_+$ (see (4.116)):*

$$R_r(\omega)^{-2} = \rho_\omega(\omega)K_\omega^{\mathfrak{E}}(\omega)/2 \quad \text{and} \quad R_\ell(\omega)^{-2} = -\rho_{\bar{\omega}}(\bar{\omega})K_{\bar{\omega}}^{\mathfrak{E}}(\bar{\omega})/2. \tag{3.186}$$

Affine generalizations of $\mathcal{C}^{p\times p}$

Let $\widetilde{\mathcal{C}}^{p\times p}$ denote the set of pairs $\{u(\lambda), v(\lambda)\}$ of $p \times p$ mvf's that are meromorphic in $\mathbb{C}_+$ and meet the following two conditions:

(1) $[u(\lambda)^* \quad v(\lambda)^*]J_p \begin{bmatrix} u(\lambda) \\ v(\lambda) \end{bmatrix} \leq 0$ for $\lambda \in \mathfrak{h}_u^+ \cap \mathfrak{h}_v^+$;

(2) $[u(\lambda)^* \quad v(\lambda)^*] \begin{bmatrix} u(\lambda) \\ v(\lambda) \end{bmatrix} > 0$ for at least one point $\lambda \in \mathfrak{h}_u^+ \cap \mathfrak{h}_v^+$;

and, for $A \in \mathcal{P}(J_p)$, let the linear fractional transformation $\widetilde{T}_A$ of such pairs be defined by the formula

$$\widetilde{T}_A[\{u, v\}] = (a_{11}(\lambda)u(\lambda) + a_{12}(\lambda)v(\lambda))(a_{21}(\lambda)u(\lambda) + a_{22}(\lambda)v(\lambda))^{-1} \quad (3.187)$$

on the set

$$\widetilde{\mathcal{C}}^{p\times p} \cap \mathcal{D}(\widetilde{T}_A) \stackrel{\text{def}}{=} \{\{u, v\} \in \widetilde{\mathcal{C}}^{p\times p} : \det(a_{21}(\lambda)u(\lambda) + a_{22}(\lambda)v(\lambda)) \not\equiv 0\}. \quad (3.188)$$

If $\det v(\lambda) \not\equiv 0$, then:

$$\{u, v\} \in \widetilde{\mathcal{C}}^{p\times p} \Longleftrightarrow uv^{-1} \in \mathcal{C}^{p\times p}$$

and

$$\{u, v\} \in \widetilde{\mathcal{C}}^{p\times p} \cap \mathcal{D}(\widetilde{T}_A) \Longleftrightarrow uv^{-1} \in \mathcal{C}^{p\times p} \cap \mathcal{D}(T_A).$$

Moreover, the set $\mathcal{C}(A)$ defined in (3.167) may be reexpressed as

$$\mathcal{C}(A) = \{\widetilde{T}_A[\{u, v\}] : \{p, q\} \in \widetilde{\mathcal{C}}^{p\times p} \cap \mathcal{D}(\widetilde{T}_A)\}. \quad (3.189)$$

3.13 Associated pairs of the second kind

Associated pairs of the second kind will be defined initially for mvf's $A \in \mathcal{U}(J_p)$ in terms of inner–outer and outer–inner factorizations of the blocks $(b_{21}^{\#})^{-1}$ and b_{22}^{-1} in the mvf $B(\lambda) = A(\lambda)\mathfrak{V}$; see (3.190) below. Subsequently, associated pairs of the second kind will be defined for mvf's $U \in \mathcal{U}(J)$ when J is unitarily equivalent to J_p. Thus, a mvf $U \in \mathcal{U}(J)$ with $p = q$, i.e., with rank$P_+ =$ rankP_-, will have associated pairs of both the first and second kind.

Let $A \in \mathcal{U}(J_p)$ and let

$$B(\lambda) = A(\lambda)\mathfrak{V} \text{ and } \mathrm{W}(\lambda) = \mathfrak{V}\mathrm{A}(\lambda)\mathfrak{V}.$$

Then $B \in \mathcal{U}(j_p, J_p)$ and $W \in \mathcal{U}(j_p)$ with blocks b_{ij} and w_{ij}, respectively. Then, in view of Lemma 3.66,

$$(b_{21}^{\#})^{-1} \in \mathcal{N}_+^{p\times p} \quad \text{and} \quad b_{22}^{-1} \in \mathcal{N}_+^{p\times p}. \quad (3.190)$$

Therefore, $(b_{21}^{\#})^{-1}$ and b_{22}^{-1} admit essentially unique inner–outer and outer–inner factorizations in $\mathcal{N}_{+}^{p\times p}$:

$$(b_{21}^{\#})^{-1} = b_3\varphi_3 \quad \text{and} \quad b_{22}^{-1} = \varphi_4 b_4, \tag{3.191}$$

where $b_j \in \mathcal{S}_{\text{in}}^{p\times p}$ and $\varphi_j = \mathcal{N}_{\text{out}}^{p\times p}$, $j = 1, 2$. The pair $\{b_3, b_4\}$ is called an **associated pair** of the mvf B, and the set of all such pairs is denoted by $ap(B)$. Thus, $\{b_3, b_4\} \in ap(B)$. Moreover, if $\{b_3, b_4\}$ is a fixed associated pair of B, then

$$ap(B) = \{\{b_3 u,\ v b_4\} : u \text{ and } v \text{ are unitary } p\times p \text{ matrices}\}.$$

In view of the relation $A(\lambda) = \mathfrak{V}W(\lambda)\mathfrak{V}$, the set $ap_I(A)$ of associated pairs of the first kind for A was defined as $ap_I(A) = ap(W)$. Analogously, we define $ap_{II}(A) = ap(B)$ as the set of **associated pairs of the second kind** for A. This definition can be extended to mvf's in the class $\mathcal{U}(J)$ based on a signature matrix J that is unitarily equivalent to J_p: If

$$J = V^* J_p V \text{ for some unitary matrix } V \in \mathbb{C}^{m\times m}, \tag{3.192}$$

then

$$U \in U(J) \Longleftrightarrow VUV^* \in U(J_p).$$

If $U \in \mathcal{U}(J)$, then the pair $\{b_3, b_4\}$ is said to be an associated pair of the second kind for $U(\lambda)$ and we write $\{b_3, b_4\} \in ap_{II}(U)$ if $\{b_3, b_4\} \in ap_{II}(VUV^*)$. The set $ap_{II}(U)$ depends upon the choice of the unitary matrix V in (3.192).

The next two lemmas follow from lemmas 4.36 and 4.38 in [ArD08b].

Lemma 3.72 *Let $A \in \mathcal{U}(J_p)$, $\{b_1, b_2\} \in ap_I(A)$ and $\{b_3, b_4\} \in ap_{II}(A)$ and let $W = \mathfrak{V}A\mathfrak{V}$ and $s_{12} = T_W[0]$. Then*

$$\frac{(I_p + s_{12})}{2} b_3 = b_1\varphi \quad \text{and} \quad b_4 \frac{(I_p + s_{12})}{2} = \psi b_2 \tag{3.193}$$

for some $\varphi \in \mathcal{S}_{\text{out}}^{p\times p}$ and $\psi \in \mathcal{S}_{\text{out}}^{p\times p}$.

Corollary 3.73 *If $U \in \mathcal{U}(J)$, J is unitarily equivalent to J_p, $\alpha \geq 0$ and $\beta \geq 0$, then*

$$\{e_\alpha I_p, e_\beta I_p\} \in ap_I(U) \Longleftrightarrow \{e_\alpha I_p, e_\beta I_p\} \in ap_{II}(U).$$

Proof This follows from (3.193), since $(I_p + s_{12})/2 \in \mathcal{S}_{\text{out}}^{p\times p}$ by Theorem 3.11. □

Lemma 3.74 *Let $A \in \mathcal{U}(J_p)$ and let $\{b_1, b_2\} \in ap_I(A)$ and let $\{b_3, b_4\} \in ap_{II}(A)$. Then the following equivalences are valid:*

(1) *$A \in \mathcal{N}_{+}^{m\times m} \Longleftrightarrow b_2$ is a constant matrix $\Longleftrightarrow b_4$ is a constant matrix.*
(2) *$A^{-1} \in \mathcal{N}_{+}^{m\times m} \Longleftrightarrow b_1$ is a constant matrix $\Longleftrightarrow b_3$ is a constant matrix.*

(3) $A \in \mathcal{U}_S(J_p) \Longleftrightarrow b_1$ *and* b_2 *are constant matrices*
$\Longleftrightarrow b_3$ *and* b_4 *are constant matrices.*

Theorem 3.75 *Let* $U \in \mathcal{U}(J)$, *where* J *is unitarily equivalent to* J_p *and let* $\{b_3, b_4\} \in ap_{II}(U)$. *Then*

(1) $\mathfrak{h}_U^+ = \mathfrak{h}_{b_4^\#}^+, \quad \mathfrak{h}_U^- = \mathfrak{h}_{b_3}^- \quad$ *and* $\quad \mathfrak{h}_U \subseteq \mathfrak{h}_{b_4^\#} \cap \mathfrak{h}_{b_3}$.
(2) *If* U *is entire, then* b_3 *and* b_4 *are entire.*
(3) *If* $U \in \mathcal{U}_{rR}(J)$, *then* $\mathfrak{h}_U = \mathfrak{h}_{b_4^\#} \cap \mathfrak{h}_{b_3}$ *and hence*

$$\text{if } U \in \mathcal{U}_{rR}(J) \text{ and } b_3 \text{ and } b_4 \text{ are entire, then } U \text{ is entire.} \tag{3.194}$$

Proof This is theorem 4.56 in [ArD08b]. □

Theorem 3.76 *Let* $A \in \mathcal{E} \cap \mathcal{U}(J_p)$ *and* $A_1 \in \mathcal{P}^\circ(J_p)$, *let* $\{b_3, b_4\} \in ap_{II}(A)$ *and suppose that*

$$\mathcal{C}(A) = \mathcal{C}(A_1).$$

Then $A_1 \in \mathcal{E} \cap \mathcal{U}(J_p)$ *and can be expressed in the form*

$$A_1(\lambda) = \frac{e_{\beta_2}(\lambda)}{e_{\beta_1}(\lambda)} A(\lambda) V,$$

where $e_{-\beta_1} b_3 \in \mathcal{S}_{\rm in}^{p\times p}$, $e_{-\beta_2} b_4 \in \mathcal{S}_{\rm in}^{p\times p}$, $\beta_1 \geq 0$, $\beta_2 \geq 0$ *and* $V \in \mathcal{U}_{\rm const}(J_p)$.

Proof See corollary 4.97 in [ArD08b]. □

Theorem 3.77 *Let* A *and* A_1 *both belong to* $\mathcal{U}(J_p)$, *let* $\{b_3, b_4\} \in ap_{II}(A)$ *and* $\{b_3^{(1)}, b_4^{(1)}\} \in ap_{II}(A_1)$. *Then the conditions*

$$(b_3^{(1)})^{-1} b_3 \in \mathcal{S}_{\rm in}^{p\times p}, \quad b_4 (b_4^{(1)})^{-1} \in \mathcal{S}_{\rm in}^{p\times p} \quad \text{and} \quad \mathcal{C}(A) \subseteq \mathcal{C}(A_1)$$

hold if and only if $A_1^{-1} A \in \mathcal{U}(J_p)$.

Proof This is theorem 4.98 in [ArD08b]. □

Remark 3.78 *Theorem 3.77 remains valid if the condition* $\mathcal{C}(A) \subseteq \mathcal{C}(A_1)$ *is replaced by the condition*

$$T_A[\mathcal{C}^{p\times p} \cap \mathcal{D}(T_A)] \subseteq T_{A_1}[\mathcal{C}^{p\times p} \cap \mathcal{D}(T_{A_1})].$$

Corollary 3.79 *Let* A *and* A_1 *both belong to* $\mathcal{U}(J_p)$. *Then*

$$\mathcal{C}(A) = \mathcal{C}(A_1) \quad \text{and} \quad ap_{II}(A) = ap_{II}(A_1)$$

if and only if

$$A_1(\lambda) = A(\lambda) U \quad \text{and} \quad U \in \mathcal{U}_{\rm const}(J_p).$$

3.14 Supplementary notes

In [Po60] V.P. Potapov considered more general injective linear fractional transformations of the class $\mathcal{P}_{\rm const}(J)$ into $\mathcal{S}_{\rm const}^{m\times m}$ than (3.98). The transformation (3.98) was introduced later by Y.Yu. Ginzburg [Gi57] for bounded linear operators in Hilbert space. The PG transform for $J = j_{pq}$ was also considered by R. Redheffer in his development of transmission line theory [Re62]. He also studied the transformation (3.189) in the more general setting of contractive linear operators between Hilbert spaces instead of contractive matrices.

Bibliographical notes on mvf's in the Nevanlinna class and the subclasses considered here (especially $\mathcal{P}(J)$ and $\mathcal{U}(J)$) may be found in the Supplementary Notes to chapters 3 and 4 in [ArD08b].

Associated pairs for mvs $W \in \mathcal{U}(j_{pq})$ and $A \in \mathcal{U}(J_p)$ and the subclasses of singular and right regular mvf's were introduced and studied by D.Z. Arov in his investigations of generalized Schur and Carathéodory interpolation problems in [Ar84], [Ar88], [Ar89], [Ar90] and [Ar93]. The supplementary classifications of right regular, left regular, right strongly regular, left strongly regular and a number of characterizations of these classes were developed later in the joint work of the authors in their study of inverse problems for canonical integral and differential systems [ArD97]–[ArD12a].

Lemma 3.61 and Theorems 3.62 and 3.76 follow from general results on meromorphic minus matrices in $\mathbb{C}_+$ by L.A. Simakova; see section 4.17 of [ArD08b], [Si74], [Si75] and [Si03]. In particular, she proved the following useful facts:

Theorem 3.80 *If W is an $m \times m$ mvf that is meromorphic in $\mathbb{C}_+$ and if $m = p+q$ with $1 \le p \le m-1$ and*

$$\mathcal{S}^{p\times q} \subset \mathcal{D}(T_W), \quad T_W[\mathcal{S}^{p\times q}] \subseteq \mathcal{S}^{p\times q} \quad \textit{and} \quad \det W(\lambda) \not\equiv 0,$$

then $\rho W \in \mathcal{P}^\circ(j_{pq})$ for some scalar function ρ that is meromorphic in $\mathbb{C}_+$. Moreover, if

$$T_W[\mathcal{S}_{\rm in}^{p\times q}] \subseteq \mathcal{S}_{\rm in}^{p\times q}, \quad \textit{then} \quad \rho W \in \mathcal{U}(j_{pq}).$$

Simakova also described the set of scalar multipliers that can arise in Theorem 3.80; see theorem 4.80 in [ArD08b].

Theorem 3.81 *If A is an $m \times m$ mvf that is meromorphic in $\mathbb{C}_+$ and if $m = 2p$ with $p \ge 1$ and*

$$T_A[\mathcal{C}^{p\times p} \cap \mathcal{D}(T_A)] \subset \mathcal{C}^{p\times p} \quad \textit{and} \quad \det A(\lambda) \not\equiv 0,$$

then $\rho A \in \mathcal{P}^\circ(J_p)$ for some scalar function ρ that is meromorphic in $\mathbb{C}_+$. Moreover, if

$$T_A[\mathcal{C}_{\rm sing}^{p\times p} \cap \mathcal{D}(T_A)] \subseteq \mathcal{C}_{\rm sing}^{p\times p}, \quad \textit{then} \quad \rho A \in \mathcal{U}(J_p).$$

The proof of Theorem 3.80 was based in part on a theorem of M.G. Krein and Yu.L. Shmuljan in [KrS66] that established analogous results for injective constant linear fractional transformations of contractive linear operators in a Hilbert space into contractive linear operators in a possibly different Hilbert space. This result of Krein and Shmuljan also follows from R. Redheffer's work on linear fractional transformations in [Re60], where the corresponding result was obtained for the Redheffer transform (3.28) even when the ε are contractive linear operators between two Hilbert spaces and the values of the transform are contractive linear operators between two Hilbert spaces. Linear fractional transformations of linear contractive operators are studied in [KrS67].

4

Interpolation problems, resolvent matrices and de Branges spaces

This chapter is devoted to the interplay between

(1) three interpolation problems;
(2) descriptions of their solutions as linear fractional transformations (of Schur class mvf's) based on appropriately restricted resolvent matrices when the interpolation problems are completely indeterminate; and
(3) formulas for these resolvent matrices in terms of the given data when the interpolation problems are strictly completely indeterminate.

The basic fact is that there is a one-to-one correspondence between completely indeterminate interpolation problems and appropriately restricted resolvent matrices. More precisely:

Interpolation problem	In the class	Class of resolvent matrices	Applicable to direct and inverse
GCIP	$\mathcal{C}^{p\times p}$	$\mathcal{U}_{rR}(J_p)$	Impedance & spectral problems
GSIP	$\mathcal{S}^{p\times q}$	$\mathcal{U}_{rR}(j_{pq})$	Input scattering problem
NP	$\mathbb{B}^{p\times q}$	$\mathfrak{M}_{rR}(j_{pq})$	Asymptotic scattering problem

The abbreviations in the first column stand for generalized Carathéodory interpolation problem, generalized Schur interpolation problem and Nehari problem, respectively

In all three of these problems uniqueness of the resolvent matrices is obtained by restricting to suitably normalized **right regular** mvf's in the indicated class and, in the first two listed problems, also requiring the two inner mvf's in the interpolation problem to be **associated pairs** of the resolvent matrix. If the interpolation problem is **strictly completely indeterminate** then the unique resolvent matrix for the problem that is determined as outlined above will be **right strongly regular**. The de Branges theory of reproducing kernel Hilbert spaces will be exploited to obtain formulas for the resolvent matrices for the first two problems in the list and also in the study of direct and inverse spectral problems.

4.1 The Nehari problem

It is convenient to begin with the following formulation of the Nehari problem:

NP(Γ): Given a bounded linear operator Γ from H_2^q into $(H_2^p)^\perp$, describe the set

$$\mathcal{N}(\Gamma) = \{f \in \mathbb{B}^{p\times q} : \Pi_- M_f|_{H_2^q} = \Gamma\}. \tag{4.1}$$

The NP(Γ) is called **determinate** if it has exactly one solution and **indeterminate** otherwise. It is called **completely indeterminate** if for every nonzero vector $\eta \in \mathbb{C}^q$ there exist solutions $f_1, f_2 \in \mathcal{N}(\Gamma)$ such that $\|f_1\eta - f_2\eta\|_\infty > 0$.

Let

$$\mathring{\mathbb{B}}^{p\times q} = \{f \in \mathbb{B}^{p\times q} : \|f\|_\infty < 1\}.$$

The NP(Γ) is called **strictly completely indeterminate** if

$$\mathcal{N}(\Gamma) \cap \mathring{\mathbb{B}}^{p\times q} \neq \emptyset.$$

It is readily checked by direct calculation that if

$$\Gamma = \Pi_- M_f|_{H_2^q}. \tag{4.2}$$

for some $f \in L_\infty^{p\times q}$, then

$$\Pi_- M_{e_t}\Gamma = \Gamma M_{e_t}|_{H_2^q} \quad \text{for every } t \geq 0. \tag{4.3}$$

A bounded linear operator Γ from H_2^q into $(H_2^p)^\perp$ that satisfies the constraints in (4.3) is called a **Hankel** operator.

Theorem 4.1 *If Γ is a bounded linear operator from H_2^q into $(H_2^p)^\perp$, then:*

(1) $\Gamma = \Pi_- M_f|_{H_2^q}$ *for some* $f \in L_\infty^{p\times q} \Longleftrightarrow \Gamma$ *is a Hankel operator.*

(2) *If Γ is a Hankel operator, then*

$$\|\Gamma\| = \min\{\|f\|_\infty : f \in L_\infty^{p\times q} \quad \textit{and} \quad \Gamma f = \Pi_- f|_{H_2^q}\}. \tag{4.4}$$

(3) $\mathcal{N}(\Gamma) \neq \emptyset \Longleftrightarrow \Gamma$ *is a Hankel operator and* $\|\Gamma\| \leq 1$.

Proof The implication $\Longrightarrow$ in (1) and the inequality

$$\|\Gamma\| \leq \|f\|_\infty \quad \text{for every } f \in L_\infty^{p\times q} \text{ for which (4.2) holds} \tag{4.5}$$

are easily checked. The rest is more complicated; see, e.g., theorem 7.2 in [ArD08b]. □

Corollary 4.2 *If Γ is a bounded linear operator from H_2^q into $(H_2^p)^\perp$, then $NP(\Gamma)$ is strictly completely indeterminate if and only if Γ is a strictly contractive Hankel operator.*

Let $\omega \in \mathbb{C}_+$ and let

$$\mathfrak{A}_+(\omega) = \left\{ \frac{\eta}{\rho_\omega} : \eta \in \mathbb{C}^q \right\} \cap (I - \Gamma^*\Gamma)^{1/2} H_2^q \tag{4.6}$$

and

$$\mathfrak{A}_-(\omega) = \left\{ \frac{\xi}{\rho_{\overline{\omega}}} : \xi \in \mathbb{C}^p \right\} \cap (I - \Gamma\Gamma^*)^{1/2} (H_2^p)^\perp. \tag{4.7}$$

Theorem 4.3 *Let Γ be a Hankel operator with $\|\Gamma\| \le 1$. Then:*

(1) *The numbers $\dim \mathfrak{A}_+(\omega)$ and $\dim \mathfrak{A}_-(\omega)$ are independent of the choice of the point $\omega \in \mathbb{C}_+$.*
(2) *The $NP(\Gamma)$ is determinate if and only if*

$$\mathfrak{A}_+(\omega) = \{0\} \quad or \quad \mathfrak{A}_-(\omega) = \{0\} \tag{4.8}$$

for at least one (and hence every) point $\omega \in \mathbb{C}_+$.
(3) *The $NP(\Gamma)$ is completely indeterminate if and only if*

$$\dim \mathfrak{A}_+(\omega) = q \quad and \quad \dim \mathfrak{A}_-(\omega) = p \tag{4.9}$$

for at least one (and hence every) point $\omega \in \mathbb{C}_+$. Moreover, the two conditions in (4.9) are equivalent.

Proof See, e.g., theorem 7.5 in [ArD08b]. □

Remark 4.4 *Assertion (3) of Theorem 4.3 implies that:*

(1) *The $NP(\Gamma)$ is completely indeterminate if and only if*

$$\lim_{r \uparrow 1} \left\langle (I - r^2\Gamma^*\Gamma)^{-1} \frac{\xi}{\rho_i}, \frac{\xi}{\rho_i} \right\rangle_{st} < \infty \quad \textit{for every } \xi \in \mathbb{C}^q.$$

(2) *If the $NP(\Gamma)$ is completely indeterminate, then $I - \Gamma^*\Gamma > 0$.*

A measurable $m \times m$ mvf $\mathfrak{A}$ on $\mathbb{R}$ with block decomposition

$$\mathfrak{A} = \begin{bmatrix} \mathfrak{a}_- & \mathfrak{b}_- \\ \mathfrak{b}_+ & \mathfrak{a}_+ \end{bmatrix}$$

belongs to the class $\mathfrak{M}_r(j_{pq})$ of **right gamma generating matrices** if

(1) $\mathfrak{A}(\mu)^* j_{pq} \mathfrak{A}(\mu) = j_{pq}$ a.e. on $\mathbb{R}$.
(2) $\mathfrak{b}_+$ and $\mathfrak{a}_+$ are nontangential boundary values of mvf's in the Nevanlinna class in $\mathbb{C}_+$ such that

$$s_{22} \stackrel{\text{def}}{=} \mathfrak{a}_+^{-1} \in \mathcal{S}_{\text{out}}^{q\times q} \quad \text{and} \quad s_{21} \stackrel{\text{def}}{=} -\mathfrak{a}_+^{-1}\mathfrak{b}_+ \in \mathcal{S}^{q\times p}. \tag{4.10}$$

(3) $\mathfrak{a}_-$ and $\mathfrak{b}_-$ are nontangential boundary values of mvf's in the Nevanlinna class in $\mathbb{C}_-$ such that

$$s_{11} \stackrel{\text{def}}{=} (\mathfrak{a}_-^\#)^{-1} \in \mathcal{S}_{\text{out}}^{p\times p}. \tag{4.11}$$

Since $\mathfrak{A}(\mu)j_{pq}\mathfrak{A}(\mu)^* = j_{pq}$ a.e. on $\mathbb{R}$ when $\mathfrak{U} \in \mathfrak{M}_r(j_{pq})$,

$$\mathfrak{a}_-(\mu)\mathfrak{b}_+(\mu)^* = \mathfrak{b}_-(\mu)\mathfrak{a}_+(\mu)^* \quad \text{a.e. on } \mathbb{R}$$

and hence

$$s_{21} = -(\mathfrak{a}_-^{-1}\mathfrak{b}_-)^\#. \tag{4.12}$$

Moreover,

$$s_{21}(\lambda)^* s_{21}(\lambda) < I_p \quad \text{for } \lambda \in \mathbb{C}_+. \tag{4.13}$$

In view of (4.10) and (4.13) it is readily seen that if $\mathfrak{A} \in \mathfrak{M}_r(j_{pq})$, then

$$\mathcal{S}^{p\times q} \subset \mathcal{D}(T_{\mathfrak{A}}) \quad \text{and} \quad T_{\mathfrak{A}}[\mathcal{S}^{p\times q}] \subseteq \mathbb{B}^{p\times q}. \tag{4.14}$$

It is easily checked that the nontangential boundary values of a mvf $W \in \mathcal{U}_S(j_{pq})$ belong to the class $\mathfrak{M}_r(j_{pq})$. Moreover,

$$\mathcal{U}_S(j_{pq}) = \mathcal{N}_{\text{out}}^{m\times m} \cap \mathfrak{M}_r(j_{pq}) = \{\mathfrak{A} \in \mathfrak{M}_r(j_{pq}) : T_{\mathfrak{A}}[\mathcal{S}^{p\times q}] \subseteq \mathcal{S}^{p\times q}\}.$$

Lemma 4.5 *If $\mathfrak{A}_1 \in \mathfrak{M}_r(j_{pq})$, $W \in \mathcal{U}_S(j_{pq})$ and $\mathfrak{A} = \mathfrak{A}_1 W$, then $\mathfrak{A} \in \mathfrak{M}_r(j_{pq})$ and*

$$T_{\mathfrak{A}}[\mathcal{S}^{p\times q}] \subseteq T_{\mathfrak{A}_1}[\mathcal{S}^{p\times q}]. \tag{4.15}$$

Conversely, if $\mathfrak{A}_1$ and $\mathfrak{A}$ belong to $\mathfrak{M}_r(j_{pq})$ and (4.15) is in force, then $\mathfrak{A} = \mathfrak{A}_1 W$ for some mvf $W \in \mathcal{U}_S(j_{pq})$.

Proof See lemma 7.17 in [ArD08b]. □

A mvf $\mathfrak{A} \in \mathfrak{M}_r(j_{pq})$ belongs to the class $\mathfrak{M}_{rR}(j_{pq})$ of **right regular gamma generating matrices** if $\mathfrak{A} = \mathfrak{A}_1 W$ with $\mathfrak{A}_1 \in \mathfrak{M}_r(j_{pq})$ and $W \in \mathcal{U}_S(j_{pq})$ implies that $W \in \mathcal{U}_{\text{const}}(j_{pq})$.

A mvf $\mathfrak{A} \in \mathfrak{M}_r(j_{pq})$ belongs to the class $\mathfrak{M}_{rsR}(j_{pq})$ of **right strongly regular gamma generating matrices** if

$$T_{\mathfrak{A}}[\mathcal{S}^{p\times q}] \cap \mathring{\mathcal{B}}^{p\times q} \neq \emptyset.$$

Thus, in view of the equivalence

$$\mathfrak{A} \in L_\infty^{m\times m} \Longleftrightarrow \|T_{\mathfrak{A}}[0]\|_\infty < 1$$

for mvf's $\mathfrak{A} \in \mathfrak{M}_r(j_{pq})$ (see, e.g., lemma 7.11 in [ArD08b]), it follows that

$$L_\infty^{m\times m} \cap \mathfrak{M}_r(j_{pq}) \subseteq \mathfrak{M}_{rsR}(j_{pq}).$$

Necessary and sufficient conditions for a mvf in the class $\mathfrak{M}_r(j_{pq})$ to belong to the classes $\mathfrak{M}_{rR}(j_{pq})$ and $\mathfrak{M}_{rsR}(j_{pq})$ may be found in sections 7.3 and 10.2 of [ArD08b], respectively. These conditions imply that

$$\mathfrak{M}_{rsR}(j_{pq}) \subset \mathfrak{M}_{rR}(j_{pq}).$$

Theorem 4.6 *If NP*(Γ) *is completely indeterminate, then:*

(1) *There exists a mvf* $\mathfrak{A} \in \mathfrak{M}_r(j_{pq})$ *such that*

$$\mathcal{N}(\Gamma) = T_{\mathfrak{A}}[\mathcal{S}^{p\times q}]. \tag{4.16}$$

Moreover, this mvf is defined by Γ *up to a constant* j_{pq}*-unitary multiplier on the right and automatically belongs to the class* $\mathfrak{M}_{rR}(j_{pq})$.

(2) *For each point* $\omega \in \mathbb{C}_+$, *there is exactly one mvf* $\mathfrak{A} \in \mathfrak{M}_r(j_{pq})$ *for which (4.16) holds that meets the normalization conditions*

$$\mathfrak{a}_-(\overline{\omega}) > 0, \quad \mathfrak{b}_+(\omega) = 0 \quad \text{and} \quad \mathfrak{a}_+(\omega) > 0. \tag{4.17}$$

(3) *The NP*(Γ) *is strictly completely indeterminate if and only if* $\mathfrak{A} \in \mathfrak{M}_{rsR}(j_{pq})$. *If*

$$\mathfrak{A} \in \mathfrak{M}_r(j_{pq}), \quad f^\circ \in T_{\mathfrak{A}}[\mathcal{S}^{p\times q}] \quad \text{and} \quad \Gamma = \Pi_- M_{f^\circ}|_{H_2^q}, \tag{4.18}$$

then:

(a) *NP*(Γ) *is completely indeterminate and*

$$T_{\mathfrak{A}}[\mathcal{S}^{p\times q}] \subseteq \mathcal{N}(\Gamma)$$

with equality if and only if $\mathfrak{A} \in \mathfrak{M}_{rR}(j_{pq})$.

(b) $\Gamma = \Pi_- M_f|_{H_2^q}$ *for every choice of* $f \in T_{\mathfrak{A}}[\mathcal{S}^{p\times q}]$.

Proof See, e.g., theorem 7.22 in [ArD08b]. □

A mvf $\mathfrak{A} \in \mathfrak{M}_r(j_{pq})$ for which the equality (4.16) holds is called a **resolvent matrix** for NP(Γ). Formulas for the unique resolvent matrix $\mathfrak{A} \in \mathfrak{M}_r(j_{pq})$ for NP(Γ) that is subject to the supplementary constraint (4.17) may be found in theorem 7.45 of [ArD08b] when NP(Γ) is strictly completely indeterminate.

Theorem 4.7 *Every mvf* $\mathfrak{A} \in \mathfrak{M}_r(j_{pq})$ *admits an essentially unique factorization*

$$\mathfrak{A} = \mathfrak{A}_1 W \quad \text{with} \quad \mathfrak{A}_1 \in \mathfrak{M}_{rR}(j_{pq}) \quad \text{and} \quad W \in \mathcal{U}_S(j_{pq}).$$

Proof See theorem 7.24 in [ArD08b]. □

There are classes $\mathfrak{M}_\ell(j_{pq})$ of left gamma generating matrices, $\mathfrak{M}_{\ell R}(j_{pq})$, of left regular gamma generating matrices and $\mathfrak{M}_{\ell sR}(j_{pq})$ of left strongly regular gamma generating matrices. They are defined in such a way that

$$\mathfrak{A} \in \mathfrak{M}_{\ell}(j_{pq}) \Longleftrightarrow \mathfrak{A}^{\sim} \in \mathfrak{M}_r(j_{pq})$$

$$\mathfrak{A} \in \mathfrak{M}_{\ell R}(j_{pq}) \Longleftrightarrow \mathfrak{A}^{\sim} \in \mathfrak{M}_{rR}(j_{pq})$$

$$\mathfrak{A} \in \mathfrak{M}_{\ell sR}(j_{pq}) \Longleftrightarrow \mathfrak{A}^{\sim} \in \mathfrak{M}_{rsR}(j_{pq}).$$

These classes play a role in the consideration of left linear fractional transformations.

Theorem 4.8 *If $W \in \mathcal{U}(j_{pq})$ and $\{b_1, b_2\} \in ap(W)$, then*

$$W(\mu) = \begin{bmatrix} b_1(\mu) & 0 \\ 0 & b_2(\mu)^{-1} \end{bmatrix} \mathfrak{A}(\mu) \quad a.e.\ on\ \mathbb{R} \tag{4.19}$$

for some mvf $\mathfrak{A} \in \Pi \cap \mathfrak{M}_r(j_{pq})$ and

$$b_1 T_{\mathfrak{A}}[\mathcal{S}^{p\times q}] b_2 \subseteq \mathcal{S}^{p\times q}. \tag{4.20}$$

Conversely, if $\mathfrak{A} \in \Pi \cap \mathfrak{M}_r(j_{pq})$, then there exists a pair of mvf's $b_1 \in \mathcal{S}_{in}^{p\times p}$ and $b_2 \in \mathcal{S}_{in}^{q\times q}$ such that

$$b_1 T_{\mathfrak{A}}[0] b_2 \in \mathcal{S}^{p\times q}.$$

Moreover, the mvf W defined by (4.19) with this choice of b_1 and b_2 belongs to the class $\mathcal{U}(j_{pq})$, $\{b_1, b_2\} \in ap(W)$ and (4.20) holds.

If a pair of mvf's $\mathfrak{A} \in \Pi \cap \mathfrak{M}_r(j_{pq})$ and $W \in \mathcal{U}(j_{pq})$ are connected by formula (4.19), then

$$\mathfrak{A} \in \mathfrak{M}_{rR}(j_{pq}) \Longleftrightarrow W \in \mathcal{U}_{rR}(j_{pq}) \tag{4.21}$$

and

$$\mathfrak{A} \in \mathfrak{M}_{rsR}(j_{pq}) \Longleftrightarrow T_W[\mathcal{S}^{p\times q}] \cap \mathring{\mathcal{S}}^{p\times q} \neq \emptyset. \tag{4.22}$$

Proof See theorems 7.26 and 7.27 in [ArD08b]. □

4.2 The generalized Schur interpolation problem

In this section we shall study the following interpolation problem for mvf's in the Schur class $\mathcal{S}^{p\times q}$, which we shall refer to as the GSIP, an acronym for the **generalized Schur interpolation problem**:

GSIP$(b_1, b_2; s^\circ)$: *Given mvf's $b_1 \in \mathcal{S}_{in}^{p\times p}$, $b_2 \in \mathcal{S}_{in}^{q\times q}$ and $s^\circ \in \mathcal{S}^{p\times q}$, describe the set*

$$\mathcal{S}(b_1, b_2; s^\circ) = \{s \in \mathcal{S}^{p\times q} : b_1^{-1}(s - s^\circ) b_2^{-1} \in H_\infty^{p\times q}\}. \tag{4.23}$$

The mvf's $s(\lambda)$ in this set are called solutions of this problem.

The set $\mathcal{S}(b_1, b_2; s^\circ)$ of solutions s of a GSIP$(b_1, b_2; s^\circ)$ is the set of mvf's $s \in \mathcal{S}^{p\times q}$ such that

$$\Pi_{\mathcal{H}(b_1)} M_s|_{H_2^q} = X_{11}, \quad \Pi_{\mathcal{H}(b_1)} M_s|_{\mathcal{H}_*(b_2)} = X_{12} \\ \text{and} \quad \Pi_- M_s|_{\mathcal{H}_*(b_2)} = X_{22}, \tag{4.24}$$

where the X_{ij} are the blocks of the linear contractive operator

$$X = \begin{bmatrix} X_{11} & X_{12} \\ 0 & X_{22} \end{bmatrix} : \begin{array}{c} H_2^q \\ \oplus \\ \mathcal{H}_*(b_2) \end{array} \longrightarrow \begin{array}{c} \mathcal{H}(b_1) \\ \oplus \\ (H_2^p)^\perp \end{array} \tag{4.25}$$

that are defined by the formulas

$$X_{11} = \Pi_{\mathcal{H}(b_1)} M_{s^\circ}|_{H_2^q}, \quad X_{12} = \Pi_{\mathcal{H}(b_1)} M_{s^\circ}|_{\mathcal{H}_*(b_2)} \\ \text{and} \quad X_{22} = \Pi_- M_{s^\circ}|_{\mathcal{H}_*(b_2)}. \tag{4.26}$$

Thus, only the part of s° that influences the operator X plays a role in the GSIP$(b_1, b_2; s^\circ)$.

Remark 4.9 *The relevance of the operator X to the GSIP$(b_1, b_2; s^\circ)$ is perhaps best understood by noting that if*

$$f^\circ = b_1^{-1} s^\circ b_2^{-1} \quad \text{and} \quad \Gamma = \Pi_- M_{f^\circ}|_{H_2^q}, \tag{4.27}$$

then

$$s \in \mathcal{S}(b_1, b_2; s^\circ) \Longleftrightarrow b_1^{-1} s b_2^{-1} \in \mathcal{N}(\Gamma). \tag{4.28}$$

Thus, the set $\mathcal{S}(b_1, b_2; s^\circ)$ depends only upon the mvf's b_1, b_2 and the Hankel operator Γ defined in (4.27).

Thus, the GSIP$(b_1, b_2; s^\circ)$ can be formulated in terms of b_1, b_2, and a given block triangular operator X of the form (4.25); the problem is to describe the set

$$\mathcal{S}(b_1, b_2; X) = \{s \in \mathcal{S}^{p\times q} : X_{11} = \Pi_{\mathcal{H}(b_1)} M_s|_{H_2^q}, \quad X_{12} = \Pi_{\mathcal{H}(b_1)} M_s|_{\mathcal{H}_*(b_2)} \\ \text{and} \quad X_{22} = \Pi_- M_s|_{\mathcal{H}_*(b_2)}\}. \tag{4.29}$$

This problem will be referred to as the **generalized Sarason problem** and will be denoted GSP$(b_1, b_2; X)$. It will prove convenient, however, to consider a relaxed version of this problem:

GSP$(b_1, b_2; X; H_\infty^{p\times q})$, which is to describe the set

$$H_\infty(b_1, b_2; X) = \{s \in H_\infty^{p\times q} : X_{11} = \Pi_{\mathcal{H}(b_1)} M_s|_{H_2^q}, \quad X_{12} = \Pi_{\mathcal{H}(b_1)} M_s|_{\mathcal{H}_*(b_2)} \\ \text{and} \quad X_{22} = \Pi_- M_s|_{\mathcal{H}_*(b_2)}\}. \tag{4.30}$$

Theorem 4.10 *Let $b_1 \in \mathcal{S}_{in}^{p\times p}$ and $b_2 \in \mathcal{S}_{in}^{q\times q}$ and let X be a bounded linear block triangular operator of the form (4.25). Then*

(1) *$H_\infty(b_1, b_2 : X) \neq \emptyset$ if and only if the operator*

$$\Gamma = M_{b_1^*}\{X_{11}\Pi_+ + (X_{22} + X_{12})P_{\mathcal{H}_*(b_2)}\}M_{b_2^*}|_{H_2^q} \tag{4.31}$$

acting from H_2^q into $(H_2^p)^\perp$ is a bounded Hankel operator.

(2) *If $H_\infty(b_1, b_2 : X) \neq \emptyset$, then*

$$\|X\| = \min\{\|s\|_\infty : s \in H_\infty(b_1, b_2; X)\}.$$

(3) *$\mathcal{S}(b_1, b_2; X) \neq \emptyset$ if and only if the operator Γ that is defined in (4.31) is a contractive Hankel operator. Moreover, if $s^\circ \in \mathcal{S}(b_1, b_2; X)$, then*

$$\mathcal{S}(b_1, b_2; s^\circ) = \mathcal{S}(b_1, b_2; X)$$

and

$$\|X\| = \min\{\|s\|_\infty : s \in \mathcal{S}(b_1, b_2; X)\}.$$

Proof This follows from Remark 4.9 and Theorem 4.1. □

Remark 4.11 *If $b_2 = I_q$, then $X = X_{11} = \Pi_{\mathcal{H}(b_1)} M_s|_{H_2^q}$ and formula (4.31) simplifies to*

$$\Gamma = M_{b_1^*}X. \tag{4.32}$$

Moreover, Γ is a Hankel operator from H_2^q into $(H_2^p)^\perp$ if and only if X is a Toeplitz operator in the sense that

$$\Pi_{\mathcal{H}(b_1)} M_{e_t} X = X M_{e_t}|_{H_2^q} \quad \text{for every } t \geq 0. \tag{4.33}$$

Thus, $\mathcal{S}(b_1, I_q; X) \neq \emptyset$ if and only if $\|X\| \leq 1$ and X meets the condition (4.33).

The GSIP$(b_1, b_2, ; s^\circ)$ is said to be **completely indeterminate** if there exists at least one point $\omega \in \mathfrak{h}_{b_1^{-1}}^+ \cap \mathfrak{h}_{b_2^{-1}}^+$ for which

$$\{s \in \mathcal{S}(b_1, b_2, ; s^\circ) : \{s(\omega) - s^\circ(\omega)\}\eta \neq 0\} \neq \emptyset \quad \text{for every nonzero vector } \eta \in \mathbb{C}^q. \tag{4.34}$$

Moreover, if the condition (4.34) holds for one point $\omega \in \mathbb{C}_+$, then it holds for every point $\omega \in \mathfrak{h}_{b_1^{-1}}^+ \cap \mathfrak{h}_{b_2^{-1}}^+$ and hence for every point $\omega \in \mathbb{C}_+$ if b_1 and b_2 are entire.

A GSIP$(b_1, b_2; s^\circ)$ is called **strictly completely indeterminate** if

$$\mathcal{S}(b_1, b_2; s^\circ) \cap \mathring{\mathcal{S}}^{p\times q} \neq \emptyset. \tag{4.35}$$

The GSIP$(b_1, b_2; s^\circ)$ with entire inner mvf's b_1, b_2 plays an essential role in the study of bitangential direct and inverse monodromy and input scattering problems.

Under this extra restriction, the inner mvf's b_1 and b_2 may be normalized by the conditions

$$b_1(0) = I_p \quad \text{and} \quad b_2(0) = I_q.$$

Then the Paley–Wiener theorem yields the integral representations

$$b_1(\lambda) - I_p = i\lambda \int_0^{a_1} e^{i\lambda t} h_1(t)dt \quad \text{and} \quad b_2(\lambda) - I_q = i\lambda \int_0^{a_2} e^{i\lambda t} h_2(t)dt, \tag{4.36}$$

where

$$h_1 \in L_2^p([0, a_1]), \quad h_2 \in L_2^q([0, a_2]), \quad a_j = \tau_{b_j}.$$

The set $\mathcal{S}(b_1, b_2; s^\circ)$ depends only upon the mvf's h_1 and h_2 and the restriction of the inverse Fourier transform (in the generalized sense) of s° to the interval $[0, a_1 + a_2]$. A more precise discussion of this dependence will be furnished in Chapter 9.

The class $\mathcal{E} \cap \mathcal{U}_{rR}(j_{pq})$ of entire right regular j_{pq}-inner mvf's may be identified with the class of resolvent matrices of such completely indeterminate interpolation problems based on entire inner mvf's b_1 and b_2:

Theorem 4.12 *If the GSIP$(b_1, b_2; s^\circ)$ is completely indeterminate and b_1 and b_2 are entire inner mvf's, then:*

(1) *There exists exactly one mvf $W \in \mathcal{E} \cap \mathcal{U}_{rR}^\circ(j_{pq})$ such that*

$$\mathcal{S}(b_1, b_2; s^\circ) = T_W[\mathcal{S}^{p\times q}] \quad \text{and} \quad \{b_1, b_2\} \in ap(W). \tag{4.37}$$

(2) *If $\widetilde{W} \in \mathcal{P}^\circ(j_{pq})$ and*

$$\mathcal{S}(b_1, b_2; s^\circ) = T_{\widetilde{W}}[\mathcal{S}^{p\times q}], \tag{4.38}$$

then $\widetilde{W} \in \mathcal{E} \cap \mathcal{U}_{rR}(j_{pg})$.

(3) *The set of all mvf's $\widetilde{W} \in \mathcal{U}^\circ(j_{pq})$ for which (4.38) holds is described by the formula*

$$\widetilde{W}(\lambda) = \frac{e_{\beta_2}(\lambda)}{e_{\beta_1}(\lambda)} W(\lambda), \tag{4.39}$$

where $\beta_j \geq 0$, $e_{\beta_1}^{-1} b_1 \in \mathcal{S}_{\text{in}}^{p\times p}$ and $e_{\beta_2}^{-1} b_2 \in \mathcal{S}_{\text{in}}^{q\times q}$.

Proof This follows from Theorems 4.6 and 4.8 and Lemma 3.65. □

Theorem 4.13 *If $W \in \mathcal{E} \cap \mathcal{U}(j_{pq})$, $\{b_1, b_2\} \in ap(W)$ and $s^\circ \in T_W[\mathcal{S}^{p\times q}]$, then b_1 and b_2 are entire mvf's and:*

(1) *The GSIP$(b_1, b_2, ; s^\circ)$ is completely indeterminate.*
(2) *$T_W[\mathcal{S}^{p\times q}] \subseteq \mathcal{S}(b_1, b_2; s^\circ)$, with equality if and only if $W \in \mathcal{U}_{rR}(j_{pq})$.*
(3) *$\mathcal{S}(b_1, b_2; s^\circ)$ is independent of the choice of $s^\circ \in T_W[\mathcal{S}^{p\times q}]$.*

(4) *If* $\widetilde{b}_1 \in \mathcal{S}_{\rm in}^{p\times p}$ *and* $\widetilde{b}_2 \in \mathcal{S}_{\rm in}^{q\times q}$ *are such that* $\widetilde{b}_1^{-1}b_1 \in \mathcal{S}_{\rm in}^{p\times p}$ *and* $b_2\widetilde{b}_2^{-1} \in \mathcal{S}_{\rm in}^{q\times q}$, *then there exists exactly one mvf* $\widetilde{W} \in \mathcal{E}\cap\mathcal{U}_{rR}^{\circ}(j_{pq})$ *such that* $\widetilde{W}^{-1}W \in \mathcal{U}(j_{pq})$ *and* $\{\widetilde{b}_1, \widetilde{b}_2\} \in ap(\widetilde{W})$.

Proof The mvf's $b_j, \widetilde{b}_j$, $j = 1, 2$, and $\widetilde{W}$ are entire by Theorem 3.55 and Lemma 3.35. (1), (2) and (4) follow from theorems 7.50 and 7.51 in [ArD08b].

Finally, if $s \in \mathcal{S}(b_1, b_2; s_1)$ for some mvf $s_1 \in T_W[\mathcal{S}^{p\times q}]$, then (2) guarantees that $s_1 \in \mathcal{S}(b_1, b_2; s^\circ)$ and hence that

$$b_1^{-1}(s - s^\circ)b_2^{-1} = b_1^{-1}(s - s_1)b_2^{-1} + b_1^{-1}(s_1 - s^\circ)b_2^{-1}$$

belongs to $H_\infty^{p\times q}$. Thus, $\mathcal{S}(b_1, b_2; s_1) \subseteq \mathcal{S}(b_1, b_2; s^\circ)$. This justifies (3), since the same inclusion holds if s° and s_1 are interchanged. □

The next result gives a full description of $T_W[\mathcal{S}^{p\times q}]$ for $W \in \mathcal{U}(j_{pq})$. It plays a useful role in interpolation problems.

Theorem 4.14 *If* $W \in \mathcal{U}(j_{pq})$, *then a mvf* $s \in \mathcal{S}^{p\times q}$ *belongs to the set* $T_W[\mathcal{S}^{p\times q}]$ *if and only if:*

(1) $\begin{bmatrix} I_p & -s \end{bmatrix} f \in H_2^p$;
(2) $\begin{bmatrix} -s^* & I_q \end{bmatrix} f \in (H_2^q)^\perp$ *and*
(3) $\left\langle \begin{bmatrix} I_p & -s \\ -s^* & I_q \end{bmatrix} f, f \right\rangle_{\rm st} \le \langle f, f\rangle_{\mathcal{H}(W)}$

for every f *in the RKHS* $\mathcal{H}(W)$ *based on* W.

Proof See [Dy03b]. □

Remark 4.15 *We focus primarily on entire* b_1, b_2 *and* W *in this monograph, because this setting suffices for the identification of matrizants of canonical systems with resolvent matrices of appropriate interpolation problems based on entire* b_1 *and* b_2. *However, to prove the next theorem, we need general versions of Theorems 4.12 and 4.13 in which* b_1, b_2 *and* W *are not necessarily entire; see theorems 7.48 and 7.50 in [ArD08b].*

Theorem 4.16 *The GSIP*$(b_1, b_2; s^\circ)$ *is completely indeterminate if and only if there exists a mvf* $s \in \mathcal{S}(b_1, b_2; s^\circ)$ *such that*

$$\ln\det(I_q - s^*s) \in \widetilde{L}_1, \tag{4.40}$$

i.e., if and only if

$$\mathcal{S}(b_1, b_2; s^\circ) \cap \mathcal{S}_{sz}^{p\times q} \neq \emptyset. \tag{4.41}$$

Proof If the GSIP$(b_1, b_2; s^\circ)$ is completely indeterminate, then by theorem 7.57 in [ArD08b] (which is essentially the same as Theorem 4.12, but without the restriction that b_1, b_2 and W are entire) there exists a mvf $W \in \mathcal{U}_{rR}(j_{pq})$ such that

$$T_W[\mathcal{S}^{p\times q}] = \mathcal{S}(b_1, b_2; s^\circ).$$

Therefore, $s_{12} = T_W[0] \in \mathcal{S}(b_1, b_2; s^\circ)$. But for this choice of s (4.40) holds, since

$$\det\{I_q - s_{12}(\mu)^* s_{12}(\mu)\} = \det\{s_{22}(\mu)^* s_{22}(\mu)\} = |\det \varphi_2(\mu)|^2$$

a.e. on $\mathbb{R}$ and $\det \varphi_{22}$ is outer in H_∞.

The proof in the other direction is divided into steps:

1. *If $s \in L_\infty^{p\times q}$ and $\|s\|_\infty \le 1$, then s satisfies the constraint (4.40) if and only if*

$$\int_{-\infty}^{\infty} \frac{\ln(1 - \|s(\mu)\|)}{1+\mu^2} d\mu > -\infty. \tag{4.42}$$

This follows easily from the observation that if $A \in \mathbb{C}^{p\times q}$ is contractive, then

$$(1 - \|A\|^2)^q \le \det(I_q - A^*A) \le 1 - \|A\|^2.$$

2. *If $s \in \mathcal{S}(b_1, b_2; s^\circ)$ meets the condition (4.42), then the GSIP$(b_1, b_2; s^\circ)$ is completely indeterminate.*

If $s \in \mathcal{S}(b_1, b_2; s^\circ)$ meets the condition (4.42), then there exists a scalar function $\varphi \in \mathcal{S}_{\text{out}}$ such that

$$1 - \|s(\mu)\| = |\varphi(\mu)| \quad \text{a.e. on } \mathbb{R}.$$

Let $E \in \mathbb{C}^{p\times q}$ be a matrix with $\|E\| = 1$ and set

$$s_1(\lambda) = s(\lambda) + \varphi(\mu) b_1(\lambda) E b_2(\lambda).$$

Then clearly $s_1 \in H_\infty^{p\times q}$ and

$$\begin{aligned} \|s_1\| &= \|s + \varphi b_1 E b_2\|_\infty \\ &= \text{ess sup}\{\|s(\mu) + \varphi(\mu) b_1(\mu) E b_2(\mu) : \mu \in \mathbb{R}\} \\ &\le \text{ess sup}\{\|s(\mu)\| + |\varphi(\mu)| : \mu \in \mathbb{R}\} = 1. \end{aligned}$$

Moreover, since $\varphi E \in H_\infty^{p\times q}$, it is readily seen that $s_1 \in \mathcal{S}(b_1, b_2; s^\circ)$ and that for any point $\omega \in \mathbb{C}_+$ at which $b_1(\omega)$ and $b_2(\omega)$ are both invertible and any non zero vector $\eta \in \mathbb{C}^q$,

$$\{s_1(\omega) - s(\omega)\}\eta = \varphi(\omega) b_1(\omega) E b_2(\omega)\eta$$

can be made nonzero by an appropriate choice of E. Therefore $\mathcal{S}(b_1, b_2; s)$ is completely indeterminate. This completes the proof, since $\mathcal{S}(b_1, b_2; s) = \mathcal{S}(b_1, b_2; s^\circ)$. □

Corollary 4.17 *If a GSIP$(b_1, b_2; s^\circ)$ is strictly completely indeterminate, then it is completely indeterminate.*

4.3 Right and left strongly regular J-inner mvf's

The classes $\mathcal{U}_{rsR}(J)$ and $\mathcal{U}_{\ell sR}(J)$ of right and left strongly regular J-inner mvf's play a significant role in our study of direct and inverse problems for canonical integral and differential systems. There are a number of different characterizations of these classes. In this section we shall focus mainly on a characterization that is particularly useful in the study of bitangential direct and inverse problems.

A mvf $W \in \mathcal{U}(j_{pq})$ is said to be a **right strongly regular** j_{pq}-inner mvf if

$$T_W[\mathcal{S}^{p\times q}] \cap \mathring{\mathcal{S}}^{p\times q} \neq \emptyset; \tag{4.43}$$

it is said to be a **left strongly regular** j_{pq}-inner mvf if

$$T_W^{\ell}[\mathcal{S}^{q\times p}] \cap \mathring{\mathcal{S}}^{q\times p} \neq \emptyset. \tag{4.44}$$

If $J \neq \pm I_m$ and $J = V^* j_{pq} V$ for some constant $m \times m$ unitary matrix V, then $U \in \mathcal{U}(J)$ is said to be a **right (left) strongly regular** J-inner mvf if $W(\lambda) = VU(\lambda)V^*$ is a right (left) strongly regular j_{pq}-inner mvf. This definition does not depend upon the choice of V: If $V_1^* j_{pq} V_1 = V_2^* j_{pq} V_2$, then the matrix $V_1V_2^*$ is both unitary and j_{pq}-unitary and therefore it must be of the form $V_1V_2^* = \mathrm{diag}\,\{u_1, u_2\}$, where u_1 and u_2 are constant unitary matrices of sizes $p \times p$ and $q \times q$, respectively. Thus, the mvf's $W_1(\lambda) = V_1U(\lambda)V_1^*$ and $W_2(\lambda) = V_2U(\lambda)V_2^*$ are related by the formula

$$W_2(\lambda) = \begin{bmatrix} u_1^* & 0 \\ 0 & u_2^* \end{bmatrix} W_1(\lambda) \begin{bmatrix} u_1 & 0 \\ 0 & u_2 \end{bmatrix}$$

and, consequently,

$$T_{W_2}[\mathcal{S}^{p\times q}] = u_1^* T_{W_1}[\mathcal{S}^{p\times q}]u_2.$$

Therefore,

$$T_{W_2}[\mathcal{S}^{p\times q}] \cap \mathring{\mathcal{S}}^{p\times q} \neq \emptyset \Longleftrightarrow T_{W_1}[\mathcal{S}^{p\times q}] \cap \mathring{\mathcal{S}}^{p\times q} \neq \emptyset.$$

The classes of left and right strongly regular J-inner mvf's will be denoted $\mathcal{U}_{\ell sR}(J)$ and $\mathcal{U}_{rsR}(J)$, respectively. The convention

$$\mathcal{U}_{\ell sR}(\pm I_m) = \mathcal{U}_{rsR}(\pm I_m) = \mathcal{U}(\pm I_m)$$

will be convenient. The preceding definitions imply that

$$U \in \mathcal{U}_{rsR}(J) \Longleftrightarrow U^{\sim} \in \mathcal{U}_{\ell sR}(J). \tag{4.45}$$

Other characterizations of the class $\mathcal{U}_{rsR}(J)$ in terms of properties of the RKHS $\mathcal{H}(U)$ and in terms of the Treil–Volberg matrix Muckenhoupt (A_2) condition are presented in (1.26) and in chapter 10 of [ArD08b], respectively.

Theorem 4.18 *The following inclusions hold:*

(1) $\mathcal{U}(J) \cap L_\infty^{m\times m}(\mathbb{R}) \subseteq \mathcal{U}_{rsR}(J) \cap \mathcal{U}_{\ell sR}(J)$.
(2) $\mathcal{U}_{rsR}(J) \cup \mathcal{U}_{\ell sR}(J) \subset \mathcal{U}(J) \cap \widetilde{L}_2^{m\times m} \subset \mathcal{U}_{\ell R}(J) \cap \mathcal{U}_{rR}(J)$.

Proof This is theorem 4.75 in [ArD08b]. □

An example of a mvf $U \in \mathcal{E} \cap \mathcal{U}_{rsR}(J_1)$ that does not belong to $L_\infty^{2\times 2}$ is presented in Section 11.6 (see Remark 11.28); another example is furnished on pp. 525–527 of [ArD08b].

Theorem 4.19 *If $U \in \mathcal{U}_{rsR}(J) \cup \mathcal{U}_{\ell sR}(J)$, then $U \in \mathcal{U}_{\ell R}(J) \cap \mathcal{U}_{rR}(J)$, i.e.,*

$$\mathcal{U}_{rsR}(J) \cup \mathcal{U}_{\ell sR}(J) \subseteq \mathcal{U}_{\ell R}(J) \cap \mathcal{U}_{rR}(J).$$

Moreover, if $U \in \mathcal{U}_{rsR}(J)$ and

$$U(\lambda) = U_1(\lambda)U_2(\lambda)U_3(\lambda)$$

with $U_i \in \mathcal{U}(J)$ for $i = 1, 2, 3$, then:

(1) $U_1 \in \mathcal{U}_{rsR}(J)$.
(2) $U_1U_2 \in \mathcal{U}_{rsR}(J)$.
(3) $U_i \in \mathcal{U}_{\ell R}(J) \cap \mathcal{U}_{rR}(J)$ *for* $i = 1, 2, 3$.

Proof This is theorem 4.76 in [ArD08b]. □

Theorem 4.19 justifies the use of the terminology **strongly regular**.

In view of the inclusion $\mathcal{U}_{rsR}(J) \subset \mathcal{U}_{rR}(J)$, the class $\mathcal{U}_{rsR}(j_{pq})$ may (and will) be identified with the subclass of mvf's in $\mathcal{U}(j_{pq})$ that are resolvent matrices of strictly completely indeterminate GSIP's. This leads to the following conclusions:

Theorem 4.20 *If $W \in \mathcal{E} \cap \mathcal{U}_{rsR}(j_{pq})$, $\{b_1, b_2\} \in ap(W)$ and $s^\circ \in T_W[\mathcal{S}^{p\times q}]$, then:*

(1) *b_1 and b_2 are entire mvf's.*
(2) *The GSIP$(b_1, b_2; s^\circ)$ is strictly completely indeterminate.*

Conversely, if (1), (2) and (4.38) hold for some choice of $b_1 \in \mathcal{S}_{in}^{p\times p}$, $b_2 \in \mathcal{S}_{in}^{q\times q}$, $s^\circ \in \mathcal{S}^{p\times q}$, and if $W \in \mathcal{P}^\circ(j_{pq})$, then $W \in \mathcal{E} \cap \mathcal{U}_{rsR}(j_{pq})$.

Proof This follows from the the definition (4.43) of the class $\mathcal{U}_{rsR}(j_{pq})$, the inclusion $\mathcal{U}_{rsR}(j_{pq}) \subset \mathcal{U}_{rR}(j_{pq})$ and Theorem 4.12. □

4.4 The generalized Carathéodory interpolation problem

The GCIP (generalized Carathéodory interpolation problem) is an analog of the GSIP that is considered in the class $\mathcal{C}^{p\times p}$ instead of the class $\mathcal{S}^{p\times q}$.

GCIP$(b_3, b_4; c^\circ)$**:** given $b_3 \in \mathcal{S}_{in}^{p\times p}$, $b_4 \in \mathcal{S}_{in}^{p\times p}$ and $c^\circ \in \mathcal{C}^{p\times p}$, describe the set

$$\mathcal{C}(b_3, b_4; c^\circ) = \{c \in \mathcal{C}^{p\times p} : b_3^{-1}(c - c^\circ)b_4^{-1} \in \mathcal{N}_+^{p\times p}\}. \tag{4.46}$$

The restriction of this problem to the setting of entire inner mvf's b_3 and b_4 plays an essential role in the study of direct and inverse input impedance and spectral problems for canonical systems. Accordingly we shall focus on this case in this monograph.

The space $\mathcal{N}_+^{p\times p}$ is introduced in the formulation of this problem (rather than $H_\infty^{p\times p}$), because there are mvf's $c \in \mathcal{C}^{p\times p}$ that do not belong to $H_\infty^{p\times p}$ (e.g., $c(\lambda) = i/\lambda$), whereas $\mathcal{C}^{p\times p} \subset \mathcal{N}_+^{p\times p}$.

The GCIP$(b_3, b_4; c^\circ)$ is called **completely indeterminate** if there exists a point $\omega \in \mathfrak{h}_{b_3^{-1}} \cap \mathfrak{h}_{b_4^{-1}}$ such that

$$\{c \in \mathcal{C}(b_3, b_4; c^\circ) : \{c(\omega) - c^\circ(\omega\}\eta\} \neq \emptyset \tag{4.47}$$

for every nonzero vector $\eta \in \mathbb{C}^p$; it is called **strictly completely indeterminate** if

$$\mathcal{C}(b_3, b_4; c^\circ) \cap \mathring{\mathcal{C}}^{p\times p} \neq \emptyset.$$

There are connections between the sets $\mathcal{S}(b_1, b_2; s^\circ)$ and $\mathcal{C}(b_3, b_4; c^\circ)$:

Theorem 4.21 *Let $c^\circ \in \mathcal{C}^{p\times p}$ and let $s^\circ = T_{\mathfrak{V}}[c^\circ]$. Then the conditions*

$$\frac{1}{2}b_1^{-1}(I_p + s^\circ)b_3 \in \mathcal{S}_{out}^{p\times p} \quad \textit{and} \quad \frac{1}{2}b_4(I_p + s^\circ)b_2^{-1} \in \mathcal{S}_{out}^{p\times p} \tag{4.48}$$

serve to define one of the pairs $\{b_1, b_2\}$ and $\{b_3, b_4\}$ of $p \times p$ inner mvf's in terms of the other, up to constant unitary multipliers. Moreover, for any two such pairs, the conditions

$$\frac{1}{2}b_1^{-1}(I_p + s)b_3 \in \mathcal{S}_{out}^{p\times p} \quad \textit{and} \quad \frac{1}{2}b_4(I_p + s)b_2^{-1} \in \mathcal{S}_{out}^{p\times p} \tag{4.49}$$

are satisfied for every mvf $s \in \mathcal{S}(b_1, b_2; s^\circ) \cap \mathcal{D}(T_{\mathfrak{V}})$ and

$$\mathcal{C}(b_3, b_4; c^\circ) = T_{\mathfrak{V}}[\mathcal{S}(b_1, b_2; s^\circ) \cap \mathcal{D}(T_{\mathfrak{V}})]. \tag{4.50}$$

Proof This theorem is the same as lemma 7.68 in [ArD08b]. □

Theorem 4.22 *The GCIP$(b_3, b_4, ; c^\circ)$ is completely indeterminate if and only if there exists a mvf $c \in \mathcal{C}(b_3, b_4, ; c^\circ)$ such that*

$$\ln \det(c + c^*) \in \widetilde{L}_1, \tag{4.51}$$

i.e., if and only if

$$\mathcal{C}(b_3, b_4; c^\circ) \cap \mathcal{C}_{sz}^{p\times p} \neq \emptyset. \tag{4.52}$$

Proof This follows from Theorem 4.16 and Theorem 4.21. □

Corollary 4.23 *If a GCIP$(b_3, b_4, ; c^\circ)$ is strictly completely indeterminate, then it is completely indeterminate.*

If $W(\lambda) = \mathfrak{V}A(\lambda)\mathfrak{V}$, then, by definition of the classes $\mathcal{U}_{rsR}(J)$,

$$A \in \mathcal{U}_{rsR}(J_p) \Longleftrightarrow W \in \mathcal{U}_{rsR}(j_p) \Longleftrightarrow T_W[\mathcal{S}^{p\times q}] \cap \mathring{\mathcal{S}}^{p\times p} \neq \emptyset. \tag{4.53}$$

Moreover, since

$$\mathring{\mathcal{C}}^{p\times p} = T_{\mathfrak{V}}[\mathring{\mathcal{S}}^{p\times p}], \tag{4.54}$$

this implies that

$$A \in \mathcal{U}_{rsR}(J_p) \Longleftrightarrow \mathcal{C}(A) \cap \mathring{\mathcal{C}}^{p\times p} \neq \emptyset. \tag{4.55}$$

Theorem 4.24 *Let $U \in \mathcal{E} \cap \mathcal{U}(J)$ with J unitarily equivalent to j_{pq} and let $\{b_1, b_2\} \in ap_I(U)$. Then:*

(1) *b_1 and b_2 are entire, $\tau_U^+ = \tau_{b_2}^-$ and $\tau_U^- = \tau_{b_1}^-$.*
(2) *$\tau_U^+ \le \delta_{b_2}^- \le q\tau_U^+$ and $\tau_U^- \le \delta_{b_1}^- \le p\tau_U^-$.*

If $q = p$ and $\{b_3, b_4\} \in ap_{II}(U)$, then b_3 and b_4 are entire and:

(3) *$\tau_{b_1}^- = \tau_{b_3}^-$ and $\delta_{b_1}^- = \delta_{b_3}^-$.*
(4) *$\tau_{b_2}^- = \tau_{b_4}^-$ and $\delta_{b_2}^- = \delta_{b_4}^-$.*

Proof This follows from theorem 4.58 in [ArD08b]. □

In the bitangential generalizations of the extension problem considered by Krein and Melik-Adamyan [KrMA86] and the Krein extension problems that will be discussed in Chapter 9, the inner mvf's b_3 and b_4 will be restricted to be entire and will be normalized by the conditions

$$b_3(0) = I_p \quad \text{and} \quad b_4(0) = I_p.$$

Then the Paley–Wiener theorem yields the integral representations

$$b_3(\lambda) - I_p = i\lambda \int_0^{a_3} e^{i\lambda t} h_3(t)dt \quad \text{and} \quad b_4(\lambda) - I_p = i\lambda \int_0^{a_4} e^{i\lambda t} h_4(t)dt, \tag{4.56}$$

where

$$h_3 \in L_2^p([0, a_3]), \quad h_4 \in L_2^p([0, a_4]) \quad \text{and} \quad a_j = \tau_{b_j}.$$

The set $\mathcal{C}(b_3, b_4; c^\circ)$ depends only upon the mvf's h_3 and h_4 and the restriction of the inverse Fourier transform (in the generalized sense) of c° to the interval $[0, a_3 + a_4]$. A more precise discussion of this dependence will be presented in Section 9.1.

The next theorem follows from theorems 7.69 and 7.70 in [ArD08b] and from Theorem 3.75.

Theorem 4.25 *The following three sets of implications are in force:*

I. *If the GCIP$(b_3, b_4; c^\circ)$ is completely indeterminate and b_3 and b_4 are entire inner mvf's, then:*

(1) *There exists exactly one mvf $\mathring{A} \in \mathcal{E} \cap \mathcal{U}_{rR}^\circ(J_p)$ such that*

$$\mathcal{C}(b_3, b_4; c^\circ) = \mathcal{C}(\mathring{A}) \quad and \quad \{b_3, b_4\} \in ap_{II}(\mathring{A}). \tag{4.57}$$

(2) *If $A \in \mathcal{P}^\circ(J_p)$ is such that*

$$\mathcal{C}(A) = \mathcal{C}(b_3, b_4; c^\circ), \tag{4.58}$$

then $A \in \mathcal{E} \cap \mathcal{U}_{rR}(J_p)$ and must be of the form

$$A(\lambda) = \frac{e_{\beta_4}(\lambda)}{e_{\beta_3}(\lambda)} \mathring{A}(\lambda) V, \tag{4.59}$$

where $\mathring{A} \in \mathcal{E} \cap \mathcal{U}_{rR}^\circ(J_p)$ satisfies (4.57), $\beta_3 \geq 0$, $\beta_4 \geq 0$,

$$e_{-\beta_4} b_4 \in \mathcal{S}_{\text{in}}^{p \times p}, \quad e_{-\beta_3} b_3 \in \mathcal{S}_{\text{in}}^{p \times p} \quad and \quad V \in \mathcal{U}_{\text{const}}(J_p). \tag{4.60}$$

(3) *Every mvf of the form (4.59) that meets the constraints in (4.60) satisfies the conditions*

$$A \in \mathcal{U}(J_p) \quad and \quad \mathcal{C}(A) = \mathcal{C}(b_3, b_4; c^\circ). \tag{4.61}$$

Moreover, $A \in \mathcal{E} \cap \mathcal{U}_{rR}(J_p)$.

II. *If $A \in \mathcal{E} \cap \mathcal{U}(J_p)$, $\{b_3, b_4\} \in ap_{II}(A)$ and $c^\circ \in \mathcal{C}(A)$, then b_3 and b_4 are entire mvf's, the GCIP$(b_3, b_4; c^\circ)$ is completely indeterminate and*

$$\mathcal{C}(A) \subseteq \mathcal{C}(b_3, b_4; c^\circ), \tag{4.62}$$

with equality if and only if $A \in \mathcal{U}_{rR}(J_p)$.

III. *If $A \in \mathcal{E} \cap \mathcal{U}(J_p)$, $\{b_3, b_4\} \in ap_{II}(A)$ and $c^\circ \in \mathcal{C}(A)$, then $A \in \mathcal{U}_{rsR}(J_p)$ if and only (4.58) holds for some strictly completely indeterminate GCIP.*

A mvf A for which (4.58) holds is called a **resolvent matrix** for the GCIP$(b_3, b_4; c^\circ)$. The class $\mathcal{E} \cap \mathcal{U}_{rR}(J_p)$ of entire right regular J_p-inner mvf's may be identified with the class of resolvent matrices of completely indeterminate generalized Carathéodory interpolation problems based on entire inner mvf's b_3 and b_4. The class $\mathcal{E} \cap \mathcal{U}_{rsR}(J_p)$ of entire right strongly regular J_p-inner mvf's may be identified with the class of resolvent matrices of such strictly completely indeterminate problems.

Remark 4.26 *If $A \in \mathcal{U}_{rsR}(J_p)$, $b(\lambda) = A(\lambda)\mathfrak{V}$, $\{b_3, b_4\} \in ap_{II}(A)$, $\chi_1(\lambda) = b_4(\lambda) b_3(\lambda)$ and*

$$I_p - \chi_1(\omega)\chi_1(\omega)^* > 0 \text{ for at least one point } \omega \in \mathbb{C}_+,$$

then

$$\mathcal{S}^{p\times p} \subseteq \mathcal{D}(T_B) \quad \text{and hence} \quad \mathcal{C}(A) = T_B[\mathcal{S}^{p\times p}].$$

In particular, $\mathcal{S}^{p\times p} \subseteq \mathcal{D}(T_B)$ *if either* $e_{-\beta}b_3 \in \mathcal{S}^{p\times p}$ *or* $e_{-\beta}b_4 \in \mathcal{S}^{p\times p}$ *for some* $\beta > 0$*; see remark 7.72 in [ArD08b].*

Lemma 4.27 *Let* $A \in \mathcal{U}(J_p)$, $\{b_1, b_2\} \in ap_I(A)$, $\{b_3, b_4\} \in ap_{II}(A)$, $B(\lambda) = A(\lambda)\mathfrak{V}$, $\chi = b_{22}^{-1}b_{21}$ *and* $W(\lambda) = \mathfrak{V}A(\lambda)\mathfrak{V}$. *Then for every* $s \in T_W[\mathcal{S}^{p\times p}] \cap \mathcal{D}(T_{\mathfrak{V}})$, *there exists a pair of mvf's* $\varphi_s \in \mathcal{S}_{out}^{p\times p}$ *and* $\psi_s \in \mathcal{S}_{out}^{p\times p}$ *such that*

$$\frac{(I_p + s)}{2} b_3 = b_1\varphi_s \quad \text{and} \quad b_4\frac{(I_p + s)}{2} = \psi_s b_2. \tag{4.63}$$

If

$$I_p - \chi(\omega)\chi(\omega)^* > 0 \quad \text{for at least one point } \omega \in \mathbb{C}_+, \tag{4.64}$$

then:

(1) $T_W[\mathcal{S}^{p\times p}] \subset \mathcal{D}(T_{\mathfrak{V}})$ *and (4.63) holds for every* $s \in T_W[\mathcal{S}^{p\times p}]$.
(2) $\mathcal{S}^{p\times p} \subset \mathcal{D}(T_B)$ *and* $\mathcal{C}(A) = T_B[\mathcal{S}^{p\times p}]$.

Proof Let $c^\circ = T_A[I_p]$. Then, by Lemma 3.72, (4.48) holds with $s^\circ = T_{\mathfrak{V}}[c^\circ]$, and, by Theorem 4.21, (4.49) holds for every $s \in \mathcal{S}(b_1, b_2; s^\circ)$. Therefore, (4.49) holds for every $s \in T_W[\mathcal{S}^{p\times p}] \cap \mathcal{D}(T_{\mathfrak{V}})$, since $T_W[\mathcal{S}^{p\times p}] \subseteq \mathcal{S}(b_1, b_2; s^\circ)$, by theorem 7.50 in [ArD08b]. □

Lemma 4.28 *Let* $A \in \mathcal{U}_{rsR}(J_p)$, $\{b_1, b_2\} \in ap_I(A)$, $\{b_3, b_4\} \in ap_{II}(A)$, $W(\lambda) = \mathfrak{V}A(\lambda)\mathfrak{V}$ *and let*

$$\gamma = \inf\{\|s\|_\infty : s \in T_W[\mathcal{S}^{p\times q}] \cap \mathring{\mathcal{S}}^{p\times q}\}.$$

Then $\gamma < 1$ *and*

$$\left(\frac{1}{2}\right)(1-\gamma)\|b_3(\lambda)\| \le \|b_1(\lambda)\| \le 2^p(1-\gamma)^{-p}\|b_3(\lambda)\| \tag{4.65}$$

and

$$\left(\frac{1}{2}\right)(1-\gamma)\|b_4(\lambda)\| \le \|b_2(\lambda)\| \le 2^p(1-\gamma)^{-p}\|b_4(\lambda)\| \tag{4.66}$$

for every point $\lambda \in \mathbb{C}_+$.

Proof See lemma 3.5 in [ArD03a]. □

Lemma 4.29 *Let* $A \in \mathcal{U}_{rsR}(J_p)$ *and let* $\{b_1, b_2\} \in ap_I(A)$ *and* $\{b_3, b_4\} \in ap_{II}(A)$. *Then:*

$$\lim_{\nu\uparrow\infty} b_1(i\nu) = 0 \iff \lim_{\nu\uparrow\infty} b_3(i\nu) = 0 \tag{4.67}$$

and

$$\lim_{\nu\uparrow\infty} b_2(i\nu) = 0 \iff \lim_{\nu\uparrow\infty} b_4(i\nu) = 0. \tag{4.68}$$

Proof This is an immediate corollary of Lemma 4.28. □

Lemma 4.30 *Let* $A \in \mathcal{U}_{rsR}(J_p)$, $\{b_3, b_4\} \in ap_{II}(A)$, $B(\lambda) = A(\lambda)\mathfrak{V}$, $W(\lambda) = \mathfrak{V}A(\lambda)\mathfrak{V}$ *and assume that*

$$\text{either} \quad \lim_{\nu\uparrow\infty} b_3(i\nu) = 0 \quad \text{or} \quad \lim_{\nu\uparrow\infty} b_4(i\nu) = 0. \tag{4.69}$$

Then every mvf $s \in T_W[\mathcal{S}^{p\times p}]$ *satisfies the inequality*

$$s(\lambda)^* s(\lambda) < I_p \quad \text{for every} \quad \lambda \in \mathbb{C}_+ \tag{4.70}$$

and hence

$$\mathcal{S}^{p\times p} \subseteq \mathcal{D}(T_B) \text{ and } \mathcal{C}(A) = T_B[\mathcal{S}^{p\times p}]. \tag{4.71}$$

Consequently, (4.71) holds, if either $e_{-\alpha}b_3 \in \mathcal{S}_{\rm in}^{p\times p}$ *or* $e_{-\alpha}b_4 \in \mathcal{S}_{\rm in}^{p\times p}$ *for some* $\alpha > 0$.

Proof See lemma 3.7 on p.24 in [ArD03a] □

4.5 Detour on scalar determinate interpolation problems

In this section we establish a few facts connected with determinate interpolation problems in $\mathcal{S} = \mathcal{S}^{1\times 1}$ and $\mathcal{C} = \mathcal{C}^{1\times 1}$ for future use. The main principle that underlies these facts is contained in the next theorem.

Theorem 4.31 *If* $s^\circ \in \mathcal{S}$, $b \in \mathcal{S}_{\rm in}$ *and* 1 *is a singular value of the operator* $X = \Pi_{\mathcal{H}(b)} M_{s^\circ}|_{H_2}$, *then* $s^\circ \in \mathcal{S}_{\rm in}$ *and*

$$\mathcal{S}(b, 1; s^\circ) = \{s^\circ\}.$$

Proof If 1 is a singular value of X, then there exists a nonzero vector $u \in H_2$ such that

$$\|u\|_{\rm st} = \|Xu\|_{\rm st} = \|\Pi_{\mathcal{H}(b)} s^\circ u\|_{\rm st} \le \|s^\circ u\|_{\rm st} \le \|u\|_{\rm st}.$$

But this implies that $|s^\circ(\mu)| = 1$ a.e. on $\mathbb{R}$ and that $s^\circ u \in \mathcal{H}(b)$. Thus, $s^\circ \in \mathcal{S}_{\rm in}$ and, if $s = s^\circ + bf$ for some $f \in H_\infty$, then

$$\Pi_{\mathcal{H}(b)} su = \Pi_{\mathcal{H}(b)} s^\circ u + \Pi_{\mathcal{H}(b)} bfu = \Pi_{\mathcal{H}(b)} s^\circ u = s^\circ u.$$

Therefore,

$$\|u\|_{\rm st} = \|s^\circ u\|_{\rm st} = \|\Pi_{\mathcal{H}(b)} s^\circ u\|_{\rm st} = \|\Pi_{\mathcal{H}(b)} su\|_{\rm st} \le \|su\|_{\rm st} \le \|u\|_{\rm st},$$

which in turn implies that $\Pi_{\mathcal{H}(b)}su = su$ and hence that

$$s^\circ u = \Pi_{\mathcal{H}(b)}s^\circ u = \Pi_{\mathcal{H}(b)}su = su.$$

Consequently, as $u \not\equiv 0$, $s^\circ = s$. □

Lemma 4.32 *If s° is a rational inner function and $a > 0$, then there exists a nonzero function $h \in \mathcal{H}(e_a)$ such that $s^\circ h \in \mathcal{H}(e_a)$.*

Proof The proof is broken into steps.

1. *If h is holomorphic at a point $\omega \in \mathbb{C}$, then*

$$(R_\omega^k h)(\lambda) = \begin{cases} \dfrac{f(\lambda) - f(\omega) - \cdots - f^{(k-1)}(\omega)\dfrac{(\lambda-\omega)^{k-1}}{(k-1)!}}{(\lambda-\omega)^k} & if \quad \lambda \neq \omega \\ \dfrac{f^{(k)}(\omega)}{k!} & if \quad \lambda = \omega \end{cases} \tag{4.72}$$

for every integer $k \geq 1$.

This is a straightforward calculation that may be verified by induction.

2. *If $\omega_1, \ldots, \omega_n$ is a given set of distinct points in $\mathbb{C}$ and $k_1, \ldots, k_n$ is a given set of nonnegative integers, then there exists a function $h \in \mathcal{H}(e_a)$ such that*

$$h^{(0)}(\omega_j) = \cdots = h^{(k_j)}(\omega_j) = 0 \quad \text{for } j = 1, \ldots, n.$$

Let

$$\varphi_\omega^{(k)}(t) = t^k e^{-i\overline{\omega}t} \quad \text{for } k = 0, 1, \ldots$$

and suppose that $f \in L_2([0, a])$ is orthogonal to $\varphi_\omega^{(k)}$ for $k = 0, \ldots, \ell$. Then $h = \widehat{f}$ belongs to $\mathcal{H}(e_a)$ and

$$0 = \langle f, \varphi_\omega^{(k)} \rangle_{\text{st}} = \int_0^a f(t) t^k e^{i\omega t} dt = (-i)^k h^{(k)}(\omega)$$

for $k = 0, \ldots, \ell$. The same calculation with $f \in L_2([0, a])$ now chosen to be orthogonal to $\varphi_{\omega_j}^{(k)}$ for $k = 0, \ldots, k_j$ and $j = 1, \ldots, n$ implies that $h = \widehat{f}$ meets the stated conditions.

3. *There exists a function $h \in \mathcal{H}(e_a)$ such that $s^\circ h \in \mathcal{H}(e_a)$.*

If s° is a rational inner function and $\gamma = \lim_{|\lambda| \uparrow \infty} s^\circ(\lambda)$, then $|\gamma| = 1$. If $s^\circ \equiv \gamma$, then the assertion is obvious. If not, then s° will have poles ω_j of order k_j for $j = 1, \ldots, n$. Thus, if $h \in \mathcal{H}(e_a)$ is constructed as in Step 2 to have a zero of order k_j at the poles ω_j, then

$$s^\circ(\lambda)h(\lambda) = \gamma h(\lambda) + \sum_{j=1}^{k_1} \frac{c_{1j}}{(\lambda - \omega_1)^j} h(\lambda) + \cdots$$

Therefore, in view of the special properties of h and the formula in Step 1,

$$\frac{h(\lambda)}{\lambda-\omega_1} = \frac{h(\lambda)-h(\omega_1)}{\lambda-\omega_1} = (R_{\omega_1}h)(\lambda)$$

$$\frac{h(\lambda)}{(\lambda-\omega_1)^2} = \frac{h(\lambda)-h(\omega_1)-(\lambda-\omega_1)h^{(1)}(\omega_1)}{(\lambda-\omega_1)^2} = (R^2_{\omega_1}h)(\lambda)$$

$$\vdots \quad \vdots$$

$$\frac{h(\lambda)}{(\lambda-\omega_1)^{k_j}} = \cdots = (R^{k_j}_{\omega_1}h)(\lambda).$$

Thus, as $\mathcal{H}(e_a)$ is invariant under the action of R_α it is readily seen that $s^\circ h \in \mathcal{H}(e_a)$. □

Theorem 4.33 *If $s^\circ \in \mathcal{S}_{\text{in}}$ is rational, then:*

(1) *1 is a singular value of the operator $X = \Pi_{\mathcal{H}(e_a)} M_{s^\circ}|_{\mathcal{H}(e_a)}$.*
(2) *$SP(s^\circ; a) = \{s^\circ\}$ for every $a > 0$.*

Proof In view of Lemma 4.32, there exists a nonzero function $h \in \mathcal{H}(e_a)$ such that $s^\circ h \in \mathcal{H}(e_a)$. Therefore, $Xh = s^\circ h$ and

$$\langle h, h\rangle - \langle Xh, Xh\rangle = \langle h, h\rangle - \langle s^\circ h, s^\circ h\rangle = 0.$$

Thus, as $\|X\| \le 1$, it follows that

$$(I - X^*X)h = 0$$

and hence that (1) holds; (2) then follows from Theorem 4.31. □

Remark 4.34 *It is also possible to prove (2) of Theorem 4.33 directly without relying on (1):*

If $s = s^\circ + e_a f$ for some $f \in H_\infty$ and h is a nonzero function in $\mathcal{H}(e_a)$ such that $Xh = s^\circ h$, then

$$\Pi_{\mathcal{H}(e_a)} sh = \Pi_{\mathcal{H}(e_a)} s^\circ h + \Pi_{\mathcal{H}(e_a)} e_a f h = \Pi_{\mathcal{H}(e_a)} s^\circ h.$$

Therefore,

$$\|h\|_{\text{st}} = \|s^\circ h\|_{\text{st}} = \|\Pi_{\mathcal{H}(e_a)} s^\circ h\|_{\text{st}} = \|\Pi_{\mathcal{H}(e_a)} sh\|_{\text{st}} \le \|sh\|_{\text{st}} \le \|h\|_{\text{st}}.$$

Thus,

$$sh = \Pi_{\mathcal{H}(e_a)} sh = \Pi_{\mathcal{H}(e_a)} s^\circ h = s^\circ h$$

and hence, as $h \not\equiv 0$, $s = s^\circ$.

Theorem 4.35 *If $c^\circ \in \mathcal{C}$ is rational and $\Re c^\circ(\mu) = 0$ at every point $\mu \in \mathbb{R}$ except at the poles of c° (i.e., $c^\circ \in \mathcal{C}_{\rm sing}$) and $a > 0$, then*

$$CP(c^\circ; a) = \{c^\circ\}.$$

Proof If $c^\circ(\mu) = 0$, then

$$s^\circ = (1 - c^\circ)/(1 + c^\circ)$$

is a rational inner function. Therefore, since $\mathrm{SP}(s^\circ; a) = \{s^\circ\}$ by Theorem 4.33, the asserted result follows. □

4.6 The reproducing kernel Hilbert space $\mathcal{H}(U)$

Theorem 4.36 *Let Ω be a subset of $\mathbb{C}$ and let the $m \times m$ matrix-valued kernel $K_\omega(\lambda)$ be positive on $\Omega \times \Omega$. Then there is a unique Hilbert space $\mathcal{H}$ of $m \times 1$ vvf's $f(\lambda)$ on Ω such that*

$$K_\omega \xi \in \mathcal{H} \quad \text{and} \quad \langle K_\omega \xi, f\rangle_{\mathcal{H}} = \xi^* f(\omega)$$

for every $\omega \in \Omega$, $\xi \in \mathbb{C}^m$ and $f \in \mathcal{H}$.

Proof This is a matrix version of a theorem of Aronszajn in [Arn50]; for a proof, see, e.g., theorem 5.2 in [ArD08b]. □

The space $\mathcal{H}$ is called an **RKHS** (**reproducing kernel Hilbert space**) with **RK** (**reproducing kernel**) $K_\omega(\lambda)$. It is readily checked that an RKHS has exactly one RK and that

$$K_\alpha(\beta)^* = K_\beta(\alpha) \quad \text{for } \alpha,\ \beta \in \Omega.$$

Lemma 4.37 *Let $\mathcal{H}$ be an RKHS of $m \times 1$ vvf's on some nonempty open subset Ω of $\mathbb{C}$ with RK $K_\omega(\lambda)$ on $\Omega \times \Omega$. Then the two conditions*

(1) *$K_\omega(\lambda)$ is a holomorphic function of λ in Ω for every point $\omega \in \Omega$;*
(2) *the function $K_\omega(\omega)$ is continuous on Ω;*
are in force if and only if
(3) *every vvf $f \in \mathcal{H}$ is holomorphic in Ω.*

Proof See lemma 5.6 and corollary 5.7 in [ArD08b] □

We shall be particularly interested in RKHS's of holomorphic vvf's on an open set Ω that are invariant under the generalized backwards shift operator R_α for $\alpha \in \Omega$. More precisely, we shall focus on the RKHS's $\mathcal{H}(b)$ and $\mathcal{H}_*(b)$ for $b \in \mathcal{E} \cap \mathcal{S}_{\rm in}^{p\times p}$, $\mathcal{H}(U)$ for $U \in \mathcal{E} \cap \mathcal{U}(J)$ and $\mathcal{B}(\mathfrak{E})$ for entire de Branges matrices $\mathfrak{E}$; and in particular for $\mathfrak{E}(\lambda) = \sqrt{2}N_2^* A(\lambda)\mathfrak{V}$ for some $A \in \mathcal{E} \cap \mathcal{U}(J_p)$. However, for the sake of added perspective, we shall begin with a more general setting:

If $U \in \mathcal{U}(J)$, then the kernel

$$K_\omega^U(\lambda) = \begin{cases} \dfrac{J - U(\lambda)JU(\omega)^*}{\rho_\omega(\lambda)} & \text{if } \lambda \neq \overline{\omega} \\ \\ \dfrac{U'(\overline{\omega})JU(\omega)^*}{2\pi i} & \text{if } \lambda = \overline{\omega} \end{cases} \tag{4.73}$$

is positive on $\mathfrak{h}_U \times \mathfrak{h}_U$. Therefore, it defines a unique RKHS that will be denoted $\mathcal{H}(U)$; see e.g., (1) of theorem 5.31 in [ArD08b].

Theorem 4.38 *If $U \in \mathcal{E} \cap \mathcal{U}(J)$ for some signature matrix $J \in \mathbb{C}^{m\times m}$, then:*

(1) *$\mathcal{H}(U) \subset \mathcal{E} \cap \Pi^p$ and every vvf in $\mathcal{H}(U)$ is of exponential type and satisfies the Cartwright condition (3.86).*
(2) *The RKHS $\mathcal{H}(U)$ is R_α invariant for every point $\alpha \in \mathbb{C}$.*
(3) *R_α is a bounded linear operator on $\mathcal{H}(U)$ for every point $\alpha \in \mathbb{C}$.*
(4) *The de Branges identity*

$$\langle R_\alpha f, g\rangle_{\mathcal{H}} - \langle f, R_\beta g\rangle_{\mathcal{H}} - (\alpha - \overline{\beta})\langle R_\alpha f, R_\beta g\rangle_{\mathcal{H}} = 2\pi i g(\beta)^* J f(\alpha) \tag{4.74}$$

in which the subscript $\mathcal{H}$ denotes $\mathcal{H}(U)$, is in force for every choice of $\alpha, \beta \in \mathbb{C}$ and $f, g \in \mathcal{H}(U)$.

Conversely, if (1)–(4) hold for a RKHS $\mathcal{H}$, then $\mathcal{H} = \mathcal{H}(U)$ for some $U \in \mathcal{E} \cap \mathcal{U}(J)$.

Proof Assertions (1)–(4) follow from theorems 5.31, 5.49 and remark 5.32 in [ArD08b] and from Theorem 3.30; for a proof of the converse statement, see, e.g., theorem 5.21 in [ArD08b]. □

The RKHS's

$$\mathcal{H}(b) = H_2^p \ominus bH_2^p \quad \text{and} \quad \mathcal{H}_*(b) = (H_2^p)^\perp \ominus b^\#(H_2^p)^\perp \tag{4.75}$$

based on $b \in \mathcal{S}_{\text{in}}^{p\times p}$ will play an important role. The RK's $k_\omega^b(\lambda)$ for $\mathcal{H}(b)$ and $\ell_\omega^b(\lambda)$ for $\mathcal{H}_*(b)$ are given by the formulas

$$k_\omega^b(\lambda) = \begin{cases} \dfrac{I_p - b(\lambda)b(\omega)^*}{-2\pi i(\lambda - \overline{\omega})} & \text{if } \lambda \neq \overline{\omega} \\ \\ \dfrac{b'(\overline{\omega})b(\omega)^*}{2\pi i} & \text{if } \lambda = \overline{\omega} \end{cases} \tag{4.76}$$

and

$$\ell_\omega^b(\lambda) = \begin{cases} \dfrac{b^\#(\lambda)b^\#(\omega)^* - I_p}{-2\pi i(\lambda - \overline{\omega})} & \text{if } \lambda \neq \overline{\omega} \\ \\ \dfrac{b'(\omega)^* b(\overline{\omega})}{2\pi i} & \text{if } \lambda = \overline{\omega} \end{cases} \tag{4.77}$$

at points λ, ω and $\overline{\omega}$ at which the indicated functions are holomorphic. If b is entire, then both RK's are defined on $\mathbb{C} \times \mathbb{C}$.

We remark that formulas (4.76) and (4.77) are both special cases of (4.73); the former corresponds to the choice $U = b$ and $J = I_p$; the latter to the choice $U = b^\#$ and $J = -I_p$.

Theorem 4.39 *If $U \in \mathcal{E} \cap \mathcal{U}^\circ(J)$, then:*

(1) *The operator R_0 is a Volterra operator, i.e., it is a compact linear operator from $\mathcal{H}(U)$ into itself with $\{0\}$ as its only point of spectrum.*
(2) *The linear operator*

$$F_0 : f \in \mathcal{H}(U) \longrightarrow \sqrt{2\pi} f(0) \in \mathbb{C}^m \tag{4.78}$$

is bounded and

$$(F_0^* v)(\lambda) = \sqrt{2\pi} K_0^U(\lambda) v \quad \textit{when } v \in \mathbb{C}^m. \tag{4.79}$$

(3) *The operators R_0, F_0 and J are connected by the relations*

$$R_0 - R_0^* = i F_0^* J F_0 \tag{4.80}$$

and

$$\bigvee_{n \geq 0} R_0^n F_0^* \mathbb{C}^m = \mathcal{H}(U). \tag{4.81}$$

(4) *The mvf $U(\lambda)$ may be recovered from R_0, F_0 and J by the formula*

$$U(\lambda) = I_m + i\lambda F_0 (I - \lambda R_0)^{-1} F_0^* J \quad \textit{for every } \lambda \in \mathbb{C} \tag{4.82}$$

and is subject to the bound

$$\|U(\lambda)\| \leq 1 + 2\pi |\lambda| \left\| \sqrt{K_0^U(0)} \right\| \left\| \sqrt{K_\lambda^U(\lambda)} \right\|. \tag{4.83}$$

Proof The fact that R_α is a bounded linear operator for every $\alpha \in \mathbb{C}$ was established in Theorem 4.38. It is also easy to check that it satisfies the resolvent identity

$$R_\alpha - R_\beta = (\alpha - \beta) R_\alpha R_\beta. \tag{4.84}$$

The identity

$$(I - \omega R_0)^{-1} = I + \omega R_w \quad \text{for } \omega \in \mathbb{C} \tag{4.85}$$

then serves to guarantee that $\{0\}$ is the only point of spectrum of R_0.

The two assertions in (2) follow from the formula

$$v^* F_0 f = \sqrt{2\pi} \langle f, K_0^U v \rangle_{\mathcal{H}(U)} \quad \text{for } v \in \mathbb{C}^m.$$

Formula (4.80) is equivalent to the de Branges identity (4.74) with $\alpha = \beta = 0$.

The formula

$$R_0 = \frac{R_0 + R_0^*}{2} + i\frac{R_0 - R_0^*}{2i} = A_R + iA_I$$

exhibits R_0 as a finite dimensional perturbation of the operator A_R, which will be shown to be compact in Theorem 4.42. Therefore, R_0 is compact.

Next, (4.80) implies that

$$\langle f, R_0^n F_0^* v\rangle_{\mathcal{H}(U)} = 0 \quad \text{for every } v \in \mathbb{C}^m \text{ and } n = 0, 1, \ldots$$

if and only if

$$\langle f, (R_0^*)^n F_0^* v\rangle_{\mathcal{H}(U)} = 0 \quad \text{for every } v \in \mathbb{C}^m \text{ and } n = 0, 1, \ldots,$$

or, equivalently, if and only if

$$\frac{v^* f^{(n)}(0)}{n!} = 0 \quad \text{for every } v \in \mathbb{C}^m \text{ and } n = 0, 1, \ldots.$$

which serves to justify (4.81).

Next, (4.82) follows from the fact that

$$U(\lambda) = \sum_{j=0}^{\infty} \lambda^j \frac{U^{(j)}(0)}{j!},$$

since

$$\sqrt{2\pi}\frac{U^{(j)}(0)}{j!}v = F_0 R_0^j U v \quad \text{and} \quad F_0^* J v = \sqrt{2\pi} K_0^U(\lambda) J v = \frac{1}{i\sqrt{2\pi}}(R_0 U)(\lambda) v.$$

On the other hand, the identity

$$U(\lambda) = \lambda\frac{U(\lambda) - I_m}{\lambda} + I_m = 2\pi i\lambda K_0^U(\lambda) J + I_m$$

implies that

$$\|U(\lambda)\| \leq 1 + 2\pi|\lambda|\|K_0^U(\lambda)\|.$$

The inequality (4.83) then follows from the fact that

$$\begin{aligned}|\eta^* K_0^U(\lambda)\xi| &= |\langle K_0^U \xi, K_\lambda^U \eta\rangle_{\mathcal{H}(U)}| \leq \sqrt{\xi^* K_0^U(0)\xi}\sqrt{\eta^* K_\lambda^U(\lambda)\eta} \\ &\leq \|\sqrt{K_0^U(0)}\|\|\xi\|\,\|\sqrt{K_\lambda^U(\lambda)}\|\|\eta\|\end{aligned}$$

for every choice of $\xi, \eta \in \mathbb{C}^m$. □

Remark 4.40 *If $U \in \mathcal{E} \cap \mathcal{U}^\circ(J)$, then the supplementary formulas*

$$\begin{aligned}K_\omega^U(\lambda)\xi &= (I - \overline{\omega}R_0)^{-1} K_0^U(\lambda) U(\omega)^*\xi \\ &= \frac{1}{\sqrt{2\pi}}(I - \overline{\omega}R_0)^{-1} F_0^* U(\omega)^*\xi, \quad \xi \in \mathbb{C}^m.\end{aligned} \tag{4.86}$$

and, with the usual identification of $m \times m$ matrices as bounded linear operators in $\mathbb{C}^m$,

$$K_\omega^U(\lambda) = \frac{1}{2\pi} F_0 (I - \lambda R_0)^{-1} (I - \overline{\omega} R_0^*)^{-1} F_0^*, \tag{4.87}$$

in which F_0^* acts on a matrix column by column, are also useful. The former yields the identity

$$\bigvee_{\omega \in \Omega} (I - \omega R_0)^{-1} F_0^* \mathbb{C}^m = \mathcal{H}(U) \tag{4.88}$$

for any open set Ω that contains the point 0, which is equivalent to (4.81).

Remark 4.41 *In view of (4.80)–(4.82), the colligation*

$$\mathfrak{N}_\circ = (R_0, F_0; \mathcal{H}(U), \mathbb{C}^m; J)$$

is the de Branges model of a simple Livsic–Brodskii J-node with characteristic mvf $U(\lambda)$; it will be discussed Section 8.4.

Theorem 4.42 *If $U \in \mathcal{E} \cap \mathcal{U}^\circ(J)$ and*

$$A_R \stackrel{\text{def}}{=} \frac{1}{2}(R_0 + R_0^*),$$

then:

(1) *A_R is a compact self-adjoint operator.*
(2) *λ_j is a nonzero eigenvalue of A_R if and only if $\det(I_m + U(1/\lambda_j)) = 0$. Moreover, the multiplicity of λ_j is equal to the dimension of the null space of the matrix $(I_m + U(1/\lambda_j))$ and, if $\omega_j = 1/\lambda_j$,*

$$(A_R f)(\lambda) = \lambda_j f(\lambda) \Longleftrightarrow f(\lambda) = \pi i \omega_j K_{\omega_j}^U(\lambda) \eta_j$$

for some nonzero vector $\eta_j \in \mathbb{C}^m$ such that $(I_m + U(\omega_j))\eta_j = 0$.
(3) *A_R belongs to the von Neumann–Schatten class $\mathcal{S}_p$ for $p > 1$, i.e., A_R is a compact operator with a countable set of singular values $s_1, s_2, \ldots$ such that $\sum |s_j|^p < \infty$ for $p > 1$.*

Proof The formula

$$R_0 - R_0^* = iF_0^* J F_0$$

implies that

$$A_R = \frac{1}{2}(R_0 + R_0^*) = R_0 - \frac{1}{2} i F_0^* J F_0$$

and hence that

$$\begin{aligned}(A_R f)(\lambda) &= (R_0 f)(\lambda) - \frac{1}{2}2\pi i K_0^U(\lambda) J f(0) \\ &= \frac{f(\lambda) - f(0)}{\lambda} + \frac{1}{2}\frac{I_m - U(\lambda)}{\lambda} f(0) \\ &= \frac{2f(\lambda) - \{I_m + U(\lambda)\}f(0)}{2\lambda}.\end{aligned}$$

The rest of the proof is broken into steps.

1. *If $\alpha \in \mathbb{R} \setminus \{0\}$, $\beta = 1/\alpha$ and $I_m + U(\beta)$ is not invertible, then α is an eigenvalue of A_R of multiplicity equal to the dimension of the null space of the matrix $I_m + U(\beta)$.*

In view of the preceding discussion,

$$(A_R f_j)(\lambda) = \lambda_j f_j(\lambda) \Longleftrightarrow f_j(\lambda) = \frac{\{I_m + U(\lambda)\}}{2(1 - \lambda\lambda_j)} f_j(0),$$

i.e., if $\lambda_j \neq 0$, $\omega_j = 1/\lambda_j$ and $\lambda \neq \omega_j$, then,

$$f_j(\lambda) = \omega_j \frac{\{I_m + U(\lambda)\}}{2(\omega_j - \lambda)} \eta_j, \quad \text{where } \{I_m + U(\omega_j)\}\eta_j = 0,$$

since $f(\lambda)$ is an entire vvf. Thus,

$$f_j(\lambda) = \omega_j \frac{U(\lambda) - U(\omega_j)}{2(\omega_j - \lambda)} \eta_j, \quad f_j(\omega_j) = -\omega_j \frac{1}{2} U'(\omega_j)\eta_j$$

and, as $\omega_j \in \mathbb{R}$,

$$\eta_j = -U(\omega_j)\eta_j \Longleftrightarrow U(\omega_j)^* J \eta_j = -J\eta_j.$$

Consequently, the formula for the eigenfunction $f_j(\lambda)$ can be rewritten in terms of the RK for $\mathcal{H}(U)$ as

$$f_j(\lambda) = \pi i \omega_j K_{\omega_j}^U(\lambda) J \eta_j.$$

Thus,

$$\langle f_j, f_k \rangle_{\mathcal{H}(U)} = \pi^2 \omega_j \omega_k \langle K_{\omega_j}^U J\eta_j, K_{\omega_k}^U J\eta_k \rangle = \eta_k^* J K_{\omega_j}^U(\omega_k) J \eta_j = 0$$

if $k \neq j$. Moreover, if also $\{I_m + U(\omega_j)\}\xi_j = 0$, then

$$\langle K_{\omega_j}^U J\eta_j, K_{\omega_j}^U J\xi_j \rangle = \xi_j^* J K_{\omega_j}^U(\omega_j) J \eta_j = \frac{i}{2\pi} \xi_j^* J U'(\omega_j)\eta_j.$$

Consequently, such a pair of vvf's in $\mathcal{H}(U)$ will be orthogonal in $\mathcal{H}(U)$ if ξ_j and η_j are orthogonal with respect to the inner product

$$\langle J K_{\omega_j}^U(\omega_j) J \eta, \xi \rangle$$

in $\mathbb{C}^m$ based on the positive definite matrix $J K_{\omega_j}^U(\omega_j) J$.

2. *If $\alpha \in \mathbb{R} \setminus \{0\}$, $\beta = 1/\alpha$ and $I_m + U(\beta)$ is invertible, then $\alpha I - A_R$ is a bounded invertible operator from $\mathcal{H}(U)$ onto itself, i.e., the nonzero spectrum of A_R consists only of eigenvalues.*

It suffices to check that for each $g \in \mathcal{H}(U)$ there exists exactly one vvf $f \in \mathcal{H}(U)$ such that $(\alpha I - A_R)f = g$. But a necessary condition for this is that

$$f(\lambda) = \frac{-2\lambda g(\lambda) + \{I_m + U(\lambda)\}f(0)}{2(1 - \lambda\alpha)}$$

and hence, since $f(\lambda)$ is an entire vvf,

$$2\beta g(\beta) = \{I_m + U(\beta)\}f(0),$$

i.e.,

$$f(0) = \{I_m + U(\beta\}^{-1}2\beta g(\beta).$$

But this in turn leads to the formula

$$\begin{aligned}
f(\lambda) &= \frac{\lambda g(\lambda) - \{I_m + U(\lambda)\}\{I_m + U(\beta)\}^{-1}\beta g(\beta)}{\alpha(\lambda - \beta)} \\
&= \frac{\lambda g(\lambda) - \beta g(\beta)}{\alpha(\lambda - \beta)} - \frac{\{I_m - \{I_m + U(\lambda)\}\{I_m + U(\beta)\}^{-1}\}\beta g(\beta)}{\alpha(\lambda - \beta)} \\
&= \beta g(\lambda) + \beta^2 (R_\beta g)(\lambda) - 2\pi i \beta K_\beta^U(\lambda)x,
\end{aligned}$$

where

$$x = \beta J U(\beta)\{I_m + U(\beta)\}^{-1} g(\beta).$$

3. *A_R is compact*

Since the reciprocals $1/\lambda_j$ of the nonzero eigenvalues λ_j of A_R are the roots of an entire function, zero is the only possible limit point of the λ_j. Therefore, they can be reindexed according to size. Let $\gamma_1, \gamma_2, \ldots$ denote a reindexing of the eigenvalues of A_R with $|\gamma_j| \geq |\gamma_{j+1}|$ and let φ_j, $j = 1, 2, \ldots$ denote a corresponding set of orthonormal vvf's in $\mathcal{H}(U)$. Then, since $A_R = A_R^*$, the spectral theorem for bounded self-adjoint operators implies that

$$A_R f = \sum_{j \geq 1} \gamma_j \langle f, \varphi_j \rangle_{\mathcal{H}(U)} \varphi_j$$

for every $f \in \mathcal{H}(U)$. If A_R has finite dimensional range, then it is automatically compact. If not, set

$$A_R^{(n)} f = \sum_{j=1}^{n} \langle A_R f, \varphi_j \rangle_{\mathcal{H}(U)} \varphi_j = \sum_{j=1}^{n} \gamma_j \langle f, \varphi_j \rangle_{\mathcal{H}(U)} \varphi_j$$

for $n = 1, 2, \ldots$. Then the bounds

$$\|A_R^{(n+k)} f - A_R^{(n)} f\|_{\mathcal{H}(U)}^2 \leq |\gamma_{n+1}|^2 \|f\|_{\mathcal{H}(U)}^2$$

guarantee that $\|A_R - A_R^{(n)}\| \leq |\gamma_{n+1}|$, and hence that A_R can be uniformly approximated in operator norm by finite rank operators and is therefore compact.

4. *A_R belongs to the von Neumann–Schatten class $\mathcal{S}_p$ for $p > 1$.*

Since $A_R = A_R^*$, the numbers $|\gamma_j| = s_j$, the singular values of A_R. Moreover,

$$\sum_{j \geq 1} s_j^{1+\delta} = \sum_{j \geq 1} |\gamma_j|^{1+\delta} < \infty,$$

since the numbers $\gamma_1^{-1}, \gamma_2^{-1}, \ldots$ are the nonzero roots (counting multiplicities) of the entire function of exponential type $\det\{I_m + U(\lambda)\}$; see e.g., p. 19 of [DMc72] for the bound. Therefore, $A_R \in \mathcal{S}_p$ for $p > 1$; pp. 91–95 of [GK69] is a good source of information on the von Neumann–Schatten class. □

Theorem 4.43 *If $U \in \mathcal{E} \cap \mathcal{U}(J)$, then:*

(1) *Every $f \in \mathcal{H}(U)$ is an entire vvf of exponential type τ_f and meets the Cartwright condition*

$$\ln(f^* f) \in \widetilde{L}_1. \tag{4.89}$$

(2) *The type $\tau_f = \max\{\tau_f^+, \tau_f^-\}$, where*

$$\tau_f^{\pm} = \lim_{\nu \uparrow \infty} \frac{\ln \|f(\pm i\nu)\|}{\nu}.$$

(3) *$\tau_f^{\pm} = \tau_{\pm}(f)$, where*

$$\tau_{\pm}(f) = \lim_{r \uparrow \infty} \frac{\max\{\|f(\lambda)\| : |\lambda| \leq r \text{ and } \lambda \in \overline{\mathbb{C}_{\pm}}\}}{r}.$$

(4) *$\tau_f \leq \tau_U$ and $\tau_f^{\pm} \leq \tau_U^{\pm}$ for every $f \in \mathcal{H}(U)$. Moreover, equality is achieved for each of these three inequalities by $K_0^U \xi$ for appropriate choices of $\xi \in \mathbb{C}^m$.*

Proof This follows from Theorem 4.44 and a theorem of M.G. Krein on entire mvf's with bounded Nevanlinna characteristics in $\mathbb{C}_+$ and $\mathbb{C}_-$; see Theorem 3.24. □

Theorem 4.44 *If $U \in \mathcal{U}(J)$, then $\mathcal{H}(U) \subset \Pi^m$ and $\cap_{f \in \mathcal{H}(U)} \mathfrak{h}_f = \mathfrak{h}_U$. Moreover, $R_\alpha U \xi \in \mathcal{H}(U)$ for every $\alpha \in \mathfrak{h}_U$ and $\xi \in \mathbb{C}^m$.*

Proof See theorem 5.49 in [ArD08b]. □

An isometry from $\mathcal{H}(S)$ onto $\mathcal{H}(U)$ when $S = PG(U)$

If $J = I_m$, then $\mathcal{U}(J) = \mathcal{S}_{\rm in}^{m\times m}$ and

$$\mathcal{H}(S) = H_2^m \ominus SH_2^m \quad \text{for } S \in \mathcal{S}_{\rm in}^{m\times m}. \tag{4.90}$$

If $J = -I_m$, then $\mathcal{U}(J) = \{U : U^\# \in \mathcal{S}_{\rm in}^{m\times m}\}$ and

$$\mathcal{H}(U) = (H_2^m)^\perp \ominus S^\#(H_2^m)^\perp = \mathcal{H}_*(S) \quad \text{for } S = U^\# \in \mathcal{S}_{\rm in}^{m\times m}. \tag{4.91}$$

If $J \neq \pm I_m$, $U \in \mathcal{U}(J)$ and $S = PG(U)$, then the description of $\mathcal{H}(U)$ will be formulated in terms of

$$L(\lambda) = (P_+ - S(\lambda)P_-)^{-1} \quad \text{for } \lambda \in \mathfrak{h}_U \cap \mathfrak{h}_S, \tag{4.92}$$

with the help of the formula

$$K_\omega^U(\lambda) = L(\lambda)K_\omega^S(\lambda)L(\omega)^* \quad \text{on } \Omega \times \Omega. \tag{4.93}$$

Theorem 4.45 *Let J be an $m \times m$ signature matrix, $U \in \mathcal{U}(J)$, $S = PG(U)$,*

$$G_\ell(\mu) = P_+ + U(\mu)P_-U(\mu)^* \quad a.e. \text{ on } \mathbb{R}$$

and let $L(\lambda)$ be defined by (4.92). Then the formula

$$f(\lambda) = L(\lambda)g(\lambda), \quad g \in \mathcal{H}(S) \tag{4.94}$$

defines a unitary operator, acting from $\mathcal{H}(S)$ onto $\mathcal{H}(U)$, i.e., $f \in \mathcal{H}(U)$ if and only if

$$(P_+ - SP_-)f \in H_2^m \quad and \quad (P_- - S^\#P_+)f \in (H_2^m)^\perp$$

and, if $f \in \mathcal{H}(U)$, then

$$\| f \|_{\mathcal{H}(U)}^2 = \| (P_+ - SP_-)f \|_{st}^2 = \|G_\ell^{-1/2} f\|_{st}^2.$$

Proof This follows from theorem 5.45 in [ArD08b]. □

4.7 de Branges' inclusion theorems

The next two theorems explore the connection between closed R_α invariant subspaces $\mathcal{L}$ of $\mathcal{H}(U)$ and the left divisors U_1 of $U \in \mathcal{U}(J)$.

Theorem 4.46 (L. de Branges) *Let $U \in \mathcal{U}(J)$ and let $\mathcal{L}$ be a closed subspace of $\mathcal{H}(U)$ that is R_α invariant for every point $\alpha \in \mathfrak{h}_U$. Then there exists an essentially unique mvf $U_1 \in \mathcal{U}(J)$ such that $\mathcal{L} = \mathcal{H}(U_1)$ and $U_1^{-1}U \in \mathcal{U}(J)$. Moreover, the space $\mathcal{H}(U_1)$ is isometrically included in $\mathcal{H}(U)$, and*

$$\mathcal{H}(U) = \mathcal{H}(U_1) \oplus U_1\mathcal{H}(U_2), \quad where \quad U_2 = U_1^{-1}U. \tag{4.95}$$

Proof See e.g., theorem 5.50 in [ArD08b]. □

Theorem 4.47 (L. de Branges) *If $U, U_1, U_2 \in \mathcal{U}(J)$ and $U = U_1U_2$, then $\mathcal{H}(U_1)$ sits contractively in $\mathcal{H}(U)$, i.e., $\mathcal{H}(U_1) \subseteq \mathcal{H}(U)$ (as linear spaces) and*

$$\|f\|_{\mathcal{H}(U)} \le \|f\|_{\mathcal{H}(U_1)} \quad \text{for every } f \in \mathcal{H}(U_1).$$

The inclusion is isometric if and only if

$$\mathcal{H}(U_1) \cap U_1\mathcal{H}(U_2) = \{0\}. \tag{4.96}$$

The condition (4.96) is in force if and only if

$$\mathcal{H}(U) = \mathcal{H}(U_1) \oplus U_1\mathcal{H}(U_2). \tag{4.97}$$

Proof See e.g., theorem 5.52 in [ArD08b]. □

Theorem 4.48 *If $U_1, U_2 \in \mathcal{U}(J)$ and $U = U_1U_2$, then $\mathfrak{h}_U = \mathfrak{h}_{U_1} \cap \mathfrak{h}_{U_2}$.*

Proof In view of Theorem 4.47, the vvf $f_{\xi,\omega} = K_\omega^{U_1}\xi$ belongs to $\mathcal{H}(U)$ for every $\xi \in \mathbb{C}^m$ and every $\omega \in \mathfrak{h}_{U_1}$. Therefore, $\mathfrak{h}_{f_{\xi,\omega}} \supseteq \mathfrak{h}_U$ by Theorem 4.38. This yields the inclusion $\mathfrak{h}_{U_1} \supseteq \mathfrak{h}_U$. Since $U \in \mathcal{U}(J_p) \Longrightarrow U^\tau \in \mathcal{U}(J_p)$, $U^\tau = U_2^\tau U_1^\tau$ and $\mathfrak{h}_U = \mathfrak{h}_{U^\tau}$, the same argument applied to U_2^τ yields the inclusion $\mathfrak{h}_{U_2} \supseteq \mathfrak{h}_U$. Consequently, $\mathfrak{h}_U \subseteq \mathfrak{h}_{U_1} \cap \mathfrak{h}_{U_2}$. Thus, as the opposite inclusion is self-evident, equality prevails. □

Corollary 4.49 *If $U_1, U_2 \in \mathcal{U}(J)$ and $U = U_1U_2$ is an entire mvf, then U_1 and U_2 are also entire mvf's.*

Theorem 4.50 *If $W \in \mathcal{U}(j_{pq})$, $\{b_1, b_2\} \in ap(W)$, $\widetilde{W} \in \mathcal{U}(j_{pq})$, and $\{\widetilde{b}_1, \widetilde{b}_2\} \in ap(\widetilde{W})$, then the following three statements are equivalent:*

(1) $\widetilde{W}^{-1}W \in \mathcal{U}(j_{pq})$.
(2) $\widetilde{b}_1^{-1}b_1 \in \mathcal{S}_{\text{in}}^{p\times p}$, $b_2\widetilde{b}_2^{-1} \in \mathcal{S}_{\text{in}}^{q\times q}$ *and* $T_{\widetilde{W}}[\mathcal{S}^{p\times q}] \supseteq T_W[\mathcal{S}^{p\times q}]$.
(3) $\mathcal{H}(\widetilde{W}) \subseteq \mathcal{H}(W)$ *and the inclusion is contractive.*

Proof The equivalence of (1) and (2) is covered by Theorem 3.64; the implication (1) $\Longrightarrow$ (3) follows from Theorem 4.47. Conversely, if (3) holds, then

$$\begin{aligned}\|K_\omega^{\widetilde{W}}u\|^2_{\mathcal{H}(\widetilde{W})} &= u^*K_\omega^{\widetilde{W}}(\omega)u = \langle K_\omega^{\widetilde{W}}u, K_\omega^{W}u\rangle_{\mathcal{H}(W)} \\ &\le \|K_\omega^{\widetilde{W}}u\|_{\mathcal{H}(W)}\,\|K_\omega^{W}u\|_{\mathcal{H}(W)} \le \|K_\omega^{\widetilde{W}}u\|_{\mathcal{H}(\widetilde{W})}\,\|K_\omega^{W}u\|_{\mathcal{H}(W)},\end{aligned}$$

which in turn leads easily to the bound

$$j_{pq} - \widetilde{W}(\omega)j_{pq}\widetilde{W}(\omega)^* \le j_{pq} - W(\omega)j_{pq}W(\omega)^* \quad \text{for } \omega \in \mathfrak{h}_W^+ \cap \mathfrak{h}_{\widetilde{W}}^+.$$

Therefore,

$$\widetilde{W}(\omega)^{-1}W(\omega)j_{pq}W(\omega)^*\widetilde{W}(\omega)^{-*} \le j_{pq}$$

for points $\omega \in \mathfrak{h}_W^+ \cap \mathfrak{h}_{\widetilde{W}}^+$ at which the indicated inverses exist, i.e., (3) $\Longrightarrow$ (1). □

Remark 4.51 *An analog of Theorem 4.50 for inner mvf's is: If $b, \widetilde{b} \in \mathcal{S}_{\text{in}}^{p\times p}$, then*

(1) $\widetilde{b}^{-1}b \in \mathcal{S}_{\text{in}}^{p\times p}$ *if and only* $\mathcal{H}(\widetilde{b}) \subseteq \mathcal{H}(b)$.
(2) $b\widetilde{b}^{-1} \in \mathcal{S}_{\text{in}}^{p\times p}$ *if and only* $\mathcal{H}_*(\widetilde{b}) \subseteq \mathcal{H}_*(b)$.

Theorem 4.52 *If $A \in \mathcal{U}(J_p)$, $\widetilde{A} \in \mathcal{U}(J_p)$, $\{b_3, b_4\} \in ap_{II}(A)$ and $\{\widetilde{b}_3, \widetilde{b}_4\} \in ap_{II}(A_1)$, then the following three statements are equivalent:*

(1) $\widetilde{A}^{-1}A \in \mathcal{U}(J_p)$.
(2) $\widetilde{b}_3^{-1}b_3 \in \mathcal{S}_{\text{in}}^{p\times p}$, $b_4\widetilde{b}_4^{-1} \in \mathcal{S}_{\text{in}}^{p\times p}$ *and* $\mathcal{C}(\widetilde{A}) \supseteq \mathcal{C}(A)$.
(3) $\mathcal{H}(\widetilde{A}) \subseteq \mathcal{H}(A)$ *and the inclusion is contractive.*

Proof The equivalence of (1) and (2) is covered by Theorem 3.77; the verification of (1) $\Longleftrightarrow$ (3) is similar to the verification of (1) $\Longleftrightarrow$ (3) in the proof of Theorem 4.50. □

We remark that the equivalence of (1) and (3) in Theorem 4.50 extends to mvf's in $\mathcal{U}(J)$ for $J = \pm I_m$.

4.8 A description of $\mathcal{H}(W) \cap L_2^m$

In this section we present a description of the space $\mathcal{H}(W) \cap L_2^m$ for $W \in \mathcal{U}(j_{pq})$ and then use it to characterize the classes $\mathcal{U}_S(J)$, $\mathcal{U}_{rR}(J)$ and $\mathcal{U}_{srR}(J)$ of singular, right regular and strongly right regular J-inner mvf's in terms of the linear space $\mathcal{L}_U = \mathcal{H}(U) \cap L_2^m$ and further to show that the inclusion $\mathcal{H}(\widetilde{U}) \subseteq \mathcal{H}(U)$ for left divisors $\widetilde{U}$ of U is isometric if $U \in \mathcal{U}_{rR}(J)$.

Since $\mathcal{H}(U) \subset L_2^m(\mathbb{R})$ if $J = \pm I_m$, only the case $J \neq \pm I_m$ is of interest and thus, it suffices to focus on $J = j_{pq}$ and $\mathcal{H}(W) \cap L_2^m(\mathbb{R})$ for $W \in \mathcal{U}(j_{pq})$.

For each $W \in \mathcal{U}(j_{pq})$, let the operators $X_{11} : H_2^q \to \mathcal{H}(b_1)$ and $X_{22} : \mathcal{H}_*(b_2) \to (H_2^p)^\perp$ be defined in terms of $\{b_1, b_2\} \in ap(W)$ and $s \in T_W\left[\mathcal{S}^{p\times q}\right]$ by the formulas

$$X_{11} = \Pi_{\mathcal{H}(b_1)} M_s|_{H_2^q}, \quad X_{22} = \Pi_- M_s|_{\mathcal{H}_*(b_2)}. \tag{4.98}$$

Theorem 4.53 *Let $W \in \mathcal{U}(j_{pq})$, $\{b_1, b_2\} \in ap(W)$, $s \in T_W[\mathcal{S}^{p\times q}]$ and let the operators X_{11} and X_{22} be defined by formula (4.98). Then:*

(1) *The operators X_{11} and X_{22} do not depend upon the choice of $s \in T_W[\mathcal{S}^{p\times q}]$.*
(2) *The subspace $\mathcal{L}_W = \mathcal{H}(W) \cap L_2^m$ is also given by the formula*

$$\mathcal{L}_W = \left\{ \begin{bmatrix} g \\ X_{11}^* g \end{bmatrix} + \begin{bmatrix} X_{22} h \\ h \end{bmatrix} : g \in \mathcal{H}(b_1) \quad \text{and} \quad h \in \mathcal{H}_*(b_2) \right\}. \tag{4.99}$$

and:

$$\|f\|_{\mathcal{H}(W)}^2 = \left\langle \begin{bmatrix} I_p & -s \\ -s^* & I_q \end{bmatrix} f, f \right\rangle_{\text{st}} \quad \text{for every } f \in \mathcal{L}_W. \tag{4.100}$$

Proof This is covered by theorem 5.81 in [ArD08b]. □

Theorem 4.54 *Let $U \in \mathcal{U}(J)$. Then:*

(1) *U admits a right regular-singular factorization*

$$U = U_1U_2 \quad \text{with} \quad U_1 \in \mathcal{U}_{rR}(J) \quad \text{and} \quad U_2 \in \mathcal{U}_S(J) \tag{4.101}$$

that is unique up to multiplication by a constant J-unitary factor V on the right of U_1 and V^* on the left of U_2.

(2) $\overline{\mathcal{L}_U} = \overline{\mathcal{L}_{U_1}} = \mathcal{H}(U_1)$.

(3) *$\mathcal{H}(U_1)$ is isometrically included in $\mathcal{H}(U)$.*

Proof If $J = \pm I_m$, then the theorem is obvious, since $\mathcal{H}(U)$ is a closed subspace of L_2^m and $\mathcal{U}_S(J) = U_{\text{const}}(J)$. If $J \neq \pm I_m$, then it follows from theorem 5.89, lemma 5.85 and corollary 5.87 in [ArD08b]. □

Theorem 4.55 *Let $U \in \mathcal{U}(J)$. Then:*

(1) $U \in \mathcal{U}_S(J) \iff \mathcal{H}(U) \cap L_2^m = \{0\}$.

(2) $U \in \mathcal{U}_{rR}(J) \iff \mathcal{H}(U) \cap L_2^m$ *is dense in* $\mathcal{H}(U)$.

(3) $U \in \mathcal{U}_{rsR}(J) \iff \mathcal{H}(U) \subset L_2^m$.

Moreover, the following are equivalent:

(a) $U \in \mathcal{U}_{rsR}(J)$.

(b) *There exist a pair of constants $\gamma_2 \geq \gamma_1 > 0$ such that*

$$\gamma_1 \|f\|_{\text{st}} \leq \|f\|_{\mathcal{H}(U)} \leq \gamma_2 \|f\|_{\text{st}} \quad \text{for every} \quad f \in \mathcal{H}(U). \tag{4.102}$$

(c) *$\mathcal{H}(U)$ is a closed subspace of L_2^m.*

(d) $\mathcal{H}(U) \subset L_2^m$.

Proof This follows from theorems 5.86 and 5.92 in [ArD08b]. □

Theorem 4.56 *If $U \in \mathcal{U}(J)$, $\widetilde{U} \in \mathcal{U}_{rR}(J)$ and $\widetilde{U}^{-1}U \in \mathcal{U}(J)$, then the inclusion $\mathcal{H}(\widetilde{U}) \subseteq \mathcal{H}(U)$ is isometric.*

Proof It suffices to verify the statement for $W \in \mathcal{U}(j_{pq})$, $\widetilde{W}^{-1}W \in \mathcal{U}(j_{pq})$ and $\widetilde{W} \in \mathcal{U}_{rR}(j_{pq})$. Let $\{b_1, b_2\} \in ap(W)$ and $\{\widetilde{b}_1, \widetilde{b}_2\} \in ap(\widetilde{W})$. Then, by Theorem 4.50,

$$\widetilde{b}_1^{-1}b_1 \in \mathcal{S}_{in}^{p\times p}, \quad b_2\widetilde{b}_2^{-1} \in \mathcal{S}_{in}^{q\times q} \quad \text{and} \quad T_W[\mathcal{S}^{p\times q}] \subseteq T_{\widetilde{W}}[\mathcal{S}^{p\times q}].$$

Therefore, Theorem 4.13 guarantees that

$$T_{\widetilde{W}}[\mathcal{S}^{p\times q}] = \mathcal{S}(\widetilde{b}_1, \widetilde{b}_2; s) \quad \text{for every } s \in T_W[\mathcal{S}^{p\times q}].$$

Now let X_{11} and X_{22} be the operators defined by formula (4.98) and let $\widetilde{X}_{11}$ and $\widetilde{X}_{22}$ be defined by the same formulas but with $\widetilde{b}_1$ in place of b_1 and $\widetilde{b}_2$ in place of

b_2 and let

$$\mathcal{L}_{\widetilde{W}} = \left\{ \begin{bmatrix} g \\ \widetilde{X}_{11}^* g \end{bmatrix} + \begin{bmatrix} \widetilde{X}_{22} h \\ h \end{bmatrix} : g \in \mathcal{H}(\widetilde{b}_1) \quad \text{and} \quad h \in \mathcal{H}_*(\widetilde{b}_2) \right\}.$$

Then, since $\mathcal{H}(\widetilde{b}_1) \subseteq \mathcal{H}(b_1)$, $\mathcal{H}_*(\widetilde{b}_2) \subseteq \mathcal{H}_*(b_2)$, it follows that

$$X_{11}^* g = \widetilde{X}_{11}^* g \text{ for } g \in \mathcal{H}(\widetilde{b}_1) \text{ and } X_{22} h = \widetilde{X}_{22} h \text{ for } h \in \mathcal{H}_*(\widetilde{b}_2).$$

Therefore, by Theorem 4.53,

$$\mathcal{H}(\widetilde{W}) \cap L_2^m = \mathcal{L}_{\widetilde{W}} \subseteq \mathcal{L}_W = \mathcal{H}(W) \cap L_2^m$$

and

$$\|f\|_{\mathcal{H}(\widetilde{W})}^2 = \left\langle \begin{bmatrix} I_p & -s \\ -s^* & I_q \end{bmatrix} f, f \right\rangle_{\text{st}} = \|f\|_{\mathcal{H}(W)}^2 \quad \text{for every } f \in \mathcal{L}_{\widetilde{W}}.$$

Thus, as the inclusion $\mathcal{H}(\widetilde{W}) \subseteq \mathcal{H}(W)$ is contractive by Theorem 4.47 and isometric on $\mathcal{L}_{\widetilde{W}}$ and $\mathcal{L}_{\widetilde{W}}$ is dense in $\mathcal{H}(\widetilde{W})$ by Theorem 4.55, the inclusion must be isometric. □

Theorem 4.57 *If $U_n \in \mathcal{U}_{rR}(J)$ for $n = 1, 2, \ldots$ is a monotone $\curvearrowright$ sequence such that*

$$U(\lambda) = \lim_{n \uparrow \infty} U_n(\lambda) \quad \textit{for every point } \lambda \in \bigcap_{n \geq 1} \mathfrak{h}_{U_n}^+$$

and $U \in \mathcal{P}^\circ(J)$, then $U \in \mathcal{U}_{rR}(J)$.

Proof Lemma 3.59 implies that the limit $U \in \mathcal{U}(J)$. Therefore, in view of Theorem 4.55, it remains only to show that if $f \in \mathcal{H}(U)$ is orthogonal to $\mathcal{H}(U) \cap L_2^m$, then $f = 0$. However, since $U_n^{-1} U \in \mathcal{U}(J)$ and $U_n \in \mathcal{U}_{rR}(J)$, Theorem 4.56 guarantees that the inclusions $\mathcal{H}(U_n) \subseteq \mathcal{H}(U)$ are isometric. Thus, if $f \in \mathcal{H}(U)$ is orthogonal to $\mathcal{H}(U) \cap L_2^m$, it is also orthogonal to $\mathcal{H}(U_n) \cap L_2^m$ and hence also to $\mathcal{H}(U_n)$ for every positive integer n, since $\mathcal{H}(U_n) \cap L_2^m$ is dense in $\mathcal{H}(U_n)$. Therefore,

$$\langle f, K_\omega^{U_n} \xi \rangle_{\mathcal{H}(U)} = 0 \quad \text{for every } \omega \in \mathfrak{h}_U \text{ and } \xi \in \mathbb{C}^m.$$

Moreover, as

$$\|K_w^U \xi - K_\omega^{U_n} \xi\|_{\mathcal{H}(U)}^2 = \xi^* K_\omega^U(\omega) \xi - \xi^* K_\omega^{U_n}(\omega) \xi \to 0 \quad \text{as } n \uparrow \infty$$

for $\omega \in \cap_{n \geq 1} \mathfrak{h}_{U_n}$, it follows that

$$\langle f, K_\omega^U \xi \rangle_{\mathcal{H}(U)} = \lim_{n \uparrow \infty} \langle f, K_\omega^{U_n} \xi \rangle_{\mathcal{H}(U)} = 0$$

for every $\omega \in \mathfrak{h}_U^+$ and $\xi \in \mathbb{C}^m$. Consequently, $f = 0$. □

4.9 The classes $\mathcal{U}_{AR}(J)$ and $\mathcal{U}_{BR}(J)$ of A-regular and B-regular J-inner mvf's

A mvf $U \in \mathcal{U}(J)$ will be said to belong to the class $\mathcal{U}_{AR}(J)$ of **A-regular J-inner mvf's** if every left divisor of U belongs to the class $\mathcal{U}_{rR}(J)$. This class has a number of different characterizations:

Lemma 4.58 *The following statements are equivalent:*

(1) $U \in \mathcal{U}_{AR}(J)$.
(2) *Every right divisor of* U *belongs to the class* $\mathcal{U}_{\ell R}(J)$.
(3) *If* $U = U_1U_2$ *with factors* $U_1, U_2 \in \mathcal{U}(J)$, *then* $U_1 \in \mathcal{U}_{rR}(J)$ *and* $U_2 \in \mathcal{U}_{\ell R}(J)$.
(4) $U^\sim \in \mathcal{U}_{AR}(J)$.
(5) *If* $U = U_1U_2U_3$ *with factors* $U_1, U_2, U_3 \in \mathcal{U}(J)$, *then* $U_i \in \mathcal{U}_S(J)$ *if and only* $U_i \in \mathcal{U}_{\rm const}(J)$.

Proof If U_2 is a right divisor of U, then $U = U_1U_2$ with $U_1, U_2 \in \mathcal{U}(J)$. Thus, if $U_2 = U_3U_4$ with $U_3 \in \mathcal{U}_S(J)$ and $U_4 \in \mathcal{U}(J)$, then the formula $U = U_1U_3U_4$ exhibits U_1U_3 as a left divisor of U. Therefore, $U_1U_3 \in \mathcal{U}_{rR}(J)$, by the definition of the class $\mathcal{U}_{AR}(J)$. Consequently, $U_3 \in \mathcal{U}_{\rm const}(J)$ and hence $U_2 \in \mathcal{U}_{\ell R}(J)$, i.e., (1) $\Longrightarrow$ (2). The remaining assertions may be verified in much the same way and are left to the reader. □

In view of Theorem 4.19 and Lemma 4.58,

$$\mathcal{U}_{rsR}(J) \cup \mathcal{U}_{\ell sR}(J) \subseteq \mathcal{U}_{AR}(J) \subseteq \mathcal{U}_{rR}(J) \cap \mathcal{U}_{\ell R}(J). \tag{4.103}$$

A mvf $U \in \mathcal{U}(J)$ will be said to belong to the class $\mathcal{U}_{BR}(J)$ of **B-regular J-inner mvf's** if every left divisor $U_1 \in \mathcal{U}(J)$ of U satisfies the de Branges condition

$$\mathcal{H}(U_1) \cap U_1\mathcal{H}(U_1^{-1}U) = \{0\}. \tag{4.104}$$

In view of Theorem 4.46, the condition in (4.104) holds if and only if the inclusion $\mathcal{H}(U_1) \subseteq \mathcal{H}(U)$ is isometric.

Lemma 4.59 *Let* $U \in \mathcal{U}(J)$ *and let* T *be the operator defined on* $\mathcal{H}(U)$ *by the formula*

$$(Tf)(\lambda) = U^\sim(\lambda)Jf(-\lambda) \quad \text{for } \lambda \in \mathfrak{h}_U \cap \mathfrak{h}_{U^\sim}. \tag{4.105}$$

Then T *is a unitary operator from* $\mathcal{H}(U)$ *onto* $\mathcal{H}(U^\sim)$.

Proof Let $\lambda, \omega \in \mathfrak{h}_{U^\sim}, -\lambda, -\omega \in \mathfrak{h}_U$ and suppose that $\lambda \neq \overline{\omega}$, $\det U^\sim(\lambda) \neq 0$ and $\det U^\sim(\omega) \neq 0$. Then

$$K_\omega^{U^\sim}(\lambda) = U^\sim(\lambda)JK_{-\omega}^U(-\lambda)JU^\sim(\omega)^*. \tag{4.106}$$

Therefore, the operator T maps the dense subspace $\mathcal{L}_1$ of vvf's $f \in \mathcal{H}(U)$ of the form

$$f(\lambda) = \sum_{j=1}^{n} K_{-\omega_j}^{U}(\lambda)JU^{\sim}(\omega_j)^*\xi_j \quad \text{with } \xi_j \in \mathbb{C}^m \text{ and } n \geq 1 \tag{4.107}$$

into the dense subspace $\mathcal{L}_2$ of vvf's

$$\begin{aligned} g(\lambda) = (Tf)(\lambda) &= U^{\sim}(\lambda)J \sum_{j=1}^{n} K_{-\omega_j}^{U}(-\lambda)JU^{\sim}(\omega_j)^*\xi_j \\ &= \sum_{j=1}^{n} K_{\omega_j}^{U^{\sim}}(\lambda)\xi_j \quad \text{with } \xi_j \in \mathbb{C}^m \text{ and } n \geq 1. \end{aligned} \tag{4.108}$$

Moreover, if f and g are defined by the above formulas, then

$$\begin{aligned} \langle f, f\rangle_{\mathcal{H}(U)} &= \sum_{j=1}^{n}\sum_{k=1}^{n} \xi_j^* U^{\sim}(\omega_j)JK_{-\omega_k}^{U}(-\omega_j)JU^{\sim}(\omega_k)^*\xi_k \\ &= \sum_{j=1}^{n}\sum_{k=1}^{n} \xi_j^* K_{\omega_k}^{U^{\sim}}(\omega_j)\xi_k = \langle g, g\rangle_{\mathcal{H}(U^{\sim})}. \end{aligned} \tag{4.109}$$

Thus, T maps $\mathcal{L}_1$ isometrically onto $\mathcal{L}_2$. Moreover, if $f \in \mathcal{H}(U)$, then there exists a sequence of vvf's $f_k \in \mathcal{L}_1$ such that $\|f - f_k\|_{\mathcal{H}(U)} \to 0$ as $k \uparrow \infty$. But, as $\mathcal{H}(U)$ is a RKHS, this implies that $f_k(\lambda) \to f(\lambda)$ at each point $\lambda \in \mathfrak{h}_U$ as $k \uparrow \infty$. Thus, if $g_k = Tf_k$ for $k = 1, 2, \ldots$, then

$$g_k(\lambda) = (Tf_k)(\lambda) = U^{\sim}(\lambda)Jf_k(-\lambda) \to U^{\sim}(\lambda)Jf(-\lambda) \quad \text{as } k \uparrow \infty$$

for each point $\lambda \in \mathfrak{h}_{U^{\sim}}$ such that $-\lambda \in \mathfrak{h}_U$. Since

$$\|g_k\|_{\mathcal{H}(U^{\sim})} = \|f_k\|_{\mathcal{H}(U)} \to \|f\|_{\mathcal{H}(U)} \quad \text{as } k \uparrow \infty$$

and

$$\|g_k - g_j\|_{\mathcal{H}(U^{\sim})} = \|f_k - f_j\|_{\mathcal{H}(U)},$$

there exists a vvf $g \in \mathcal{H}(U^{\sim})$ such that $\|g_k - g\|_{\mathcal{H}(U^{\sim})} \to 0$ as $k \uparrow \infty$. Therefore, since $\mathcal{H}(U^{\sim})$ is a RKHS and

$$\mathfrak{h}_{U^{\sim}} = \bigcap_{g \in \mathcal{H}(U^{\sim})} \mathfrak{h}_g,$$

$g_k(\lambda) \to g(\lambda)$ at each point $\lambda \in \mathfrak{h}_{U^{\sim}}$ as $k \uparrow \infty$. Consequently,

$$g(\lambda) = U^{\sim}(\lambda)Jf(-\lambda) = (Tf)(\lambda) \quad \text{for } f \in \mathcal{H}(U),$$

i.e., T maps $\mathcal{H}(U)$ into $\mathcal{H}(U^{\sim})$. Therefore, since T is an isometry on the full space $\mathcal{H}(U)$ and $\mathcal{L}_2$ is dense in $\mathcal{H}(U^{\sim})$, T maps $\mathcal{H}(U)$ onto $\mathcal{H}(U^{\sim})$. □

Theorem 4.60 *If T is the operator defined on $\mathcal{H}(U)$ by formula (4.105), then*

$$\mathcal{U}_{\ell sR}(J) = \{U \in \mathcal{U}(J) : Tf \in L_2^m(\mathbb{R}) \quad \text{for every } f \in \mathcal{H}(U)\}. \tag{4.110}$$

Proof This follows from Lemma 4.59 and formulas (1.26) and (4.45). □

Remark 4.61 *Since $g(\mu)$ belongs to $L_2^m(\mathbb{R})$ if and only if $g(-\mu)$ belongs to $L_2^m(\mathbb{R})$, the equality (4.110) is equivalent to the following equality*

$$\begin{aligned} \mathcal{U}_{\ell sR}(J) &= \{U \in \mathcal{U}(J) : U^{\#}Jf \in L_2^m(\mathbb{R}) \text{ for every } f \in \mathcal{H}(U)\} \\ &= \{U \in \mathcal{U}(J) : U^{-1}f \in L_2^m(\mathbb{R}) \text{ for every } f \in \mathcal{H}(U)\}. \end{aligned} \tag{4.111}$$

Theorem 4.62 $U \in \mathcal{U}_{BR}(J) \Longleftrightarrow U^{\sim} \in \mathcal{U}_{BR}(J)$.

Proof If $U \in \mathcal{U}_{BR}(J)$ and $U = U_1U_2$ is a factorization of U with factors $U_1,\ U_2 \in \mathcal{U}(J)$, then

$$U^{\sim} = U_2^{\sim}U_1^{\sim}.$$

Let $f \in \mathcal{H}(U_2^{\sim}) \cap U_2^{\sim}\mathcal{H}(U_1^{\sim})$. Then, by Lemma 4.59,

$$f(\lambda) = U_2^{\sim}(\lambda)Jf_2(-\lambda) = U_2^{\sim}(\lambda)U_1^{\sim}(\lambda)Jf_1(-\lambda),$$

where $f_j \in \mathcal{H}(U_j)$ for $j = 1, 2$. Therefore,

$$Jf_2(-\lambda) = U_1^{\sim}(\lambda)Jf_1(-\lambda),$$

i.e.,

$$f_2(\lambda) = JU_1^{\#}(\lambda)Jf_1(\lambda) = U_1(\lambda)^{-1}f_1(\lambda).$$

Thus,

$$f_1 = U_1f_2,$$

and hence

$$f_1 \in \mathcal{H}(U_1) \cap U_1\mathcal{H}(U_2) = \{0\}.$$

Consequently, $f = 0$, i.e.,

$$\mathcal{H}(U_1) \cap U_1\mathcal{H}(U_2) = \{0\} \Longrightarrow \mathcal{H}(U_2^{\sim}) \cap U_2^{\sim}\mathcal{H}(U_1^{\sim}) = \{0\}.$$

The converse implication then follows from the fact that $(f^{\sim})^{\sim} = f$. □

Theorem 4.63 *If $J \neq \pm I_m$, then*

$$\mathcal{U}_{\ell sR} \cup \mathcal{U}_{rsR}(J) \subseteq \mathcal{U}_{AR}(J) \subseteq \mathcal{U}_{BR}(J). \tag{4.112}$$

Proof If $U \in \mathcal{U}_{rsR}(J)$ and $U_1 \in \mathcal{U}(J)$ is a left divisor of U, then Theorem 4.19 guarantees that $U_1 \in \mathcal{U}_{rR}(J)$. On the other hand, if $U \in \mathcal{U}_{\ell sR}(J)$ and $U_1 \in \mathcal{U}(J)$ is a left divisor of U, then $U^{\sim} \in \mathcal{U}_{rsR}(J)$, $U_1^{\sim} \in \mathcal{U}(J)$ and $U_1^{\sim}$ is a right divisor of $U^\sim$. Therefore, by another application of Theorem 4.19, $U_1^\sim \in \mathcal{U}_{\ell R}(J)$ and hence $U_1 \in \mathcal{U}_{rR}(J)$. Therefore, the first inclusion holds. The second follows from Theorem 4.56. □

Theorem 4.64 *If $U_1, \ldots, U_n \in \mathcal{U}(J)$ and $U = U_1 \cdots U_n$, then*

$$U \in \mathcal{U}_{BR}(J) \Longrightarrow U_k \in \mathcal{U}_{BR}(J) \quad \text{for } k = 1, \ldots, n.$$

Proof It suffices to consider the case $n = 2$. Then if $U = U_1U_2$, $U_1 = U_aU_b$, with $U_a, U_b, U_2 \in \mathcal{U}(J)$ and $U \in \mathcal{U}_{BR}(J)$, the two factorizations $U = U_1U_2$ and $U = U_a(U_bU_2)$ imply that

$$\|f\|_{\mathcal{H}(U_1)} = \|f\|_{\mathcal{H}(U)} \quad \text{for every } f \in \mathcal{H}(U_1)$$

and

$$\|f\|_{\mathcal{H}(U_a)} = \|f\|_{\mathcal{H}(U)} \quad \text{for every } f \in \mathcal{H}(U_a),$$

respectively. Therefore,

$$\|f\|_{\mathcal{H}(U_a)} = \|f\|_{\mathcal{H}(U_1)} \quad \text{for every } f \in \mathcal{H}(U_a),$$

which proves that $U_1 \in \mathcal{U}_{BR}(J)$.

The proof that $U_2 \in \mathcal{U}_{BR}(J)$ follows from formula $U^\sim = U_2^\sim U_1^\sim$ and Theorem 4.62. □

4.10 de Branges matrices $\mathfrak{E}$ and de Branges spaces $\mathcal{B}(\mathfrak{E})$

In this section we summarize a number of results from the theory of de Branges matrices $\mathfrak{E}(\lambda)$ and de Branges spaces $\mathcal{B}(\mathfrak{E})$ in the special case that $\mathfrak{E}$ is an entire $p \times 2p$ mvf and hence $\mathcal{B}(\mathfrak{E})$ is a RKHS of entire $p \times 1$ vvf's. These results are adapted from the treatment on pp. 295–305 of [ArD08b] (which deals with the more general setting of meromorphic de Branges matrices) and are presented without proof.

The mvf

$$\mathfrak{E}(\lambda) = [E_-(\lambda) \quad E_+(\lambda)], \tag{4.113}$$

with $p \times p$ blocks $E_\pm$ that are entire mvf's and meet the conditions

$$\det E_+(\lambda) \not\equiv 0 \quad \text{and} \quad \chi \stackrel{\text{def}}{=} E_+^{-1}E_- \in \mathcal{S}_{\text{in}}^{p\times p} \tag{4.114}$$

will be called an entire **de Branges matrix**. In view of (4.114),

$$E_+(\lambda)E_+^\#(\lambda) \equiv E_-(\lambda)E_-^\#(\lambda) \quad \text{in } \mathbb{C}. \tag{4.115}$$

If $\mathfrak{E}$ is an entire de Branges matrix, then the kernel

$$K_\omega^{\mathfrak{E}}(\lambda) = \begin{cases} \dfrac{E_+(\lambda)E_+(\omega)^* - E_-(\lambda)E_-(\omega)^*}{\rho_\omega(\lambda)} & \text{if } \lambda \neq \overline{\omega} \\ -\dfrac{1}{2\pi i}\{E_+'(\overline{\omega})E_+(\omega)^* - E_-'(\overline{\omega})E_-(\omega)^*\} & \text{if } \lambda = \overline{\omega}. \end{cases} \tag{4.116}$$

is positive on $\mathbb{C} \times \mathbb{C}$, since $\chi \in \mathcal{S}_{\text{in}}^{p\times p}$ and

$$K_\omega^{\mathfrak{E}}(\lambda) = E_+(\lambda)k_\omega^\chi(\lambda)E_+(\omega)^* \quad \text{on } \mathfrak{h}_\chi \times \mathfrak{h}_\chi. \tag{4.117}$$

Therefore, by Theorem 4.36, there is exactly one RKHS $\mathcal{B}(\mathfrak{E})$ with RK $K_\omega^{\mathfrak{E}}(\lambda)$ associated with each de Branges matrix $\mathfrak{E}$; it will be called a **de Branges space**. Moreover, since the kernel $K_\omega^{\mathfrak{E}}(\lambda)$ is an entire function of λ for every fixed $\omega \in \mathbb{C}$ and $K_\omega^{\mathfrak{E}}(\omega)$ is continuous on $\mathbb{C}$, Lemma 4.37 guarantees that every vvf $f \in \mathcal{B}(\mathfrak{E})$ is entire.

Theorem 4.65 *If an entire de Branges matrix $\mathfrak{E} = [E_- \quad E_+]$ belongs to $\Pi^{p\times 2p}$ and $\chi = E_+^{-1}E_-$, then:*

(1) $f \in \mathcal{B}(\mathfrak{E}) \Longleftrightarrow E_+^{-1}f \in \mathcal{H}(\chi) \Longleftrightarrow E_-^{-1}f \in \mathcal{H}_*(\chi)$.

(2) $f \in \mathcal{E} \cap \Pi^p$ *for every* $f \in \mathcal{B}(\mathfrak{E})$, *i.e.,*

$$\mathcal{B}(\mathfrak{E}) \subset \mathcal{E} \cap \Pi^p. \tag{4.118}$$

(3) *If* $f \in \mathcal{B}(\mathfrak{E})$, *then*

$$\|f\|_{\mathcal{B}(\mathfrak{E})}^2 = \|E_+^{-1}f\|_{\text{st}}^2 = \int_{-\infty}^{\infty} f(\mu)^*\Delta_{\mathfrak{E}}(\mu)f(\mu)d\mu, \tag{4.119}$$

where

$$\Delta_{\mathfrak{E}}(\mu) = E_+(\mu)^{-*}E_+(\mu)^{-1} = E_-(\mu)^{-*}E_-(\mu)^{-1} \tag{4.120}$$

is well defined on $\mathbb{R}$, except at the real zeros of $\det E(\lambda)$.

Remark 4.66 *In view of formula (4.117), the mapping*

$$g_+ \in \mathcal{H}(\chi) \longrightarrow E_+g_+ \tag{4.121}$$

defines a unitary operator from $\mathcal{H}(\chi)$ onto $\mathcal{B}(\mathfrak{E})$ and the mapping

$$g_- \in \mathcal{H}_*(\chi) \longrightarrow E_-g_- \tag{4.122}$$

defines a unitary operator from $\mathcal{H}_*(\chi)$ onto $\mathcal{B}(\mathfrak{E})$.

Lemma 4.67 *Let $\mathfrak{E} = [E_- \quad E_+]$ be an entire de Branges matrix and let $\chi = E_+^{-1}E_-$. Then the following conditions are equivalent:*

(1) *The inequality $K_\omega^{\mathfrak{E}}(\omega) > 0$ holds for at least one point $\omega \in \mathbb{C}_+$.*

(2) *The inequality $K_\omega^{\mathfrak{E}}(\omega) > 0$ holds for every point $\omega \in \mathbb{C}_+$.*

(3) *The equality $\{f(\omega) : f \in \mathcal{B}(\mathfrak{E})\} = \mathbb{C}^p$ holds for at least one point $\omega \in \mathbb{C}_+$.*

(4) *The equality* $\{f(\omega) : f \in \mathcal{B}(\mathfrak{E})\} = \mathbb{C}^p$ *holds for every point* $\omega \in \mathbb{C}_+$.
(5) *The inequality* $k_\omega^\chi(\omega) > 0$ *holds for at least one point* $\omega \in \mathfrak{h}_\chi$.
(6) *The inequality* $k_\omega^\chi(\omega) > 0$ *holds for every point* $\omega \in \mathfrak{h}_\chi$.
(7) *The equality* $\{f(\omega) : f \in \mathcal{H}(\chi)\} = \mathbb{C}^p$ *holds for at least one point* $\omega \in \mathfrak{h}_\chi$.
(8) *The equality* $\{f(\omega) : f \in \mathcal{H}(\chi)\} = \mathbb{C}^p$ *holds for every point* $\omega \in \mathfrak{h}_\chi$.

Moreover, if $\mathfrak{E} \in \Pi^{p\times 2p}$ *(and hence if* $\mathfrak{E} = \sqrt{2}N_2^*A\mathfrak{V}$ *for some* $A \in \mathcal{E} \cap \mathcal{U}(J_p)$*), then the equivalences in (1)–(4) hold with* $\mathbb{C}$ *in place of* $\mathbb{C}_+$.

Regular de Branges matrices $\mathfrak{E}$ and spaces $\mathcal{B}(\mathfrak{E})$

Let $\mathfrak{E}$ be an entire de Branges matrix. Then the space $\mathcal{B}(\mathfrak{E})$ will be called a **regular de Branges space** if it is R_α invariant for every point $\alpha \in \mathbb{C}_+$; $\mathfrak{E}$ will be called a **regular de Branges matrix** if

$$\rho_\alpha^{-1}E_+^{-1} \in H_2^{p\times p} \quad \text{and} \quad \rho_{\overline{\alpha}}^{-1}E_-^{-1} \in (H_2^{p\times p})^\perp \tag{4.123}$$

for at least one (and hence every) point $\alpha \in \mathbb{C}_+$. Since $E_- = \chi E_+$, the constraints in (4.123) guarantee that $\mathfrak{E} \in \Pi^{p\times 2p}$ and hence, by Theorem 3.30, that $\mathfrak{E}$ is an entire mvf of exponential type.

Lemma 4.68 *If* $\mathfrak{E} = [E_- \quad E_+]$ *is an entire de Branges matrix, then the following implications are in force:*

(a) $\mathfrak{E}$ *is a regular de Branges matrix* $\Longrightarrow$
(b) $\mathcal{B}(\mathfrak{E})$ *is a regular de Branges space* $\Longrightarrow$
(c) $\mathcal{B}(\mathfrak{E})$ *is* R_α *invariant for at least one point* $\alpha \in \mathbb{C}_+$;
i.e., (a) $\Longrightarrow$ (b) $\Longrightarrow$ (c). *If*

$$K_\omega^{\mathfrak{E}}(\omega) > 0 \quad \textit{for at least one (and hence every) point } \omega \in \mathbb{C}_+, \tag{4.124}$$

then (c) $\Longrightarrow$ (a) *and hence* (a) $\Longleftrightarrow$ (b) $\Longleftrightarrow$ (c). *Moreover, if* $\mathfrak{E}(\lambda)$ *is an entire regular de Branges matrix, then* $\mathcal{B}(\mathfrak{E})$ *is a RKHS of* $p \times 1$ *entire vvf's, its RK,* $K_\omega^{\mathfrak{E}}(\lambda)$*, is defined on* $\mathbb{C} \times \mathbb{C}$ *by formula (4.116) and* $\mathcal{B}(\mathfrak{E})$ *is* R_α *invariant for every point* $\alpha \in \mathfrak{h}_\chi$.

Theorem 4.69 *Let* $\mathfrak{E} = [E_- \quad E_+]$ *be an entire de Branges matrix, let* $\chi = E_+^{-1}E_-$ *and suppose that*

$$E_+(0) = E_-(0) = I_p. \tag{4.125}$$

Then:

(1) $\mathcal{B}(\mathfrak{E})$ *is a regular de Branges space if and only if it is* R_0 *invariant.*
(2) $\mathfrak{E}$ *is a regular de Branges matrix if and only if* $\mathfrak{E} \in \Pi^{p\times 2p}$ *and* $R_0E_+\xi \in \mathcal{B}(\mathfrak{E})$ *and* $R_0E_-\xi \in \mathcal{B}(\mathfrak{E})$ *for every* $\xi \in \mathbb{C}^p$.
(3) *If* $\mathfrak{E}$ *is a regular de Branges matrix, then* $\mathcal{B}(\mathfrak{E})$ *is* R_0 *invariant.*

(4) *If $\mathcal{B}(\mathfrak{E})$ is R_0 invariant and $-i\chi'(0) > 0$, then $\mathfrak{E}$ is a regular de Branges matrix,*
(5) $-i\chi'(0) > 0$ *if and only if each of the eight equivalent conditions in Lemma 4.67 is in force.*

Connections between mvf's $A \in \mathcal{E} \cap \mathcal{U}(J_p)$ and entire de Branges matrices $\mathfrak{E}$

If $A \in \mathcal{E} \cap \mathcal{U}(J_p)$, then Lemma 3.66 guarantees that

$$\begin{aligned}\mathfrak{E}(\lambda) = [E_-(\lambda) \quad E_+(\lambda)] &= \sqrt{2}N_2^* A(\lambda)\mathfrak{V} \\ &= [a_{22}(\lambda) - a_{21}(\lambda) \quad a_{22}(\lambda) + a_{21}(\lambda)]\end{aligned} \tag{4.126}$$

is an entire regular de Branges matrix. Moreover,

$$K_\omega^{\mathfrak{E}}(\lambda) = 2N_2^* K_\omega^A(\lambda) N_2 \quad \text{for } \lambda, \omega \in \mathbb{C}, \tag{4.127}$$

$\det E_\pm(\mu) \neq 0$ on $\mathbb{R}$,

$$\Delta_{\mathfrak{E}}(\mu) \stackrel{\text{def}}{=} E_+(\mu)^{-*}E_+(\mu)^{-1} = E_-(\mu)^{-*}E_-(\mu)^{-1} \quad \text{on } \mathbb{R} \tag{4.128}$$

and

$$\Delta_{\mathfrak{E}} \in \widetilde{L}_1^{p\times p}. \tag{4.129}$$

A mvf $A \in \mathcal{E} \cap \mathcal{U}(J_p)$ is called **perfect** if the mvf $c_0 = T_A[I_p]$ belongs to the class $\mathcal{C}_a^{p\times p}$, i.e., if the matrix $\beta_{c_0} = 0$ in the representation formula (1.65) for $c_0(\lambda)$. (In this case the spectral function $\sigma_{c_0}(\mu)$ in that formula is automatically locally absolutely continuous.) This condition is equivalent to the condition

$$\Re c_0(i) = \frac{1}{\pi}\int_{-\infty}^{\infty} \frac{\Re c_0(\mu)}{1+\mu^2} d\mu. \tag{4.130}$$

Additional information on the connections between entire regular de Branges matrices $\mathfrak{E}(\lambda)$ and perfect mvf's $A \in \mathcal{E} \cap \mathcal{U}(J_p)$ is provided by the next two theorems, which follow from the more general formulations (that do not assume that $\mathfrak{E}$ and A are entire) on pp. 302–305 in [ArD08b].

Theorem 4.70 *If $A \in \mathcal{E} \cap \mathcal{U}(J_p)$, $\mathfrak{E}$ is given by (4.126) and $c = T_A[I_p]$, then:*

(1) *$\mathfrak{E}$ is a regular de Branges matrix and $c \in \Pi \cap \mathcal{C}^{p\times p}$ is a meromorphic mvf that is holomorphic in $\overline{\mathbb{C}_+}$.*
(2) *The mvf A can be recovered from $\mathfrak{E}$ and c by the formula*

$$A = \frac{1}{\sqrt{2}}\begin{bmatrix} -c^\# E_- & cE_+ \\ E_- & E_+ \end{bmatrix}\mathfrak{V}. \tag{4.131}$$

(3) *The mvf c admits a unique decomposition of the form*

$$c = c_s + c_a \tag{4.132}$$

with components $c_s = -i\beta\lambda$ for some $\beta \in \mathbb{C}^{p\times p}$ that is nonnegative and

$$c_a(\lambda) = i\alpha + \frac{1}{\pi i}\int_{-\infty}^{\infty}\left\{\frac{1}{\mu-\lambda} - \frac{\mu}{1+\mu^2}\right\}E_+(\mu)^{-*}E_+(\mu)^{-1}d\mu \quad \text{for } \lambda \in \mathbb{C}_+ \quad (4.133)$$

for some Hermitian matrix $\alpha \in \mathbb{C}^{p\times p}$.

(4) *The given mvf $A \in \mathcal{E}\cap\mathcal{U}(J_p)$ admits a factorization of the form*

$$A(\lambda) = A_s(\lambda)A_a(\lambda), \quad (4.134)$$

where

$$A_s(\lambda) = \begin{bmatrix} I_p & -i\beta\lambda \\ 0 & I_p \end{bmatrix} \quad (4.135)$$

and

$$A_a(\lambda) = \frac{1}{\sqrt{2}}\begin{bmatrix} -c_a^{\#}(\lambda)E_-(\lambda) & c_a(\lambda)E_+(\lambda) \\ E_-(\lambda) & E_+(\lambda) \end{bmatrix}\mathfrak{V}. \quad (4.136)$$

Moreover, $A_s \in \mathcal{E}\cap\mathcal{U}_S(J_p)$, $A_a \in \mathcal{E}\cap\mathcal{U}(J_p)$ and the mvf's c_a and A_a are uniquely determined by $\mathfrak{E}$ up to an additive constant $i\alpha$ in (4.133) and a corresponding left constant multiplicative factor in (4.136) that is of the form

$$\begin{bmatrix} I_p & i\alpha \\ 0 & I_p \end{bmatrix} \quad \text{with} \quad \alpha = \alpha^* \in \mathbb{C}^{p\times p}. \quad (4.137)$$

Conversely, if $\mathfrak{E} = [E_- \quad E_+]$ is an entire regular de Branges matrix and if $c = c_s + c_a$, where $c_s(\lambda) = -i\beta\lambda$ for some $\beta \in \mathbb{C}^{p\times p}$, $\beta \geq 0$ c_a is defined by formula (4.133), then $c_a \in \Pi^{p\times p}$; the mvf A that is defined by formulas (4.131)–(4.133) belongs to the class $\mathcal{E}\cap\mathcal{U}(J_p)$; and (4.126) holds.

Theorem 4.71 *Let $\mathfrak{E} = [E_- \quad E_+]$ be an entire regular de Branges matrix and let $\Delta_{\mathfrak{E}}(\mu) = E_+(\mu)^{-*}E_+(\mu)^{-1}$ on $\mathbb{R}$. Then:*

(1) *There exists a perfect mvf $A \in \mathcal{E}\cap\mathcal{U}(J_p)$ such that (4.126) holds. Moreover, A is uniquely defined by $\mathfrak{E}$ up to a left constant J_p-unitary factor of the form (4.137) by formulas (4.133) and (4.136). There is only one such perfect mvf A_a for which $c_a(i) > 0$, i.e., for which $\alpha = 0$ in (4.133).*

(2) *If $\mathfrak{E}$ satisfies the conditions (4.125), then there is exactly one perfect mvf $A \in \mathcal{E}\cap\mathcal{U}^\circ(J_p)$ for which (4.126) holds. It is given by formula (4.131), where*

$$c(\lambda) = I_p + \frac{\lambda}{\pi i}\int_{-\infty}^{\infty}\frac{1}{\mu(\mu-\lambda)}\{\Delta_{\mathfrak{E}}(\mu) - I_p\}d\mu. \quad (4.138)$$

4.11 A coisometry from $\mathcal{H}(A)$ onto $\mathcal{B}(\mathfrak{E})$

The next theorem shows that if $\mathfrak{E}(\lambda)$ is a regular de Branges matrix that is related to an entire mvf $A \in \mathcal{U}(J_p)$ by (4.126), then there is a simple formula that defines a coisometric map from $\mathcal{H}(A)$ onto $\mathcal{B}(\mathfrak{E})$.

Theorem 4.72 *Let $A \in \mathcal{E} \cap \mathcal{U}(J_p)$ and let $\mathfrak{E}$, c_a, c_s, A_a and A_s be defined by A as in Theorem 4.70. Let U_2 denote the operator that is defined on $\mathcal{H}(A)$ by the formula*

$$(U_2 f)(\lambda) = \sqrt{2}[0 \quad I_p]f(\lambda) \quad for \quad f \in \mathcal{H}(A). \tag{4.139}$$

Then:

(1) $\mathcal{H}(A_s) = \{f \in \mathcal{H}(A) : (U_2 f)(\lambda) \equiv 0\}$. *Moreover,*

$$\mathcal{H}(A_s) = \ker U_2 = \left\{ \begin{bmatrix} \beta u \\ 0 \end{bmatrix} : u \in \mathbb{C}^p \right\} \tag{4.140}$$

with inner product

$$\left\langle \begin{bmatrix} \beta u \\ 0 \end{bmatrix}, \begin{bmatrix} \beta v \\ 0 \end{bmatrix} \right\rangle_{\mathcal{H}(A_s)} = 2\pi v^* \beta u.$$

(2) *The orthogonal complement of $\mathcal{H}(A_s)$ in $\mathcal{H}(A)$ is equal to $A_s\mathcal{H}(A_a)$, i.e.,*

$$\mathcal{H}(A) = \mathcal{H}(A_s) \oplus A_s\mathcal{H}(A_a). \tag{4.141}$$

(3) *The operator U_2 is a partial isometry from $\mathcal{H}(A)$ onto $\mathcal{B}(\mathfrak{E})$ with kernel $\mathcal{H}(A_s)$, i.e., U_2 maps $\mathcal{H}(A) \ominus \mathcal{H}(A_s)$ isometrically onto $\mathcal{B}(\mathfrak{E})$.*

(4) *The operator U_2 is unitary from $\mathcal{H}(A)$ onto $\mathcal{B}(\mathfrak{E})$ if and only if the mvf A is perfect, i.e., if and only if $\beta_{c_0} = 0$, where $c_0 = T_A[I_p]$.*

Proof This follows from theorem 5.76 in [ArD08b]. □

Theorem 4.73 *Let $\mathfrak{E} = [E_- \quad E_+]$ be an entire regular de Branges matrix such that (4.125) holds and let $A(\lambda)$ be the unique perfect matrix in $\mathcal{U}^\circ(J_p)$ such that (4.126) holds. Let $\Delta_{\mathfrak{E}}(\mu)$ be defined by formula (4.128),*

$$G_\pm(\lambda) = (R_0 E_\pm)(\lambda) \quad and \quad G = [G_+ + G_- \quad G_+ - G_-]. \tag{4.142}$$

Then $G_\pm \xi \in \mathcal{B}(\mathfrak{E})$ for every $\xi \in \mathbb{C}^p$ and the adjoint U_2^* of the unitary operator U_2 is given by the formula

$$(U_2^* g)(\lambda) = \frac{1}{\sqrt{2}2\pi i} \int_{-\infty}^{\infty} G(\mu)^* \Delta_{\mathfrak{E}}(\mu) \frac{\lambda g(\lambda) - \mu g(\mu)}{\lambda - \mu} d\mu. \tag{4.143}$$

Proof See theorem 6.15 in [ArD08b]. □

4.12 Formulas for resolvent matrices $W \in \mathcal{E} \cap \mathcal{U}^{\circ}_{rsR}(j_{pq})$

In view of Theorem 4.20, there exists exactly one resolvent matrix $W \in \mathcal{E} \cap \mathcal{U}^{\circ}_{rsR}(j_{pq})$ for each strictly completely indeterminate GSIP$(b_1, b_2; s^{\circ})$ with $b_1 \in \mathcal{E} \cap \mathcal{S}^{p\times p}_{\text{in}}$, $b_2 \in \mathcal{E} \cap \mathcal{S}^{q\times q}_{\text{in}}$ and $\{b_1, b_2\} \in ap(W)$. In this section we shall present a formula for this resolvent matrix in terms of b_1, b_2 and the block triangular operator X of the form (4.25) for which

$$\mathcal{S}(b_1, b_2; s^{\circ}) = \mathcal{S}(b_1, b_2; X),$$

the set defined in (4.29). Moreover, by Theorem 4.10, the GSIP$(b_1, b_2; s^{\circ})$ is strictly completely indeterminate if and only if $\|X\| < 1$.

Let

$$\mathcal{H}(b_1, b_2) = \begin{matrix} \mathcal{H}(b_1) \\ \oplus \\ \mathcal{H}_*(b_2) \end{matrix} .$$

The condition $\|X\| < 1$ is equivalent to the condition that the bounded linear operator

$$\Delta_X = \begin{bmatrix} I - X_{11}X_{11}^* & -X_{12} \\ -X_{12}^* & I - X_{22}^*X_{22} \end{bmatrix} : \mathcal{H}(b_1, b_2) \to \mathcal{H}(b_1, b_2) \tag{4.144}$$

is strictly positive, i.e.,

$$\|X\| < 1 \Longleftrightarrow \Delta_X > \varepsilon I \quad \text{for some } \varepsilon > 0. \tag{4.145}$$

To verify this, first use Schur complements to verify that

$$\|X\| < 1 \Longleftrightarrow \begin{bmatrix} I - X_{11}X_{11}^* - X_{12}(I - X_{22}^*X_{22})^{-1}X_{12}^* & 0 \\ 0 & I - X_{22}X_{22}^* \end{bmatrix} > \varepsilon I$$

for some $\varepsilon > 0$, and then replace the 22 block entry by $I - X_{22}^*X_{22}$ and compare the resulting operator matrix with the corresponding Schur complement formula for Δ_X. If $\|X\| = 1$, then applying (4.145) to ρX with $0 < \rho < 1$ and then letting $\rho \uparrow 1$ yields the supplementary equivalence

$$\|X\| \leq 1 \Longleftrightarrow \Delta_X \geq 0. \tag{4.146}$$

Theorem 4.74 *Let $W \in \mathcal{E} \cap \mathcal{U}^{\circ}_{rsR}(j_{pq})$, $\{b_1, b_2\} \in ap(W)$ and $s_{12} = T_W[0]$. Let the operators X_{11}, X_{22} and X_{12} be defined by the formulas in (4.26) with $s = s_{12}$ and let the operator Δ_X be defined by formula (4.144). Then:*

(1) *The formula*

$$f = \begin{bmatrix} I & X_{22} \\ X_{11}^* & I \end{bmatrix} \begin{bmatrix} g \\ h \end{bmatrix}, \quad \begin{bmatrix} g \\ h \end{bmatrix} \in \mathcal{H}(b_1, b_2), \tag{4.147}$$

defines a bounded bijective operator

$$L_X = \begin{bmatrix} I & X_{22} \\ X_{11}^* & I \end{bmatrix} : \mathcal{H}(b_1, b_2) \to \mathcal{H}(W) \tag{4.148}$$

from $\mathcal{H}(b_1, b_2)$ *onto* $\mathcal{H}(W)$ *with bounded inverse and hence*

$$\Delta_X = L_X^* L_X > \varepsilon I \quad \text{for some } \varepsilon > 0. \tag{4.149}$$

(2) *If* f *is as in (4.147), then*

$$\|f\|^2_{\mathcal{H}(W)} = \left\langle \Delta_X \begin{bmatrix} g \\ h \end{bmatrix}, \begin{bmatrix} g \\ h \end{bmatrix} \right\rangle_{\rm st} \tag{4.150}$$

(3) *If* $s \in T_W[\mathcal{S}^{p\times q}]$, *or, equivalently, if* $s \in \mathcal{S}(b_1, b_2; X)$, *then*

$$\|f\|^2_{\mathcal{H}(W)} = \left\langle \begin{bmatrix} I_p & -s \\ -s^* & I_q \end{bmatrix} f, f \right\rangle_{\rm st} \tag{4.151}$$

for every $f \in \mathcal{H}(W)$.

Proof This follows from Theorem 4.53, since $\mathcal{H}(W) \subset L_2^m$ when $W \in \mathcal{U}_{rsR}(j_{pq})$. □

If $d_j(\lambda) = \det b_j(\lambda)$ for $j = 1, 2$, then $\overline{X_{11}^* \mathcal{H}(b_1)}$ is a closed subspace of $\mathcal{H}(d_1 I_p)$ that is invariant under the action of R_α for every $\alpha \in \mathbb{C}_+$. Therefore, by Lemma 3.6 and Corollary 3.8, and their counterparts in $(H_2^p)^\perp$,

$$\overline{X_{11}^* \mathcal{H}(b_1)} = \mathcal{H}(\mathring{b}_1) \quad \text{and} \quad \overline{X_{22} \mathcal{H}_*(b_2)} = \mathcal{H}_*(\mathring{b}_2) \tag{4.152}$$

for some pair of mvf's $\mathring{b}_1 \in \mathcal{S}_{\rm in}^{p\times p}$ and $\mathring{b}_2 \in \mathcal{S}_{\rm in}^{q\times q}$ such that $d_1 \mathring{b}_1^{-1} \in \mathcal{S}_{\rm in}^{p\times p}$ and $d_2 \mathring{b}_2^{-1} \in \mathcal{S}_{\rm in}^{q\times q}$, respectively. If b_1 and b_2 are entire, as in the present case, then d_1 and d_2 are entire and hence, in view of Lemma 3.35, $\mathring{b}_1$ and $\mathring{b}_2$ are also entire.

Consider the set $\mathcal{H}^m(b_1, b_2)$ of $m \times m$ mvf's $F = \begin{bmatrix} f_1 & f_2 & \cdots & f_m \end{bmatrix}$ with columns $f_j \in \mathcal{H}(b_1, b_2)$ for $1 \le j \le m$, as the orthogonal sum of m copies of the space $\mathcal{H}(b_1, b_2)$ and the set $\mathcal{H}^m(W)$ of $m \times m$ mvf's $K = \begin{bmatrix} k_1 & k_2 & \cdots & k_m \end{bmatrix}$ with columns $k_j \in \mathcal{H}(W)$, for $1 \le j \le m$, as the orthogonal sum of m copies of the space $\mathcal{H}(W)$. Let the operators

$$\Delta_X : \mathcal{H}^m(b_1, b_2) \to \mathcal{H}^m(b_1, b_2) \text{ and } L_X : \mathcal{H}^m(b_1, b_2) \to \mathcal{H}^m(W)$$

act on these spaces of $m \times m$ mvf's column by column:

$$\Delta_X \begin{bmatrix} f_1 & f_2 & \cdots & f_m \end{bmatrix} = \begin{bmatrix} \Delta_X f_1 & \Delta_X f_2 & \cdots & \Delta_X f_m \end{bmatrix}$$

and

$$L_X \begin{bmatrix} f_1 & f_2 & \cdots & f_m \end{bmatrix} = \begin{bmatrix} L_X f_1 & L_X f_2 & \cdots & L_X f_m \end{bmatrix}.$$

Analogously, let the operators X_{11} and X_{22}^* act on $p \times p$ and $q \times q$ mvf's respectively, column by column.

Theorem 4.75 *If the GSIP*$(b_1, b_2; s^\circ)$ *with entire inner mvf's* b_1 *and* b_2 *is strictly completely indeterminate and the mvf* W *is the unique resolvent matrix of this problem in the class* $\mathcal{E} \cap \mathcal{U}_{rsR}^{\circ}(j_{pq})$ *with* $\{b_1, b_2\} \in ap(W)$, *then the RK of the RKHS* $\mathcal{H}(W)$ *is given by the formula*

$$K_\omega^W(\lambda) = (L_X \Delta_X^{-1} F_\omega^X)(\lambda), \tag{4.153}$$

where

$$F_\omega^X(\lambda) = \begin{bmatrix} k_\omega^{b_1}(\lambda) & (X_{11} k_\omega^{\widehat{b}_1})(\lambda) \\ (X_{22}^* \ell_\omega^{\widehat{b}_2})(\lambda) & \ell_\omega^{b_2}(\lambda) \end{bmatrix} \quad on\ \mathbb{C} \times \mathbb{C}, \tag{4.154}$$

$\widehat{b}_1 \in \mathcal{E} \cap \mathcal{S}_{\text{in}}^{p\times p}$ *and* $\widehat{b}_2 \in \mathcal{E} \cap \mathcal{S}_{\text{in}}^{q\times q}$ *are such that* $\mathring{b}_1^{-1}\widehat{b}_1 \in \mathcal{S}_{\text{in}}^{p\times p}$ *and* $\widehat{b}_2 \mathring{b}_2^{-1} \in \mathcal{S}_{\text{in}}^{q\times q}$ *and* $\mathring{b}_1$ *and* $\mathring{b}_2$ *are defined in (4.152). Thus,*

$$W(\lambda) = I_m + 2\pi i \lambda\, K_0^W(\lambda) j_{pq}, \tag{4.155}$$

where $K_0^W(\lambda)$ *is obtained from formula (4.153) with* $\omega = 0$. *In these formulas, the entire inner mvf's* b_j *and* $\widehat{b}_j$ *may be normalized at the point* $\lambda = 0$ *to be* I_p *if* $j = 1$ *and* I_q *if* $j = 2$, *in which case*

$$\begin{aligned} k_0^{b_1}(\lambda) &= \frac{I_p - b_1(\lambda)}{-2\pi i \lambda}, & k_0^{\widehat{b}_1}(\lambda) &= \frac{I_q - \widehat{b}_1(\lambda)}{-2\pi i \lambda}, \\ \ell_0^{b_2}(\lambda) &= \frac{b_2^\#(\lambda) - I_q}{-2\pi i \lambda} \quad \text{and} & \ell_0^{\widehat{b}_2}(\lambda) &= \frac{\widehat{b}_2^\#(\lambda) - I_p}{-2\pi i \lambda}. \end{aligned} \tag{4.156}$$

Proof This follows from theorem 5.95 in [ArD08b], which is obtained from Theorem 4.74. □

Remark 4.76 *In formulas (4.154) and (4.156) it is possible to choose* $\widehat{b}_1(\lambda) = \det b_1(\lambda) I_p$ *and* $\widehat{b}_2(\lambda) = \det b_2(\lambda) I_q$.

4.13 Formulas for resolvent matrices $A \in \mathcal{E} \cap \mathcal{U}_{rsR}^{\circ}(J_p)$

In view of Theorem 4.25, there exists exactly one resolvent matrix $A \in \mathcal{E} \cap \mathcal{U}_{rsR}^{\circ}(J_p)$ for each strictly completely indeterminate GCIP$(b_3, b_4; c^\circ)$ with $b_3 \in \mathcal{E} \cap \mathcal{S}_{in}^{p\times p}$, $b_4 \in \mathcal{E} \cap \mathcal{S}_{in}^{p\times p}$ and $\{b_3, b_4\} \in ap_{II}(A)$. In this section we shall present a formula for this resolvent matrix in terms of the data of the interpolation problem.

Let $A \in \mathcal{U}_{rsR}(J_p)$, $\{b_3, b_4\} \in ap_{II}(A)$, $c^\circ \in \mathcal{C}(A) \cap H_\infty^{p\times p}$,

$$\begin{aligned} \Phi_{11} &= \Pi_{\mathcal{H}(b_3)} M_{c^\circ}|_{H_2^p}, \quad \Phi_{22} = \Pi_- M_{c^\circ}|_{\mathcal{H}_*(b_4)} \\ \text{and} \quad \Phi_{12} &= \Pi_{\mathcal{H}(b_3)} M_{c^\circ}|_{\mathcal{H}_*(b_4)}. \end{aligned} \tag{4.157}$$

The operators Φ_{ij} do not change if c° is replaced by any mvf $c \in \mathcal{C}(A) \cap H_\infty^{p\times p}$. In fact,

$$\begin{aligned} \mathcal{C}(b_3, b_4; c^\circ) \cap H_\infty^{p\times p} = \{c \in \mathcal{C}^{p\times p} \cap H_\infty^{p\times p} :\ &(4.157) \text{ is in force} \\ &\text{when } c^\circ \text{ is replaced by } c\}. \end{aligned} \tag{4.158}$$

Next, formulas for the resolvent matrix A will be given in terms of the RK $K_\omega^A(\lambda)$ of the RKHS $\mathcal{H}(A)$, which will be expressed in terms of the operators

$$\Delta_\Phi = 2\mathfrak{R} \begin{bmatrix} \Phi_{11}|_{\mathcal{H}(b_3)} & \Phi_{12} \\ 0 & \Pi_{\mathcal{H}_*(b_4)}\Phi_{22} \end{bmatrix} : \begin{array}{c} \mathcal{H}(b_3) \\ \oplus \\ \mathcal{H}_*(b_4) \end{array} \begin{array}{c} \mathcal{H}(b_3) \\ \longrightarrow \\ \mathcal{H}_*(b_4) \end{array} \oplus \tag{4.159}$$

and

$$L_\Phi \begin{bmatrix} g \\ h \end{bmatrix} = \begin{bmatrix} -\Phi_{11}^* & \Phi_{22} \\ I & I \end{bmatrix} \begin{bmatrix} g \\ h \end{bmatrix} \quad \text{for} \begin{bmatrix} g \\ h \end{bmatrix} \in \begin{array}{c} \mathcal{H}(b_3) \\ \oplus \\ \mathcal{H}_*(b_4) \end{array}. \tag{4.160}$$

The orthogonal sum considered in (4.159) and (4.160) will be denoted $\mathcal{H}(b_3, b_4)$. The rest of this section follows from the results that are presented with proofs on pp. 523–531 in [ArD08b].

We first present a description of the space $\mathcal{H}(A)$.

Theorem 4.77 *Let $A \in \mathcal{E} \cap \mathcal{U}_{rsR}(J_p)$, $B(\lambda) = A(\lambda)\mathfrak{V}$ and let the operators Φ_{11}, Φ_{22} and Φ_{12} be defined by formula (4.157), where $c \in \mathcal{C}(A) \cap H_\infty^{p\times p}$ and $\{b_3, b_4\} \in ap_{II}(A)$. Furthermore, let L_Φ and Δ_Φ be defined by formulas (4.160) and (4.159), respectively. Then*

$$\mathcal{H}(A) = \left\{ L_\Phi \begin{bmatrix} g \\ h \end{bmatrix} : g \in \mathcal{H}(b_3) \text{ and } h \in \mathcal{H}_*(b_4) \right\}. \tag{4.161}$$

Moreover, L_Φ is a bounded linear operator from $\mathcal{H}(b_3, b_4)$ onto $\mathcal{H}(A)$ with bounded inverse,

$$\Delta_\Phi = L_\Phi^* L_\Phi$$

is a bounded linear positive operator from $\mathcal{H}(b_3, b_4)$ onto itself with bounded inverse and, if

$$f = L_\Phi \begin{bmatrix} g \\ h \end{bmatrix} \quad \text{for some } g \in \mathcal{H}(b_3) \text{ and } h \in \mathcal{H}_*(b_4), \tag{4.162}$$

then

$$\|f\|^2_{\mathcal{H}(A)} = \langle \Delta_\Phi f, f\rangle_{\rm st} = \langle (c+c^*)(g+h), (g+h)\rangle_{\rm st} \tag{4.163}$$

for every $c \in \mathcal{C}(A) \cap H^{p\times p}_\infty$.

Theorem 4.78 *Let $A \in \mathcal{E} \cap \mathcal{U}_{rsR}(J_p)$, let $\mathfrak{E}(\lambda) = \sqrt{2}N_2^*A(\lambda)\mathfrak{V}$ and $\{b_3, b_4\} \in ap_{II}(A)$. Then*

$$\mathcal{B}(\mathfrak{E}) = \mathcal{H}_*(b_4) \oplus \mathcal{H}(b_3) \tag{4.164}$$

as linear spaces of vvf's, but not as Hilbert spaces (unless $E_+E_+^\# = I_p$) and there exist a pair of positive constants γ_1 and γ_2 such that

$$\gamma_1\|f\|_{\rm st} \le \|f\|_{\mathcal{B}(\mathfrak{E})} \le \gamma_2\|f\|_{\rm st} \tag{4.165}$$

for every $f \in \mathcal{B}(\mathfrak{E})$.

Lemma 4.79 *Let $A \in \mathcal{E} \cap \mathcal{U}_{rsR}(J_p)$, $\mathring{A}(\lambda) = J_pA(\lambda)J_p$ and let $\{b_3, b_4\} \in ap_{II}(A)$ and $\{\mathring{b}_3, \mathring{b}_4\} \in ap_{II}(\mathring{A})$. Then*

$$\Phi_{11}^*\mathcal{H}(b_3) = \mathcal{H}(\mathring{b}_3) \quad \text{and} \quad \Phi_{22}\mathcal{H}_*(b_4) = \mathcal{H}_*(\mathring{b}_4). \tag{4.166}$$

Remark 4.80 *If $A \in \mathcal{E} \cap \mathcal{U}(J_p)$, $B = A\mathfrak{V}$ and $\mathring{A} = J_pAJ_p$, then $\mathring{b}_3$ and $\mathring{b}_4$ are $p \times p$ inner mvf's in the factorizations*

$$(b_{11}^\#)^{-1} = \mathring{b}_3\mathring{\varphi}_3 \quad \text{and} \quad b_{12}^{-1} = \mathring{\varphi}_4\mathring{b}_4 \quad \text{with } \mathring{\varphi}_j \in \mathcal{N}^{p\times p}_{\rm out} \text{ for } j = 1, 2. \tag{4.167}$$

Moreover, if $A \in \mathcal{E} \cap \mathcal{U}(J_p)$ and $B = A\mathfrak{V}$, then

$$B = \frac{1}{\sqrt{2}}\begin{bmatrix} -\mathring{E}_- & \mathring{E}_+ \\ E_- & E_+ \end{bmatrix} \tag{4.168}$$

where

$$\mathfrak{E} = [E_- \quad E_+] = [a_{22} - a_{21} \quad a_{22} + a_{21}]$$

and

$$\mathring{\mathfrak{E}} = [\mathring{E}_- \quad \mathring{E}_+] = [a_{11} - a_{12} \quad a_{11} + a_{12}]$$

are entire de Branges matrices and $\mathring{b}_3$ and $\mathring{b}_4$ are entire inner $p \times p$ mvf's. The formula

$$U_1 f = \sqrt{2}N_1^* f \overset{\rm def}{=} \sqrt{2}[I_p \quad 0] \quad \text{for } f \in \mathcal{H}(A) \tag{4.169}$$

defines a coisometric operator from $\mathcal{H}(A)$ onto $\mathcal{B}(\mathring{\mathfrak{E}})$, which is unitary if and only if $\mathring{A}$ is a perfect matrix.

Theorem 4.81 *If the GCIP$(b_3, b_4; c^\circ)$ with entire inner mvf's b_3 and b_4 is strictly completely indeterminate and the mvf A is the unique resolvent matrix of this problem in the class $\mathcal{E} \cap \mathcal{U}^\circ_{rsR}(J_p)$ with $\{b_3, b_4\} \in ap_{II}(A)$, then the RK of the RKHS $\mathcal{H}(A)$ is given by the formula*

$$K_\omega^A(\lambda) = (L_\Phi \Delta_\Phi^{-1} F_\omega^\Phi)(\lambda), \tag{4.170}$$

where the operators

$$L_\Phi : \mathcal{H}^m(b_3, b_4) \to \mathcal{H}^m(A) \quad \text{and} \quad \Delta_\Phi : \mathcal{H}^m(b_3, b_4) \to \mathcal{H}^m(b_3, b_4)$$

act on the columns of $m \times m$ mvf's and

$$F_\omega^\Phi(\lambda) = \begin{bmatrix} -(\Phi_{11} k_\omega^{\widetilde{b}_3})(\lambda) & k_\omega^{b_3}(\lambda) \\ (\Phi_{22}^* \ell_\omega^{\widetilde{b}_4})(\lambda) & \ell_\omega^{b_4})(\lambda) \end{bmatrix} \in \mathcal{H}^m(b_3, b_4).$$

$\widetilde{b}_3 \in \mathcal{E} \cap \mathcal{S}_{\text{in}}^{p\times p}$ and $\widetilde{b}_4 \in \mathcal{E} \cap \mathcal{S}_{\text{in}}^{p\times p}$ are such that $\mathring{b}_3^{-1}\widetilde{b}_3 \in \mathcal{S}_{\text{in}}^{p\times p}$ and $\widetilde{b}_4\mathring{b}_4^{-1} \in \mathcal{S}_{\text{in}}^{p\times p}$ and $\mathring{b}_3$ and $\mathring{b}_4$ are the entire inner mvf's defined in (4.166). Thus,

$$A(\lambda) = I_m + 2\pi i \lambda K_0^A(\lambda) J_p, \tag{4.171}$$

where $K_0^A(\lambda)$ is obtained from formula (4.170) by setting $\omega = 0$. In these formulas the entire mvf's $b_j(\lambda)$ and $\widetilde{b}_j(\lambda)$ may be normalized by setting $b_j(0) = \widetilde{b}_j(0) = I_p$.

Theorem 4.82 *Let $A \in \mathcal{E} \cap \mathcal{U}_{rsR}(J_p)$, $\{b_3, b_4\} \in ap_{II}(A)$, $\mathfrak{E}(\lambda) = \sqrt{2} N_2^* A(\lambda)\mathfrak{V}$ and let the operator Δ_Φ be defined by formula (4.159). Then*

$$\eta_2^* K_\omega^{\mathfrak{E}}(\lambda)\eta_1 = 2\left\langle \Delta_\Phi \begin{bmatrix} k_\omega^{b_3}\eta_1 \\ \ell_\omega^{b_4}\eta_1 \end{bmatrix}, \begin{bmatrix} k_\lambda^{b_3}\eta_2 \\ \ell_\lambda^{b_4}\eta_2 \end{bmatrix} \right\rangle_{st} \tag{4.172}$$

for every pair of points $\lambda, \omega \in \mathbb{C}$ and every pair of vectors $\eta_1, \eta_2 \in \mathbb{C}^p$. Moreover, if $c \in \mathcal{C}(A) \cap \mathring{\mathcal{C}}^{p\times p}$, then there exist numbers $\gamma_1 > 0$, $\gamma_2 > \gamma_1$ such that

$$\gamma_1 I_p \leq (\Re c)(\mu) \leq \gamma_2 I_p \tag{4.173}$$

for almost all points $\mu \in \mathbb{R}$, and for every such choice of γ_1 and γ_2, and every point $\omega \in \mathbb{C}$,

$$\gamma_2^{-1}\{k_\omega^{b_3}(\omega) + \ell_\omega^{b_4}(\omega)\} \leq K_\omega^{\mathfrak{E}}(\omega) \leq \gamma_1^{-1}\{k_\omega^{b_3}(\omega) + \ell_\omega^{b_4}(\omega)\}. \tag{4.174}$$

With a slight abuse of notation, formula (4.170) can be reexpressed in the following more convenient form:

Theorem 4.83 *In the setting of Theorem 4.81,*

$$K_0^A = L_\Phi \begin{bmatrix} \widehat{u}_{11} & \widehat{u}_{12} \\ \widehat{u}_{21} & \widehat{u}_{22} \end{bmatrix}, \tag{4.175}$$

where the $\widehat{u}_{ij} = \widehat{u}_{ij}(\lambda)$ are $p \times p$ mvf's that are obtained as the solutions of the system of equations

$$\Delta_\Phi \begin{bmatrix} \widehat{u}_{11} & \widehat{u}_{12} \\ \widehat{u}_{21} & \widehat{u}_{22} \end{bmatrix} = \begin{bmatrix} -\Phi_{11} k_0^{\widetilde{b}_3} & k_0^{b_3} \\ \Phi_{22}^* \ell_0^{\widetilde{b}_4} & \ell_0^{b_4} \end{bmatrix} \tag{4.176}$$

and the operators in formulas (4.175) and (4.176) act on the indicated matrix arrays column by column. In particular, the columns of $\widehat{u}_{11}(\lambda)$ and $\widehat{u}_{12}(\lambda)$ belong to $\mathcal{H}(b_3)$ and the columns of $\widehat{u}_{21}(\lambda)$ and $\widehat{u}_{22}(\lambda)$ belong to $\mathcal{H}_*(b_4)$.

4.14 Supplementary notes

Nehari [Ne57] formulated a scalar version of the NP(Γ) on the unit circle $\mathbb{T}$ and obtained a criterion for $\mathcal{N}(\Gamma) \neq \emptyset$. Versions of this problem for mvf's on $\mathbb{T}$ were studied in [AAK71a] and [AAK71b], wherein criteria for the problem to be determinate and completely indeterminate were formulated and parametrization formulas for the set of solutions analogous to (4.16) were obtained in the completely indeterminate case. Operator-valued versions of NP(Γ) on $\mathbb{T}$ were studied by L.A. Page [Pa70]; see V.V. Peller [Pe03] for a readable comprehensive treatise on Nehari problems and related issues. Continuous analogs on NP(Γ) for mvf's in the Wiener class were studied in [AAK71b], [KrMA86] and [Dy89b]. The parametrization in Theorem 4.6 was obtained in [AAK68] and [Ad73]. The proof was based on the generalization of the Lax–Phillips scheme in [AdAr66] and [Ad73] and on parametrization formulas for unitary extensions of isometric operators.

Generalized Schur interpolation problems and generalized Carathéodory interpolation problems were studied in [Ar84], [Ar88], [Ar89] and [Ar95a]. In these papers a parametrization of the set of solutions to each of these problems in terms of linear fractional transformations of $\mathcal{S}^{p\times q}$ were given in the completely indeterminate case and the classes $\mathcal{U}_{rR}(j_{pq})$ and $\mathcal{U}_{rR}(J_p)$ were identified with classes of resolvent matrices for these problems. Later, in [ArD02a] and [ArD05a], the formulas that are presented for these resolvent matrices in Sections 4.12 and 4.13 in the strictly completely indeterminate case were obtained by methods based on the theory of RKHS's of L. de Branges that were used earlier to parametrize the solutions of a number of bitangential interpolation problems for mvf's in [Dy89a], [Dy89b], [Dy89c], [Dy90], [Dy98], [Dy03a], [Dy03b], [BoD98] and [BoD06].

Theorem 4.16 is taken from [ArD97]; the argument used to justify Step 2 in the proof of the theorem is adapted from an argument used on pp. 137–138 of the book by Hoffman [Ho62] to characterize extreme points of the unit ball in H_∞.

The theory of the RKHS's $\mathcal{H}(U)$ and $\mathcal{B}(\mathfrak{E})$ originates in the work of L. de Branges. In particular, the characterization of RKHS's $\mathcal{H}(U)$ in terms of R_α invariance and the identity (4.74) is presented in [Br63] and [Br65] for vector-valued functions that are holomorphic in an open connected set Ω that contains the origin; an important technical improvement by J. Rovnyak extended the result to open connected sets Ω, even if $\Omega \cap \mathbb{R} = \emptyset$ [Rov68]; some extensions to Krein spaces and general classes of domains are presented in [AlD93]. A more general version of Theorem 4.38 for RKHS's $\mathcal{H}(U)$ with $U \in \mathcal{P}^\circ(J)$ is presented in theorem 5.31 of [ArD08b]. A readable introduction to the spaces $\mathcal{H}(U)$ and $\mathcal{B}(\mathfrak{E})$ for $m = 2$ is [GoM97].

Additional insight into the role of R_α invariance and the de Branges identity (4.74) is obtained by considering finite dimensional RKHS's: A finite dimensional RKHS of vvf's that is R_α invariant is automatically a space of rational vvf's of the form $\{F(\lambda)v : v \in \mathbb{C}^n\}$ with $F(\lambda) = C(A_1 - \lambda A_2)^{-1}$ and an inner product that is defined in terms of a positive semidefinite matrix P. The de Branges identity is then equivalent to a Lyapunov–Stein equation for P if it is invertible and to a Riccati equation if not; see, e.g., [Dy01a], [Dy01b] and [Dy03a].

Sections 4.8 and 4.13 extend the descriptions of the spaces $\mathcal{H}(W)$ and $\mathcal{H}(A)$ in Chapter 2 of [Dy89b] and [Dy90].

The class $\mathcal{U}_{BR}(J)$ was introduced and studied recently in [ArD12a]; the class $\mathcal{U}_{AR}(J)$, which plays a significant role in the study of the bitangential inverse monodromy problem, was introduced even more recently, in the final stages of preparation of this book for publication. As of today we do not have alternate descriptions of either of these classes that are analogous to the Treil–Volberg matrix version of the Muckenhoupt (A_2) condition in [TrV97] that is used to characterize the class $\mathcal{U}_{rsR}(J)$ in theorem 10.12 of [ArD08b], (see also [ArD01b], [ArD03b]), or even to the characterizations of $\mathcal{U}_{rsR}(J)$ in (1.26) and (1.36). The inclusions in (4.112) imply that in order for U to be in $\mathcal{U}_{AR}(J)$ it is necessary that $U \in \mathcal{U}_{BR}(J)$ and sufficient that $U \in \mathcal{U}_{rsR}(J) \cup \mathcal{U}_{\ell sR}(J)$. However, as $\mathcal{U}_{rsR}(J) \cup \mathcal{U}_{\ell sR}(J)$ is a proper subclass of $\mathcal{U}_{BR}(J)$, these two conditions are not the same.

The GSIP$(b_1, b_2; s^\circ)$ and the GCIP$(b_3, b_4; c^\circ)$ with entire inner mvf's $b_1, \ldots, b_4$ were studied in [ArD98] as bitangential generalizations of a pair of extension problems that were studied by M.G. Krein. They will be used extensively in the formulation and analysis of bitangential direct and inverse problems in the next several chapters. The continuous analogs of the classical Schur and Carathéodory extension problems with solutions in a Wiener class of mvf's that were studied by M.G. Krein and F.E. Melik-Adamyan in [KrMA86] can be viewed as special cases of the GSIP$(e_a I_p, I_q; s^\circ)$ and the GCIP$(e_a I_p, I_p; c^\circ)$, respectively. This will be discussed in more detail in Chapter 9.

Time domain versions of the GCIP($b_3, b_4; c^\circ$) with entire inner mvf's b_3, b_4 were considered in section 8.5 of [ArD08b] and will be reviewed briefly in Chapter 9 below. Time domain versions of the GSIP($b_1, b_2; s^\circ$) with entire inner mvf's b_1, b_2 will also be considered in this chapter.

The problem of describing the set $\mathcal{S}(b_1, b_2; X)$ that is considered in Section 4.12 is a bitangential generalization of the scalar Sarason problem, which is to describe the set

$$\mathcal{S}(b; X) = \{s \in \mathcal{S} : \Pi_{\mathcal{H}(b)} M_s|_{H_2} = X\},$$

given a scalar inner function b and a bounded linear operator X from H_2 into $\mathcal{H}(b)$; see [Sar67] and, for further generalizations with nonsquare inner mvf's, [Kh90] and [Kh95].

5

Chains that are matrizants and chains of associated pairs

Recall that a family of mvf's $U_t \in \mathcal{E} \cap \mathcal{U}^\circ(J)$, $0 \le t < d$, is said to be a normalized nondecreasing $\curvearrowright$ (resp., $\curvearrowleft$) chain of entire J-inner mvfs if $U_0(\lambda) \equiv I_m$ and

$$U_{t_1}^{-1}U_{t_2} \in \mathcal{E} \cap \mathcal{U}^\circ(J) \quad (\text{resp., } U_{t_2}U_{t_1}^{-1} \in \mathcal{E} \cap \mathcal{U}^\circ(J) \quad \text{for } 0 \le t_1 \le t_2 < d.$$

The chain is said to be strictly increasing if $U_{t_1}^{-1}U_{t_2} \not\equiv I_m$ (resp., $U_{t_2}U_{t_1}^{-1} \not\equiv I_m$) when $t_1 < t_2$.

In view of Theorem 2.5, the matrizant $U_t(\lambda)$, $0 \le t < d$, of a canonical integral system (2.1) with a continuous nondecreasing mass function $M(t)$ on the interval $[0, d)$ is a normalized nondecreasing $\curvearrowright$ continuous chain of entire J-inner mvf's. However, the converse is not true: some sufficient conditions for the converse to hold are presented in Theorem 5.6.

A chain of pairs $\{b_1^t, b_2^t\}$, $0 \le t < d$, of entire inner mvf's is said to be a **normalized nondecreasing continuous chain of pairs** if $\{b_1^t\}$, $0 \le t < d$, is a normalized nondecreasing $\curvearrowright$ continuous chain and $\{b_2^t\}$, $0 \le t < d$, is a normalized nondecreasing $\curvearrowleft$ continuous chain. It is said to be a **strictly increasing chain of pairs** if at least one of the ratios $(b_1^s)^{-1}b_1^t$, $b_2^t(b_2^s)^{-1}$ is a nonconstant mvf for each choice of $0 \le s < t < d$.

In Section 5.3, it will be shown that normalized nondecreasing chains of associated pairs of the first and second kind of a normalized motonic $\curvearrowright$ continuous chain of entire A-regular J-inner mvf's (with $J \ne \pm I_m$) are continuous.

5.1 Continuous chains of entire J-inner mvf's

This section reviews some properties of normalized nondecreasing $\curvearrowright$ and $\curvearrowleft$ continuous chains of entire J-inner mvf's that will be needed in the study of bitangential direct and inverse problems.

Theorem 5.1 *Let $\{U_t(\lambda): 0 \le t \le d\}$ be a normalized nondecreasing $\curvearrowright$ chain of entire J-inner $m \times m$ mvf's such that the Hilbert spaces $\mathcal{H}(U_t)$ are included isometrically in $\mathcal{H}(U_d)$ and $K_\omega^t(\lambda) = K_\omega^{U_t}(\lambda)$ for $0 \le t \le d$. Let Π_t denote the orthogonal projection from $\mathcal{H}(U_d)$ onto $\mathcal{H}(U_t)$. Then*

$$\Pi_t K_\omega^d \xi = K_\omega^t \xi \quad \textit{for every } \xi \in \mathbb{C}^m,\ \omega \in \mathbb{C} \textit{ and } t \in [0, d] \tag{5.1}$$

and the following statements are equivalent:

(1) *$K_\omega^t(\omega)$ is a continuous mvf of t on the interval $[0, d]$ for some point $\omega \in \mathbb{C}$.*
(2) *$U_t(\lambda)$ is a continuous mvf of t on the interval $[0, d]$ for every point $\lambda \in \mathbb{C}$.*
(3) *$K_\omega^t(\lambda)$ is a continuous mvf of t on the interval $[0, d]$ for every pair of points λ, ω in $\mathbb{C}$.*
(4) *$K_\omega^t \xi$ is a continuous vvf of t in the Hilbert space $\mathcal{H}(U_d)$ on the interval $[0, d]$ for every $\xi \in \mathbb{C}^m$ and $\omega \in \mathbb{C}$.*
(5) *The orthogonal projections Π_t are strongly continuous in the Hilbert space $\mathcal{H}(U_d)$.*
(6) *The two equalities*

$$\bigcap_{0<\varepsilon<d-t} \mathcal{H}(U_{t+\varepsilon}) = \mathcal{H}(U_t) \quad \textit{for every } t \in [0, d) \tag{5.2}$$

and

$$\bigvee_{0<\varepsilon\le t} \mathcal{H}(U_{t-\varepsilon}) = \mathcal{H}(U_t) \quad \textit{for every } t \in (0, d]. \tag{5.3}$$

hold.

Proof The formula

$$\langle f, \Pi_t K_\omega^d \xi - K_\omega^t \xi \rangle_{\mathcal{H}(U_d)} = \xi^* f(\omega) - \xi^* f(\omega) = 0,$$

which is valid for every choice of $f \in \mathcal{H}(U_t)$ with $0 \le t \le d$, $\xi \in \mathbb{C}^m$ and $\omega \in \mathbb{C}$, clearly justifies (5.1). The rest of the proof is broken into steps.

1. (1) $\Longrightarrow$ (2): If (1) is in force for some point $\omega \in \mathbb{C}$, then the inequalities

$$\begin{aligned}
|\xi^*(K_\omega^{t_2}(\lambda) - K_\omega^{t_1}(\lambda))\eta|^2 &= |\langle (K_\omega^{t_2} - K_\omega^{t_1})\eta, (K_\lambda^{t_2} - K_\lambda^{t_1})\xi \rangle_{\mathcal{H}(U_d)}|^2 \\
&\le \|(K_\omega^{t_2} - K_\omega^{t_1})\eta\|_{\mathcal{H}(U_d)}^2 \, \|(K_\lambda^{t_2} - K_\lambda^{t_1})\xi\|_{\mathcal{H}(U_d)}^2 \\
&= \eta^*(K_\omega^{t_2}(\omega) - K_\omega^{t_1}(\omega))\eta \, \xi^*(K_\lambda^{t_2}(\lambda) - K_\lambda^{t_1}(\lambda))\xi \\
&\le \eta^*(K_\omega^{t_2}(\omega) - K_\omega^{t_1}(\omega))\eta \, \xi^*(K_\lambda^d(\lambda))\xi
\end{aligned}$$

imply that $K_\omega^t(\lambda)$ is a continuous mvf of t on the interval $[0, d]$ for every $\lambda \in \mathbb{C}$. Thus, the formulas

$$K_\omega^t(0) = \frac{J - JU_t(\omega)^*}{2\pi i\overline{\omega}} \quad \text{for } \omega \ne 0 \text{ and } U_t(0) = I_m$$

imply that $U_t(\omega)$ is a continuous function of t on the interval $[0, d]$ for the point ω specified in (1). The continuity of $U_t(\lambda)$ for arbitrary points $\lambda \in \mathbb{C}$ then follows from formula (2.21). This completes the proof of Step 1.

2. (2)$\Longrightarrow$ (3): Formula (4.73) clearly implies the desired implication if $\lambda \neq \overline{\omega}$. On the other hand, if $\lambda = \overline{\omega}$, then

$$K_\omega^t(\overline{\omega}) = \frac{U_t'(\overline{\omega})JU_t(\omega)^*}{2\pi i}$$

which is also a continuous function of t as is perhaps most easily seen by noting that

$$\frac{U_t'(\overline{\omega})JU_t(\omega)^*}{2\pi i} = \frac{1}{(2\pi i)^2}\int_{\Gamma_r(\overline{\omega})} \frac{U_t(\zeta)}{(\zeta - \overline{\omega})^2} d\zeta JU_t(\omega)^*$$

where $\Gamma_r(\overline{\omega})$ denotes a circle of radius r centered at $\overline{\omega}$ and directed counterclockwise. In view of the bound (4.83),

$$\|U_t(\lambda)\| \leq 1 + 2\pi|\lambda| \left\|\sqrt{K_0^t(0)}\right\| \left\|\sqrt{K_\lambda^t(\lambda)}\right\| \leq 1+2\pi|\lambda| \left\|\sqrt{K_0^d(0)}\right\| \left\|\sqrt{K_\lambda^d(\lambda)}\right\|,$$

and the fact that $\|K_\lambda^d(\lambda)\|$ is a continuous function of λ on $\mathfrak{h}_U$ by Lemma 4.37, limits in t can be brought inside the integral by Lebesgue's theorem on dominated convergence.

3. (3)$\Longrightarrow$ (4): This is an immediate consequence of the formulas

$$\|K_\omega^{t_2}\xi - K_\omega^{t_1}\xi\|_{\mathcal{H}(U_d)}^2 = \xi^*(K_\omega^{t_2}(\omega) - K_\omega^{t_1}(\omega))\xi \text{ for } 0 \leq t_1 \leq t_2 \leq d.$$

4. (4)$\Longrightarrow$ (5): In view of (5.1),

$$\|\Pi_{t_2}K_\omega^d\xi - \Pi_{t_1}K_\omega^d\xi\|_{\mathcal{H}(U_d)}^2 = \|K_\omega^{t_2}\xi - K_\omega^{t_1}\xi\|_{\mathcal{H}(U_d)}^2$$

for $0 \leq t_1 \leq t_2 \leq d$, and tends to zero as $t_2 - t_1 \to 0$, thanks to (4). Now for any $f \in \mathcal{H}(U_d)$ and any $\varepsilon > 0$, there exists a finite linear combination

$$f_n = \sum K_{\omega_j}^d \xi_j \quad \text{such that} \quad \|f - f_n\|_{\mathcal{H}(U_d)} \leq \varepsilon/3.$$

Therefore,

$$\begin{aligned}
\|\Pi_{t_2}f - \Pi_{t_1}f\|_{\mathcal{H}(U_d)} &\leq \|\Pi_{t_2}(f - f_n)\|_{\mathcal{H}(U_d)} + \|\Pi_{t_2}f_n - \Pi_{t_1}f_n\|_{\mathcal{H}(U_d)} \\
&\quad + \|\Pi_{t_1}(f_n - f)\|_{\mathcal{H}(U_d)} \\
&\leq 2\|f - f_n\|_{\mathcal{H}(U_d)} + \|\Pi_{t_2}f_n - \Pi_{t_1}f_n\|_{\mathcal{H}(U_d)} \\
&\leq 2\varepsilon/3 + \|\Pi_{t_2}f_n - \Pi_{t_1}f_n\|_{\mathcal{H}(U_d)}
\end{aligned}$$

which serves to justify the asserted continuity by standard arguments.

5. (5)$\Longleftrightarrow$(6): Let $f \in \mathcal{H}(U_{t+\varepsilon})$ for every ε such that $0 < \varepsilon \leq d - t$. Then $f = \Pi_{t+\varepsilon}f$ for all such ε and hence, in view of (5),

$$f = \lim_{\varepsilon \downarrow 0} \Pi_{t+\varepsilon}f = \Pi_t f,$$

which belongs to $\mathcal{H}(U_t)$. This proves that $\cap_{0<\varepsilon<d-t}\mathcal{H}(U_{t+\varepsilon}) \subseteq \mathcal{H}(U_t)$ and hence, as the opposite inclusion is self-evident, justifies (5.2). The verification of (5.3) is similar. Thus (5) $\Rightarrow$ (6); the converse implication is obvious.

6. (5)$\Longrightarrow$(1): If $0 \le t \le d$, $\omega \in \mathbb{C}$, $\xi \in \mathbb{C}^m$ and $\eta \in \mathbb{C}^m$, then

$$\begin{aligned}\eta^* K^t_\omega(\omega)\xi &= \langle K^t_\omega \xi, K^t_\omega \eta\rangle_{\mathcal{H}(U_d)} \\ &= \langle \Pi_t K^d_\omega \xi, K^d_\omega \eta\rangle_{\mathcal{H}(U_d)},\end{aligned}$$

which is a continuous function of t on $[0, d]$, as claimed. □

Theorem 5.2 *Let $\{U_t(\lambda)\}$, $0 \le t < d$, be a normalized nondecreasing $\curvearrowright$ chain of entire J-inner mvf's. Then*

$$U_t \text{ continuous on } [0, d) \Longrightarrow 2\pi K_0^{U_t}(0) \text{ is continuous on } [0, d).$$

Proof Let $M(t) = 2\pi K_0^{U_t}(0)$. Then $M(0) = 0$ because $U_0(\lambda) \equiv I_m$ and

$$\begin{aligned} M(t_2) - M(t_1) &= \lim_{\lambda\to 0} -i\left\{\frac{U_{t_2}(\lambda) - U_{t_1}(\lambda)}{\lambda}\right\} J \\ &= \lim_{\lambda\to 0} U_{t_1}(\lambda) \lim_{\lambda\to 0}\left\{\frac{U_{t_1}(\lambda)^{-1}U_{t_2}(\lambda) - I_m}{i\lambda}\right\} J \\ &\ge 0 \ \text{ for } 0 \le t_1 \le t_2 < d, \end{aligned}$$

since $U = U_{t_1}^{-1}U_{t_2}$ belongs to the class $\mathcal{E} \cap \mathcal{U}^\circ(J)$ and hence

$$\begin{aligned} -2iU'(0)J &= -iU'(0)J + iJU'(0)^* \\ &= \lim_{\nu\downarrow 0}\left\{-iJ\frac{U(i\nu) - I_m}{i\nu}JU(i\nu)^* + iJ\left(\frac{U(i\nu) - I_m}{i\nu}\right)^*\right\} \\ &= \lim_{\nu\downarrow 0}\frac{J - U(i\nu)JU(i\nu)^*}{\nu} \ge 0. \end{aligned}$$

Finally, continuity follows from the Cauchy formula,

$$\frac{\partial}{\partial\lambda}U_t(\lambda)|_{\lambda=0} = \frac{1}{2\pi i}\int_{|\rho|=1}\frac{U_t(\rho)}{\rho^2}d\rho$$

and the fact that $U_t(\lambda)$ is a continuous mvf of t on the interval $[0, d)$ for each fixed $\lambda \in \mathbb{C}$ that is subject to the bound (2.50) with $U = U_t$. □

Theorem 5.3 *If $M(t)$ is a continuous nondecreasing $m \times m$ mvf on the interval $[0, d]$ with $M(0) = 0$, then the matrizant U_t of the corresponding canonical integral system (2.1) is a normalized nondecreasing $\curvearrowright$ continuous chain of entire J-inner mvf's and*

$$\mathcal{H}(U_s) \quad \text{is included contractively in } \mathcal{H}(U_t) \text{ when } 0 \le s \le t \le d.$$

Proof Theorem 2.5 guarantees that the matrizant $U_t(\lambda)$, $0 \le t < d$, is a normalized nondecreasing continuous chain of entire J-inner mvf's. The rest follows from Theorem 4.47. □

Theorem 5.4 *Let $\{U_t\}$, $0 \le t \le d$, be a normalized nondecreasing $\curvearrowright$ chain of entire J-inner mvf's such that $U_t \in \mathcal{U}_{rR}(J)$ for every $t \in [0, d]$. Then*

$$U_t \text{ is continuous on } [0, d] \Longleftrightarrow 2\pi K_0^{U_t}(0) \text{ is continuous on } [0, d].$$

Proof In view of Theorem 4.56, the spaces $\mathcal{H}(U_t)$ are included isometrically in $\mathcal{H}(U_d)$. The rest then follows from Theorem 5.1. □

Corollary 5.5 *If $\{U_t\}$, $0 \le t \le d$, is a normalized nondecreasing $\curvearrowright$ chain of entire J-inner mvf's such that $U_d \in \mathcal{U}_{AR}(J)$, then the conclusions of Theorem 5.4 are in force.*

5.2 Chains that are matrizants

The next theorem presents a sufficient condition for a chain of mvf's to be the matrizant of a canonical integral system (2.1).

Theorem 5.6 *Let $U_t(\lambda)$, $0 \le t \le d$, be a normalized nondecreasing $\curvearrowright$ continuous chain of entire J-inner mvf's such that the spaces $\mathcal{H}(U_t)$ are isometrically included in $\mathcal{H}(U_d)$ for $0 \le t \le d$ and let $K_\omega^t(\lambda) = K_\omega^{U_t}(\lambda)$. Then*

$$M(t) = 2\pi K_0^t(0) \tag{5.4}$$

is a continuous nondecreasing mvf on the interval $0 \le t \le d$ with $M(0) = 0$ and $U_t(\lambda)$ is the matrizant of the canonical integral system (2.1) with mass function $M(t)$, i.e.,

$$U_t(\lambda) = I_m + i\lambda \int_0^t U_s(\lambda) dM(s) J \quad \textit{for } \le t \le d. \tag{5.5}$$

Proof The asserted properties of $M(t)$ are immediate from Theorem 5.1. Next, in order to establish (5.5), it is convenient to let Π_t denote the orthogonal projection of $\mathcal{H}(U_d)$ onto $\mathcal{H}(U_t)$ for $0 \le t \le d$. Then, since $R_0 \Pi_t = \Pi_t R_0 \Pi_t$ and the Π_t are continuous and R_0 is compact by Theorem 4.39, it follows from lemma 5.1 in chapter 1 of Gohberg and Krein [GK69] that

$$\left\| \sum_{j=1}^n \left(\Pi_{t_j} - \Pi_{t_{j-1}}\right) R_0^* \left(\Pi_{t_j} - \Pi_{t_{j-1}}\right) \right\|_{\mathcal{H}(U_d)} \longrightarrow 0$$

as the mesh size of the partition $0 = t_0 < t_1 < \cdots < t_n = t$ tends to zero. Thus, in view of (5.1), the sum

$$\sum_{j=1}^n \left\langle R_0^* \left(K_0^{t_j} - K_0^{t_{j-1}}\right) u\,,\; \left(K_\omega^{t_j} - K_\omega^{t_{j-1}}\right) v \right\rangle_{\mathcal{H}(U_d)}$$

also tends to zero for every choice of $\omega \in \mathbb{C}$ and u, v in $\mathbb{C}^m$ as the mesh size shrinks to zero. But as $R_0 K_\omega^{t_j-1} v \in \mathcal{H}(U_{t_{j-1}})$ and

$$\left\langle R_0^* \left(K_0^{t_j} - K_0^{t_{j-1}}\right) u, f\right\rangle_{\mathcal{H}(U_d)} = f'(0)^* u - f'(0)^* u = 0 \quad \text{for } f \in \mathcal{H}(U_{t_{j-1}}),$$

it is readily checked that ω times the preceding sum is equal to

$$\begin{aligned}
&\omega \sum_{j=1}^{n} \langle (K_0^{t_j} - K_0^{t_{j-1}})u, R_0 K_\omega^{t_j} v\rangle_{\mathcal{H}(U_d)} \\
&= \sum_{j=1}^{n} \langle (K_0^{t_j} - K_0^{t_{j-1}})u, K_\omega^{t_j} v - K_0^{t_j} U_{t_j}(\omega)^* v\rangle_{\mathcal{H}(U_d)} \\
&= v^* \sum_{j=1}^{n} \left\{ K_0^{t_j}(\omega) - K_0^{t_{j-1}}(\omega) - U_{t_j}(\omega) \left[K_0^{t_j}(0) - K_0^{t_{j-1}}(0) \right] \right\} u \\
&= v^* K_0^t(\omega) u - v^* \sum_{j=1}^{n} U_{t_j}(\omega) \left[K_0^{t_j}(0) - K_0^{t_{j-1}}(0) \right] u.
\end{aligned}$$

In the limit, as the mesh size of the partition tends to zero, this supplies the identity

$$\begin{aligned}
K_0^t(\omega) &= \int_0^t U_s(\omega) dK_0^s(0) \\
&= \frac{1}{2\pi} \int_0^t U_s(\omega) dM(s),
\end{aligned}$$

which is equivalent to (5.5). □

The following simple example shows that the inclusion $\mathcal{H}(U_t) \subseteq \mathcal{H}(U)$ may be contractive but not isometric for the matrizant U_t, $0 \le t \le d$, of a canonical system (1.1).

Example 5.7 *Let* $U(\lambda) = I_m + i\lambda VV^*J$, *where* $V \in \mathbb{C}^{m\times k}$, $V^*JV = 0$ *and* $V^*V = I_k$. *Then* $U_t(\lambda) = I_m + i\lambda t VV^*J$, $0 \le t \le 1$, *is the matrizant of the differential system*

$$y'(t, \lambda) = i\lambda y(t, \lambda) H(t) J, \quad 0 \le t \le 1,$$

with Hamiltonian

$$H(t) = VV^* \quad for\ 0 \le t \le 1.$$

The RK $K_\omega^t(\lambda)$ *of* $\mathcal{H}(U_t)$ *is*

$$K_\omega^t(\lambda) = \frac{1}{2\pi} tVV^* \quad for\ 0 \le t \le 1$$

and if $t > 0$, then $\mathcal{H}(U_t)$ is the k-dimensional space spanned by the columns of V. Moreover,

$$\|K_\omega^s\xi\|^2_{\mathcal{H}(U_s)} = \xi^* K_\omega^s(\omega)\xi = \frac{s}{2\pi}\xi^* V^* V\xi,$$

whereas

$$\|K_\omega^s\xi\|^2_{\mathcal{H}(U_t)} = \frac{s^2}{t^2}\|K_\omega^t\xi\|^2_{\mathcal{H}(U_t)} = \frac{s^2}{t2\pi}\xi^* V^* V\xi = \frac{s}{t}\|K_\omega^s\xi\|^2_{\mathcal{H}(U_s)}$$

for $0 < s \le t \le 1$. Thus, $\mathcal{H}(U_t)$ coincides with the space

$$\mathcal{H}(U_1) = \{Vx : x \in \mathbb{C}^k\}$$

as vector spaces for every $0 < t \le 1$, but the inclusions $\mathcal{H}(U_{t_1}) \subseteq \mathcal{H}(U_{t_2})$ are strictly contractive for $0 < t_1 < t_2 \le 1$.

Notice that in this example $U_t \in \mathcal{E} \cap \mathcal{U}_S(J)$ for every $t \in [0, 1]$.

Theorem 5.8 *Let $\{U_t(\lambda)\}$, $0 \le t < d$, be a normalized nondecreasing continuous chain of entire J-inner mvf's such that either*

$$U_t \in \mathcal{U}_{BR}(J) \text{ for each } t \in [0, d) \tag{5.6}$$

or

$$U_t \in \mathcal{U}_{rR}(J) \text{ for each } t \in [0, d). \tag{5.7}$$

Then, $U_t(\lambda)$, $0 \le t < d$, is the matrizant of exactly one canonical integral system (2.1) with continuous nondecreasing mass function $M(t)$, $0 \le t < d$, with $M(0) = 0$. Moreover, this mass function is defined by formula (5.4).

Proof If (5.6) (resp., (5.7)) is in force, then the RKHS's $\mathcal{H}(U_t)$, $0 \le t \le d_1$, are isometrically included in $\mathcal{H}(U_{d_1})$ for $0 < d_1 < d$ by Theorem 4.46 (resp., Theorem 4.56). Theorem 5.6 guarantees that the family $\{U_t\}$ is the matrizant of the canonical system (2.1) on the closed interval $[0, d_1]$ for every choice of $0 < d_1 < d$. □

We shall say that a normalized nondecreasing $\curvearrowright$ chain of entire J-inner mvf's $U_t(\lambda)$, $0 \le t \le d$, is absolutely continuous with respect to t on the interval $[0, d]$ uniformly in Ω if for every $\varepsilon > 0$, there exists a $\delta > 0$ such that

$$\sum_j \|U_{t_j}(\lambda) - U_{s_j}(\lambda)\| \le \varepsilon$$

for any set of disjoint intervals $[s_j, t_j]$ in $[0, d]$ with $\sum_j (t_j - s_j) < \delta$ and every $\lambda \in \Omega$.

Lemma 5.9 *If $U_t(\lambda)$, $0 \le t \le d$, is the matrizant of a canonical differential system (2.14) with a summable nonnegative Hamiltonian $H(t)$, then $U_t(\lambda)$, $0 \le t \le d$, is absolutely continuous with respect to t on the interval $[0, d]$ uniformly in Ω for every bounded subset Ω of $\mathbb{C}$.*

Proof Since $\|U_t(\lambda)\| \leq M$ for every $t \in [0, d]$ and $\lambda \in \Omega$, it follows that

$$\|U_{t_j}(\lambda) - U_{s_j}(\lambda)\| = \|\lambda \int_{s_j}^{t_j} U_s(\lambda)H(s)dsJ\| \leq RM \int_{s_j}^{t_j} \|H(s)\|ds$$

for $|\lambda| \leq R$ and $0 \leq s_j \leq t_j \leq d$. Therefore,

$$\sum_j \|U_{t_j}(\lambda) - U_{s_j}(\lambda)\| \leq RM \sum_j \int_{s_j}^{t_j} \|H(s)\|ds$$

and the asserted inequality now follows easily from the fact that

$$\int_0^t \|H(s)\|ds \quad \text{is an absolutely continuous function of } t \text{ on } [0, d].$$

□

Theorem 5.10 *If $U_t(\lambda)$, $0 \leq t \leq d$, is a normalized nondecreasing $\curvearrowright$ chain of entire J-inner mvf's that is absolutely continuous with respect to t on $[0, d]$ uniformly with respect to some circle $\Omega_r = \{\lambda \in \mathbb{C} : |\lambda| = r\}$, and the spaces $\mathcal{H}(U_t)$ are isometrically included in $\mathcal{H}(U_d)$, for every $t \in [0, d]$, then it is the matrizant of a regular canonical differential system (2.14) with a summable Hamiltonian*

$$H(t) = \frac{1}{i\lambda} U_t(\lambda)^{-1} \left(\frac{\partial U_t}{\partial t} \right) (\lambda) J \quad \text{a.e. on } [0, d]$$

that is nonnegative a.e. on $[0, d]$.

Proof By Theorem 5.6, $U_t(\lambda), 0 \leq t \leq d$, is the matrizant of a canonical integral system (2.1) with mass function

$$M(t) = -i \left(\frac{\partial U_t}{\partial \lambda} \right) (0) J = -\frac{1}{2\pi} \int_{\Omega_r} \frac{U_t(\zeta)}{\zeta^2} J d\zeta.$$

Therefore,

$$\|M(t) - M(s)\| \leq \frac{1}{2\pi r} \int_0^{2\pi} \|U_t(re^{i\theta}) - U_s(re^{i\theta})\| d\theta,$$

and hence, in view of the assumptions on $U_t(\lambda)$, $M(t)$ is absolutely continuous on $[0, d]$, $H(t) = M'(t) \geq 0$ a.e. on $[0, d]$ and $H \in L_1^{m \times m}([0, d])$. □

Lemma 5.11 *Let $\{b^t(\lambda) : 0 \leq t < d\}$ be a normalized nondecreasing $\curvearrowright$ [resp. $\curvearrowleft$] continuous chain of entire inner $p \times p$ mvf's. Then the following conditions are equivalent:*

(1) $\lim_{t \uparrow d} |\det b^t(\omega)| > 0$ *for at least one point* $\omega \in \mathbb{C}_+$.
(2) $\lim_{t \uparrow d} |\det b^t(\lambda)| > 0$ *for every point* $\lambda \in \mathbb{C}$.
(3) *The limit*

$$b^d(\lambda) = \lim_{t \uparrow d} b^t(\lambda)$$

exists and the convergence is uniform on every compact subset of $\mathbb{C}$.

Moreover, if any one (and hence every one) of these conditions is in force, then

$$b^d \in \mathcal{E} \cap \mathcal{S}_{\text{in}}^{p\times p} \quad \text{and} \quad b^d(0) = I_p.$$

Furthermore, if $d < \infty$, then $b^d(\lambda)$ is the monodromy matrix of the canonical integral system (2.1) with $J = I_p$, a mass function that is subject to the constraints (1.22) with $m = p$ and matrizant b^t, $0 \le t \le d$.

Proof Since $b^t \in \mathcal{E} \cap \mathcal{S}_{\text{in}}^{p\times p}$ and $b^t(0) = I_p$ the scalar function $\det\{b^t(\lambda)\} = \exp\{i\lambda\delta(t)\}$ is an entire inner function such that $\det\{b^t(0)\} = 1$ for $0 \le t < d$. Moreover, in view of the assumptions on the chain, it is clear that $\delta(t)$ is a continuous nondecreasing nonnegative function on the interval $[0, d)$. Thus (1) implies that

$$\delta(d) := \lim_{t\uparrow d} \delta(t) < \infty$$

and hence that $\det\{b^t(\lambda)\} \longrightarrow \exp\{i\lambda\delta(d)\}$ uniformly on compact subsets of $\mathbb{C}$ as $t \uparrow d$. This serves to establish the equivalence of (1) and (2), since the implication (2) $\Longrightarrow$ (1) is self-evident.

Next, with the help of the identity $(b^t)^\#(\lambda)b^t(\lambda) = I_p$, it is readily checked that if (1) is in force, then the numbers $\|b^t(\lambda)\|$ are bounded on every compact subset of $\mathbb{C}$. Therefore, there exists a sequence of points $t_n \uparrow d$ such that $b^{t_n}(\lambda)$ converges uniformly to a limit $b^d(\lambda)$ on every compact subset of $\mathbb{C}$. Thus, $b^d \in \mathcal{E} \cap \mathcal{S}_{in}^{p\times p}$ and $b^d(0) = I_p$.

Moreover, since

$$b^d(\omega)b^d(\omega)^* \le b^t(\omega)b^t(\omega)^* \le b^{t_n}(\omega)b^{t_n}(\omega)^*$$

and

$$b^{t_n}(\overline{\omega})^* b^{t_n}(\overline{\omega}) \le b^t(\overline{\omega})^* b^t(\overline{\omega}) \le b^d(\overline{\omega})^* b^d(\overline{\omega})$$

for $\omega \in \mathbb{C}_+$ and $t > t_n$, it follows that

$$b^t(\lambda)b^t(\lambda)^* \longrightarrow b^d(\lambda)b^d(\lambda)^*$$

as $t \uparrow d$ for every point $\lambda \in \mathbb{C}$. Thus, the reproducing kernel

$$k_\omega^t(\lambda) = \frac{I_p - b^t(\lambda)b^t(\omega)^*}{\rho_\omega(\lambda)}$$

of the space $\mathcal{H}(b^t) = H_2^p \ominus b^t H_2^p$ satisfies the condition

$$\lim_{t\uparrow d} k_\lambda^t(\lambda) = k_\lambda^d(\lambda)$$

for every point $\lambda \notin \mathbb{R}$ and therefore, in view of Theorem 5.1, $b^t(\lambda) \to b^d(\lambda)$ as $t \uparrow d$ for every point $\lambda \in \mathbb{C}$. Moreover, this convergence is uniform on compact subsets of $\mathbb{C}$, because $b_t(\lambda)$ are uniformly bounded on such sets, in view of the bounds

$$k_\lambda^t(\lambda) \le k_\lambda^d(\lambda) \quad \text{for } 0 \le t \le d \text{ and } \lambda \in \mathbb{C}$$

and (4.83) applied to $b^t(\lambda)$ (in place of $U_t(\lambda)$).

Thus, (1) $\Longrightarrow$ (3). The implications (3) $\Longrightarrow$ (2) and (2) $\Longrightarrow$ (1) are self-evident. The rest follows from Theorem 5.6. □

Lemma 5.12 *If $d < \infty$ and U_t, $0 \le t < d$, is a normalized nondecreasing $\curvearrowright$ continuous chain of entire right regular J-inner mvfs such that*

$$U_d(\lambda) \stackrel{\text{def}}{=} \lim_{t \uparrow d} U_t(\lambda) \quad \text{for } \lambda \in \mathbb{C}_+ \quad \text{and} \quad U_d \in \mathcal{P}^\circ(J),$$

then $U_d \in \mathcal{E} \cap \mathcal{U}_{rR}^\circ(J)$ and $U_t^{-1}U_d \in \mathcal{E} \cap \mathcal{U}^\circ(J)$, i.e., U_t, $0 \le t \le d$, is a normalized nondecreasing $\curvearrowright$ continuous chain of right regular J-inner mvfs on the closed interval $[0, d]$. Moreover, U_d is the monodromy matrix of the canonical integral system (2.1) with mass function that is subject to the constraints (1.22) and matrizant U_t, $0 \le t \le d$.

Proof The first assertion is justified by Theorem 4.57. The rest follows from Lemma 5.11 if $J = \pm I_m$ and from Theorem 5.8 if $J \neq \pm I_m$. □

5.3 Continuity of chains of associated pairs

Theorem 5.13 *If $U_t \in \mathcal{E} \cap \mathcal{U}_{rR}^\circ(J)$ for $0 \le t < d$ is a normalized nondecreasing $\curvearrowright$ chain, and if $J \neq \pm I_m$ and $\{b_1^t, b_2^t\} \in ap_I(U_t)$ for $0 \le t < d$ is a normalized chain of pairs, then*

$$U_t \text{ continuous on } [0, d) \Longrightarrow \{b_1^t, b_2^t\} \text{ is a normalized nondecreasing}$$
$$\text{continuous chain of pairs of entire inner mvf's on } [0, d).$$

Moreover, if $d < \infty$ and U_t is continuous on $[0, d]$, then the chain$\{b_1^t, b_2^t\}$ is also continuous on $[0, d]$.

If J is unitarily equivalent to J_p, then the same implications hold for $\{b_3^t, b_4^t\} \in ap_{II}(U_t)$.

Proof Theorem 3.55 guarantees that $b_1^t(\lambda)$ and $b_2^t(\lambda)$ are entire mvf's of λ for each choice of t in the interval $[0, d)$ and hence that they may be presumed to be normalized with $b_1^0(\lambda) \equiv I_p$ and $b_2^0(\lambda) \equiv I_q$, since $U_0(\lambda) \equiv I_m$. Theorem 3.64 insures that the normalized chain $\{b_1^t, b_2^t\}$, $0 \le t < d$, is nondecreasing. It remains only to prove the continuity. In view of Theorem 5.4 and its dual version (applied to $b_1^t \in \mathcal{E} \cap \mathcal{U}(I_p)$ and $(b_2^t)^\# \in \mathcal{E} \cap \mathcal{U}(-I_q)$), this is equivalent to verifying the following four identities for each choice of d_1 with $0 < d_1 < d$:

$$(a) \bigcap_{0<\varepsilon<d_1-t} \mathcal{H}(b_1^{t+\varepsilon}) = \mathcal{H}(b_1^t), \quad (b) \bigvee_{0<\varepsilon<t} \mathcal{H}(b_1^{t-\varepsilon}) = \mathcal{H}(b_1^t),$$

$$(c) \bigcap_{0<\varepsilon<d_1-t} \mathcal{H}_*(b_2^{t+\varepsilon}) = \mathcal{H}_*(b_2^t) \quad \text{and} \quad (d) \bigvee_{0<\varepsilon<t} \mathcal{H}_*(b_2^{t-\varepsilon}) = \mathcal{H}_*(b_2^t)$$

for $t \in [0, d_1)$ in (a) and (c) and $t \in (0, d_1]$ in (b) and (d). The inclusions $\supseteq$ in (a) and (c) are obvious. On the other hand, if $t < d_1 < d$, then the space on the left-hand side of (a) is a closed subspace of $\mathcal{H}(b_1^{d_1})$ that is R_α-invariant for every $\alpha \in \mathbb{C}$. Therefore, by Lemmas 3.6 and 3.7, there exists an essentially unique mvf $\widetilde{b}_1^t$ such that

$$\bigcap_{0<\varepsilon<d_1-t} \mathcal{H}(b_1^{t+\varepsilon}) = \mathcal{H}(\widetilde{b}_1^t).$$

Moreover, since

$$(\widetilde{b}_1^t)^{-1} b_1^{t+\varepsilon} \in \mathcal{E} \cap \mathcal{S}_{in}^{p\times p}, \tag{5.8}$$

the mvf $\widetilde{b}_1^t \in \mathcal{E} \cap \mathcal{S}_{in}^{p\times p}$ and may be uniquely specified by imposing the normalization condition $\widetilde{b}_1^t(0) = I_p$. Much the same sort of argument yields the identities

$$\bigcap_{0<\varepsilon<d_1-t} \mathcal{H}_*(b_2^{t+\varepsilon}) = \mathcal{H}_*(\widetilde{b}_2^t)$$

for some essentially unique mvf $\widetilde{b}_2^t \in \mathcal{S}_{in}^{q\times q}$ and

$$b_2^{t+\varepsilon}(\widetilde{b}_2^t)^{-1} \in \mathcal{E} \cap \mathcal{S}_{in}^{q\times q}. \tag{5.9}$$

Thus, $\widetilde{b}_2^t \in \mathcal{E} \cap \mathcal{S}_{in}^{q\times q}$ and may be uniquely specified by imposing the normalization $\widetilde{b}_2^t(0) = I_q$.

Now let $W_t(\lambda) = VU_t(\lambda)V^*$ and choose $s^\circ \in T_{W_{d_1}}[\mathcal{S}^{p\times q}]$. Then, in view of Theorem 4.13, the GSIP$(b_1^{d_1}, b_2^{d_1}; s^\circ)$ is completely indeterminate. Therefore, since $\mathcal{S}(\widetilde{b}_1^t, \widetilde{b}_2^t : s^\circ) \supseteq \mathcal{S}(b_1^{d_1}, b_2^{d_1}; s^\circ)$, the GSIP$(\widetilde{b}_1^t, \widetilde{b}_2^t : s^\circ)$ is also completely indeterminate. Consequently, Theorem 4.13 guarantees the existence of a unique resolvent matrix $\widetilde{W}_t \in \mathcal{E} \cap \mathcal{U}_{rR}^\circ(j_{pq})$ such that

$$\{\widetilde{b}_1^t, \widetilde{b}_2^t\} \in ap(\widetilde{W}_t).$$

Thus, as (5.8) and (5.9) imply that

$$\mathcal{S}(b_1^t, b_2^t; s^\circ) \subseteq \mathcal{S}(\widetilde{b}_1^t, \widetilde{b}_2^t : s^\circ) \subseteq \mathcal{S}(b_1^{t+\varepsilon}, b_2^{t+\varepsilon}; s^\circ),$$

it follows that

$$T_{W_t}[\mathcal{S}^{p\times q}] \subseteq T_{\widetilde{W}_t}[\mathcal{S}^{p\times q}] \subseteq T_{W_{t+\varepsilon}}[\mathcal{S}^{p\times q}]$$

and hence, by Theorem 3.64, that

$$W_t^{-1}\widetilde{W}_t \in \mathcal{U}(j_{pq}) \quad \text{and} \quad \widetilde{W}_t^{-1}W_{t+\varepsilon} \in \mathcal{U}(j_{pq}) \quad \text{for } 0 < \varepsilon < d_1 - t.$$

Therefore,

$$\mathcal{H}(W_t) \subseteq \mathcal{H}(\widetilde{W}_t) \subseteq \bigcap_{0<\varepsilon<t-d_1} \mathcal{H}(W_{t+\varepsilon}),$$

which in turn implies that

$$\mathcal{H}(W_t) = \mathcal{H}(\widetilde{W}_t),$$

thanks to Theorems 5.1 and 4.56. Thus, $\widetilde{b}_1^t = b_1^t$ and $\widetilde{b}_2^t = b_2^t$ and (a) and (c) must hold.

Formulas (b) and (d) may be verified in much the same way (though in these two cases it is the inclusions $\subseteq$ that are obvious).

If $d < \infty$ and U_t is continuous on $[0, d]$, then $U_d(\lambda) = \lim_{t\uparrow d} U_t(\lambda)$ at each point $\lambda \in \mathbb{C}$. Therefore, by Lemma 5.12, U_d also belongs to the class $\mathcal{U}_{rR}(J)$. The verification of the continuity of $b_1^t(\lambda)$ and $b_2^t(\lambda)$ on $[0, d]$ for each point $\lambda \in \mathbb{C}$ is the same as for $t < d$.

Finally, the proof of the implications for $\{b_3^t, b_4^t\}$ when $q = p$ is easily adapted from the proof for $\{b_1^t, b_2^t\}$ by considering the GCIP$(b_3^t, b_4^t; c^\circ)$ for $c^\circ \in \mathcal{C}(A_{d_1})$, $0 < d_1 < d$ and $0 \le t \le d_1$. □

Lemma 5.14 *If U_t, $0 \le t < d$, is a normalized nondecreasing $\curvearrowright$ chain of entire J-inner mvf's and if*

$$x(t) = \text{trace}\,\{2\pi\, K_0^{U_t}(0)\} \quad \textit{and} \quad \{b_1^t, b_2^t\} \in ap_I(U_t) \quad \textit{for } 0 \le t < d,$$

then

(1) *U_t is strictly increasing on $[0, d]$ $\Longleftrightarrow$ $x(t)$ is strictly increasing on $[0, d]$.*
(2) *$\{b_1^t, b_2^t\}$ is strictly increasing on $[0, d]$ $\Longrightarrow$ U_t is strictly increasing on $[0, d]$.*
(3) *If also $U_t \in \mathcal{U}_{rR}(J)$ for every $t \in [0, d)$, then U_t is strictly increasing on $[0, d]$ $\Longrightarrow$ $\{b_1^t, b_2^t\}$ is strictly increasing on $[0, d]$.*

Proof Let

$$M(t) = 2\pi\, K_0^{U_t}(0)\} = -i\left(\frac{\partial}{\partial\lambda}U_t\right)(0)J$$

and suppose that $0 \le t_1 < t_2 < d$. Then the mvf $\Theta = U_{t_1}^{-1}U_{t_2}$ belongs to the class $\mathcal{E} \cap \mathcal{U}^\circ(J)$ and $U_{t_2} = U_{t_1}\Theta$. Therefore, since

$$M(t) = -i\left(\frac{\partial}{\partial\lambda}U_t\right)(0)J \quad \text{and} \quad -\, i\Theta'(0)J = 2\pi K_0^\Theta(0),$$

it is readily seen that

$$M(t_2) = M(t_1) + 2\pi K_0^\Theta(0) \ge M(t_1)$$

and hence that

$$x(t_2) = x(t_1) \Longleftrightarrow M(t_2) = M(t_1) \Longleftrightarrow K_0^\Theta(0) = 0.$$

Thus, as

$$v^* K_0^\Theta(\lambda) u = \langle K_0^\Theta u, K_\lambda^\Theta v\rangle_{\mathcal{H}(\Theta)}$$

for every choice of $u, v \in \mathbb{C}^m$ and $\lambda \in \mathbb{C}$, it follows readily from the Cauchy–Schwarz inequality in $\mathcal{H}(\Theta)$ that

$$K_0^{\Theta}(0) = 0 \Longleftrightarrow K_0^{\Theta}(\lambda) = 0 \quad \text{for every } \lambda \in \mathbb{C} \Longleftrightarrow \Theta(\lambda) \equiv I_m,$$

which in turn holds if and only if $U_{t_2} = U_{t_1}$, contrary to assumption. This serves to justify (1).

Assertions (2) and (3) follow from Theorem 3.64. □

Theorem 5.15 *Let $U_t(\lambda)$, $0 \le t \le d$, be the matrizant of a canonical integral system (2.1) with $J = V^* j_{pq} V$ for some unitary matrix $V \in \mathbb{C}^{m\times m}$ and mass function $M(t)$ that meets the constraints (1.22), let $\{b_1^t, b_2^t\} \in ap_I(U_t)$ with $b_1^t(0) = I_p$ and $b_2^t(0) = I_q$ for $0 \le t \le d$, $b_1^0(\lambda) \equiv I_p$ and $b_2^0(\lambda) \equiv I_q$, and suppose further that $U_d \in \mathcal{U}_{AR}(J)$. Then $\{b_1^t, b_2^t\}$, $0 \le t \le d$, is a normalized nondecreasing continuous chain of pairs of entire inner mvf's.*

Proof This follows from Theorem 5.13. □

5.4 Type functions for chains

The functions

$$\tau(t) = \tau_{b_t}^- \tag{5.10}$$

and

$$\varphi(t) = \text{trace}\left\{-i\left(\frac{\partial}{\partial\lambda} b_t\right)(0)\right\} = \text{trace}\left\{2\pi k_0^{b_t}(0)\right\} = \delta_{b_t}^- \tag{5.11}$$

will play a useful role in the study of normalized nondecreasing $\curvearrowright$ continuous chains $\{b_t\}$, $0 \le t$, of entire inner mvf's. Analogously, for normalized nondecreasing continuous chains of pairs $\{b_1^t, b_2^t\}$, $0 \le t < d$, of entire inner mvf's, we set

$$\tau_1(t) = \tau_{b_1^t}^-, \quad \tau_2(t) = \tau_{b_2^t}^-, \quad \varphi_1(t) = \delta_{b_1^t}^- \quad \text{and} \quad \varphi_2(t) = \delta_{b_2^t}^-.$$

Lemma 5.16 *Let $b \in \mathcal{E} \cap \mathcal{S}_{\text{in}}^{p\times p}$, and let $\{b_t : 0 \le t \le d\}$ be a normalized nondecreasing $\curvearrowright$ continuous chain of entire inner mvf's such that $b_d(\lambda) = b(\lambda)$ (and $b_0(\lambda) \equiv I_p$, since the chain is normalized). Then the functions $\tau(t)$ and $\varphi(t)$ defined in (5.10) and (5.11) have the following properties:*

1. $\tau(t) - \tau(s) \le \text{type}\{b_s^{-1} b_t\}$ *for* $0 \le s \le t \le d$.
2. *The functions $\tau(t)$ and $\varphi(t)$ are continuous and nondecreasing on the interval $[0, d]$ and $\tau(0) = \varphi(0) = 0$.*
3. $\tau(t) - \tau(s) \le \varphi(t) - \varphi(s) \le p\,\text{type}\{b_s^{-1} b_t\}$ *for* $0 \le s \le t \le d$.
4. *$\tau(t)$ and $\varphi(t)$ are strictly increasing on $[0, d]$ if and only if the chain b_t, $0 \le t \le d$, is strictly increasing $\curvearrowright$.*

Proof If $s \leq t$, then the inequalty

$$\ln \|b_t\| = \ln \|b_s b_s^{-1} b_t\| \leq \ln \|b_s\| + \ln \|b_s^{-1} b_t\|$$

leads easily to (1). Next, Theorem 3.36 implies that

$$\det b_t(\lambda) = e^{i\lambda\varphi(t)}. \tag{5.12}$$

The function $\varphi(t)$ is a continuous nondecreasing function on $[0, d]$, since $k_0^t(0) = k_0^{b_t}(0)$ is continuous on the interval $[0, d]$, by Theorem 5.1, and

$$k_0^t(0) - k_0^s(0) = k_0^{b_s^{-1} b_t}(0) \geq 0 \quad \text{for } 0 \leq s \leq t \leq d.$$

Thus, (2)–(4) follow from the following string of inequalities that are obtained from Theorem 3.36:

$$\begin{aligned} 0 \leq \tau(t) - \tau(s) = \text{type}(b_t) - \text{type}(b_s) &\leq \text{type}(b_s^{-1} b_t) \leq \delta_{b_s^{-1} b_t} \\ &\leq \text{ptype}(b_s^{-1} b_t), \end{aligned}$$

and the equality

$$\delta^-_{b_s^{-1} b_t} = \varphi(t) - \varphi(s).$$

□

Lemma 5.17 *Let b_x, $0 \leq x \leq \ell$, be the matrizant of the canonical differential system (2.8) with $J = I_p$ and Hamiltonian $H(x)$ subject to the normalization condition in (1.19). Then*

$$\text{type} \det b(x) = x \quad \textit{for } 0 \leq x \leq \ell$$

and

$$\tau(x) = \text{type}\,(b_x) = \tau_{b_x}$$

is strictly increasing and absolutely continuous with $\tau'(x) \leq 1$ a.e. on the interval $[0, \ell]$.

Proof The first assertion follows from (5) of Theorem 2.5. The second then follows from (3) of Lemma 5.16, since $\varphi(x) = x$ for $0 \leq x \leq \ell$ in this setting. □

The **notation**

$$\mathcal{E}(a_1, a_2) \cap \mathcal{X} = \{U \in \mathcal{E} \cap \mathcal{X} : \tau_U^- = a_1 \quad \text{and} \quad \tau_U^+ = a_2\}, \tag{5.13}$$

for $\mathcal{X} \subseteq \mathcal{U}(J)$,

$$U_P(\lambda) = Q + PU(\lambda)P \quad \text{and} \quad U_Q(\lambda) = P + QU(\lambda)Q, \quad \text{with}$$

(5.14)

$$P = P_+ = \frac{(I_m + J)}{2} \quad \text{and} \quad Q = P_- = \frac{(I_m - J)}{2} \quad \text{for } U \in \mathcal{U}(J),$$

will be useful. If $W \in \mathcal{U}(j_{pq})$, then

$$W_P(\lambda) = \begin{bmatrix} w_{11}(\lambda) & 0 \\ 0 & I_q \end{bmatrix} \quad \text{and} \quad W_Q(\lambda) = \begin{bmatrix} I_p & 0 \\ 0 & w_{22}(\lambda) \end{bmatrix}. \tag{5.15}$$

The next theorem presents a number of equalities and inequalities connected with type; the simple example

$$U(\lambda) = W(\lambda) = \begin{bmatrix} e^{i\lambda a_1} I_p & 0 \\ 0 & e^{-i\lambda a_2} I_q \end{bmatrix}$$

for $J = j_{pq}$ may help to orient the reader.

Theorem 5.18 *Let $U \in \mathcal{E}(a_1, a_2) \cap \mathcal{U}(J)$ with $J \neq \pm I_m$. Let P, Q and $U_P(\lambda), U_Q(\lambda)$ be defined by (5.14). Then:*

(1) $a_1 = \tau_U^- = \tau_{U_P}^- \geq 0$ *and* $a_2 = \tau_U^+ = \tau_{U_Q}^+ \geq 0$.
(2) $e_{a_2} U \in \mathcal{N}_+^{m\times m}$ *and* $e_{a_1} U^\# \in \mathcal{N}_+^{m\times m}$.
(3) $U \in \mathcal{E} \cap \mathcal{U}_S(J) \Longleftrightarrow U \in \mathcal{E}(0,0) \cap \mathcal{U}(J) \Longleftrightarrow \tau_U^- = \tau_U^+ = 0$.
(4) $\tau_{U_P}^- \leq \delta_{U_P}^- \leq p\tau_{U_P}^-$ *and* $\tau_{U_Q}^+ \leq \delta_{U_Q}^+ \leq q\tau_{U_Q}^+$.
(5) $e_{-a_1} U \in \mathcal{U}(J) \Longleftrightarrow \delta_{U_P}^- = p\tau_{U_P}^-$.
(6) $e_{a_2} U \in \mathcal{U}(J) \Longleftrightarrow \delta_{U_Q}^+ = q\tau_{U_Q}^+$.
(7) $\delta_U^+ = \delta_{U_Q}^+ - \delta_{U_P}^- = -\delta_U^-$.
(8) $\tau_{U_P}^- = 0 \Longleftrightarrow \delta_{U_P}^- = 0 \Longleftrightarrow \delta_U^+ = \delta_{U_Q}^+ \Longleftrightarrow \tau_U^- = 0 \Longleftrightarrow U^\# \in \mathcal{N}_+^{m\times m}$
(9) $\tau_{U_Q}^+ = 0 \Longleftrightarrow \delta_{U_Q}^+ = 0 \Longleftrightarrow \delta_U^- = \delta_{U_P}^- \Longleftrightarrow \tau_U^+ = 0 \Longleftrightarrow U \in \mathcal{N}_+^{m\times m}$.
(10) *If* $\tau_U^+ = 0$, *then* $e_{-a_1} U \in \mathcal{U}(J) \Longleftrightarrow \delta_U^- = p\tau_U^-$.
(11) *If* $\tau_U^- = 0$, *then* $e_{a_2} U \in \mathcal{U}(J) \Longleftrightarrow \delta_U^+ = q\tau_U^+$.
Moreover, if $U_1, U_2 \in \mathcal{U}(J)$ and $U = U_1 U_2$, then
(12) $\delta_{U_P}^- = \delta_{(U_1)_P}^- + \delta_{(U_2)_P}^-$ *and* $\delta_{U_Q}^+ = \delta_{(U_1)_Q}^+ + \delta_{(U_2)_Q}^+$.

Proof It suffices to check these assertions for $J = j_{pq}$ and $W \in \mathcal{E}(a_1, a_2) \cap \mathcal{U}(j_{pq})$. Then the PG transform $S = PG(W)$ and each of its four blocks s_{ij}, $i, j = 1, 2$, are holomorphic and contractive in $\mathbb{C}_+$. The formulas

$$s_{11} = (w_{11}^\#)^{-1} \quad \text{and} \quad s_{22} = w_{22}^{-1}$$

lead readily to (1) and (4) with the help of (3.121), (3.123) and (2) of Theorem 3.56, respectively.

Let $\{b_1, b_2\} \in ap(W)$. Then, in view of Theorem 3.56, $b_1 \in \mathcal{E} \cap \mathcal{S}_{\text{in}}^{p\times p}$, $b_2 \in \mathcal{E} \cap \mathcal{S}_{\text{in}}^{q\times q}$, $a_1 = \tau_U^- = \tau_{b_1}^-$, $a_2 = \tau_U^+ = \tau_{b_2}^-$ and

$$U \in \mathcal{U}_S(J) \Longleftrightarrow a_1 = a_2 = 0 \Longleftrightarrow \tau_{b_1} = \tau_{b_2} = 0$$
$$\Longleftrightarrow b_1(\lambda) \text{ and } b_2(\lambda) \text{ are constant mvf's.}$$

Thus, (3) holds. Moreover, since $e_{a_2}b_2^{-1} \in \mathcal{S}^{p\times p}$ by (1) of Lemma 3.41, the formulas

$$w_{21} = -w_{22}s_{21}, \quad w_{12} = s_{12}w_{22}, \quad w_{11} = s_{11} + w_{12}w_{22}^{-1}w_{21} = s_{11} - w_{12}s_{21}$$

imply that $e_{a_2}W \in \mathcal{N}_+^{m\times m}$. A similar argument shows that $e_{a_1}W^{\#} \in \mathcal{N}_+^{m\times m}$. Therefore, (2) holds.

Next, (5) and (6) follow from the equivalences

$$\delta_{U_P}^- = p\tau_U^- \Longleftrightarrow \delta_{b_1}^- = p\tau_{b_1}^- \Longleftrightarrow b_1(\lambda) = e_{\alpha_1}b_1(0) \quad \text{with } \alpha_1 = \delta_{b_1}^-/p$$

and

$$\delta_{U_Q}^+ = q\tau_U^+ \Longleftrightarrow \delta_{b_2}^- = q\tau_{b_2}^- \Longleftrightarrow b_2(\lambda) = e_{\alpha_2}b_2(0) \quad \text{with } \alpha_2 = \delta_{b_2}^-/q,$$

which depend in part on (4) of Theorem 3.36, and Lemma 3.61.

The equalities in (7) follow from (3.119), which yields the formula

$$\det W = \frac{\det s_{11}}{\det s_{22}} = \gamma\frac{\det b_1}{\det b_2} \tag{5.16}$$

with $\gamma \in \mathbb{C}$ and $|\gamma| = 1$, since

$$\begin{aligned}\det\{\varphi_2(\mu)\varphi_2(\mu)^*\} &= \det\{s_{22}(\mu)s_{22}(\mu)^*\} = \det\{I_q - s_{21}(\mu)s_{21}(\mu)^*\}\\ &= \det\{I_p - s_{21}^*(\mu)s_{21}(\mu)\} = \det\{s_{11}(\mu)^*s_{11}(\mu)\}\\ &= \det\{\varphi_1(\mu)^*\varphi_1(\mu)\}\end{aligned}$$

for almost all points $\mu \in \mathbb{R}$. Therefore, since $\det\varphi_1$ and $\det\varphi_2$ are outer functions such that $|\det\varphi_1(\mu)| = |\det\varphi_2(\mu)|$ a.e. on $\mathbb{R}$, there exists a constant $\gamma \in \mathbb{T}$ such that $\det\varphi_1(\lambda) = \gamma\det\varphi_2(\lambda)$.

Assertions (8)–(11) follow easily from (1)–(6); (12) follows from (3.150), (3.151) and (4) of Theorem 3.11. □

We remark that if the mvf's $W(\lambda)$, $b_1(\lambda)$ and $b_2(\lambda)$ are normalized in the usual way at $\lambda = 0$, then $\gamma = 1$ in formula (5.16).

Theorem 5.19 *Let $\{U_t\}$, $0 \le t \le d$, be the matrizant of a canonical integral system (2.1) with J unitarily equivalent to j_{pq}, let $\{b_1^t, b_2^t\} \in ap_I(U_t)$ with $b_1^t(0) = I_p$ and $b_2^t(0) = I_q$ for $0 \le t \le d$ and let*

$$\tau_1(t) = \tau_{U_t}^-, \quad \tau_2(t) = \tau_{U_t}^+ \quad \text{and } \tau(t) = \tau_1(t) + \tau_2(t) \text{ for } 0 \le t \le d$$

and

$$\varphi_1(t) = \delta_{(U_t)_P}^-, \quad \varphi_2(t) = \delta_{(U_t)_Q}^+ \quad \text{and } \varphi(t) = \varphi_1(t) + \varphi_2(t) \text{ for } 0 \le t \le d.$$

Then:

(1) *$\tau_j(0) = \varphi_j(0) = 0$ and $\tau_j(t)$ and $\varphi_j(t)$ are continuous nondecreasing functions on the interval $[0, d]$ for $j = 1, 2$.*

(2) $\tau_1(t) - \tau_1(s) \le \varphi_1(t) - \varphi_1(s) \le \text{ptype}\{b_s^{-1}b_t\}$ *and* $\tau_2(t) - \tau_2(s) \le \varphi_2(t) - \varphi_2(s) \le q\{\tau_2(t) - \tau_2(s)\}$ *for* $0 \le s \le t \le d$.

(3) $\tau_j(t) = \tau_{b_j^t}^-$ *and* $\varphi_j(t) = \delta_{b_j^t}^-$ *for* $j = 1, 2$.

(4) *If* $q = p$ *and* $\{b_3^t, b_4^t\} \in ap_{II}(U_t)$ *with* $b_3^t(0) = b_4^t(0) = I_p$ *for* $0 \le t \le d$, *then* $\tau_1(t) = \tau_{b_3^t}^-$, $\varphi_1(t) = \delta_{b_3^t}^-$, $\tau_2(t) = \tau_{b_4^t}^-$ *and* $\varphi_2(t) = \delta_{b_4^t}^-$.

(5) *If* $\{U_t\}$ *is a strictly increasing chain, then* $\tau(t)$ *and* $\varphi(t)$ *are both strictly increasing functions of* t *on* $[0, d]$.

Proof (1)–(4) follow from Theorem 4.24 and Lemma 5.16. (5) follows from the inequalities

$$0 \le \tau(t) - \tau(s) \le \varphi(t) - \varphi(s) = \delta_{(b_1^s)^{-1}b_1^t} + \delta_{b_2^t(b_2^s)^{-1}} \le c(\tau(t) - \tau(s))$$

where $c = \max\{p, q\}$. □

Lemma 5.20 *Let* $M(t)$ *be a continuous nondecreasing* $m \times m$ *mvf on the interval* $[0, d]$ *with* $M(d) \ne M(0) = 0$, *let* $U_t(\lambda) = U(t, \lambda)$, $0 \le t \le d$, *denote the matrizant of the corresponding integral equation (2.1) and let*

$$\tau_1(t) = \tau_{U_t}^- \quad \text{and} \quad \tau_2(t) = \tau_{U_t}^+ \quad \text{for } 0 \le t \le d.$$

Then the following conclusions hold for $0 \le s \le t \le d$:

(1) *If* $U_d \in \mathcal{E}(a, 0)$ *for some* $a > 0$, *then* $J \ne \pm I_m$, $\tau_2(t) = 0$ *and*

$$\tau_1(t) - \tau_1(s) \le \text{trace}\{[M(t) - M(s)]J\} \le \text{trace}\{M(t) - M(s)\} \tag{5.17}$$

and hence if $M(t)$ *is absolutely continuous on* $[0, d]$, *then* $\tau_1(t)$ *is absolutely continuous on* $[0, d]$.

(2) *If* $U_d \in \mathcal{E}(0, a)$ *for some* $a > 0$, *then* $J \ne \pm I_m$, $\tau_1(t) = 0$ *and*

$$\tau_2(t) - \tau_2(s) \le \text{trace}\{[M(t) - M(s)](-J)\} \le \text{trace}\{M(t) - M(s)\} \tag{5.18}$$

and hence if $M(t)$ *is absolutely continuous on* $[0, d]$, *then* $\tau_2(t)$ *is absolutely continuous on* $[0, d]$.

Proof Let $\Omega = U_s^{-1}U_t$ for $0 \le s \le t \le d$. Then $\Omega(0) = I_m$ and, by Theorem 2.5, $\Omega \in \mathcal{E} \cap \mathcal{U}(J)$. Therefore, the inequalities

$$\ln \|U_t(\lambda)\| - \ln \|U_s(\lambda)\| \le \ln \|\Omega(\lambda)\|$$

and

$$\ln \|U_t^\#(\lambda)\| - \ln \|U_s^\#(\lambda)\| \le \ln \|\Omega^\#(\lambda)\|$$

hold for every point $\lambda \in \mathbb{C}$. Thus it is readily seen that

$$\tau_1(t) - \tau_1(s) \le \tau_\Omega^- \quad \text{and} \quad \tau_2(t) - \tau_2(s) \le \tau_\Omega^+.$$

Moreover, by Theorem 5.18,

$$\tau_\Omega^- = \tau_{\Omega_P}^- \le \delta_{\Omega_P}^- \quad \text{and} \quad \tau_\Omega^+ = \tau_{\Omega_Q}^+ \le \delta_{\Omega_Q}^+.$$

The rest of the proof splits into two cases.

Suppose first that $U_d \in \mathcal{E}(a,0)$ for some $a > 0$. Then $U_d \in \mathcal{N}_+^{m\times m}$ and hence, by Lemma 3.54, the same holds true for Ω. Therefore, $\tau_\Omega^+ = 0$ and hence, by (9) of Theorem 5.18, $\delta_{\Omega_P}^- = \delta_\Omega^-$. However, by (5) of Theorem 2.5,

$$\delta_\Omega^- = \mathrm{trace}\{[M(t) - M(s)]J\}$$

and the inequality asserted in (1) drops out easily by combining estimates.

The proof of (2) goes through in much the same way. □

Theorem 5.21 *Let $M(t)$ be a continuous nondecreasing $m \times m$ mvf on the interval $[0,d]$ with $M(d) \ne M(0) = 0$, let $U_t(\lambda) = U(t,\lambda)$, $0 \le t \le d$, denote the matrizant of the corresponding integral equation (2.1) with J unitarily equivalent to j_{pq} and let*

$$\tau_1(t) = \tau_{U_t}^-, \quad \tau_2(t) = \tau_{U_t}^+ \quad \text{and} \quad \tau(t) = \tau_1(t) + \tau_2(t) \quad \text{for } 0 \le t \le d.$$

Then for $0 \le s \le t \le d$, the following sets of inequalities prevail:

(1) *If $\delta_{(U_t)_Q}^+ = q\tau_{U_t}^+$ for every $t \in [0,d]$, then*

$$\begin{aligned}\tau(t) - \tau(s) &\le \mathrm{trace}\{[M(t) - M(s)]J\} + \{\tau_2(t) - \tau_2(s)\}m \\ &\le \mathrm{trace}\{M(t) - M(s)\} + [\tau_2(t) - \tau_2(s)](p - q).\end{aligned}$$

(2) *If $\delta_{(U_t)_P}^- = p\tau_{U_t}^-$ for every $t \in [0,d]$, then*

$$\begin{aligned}\tau(t) - \tau(s) &\le \mathrm{trace}\{[M(t) - M(s)](-J)\} + \{\tau_1(t) - \tau_1(s)\}m \\ &\le \mathrm{trace}\{M(t) - M(s)\} + [\tau_1(t) - \tau_1(s)](q - p).\end{aligned}$$

(3) *If $\delta_{(U_t)_Q}^+ = q\tau_{U_t}^+$ and $\delta_{(U_t)_P}^- = p\tau_{U_t}^-$ for every $t \in [0,d]$, then*

$$\tau(t) - \tau(s) \le \mathrm{trace}\{M(t) - M(s)\}$$

and hence if $M(t)$ is absolutely continuous on $[0,d]$ then $\tau(t)$, $\tau_1(t)$ and $\tau_2(t)$ are absolutely continuous on $[0,d]$.

Proof Suppose first that condition (1) is in force. Then, by (6) of Theorem 5.18, the mvf $\widetilde{U}_t = e_{\tau_2(t)}U_t$ belongs to $\mathcal{E}(\tau(t),0) \cap \mathcal{U}(J)$. Moreover,

$$\widetilde{\tau}_1(t) = \tau_{\widetilde{U}_t}^- = \tau(t)$$

and

$$\widetilde{M}(t) = -i\left(\frac{\partial \widetilde{U}_t}{\partial \lambda}\right)(0)J = M(t) + \tau_2(t)J$$

are both nondecreasing functions on $[0, d]$, since

$$\widetilde{U}_s^{-1}\widetilde{U}_t \in \mathcal{U}(J) \quad \text{for } 0 \le s \le t \le d$$

(see Lemma 8.49). Therefore,

$$\begin{aligned}\widetilde{\tau}_1(t) - \widetilde{\tau}_1(s) &\le \operatorname{trace}\{[\widetilde{M}(t) - \widetilde{M}(s)]J\} \quad \text{(by Lemma 5.20)} \\ &\le \operatorname{trace}\{\widetilde{M}(t) - \widetilde{M}(s)\}\end{aligned}$$

and the asserted inequalities now follow from the preceding two inequalities by direct substitution.

Suppose next that the condition in (2) is in force. Then, by (5) of Theorem 5.18, the mvf $\widetilde{U}_t = e_{-\tau_1(t)}U_t$ belongs to $\mathcal{E}(0, \tau(t)) \cap \mathcal{U}(J)$. Moreover,

$$\widetilde{\tau}_2(t) = \tau_{\widetilde{U}_t}^+ = \tau(t)$$

and

$$\widetilde{M}(t) = -i\left(\frac{\partial \widetilde{U}_t}{\partial \lambda}\right)(0)J = M(t) - \tau_1(t)J$$

are both nondecreasing functions on $[0, d]$. The rest now follows from (2) of Lemma 5.20, much as before.

Finally, if both conditions are in force, then the inequality in (3) follows from the second inequality in (1) when $p \le q$ and from the second inequality in (2) when $q \le p$. □

Corollary 5.22 *Let U_t, $0 \le t < d$, be the matrizant of the canonical differential system (1.1) with J unitarily equivalent to j_{pq} and a locally summable Hamiltonian $H(t)$ that is subject to the constraints*

$$H(t) \ge 0 \quad \textit{and} \quad \operatorname{trace} H(t) = 1 \ \textit{a.e. in } [0, d),$$

and suppose further that

$$\delta^+_{(U_t)_Q} = q\tau^+_{U_t} \quad \textit{and} \quad p \le q \quad \textit{or} \quad \delta^-_{(U_t)_P} = p\tau^-_{U_t} \quad \textit{and} \quad q \le p.$$

Then the functions $\tau_1(t)$ and $\tau_2(t)$ that are defined in Theorem 5.21 are locally absolutely continuous nondecreasing functions on $[0, d)$ with $\tau_1(0) = \tau_2(0) = 0$ and

$$\tau_1'(t) + \tau_2'(t) \le 1 \quad \textit{a.e. on } [0, d).$$

If the system is regular, then $\tau_1(t)$ and $\tau_2(t)$ are absolutely continuous on $[0, d]$.

Proof This follows from Theorems 5.21 and 5.19. □

5.5 Supplementary notes

Most of the results in this chapter are adapted from the papers [ArD97], [ArD00a] and [ArD00b], which contain much additional information. The stronger versions of some of these theorems rest largely on Theorem 4.56 and the class $U_{AR}((J)$ that was introduced in this monograph. Lemma 5.14 is much stronger than its counterpart lemma 3.3 in [ArD00a]; it will play a very useful role in the study of the inverse monodromy problem in Chapter 8; see, e.g., Theorem 8.16. The applicability of the class $\mathcal{U}_{AR}(J)$ to the inverse problems that will be considered in later chapters is guaranteed by Corollary 5.5 and Theorem 5.15. The class $\mathcal{U}_{BR}(J)$ that was introduced in the recent paper [ArD12a] enters in Theorem 5.8, which is also one of the main results of this chapter.

6

The bitangential direct input scattering problem

This chapter focuses on the set $\mathcal{S}_{\text{scat}}^d(dM)$ of input scattering matrices for canonical integral systems (2.1) with $J = j_{pq}$ and matrizant $W_t(\lambda)$ for $0 \leq t < d$. The set $\{s(\omega) : s \in \mathcal{S}_{\text{scat}}^d(dM)\}$ is identified as a matrix ball and bounds are obtained on left and right semiradii of this ball, as well as formulas for the ranks of these semiradii. A limit ball/limit point classification of these ranks analogous to the classical Weyl–Titchmarsh classification for spectral problems for differential equations is presented. A characterization of the unique input scattering matrix when the left semiradius of this ball is zero is also given.

6.1 The set $\mathcal{S}_{\text{scat}}^d(dM)$ of input scattering matrices

Let W_t, $0 \leq t < d$, be the matrizant of the canonical integral system

$$y(t, \lambda) = y(0, \lambda) + i\lambda \int_0^t y(s, \lambda) dM(s) j_{pq} \quad \text{for } 0 \leq t < d, \tag{6.1}$$

in which the $m \times m$ mvf (the mass function)

$$M(t) \text{ is nondecreasing and continuous on } [0, d) \text{ with } M(0) = 0. \tag{6.2}$$

Since $W_t \in \mathcal{U}(j_{pq})$ and $W_{t_1}^{-1} W_{t_2} \in \mathcal{U}(j_{pq})$ when $0 \leq t_1 \leq t_2 < d$, the inclusions

$$\mathcal{S}^{p\times q} \subseteq \mathcal{D}(T_{W_t}) \quad \text{and} \quad T_{W_t}[\mathcal{S}^{p\times q}] \subseteq \mathcal{S}^{p\times q}, \quad 0 \leq t < d, \tag{6.3}$$

are in force and

$$T_{W_{t_2}}[\mathcal{S}^{p\times q}] \subseteq T_{W_{t_1}}[\mathcal{S}^{p\times q}] \quad \text{when} \quad 0 \leq t_1 \leq t_2 < d, \tag{6.4}$$

by Lemma 3.58 and Theorem 2.5.

The set of **input scattering matrices** of the canonical integral system (6.1) with mass function $M(t)$ on the interval $[0, d)$ that is subject to the conditions (6.2) is defined as the intersection

$$\mathcal{S}_{\text{scat}} = \mathcal{S}_{\text{scat}}^d(dM) = \bigcap_{0 \le t < d} T_{W_t}[\mathcal{S}^{p \times q}]. \tag{6.5}$$

The nesting property (6.4) and the sequential compactness of the class $\mathcal{S}^{p\times q}$ insure that

$$\mathcal{S}_{\text{scat}}^d(dM) \neq \emptyset. \tag{6.6}$$

We shall refer to the problem of describing the set $\mathcal{S}_{\text{scat}}$, given a canonical integral system of the form (6.1), as the **direct input scattering problem**. This problem is simple for regular canonical systems:

$$\mathcal{S}_{\text{scat}}^d(dM) = T_{W_d}[\mathcal{S}^{p\times q}], \tag{6.7}$$

where $W_d(\lambda)$ is the monodromy matrix of the given system. The terminology **input scattering matrix** originates from this case: Let

$$y(t,\lambda) = [\varphi^-(t,\lambda) \quad \varphi^+(t,\lambda)]$$

be a $1 \times m$ mvf solution of the canonical integral system (6.1) with components $\varphi^-(t,\lambda)$ of size $1 \times p$ and $\varphi^+(t,\lambda)$ of size $1 \times q$ and suppose that $y_t(\lambda) = y(t,\lambda)$, $0 \le t \le d$, is subject to the boundary condition

$$y(d,\lambda)\begin{bmatrix} \varepsilon(\lambda) \\ I_q \end{bmatrix} = 0$$

for some *passive load* $\varepsilon \in \mathcal{S}^{p\times q}$ at the right end point. Then, since

$$y(t,\lambda) = y(0,\lambda)W_t(\lambda)$$

for $0 \le t \le d$, it follows that

$$0 = y(0,\lambda)W_d(\lambda)\begin{bmatrix} \varepsilon(\lambda) \\ I_q \end{bmatrix}$$

and hence that

$$0 = y(0,\lambda)\begin{bmatrix} s(\lambda) \\ I_q \end{bmatrix}, \quad \text{with} \quad s(\lambda) = T_{W_d}[\varepsilon],$$

since the bottom block row of

$$W_d(\lambda)\begin{bmatrix} \varepsilon(\lambda) \\ I_q \end{bmatrix}$$

is invertible in $\mathbb{C}_+$. Thus,

$$\varphi^+(d,\lambda) = -\varphi^-(d,\lambda)\varepsilon(\lambda) \quad \Longrightarrow \quad \varphi^+(0,\lambda) = -\varphi^-(0,\lambda)s(\lambda).$$

These relations reinforce the interpretation of $\varphi^-(t,\lambda)$ and $\varphi^+(t,\lambda)$ as the Fourier transforms of the complex amplitudes of the incoming and outgoing waves, respectively, and exhibit $s(\lambda)$ as an input scattering matrix.

The sets of input scattering matrices for the differential systems (2.72) and (2.8) with $J = j_{pq}$ are also defined in terms of the corresponding matrizants $Y_x(\lambda)$ and $U_x(\lambda)$, respectively:

$$\mathcal{S}^d_{\text{scat}}(N,\mathcal{V}) = \bigcap_{0\le x<d} T_{Y_x}[\mathcal{S}^{p\times q}] \quad\text{and}\quad \mathcal{S}^d_{\text{scat}}(Hdx) = \bigcap_{0\le x<d} T_{U_x}[\mathcal{S}^{p\times q}].$$

Moreover, if the system (2.8) corresponds to the system (2.72) via (2.79) and (2.74), then

$$\mathcal{S}^d_{\text{scat}}(N,\mathcal{V}) = \mathcal{S}^d_{\text{scat}}(Hdx),$$

since

$$Y_x(\lambda) = U_x(\lambda)Y(x)\,, \quad \text{where} \quad Y(x) \in \mathcal{U}_{\text{const}}(j_{pq}), \quad 0 \le x < \ell,$$

and, for such $Y(x)$,

$$T_{Y(x)}[\mathcal{S}^{p\times q}] = \mathcal{S}^{p\times q}\,.$$

Input scattering for arbitrary $J \neq \pm I_m$

If $U_t(\lambda)$ is the matrizant of a system (2.1) with mass function $M(t)$ subject to (1.21) and signature matrix J with

$$p = \operatorname{rank}(I_m + J) \ge 1 \quad\text{and}\quad q = \operatorname{rank}(I_m - J) \ge 1 \quad\text{but}\quad J \neq j_{pq},$$

then

$$J = V^* j_{pq} V \;\; \text{for some unitary } m\times m \text{ matrix } V$$

and $W_t(\lambda) = VU_t(\lambda)V^*$ is the matrizant of a system of the form (6.1) with mass function $VM(t)V^*$ and signature matrix j_{pq}. Thus, it is natural to view the set

$$\mathcal{S}^d_{\text{scat}}(dM) = \bigcap_{0\le t<d} T_{VU_tV^*}[\mathcal{S}^{p\times q}]$$

as the set of input scattering matrices for the original system. With this convention, all the results that are obtained for the direct and bitangential inverse input scattering problem for systems (6.1) with $J = j_{pq}$ are easily reformulated for systems (6.1) with j_{pq} replaced by an arbitrary signature matrix $J \neq \pm I_m$. Moreover, in this case, there is exactly one normalized nondecreasing chain of pairs $\{b_1^t, b_2^t\}$ of entire inner mvf's of sizes $p\times p$ and $q \times q$ such that

$$\{b_1^t, b_2^t\} \in ap(VU_tV^*), \quad 0 \le t < d,$$

associated with each matrizant U_t, i.e., $\{b_1^t, b_2^t\} \in ap_I(U_t)$. Theorem 5.15 guarantees that this chain of pairs will be continuous in the variable t when

$$U_t \in \mathcal{U}_{rR}(J) \quad \text{for } 0 \le t < d. \tag{6.8}$$

The **bitangential direct input scattering problem** is to study the connections between the properties of $M(t)$, $U_t(\lambda)$, $\{b_1^t, b_2^t\}$ and $\mathcal{S}^d_{\text{scat}}(dM)$.

6.2 Parametrization of $\mathcal{S}^d_{\text{scat}}(dM)$ in terms of Redheffer transforms

It is advantageous to work in the more general setting of chains, i.e., we shall assume the given data is a nondecreasing $\curvearrowright$ continuous chain of entire j_{pq}-inner mvf's $\{W_t(\lambda):\ 0 \le t < d\}$ and shall obtain a description of the set $\bigcap_{0\le t<d} T_{W_t}[\mathcal{S}^{p\times q}]$ by reexpressing the linear fractional transformations $T_{W_t}[\mathcal{S}^{p\times q}]$ in terms of the Potapov–Ginzburg transform

$$S_t(\lambda) = \begin{bmatrix} w_{11}^t & w_{12}^t(\lambda) \\ 0 & I_q \end{bmatrix} \begin{bmatrix} I_p & 0 \\ w_{21}^t & w_{22}^t(\lambda) \end{bmatrix}^{-1} \tag{6.9}$$

of $W_t(\lambda)$. The blocks $s_{ij}^t(\lambda)$ of $S_t(\lambda)$ are given by the formula

$$\begin{bmatrix} s_{11}^t & s_{12}^t \\ s_{21}^t & s_{22}^t \end{bmatrix} = \begin{bmatrix} w_{11}^t - w_{12}^t(w_{22}^t)^{-1}w_{21}^t & w_{12}^t(w_{22}^t)^{-1} \\ -(w_{22}^t)^{-1}w_{21}^t & (w_{22}^t)^{-1} \end{bmatrix}; \tag{6.10}$$

and thus, by Lemma 3.17,

$$T_{W_t}[\varepsilon] = R_{S_t}[\varepsilon], \tag{6.11}$$

where

$$R_{S_t}[\varepsilon] = s_{12}^t(\lambda) + s_{11}^t(\lambda)\varepsilon(\lambda)\{I_q - s_{21}^t(\lambda)\varepsilon(\lambda)\}^{-1}s_{22}^t(\lambda). \tag{6.12}$$

This formula is more convenient than the left-hand side of (6.11) for analyzing the limit as $t \uparrow d$ because $S_t \in \mathcal{S}^{m\times m}$. The term $\{I_q - s_{21}^t(\lambda)\varepsilon(\lambda)\}^{-1}$ is well defined for $W_t \in \mathcal{E} \cap \mathcal{U}(j_{pq})$ because in that case $\|s_{21}^t(\lambda)\| < 1$ for $\lambda \in \mathbb{C}_+$. Since $S_t \in \mathcal{S}^{m\times m}$, we can always choose a sequence $t_j \uparrow d$ such that the limit

$$S_*(\lambda) = \lim_{j\uparrow\infty} S_{t_j}(\lambda)$$

exists. However, although $S_*(\lambda) \in \mathcal{S}^{m\times m}$, we cannot guarantee that $R_{S_*}[\varepsilon]$ is well defined for all $\varepsilon \in \mathcal{S}^{p\times q}$. To overcome this difficulty we shall renormalize the family $W_t(\lambda)$, $0 \le t < d$, by multiplying on the right by appropriately chosen matrices $U_t \in \mathcal{U}_{\text{const}}(j_{pq})$, $0 \le t < d$. By Lemma 3.18 there exists a unique $U_t \in \mathcal{U}_{\text{const}}(j_{pq})$ such that the blocks $\Omega_{ij}^t(\lambda)$ of

$$\Omega_t(\lambda) = W_t(\lambda)U_t \tag{6.13}$$

meet the following normalization conditions at a point $\omega \in \mathbb{C}_+$:

$$\Omega_{21}^t(\omega) = 0, \quad \Omega_{11}^t(\omega) > 0 \quad \text{and} \quad \Omega_{22}^t(\omega) > 0. \tag{6.14}$$

(Here $\Omega_{11}^t(\omega) > 0$, since $\det W_t(\omega) \neq 0$, because W_t is an entire j_{pq}-inner mvf.)

Notice that the blocks of the PG transform $\Sigma_t(\lambda) = [\sigma_{ij}^t(\lambda)]$ of $\Omega_t(\lambda)$ also meet the same conditions:

$$\sigma_{21}^t(\omega) = 0, \;\; \sigma_{11}^t(\omega) > 0, \;\; \text{and} \;\; \sigma_{22}^t(\omega) > 0. \tag{6.15}$$

Let

$$\mathcal{L}_d^\omega = \mathcal{L}_d^\omega(\{\Sigma_t : 0 \le t < d\})$$

denote the set of *limit points* of the indicated family Σ_t of $m \times m$-inner mvf's that are subject to the normalization (6.15), i.e., $\Sigma \in \mathcal{L}_d^\omega$ if and only if there exists a sequence $\{t_j\}$, $t_j \in [0, d)$, such that

$$\Sigma(\lambda) = \lim_{j\uparrow\infty} \Sigma_{t_j}(\lambda)$$

uniformly in compact subsets of $\mathbb{C}_+$.

Theorem 6.1 *Let $\{W_t : 0 \le t < d\}$ be a nondecreasing $\curvearrowright$ continuous chain of entire j_{pq}-inner mvf's, let $\Omega_t(\lambda) = W_t(\lambda)U_t$ be the corresponding chain that meets the normalization conditions (6.14) at a fixed point $\omega \in \mathbb{C}_+$ and let $\Sigma_t(\lambda)$ denote the Potapov–Ginzburg transform of $\Omega_t(\lambda)$ for $0 \le t < d$. Then the set of "limit points"*

$$\mathcal{L}_d^\omega(\{\Sigma_t(\lambda) : 0 \le t < d\}) \neq \emptyset.$$

If $\Sigma = [\sigma_{ij}]$ belongs to $\mathcal{L}_d^\omega$, then:

(1)

$$\sigma_{21}(\omega) = 0, \quad \sigma_{11}(\omega) \ge 0, \quad \sigma_{22}(\omega) \ge 0. \tag{6.16}$$

(2) *The Redheffer transform R_Σ is well defined on $\mathcal{S}^{p\times q}$, i.e.,*

$$\det\{I_q - \sigma_{21}(\lambda)\varepsilon(\lambda)\} \not\equiv 0 \, \textit{for every} \; \varepsilon \in \mathcal{S}^{p\times q}.$$

(3) *The set of input scattering matrices*

$$\bigcap_{0\le t<d} T_{W_t}[\mathcal{S}^{p\times q}] = R_\Sigma[\mathcal{S}^{p\times q}]. \tag{6.17}$$

Proof Assertions (1) and (2) are self-evident. To verify (3), let

$$s \in \bigcap_{0\le t<d} T_{W_t}[\mathcal{S}^{p\times q}]\,.$$

Then, for every $t \in [0, d)$, there exists a unique mvf $\varepsilon_t \in \mathcal{S}^{p\times q}$ such that

$$s = T_{\Omega_t}[\varepsilon_t] \;=\; R_{\Sigma_t}[\varepsilon_t]\,.$$

Let $\Sigma \in \mathcal{L}_d^\omega$. Then there exists a sequence of points $0 \le t_0 \le t_1 \le \cdots$ tending to d such that

$$\Sigma_{t_n}(\lambda) \longrightarrow \Sigma(\lambda)$$

uniformly on compact subsets of $\mathbb{C}_+$. Let $\{t_{n_j}\}$ be a subsequence of the $\{t_n\}$ such that $\varepsilon_{t_{n_j}}(\lambda)$ converges uniformly to a limit $\varepsilon(\lambda)$ on compact subsets of $\mathbb{C}_+$ as $j \uparrow \infty$. Then

$$s = \lim_{j\uparrow\infty} R_{\Sigma_{t_{n_j}}}[\varepsilon_{t_{n_j}}] = R_\Sigma[\varepsilon] .$$

This proves that

$$\bigcap_{0\le t<d} T_{W_t}[\mathcal{S}^{p\times q}] \subseteq R_\Sigma[\mathcal{S}^{p\times q}] .$$

Now suppose conversely that $s = R_\Sigma[\varepsilon]$ for some $\varepsilon \in \mathcal{S}^{p\times q}$ and let $s_{t_n} = R_{\Sigma_{t_n}}[\varepsilon]$ for $n = 1, 2 \ldots$. Then $s_{t_n} \in T_{W_t}[\mathcal{S}^{p\times q}]$ for $t_n \ge t$ and hence

$$s = \lim_{n\uparrow\infty} s_{t_n} \in T_{W_t}[\mathcal{S}^{p\times q}]$$

for every $t \in [0, d)$. □

We remark that the Redheffer transform $R_\Sigma[\varepsilon]$ based on $\Sigma \in \mathcal{L}_d^\omega$ (that is defined by (6.12)) is an injective map of $\mathcal{S}^{p\times q}$ into itself if and only if its diagonal blocks $\sigma_{11}(\omega)$ and $\sigma_{22}(\omega)$ are both invertible. This will follow from (3) of theorem 6.4.

6.3 Regular canonical integral systems

Recall that the system (6.1) is said to be a regular canonical integral system if $d < \infty$ and $M(t)$ is nondecreasing and continuous on $[0, d]$ with $M(0) = 0$. Then the matrizant $W_t(\lambda)$ is defined on $[0, d]$, as is the mvf

$$\Omega_t(\lambda) = W_t(\lambda)U_t \tag{6.18}$$

that is normalized by the conditions (6.14).

Theorem 6.2 *If (6.1) is a regular canonical integral system on* $[0, d]$ *and* $\Sigma_t(\lambda)$ *denotes the PG transform of the renormalized matrizant* $\Omega_t(\lambda) = W_t(\lambda)U_t$ *at a point* $\omega \in \mathbb{C}_+$ *(by (6.14)) for* $0 \le t \le d$*, then:*

1. $\Sigma_d(\lambda) = \lim_{t\uparrow d} \Sigma_t(\lambda)$ *for every point* $\lambda \in \overline{\mathbb{C}_+}$.
2. $\mathcal{L}_d^\omega(\{\Sigma_t(\lambda) : 0 \le t < d\}) = \{\Sigma_d(\lambda)\}$ *, i.e., this set consists of exactly one mvf* $\Sigma_d(\lambda)$.
3. $\sigma_{11}^d(\omega) > 0$, $\sigma_{22}^d(\omega) > 0$ *and* $\sigma_{21}^d(\omega) = 0$.
4. $R_{\Sigma_d}[\mathcal{S}^{p\times q}] = T_{W_d}[\mathcal{S}^{p\times q}]$.
5. R_{Σ_d} *is an injective map of* $\mathcal{S}^{p\times q}$ *into itself.*

Proof The monodromy matrix $W_d(\lambda)$ of a regular canonical integral system on $[0, d]$ belongs to the class $\mathcal{E} \cap \mathcal{U}(j_{pq})$. Moreover, as the limit

$$W_d(\lambda) = \lim_{t \uparrow d} W_t(\lambda)$$

exists for every point $\lambda \in \mathbb{C}$, it follows from the proof of Lemma 3.18 that the constant normalizing factors U_t also tend to a limit U_d as $t \uparrow d$ and hence that the limit

$$\Omega_d(\lambda) = \lim_{t \uparrow d} \Omega_t(\lambda)$$

exists for every point $\lambda \in \mathbb{C}$. Therefore, the Potapov–Ginzburg transforms $\Sigma_t(\lambda)$ of $\Omega_t(\lambda)$ tend to the Potapov–Ginzburg transform $\Sigma_d(\lambda)$ of $\Omega_d(\lambda)$:

$$\Sigma_d(\lambda) = \lim_{t \uparrow d} \Sigma_t(\lambda) \quad \text{for} \quad \lambda \in \bigcap_{0 \leq t < d} \mathfrak{h}_{\Sigma_t},$$

which serves to justify (1), since $\overline{\mathbb{C}_+}$ is included in the latter set.

Now let $\Sigma \in \mathcal{L}_d^{\omega}$. Then there exists a sequence of points $0 \leq t_1 \leq t_2 \leq \cdots$ tending to d such that

$$\Sigma(\lambda) = \lim_{n \uparrow \infty} \Sigma_{t_n}(\lambda).$$

Then

$$\Sigma(\lambda) = \Sigma_d(\lambda),$$

i.e., (2) holds. Moreover, since $\Sigma_d(\lambda) = [\sigma_{ij}^d(\lambda)]$ is the Potapov–Ginzburg transform of $\Omega_d \in \mathcal{E} \cap \mathcal{U}(j_{pq})$ and

$$\det\, \Omega_d(\omega) = \det\, \sigma_{11}^d(\omega) \cdot \det\, \sigma_{22}^d(\omega) \neq 0,$$

it follows that the diagonal blocks $\sigma_{11}^d(\omega)$ and $\sigma_{22}^d(\omega)$ of $\Sigma_d(\omega)$ are invertible matrices. This proves (3).

Finally, (4) is a straightforward calculation and (5) is a consequence of the fact that T_{W_d} is injective, since W_d is invertible. □

6.4 Limit balls for input scattering matrices

Our next objective is to study the set

$$\mathcal{B}_d(\omega) = \{s(\omega) : s \in \mathcal{S}_{\text{scat}}^d(dM)\} \quad \text{for } \omega \in \mathbb{C}_+ \tag{6.19}$$

in the general case, when $M(t)$ is not assumed to be bounded and $d \leq \infty$.

Lemma 6.3 *Let $W \in \mathcal{U}(j_{pq})$, $S = PG(W)$ and $\{b_1, b_2\} \in ap(W)$. Then the set*

$$\mathcal{B}(\omega) = \{s(\omega) : s \in T_W[\mathcal{S}^{p \times q}]\}, \quad \omega \in \mathbb{C}_+ \tag{6.20}$$

is a matrix ball with left and right semiradii

$$R_\ell(\omega)^2 = \{I_p - \rho_{\overline{\omega}}(\overline{\omega})K_{\overline{\omega}}^{11}(\overline{\omega})\}^{-1} \tag{6.21}$$

and

$$R_r(\omega)^2 = \{I_q + \rho_\omega(\omega)K_\omega^{22}(\omega)\}^{-1} \tag{6.22}$$

and center

$$\gamma_c(\omega) = \rho_\omega(\omega)K_\omega^{12}(\omega)\{I_q + \rho_\omega(\omega)K_\omega^{22}(\omega)\}^{-1}, \tag{6.23}$$

where $K_\omega^{ij}(\lambda)$ *denotes the* ij *block of* $K_\omega^W(\lambda)$. *The semiradii are subject to the bounds*

$$s_{11}(\omega)s_{11}(\omega)^* \leq R_\ell(\omega)^2 \leq b_1(\omega)b_1(\omega)^* \tag{6.24}$$

and

$$s_{22}(\omega)^*s_{22}(\omega) \leq R_r(\omega)^2 \leq b_2(\omega)^*b_2(\omega). \tag{6.25}$$

Proof Formulas (6.21)–(6.23) for points ω in $\omega \in \mathfrak{h}_W^+ \cap \mathfrak{h}_{W^{-1}}^+$ follow easily from formulas (3.35)–(3.37) with U set equal to $W \in \mathcal{U}(j_{pq})$ and formula (4.73) for the RK $K_\omega^W(\lambda)$ of $\mathcal{H}(W)$. These formulas may be rewritten in terms of the blocks s_{ij} of $S = PG(W)$ as:

$$R_r(\omega)^2 = s_{22}(\omega)^*\{I_q - s_{21}(\omega)s_{21}(\omega)^*\}^{-1}s_{22}(\omega) \tag{6.26}$$

and

$$R_\ell(\omega)^2 = s_{11}(\omega)\{I_p - s_{21}(\omega)^*s_{21}(\omega)\}^{-1}s_{11}(\omega)^*. \tag{6.27}$$

Formulas (6.26) and (6.27) are valid for every point $\omega \in \mathbb{C}_+$ and yield the lower bounds in (6.24) and (6.25).

Next, to obtain the upper bounds, fix a point $\omega \in \mathfrak{h}_W^+ \cap \mathfrak{h}_{W^{-1}}^+$ and choose a matrix $U \in \mathcal{U}_{\text{const}}$ such that the blocks of $\Omega(\lambda) = W(\lambda)U$ meet the conditions $\Omega_{11}(\omega) > 0$, $\Omega_{22}(\omega) > 0$ and $\Omega_{21}(\omega) = 0$. Then $\Omega \in \mathcal{U}(j_{pq})$ and, by Lemma 3.54, $\{b_1, b_2\} \in ap(\Omega)$. Moreover, the semiradii can be expressed in terms of the blocks σ_{ij} in $\Sigma = PG(\Omega)$:

$$R_\ell(\omega)^2 = \sigma_{11}(\omega)^2 \quad \text{and} \quad R_r(\omega)^2 = \sigma_{22}(\omega)^2.$$

Thus, as

$$\sigma_{11} = b_1\varphi_1 \quad \text{and} \quad \sigma_{22} = \varphi_2 b_2, \quad \text{where } \varphi_1 \in \mathcal{S}_{\text{out}}^{p\times p} \text{ and } \varphi_2 \in \mathcal{S}_{\text{out}}^{q\times q},$$

it is readily seen that

$$R_\ell(\omega)^2 \leq b_1(\omega)b_1(\omega)^* \quad \text{and} \quad R_r(\omega)^2 \leq b_2(\omega)^*b_2(\omega). \qquad \square$$

If $\Sigma \in \mathcal{L}_d^\omega$, then

$$\mathcal{S}_{\text{scat}}^d(dM) = R_\Sigma[\mathcal{S}^{p\times q}] \tag{6.28}$$

by Theorem 6.1 and the blocks $\sigma_{ij}(\omega)$ of $\Sigma(\omega)$ meet the conditions (6.16). Thus, as $\sigma_{21}(\omega) = 0$,

$$\mathcal{B}_d(\omega) = \{\sigma_{12}(\omega) + \sigma_{11}(\omega)\varepsilon(\omega)\sigma_{22}(\omega) : \varepsilon \in \mathcal{S}^{p\times q}\}. \tag{6.29}$$

Since $\varepsilon(\omega)$ may be any contactive matrix in $\mathbb{C}^{p\times q}$ in (4.35), the last formula may be rewritten as

$$\mathcal{B}_d(\omega) = \{\sigma_{12}(\omega) + \sigma_{11}(\omega)Q\sigma_{22}(\omega) : Q \in \mathcal{S}^{p\times q}_{\text{const}}\}. \tag{6.30}$$

Formula (6.30) serves to identify $\mathcal{B}_d(\omega)$ as a matrix ball in the vector space $\mathbb{C}^{p\times q}$ with center

$$\gamma_c^d(\omega) = \sigma_{12}(\omega) \tag{6.31}$$

and left and right semiradii

$$R_\ell^d(\omega) = \sigma_{11}(\omega) \geq 0 \quad \text{and} \quad R_r^d(\omega) = \sigma_{22}(\omega) \geq 0. \tag{6.32}$$

The matrix ball $\mathcal{B}_d(\omega)$ is called the **limit ball** of the canonical integral system under consideration, since

$$\mathcal{B}_d(\omega) = \cap_{0\leq t<d}\mathcal{B}_t(\omega), \tag{6.33}$$

where

$$\begin{aligned}\mathcal{B}_t(\omega) &= \{s(\omega) : s \in T_{W_t}[\mathcal{S}^{p\times q}]\} = \{s(\omega) : s \in R_{\Sigma_t}[\mathcal{S}^{p\times q}]\} \\ &= \{\sigma_{12}^t(\omega) + \sigma_{11}^t(\omega)Q\sigma_{22}^t(\omega) : Q \in \mathcal{S}^{p\times q}\}\end{aligned} \tag{6.34}$$

are the matrix balls defined by $W_t(\omega)$, $\sigma_{ij}^t(\omega)$ are the blocks of $\Sigma_t(\omega)$ that are considered in Theorem 6.2 and

$$R_\ell^t(\omega) = \sigma_{11}^t(\omega) > 0 \quad \text{and} \quad R_r^t(\omega) = \sigma_{22}^t(\omega) > 0. \tag{6.35}$$

Theorem 6.4 *Let $W_t(\lambda)$, $0 \leq t < d$, be the matrizant of a canonical integral system (6.1) for which (6.2) holds and let $\omega \in \mathbb{C}_+$. Then:*

(1) *The matrix balls*

$$\mathcal{B}_t(\omega) = \left\{s(\omega) : s \in T_{W_t}[\mathcal{S}^{p\times q}]\right\} \tag{6.36}$$

are nested:

$$\mathcal{B}_{t_2}(\omega) \subseteq \mathcal{B}_{t_1}(\omega) \text{ if } 0 \leq t_1 \leq t_2 < d\,. \tag{6.37}$$

(2) *The limit ball*

$$\mathcal{B}_d(\omega) := \bigcap_{0\leq t<d} \mathcal{B}_t(\omega) = \left\{\gamma_c^d(\omega) + \mathfrak{r}_\ell^d(\omega)Q\mathfrak{r}_r^d(\omega) : Q \in \mathcal{S}^{p\times q}_{\text{const}}\right\} \tag{6.38}$$

is a matrix ball with center

$$\gamma_c^d(\omega) = \lim_{t \uparrow d} \gamma_c^t(\omega) \tag{6.39}$$

and left and right semiradii

$$\mathfrak{r}_\ell^d(\omega) = \lim_{t \uparrow d} R_\ell^t(\omega) \ \ and \ \ \mathfrak{r}_r^d(\omega) = \lim_{t \uparrow d} R_r^t(\omega)\,, \tag{6.40}$$

where $R_\ell^t(\omega)$, $R_r^t(\omega)$ and $c^t(\omega)$ are the left and right semiradii and center of the ball $\mathcal{B}_t(\omega)$ that are defined in terms of the blocks of the reproducing kernel $K_\omega^{W_t}(\lambda)$ by the formulas

$$R_r^t(\omega) = \{I_q + \rho_\omega(\omega)\{K_\omega^t(\omega)\}_{22}\}^{-1/2}, \tag{6.41}$$

$$R_\ell^t(\omega) = \{I_p - \rho_{\overline{\omega}}(\overline{\omega})\{K_{\overline{\omega}}^t(\overline{\omega})\}_{11}\}^{-1/2} \tag{6.42}$$

and

$$\gamma_c^t(\omega) = \rho_\omega(\omega)\{K_\omega^t(\omega)\}_{12}\{I_q + \rho_\omega(\omega)\{K_\omega^t(\omega)\}_{22}\}^{-1}. \tag{6.43}$$

(3) *The ranks*

$$\mathfrak{n}_\ell = \operatorname{rank} \mathfrak{r}_\ell^d(\omega) \ \ and \ \ \mathfrak{n}_r = \operatorname{rank} \mathfrak{r}_r^d(\omega)$$

are independent of the choice of the point $\omega \in \mathbb{C}_+$.

Proof Since the mvf $\Omega_t(\lambda) = W_t(\lambda)U_t$ introduced in (6.13) meets the conditions

$$\Omega_t(\overline{\omega})^* j_{pq} \Omega_t(\omega) = j_{pq} \quad \text{and} \quad \Omega_{21}^t(\omega) = 0, \tag{6.44}$$

it is readily checked that

$$\Omega_{11}^t(\omega) = \{\Omega_{11}^t(\overline{\omega})^*\}^{-1} \quad \text{and} \quad \Omega_{12}^t(\overline{\omega}) = 0, \tag{6.45}$$

and hence that the following formulas hold for the blocks σ_{ij}^t of $\Sigma_t = PG(\Omega_t)$:

$$\sigma_{11}^t(\omega) = \{\Omega_{11}^t(\overline{\omega})^*\}^{-1} = \Omega_{11}^t(\omega).$$

Let $\{K_\alpha^t(\beta)\}_{ij}$ denote the ij block of

$$K_\alpha^t(\beta) \stackrel{\text{def}}{=} K_\alpha^{W_t}(\beta) = K_\alpha^{\Omega_t}(\beta).$$

Then, since $\Omega_{21}^t(\omega) = 0$ and $\Omega_{22}^t(\omega) > 0$,

$$\{K_\omega^t(\omega)\}_{22} = \frac{\Omega_{22}^t(\omega)^2 - I_q}{\rho_\omega(\omega)}$$

and hence

$$R_r^t(\omega) = \sigma_{22}^t(\omega) = \Omega_{22}^t(\omega)^{-1} = \{I_q + \rho_\omega(\omega)\{K_\omega^t(\omega)\}_{22}\}^{-1/2}.$$

Similarly, since $\Omega^t_{12}(\overline{\omega}) = 0$ and $\Omega^t_{11}(\overline{\omega}) > 0$,

$$\{K^t_{\overline{\omega}}(\overline{\omega})\}_{11} = \frac{I_p - \Omega^t_{11}(\overline{\omega})^2}{\rho_{\overline{\omega}}(\overline{\omega})}$$

and therefore,

$$R^t_\ell(\omega) = \sigma^t_{11}(\omega) = \Omega^t_{11}(\overline{\omega})^{-1} = \{I_p - \rho_{\overline{\omega}}(\overline{\omega})\{K^t_{\overline{\omega}}(\overline{\omega})\}_{11}\}^{-1/2}.$$

Finally, since $\Omega^t_{22}(\omega) > 0$, it follows that

$$\sigma^t_{12}(\omega) = \Omega^t_{12}(\omega)\Omega^t_{22}(\omega)^{-1} = \rho_\omega(\omega)\{K^t_\omega(\omega)\}_{12}\Omega^t_{22}(\omega)^{-2}$$

and consequently that

$$\sigma^t_{12}(\omega) = \rho_\omega(\omega)\{K^t_\omega(\omega)\}_{12}\{I_q + \rho_\omega(\omega)\{K^t_\omega(\omega)\}_{22}\}^{-1}.$$

It is readily checked that the monotonicity of the family W_t, $0 \le t < d$, of entire j_{pq}-inner mvf's implies that

$$0 \le K^{t_1}_\omega(\omega) \le K^{t_2}_\omega(\omega) \quad \text{for } 0 \le t_1 \le t_2 < d. \tag{6.46}$$

Thus,

$$0 \le R^{t_2}_\ell(\omega)^2 \le R^{t_1}_\ell(\omega)^2 \quad \text{and} \quad 0 \le R^{t_2}_r(\omega)^2 \le R^{t_1}_r(\omega)^2 \quad \text{for } 0 \le t_1 \le t_2 < d. \tag{6.47}$$

This in turn guarantees the existence of the positive semidefinite square roots $\mathfrak{r}^d_\ell(\omega) \ge 0$ and $\mathfrak{r}^d_r(\omega) \ge 0$ of the limits

$$\mathfrak{r}^d_r(\omega)^2 = \lim_{t\uparrow d} R^t_r(\omega)^2 \quad \text{and} \quad \mathfrak{r}^d_\ell(\omega)^2 = \lim_{t\uparrow d} R^t_\ell(\omega)^2. \tag{6.48}$$

Therefore, since a positive semidefinite matrix has a unique positive semidefinite square root, and it is a continuous function of its argument, assertion (6.40) follows. (The point is that if, say, B and C are limit points of the matrices $R^t_\ell(\omega)$ as $t \uparrow d$, then $B \ge 0$, $C \ge 0$ and, in view of (6.48), $B^2 = C^2$. The uniqueness of the positive semidefinite square root guarantees that $B = C$ and hence, as all limit points coincide, that the limit exists.)

The existence of the limit in (6.39) follows from Yu.L. Shmuljan [Shm68]. Item (3) follows from a result of S.A. Orlov [Or76]; see Theorem 7.39 below. □

Remark 6.5 *The semiradii that are defined by formulas (6.41) and (6.42) are connected:*

$$\frac{\det R^t_\ell(\omega)}{\det R^t_r(\omega)} = \frac{|\det b^t_1(\omega)|}{|\det b^t_2(\omega)|} = |\det W_t(\omega)|, \tag{6.49}$$

where $\{b^t_1, b^t_2\} \in ap(W_t)$. Moreover, the semiradii $R^t_\ell(\omega)$ and $R^t_r(\omega)$ of the ball $\mathcal{B}_t(\omega)$ are uniquely defined by the relation (6.49); see theorem 5.59 on pp. 291–292 of [ArD08b] and the Supplementary Notes to Chapter 7.

Remark 6.6 *A result of Shmuljan [Shm68] guarantees that the left and right semiradii $R_\ell^d(\omega)$ and $R_r^d(\omega)$ of a matrix ball $\mathcal{B}_d(\omega)$ that contains at least two distinct points P_1, $P_2 \in \mathbb{C}^{p\times q}$ are uniquely determined up to a positive scalar factor by the formulas*

$$R_\ell^d(\omega) = \rho \mathfrak{r}_\ell^d(\omega) \quad \textit{and} \quad R_r^d(\omega) = \rho^{-1}\mathfrak{r}_r^d(\omega) \quad \textit{for some } \rho > 0.$$

However, this essential uniqueness fails if the matrix ball contains only one point. Thus, for example, if $\mathfrak{r}_\ell^d(\omega) = 0_{p\times p}$, then $R_r^d(\omega)$ may be set equal to any $q \times q$ positive semidefinite matrix. Similarly, if $\mathfrak{r}_r^d(\omega) = 0_{q\times q}$, then $R_\ell^d(\omega)$ may be set equal to any $p \times p$ positive semidefinite matrix. To avoid this ambiguity, the left and right semiradii of the limit ball will always be understood as the limits $\mathfrak{r}_\ell^d(\omega)$ and $\mathfrak{r}_r^d(\omega)$ in (6.40), where $R_\ell^t(\omega)$ and $R_r^t(\omega)$ denote the left and right semiradii of the ball $\mathcal{B}_t(\omega)$, $0 < t < d$, that meet the constraints in (6.49).

Lemma 6.7 *If $J \neq \pm I_m$, then the set $\mathcal{S}_{\text{scat}}^d(dM)$ for the canonical integral system (2.1) is convex.*

Proof It suffices to consider the case $J = j_{pq}$. Let $\mathcal{B}_d(\omega)$ be the limit ball for a system of the form (6.1). By Theorem 6.4,

$$\mathcal{B}_d(\omega) = \{s(\omega) : s \in \mathcal{S}_{\text{scat}}\}.$$

Let $s_1, s_2 \in \mathcal{S}_{\text{scat}}$ and let $s^{(a)} = as_1 + (1-a)s_2$ for $0 < a < 1$. Then, for every $\omega \in \mathbb{C}_+$ the mvf $s^{(a)}(\omega) \in \mathcal{B}_d(\omega)$, since $\mathcal{B}_d(\omega)$ is a convex set. Consequently,

$$s^{(a)}(\omega) = T_{W_t(\omega)}[\varepsilon^t(\omega)],$$

for some choice of $\varepsilon^t(\omega) \in \mathcal{S}_{\text{const}}^{p\times p}$. Repeating this procedure for every point $\lambda \in \mathbb{C}_+$, we obtain the formula

$$s^{(a)}(\lambda) = T_{W_t(\lambda)}[\varepsilon^t(\lambda)],$$

for all such points and hence that

$$\varepsilon^t(\lambda) = T_{W_t(\lambda)^{-1}}[s^{(a)}(\lambda)]$$

is a meromorphic function that is contractive in $\mathfrak{h}_{\varepsilon^t}^+$. Therefore, $\varepsilon^t \in \mathcal{S}^{p\times q}$. □

6.5 The full rank case

We shall say that the right end point d of a canonical integral system (6.1) is a **full rank point** if the ranks of the limiting semiradii are maximal, i.e., if

$$\mathfrak{n}_\ell = p \quad \text{and} \quad \mathfrak{n}_r = q. \tag{6.50}$$

Theorem 6.8 *Let $W_t(\lambda) = W(t,\lambda)$, $0 \le t < d$, be the matrizant for the canonical integral system (6.1) with mass function $M(t)$ that is subject to (6.2). Let*

$\{b_1^t, b_2^t\} \in ap(W_t)$ and assume that $b_1^t(0) = I_p$ and $b_2^t(0) = I_q$ for $0 \leq t < d$. Then the following conclusions are valid:

(1) *If $M(t)$ is bounded on $[0, d)$, then d is a full rank point.*
(2) *If d is a full rank point and $W_{t_k} \in \mathcal{U}_{rR}(j_{pq})$ for a sequence of points $t_k \in [0, d)$ that tends to d, then $M(t)$ is bounded on $[0, d)$ and*

$$W_d(\lambda) = \lim_{t \uparrow \infty} W_t(\lambda) \quad \text{exists for every } \lambda \in \mathbb{C} \text{ and is in } \mathcal{U}_{rR}(j_{pq}).$$

Moreover, if $d < \infty$, then the system is regular and W_d is its monodromy matrix.

Proof If $M(t)$ is bounded on $[0, d)$, then $W_t(\lambda)$ converges uniformly to a limit $W_d(\lambda) \in \mathcal{E} \cap \mathcal{U}(j_{pq})$ on compact subsets of $\mathbb{C}$ and hence

$$\mathfrak{r}_\ell^d(\lambda) = R_\ell^d(\lambda) \text{ and } \mathfrak{r}_r^d(\lambda) = R_r^d(\lambda)$$

for every point $\lambda \in \mathbb{C}_+$. Therefore, by formulas (6.42) and (6.41), $\mathfrak{n}_\ell = p$ and $\mathfrak{n}_r = q$, i.e., d is a full rank point.

Suppose next that d is a full rank point. Then Lemma 6.3 supplies the inequalities

$$s_{11}^t(\lambda)s_{11}^t(\lambda)^* \leq R_\ell^t(\lambda)^2 \leq b_1^t(\lambda)b_1^t(\lambda)^* \tag{6.51}$$

and

$$s_{22}^t(\lambda)^*s_{22}^t(\lambda) \leq R_r^t(\lambda)^2 \leq b_2^t(\lambda)^*b_2^t(\lambda) \tag{6.52}$$

for every point $\lambda \in \mathbb{C}_+$ and every $t \in [0, d)$. But this in turn implies that

$$|\det\ b_1^t(\lambda)| \geq |\det\ R_\ell^t(\lambda)| \geq |\det\ \mathfrak{r}_\ell^d(\lambda)|$$

and hence that

$$\mathfrak{n}_\ell = p \Longrightarrow \lim_{t \uparrow d} |\det\ b_1^t(\lambda)| > 0$$

for at least one point $\lambda \in \mathbb{C}_+$. Therefore, by Lemma 5.11, $b_1^t(\lambda)$ tends uniformly to a limit $b_1^d(\lambda) \in \mathcal{E} \cap \mathcal{S}_{\text{in}}^{p\times p}$ on compact subsets of $\mathbb{C}$. Similar considerations based on the second set of inequalities lead to the conclusion that if $\mathfrak{n}_r = q$, then $b_2^t(\lambda)$ tends uniformly to a limit $b_2^d(\lambda) \in \mathcal{E} \cap \mathcal{S}_{in}^{q\times q}$ on compact subsets of $\mathbb{C}$ as t tends to d. Moreover,

$$\Sigma_d(\lambda) = \lim_{j \uparrow \infty} \Sigma_{t_j}(\lambda) \text{ in } \mathbb{C}_+,$$

where the sequence t_j may be chosen such that $W_{t_j} \in \mathcal{U}_{rR}(j_{pq})$. Thus, the diagonal blocks must also converge, i.e.,

$$\sigma_{11}^d(\lambda) = \lim_{j \uparrow \infty} \sigma_{11}^{t_j}(\lambda) \text{ and } \sigma_{22}^d(\lambda) = \lim_{j \uparrow \infty} \sigma_{22}^{t_j}(\lambda)$$

for every $\lambda \in \mathbb{C}_+$. If $\lambda = \omega$, the special point in $\mathbb{C}_+$ at which the normalization (6.14) is imposed, then

$$\sigma_{11}^t(\omega) = R_\ell^t(\omega) \text{ and } \sigma_{22}^t(\omega) = R_r^t(\omega).$$

Therefore, the limits $\sigma_{11}^d(\omega)$ and $\sigma_{22}^d(\omega)$ are invertible.

The latter condition permits us to define the PG transform $\Omega_d(\lambda)$ of $\Sigma_d(\lambda)$ and to check that

$$\lim_{n\uparrow\infty} \Omega_{t_n}(\lambda) = \Omega_d(\lambda)$$

for $\lambda \in \mathbb{C}_+$. Moreover, at the special normalization point ω,

$$\Omega_d(\omega) = \begin{bmatrix} \sigma_{11}^d(\omega) & \sigma_{12}^d(\omega) \\ 0 & I_q \end{bmatrix} \begin{bmatrix} I_p & 0 \\ \sigma_{21}^d(\omega) & \sigma_{22}^d(\omega) \end{bmatrix}^{-1}$$

is invertible. Therefore, since $\Omega_{t_n} \in \mathcal{U}(j_{pq})$ and

$$j_{pq} - \Omega_{t_n}(\lambda) j_{pq} \Omega_{t_n}(\lambda)^* \le j_{pq} - \Omega_d(\lambda) j_{pq} \Omega_d(\lambda)^*$$

for $\lambda \in \mathbb{C}_+$ and $\det \Omega_d(\lambda) \not\equiv 0$, it follows from Lemma 5.12 that $\Omega_d \in \mathcal{U}_{rR}(j_{pq})$, and from Lemma 3.59 that $\{b_1^d, b_2^d\} \in ap(\Omega_d)$. Therefore, since $\Omega_d \in \mathcal{U}_{rR}(j_{pq})$ and $b_1^d(\lambda)$ and $b_2^d(\lambda)$ are both entire mvf's, it follows from Theorem 3.55 that $\Omega_d(\lambda)$ is entire. Next, because the reproducing kernels

$$K_\omega^t(\lambda) = \frac{j_{pq} - W_t(\lambda) j_{pq} W_t(\omega)^*}{\rho_\omega(\lambda)} = \frac{j_{pq} - \Omega_t(\lambda) j_{pq} \Omega_t(\omega)^*}{\rho_\omega(\lambda)}$$

are nondecreasing functions of t when $\lambda = \omega$, we see that

$$\begin{aligned} \frac{j_{pq} - W_t(\lambda) j_{pq} W_t(\lambda)^*}{\rho_\lambda(\lambda)} &= \frac{j_{pq} - \Omega_t(\lambda) j_{pq} \Omega_t(\lambda)^*}{\rho_\lambda(\lambda)} \le \frac{j_{pq} - \Omega_{t_n}(\lambda) j_{pq} \Omega_{t_n}(\lambda)^*}{\rho_\lambda(\lambda)} \\ &\le \frac{j_{pq} - \Omega_d(\lambda) j_{pq} \Omega_d(\lambda)^*}{\rho_\lambda(\lambda)} \end{aligned}$$

for $\lambda \in \mathbb{C}_+$ and $t \le t_n$. Thus upon letting $\lambda \to 0$ (through points in $\mathbb{C}_+$), we obtain the bound

$$M(t) = 2\pi \lim_{\lambda\to 0} \frac{j_{pq} - W_t(\lambda) j_{pq} W_t(\lambda)^*}{\rho_\lambda(\lambda)} \le 2\pi \lim_{\lambda\to 0} \frac{j_{pq} - \Omega_d(\lambda) j_{pq} \Omega_d(\lambda)^*}{\rho_\lambda(\lambda)}.$$

Therefore, $M(t)$ is bounded on $[0, d)$. □

Corollary 6.9 *Let $s \in \mathring{\mathcal{S}}^{p\times q}$ be an input scattering matrix for the canonical integral system (6.1) with mass function $M(t)$ subject to (6.2) on $[0, d)$. Then $M(t)$ is bounded on $[0, d)$ if and only if d is a full rank point.*

Proof This is immediate from the preceding theorem and the fact that if $s \in \mathring{\mathcal{S}}^{p\times q}$, then the matrizants $W_t(\lambda)$, $0 \le t < d$, of the canonical integral system are all strongly regular and hence a fortiori right regular. □

Remark 6.10 *Corollary 6.9 is still valid if the condition* $s \in \mathring{\mathcal{S}}^{p\times q}$ *is replaced by the condition* $s \in \mathcal{W}_+^{p\times q}(\gamma) \cap \mathcal{S}^{p\times q}$ *with* $\|\gamma\| < 1$ *because in this case too* $W_t \in \mathcal{U}_{rsR}(j_{pq})$ *for every* $t \in [0, d)$ *by Theorem 9.47.*

6.6 Rank formulas

Our next objective is to apply the bounds (6.24) and (6.25) to the semiradii $R_\ell^t(\omega)$ and $R_r^t(\omega)$ of the matrix balls

$$\mathcal{B}_t(\omega) = \{s(\omega) : s \in T_{W_t}[\mathcal{S}^{p\times q}]\}$$

corresponding to the matrizant $W_t(\lambda) = W(t, \lambda)$, $0 \le t < d$, of the canonical system (6.1). A passage to the limit as $t \uparrow d$ will then yield information on the semiradii $\mathfrak{r}_\ell^d(\omega)$ and $\mathfrak{r}_r^d(\omega)$ of the matrix ball

$$\mathcal{B}_d(\omega) = \bigcap_{0\le t<d} \mathcal{B}_t(\omega) .$$

Theorem 6.11 *Let* W_t, $0 \le t < d$, *be the matrizant of the canonical integral system (6.1), let* $\{b_1^t, b_2^t\} \in ap(W_t)$ *and let* $s \in \mathcal{S}_{scat}^d(dM)$, $\gamma = \|s\|_\infty$ *and* $\omega \in \mathbb{C}_+$. *Then the left and right semiradii of the limit ball* $\mathcal{B}_d(\omega)$ *are subject to the bounds*

$$(1-\gamma^2)\lim_{t\uparrow d} b_1^t(\omega)b_1^t(\omega)^* \le \mathfrak{r}_\ell^d(\omega)^2 \le \lim_{t\uparrow d} b_1^t(\omega)b_1^t(\omega)^* \tag{6.53}$$

$$(1-\gamma^2)\lim_{t\uparrow d} b_2^t(\omega)^*b_2^t(\omega) \le \mathfrak{r}_r^d(\omega)^2 \le \lim_{t\uparrow d} b_2^t(\omega)^*b_2^t(\omega). \tag{6.54}$$

Moreover, if

$$\mathring{\mathcal{S}}_{\text{scat}}^d(dM) \stackrel{\text{def}}{=} \mathcal{S}_{\text{scat}}^d(dM) \cap \mathring{\mathcal{S}}^{p\times q} \ne \emptyset, \tag{6.55}$$

then (6.53) and (6.54) hold with

$$\gamma = \inf\{\|s\|_\infty : s \in \mathring{\mathcal{S}}_{\text{scat}}^d(dM)\} \tag{6.56}$$

and the ranks of the semiradii $\mathfrak{r}_\ell^d(\omega)$ *and* $\mathfrak{r}_r^d(\omega)$ *are given by the formulas*

$$\mathfrak{n}_\ell = \text{rank}\left\{\lim_{t\uparrow d} b_1^t(\omega)b_1^t(\omega)^*\right\} \quad \textit{and} \quad \mathfrak{n}_r = \text{rank}\left\{\lim_{t\uparrow d} b_2^t(\omega)^*b_2^t(\omega)\right\}. \tag{6.57}$$

Proof The upper bounds in (6.53) and (6.54) follow from (6.51) and (6.52). The lower bounds follow from the formulas

$$R_\ell^t(\omega)^2 = \sigma_{11}^t(\omega)^2 = b_1^t(\omega)\varphi_1^t(\omega)\varphi_1^t(\omega)^*b_1^t(\omega)^*$$

and

$$R_r^t(\omega)^2 = \sigma_{22}^t(\omega)^2 = b_2^t(\omega)^*\varphi_2^t(\omega)^*\varphi_2^t(\omega)b_2^t(\omega)$$

and the inequalities

$$\varphi_1^t(\omega)\varphi_1^t(\omega)^* \geq 1 - \|s^\circ\|_\infty \quad \text{and} \quad \varphi_2^t(\omega)^*\varphi_2^t(\omega) \geq 1 - \|s^\circ\|_\infty, \tag{6.58}$$

which hold for every mvf $s^\circ \in \mathring{\mathcal{S}}_{\text{scat}}^d(dM)$. Proofs of the last inequalities may be found on pp. 30–31 of [ArD02a]. □

6.7 Regular systems (= full rank) case

Theorem 6.12 *In the setting of Theorem 6.11, under condition (6.55), the mass function $M(t)$ is bounded on $[0, d)$ if and only if*

$$\lim_{t\uparrow d} |\det b_1^t(\omega)| > 0 \quad \text{and} \quad \lim_{t\uparrow d} |\det b_2^t(\omega)| > 0 \tag{6.59}$$

for at least one (and hence for every) point $\omega \in \mathbb{C}_+$.

Proof Condition (6.59) guarantees that the lower limits in the inequalities (6.53) and (6.54) are positive definite matrices. Therefore, the semiradii $\mathfrak{r}_\ell^d(\omega)$ and $\mathfrak{r}_r^d(\omega)$ of the limit ball are both invertible matrices. Thus, d is a full rank point. Moreover, since $s^\circ \in T_{W_t}[\mathcal{S}^{p\times q}] \cap \mathring{\mathcal{S}}^{p\times q}$ for every t in the interval $0 \leq t < d$, Theorem 6.8 guarantees that the mass function $M(t)$ is bounded on $[0, d)$. Conversely, if the mass function $M(t)$ of (6.1) is bounded on $[0, d)$, then, by another application of Theorem 6.8, d is a full rank point. Therefore condition (6.59) is in force. □

6.8 The limit point case

The canonical integral system (6.1) is said to be in the **limit point** case if it has only one input scattering matrix:

$$\bigcap_{0\leq t<d} T_{W_t}[\mathcal{S}^{p\times q}] = \{s^\circ\}, \tag{6.60}$$

i.e., if there is only one mvf $s^\circ \in \mathcal{S}^{p\times q}$ which belongs to $T_{W_t}[\mathcal{S}^{p\times q}]$ for every $t \in [0, d)$.

Theorem 6.13 *The canonical integral system (6.1) is in the limit point case if and only if either $\mathfrak{r}_\ell^d(\omega)$ or $\mathfrak{r}_r^d(\omega)$ is equal to zero for at least one point $\omega \in \mathbb{C}_+$.*

Proof In view of the result of Orlov [Or76] that was referred to earlier, it is clear that if $\mathfrak{r}_\ell^d(\omega) = 0$ (resp. $\mathfrak{r}_r^d(\omega) = 0$) for a point $\omega \in \mathbb{C}_+$, then it is equal to zero for every point $\omega \in \mathbb{C}_+$ and hence, by the definition of the ball $\mathcal{B}_d(\omega)$, there is only one input scattering matrix.

Suppose next that there is only one input scattering matrix and let $\omega \in \mathbb{C}_+$. Then by formula (6.38),

$$\mathfrak{r}_\ell^d(\omega)Q\mathfrak{r}_r^d(\omega) = 0_{p\times q}$$

for every constant matrix $Q \in \mathcal{S}^{p\times q}$. But this in turn implies that at least one of the semiradii is identically equal to zero. □

Theorem 6.14 *Let $W_t(\lambda) = W(t,\lambda)$, $0 \le t < d$, be the matrizant of the canonical integral system (6.1), let $\{b_1^t, b_2^t\} \in ap(W_t)$ for $0 \le t < d$ and suppose that*

$$\lim_{t\uparrow d} \left\{ \|b_1^t(\omega)\| \, \|b_2^t(\omega)\| \right\} = 0 \tag{6.61}$$

for some point $\omega \in \mathbb{C}_+$. Then the system is in the limit point case. Moreover, if

$$\begin{aligned} &\lim_{t\uparrow d} b_1^t(\omega) = 0 \quad \textit{for some } \omega \in \mathbb{C}_+, \textit{ then } \mathfrak{n}_\ell = 0, \\ &\textit{i.e.,} \quad \mathfrak{r}_\ell^d(\omega) = 0 \textit{ for every } \omega \in \mathbb{C}_+; \end{aligned} \tag{6.62}$$

similarly, if

$$\begin{aligned} &\lim_{t\uparrow d} b_2^t(\omega) = 0 \quad \textit{for some } \omega \in \mathbb{C}_+, \textit{ then } \mathfrak{n}_r = 0, \\ &\textit{i.e.,} \quad \mathfrak{r}_r^d(\omega) = 0 \textit{ for every } \omega \in \mathbb{C}_+. \end{aligned} \tag{6.63}$$

Proof Since $\|b_1^t(\omega)\|$ and $\|b_2^t(\omega)\|$ are both monotone nonincreasing functions of t for each fixed $\omega \in \mathbb{C}_+$, the condition (6.61) implies that either $\|b_1^t(\omega)\| \longrightarrow 0$ or $\|b_2^t(\omega)\| \longrightarrow 0$ (or both) as $t \uparrow d$. In the first case, the inequality (6.51) guarantees that $\mathfrak{r}_\ell^d(\omega) = 0$; in the second case, the inequality (6.52) guarantees that $\mathfrak{r}_r^d(\omega) = 0$. The result now follows by the preceding theorem. □

Corollary 6.15 *Let $W_t(\lambda) = W(t,\lambda)$, $0 \le t < d$, be the matrizant of the canonical integral system (6.1), let $\{b_1^t, b_2^t\} \in ap(W_t)$ for $0 \le t < d$ and suppose that $d = \infty$. Then*

(1) *If $e_{-\alpha t} b_1^t \in \mathcal{S}_{in}^{p\times p}$ for some $\alpha > 0$ and every $t \ge t_0$, then $\mathfrak{n}_\ell = 0$.*
(2) *If $e_{-\beta t} b_2^t \in \mathcal{S}_{in}^{q\times q}$ for some $\beta > 0$ and every $t \ge t_0$, then $\mathfrak{n}_r = 0$.*

Theorem 6.14 admits a converse in the setting of Theorem 6.11 under condition (6.55).

Theorem 6.16 *In the setting of Theorem 6.11, under condition (6.55), the canonical system (6.1) is in the limit point case if and only if*

$$\lim_{t\uparrow d} b_1^t(\omega) = 0 \quad \textit{or} \quad \lim_{t\uparrow d} b_2^t(\omega) = 0 \tag{6.64}$$

for at least one (and hence every) point $\omega \in \mathbb{C}_+$.

Proof This is an immediate consequence of formulas (6.53) and (6.54). □

6.9 The diagonal case

In this section we shall specialize a number of the statements of the preceding theorems to the case where:

$$b_1^t(\lambda) = \mathrm{diag}\{e_{\alpha_1(t)}, \ldots, e_{\alpha_p(t)}\}, \quad b_2^t(\lambda) = \mathrm{diag}\{e_{\beta_1(t)}, \ldots, e_{\beta_q(t)}\}$$

and the $\alpha_j(t)$ and $\beta_k(t)$ are continuous nondecreasing functions on the interval $[0, d)$ with $\alpha_j(0) = \beta_k(0) = 0$ for $j = 1, \ldots, p$ and $k = 1, \ldots, q$. (Recall that $e_a = e_a(\lambda) = e^{ia\lambda}$.)

Theorem 6.17 *In the setting of Theorem 6.11 under condition (6.55), with $b_1^t(\lambda)$ and $b_2^t(\lambda)$ of the diagonal form described above, the following conclusions hold for the canonical integral system (6.1) on the interval $[0, d)$.*

1. *$M(t)$ is bounded on $[0, d)$ if and only if the functions $\alpha_j(t)$ and $\beta_k(t)$ are bounded on $[0, d)$.*
2. *(6.1) is in the limit point case if and only if either all the functions $\alpha_j(t)$, $j = 1, \ldots, p$, are unbounded on $[0, d)$ or all the functions $\beta_k(t)$, $k = 1, \ldots, q$, are unbounded on $[0, d)$.*
3. *The ranks $\mathfrak{n}_\ell$ and $\mathfrak{n}_r$ of the left and right semiradii $\mathfrak{r}_\ell^d(\omega)$ and $\mathfrak{r}_r^d(\omega)$ of the limit ball $\mathcal{B}_d(\omega) = \bigcap_{0 \le t < d} \mathcal{B}_t(\omega)$ are given by the formulas*

$$\mathfrak{n}_\ell = \left\{ \#\{j\} : \lim_{t \uparrow d} \alpha_j(t) < \infty \right\} \quad \textit{and}$$

$$\mathfrak{n}_r = \left\{ \#\{k\} : \lim_{t \uparrow d} \beta_k(t) < \infty \right\}.$$

Proof Statement (1) is immediate from Theorem 6.12, (2) from Theorem 6.16 and (3) from Theorem 6.11, respectively. □

6.10 A Weyl–Titchmarsh like characterization for input scattering matrices

In this section we will consider the canonical integral system

$$u(t, \lambda) = u(0, \lambda) - i\lambda j_{pq} \int_0^t dM(s) u(s, \lambda) \tag{6.65}$$

that is adjoint to the system (6.1): $u_t(\lambda) = u(t, \lambda)$ is the solution of the system (6.65) if and only if $y_t(\lambda) = u_t^\#(\lambda)$ is the solution of the system (6.1). We will investigate the following problem: Given a point $\omega \notin \mathbb{R}$, describe the linear space in $\mathbb{C}^m$ of initial data $u(0, \omega)$ for which the $m \times 1$ vector-valued solution $u_t(\omega)$ of the system (6.65) (with $\lambda = \omega$) satisfies the condition

$$\int_0^d u(t, \omega)^* dM(t) u(t, \omega) \; < \; \infty. \tag{6.66}$$

The main conclusion is that if $\omega \in \mathbb{C}_+$, then (6.66) holds if and only if condition (6.76) below holds for some input scattering matrix $s(\lambda)$ of the system (6.1). Enroute some auxiliary results are obtained for $\omega \in \mathbb{C}_-$. A characterization of input scattering matrices of (6.1) in the *left limit point* case, i.e., when $\mathfrak{n}_\ell = 0$, is also established.

A property of the semiradii of the limit ball

Lemma 6.18 *Let $A(t)$ be a $k \times k$ mvf on the interval $[0, d)$ such that $A(t) > 0$ and*

$$A(t_2)^2 \leq A(t_1)^2 \quad for \quad 0 \leq t_1 \leq t_2 < d. \tag{6.67}$$

Let $A_d = \lim_{t \uparrow d} A(t)$. Then

$$\xi \in A_d \mathbb{C}^k \iff \lim_{t \uparrow d} \xi^* A(t)^{-2} \xi < \infty. \tag{6.68}$$

Proof Condition (6.67) implies that

$$(A_d)^2 \leq A(t)^2, \quad 0 \leq t < d,$$

and hence that

$$\|A(t)^{-1} A_d\| \leq 1, \quad 0 \leq t < d.$$

Thus, if $\xi = A_d \eta$ for some $\eta \in \mathbb{C}^k$, then

$$\|A(t)^{-1} \xi\| = \|A(t)^{-1} A_d \eta\| \leq \|\eta\|,$$

i.e., the implication $\Longrightarrow$ holds in (6.68).

On the other hand, if the right-hand side of (6.68) holds, then the vectors $\xi_t = A(t)^{-1}\xi$ are bounded. Therefore, a subsequence ξ_{t_j} tends to a limit η as $t_j \uparrow d$. Consequently,

$$\xi = A(t_j)\xi_{t_j} = \lim_{t_j \uparrow d} A(t_j)\xi_{t_j} = A_d \eta,$$

i.e., the implication $\Longleftarrow$ holds in (6.68). □

Let $W_t(\lambda) = W(t, \lambda)$ be the matrizant of the system (6.1). Then

$$y(t, \lambda) = y(0, \lambda) W(t, \lambda)$$

is the solution of the system (6.1) with initial data $y(0, \lambda)$ and

$$u(t, \lambda) = W^{\#}(t, \lambda) u(0, \lambda)$$

is the solution of the system (6.65) with initial data $u(0, \lambda)$. Thus, the identity

$$\frac{1}{2\pi} \int_0^t W(s, \lambda) dM(s) W(s, \omega)^* = K_\omega^t(\lambda)$$

for the RK $K_\omega^t(\lambda)$ for the RKHS $\mathcal{H}(W_t)$ based on the j_{pq}-inner mvf $W_t(\lambda) = W(t, \lambda)$, implies that

$$\frac{1}{2\pi}\int_0^t u(s,\omega)^* dM(s) u(s,\omega) = u(0,\omega)^* K_{\overline{\omega}}^t(\overline{\omega}) u(0,\omega). \tag{6.69}$$

Thus, condition (6.66) is equivalent to the condition

$$\lim_{t\uparrow d} u(0,\omega)^* K_{\overline{\omega}}^t(\overline{\omega}) u(0,\omega) < \infty. \tag{6.70}$$

Let $\xi = [I_p \quad 0]u(0,\omega)$ and $\eta = [0 \quad I_q]u(0,\omega)$. Then $\xi \in \mathbb{C}^p$, $\eta \in \mathbb{C}^q$ and

$$\begin{aligned} u(0,\omega)^* K_{\overline{\omega}}^t(\overline{\omega}) u(0,\omega) &= \left\| K_{\overline{\omega}}^t \begin{bmatrix} \xi \\ \eta \end{bmatrix} \right\|_{\mathcal{H}(W_t)}^2 \\ &= \left\| K_{\overline{\omega}}^t \begin{bmatrix} \xi \\ 0 \end{bmatrix} + K_{\overline{\omega}}^t \begin{bmatrix} 0 \\ \eta \end{bmatrix} \right\|_{\mathcal{H}(W_t)}^2 \\ &\leq 2 \left\| K_{\overline{\omega}}^t \begin{bmatrix} \xi \\ 0 \end{bmatrix} \right\|_{\mathcal{H}(W_t)}^2 + 2 \left\| K_{\overline{\omega}}^t \begin{bmatrix} 0 \\ \eta \end{bmatrix} \right\|_{\mathcal{H}(W_t)}^2 \\ &= 2\xi^* \{K_{\overline{\omega}}^t(\overline{\omega})\}_{11}\xi + 2\eta^* \{K_{\overline{\omega}}^t(\overline{\omega})\}_{22}\eta. \end{aligned}$$

Thus, in addition to the obvious bound, it is readily seen that

$$\text{if } \xi = 0, \text{ then (6.70) holds} \Longleftrightarrow \lim_{t\uparrow d} \eta^* \{K_{\overline{\omega}}^t(\overline{\omega})\}_{22}\eta < \infty, \tag{6.71}$$

whereas,

$$\text{if } \eta = 0, \text{ then (6.70) holds} \Longleftrightarrow \lim_{t\uparrow d} \xi^* \{K_{\overline{\omega}}^t(\overline{\omega})\}_{11}\xi < \infty. \tag{6.72}$$

Lemma 6.19 *Let $\omega \in \mathbb{C}_+$ and let $u(0,\omega) = \begin{bmatrix} \xi \\ 0 \end{bmatrix}$. Then the corresponding solution $u(t,\omega)$ of the system (6.65) (with $\lambda = \omega$) satisfies the condition (6.66) if and only if*

$$\xi \in \mathfrak{r}_\ell^d(\omega)\mathbb{C}^p. \tag{6.73}$$

Proof If $\eta = 0$, then, in view of the preceding discussion, (6.66) holds if and only if the constraint on the limit in (6.72) is met. If $\omega \in \mathbb{C}_+$, then, by formula (6.21), this condition will be met if and only if

$$\lim_{t\uparrow d} \xi^* R_\ell^t(\omega)^{-2}\xi < \infty.$$

However, by Lemma 6.18, this will hold if and only if (6.73) holds. □

Similar considerations lead to the following supplementary conclusion for $\omega \in \mathbb{C}_-$.

Lemma 6.20 *Let $\omega \in \mathbb{C}_-$ and let $u(0,\omega) = \begin{bmatrix} 0 \\ \eta \end{bmatrix}$. Then the corresponding solution $u(t,\omega)$ of the system (6.65) (with $\lambda = \omega$) satisfies the condition (6.66) if and only if*

$$\eta \in \mathfrak{r}_r^d(\overline{\omega})\mathbb{C}^q. \tag{6.74}$$

Proof If $\xi = 0$, then, in view of the preceding discussion, (6.66) holds if and only if the constraint on the limit in (6.71) is met. If $\omega \in \mathbb{C}_-$, then, by formula (6.22), this condition will be met if and only if

$$\lim_{t \uparrow d} \eta^* R_r^t(\overline{\omega})^{-2}\eta < \infty.$$

Therefore, by Lemma 6.18, this will hold if and only if (6.74) holds. □

The Weyl property of the input scattering matrices

Lemma 6.21 *Let $s(\lambda)$ be an input scattering matrix for the system (6.1), let $\omega \in \mathbb{C}_+$ and $\eta \in \mathbb{C}^q$. Then the solution $u(t,\omega)$ of the system (6.65) (with $\lambda = \omega$) and initial data*

$$u(0,\omega) = \begin{bmatrix} -s(\omega)\eta \\ \\ \eta \end{bmatrix}$$

satisfies the bound

$$\frac{1}{2\pi}\int_0^d u(t,\omega)^* dM(t) u(t,\omega) \le \eta^* \frac{I_q - s(\omega)^* s(\omega)}{\rho_\omega(\omega)}\eta. \tag{6.75}$$

Proof The inequality (6.75) is obtained by estimating the right-hand side of (6.69). In particular, since

$$-\rho_{\overline{\omega}}(\overline{\omega}) = \rho_\omega(\omega) > 0 \quad \text{for} \quad \omega \in \mathbb{C}_+,$$

we have

$$\rho_\omega(\omega) u(0,\omega)^* K_{\overline{\omega}}^t(\overline{\omega}) u(0,\omega) = ① + ②,$$

where

$$① = u(0,\omega)^* W_t(\overline{\omega}) j_{pq} W_t(\overline{\omega})^* u(0,\omega)$$

and

$$② = -u(0,\omega)^* j_{pq} u(0,\omega) = \eta^*\{I_q - s(\omega)^* s(\omega)\}\eta.$$

Therefore, in order to complete the proof, it suffices to show that ① ≤ 0. But if $s(\lambda)$ is an input scattering matrix for the system (6.1), then for every $t \in [0,d)$ there exists an mvf $\varepsilon_t \in \mathcal{S}^{p\times q}$ such that

$$\begin{bmatrix} -s(\omega) \\ I_q \end{bmatrix} = -j_{pq} W_t(\omega) \begin{bmatrix} \varepsilon_t(\omega) \\ I_q \end{bmatrix} \psi_t(\omega),$$

where

$$\psi_t(\omega) = \{w_{21}^t(\omega)\varepsilon_t(\omega) + w_{22}^t(\omega)\}^{-1}.$$

Thus, as

$$W_t(\overline{\omega})^* j_{pq} W_t(\omega) = j_{pq},$$

it now follows easily that

$$① = \eta^* \psi_t(\omega)^* \{\varepsilon_t(\omega)^* \varepsilon_t(\omega) - I_q\} \psi_t(\omega)\eta \leq 0,$$

as needed. □

Theorem 6.22 *Let $\omega \in \mathbb{C}_+$. Then the solution $u(t, \omega)$ of the system (6.65) (with $\lambda = \omega$) satisfies the condition (6.66) if and only if the initial condition $u(0, \omega)$ meets the condition*

$$[I_p \quad s(\omega)]\, u(0, \omega) \in \mathfrak{r}_\ell^d(\omega)\mathbb{C}^p \tag{6.76}$$

for some input scattering matrix $s(\lambda)$ of the system (6.1).

Proof Let $s(\lambda)$ be an input scattering matrix for the system (6.1). Then

$$u(0, \omega) = \begin{bmatrix} \xi \\ \eta \end{bmatrix} = \begin{bmatrix} \xi + s(\omega)\eta \\ 0 \end{bmatrix} + \begin{bmatrix} -s(\omega)\eta \\ \eta \end{bmatrix}$$

and hence, in view of Lemma 6.21, the corresponding solution $u(t, \omega)$ will satisfy condition (6.66) if and only if the solution corresponding to the first vector on the right satisfies condition (6.66). The stated conclusion is now immediate from Lemma 6.19. □

Corollary 6.23 *If $\mathfrak{n}_\ell = 0$, and $s(\lambda)$ is the unique input scattering matrix for the system (6.1), then the solution $u(t, \omega)$ of the system (6.65) with $\lambda = \omega$ and initial data $u(0, \omega)$ will satisfy the condition (6.66) if and only if*

$$[I_p \quad s(\omega)]u(0, \omega) = 0. \tag{6.77}$$

If $\mathfrak{r}_\ell^d(\lambda) = 0$ for any one (and hence every) point $\lambda \in \mathbb{C}_+$, then the description (6.77) of the initial data that corresponds to condition (6.66) serves to characterize the input scattering matrix of the system (6.1).

Theorem 6.24 *Let the left limit semiradius $\mathfrak{r}_\ell^d(\lambda)$ for the system (6.1) be equal to zero for at least one (and hence every) point $\lambda \in \mathbb{C}_+$. Let $s(\lambda)$ be a $p \times q$ mvf such that for every $\lambda \in \mathbb{C}_+$ and every $\eta \in \mathbb{C}^q$ the solution $u(t, \lambda)$ of the system (6.65) with initial data*

$$u(0, \lambda) = \begin{bmatrix} -s(\lambda)\eta \\ \eta \end{bmatrix}$$

satisfies the condition (6.66). Then $s(\lambda)$ is the input scattering matrix for the system (6.1).

Proof Let $s^\circ(\lambda)$ be the input scattering matrix for the system (6.1). Then since the solutions $u^\circ(t,\lambda)$ and $u(t,\lambda)$ of (6.65) with initial data

$$u^\circ(0,\lambda) = \begin{bmatrix} -s^\circ(\lambda)\eta \\ \eta \end{bmatrix} \quad \text{and} \quad u(0,\lambda) = \begin{bmatrix} -s(\lambda)\eta \\ \eta \end{bmatrix},$$

respectively, both meet (6.66), the same holds true for the difference $u^\circ(t,\lambda) - u(t,\lambda)$. Therefore, by Lemma 6.19, $\{s(\lambda) - s^\circ(\lambda)\}\eta \in \mathfrak{r}_\ell^d(\lambda)\mathbb{C}^q$ for every point $\lambda \in \mathbb{C}_+$ and every $\eta \in \mathbb{C}^q$. Thus, as $\mathfrak{r}_\ell^d(\lambda) = 0$, it follows that $s(\lambda) = s^\circ(\lambda)$, as claimed. □

6.11 Supplementary notes

Most of the results presented in this chapter are adapted from [ArD02a], where bitangential generalizations of direct and inverse input scattering problems for canonical integral and differential equations were considered for the first time.

Input scattering matrices play a significant role in mathematical physics [Fa59], [Vl02] and in the theory of passive networks with lumped and distributed parameters; see, e.g., [YCC59], [Bel68], [Wo69], [Hel73], [DeD84], [ADD89], [Ar95b] and the references cited therein.

In the literature, there are two different definitions of input scattering matrix (or, as it is more commonly referred to, scattering matrix, or reflection coefficient): The stationary definition views the scattering matrix as a transfer function between the Fourier transforms of incoming and outgoing waves. The nonstationary, or dynamic definition is based on wave operators $W_\pm$ and the representation of the scattering operator $S = W_+^* W_-$ as the operator of multiplication by the scattering matrix acting between two L_2 spaces.

The equivalence of these two definitions for the Lax–Phillips scattering scheme is established in [AdAr66] and [LP67]. In [LP67] the scattering matrix is considered for the wave equation, hyperbolic systems and the Schrödinger equation. The equivalence of the two definitions of scattering matrix for DK-systems is established in [Den06]; see also [Ar71], [Hel74] and [Ar79] for connections with system theory.

The classification of the singular canonical integral system (6.1) by the ranks $\mathfrak{n}_\ell(\omega)$ and $\mathfrak{n}_r(\omega)$ of the left and right semiradii of the limit ball $\{s(\omega) : s \in \mathcal{S}_{\rm scat}\}$ is similar in spirit to the classification by H. Weyl [We10] and E.C. Titchmarsh [Ti46] for the limit balls defined by the input impedances (Weyl–Titchmarsh functions) that will be considered in Chapter 7. The Supplementary Notes to that chapter contain a more extensive discussion of Orlov's results that guarantee that the ranks of $\mathfrak{n}_\ell(\omega)$ and $\mathfrak{n}_r(\omega)$ are independent of the choice of the point $\omega \in \mathbb{C}_+$.

The matching of the ranks of the semiradii of the limit ball with the ranks of limits based on associated pairs that is presented in Theorem 6.11 is adapted from [ArD02a]. Formulas for the semiradii of matrix balls in terms of reproducing kernels are developed in section 8 of [DI84].

7

Bitangential direct input impedance and spectral problems

This chapter focuses on the bitangential direct problem for input impedances and spectral functions for canonical integral systems (2.1) with $J = J_p$. Most of the results on the direct input impedance problem that will be presented below are analogs of the results for bitangential direct input scattering problems that were discussed in Chapter 6. The change from $J = j_{pq}$ to $J = J_p$ introduces some complications in the domain of definition of linear fractional transformations based on mvf's in $\mathcal{U}(J_p)$. The symbol $A(\lambda)$ will be used to denote mvf's in this class.

7.1 Input impedance matrices

Recall that the matrizant of the canonical integral system

$$y(t,\lambda) = y(0,\lambda) + i\lambda \int_0^t y(s,\lambda)dM(s)J_p, \quad 0 \le t < d, \tag{7.1}$$

with mass function $M(t)$ subject to (6.2) is denoted $A_t(\lambda) = A(t,\lambda)$ for $0 \le t < d$. It will also be convenient to consider the modified matrizant

$$B_t(\lambda) = B(t,\lambda) = A(t,\lambda)\mathfrak{V}, \quad 0 \le t < d, \tag{7.2}$$

that belongs to the class $\mathcal{E} \cap \mathcal{U}(j_p, J_p)$ of entire (j_p, J_p)-inner mvf's for each fixed $t \in [0, d)$ and can be identified as the only continuous solution of the canonical integral equation

$$y(t,\lambda) = u(0,\lambda) + i\lambda \int_0^t u(s,\lambda)d(\mathfrak{V}M(s)\mathfrak{V})j_p, \quad 0 \le t < d, \tag{7.3}$$

that satisfies the initial condition $u(0,\lambda) = \mathfrak{V}$. Moreover, since

$$B_{t_1}^{-1}B_{t_2} \in \mathcal{E} \cap \mathcal{U}(j_p) \text{ for } 0 \le t_1 \le t_2 < d,$$

the inclusions

$$\mathcal{S}^{p\times p} \cap \mathcal{D}(T_{B_{t_2}}) \subseteq \mathcal{S}^{p\times p} \cap \mathcal{D}(T_{B_{t_1}}) \tag{7.4}$$

and

$$\mathcal{C}(A_{t_2}) \subseteq \mathcal{C}(A_{t_1}) \subseteq \mathcal{C}^{p\times p} \tag{7.5}$$

hold; see, e.g., Lemma 3.68.

The set of **input impedance matrices** $\mathcal{C}_{\rm imp}$ of the canonical system (7.1) is defined by the formula

$$\mathcal{C}_{\rm imp} = \mathcal{C}_{\rm imp}^{d}(dM) = \bigcap_{0\le t<d} \mathcal{C}(A_t). \tag{7.6}$$

Clearly, $\mathcal{C}_{\rm imp} \subseteq \mathcal{C}^{p\times p}$. However, this statement may be vacuous, since the set $\mathcal{C}_{\rm imp}$ may be empty. The formulas

$$\mathcal{S}_{\rm scat} = \bigcap_{0\le t<d} T_{\mathfrak{V}A_t\mathfrak{V}}[\mathcal{S}^{p\times q}] \quad \text{and} \quad \mathcal{C}_{\rm imp} = T_{\mathfrak{V}}[\mathcal{S}_{\rm scat} \cap \mathcal{D}(T_{\mathfrak{V}})] \tag{7.7}$$

exhibit the fact that

$$\mathcal{C}_{\rm imp} \neq \emptyset \Longleftrightarrow \mathcal{S}_{\rm scat} \cap \mathcal{D}(T_{\mathfrak{V}}) \neq \emptyset. \tag{7.8}$$

Both extremes are possible: It can happen that

$$\mathcal{S}_{\rm scat} \subset \mathcal{D}(T_{\mathfrak{V}}) \quad \text{and} \quad \mathcal{C}_{\rm imp} = T_{\mathfrak{V}}[\mathcal{S}_{\rm scat}],$$

or that

$$\mathcal{S}_{\rm scat} \cap \mathcal{D}(T_{\mathfrak{V}}) = \emptyset \quad \text{and} \quad \mathcal{C}_{\rm imp} = \emptyset.$$

This comes about because the Cayley transform

$$T_{\mathfrak{V}}[c] = \{I_p - c(\lambda)\}\{I_p + c(\lambda)\}^{-1}$$

maps the class $\mathcal{C}^{p\times p}$ onto

$$\mathcal{S}^{p\times p} \cap \mathcal{D}(T_{\mathfrak{V}}) = \left\{s \in \mathcal{S}^{p\times p} : \det(I_p + s(\lambda)) \not\equiv 0\right\},$$

and hence,

$$\mathcal{S}_{\rm scat} \cap \mathcal{D}(T_{\mathfrak{V}}) = \left\{s \in \mathcal{S}_{\rm scat} : \det(I_p + s(\lambda)) \not\equiv 0\right\}.$$

The last intersection may be empty. To overcome this difficulty, one can introduce an affine generalization $\widetilde{\mathcal{C}}^{p\times p}$ of the Carathéodory class $\mathcal{C}^{p\times p}$; see, e.g., the discussion at the end of Section 3.12. The set of generalized input impedances $\widetilde{\mathcal{C}}_{\rm imp}$ based on $\widetilde{\mathcal{C}}^{p\times p}$ is nonempty. In this monograph, we will not use this generalization. Instead, we will consider systems (7.1) for which

$$\mathcal{S}^{p\times p} \subseteq \mathcal{D}(T_{B_{t_0}}) \text{ for some } t_0 \in (0, d). \tag{7.9}$$

Thus, if $W_t = \mathfrak{V}B_t$ for $0 \le t < d$, the inclusion (7.9) implies that

$$T_{W_{t_0}}[\mathcal{S}^{p\times q}] \subseteq \mathcal{D}(T_{\mathfrak{V}}).$$

Therefore,

$$T_{W_t}[\mathcal{S}^{p\times q}] \subseteq T_{W_{t_0}}[\mathcal{S}^{p\times q}] \subseteq \mathcal{D}(T_{\mathfrak{V}}) \quad \text{for } t_0 \le t < d$$

and hence

$$\mathcal{S}^{p\times p} \subseteq \mathcal{D}(T_{B_t}) \ \text{ for every } \ t \in [t_0, d). \tag{7.10}$$

Under the extra condition (7.9),

$$\mathcal{S}_{\text{scat}} \subseteq T_{\mathfrak{V}B_{t_0}}[\mathcal{S}^{p\times p}].$$

Consequently, under condition (7.9),

$$\mathcal{S}_{\text{scat}} \subseteq \mathcal{D}(T_{\mathfrak{V}}) \quad \text{and} \quad \mathcal{C}_{\text{imp}} = T_{\mathfrak{V}}[\mathcal{S}_{\text{scat}}]. \tag{7.11}$$

If (7.1) is a regular canonical integral system with right-hand end point d, then the set $\mathcal{C}_{\text{imp}}$ of input impedances is simply described by the formula

$$\mathcal{C}_{\text{imp}} = \mathcal{C}(A_d). \tag{7.12}$$

If (7.1) is not a regular canonical integral system, then the description of $\mathcal{C}_{\text{imp}}$ is more complicated and will be discussed in the next several sections. The physical interpretation of the input impedance of a regular canonical integral system of the form (7.1) is similar to the interpretation of the input scattering matrix that was discussed in the previous chapter. The only difference is that now the components of the solution $y(t, \lambda) = [v(t, \lambda) \quad i(t, \lambda)]$ are interpreted as the Fourier transforms of voltage and current in p-line networks.

The **direct input impedance problem** is to describe the set $\mathcal{C}^d_{\text{imp}}(dM)$. If (7.1) is a regular canonical integral system, then the solution is given by formula (7.12). If condition (7.9) is in force, then a solution of this problem may be expressed in terms of the solution of the direct input scattering problem via formula (7.11).

The sets of input impedance matrices $\mathcal{C}^\ell_{\text{imp}}(N, \mathcal{V})$ and $\mathcal{C}^d_{\text{imp}}(Hdx; [0, d))$ for the canonical differential systems (2.72) and (2.14) with $J = J_p$ are also defined in terms of the corresponding matrizants, in much the same way as for the canonical integral system (7.1). If $M(x) = \int_0^x H(s)ds$, then

$$\mathcal{C}^d_{\text{imp}}(dM) = \mathcal{C}^d_{\text{imp}}(Hdx).$$

Moreover, if the Hamiltonian $H(x)$ of the canonical differential system (2.14) is defined in terms of the matrix N and the potential $\mathcal{V}(x)$ of the differential system (2.72) with potential via formulas (2.79) and (2.74), then

$$\mathcal{C}^\ell_{\text{imp}}(N, \mathcal{V}) = \mathcal{C}^\ell_{\text{imp}}(Hdx),$$

since the corresponding matrizants $Y_x(\lambda)$ and $U_x(\lambda)$ are connected by the formula

$$Y_x(\lambda)\mathfrak{V} = U_x(\lambda)\mathfrak{V}\mathfrak{V}Y(x)\mathfrak{V}, \quad \text{where} \quad \mathfrak{V}Y(x)\mathfrak{V} \in \mathcal{U}_{\text{const}}(j_p) \quad \text{for } 0 \le x < \ell.$$

Lemma 7.1 *Let $A_t(\lambda) = A(t, \lambda)$, $0 \le t < d$, be the matrizant of a canonical integral system of the form (7.1), let $\{b_3^t, b_4^t\} \in ap_{II}(A_t)$ with $b_3^t(0) = b_4^t(0) = I_p$ for every $t \in [0, d)$ and suppose that the following two conditions are met:*

(1) $A_t \in \mathcal{U}_{rsR}(J_p)$ *for every* $t \in [0, d)$.
(2) *The mvf* $\chi_1^t(\lambda) = b_4^t(\lambda)b_3^t(\lambda)$ *is subject to the bound*

$$\chi_1^{t_0}(\omega)\chi_1^{t_0}(\omega)^* < I_p \tag{7.13}$$

for some point $\omega \in \mathbb{C}_+$ and some point $t_0 \in (0, d)$.

Then

$$\mathcal{C}_{\text{imp}} = \bigcap_{t_0 \le t < d} T_{B_t}[\mathcal{S}^{p\times p}] = T_{\mathfrak{V}}[\mathcal{S}_{\text{scat}}] \neq \emptyset. \tag{7.14}$$

Moreover,

$$c \in \mathcal{C}_{\text{imp}} \Longrightarrow \Re c(\lambda) > 0 \textit{ for every point } \lambda \in \mathbb{C}_+. \tag{7.15}$$

Proof Under the given assumptions, $\mathcal{S}^{p\times p} \subseteq \mathcal{D}(T_{B_t})$ for $t = t_0$ by Remark 4.26. Therefore, this inclusion is valid for all t in the interval $t_0 \le t < d$, and (4.70) holds for $s \in \mathcal{S}_{\text{scat}}$. □

It is clear that condition (2) in the last lemma holds if

$$\text{either} \quad \lim_{\nu\uparrow\infty} b_3^{t_0}(i\nu) = 0 \quad \text{or} \quad \lim_{\nu\uparrow\infty} b_4^{t_0}(i\nu) = 0 \quad \text{for some point } t \in [0, d).$$

Moreover, if condition (2) in the last lemma holds for $t = t_0$, then it also holds for $t_0 \le t < d$.

Lemma 7.2 *Let $A_t(\lambda) = A(t, \lambda)$, $0 \le t < d$, be the matrizant of a canonical integral system (7.1) with input impedance matrix $c \in \mathcal{C}_{\text{imp}} \cap \mathring{\mathcal{C}}^{p\times p}$. Let $\{b_1^t, b_2^t\} \in ap_I(A_t)$ and $\{b_3^t, b_4^t\} \in ap_{II}(A_t)$ for $0 \le t < d$. Then the following equivalences hold for every point $\lambda \in \mathbb{C}_+$:*

$$\lim_{t\uparrow d} b_1^t(\lambda) = 0 \Longleftrightarrow \lim_{t\uparrow d} b_3^t(\lambda) = 0 \tag{7.16}$$

$$\lim_{t\uparrow d} b_2^t(\lambda) = 0 \Longleftrightarrow \lim_{t\uparrow d} b_4^t(\lambda) = 0. \tag{7.17}$$

Proof The conclusions are obtained by adapting the bounds of Lemma 4.28 to the present setting. □

7.2 Limit balls for input impedance matrices

In this section we will consider a family of matrix balls $\mathcal{B}_t(\omega)$ defined by formula (3.181) for $B(\lambda) = B_t(\lambda)$ and $t \geq t_0$ under assumption (7.19) (that is given below), where

$$B_t(A) = A_t(\lambda)\mathfrak{V} = \begin{bmatrix} b_{11}^t(\lambda) & b_{12}^t(\lambda) \\ b_{21}^t(\lambda) & b_{22}^t(\lambda) \end{bmatrix}$$

and $A_t(\lambda) = A(t, \lambda),\ 0 \leq t < d$, is the matrizant of the canonical integral system (7.1). Correspondingly, set

$$\begin{aligned}
\mathfrak{E}_t(\lambda) &= [E_-^t(\lambda) \quad E_+^t(\lambda)] = \sqrt{2}[b_{21}^t(\lambda) \quad b_{22}^t(\lambda)] \\
\chi_t(\lambda) &= E_+^t(\lambda)^{-1}E_-^t(\lambda) \\
K_\omega^{A_t}(\lambda) &= \rho_\omega(\lambda)^{-1}\{J_p - A_t(\lambda)J_pA_t(\omega)^*\} \\
K_\omega^{\mathfrak{E}_t}(\lambda) &= \rho_\omega(\lambda)^{-1}\{E_+^t(\lambda)E_+^t(\omega)^* - E_-^t(\lambda)E_-^t(\omega)^*\} \\
2r_t(\lambda) &= -\mathfrak{E}_t(\lambda)j_p\mathfrak{E}_t(\lambda)^* = E_+^t(\lambda)E_+^t(\lambda)^* - E_-^t(\lambda)E_-^t(\lambda)^*
\end{aligned}$$

and note that $\mathfrak{E}_t$ is the unique continuous solution of the equation

$$\mathfrak{E}_t'(\lambda) = \begin{bmatrix} I_p & I_p \end{bmatrix} + i\lambda \int_0^t \mathfrak{E}_s(\lambda)d(\mathfrak{V}M(s)\mathfrak{V})j_p \quad \text{for } t \in [0, d). \tag{7.18}$$

We shall assume that there exists a point $t_0 \in (0, d)$ such that

$$K_\omega^{\mathfrak{E}_{t_0}}(\omega) > 0 \text{ for at least one point } \omega \in \mathbb{C}_+. \tag{7.19}$$

This is equivalent to the condition that

$$\chi_{t_0}(\omega)\chi_{t_0}(\omega)^* < I_p \text{ for at least one point } \omega \in \mathbb{C}_+, \tag{7.20}$$

since $E_+^t(\lambda)$ is invertible in $\mathbb{C}_+$ by Lemma 3.67. Moreover, as $\chi \in \mathcal{S}^{p\times p}$, if either (7.19) or (7.20) holds for a point $\omega \in \mathbb{C}_+$, then, by Lemma 4.67, it holds for every point $\omega \in \mathbb{C}_+$.

Let

$$N_2^* = [0 \ I_p]. \tag{7.21}$$

Then, if $0 \leq t_1 \leq t_2 < d$,

$$K_\omega^{\mathfrak{E}_{t_1}}(\omega) = 2N_2^*K_\omega^{A_{t_1}}(\omega)N_2 \leq 2N_2^*K_\omega^{A_{t_2}}(\omega)N_2 = K_\omega^{\mathfrak{E}_{t_2}}(\omega),$$

whereas $\rho_\omega(\omega) > 0$ for $\omega \in \mathbb{C}_+$ and $\rho_\omega(\omega) < 0$ for $\omega \in \mathbb{C}_-$. Thus,

$$r_{t_1}(\omega) \leq r_{t_2}(\omega) \text{ and } - r_{t_1}(\bar{\omega}) \leq -r_{t_2}(\bar{\omega}) \text{ for } \omega \in \mathbb{C}_+. \tag{7.22}$$

Consequently, by formulas (3.183)–(3.185),

$$R_r^t(\omega)^2 = r_t(\omega)^{-1} = 2E_+^t(\omega)^{-*}\{I_p - \mathcal{X}_t(\omega)\mathcal{X}_t(\omega)^*\}^{-1}E_+^t(\omega)^{-1}, \tag{7.23}$$

$$R_\ell^t(\omega)^2 = -r_t(\overline{\omega})^{-1} = 2E_-^t(\overline{\omega})^{-*}\{I_p - \mathcal{X}_t(\omega)^*\mathcal{X}_t(\omega)\}^{-1}E_-^t(\overline{\omega})^{-1}, \tag{7.24}$$

$$\gamma_c^t(\omega)^2 = T_{B_t(\omega)}[-\mathcal{X}_t(\omega)^*] \tag{7.25}$$

and the inequalities in (7.22), the squares of the semiradii are positive definite and nonincreasing:

$$0 < R_\ell^{t_2}(\omega)^2 \le R_\ell^{t_1}(\omega)^2 \text{ and } 0 < R_r^{t_2}(\omega)^2 \le R_r^{t_1}(\omega)^2 \text{ for } t_0 \le t_1 \le t_2 < d. \tag{7.26}$$

Moreover, the matrix balls

$$\mathcal{B}_t(\omega) = \{c(\omega) : c \in T_{B_t}[\mathcal{S}^{p\times p}]\} \tag{7.27}$$

are well defined for $t_0 \le t < d$ and they are nested sets:

$$\mathcal{B}_{t_2}(\omega) \subseteq \mathcal{B}_{t_1}(\omega) \text{ for } t_0 \le t_1 \le t_2 < d. \tag{7.28}$$

It is known that the intersection of a nested family of matrix balls is a matrix ball and that the center of this ball is the limit of the centers of the considered family of matrix balls, see [Shm68]. Moreover, for $\omega \in \mathbb{C}_+$, the intersection

$$\mathcal{B}_d(\omega) = \bigcap_{0\le t<d} \mathcal{B}_t(\omega) \tag{7.29}$$

is a matrix ball with center

$$\begin{aligned}\gamma_c^d(\omega) &= \lim_{t\uparrow d} \gamma_c^t(\omega) = \lim_{t\uparrow d} T_{B_t(\omega)}[-\mathcal{X}_t(\omega)^*] \\ &= \lim_{t\uparrow d}\{[I_p + \rho_\omega(\omega)\{K_\omega^{A_t}(\omega)\}_{12}][\rho_\omega(\omega)\{K_\omega^{A_t}(\omega)\}_{22}]^{-1}\}\end{aligned} \tag{7.30}$$

and semiradii

$$\mathfrak{r}_\ell^d(\omega) = \lim_{t\uparrow d} R_\ell^t(\omega) = \sqrt{2}\,\lim_{t\uparrow d}\{-\rho_{\overline{\omega}}(\bar{\omega})K_{\overline{\omega}}^{\mathfrak{E}_t}(\overline{\omega})\}^{-1/2} \tag{7.31}$$

and

$$\mathfrak{r}_r^d(\omega) = \lim_{t\uparrow d} R_r^t(\omega) = \sqrt{2}\,\lim_{t\uparrow d}\{\rho_\omega(\omega)K_\omega^{\mathfrak{E}_t}(\omega)\}^{-1/2}. \tag{7.32}$$

Let

$$\mathfrak{n}_\ell = \operatorname{rank} \mathfrak{r}_\ell^d(\omega) \text{ and } \mathfrak{n}_r = \operatorname{rank} \mathfrak{r}_r^d(\omega). \tag{7.33}$$

By a general result of S.A. Orlov [Or76], the numbers $\mathfrak{n}_\ell$ and $\mathfrak{n}_r$ are independent of the choice of the point $\omega \in \mathbb{C}_+$. Thus, the canonical integral system (7.1) is in the full rank case (i.e., $\mathfrak{n}_\ell = \mathfrak{n}_r = p$) if and only if $\mathfrak{r}_\ell^d(\omega) > 0$ and $\mathfrak{r}_r^d(\omega) > 0$ for at least one (and hence every) point $\omega \in \mathbb{C}_+$. Similarly, the system (7.1) is in the limit point case if either $\mathfrak{r}_\ell^d(\omega) = 0$ or $\mathfrak{r}_r^d(\omega) = 0$ for at least one (and hence every) point $\omega \in \mathbb{C}_+$.

Remark 7.3 *The semiradii that are defined by formulas (7.23) and (7.24) are connected:*

$$\frac{\det R_\ell^t(\omega)}{\det R_r^t(\omega)} = \frac{|\det b_3^t(\omega)|}{|\det b_4^t(\omega)|} = |\det B_t(\omega)| = |\det A_t(\omega)|, \; 0 \le t < d, \tag{7.34}$$

where $\{b_3^t, b_4^t\} \in ap(B_t)$. *Moreover, the semiradii* $R_\ell^t(\omega)$ *and* $R_r^t(\omega)$ *of the ball* $\mathcal{B}_t(\omega)$ *are uniquely defined by the relation (7.34); see theorem 5.59 on pp. 291–292 of [ArD08b].*

Remark 7.4 *Remark 6.6 is applicable to the matrix balls considered in this chapter also.*

Theorem 7.5 *Let* $\mathcal{C}_{\rm imp}$ *be the set of input impedance matrices for the system (7.1) and* $\mathcal{S}_{\rm scat}$ *the set of input scattering matrices for the corresponding system (6.1) (with matrizant* $W_t(\lambda) = \mathfrak{V}A_t(\lambda)\mathfrak{V}$ *and mass function* $\mathfrak{V}M(t)\mathfrak{V}$*), and suppose that assumption (7.19) is in force. Then:*

$$\mathcal{C}_{\rm imp} = \bigcap_{t_0 \le t < d} T_{B_t}[\mathcal{S}^{p\times p}] = T_{\mathfrak{V}}[\mathcal{S}_{\rm scat}] \neq \emptyset \tag{7.35}$$

and

$$\{c(\omega) : c \in \mathcal{C}_{\rm imp}\} = \mathcal{B}_d(\omega) \quad \textit{for every point } \omega \in \mathbb{C}_+. \tag{7.36}$$

Moreover, $\mathcal{C}_{\rm imp}$ *contains exactly one input impedance matrix if and only if the canonical integral system (7.1) is in the limit point case.*

Proof The first assertion is immediate from Lemma 3.67 and (7.11). To verify the second assertion, recall first that there exists a unique j_p-unitary mvf $U(t)$ such that the blocks of the j_p-inner mvf's

$$\Omega_t(\lambda) = W_t(\lambda)U(t), \quad 0 \le t < d,$$

meet the conditions

$$(\Omega_t)_{21}(\omega) = 0, \quad (\Omega_t)_{11}(\omega) > 0 \quad \text{and} \quad (\Omega_t)_{22}(\omega) > 0$$

at some fixed point $\omega \in \mathbb{C}_+$. Let $\Sigma_t(\lambda)$ denote the Potapov–Ginzburg transform of $\Omega_t(\lambda)$. Then $\Sigma_t \in \mathcal{S}_{\rm in}^{m\times m}$,

$$(\Sigma_t)_{21}(\omega) = 0, \quad (\Sigma_t)_{11}(\omega) > 0, \quad (\Sigma_t)_{22}(\omega) > 0$$

and

$$\begin{aligned} T_{W_t}[\mathcal{S}^{p\times p}] &= R_{\Sigma_t}[\mathcal{S}^{p\times p}] \\ &= \{(\Sigma_t)_{12} + (\Sigma_t)_{11}\varepsilon\{I_p - (\Sigma_t)_{21}\varepsilon\}^{-1}(\Sigma_t)_{22} : \varepsilon \in \mathcal{S}^{p\times p}\}. \end{aligned}$$

The advantage of this formulation is that it makes it clear that there exists a sequence of points $t_n \uparrow d$ such that $\Sigma_{t_n} \longrightarrow \Sigma_* \in \mathcal{S}^{m\times m}$ uniformly on compact subsets of $\mathbb{C}_+$. Now fix $\omega \in \mathbb{C}_+$ and let $\alpha \in \mathcal{B}_d(\omega)$. Then $\alpha \in \mathcal{C}_{\rm const}^{p\times p}$ and, since $\mathcal{B}_d(\omega) \subseteq \mathcal{B}_t(\omega)$ for every $t \in [0, d)$, there exists a mvf $\varepsilon_t \in \mathcal{S}^{p\times p}$ such that

$$\beta = T_{\mathfrak{V}}[\alpha] = R_{\Sigma_t(\omega)}[\varepsilon_t(\omega)].$$

The next step is to choose a sequence of points $t_n \in (0, d)$ tending to d so that $\Sigma_{t_n}(\lambda) \longrightarrow \Sigma_*(\lambda)$ and $\varepsilon_{t_n}(\lambda) \longrightarrow \varepsilon_*(\lambda)$ locally uniformly on $\mathbb{C}_+$. Then

$$s(\lambda) = R_{\Sigma_*(\lambda)}[\varepsilon_*(\lambda)] \in \mathcal{S}_{\rm scat} \quad \text{and} \quad s(\omega) = \beta = T_{\mathfrak{V}}[\alpha].$$

The latter implies in particular that

$$I_p + s(\omega) = 2(I_p + \alpha)^{-1}$$

and hence, as $\det\{I_p + s(\lambda)\} \not\equiv 0$, that

$$s \in \mathcal{D}(T_{\mathfrak{V}}) \text{ and } T_{\mathfrak{V}}[s(\omega)] = \alpha.$$

Thus,

$$c = T_{\mathfrak{V}}[s] \in \mathcal{C}_{\rm imp} \text{ and } c(\omega) = \alpha,$$

i.e.,

$$\mathcal{B}_d(\omega) \subseteq \{c(\omega) : c \in \mathcal{C}_{\rm imp}\}.$$

Therefore, since the opposite inclusion is self-evident, the proof of (7.36) is complete. The final assertion follows from (7.36). □

7.3 Formulas for the ranks of semiradii of the limit ball

Lemma 7.6 *Let $A_t(\lambda) = A(t, \lambda)$, $0 \le t < d$, be the matrizant of the system (7.1), let $\{b_3^t, b_4^t\} \in ap_{II}(A_t)$ and $\chi_1^t(\lambda) = b_4^t(\lambda)b_3^t(\lambda)$ for every $t \in [0, d)$ and suppose that $c \in \mathcal{C}_{\rm imp} \cap \mathring{\mathcal{C}}^{p\times p}$ is such that*

$$0 < \gamma_1 I_p \le \Re c(\mu) \le \gamma_2 I_p \ \ for\ a.e.\ \ \mu \in \mathbb{R}$$

and

$$\|\chi_1^{t_0}(\omega)\| = \gamma < 1$$

for some $t_0 \in (0, d)$ and some point $\omega \in \mathbb{C}_+$. Then, the left and right semiradii $R_\ell^t(\omega)$ and $R_r^t(\omega)$ of the ball

$$\mathcal{B}_t(\omega) = \{c(\omega) : c \in T_{B_t}[\mathcal{S}^{p\times p}]\}$$

are subject to the bounds

$$2\gamma_1 b_4^t(\omega)^* b_4^t(\omega) \le R_r^t(\omega)^2 \le (1-\gamma^2)^{-1} 2\gamma_2 b_4^t(\omega)^* b_4^t(\omega) \tag{7.37}$$

and

$$2\gamma_1 b_3^t(\omega) b_3^t(\omega)^* \le R_\ell^t(\omega)^2 \le (1-\gamma^2)^{-1} 2\gamma_2 b_3^t(\omega) b_3^t(\omega)^* \tag{7.38}$$

for every $t \in [t_0, d)$.

Proof The condition on $c(\lambda)$ guarantees that $A_t \in \mathcal{U}_{rsR}(J_p)$ for every $t \in [0, d)$. Thus, upon setting $B_t(\lambda) = A_t(\lambda)\mathfrak{V}$, $\mathfrak{E}_t(\lambda) = \sqrt{2}N_2^* B_t(\lambda)$ and invoking Remark 4.26 and Lemma 3.67, we obtain the inequality

$$K_\omega^{\mathfrak{E}_t}(\omega) > 0 \tag{7.39}$$

for every point $\omega \in \mathbb{C}_+$ and every $t \in [t_0, d)$. Since $\mathfrak{E}_t(\lambda)$ is an entire mvf, the inequality (7.39) is in fact valid for every point $\omega \in \mathbb{C}$ by Lemma 4.67. Lemma 3.67 also guarantees that $\mathcal{S}^{p\times p} \subseteq \mathcal{D}(T_{B_t})$ for $t \in [t_0, d)$ and hence that the ball $\mathcal{B}_t(\omega)$ is given by the formula

$$\mathcal{B}_t(\omega) = \{c(\omega) : \ c \in T_{B_t}[\mathcal{S}^{p\times p}]\} \ \text{ for } \ t \in [t_0, d).$$

The left and right semiradii $R_\ell^t(\omega)$ and $R_r^t(\omega)$ of this ball are given by formulas (3.183) and (3.184):

$$R_r^t(\omega)^2 = -2\{\mathfrak{E}_t(\omega) j_p \mathfrak{E}_t(\omega)^*\}^{-1} \ \text{ and } \ R_\ell^t(\omega)^2 = 2\{\mathfrak{E}_t(\bar{\omega}) j_p \mathfrak{E}_t(\bar{\omega})^*\}^{-1}, \tag{7.40}$$

for $\omega \in \mathbb{C}_+$. Next, in view of the assumption on $c(\lambda)$ and Theorem 4.82,

$$\gamma_2^{-1}\{k_\omega^{b_3^t}(\omega) + \ell_\omega^{b_4^t}(\omega)\} \leq K_\omega^{\mathfrak{E}_t}(\omega) \leq \gamma_1^{-1}\{k_\omega^{b_3^t}(\omega) + \ell_\omega^{b_4^t}(\omega)\}, \tag{7.41}$$

for every point $\omega \in \mathbb{C}$ and every $t \in [t_0, d)$. Thus, as

$$\rho_\omega(\omega)\{k_\omega^{b_3^t}(\omega) + \ell_\omega^{b_4^t}(\omega)\} = b_4^t(\omega)^{-1} b_4^t(\omega)^{-*} - b_3^t(\omega) b_3^t(\omega)^*, \tag{7.42}$$

it follows from (7.39) and (7.41) that

$$\begin{aligned} &2\gamma_1 b_4^t(\omega)^*\{I_p - \chi_1^t(\omega)\chi_1^t(\omega)^*\}^{-1} b_4^t(\omega) \\ &\leq R_r^t(\omega)^2 \leq 2\gamma_2 b_4^t(\omega)^*\{I_p - \chi_1^t(\omega)\chi_1^t(\omega)^*\}^{-1} b_4^t(\omega) \end{aligned}$$

for $t \in [t_0, d)$ and $\omega \in \mathbb{C}_+$. Therefore, since

$$\|\chi_1^t(\omega)\| \leq \|\chi_1^{t_0}(\omega)\| = \gamma \ \text{ for } \ t \in [t_0, d) \quad \text{and } \gamma < 1,$$

it follows that

$$I_p \leq (I_p - \chi_1^t(\omega)\chi_1^t(\omega)^*)^{-1} \leq (1 - \gamma^2)^{-1} I_p, \tag{7.43}$$

and hence also that (7.37) holds.

The bounds in (7.38) are established in much the same way, from (7.40), (7.41), (7.42) and the observation that

$$\begin{aligned} -\rho_{\bar{\omega}}(\bar{\omega})\{k_{\bar{\omega}}^{b_3^t}(\bar{\omega}) + \ell_{\bar{\omega}}^{b_4^t}(\bar{\omega})\} &= -b_4^t(\omega)^* b_4^t(\omega) + b_3^t(\omega)^{-*} b_3^t(\omega)^{-1} \\ &= b_3^t(\omega)^{-*}\{I_p - \chi_1^t(\omega)^*\chi_1^t(\omega)\} b_3^t(\omega)^{-1}. \end{aligned}$$

□

Theorem 7.7 *Let $A_t(\lambda) = A(t, \lambda)$, $0 \leq t < d$, be the matrizant for the system (7.1), let $\{b_3^t, b_4^t\} \in ap_{II}(A_t)$ and $\chi_1^t(\lambda) = b_4^t(\lambda) b_3^t(\lambda)$ for every $t \in [0, d)$ and suppose that*

$$\mathcal{C}_{\rm imp} \cap \mathring{\mathcal{C}}^{p\times p} \neq \emptyset \ \textit{ and } \ \|\chi_1^{t_0}(\omega)\| < 1 \ \textit{ for some } \ t_0 \in (0, d) \ \textit{ and some } \ \omega \in \mathbb{C}_+.$$

Then the ranks $\mathfrak{n}_\ell$ *and* $\mathfrak{n}_r$ *of the left and right semiradii of the limit ball*

$$\mathcal{B}_d(\omega) = \{c(\omega) : c \in \mathcal{C}_{\rm imp}\}$$

(that are independent of $\omega \in \mathbb{C}_+$*) may be computed by the formulas*

$$\mathfrak{n}_\ell = \text{rank } \lim_{t \uparrow d} b_3^t(\omega) b_3^t(\omega)^* \quad \text{and} \quad \mathfrak{n}_r = \text{rank } \lim_{t \uparrow d} b_4^t(\omega)^* b_4^t(\omega). \tag{7.44}$$

Proof The proof is immediate from the the bounds in (7.37) and (7.38) and the fact that the limits in the statement of theorem both exist. □

7.4 Bounded mass functions and full rank end points

In this section we first explore the connections between bounded mass functions and full rank end points for canonical integral systems (7.1). We then consider some connections with a limit property of the corresponding chain of associated pairs $\{b_3^t, b_4^t\}$ as $t \uparrow d$.

Theorem 7.8 *Let* $A_t(\lambda) = A(t, \lambda)$, $0 \le t < d$, *be the matrizant of the canonical integral system (7.1) with mass function* $M(t)$ *subject to (6.2), let* $B_t(\lambda) = A_t(\lambda)\mathfrak{V}$ *and let condition (7.20) be in force. Then:*

1. *If* $M(t)$ *is bounded on* $[0, d)$, *then* d *is a full rank point.*
2. *If* d *is a full rank point and* $A_t \in \mathcal{U}_{rR}(J_p)$ *for* $0 \le t < d$, *then* $M(t)$ *is bounded on* $[0, d)$, *the limit* $A_d(\lambda) = \lim_{t\uparrow d} A_t(\lambda)$ *exists for every* $\lambda \in \mathbb{C}$ *and* $A_d \in \mathcal{E} \cap \mathcal{U}_{rR}^\circ(J)$. *If* $d < \infty$, *then the system is regular and* A_d *is its monodromy matrix.*

Proof It suffices to restrict attention to the case $d < \infty$, because this can always be achieved by appropriate change of the independent variable. Then under the assumption in (1), the system (7.1) is regular on $[0, d]$ with monodromy matrix $A_d(\lambda)$.

The first statement follows from the identities $B_d = A_d\mathfrak{V}$,

$$\mathcal{B}_d(\omega) = \bigcap_{0 \le t < d} \mathcal{B}_t(\omega) = T_{B_d(\omega)}[\mathcal{S}_{\rm const}^{p\times p}] \tag{7.45}$$

and the fact that for regular canonical systems,

$$R_r^d(\omega)^2 = r_d(\omega)^{-1} = \{b_{22}^d(\omega)b_{22}^d(\omega)^* - b_{21}^d(\omega)b_{21}^d(\omega)^*\}^{-1} = \mathfrak{r}_r^d(\omega)^2 \tag{7.46}$$

$$R_\ell^d(\omega)^2 = -r_d(\bar\omega)^{-1} = -\{b_{22}^d(\bar\omega)b_{22}^d(\bar\omega)^* - b_{21}^d(\bar\omega)b_{21}^d(\bar\omega)^*\}^{-1} = \mathfrak{r}_\ell^d(\omega)^2. \tag{7.47}$$

Suppose next that d is a full rank point and that $A_t(\lambda)$ is right regular for every $t \in [0, d)$. Then $W_t(\lambda) = \mathfrak{V}A_t(\lambda)\mathfrak{V}$ is the matrizant of a canonical system of the form (6.1) with mass function

$$\widetilde{M}(t) = -i\left(\frac{\partial W_t}{\partial \lambda}\right)(0)j_p = -i\mathfrak{V}\left(\frac{\partial A_t}{\partial \lambda}\right)(0)J_p\mathfrak{V} = \mathfrak{V}M(t)\mathfrak{V}$$

for $0 \le t < d$. Moreover, the matrix balls

$$\widetilde{\mathcal{B}}_t(\omega) = \{s(\omega) : s \in T_{W_t}[\mathcal{S}^{p\times p}]\} \text{ and } \mathcal{B}_t(\omega) = \{c(\omega) : c \in \mathcal{C}(A_t)\} \quad (7.48)$$

are connected by the formula

$$\mathcal{B}_t(\omega) = T_{\mathfrak{V}}[\widetilde{\mathcal{B}}_t(\omega) \cap \mathcal{D}(T_{\mathfrak{V}})], \quad \text{for } 0 \le t < d.$$

Therefore, the same formula prevails for the limit balls:

$$\mathcal{B}_d(\omega) = T_{\mathfrak{V}}[\widetilde{\mathcal{B}}_d(\omega) \cap \mathcal{D}(T_{\mathfrak{V}})]. \quad (7.49)$$

By assumption, the ball $\mathcal{B}_d(\omega)$ has positive semiradii. Therefore, the set $\widetilde{\mathcal{B}}_d(\omega) \cap \mathcal{D}(T_{\mathfrak{V}})$ must contain an open set and hence the semiradii of the matrix ball $\widetilde{\mathcal{B}}_d(\omega)$ must also be positive, i.e., d is a full rank point for the canonical integral system (6.1). Thus, as $W_t \in \mathcal{U}_{rR}(j_p)$ for every $t \in [0, d)$, Theorem 6.8 guarantees that

$$\ell_{\widetilde{M}} = \lim_{t\uparrow d} \text{trace}\, \widetilde{M}(t) < \infty,$$

and hence as trace $\widetilde{M}(t) = $ trace $M(t)$ that $\ell_M < \infty$ also. Moreover, $A_d \in \mathcal{U}_{rR}(J_p)$, since $W_d \in \mathcal{U}_{rR}(j_p)$ by Theorem 6.8. □

Corollary 7.9 *Let $\mathcal{C}^d_{\text{imp}}(dM)$ denote the set of input impedance matrices for a canonical integral system (7.1) on the interval $[0, d)$ and assume that $\mathcal{C}^d_{\text{imp}}(dM) \cap \mathring{\mathcal{C}}^{p\times p} \neq \emptyset$ and that condition (7.20) is in force. Then $M(t)$ is bounded on $[0, d)$ if and only if d is a full rank point.*

Proof This is immediate from Theorem 7.8 and the observation that under the given assumption $A_t \in \mathcal{U}^{\circ}_{rsR}(J_p)$ for every $t \in [0, d)$. □

Theorem 7.10 *Let $A_t(\lambda) = A(t, \lambda)$, $0 \le t < d$, be the matrizant of the canonical integral system (7.1) with mass function $M(t)$ subject to (6.2), let $\{b_3^t, b_4^t\} \in ap_{II}(A_t)$ and assume that $b_3^t(0) = b_4^t(0) = I_p$ for $0 \le t < d$ and that $\mathcal{C}_{\text{imp}} \cap \mathring{\mathcal{C}}^{p\times p} \neq \emptyset$. Then the following three conditions are equivalent:*

1. *$M(t)$ is bounded on $[0, d)$.*
2. *There exists at least one point $\omega \in \mathbb{C}_+$ such that*

$$\lim_{t\uparrow d} |\det b_3^t(\omega)| > 0 \quad \text{and} \quad \lim_{t\uparrow d} |\det b_4^t(\omega)| > 0. \quad (7.50)$$

3. *(7.50) holds for every point $\omega \in \mathbb{C}_+$.*

The implication (1) $\Longrightarrow$ (2) is valid even if $\mathcal{C}^d_{\text{imp}} \cap \mathring{\mathcal{C}}^{p\times p} = \emptyset$.

Proof Again, without loss of generality, we can assume that $d < \infty$.

Suppose first that (1) holds. Then the matrizant $A_t(\lambda)$ is well defined on the closed interval $[0, d]$. Moreover, $A_d \in \mathcal{E} \cap \mathcal{U}(J_p)$. Thus, we can assume that the associated pair $\{b_3^d, b_4^d\} \in ap_{II}(A_d)$ is normalized at $\lambda = 0$ in the usual way: $b_3^d(0) = b_4^d(0) = I_p$. Then,

$$(b_3^t)^{-1} b_3^d \in \mathcal{E} \cap \mathcal{S}_{\rm in}^{p \times p} \quad \text{and} \quad b_4^d (b_4^t)^{-1} \in \mathcal{E} \cap \mathcal{S}_{\rm in}^{p \times p} \tag{7.51}$$

for every $t \in [0, d)$. Consequently,

$$|\det\{b_3^t(\omega)\}| \geq |\det\{b_3^d(\omega)\}| > 0 \text{ and } |\det b_4^t(\omega)| \geq |\det\{b_4^d(\omega)| > 0$$

for every $t \in [0, d)$ and every point $\omega \in \mathbb{C}_+$. Thus, (1)$\Longrightarrow$ (3).

Conversely, if (3) holds and $\mathcal{C}_{\rm imp} \cap \mathring{\mathcal{C}}^{p \times p} \neq \emptyset$, then by Lemma 5.11, the limits

$$b_3^d(\lambda) = \lim_{t \uparrow d} b_3^t(\lambda) \quad \text{and} \quad b_4^d(\lambda) = \lim_{t \uparrow d} b_4^t(\lambda)$$

exist, the convergence is uniform on every compact subset of $\mathbb{C}$ and the limits are entire inner mvf's, i.e.,

$$b_3^d \in \mathcal{E} \cap \mathcal{S}_{\rm in}^{p \times p} \quad \text{and} \quad b_4^d \in \mathcal{E} \cap \mathcal{S}_{\rm in}^{p \times p}.$$

Moreover, (7.51) holds for every $t \in [0, d)$. Now let $c \in \mathcal{C}_{\rm imp} \cap \mathring{\mathcal{C}}^{p \times p}$. Then the GCIP $(b_3^d, b_4^d; c)$ is strictly completely indeterminate and hence, by Theorem 4.25, there exists exactly one mvf $A_d \in \mathcal{E} \cap \mathcal{U}^\circ(J_p)$ such that

$$\mathcal{C}(b_3^d, b_4^d; c) = \mathcal{C}(A_d) \quad \text{and} \quad \{b_3^d, b_4^d\} \in ap_{II}(A_d).$$

In view of (7.51),

$$\mathcal{C}(A_d) = \mathcal{C}(b_3^d, b_4^d; c) \subseteq \mathcal{C}(b_3^t, b_4^t; c) = \mathcal{C}(A_t)$$

for $0 \leq t < d$. Therefore, $A_t^{-1} A_d \in \mathcal{E} \cap \mathcal{U}(J_p)$ for $0 \leq t < d$, by Theorem 3.77. But this implies that

$$K_\omega^{A_t}(\omega) \leq K_\omega^{A_d}(\omega)$$

for every point $\omega \in \mathbb{C}$ and hence that

$$M(t) = 2\pi K_0^{A_t}(0) \leq 2\pi K_0^{A_d}(0).$$

Therefore, (3)$\Longrightarrow$(1). The equivalence of (2) and (3) is covered by Lemma 5.11. □

7.5 The limit point case

In this section we discuss the limit point case when $\mathcal{C}_{\rm imp} \cap \mathring{\mathcal{C}}^{p \times p} \neq \emptyset$. This serves to guarantee that the matrizant $A_t(\lambda)$, $0 \leq t < d$, of the corresponding integral system (7.1) is right strongly regular for every t, $0 \leq t < d$.

Theorem 7.11 *Let $A_t(\lambda) = A(t, \lambda)$, $0 \le t < d$, be the matrizant of a canonical integral system of the form (7.1). Let $B_t(\lambda) = A_t(\lambda)\mathfrak{V}$, $W_t(\lambda) = \mathfrak{V}A_t(\lambda)\mathfrak{V}$, $\{b_1^t, b_2^t\} \in ap_I(A_t)$ and $\{b_3^t, b_4^t\} \in ap_{II}(A_t)$ for $0 \le t < d$ and let $c^\circ \in \mathcal{C}_{\rm imp}$ for this system. Then the following two conditions are equivalent:*

(a) *c° is the only impedance matrix for the system (7.1).*
(b) *$s^\circ = T_{\mathfrak{V}}[c^\circ]$ is the only input scattering matrix for the system (6.1) with mass function $\mathfrak{V}M(t)\mathfrak{V}$, where $M(t)$ is taken from (7.1).*

Moreover, if $\omega \in \mathbb{C}_+$, then

$$\lim_{t\uparrow d} b_3^t(\omega) = 0 \Longrightarrow \lim_{t\uparrow d} b_1^t(\omega) = 0 \quad \text{and} \quad \lim_{t\uparrow d} b_4^t(\omega) = 0 \Longrightarrow \lim_{t\uparrow d} b_2^t(\omega) = 0. \tag{7.52}$$

Thus, if either

$$\lim_{t\uparrow d} b_3^t(\omega) = 0 \quad \text{or} \quad \lim_{t\uparrow d} b_4^t(\omega) = 0,$$

then there exists exactly one mvf $c^\circ \in \mathcal{C}_{\rm imp}^d(dM)$.

If it is also assumed that $c^\circ \in \mathring{\mathcal{C}}^{p\times p}$, then conditions (a) and (b) are also equivalent to each of the following two conditions:

(c) *There exists at least one point $\omega \in \mathbb{C}_+$ such that*

$$\text{either} \quad \lim_{t\uparrow d} b_3^t(\omega) = 0 \quad \text{or} \quad \lim_{t\uparrow d} b_4^t(\omega) = 0. \tag{7.53}$$

(d) *There exists at least one point $\omega \in \mathbb{C}_+$ such that*

$$\text{either} \quad \lim_{t\uparrow d} b_1^t(\omega) = 0 \quad \text{or} \quad \lim_{t\uparrow d} b_2^t(\omega) = 0. \tag{7.54}$$

Proof If $c^\circ \in \mathcal{C}_{\rm imp}$ and $s^\circ = T_{\mathfrak{V}}[c^\circ]$, then Lemma 4.27 guarantees the existence of a pair of mvf's $\varphi^t \in \mathcal{S}_{\rm out}^{p\times p}$ and $\psi^t \in \mathcal{S}_{\rm out}^{p\times p}$ such that

$$\frac{I_p + s^\circ}{2} b_3^t = b_1^t \varphi^t \quad \text{and} \quad b_4^t \frac{I_p + s^\circ}{2} = \psi^t b_2^t. \tag{7.55}$$

Since $\frac{I_p+s^\circ}{2} \in \mathcal{S}_{\rm out}^{p\times p}$ by Theorem 3.11, these equalities justify (7.52) because the determinants of the outer factors have the same modulus at ω. If $c^\circ \in \mathring{\mathcal{C}}^{p\times p}$, then $A_t \in \mathcal{U}_{rsR}(J_p)$ for $t \in [0, d)$ and the equivalence of (c) and (d) is immediate from Lemma 4.28. Moreover, since $s^\circ \in \mathring{\mathcal{S}}^{p\times p}$, it follows from Theorem 6.16 that (b) and (d) are equivalent. It remains only to show that (a) is equivalent to (b).

The implication (b) $\Longrightarrow$ (a) is self-evident from the second formula in (7.7). Conversely, if (a) is in force, i.e., if $\mathcal{C}_{\rm imp} = \{c^\circ\}$, then $s^\circ = T_{\mathfrak{V}}[c^\circ]$ is an input scattering matrix for the canonical integral system (6.1) with matrizant $W_t(\lambda) = W(t, \lambda) = \mathfrak{V}A_t(\lambda)\mathfrak{V}$, $0 \le t < d$. If there is a second input scattering matrix $s_1(\lambda)$, $s_1(\lambda) \not\equiv s^\circ(\lambda)$, then, by Lemma 6.7, all the mvf's $s_\alpha(\lambda) = \alpha s_1(\lambda) + (1 - \alpha)s^\circ(\lambda)$, $0 < \alpha < 1$, are input scattering matrices for the system (6.1), and at least one of these must belong to $\mathcal{D}(T_{\mathfrak{W}})$ because $\det\{I_p + s_\alpha(\lambda)\} \not\equiv 0$ in $\mathbb{C}_+$ for small enough $\alpha > 0$. The corresponding $c_\alpha = T_{\mathfrak{W}}[s_\alpha]$ belongs to $\mathcal{C}_{\rm imp}$ and differs from $c^\circ(\lambda)$, which contradicts the presumed validity of (a). □

Lemma 7.12 *Let $A_t(\lambda)$, $0 \le t < d$, be the matrizant of the canonical integral system (7.1), let $\{b_1^t, b_2^t\} \in ap_I(A_t)$ and $\{b_3^t, b_4^t\} \in ap_{II}(A_t)$ for $0 \le t < d$ and suppose that $\mathcal{C}_{\rm imp} \neq \emptyset$, $\alpha > 0$ and $\beta > 0$. Then:*

(1) $e_{-\alpha t} b_1^t \in \mathcal{S}_{\rm in}^{p\times p} \iff e_{-\alpha t} b_3^t \in \mathcal{S}_{\rm in}^{p\times p}$.
(2) $e_{-\beta t} b_2^t \in \mathcal{S}_{\rm in}^{p\times p} \iff e_{-\beta t} b_4^t \in \mathcal{S}_{\rm in}^{p\times p}$.

Moreover, if $d = \infty$, then

(3) *If $e_{-\alpha t} b_3^t \in \mathcal{S}_{\rm in}^{p\times p}$ for some $\alpha > 0$ and every $t > t_0$, then $\mathfrak{n}_\ell = 0$.*
(4) *If $e_{-\beta t} b_4^t \in \mathcal{S}_{\rm in}^{q\times q}$ for some $\beta > 0$ and every $t > t_0$, then $\mathfrak{n}_r = 0$.*

Proof Since $\frac{I_p + s^\circ}{2} \in \mathcal{S}_{\rm out}^{p\times p}$ by Theorem 3.11, the equalities (7.55) serve to justify (1) and (2). The remaining assertions follow from Theorem 7.11. □

Remark 7.13 *Matrizants with associated pairs $\{e_{-\alpha t} I_p, e_{-\beta t} I_q\}$ will be considered in the study of homogeneous matrizants in Section 8.8 and Dirac–Krein systems with locally summable potential on the half line $[0, \infty)$ in Chapter 12. Chains of such pairs automatically meet the conditions referred to in Lemma 7.12.*

7.6 The Weyl–Titchmarsh characterization of the input impedance

In this section we shall first describe the set of initial conditions $y(0, \lambda) \in \mathbb{C}^{1\times m}$ for each point $\lambda \notin \mathbb{R}$ for which the solution $y(s, \lambda)$, $0 \le s < d$, of the canonical system (7.1) satisfies the constraint

$$\int_0^d y(s, \lambda) dM(s) y(s, \lambda)^* < \infty. \tag{7.56}$$

Analogous results for input scattering matrices were presented in Section 6.10. We shall then present another characterization of the input impedance matrix of a canonical integral system in terms of the square summability (with respect to dM) of its solutions in the limit point case.

Initial data that generate square summable solutions

Since

$$y(s, \lambda) = y(0, \lambda) A(s, \lambda), \quad 0 \le s < d,$$

it suffices to focus on the integral

$$I_\omega^t(\xi,\eta)=\int_0^t [\xi^* \;\; \eta^*]A(s,\omega)dM(s)A(s,\omega)^*\begin{bmatrix}\xi\\ \eta\end{bmatrix}, \qquad 0\le t<d, \tag{7.57}$$

for each choice of $\omega\in\mathbb{C},\ \xi\in\mathbb{C}^p,\ \eta\in\mathbb{C}^p$ and the limits

$$I_\omega^d(\xi,\eta)=\underset{t\uparrow d}{\text{limit}}\, I_\omega^t(\xi,\eta).$$

The latter may be finite or infinite. The formulas

$$\frac{1}{2\pi}\int_0^t A(s,\omega)dM(s)A(s,\omega)^*=\frac{J_p-A_t(\omega)J_pA_t(\omega)^*}{\rho_\omega(\omega)} \tag{7.58}$$

and

$$I_\omega^t(0,\eta)=\pi\eta^*K_\omega^{\mathfrak{E}_t}(\omega)\eta \quad \text{for } 0\le t<d \text{ and } \omega\in\mathbb{C}$$

will be useful, as will the following relations for the semiradii of the matrix balls, which follow from formulas (7.31) and (7.32):

$$\mathfrak{r}_\ell^d(\omega)=\lim_{t\uparrow d}R_\ell^t(\omega) \text{ and } \mathfrak{r}_\ell^d(\omega)^2\le R_\ell^t(\omega)^2 \text{ for } 0\le t<d \text{ and } \omega\in\mathbb{C}_+. \tag{7.59}$$

$$\mathfrak{r}_r^d(\omega)=\lim_{t\uparrow d}R_r^t(\omega) \text{ and } \mathfrak{r}_r^d(\omega)^2\le R_r^t(\omega)^2 \text{ for } 0\le t<d \text{ and } \omega\in\mathbb{C}_+. \tag{7.60}$$

Lemma 7.14 *Let $\eta\in\mathbb{C}^p$ and $\lambda\in\mathbb{C}_+$. Then*

$$\eta\in\mathfrak{r}_\ell^d(\lambda)\mathbb{C}^p \Longleftrightarrow \lim_{t\uparrow d}\eta^*\{R_\ell^t(\lambda)\}^{-2}\eta<\infty \tag{7.61}$$

and

$$\eta\in\mathfrak{r}_r^d(\lambda)\mathbb{C}^p \Longleftrightarrow \lim_{t\uparrow d}\eta^*\{R_r^t(\lambda)\}^{-2}\eta<\infty. \tag{7.62}$$

Proof This follows from (7.59), (7.60) and Lemma 6.18. □

Lemma 7.15 *Let $A\in\mathcal{U}(J_p)$. Then $c\in\mathcal{C}(A)$ if and only if $c(\lambda)$ is a holomorphic $p\times p$ mvf in $\mathbb{C}_+$ such that*

$$[I_p \;\; c(\omega)^*]A(\bar\omega)J_pA(\bar\omega)^*\begin{bmatrix}I_p\\ c(\omega)\end{bmatrix}\le 0 \tag{7.63}$$

for every point $\omega\in\mathfrak{h}_A^+\cap\mathfrak{h}_{A^\#}^+$.

Proof Suppose first that $c(\lambda)$ is a holomorphic $p\times p$ mvf in $\mathbb{C}_+$ such that (7.63) holds and let

$$\begin{bmatrix}\alpha(\omega)\\ \beta(\omega)\end{bmatrix}=\begin{bmatrix}b_{11}^\#(\omega) & b_{21}^\#(\omega)\\ b_{12}^\#(\omega) & b_{22}^\#(\omega)\end{bmatrix}\begin{bmatrix}I_p\\ c(\omega)\end{bmatrix}$$

for every point $\omega \in \mathfrak{h}_{A^\#}^+$, where $B(\lambda) = A(\lambda)\mathfrak{V}$, as usual. Then

$$[\alpha(\omega)^* \;\; \beta(\omega)^*] j_p \begin{bmatrix} \alpha(\omega) \\ \beta(\omega) \end{bmatrix} \le 0 \text{ and rank } \begin{bmatrix} \alpha(\omega) \\ \beta(\omega) \end{bmatrix} = p$$

for every point $\omega \in \mathfrak{h}_A^+ \cap \mathfrak{h}_{A^\#}^+$, since $A(\omega)J_pA^\#(\omega) = J_p$ at all such points. But this in turn implies that $\beta(\omega)$ is invertible at all such points and hence that there exists a mvf $\varepsilon \in \mathcal{S}^{p\times p}$ such that $\varepsilon(\omega) = -\alpha(\omega)\beta(\omega)^{-1}$ for every point $\omega \in \mathbb{C}_+ \cap \mathfrak{h}_A \cap \mathfrak{h}_{A^\#}$.

Thus, as

$$Bj_p \begin{bmatrix} -\varepsilon \\ I_p \end{bmatrix} \beta = Bj_p \begin{bmatrix} \alpha \\ \beta \end{bmatrix} = Bj_pB^\# \begin{bmatrix} I_p \\ c \end{bmatrix} = J_p \begin{bmatrix} I_p \\ c \end{bmatrix},$$

it follows that

$$\begin{bmatrix} b_{11}\varepsilon + b_{12} \\ b_{21}\varepsilon + b_{22} \end{bmatrix} \beta = \begin{bmatrix} c \\ I_p \end{bmatrix}$$

and hence that $\varepsilon \in \mathcal{D}(T_B)$ and $c = T_B[\varepsilon]$, as claimed.

The converse is a straightforward calculation. □

Theorem 7.16 *Let $c(\lambda)$ be an impedance matrix for a canonical integral system (7.1) that meets condition (7.20) and let $\mathfrak{r}_\ell^d(\omega)$ and $\mathfrak{r}_r^d(\omega)$ denote the left and right semiradii of the Weyl–Titchmarsh limit ball $\mathcal{B}_d(\omega)$ for this system for every point $\omega \in \mathbb{C}_+$. Then the following conclusions hold for every point $\omega \in \mathbb{C}_+$:*

(1) $I_\omega^d(0, \eta) < \infty \iff \eta \in \mathfrak{r}_r^d(\omega)\mathbb{C}^p$.
(2) $I_{\overline{\omega}}^d(0, \eta) < \infty \iff \eta \in \mathfrak{r}_\ell^d(\omega)\mathbb{C}^p$.
(3) $I_{\overline{\omega}}^d(\xi, c(\omega)\xi) < \infty$ *for every* $\xi \in \mathbb{C}^p$.
(4) $I_{\overline{\omega}}^d(\xi, \eta) < \infty \iff \eta - c(\omega)\xi \in \mathfrak{r}_\ell^d(\omega)\mathbb{C}^p$.

Proof The first two statements are immediate from (7.40), Lemma 7.14 and the formula

$$I_\omega^t(0, \eta) = \begin{cases} 2\pi\rho_\omega(\omega)^{-1}\eta^*\{R_r^t(\omega)\}^{-2}\eta & \text{if } \omega \in \mathbb{C}_+ \\ 2\pi\rho_{\overline{\omega}}(\bar{\omega})^{-1}\eta^*\{R_\ell^t(\bar{\omega})\}^{-2}\eta & \text{if } \omega \in \mathbb{C}_-. \end{cases} \tag{7.64}$$

Next, to obtain (3), we first observe that Lemma 7.15 is applicable because the matrizant $A_t(\lambda) = A(t, \lambda)$ of the canonical integral system (7.1) belongs to the class $\mathcal{E} \cap \mathcal{U}(J_p)$ for every $0 \le t < d$. Moreover, by assumption (7.20),

$$\beta^t(\omega) = b_{21}^t(\omega)\varepsilon(\omega) + b_{22}^t(\omega)$$

is invertible for every point $\omega \in \mathbb{C}_+$ and every $t_0 \le t < d$. Thus, by formulas (7.58) and (7.63),

$$\begin{aligned} I^t_{\overline{\omega}}(\xi, \eta) &= 2\pi \rho_{\overline{\omega}}(\bar{\omega})^{-1} v^* \{J_p - A_t(\bar{\omega}) J_p A_t(\bar{\omega})^*\} v \\ &= -2\pi \rho_{\overline{\omega}}(\bar{\omega})^{-1} v^* \{A_t(\bar{\omega}) J_p A_t(\bar{\omega})^* - J_p\} v \\ &\le -2\pi \rho_{\overline{\omega}}(\bar{\omega})^{-1} \xi^* \{c(\omega) + c(\omega)^*\} \xi \end{aligned}$$

for $\omega \in \mathbb{C}_+,\ v^* = [\xi^* \quad \eta^*]$ and $\eta = c(\omega)\xi$. Therefore, (3) holds.

Finally, the inequality

$$I^d_{\overline{\omega}}(\xi_1 + \xi_2, \eta_1 + \eta_2) \le 2 I^d_{\overline{\omega}}(\xi_1, \eta_1) + 2 I^d_{\overline{\omega}}(\xi_2, \eta_2), \tag{7.65}$$

which is valid for every choice of $\omega \in \mathbb{C}$, $\xi_j \in \mathbb{C}^p$ and $\eta_j \in \mathbb{C}^p$, $j = 1, 2$, implies that

$$I^d_{\overline{\omega}}(\xi - \xi, \eta - c(\omega)\xi) \le 2 I^d_{\overline{\omega}}(\xi, \eta) + 2 I^d_{\overline{\omega}}(-\xi, -c(\omega)\xi)$$

and hence, in view of (3), that

$$I^d_{\overline{\omega}}(\xi, \eta) < \infty \Longrightarrow I^d_{\overline{\omega}}(0, \eta - c(\omega)\xi) < \infty.$$

On the other hand, an application of (3) to the inequality

$$I^d_{\overline{\omega}}(\xi, \eta) = I^d_{\overline{\omega}}(\xi, \eta - c(\omega)\xi + c(\omega)\xi) \le 2 I^d_{\overline{\omega}}(\xi, c(\omega)\xi) + 2 I^d_{\overline{\omega}}(0, \eta - c(\omega)\xi).$$

yields the converse implication, i.e.,

$$I^d_{\overline{\omega}}(0, \eta - c(\omega)\xi) < \infty \Longrightarrow I^d_{\overline{\omega}}(\xi, \eta) < \infty.$$

The rest is immediate from (2). □

The limit point case again

Theorem 7.17 *Let a canonical integral system (7.1) meet the constraint (7.20). Then the following statements are equivalent:*

(1) *The given system is in the limit point case, i.e., it admits only one input impedance matrix.*
(2) *There exists a point $\omega \notin \mathbb{R}$ such that $I^d_\omega(0, \eta) = \infty$ for every nonzero vector $\eta \in \mathbb{C}^p$.*
(3) *$I^d_\lambda(0, \eta) = \infty$ for every nonzero vector $\eta \in \mathbb{C}^p$ and every point λ in the same half plane ($\mathbb{C}_+$ or $\mathbb{C}_-$) as the point ω in (2).*

Moreover,

(a) $\mathfrak{n}_\ell = 0 \Longleftrightarrow I^d_{\overline{\omega}}(0, \eta) = \infty$ *for every nonzero vector $\eta \in \mathbb{C}^p$ and at least one (and hence every) point $\omega \in \mathbb{C}_+$.*

(b) *If* $\mathfrak{n}_\ell = 0$, *then, for each point* $\lambda \in \mathbb{C}_+$,

$$I_\lambda^d(\xi, \eta) < \infty \iff \eta = c(\lambda)\xi, \tag{7.66}$$

where $c(\lambda)$ *coincides with the unique impedance matrix of the given system.*

Proof This theorem is an immediate corollary of Theorem 7.16 and the fact (due to Orlov [Or76]) that the ranks $\mathfrak{n}_\ell^d$ and $\mathfrak{n}_r^d$ of the semiradii $\mathfrak{r}_\ell^d(\omega)$ and $\mathfrak{r}_r^d(\omega)$ are independent of the choice of $\omega \in \mathbb{C}_+$. □

7.7 Spectral functions for canonical systems

With each canonical integral system (7.1) we define the Hilbert space

$$L_2^m(dM) = L_2^m(dM; [0, d))$$

of $m \times 1$ measurable vvf's $f(t)$ on $[0, d)$ with

$$\|f\|_{L_2^m(dM)}^2 = \int_0^d f(t)^* dM(t) f(t) < \infty, \tag{7.67}$$

and scalar product

$$\langle f, g \rangle_{L_2^m(dM)} = \int_0^d g(t)^* dM(t) f(t). \tag{7.68}$$

As usual, a vvf $f \in L_2^m(dM)$ will be identified with the equivalence class of vvf's $\widetilde{f} \in L_2^m(dM)$ such that $\|f - \widetilde{f}\|_{L_2^m(dM)} = 0$.

The spectral function of a canonical integral system (7.1) is defined in terms of the generalized Fourier transforms $\mathcal{F}$ and $\mathcal{F}_2$ based on the matrizant A_t that are defined on the linear manifold $\mathcal{L}$ of the functions $f \in L_2^m(dM)$ that vanish in a neighborhood of the right end point d of the interval $[0, d)$ by the formulas

$$(\mathcal{F}f)(\lambda) = \frac{1}{\sqrt{2\pi}} \int_0^d A(t, \lambda) dM(t) f(t), \tag{7.69}$$

and

$$\begin{aligned}(\mathcal{F}_2 f)(\lambda) &= \sqrt{2} N_2^* (\mathcal{F}f)(\lambda) \\ &= \frac{1}{\sqrt{\pi}} \int_0^d [a_{21}(t, \lambda) \;\; a_{22}(t, \lambda)] dM(t) f(t). \end{aligned} \tag{7.70}$$

The operators $\mathcal{F}$ and $\mathcal{F}_2$ map $\mathcal{L}$ into linear spaces of vector-valued entire functions of sizes $m \times 1$ and $p \times 1$, respectively.

A nondecreasing $p \times p$ mvf $\sigma(\mu)$ on $\mathbb{R}$ is said to be a **spectral function** for the system (7.1) if $\mathcal{F}_2$ extends to an isometry from $L_2^m(dM)$ into $L_2^p(d\sigma)$, i.e.

(upon denoting the extension by $\mathcal{F}_2$),

$$\|f\|_{L_2^m(dM)} = \|\mathcal{F}_2 f\|_{L_2^p(d\sigma)} \quad \text{for every } f \in L_2^m(dM). \tag{7.71}$$

Let $\Sigma_{sf}^d(dM)$ denote the set of spectral functions for the system (7.1). If $\Sigma_{sf}^d(dM) \neq \emptyset$, then $\ker\ \mathcal{F}_2 = \{0\}$.

Remark 7.18 *If $J = J_1$ and*

$$M(x) = \begin{bmatrix} x & 0 \\ 0 & m(x) \end{bmatrix}$$

for some continuous nondecreasing function $m(x)$ on $[0, d)$, then the canonical integral system with mass function $M(x)$ corresponds to the Feller–Krein string equation with mass function $m(x)$ that was considered in Section 2.6. Thus, if $g \in L_2([0, d); dm)$, then $f(x)^\tau = [0 \quad g(x)]$ belongs to $L_2^2(dM : [0, d))$ and in view of the formulas (2.62)–(2.64), (7.70) and (7.71),

$$(\mathcal{F}_2 f)(\lambda) = \frac{1}{\sqrt{\pi}} \int_0^d a_{22}(s, \lambda) g(s) dm(s) = \frac{1}{\sqrt{\pi}} (\mathcal{F}_{st} g)(\lambda^2).$$

Therefore, the corresponding Parseval–Plancherel formulas imply that

$$\int_{-\infty}^{\infty} \left| \frac{1}{\sqrt{\pi}} (\mathcal{F}_{st} g)(\mu^2) \right|^2 d\sigma(\mu) = \int_{-\infty}^{\infty} |(\mathcal{F}_{st} g)(\mu)|^2 d\tau(\mu).$$

Thus, if $\sigma \in \Sigma_{sf}^d(dM)$ and $\sigma(-\mu) = -\sigma(\mu)$, then the function

$$\tau(\mu) = \begin{cases} (2/\pi)\sigma(\sqrt{\mu}) & \text{if } \mu \geq 0 \\ 0 & \text{if } \mu < 0 \end{cases} \tag{7.72}$$

is a spectral function with support in $\mathbb{R}_+$ for the corresponding S_1-string.

If $\ker\ \mathcal{F}_2 \neq \{0\}$, then we consider pseudospectral functions instead of spectral functions. If (7.1) is a regular canonical integral system, i.e., if the mass function $M(t)$ is bounded on the interval $[0, d)$ and $d < \infty$, then the transforms $\mathcal{F}$ and $\mathcal{F}_2$ are defined on the full space $L_2^m(dM)$ by formulas (7.69) and (7.70) and a nondecreasing $p \times p$ mvf $\sigma(\mu)$ on $\mathbb{R}$ is called a **pseudospectral function** for the system (7.1) if

$$\|f\|_{L_2^m(dM)} = \|\mathcal{F}_2 f\|_{L_2^p(d\sigma)} \quad \text{for every } f \in L_2^m(dM) \ominus \ker\ \mathcal{F}_2.$$

If the end point d is not subject to these restrictions, then the notion of a pseudospectral function is more complicated; it will be introduced at the end of the next section, where it will also be shown that every pseudospectral function of a system that meets the constraint (5.6) is automatically a spectral function, i.e., $\ker \mathcal{F}_2 = \{0\}$.

The symbol $\Sigma_{\text{psf}}^d(dM)$ will be used to denote the set of pseudospectral functions of a given canonical integral system (7.1). The **direct spectral problem** for a given a canonical integral system (7.1) with mass function $M(t)$ is to describe the sets $\Sigma_{\rm psf}^d(M)$ and $\Sigma_{\rm sf}^d(M)$.

It is convenient to let

$$(\Omega)_{sf} = \{\sigma_c : c \in \Omega\} \quad \text{for every set } \Omega \subset \mathcal{C}^{p\times p},$$

in which σ_c denotes the unique spectral function of c that is subject to the normalizations in (3.45). We shall show that, under appropriate constraints, the set $\Sigma_{\rm psf}^d(dM)$ coincides with the set $(\mathcal{C}_{\rm imp}^d(dM))_{\rm sf}$ of spectral functions of the input impedance matrices of the system and hence that for regular canonical integral systems, $\Sigma_{\rm psf}^d(dM) = (\mathcal{C}(A_d))_{\rm sf}$. If $dM(t) = H(t)dt$, where $H \in L_1^{m\times m}([0, d])$ and $H(t) > 0$ a.e. on $[0, d]$, then the constraints alluded to above are satisfied and $(\mathcal{C}_{\rm imp}^d(dM))_{\rm sf} = \Sigma_{\rm sf}^d(dM)$.

Regular canonical integral systems

Assume now that (7.1) is a regular canonical integral system and hence that the mass function $M(t)$ and the matrizant $A_t(\lambda) = A(t, \lambda)$ of this system are defined on the closed interval $[0, d]$. The generalized Fourier transforms $\mathcal{F}$ and $\mathcal{F}_2$ are defined on the space $L_2^m(dM; [0, d])$ by the formulas (7.69) and (7.70) and the transforms $(\mathcal{F}f)(\lambda)$ and $(\mathcal{F}_2 f)(\lambda)$ are entire vvf's for each choice of $f \in L_2^m(dM; [0, d])$. Consequently,

$$\ker \mathcal{F} = \{f \in L_2^m(dM; [0, d]) : (\mathcal{F}f)(\lambda) \equiv 0\}$$

and

$$\ker \mathcal{F}_2 = \{f \in L_2^m(dM; [0, d]) : (\mathcal{F}_2 f)(\lambda) \equiv 0\} \tag{7.73}$$

are closed subspaces of $L_2^m(dM; [0, d])$. The linear spaces

$$\mathcal{L}_d = \left\{ f : f(t) = \sum_{j=1}^n A(t, \omega_j)^* u_j, \quad \omega_j \in \mathbb{C}, \quad u_j \in \mathbb{C}^m \text{ and } n \geq 1 \right\} \tag{7.74}$$

and

$$(\mathcal{L}_d)_2 = \left\{ f : f(t) = \sum_{j=1}^n A(t, \omega_j)^* N_2 \xi_j, \quad \omega_j \in \mathbb{C}, \quad \xi_j \in \mathbb{C}^p \text{ and } n \geq 1 \right\} \tag{7.75}$$

and their closures $\overline{\mathcal{L}_d}$ and $\overline{(\mathcal{L}_d)_2}$ in the space $L_2^m(dM; [0, d])$ intervene in the next theorem.

Theorem 7.19 *If (7.1) is a regular canonical integral system with matrizant $A_t(\lambda) = A(t, \lambda)$, $0 \leq t \leq d$, and if $\mathfrak{E}_d(\lambda) = \sqrt{2} N_2^* A_d(\lambda) \mathfrak{V}$ and $\mathcal{F}$ and $\mathcal{F}_2$ are the generalized Fourier transforms defined by formulas (7.69) and (7.70), respectively,* *then:*

$$L_2^m(dM;[0,d]) = \overline{\mathcal{L}_d} \oplus \ker \mathcal{F} \tag{7.76}$$

and $\mathcal{F}$ is a coisometry from $L_2^m(dM;[0,d])$ onto $\mathcal{H}(A_d)$. Moreover,

$$L_2^m(dM;[0,d]) = \overline{(\mathcal{L}_d)_2} \oplus \ker \mathcal{F}_2 \tag{7.77}$$

and $\mathcal{F}_2$ is a coisometry from $L_2^m(dM;[0,d])$ onto $\mathcal{B}(\mathfrak{E}_d)$.

Proof In view of the formula

$$K_\omega^{A_d}(\lambda) = \frac{J_p - A_d(\lambda)J_pA_d(\omega)^*}{\rho_\omega(\lambda)} = \frac{1}{2\pi}\int_0^d A_t(\lambda)dM(t)A_t(\omega)^*, \tag{7.78}$$

it is readily seen that

$$(\mathcal{F}f)(\lambda) = \sqrt{2\pi}\sum_{j=1}^n K_{\omega_j}^{A_d}(\lambda)u_j$$

for $f \in \mathcal{L}_d$, as in (7.74) and hence that $\mathcal{F}f \in \mathcal{H}(A_d)$ for $f \in \mathcal{L}_d$. Moreover,

$$\begin{aligned}\|\mathcal{F}f\|_{\mathcal{H}(A_d)}^2 &= 2\pi\left\langle \textstyle\sum_{j=1}^n K_{\omega_j}^{A_d}u_j, \sum_{i=1}^n K_{\omega_i}^{A_d}u_i\right\rangle_{\mathcal{H}(A_d)} \\ &= 2\pi \textstyle\sum_{i,j=1}^n u_i^* K_{\omega_j}^{A_d}(\omega_i)u_j \\ &= \textstyle\int_0^d f(t)^*dM(t)f(t).\end{aligned}$$

Thus, the operator $\mathcal{F}$ maps the linear space $\mathcal{L}_d$ isometrically onto the linear space

$$\mathcal{L}_{A_d} = \left\{\sqrt{2\pi}\sum_{j=1}^n K_{\omega_j}^{A_d}(\lambda)u_j \ : \ \omega_j \in \mathbb{C}\,, \ \ u_j \in \mathbb{C}^m\,, \ \ n \geq 1\right\}$$

in $\mathcal{H}(A_d)$.

Suppose next that $f \in \overline{\mathcal{L}_d}$ in $L_2^m(dM)$. Then there exists a sequence $f_n \in \mathcal{L}_d$ such that $f_n \longrightarrow f$ in $L_2^m(dM)$ as $n \uparrow \infty$. Thus, in view of the already established isometry, $g_n = \mathcal{F}f_n$ is a Cauchy sequence in $\mathcal{H}(A_d)$. Consequently, g_n tends to a limit g in $\mathcal{H}(A_d)$ as $n \uparrow \infty$, and hence

$$\begin{aligned}g(\lambda) &= \lim_{n\uparrow\infty}\int_0^d A(t,\lambda)dM(t)f_n(t) \\ &= \int_0^d A(t,\lambda)dM(t)f(t) = (\mathcal{F}f)(\lambda) \quad \text{for } \lambda \in \mathbb{C}.\end{aligned}$$

Thus, $\mathcal{F}f \in \mathcal{H}(A_d)$ for every $f \in \overline{\mathcal{L}_d}$. Moreover,

$$\begin{aligned}\|\mathcal{F}f\|_{\mathcal{H}(A_d)} = \|g\|_{\mathcal{H}(A_d)} &= \lim_{n\uparrow\infty}\|g_n\|_{\mathcal{H}(A_d)} \\ &= \lim_{n\uparrow\infty}\|f_n\|_{L_2^m(dM)} = \|f\|_{L_2^m(dM)},\end{aligned}$$

i.e., $\mathcal{F}$ maps $\overline{\mathcal{L}_d}$ isometrically into $\mathcal{H}(A_d)$. However, since $\mathcal{L}_{A_d}$ is dense in $\mathcal{H}(A_d)$, the operator $\mathcal{F}$ maps the closure of $\mathcal{L}_d$ in $L_2^m(dM; [0, d])$ isometrically onto $\mathcal{H}(A_d)$. Now, if $f \in L_2^m(dM; [0, d])$ is orthogonal to $\mathcal{L}_d$ in $L_2^m(dM; [0, d])$, then

$$u^* \int_0^d A(t, \omega) dM(t) f(t) = 0$$

for every vector $u \in \mathbb{C}^m$ and every point $\omega \in \mathbb{C}$. Therefore, $(\mathcal{F}f)(\lambda) \equiv 0$ for all such f. Conversely, if $(\mathcal{F}f)(\lambda) \equiv 0$ for every point $\lambda \in \mathbb{C}$, then f is orthogonal to $\mathcal{L}_d$ in $L_2^m(dM; [0, d])$. Thus, we have established the Hilbert space identity (7.76) and have identified $\mathcal{F}$ as a coisometry from $L_2^m(dM; [0, d])$ onto $\mathcal{H}(A_d)$ that maps $L_2^m(dM : [0, d]) \ominus \ker \mathcal{F}$ isometrically onto $\mathcal{H}(A_d)$.

Much the same sort of analysis shows that the operator $\mathcal{F}_2$ maps the linear manifold $(\mathcal{L}_d)_2$ isometrically onto the linear manifold

$$\mathcal{L}_{\mathfrak{E}_d} = \left\{ \sqrt{\pi} \sum_{j=1}^{n} K_{\omega_j}^{\mathfrak{E}_d}(\lambda)\xi_j : \omega_j \in \mathbb{C}, \quad \xi_j \in \mathbb{C}^p \text{ and } n \geq 1 \right\}$$

in the RKHS $\mathcal{B}(\mathfrak{E}_d)$ based on $\mathfrak{E}_d(\lambda) = \sqrt{2} N_2^* A_d(\lambda) \mathfrak{V}$:

$$\|\mathcal{F}_2 f\|_{\mathcal{B}(\mathfrak{E}_d)} = \|f\|_{L_2^m(dM)} \text{ for } f \in (\mathcal{L}_d)_2;$$

and that (7.77) holds. □

Spectral functions for the spaces $\mathcal{H}(A)$ and $\mathcal{B}(\mathfrak{E})$

For the rest of this section we shall assume that the de Branges matrix $\mathfrak{E}(\lambda) = [E_-(\lambda) \quad E_+(\lambda)]$ satisfies the constraint

$$\mathfrak{E} \in \mathcal{E} \cap \Pi^{p \times 2p}. \tag{7.79}$$

Then, by Theorem 4.65, $\mathcal{B}(\mathfrak{E}) \subset \mathcal{E} \cap \Pi^p$ and hence

$$g = 0 \text{ in } \mathcal{B}(\mathfrak{E}) \Longleftrightarrow g(\lambda) \equiv 0 \text{ in } \mathbb{C}.$$

We shall be particularly interested in the case

$$\mathfrak{E} = \sqrt{2} N_2^* A(\lambda) \mathfrak{V} \quad \text{for } A \in \mathcal{E} \cap \mathcal{U}(J_p).$$

Such mvf's $\mathfrak{E}$ automatically meet the constraint (7.79). A nondecreasing $p \times p$ mvf $\sigma(\mu)$ on $\mathbb{R}$ is said to be a **spectral function** for the RKHS $\mathcal{B}(\mathfrak{E})$ based on a de Branges matrix $\mathfrak{E}(\lambda) = [E_-(\lambda) \quad E_+(\lambda)]$ that meets the constraint (7.79) if

$$\int_{-\infty}^{\infty} g(\mu)^* d\sigma(\mu) g(\mu) = \langle g, g \rangle_{\mathcal{B}(\mathfrak{E})} \text{ for every } g \in \mathcal{B}(\mathfrak{E}). \tag{7.80}$$

The set of such spectral functions will be denoted $(\mathcal{B}(\mathfrak{E}))_{\mathrm{sf}}$.

Let $A \in \mathcal{E} \cap \mathcal{U}(J_p)$ be such that $\sqrt{2}N_2^*A(\lambda))\mathfrak{V} = \mathfrak{E}(\lambda)$. Then, by Theorem 4.72, the operator U_2 that is defined by the formula

$$(U_2 f)(\lambda) = \sqrt{2}N_2^* f(\lambda)$$

is a coisometry from $\mathcal{H}(A)$ onto $\mathcal{B}(\mathfrak{E})$.

A nondecreasing $p \times p$ mvf $\sigma(\mu)$ on $\mathbb{R}$ will be called a **spectral function** for the space $\mathcal{H}(A)$ if

$$\int_{-\infty}^{\infty} (U_2 f)(\mu)^* d\sigma(\mu)(U_2 f)(\mu) = \langle f, f\rangle_{\mathcal{H}(A)} \text{ for every } f \in \mathcal{H}(A) \tag{7.81}$$

and the set of such spectral functions will be denoted $(\mathcal{H}(A))_{\rm sf}$. If the equality holds for $f \in \mathcal{H}(A) \ominus \ker U_2$, then $\sigma(\mu)$ will be called a **pseudospectral function** for the space $\mathcal{H}(A)$ and the set of such $\sigma(\mu)$ will be denoted $(\mathcal{H}(A))_{\rm psf}$. It is automatically true that

$$\mathfrak{E} = \sqrt{2}N_2^* A\mathfrak{V} \Longrightarrow (\mathcal{H}(A))_{\rm psf} = (\mathcal{B}(\mathfrak{E}))_{\rm sf} \neq \emptyset. \tag{7.82}$$

However,

$$(\mathcal{H}(A))_{\rm sf} \neq \emptyset \Longleftrightarrow \ker U_2 = \{0\}$$

and

$$\ker U_2 = \{0\} \Longrightarrow (\mathcal{H}(A))_{\rm sf} = (\mathcal{H}(A))_{\rm psf} = (\mathcal{B}(\mathfrak{E}))_{\rm sf}.$$

By Theorem 4.72,

$$\ker U_2 = \{0\} \Longleftrightarrow \text{the mvf } A \in \mathcal{U}(J_p) \text{ is perfect,}$$

i.e., if and only if $c_0 = T_A[I_p]$ belongs to $\mathcal{C}_a^{p\times p}$. The condition $c_0 \in \mathcal{C}_a^{p\times p}$ holds if and only if $\beta_{c_0} = \lim_{\nu\uparrow\infty} \nu^{-1}\Re c_0(i\nu) = 0$.

Lemma 7.20 *If $A_d(\lambda)$ is the monodromy matrix of a regular canonical integral system (7.1),*

$$\mathfrak{E}_d(\lambda) = [(E_d)_-(\lambda) \quad (E_d)_+(\lambda)] = \sqrt{2}\, N_2^* A_d(\lambda)\mathfrak{V}$$

and $U_2 : f \in \mathcal{H}(A_d) \longrightarrow \sqrt{2}\, N_2^* f \in \mathcal{B}(\mathfrak{E}_d)$, then:

(1) $\Sigma_{\rm psf}^d(dM) \neq \emptyset$: *The mvf*

$$\sigma_0(\mu) = \int_0^{\mu} (E_d)_+(\nu)^{-*}(E_d)_+(\nu)^{-1} d\nu \ \textit{belongs to}\ \Sigma_{\rm psf}^d(dM).$$

(2) $\Sigma_{\rm psf}^d(dM) = (\mathcal{B}(\mathfrak{E}))_{\rm sf}$.
(3) $\ker \mathcal{F} = \ker \mathcal{F}_2 \Longleftrightarrow \ker U_2 = \{0\}$.

Proof In view of the decomposition (7.77),

$$\sigma \in \Sigma_{\rm psf}^d(dM) \Longleftrightarrow \|\mathcal{F}_2 f\|_{L_2^p(d\sigma)} = \|f\|_{L_2^m(dM)} \text{ for every } f \in (\mathcal{L}_d)_2.$$

Therefore, since $\mathcal{F}_2$ maps $\overline{(\mathcal{L}_d)_2}$ isometrically onto $\mathcal{B}(\mathfrak{E}_d)$, (2) must follow. Next, (1) is immediate from (2), and (3) follows from the observation that $U_2\mathcal{F} = \mathcal{F}_2$. □

The last lemma guarantees that the set $\Sigma_{\rm psf}^d(dM)$ of pseudospectral functions of a regular canonical integral system is nonempty. However, the set $\Sigma_{\rm sf}^d(dM)$ of spectral functions may be empty:

$$\text{either } \ker \mathcal{F}_2 \neq \{0\} \quad \text{and} \quad \Sigma_{\rm sf}^d(dM) = \emptyset$$
$$\text{or } \ker \mathcal{F}_2 = \{0\} \quad \text{and} \quad \Sigma_{\rm sf}^d(dM) = \Sigma_{\rm psf}^d(dM) \neq \emptyset.$$

Thus, it is of interest to: (1) describe the set $\Sigma_{\rm psf}^d(dM)$ and, (2) find conditions under which $\ker \mathcal{F}_2 = \{0\}$. If (7.1) is a regular canonical integral system, then the monodromy matrix $A_d \in \mathcal{E} \cap \mathcal{U}(J_p)$ and $A_d(0) = I_m$. Thus, in view of Lemma 7.20 and Theorem 7.19, the sets $\Sigma_{\rm sf}^d(dM)$ and $\Sigma_{\rm psf}^d(dM)$ of spectral and pseudospectral functions of regular canonical integral systems (7.1) with mass function $M(t)$ are related to the sets $(\mathcal{H}(A_d))_{\rm sf}$, $(\mathcal{H}(A_d))_{\rm psf}$ and $(\mathcal{B}(\mathfrak{E}_d))_{\rm sf}$ as follows:

$$\Sigma_{\rm psf}^d(dM) = (\mathcal{H}(A_d))_{\rm psf} = (\mathcal{B}(\mathfrak{E}_d))_{\rm sf}.$$
$$(\mathcal{H}(A_d))_{\rm sf} = (\mathcal{H}(A_d))_{\rm psf} \Longleftrightarrow \ker \mathcal{F} = \ker \mathcal{F}_2.$$
$$\Sigma_{\rm sf}^d(dM) = (\mathcal{H}(A_d))_{\rm sf} \Longleftrightarrow \ker \mathcal{F} = \{0\}.$$
$$\Sigma_{\rm sf}^d(dM) = \Sigma_{\rm sf}(\mathfrak{E}_d) \Longleftrightarrow \ker \mathcal{F}_2 = \{0\}.$$

7.8 Parametrization of the set $(\mathcal{H}(A))_{\rm psf}$

In this section we shall obtain a parametrization of the set $(\mathcal{H}(A))_{\rm psf}$ for mvf's $A \in \mathcal{E} \cap \mathcal{U}(J_p)$ that are subject to one or more of the constraints (C1)–(C4) that are given below in terms of

$$\{b_3, b_4\} \in ap_{II}(A),\ \chi_1(\lambda) = b_4(\lambda)b_3(\lambda),\ B(\lambda) = A(\lambda)\mathfrak{V},\ \mathfrak{E}(\lambda) = \sqrt{2}\, N_2^* B(\lambda)$$
$$c_0(\lambda) = T_A[I_p] \ (= T_B[0] = b_{12}b_{22}^{-1}), \quad \beta_{c_0} = \lim_{\nu\uparrow\infty} \nu^{-1}\Re c_0(i\nu)$$

and the domain of definition of the operator of multiplication by λ in $\mathcal{B}(\mathfrak{E})$

$$\mathcal{M}_{\mathfrak{E}} = \{f \in \mathcal{B}(\mathfrak{E}) : \lambda f(\lambda) \text{ belongs to } \mathcal{B}(\mathfrak{E})\}: \tag{7.83}$$

(C1) $\beta_{c_0} = 0$, i.e., the mvf A is perfect.
(C2) $K_\omega^{\mathfrak{E}}(\omega) > 0$ for at least one (and hence every) point $\omega \in \mathbb{C}$.
(C3) $\overline{\mathcal{M}_{\mathfrak{E}}} = \mathcal{B}(\mathfrak{E})$.
(C4) $k_\omega^{\chi_1}(\omega) > 0$ for at least one (and hence every) point $\omega \in \mathbb{C}_+$.

(If $\chi_1(0) = I_p$, then (C4) is equivalent to $-i\chi_1'(0) > 0$.)

The next theorem summarizes the role of these constraints in the description of $(\mathcal{H}(A))_{\rm psf}$.

Theorem 7.21 *If $A \in \mathcal{E} \cap \mathcal{U}(J_p)$, then:*

(1) *(C1) holds if and only if the operator U_2 from $\mathcal{H}(A)$ onto $\mathcal{B}(\mathfrak{E})$ that is defined in (4.139) is unitary, i.e., if and only if*

$$(\mathcal{H}(A))_{\rm sf} = (\mathcal{B}(\mathfrak{E}))_{\rm sf} \; (= (\mathcal{H}(A))_{\rm psf}).$$

(2) *(C2) holds if and only if*

$$\mathcal{S}^{p\times p} \subset \mathcal{D}(T_B) \quad \textit{and} \quad \mathcal{C}(A) = T_B[\mathcal{S}^{p\times p}], \tag{7.84}$$

or, equivalently, if and only if any one (and hence every one) of the conditions (2)–(8) in Lemma 4.67 hold. Moreover, if $A \in \mathcal{E} \cap \mathcal{U}^\circ(J_p)$, then (C2) holds if and only if $ia'_{21}(0) > 0$.

(3) *If (C2) is in force, then*

$$((\mathcal{B}(\mathfrak{E}))_{\rm sf} =) \quad (\mathcal{H}(A))_{\rm psf} \subseteq (\mathcal{C}(A))_{\rm sf}$$

and hence every $\sigma \in (\mathcal{H}(A))_{\rm psf}$ satisfies the condition (3.43). Thus, if $\sigma \in (\mathcal{H}(A))_{\rm psf}$, then there exists at least one mvf $c \in \mathcal{C}(A)$ such that

$$c(\lambda) = i\alpha_c - i\beta_c\lambda + \frac{1}{\pi i}\int_{-\infty}^{\infty}\left\{\frac{1}{\mu-\lambda} - \frac{\mu}{1+\mu^2}\right\} d\sigma_c(\mu), \quad \lambda \in \mathbb{C}_+ \tag{7.85}$$

with spectral function $\sigma_c = \sigma$ and some Hermitian matrix $\alpha_c \in \mathbb{C}^{p\times p}$ and positive semidefinite $\beta_c \in \mathbb{C}^{p\times p}$. Moreover, one can choose such $c \in \mathcal{C}(A)$ with $\beta_c = \beta_{c_0}$.

(4) *If (C3) is in force, then*

$$(\mathcal{C}(A))_{\rm sf} \subseteq (\mathcal{B}(\mathfrak{E}))_{\rm sf} \; (= (\mathcal{H}(A))_{\rm psf}).$$

(5) *If (C2) and (C3) are in force, then*

$$(\mathcal{H}(A))_{\rm psf} = (\mathcal{C}(A))_{\rm sf}. \tag{7.86}$$

(6) *If (C2) and (C4) are in force and $\sigma \in (\mathcal{H}(A))_{\rm psf}$, then there exists exactly one mvf $c \in \mathcal{C}(A)$ with spectral function σ and $\beta_c = \beta_{c_0}$.*

(7) *If (C2)–(C4) are in force, then the formulas*

$$\varepsilon \in \mathcal{S}^{p\times p}, \quad T_B[\varepsilon] = c$$

and (7.85) define a one-to-one correspondence between the sets $\mathcal{S}^{p\times p}$, $\mathcal{C}(A)$ and $(\mathcal{C}(A))_{\rm sf}$. Moreover, $(\mathcal{C}(A))_{\rm sf} = (\mathcal{H}(A))_{\rm psf}$ and $\beta_c = \beta_{c_0}$ in (7.85).

(8) *If (C1)–(C4) are in force, then in the one-to-one correspondences considered in (7) $(\mathcal{C}(A))_{\rm sf} = (\mathcal{H}(A))_{\rm sf}$, and $\beta_c = 0$ for every $c \in \mathcal{C}(A)$.*

Proof Assertion (1) coincides with (4) of Theorem 4.72. Assertion (2) follows from Lemmas 4.67 and 3.16. Assertions (3)–(7) are covered by theorem 2.14 in [ArD04b]. Finally, (8) follows from (1) and (7). □

Item (4) is significant in the analysis of the inverse problem because $\chi_1(\lambda)$ is directly available from the given data, whereas $\mathfrak{E}(\lambda)$ is not.

Let $b \in \mathcal{S}_{\rm in}^{p\times p}$ and let

$$\mathcal{M}_b = \{g \in \mathcal{H}(b) : \lambda g(\lambda) \in \mathcal{H}(b)\} \tag{7.87}$$

denote the domain of multiplication by λ in the RKHS $\mathcal{H}(b) = H_2^p \ominus bH_2^p$.

Lemma 7.22 *If* $b \in \mathcal{S}_{\rm in}^{p\times p}$, *then the following statements are equivalent:*

(1) $\overline{\mathcal{M}_b} \neq \mathcal{H}(b)$.
(2) *There exist a pair of nonzero vectors* $\xi, \eta \in \mathbb{C}^p$ *such that* $\xi - b\eta \in \mathcal{H}(b)$.
(3) *There exist a pair of nonzero vectors* $\xi, \eta \in \mathbb{C}^p$ *such that* $\xi - b\eta \in L_2^p(\mathbb{R})$.

Moreover, if either (2) or (3) is in force (and hence in fact both are in force), then $\|\xi\| = \|\eta\|$.

Proof Fix a point $\alpha \in \mathbb{C}_+$ at which $b(\alpha)$ is invertible. Then $\overline{\alpha} \in \mathfrak{h}_b$ and the de Branges identity (4.74) specialized to $\mathcal{H}(b)$ spaces implies that

$$\langle R_{\overline{\alpha}} f, g\rangle_{\rm st} - \langle R_\alpha^* f, g\rangle_{\rm st} = 2\pi i g(\alpha)^* f(\overline{\alpha})$$

for every choice of f and g in $\mathcal{H}(b)$ and hence upon choosing $g = k_\lambda^b \xi$, that

$$(R_\alpha^* f)(\lambda) = (R_{\overline{\alpha}} f)(\lambda) - 2\pi i k_\alpha^b(\lambda) f(\overline{\alpha}) = \frac{f(\lambda) - b(\lambda)b(\alpha)^* f(\overline{\alpha})}{\lambda - \overline{\alpha}}. \tag{7.88}$$

At the same time, it is readily checked that

$$\mathcal{M}_b = \{R_\alpha f : f \in \mathcal{H}(b) \text{ and } f(\alpha) = 0\}.$$

Thus, if a nonzero vvf $g \in \mathcal{H}(b)$ is orthogonal to $\mathcal{M}_b$, then

$$\langle g, R_\alpha f\rangle_{\rm st} = 0$$

for every $f \in \mathcal{H}(b)$ with $f(\alpha) = 0$. Therefore, since

$$\{f \in \mathcal{H}(b) : f(\alpha) = 0\} \text{ is a RKHS with RK } k_\omega^b(\lambda) - k_\alpha^b(\lambda) k_\alpha^b(\alpha)^{-1} k_\omega^b(\alpha),$$

it follows that

$$(R_\alpha^* g)(\lambda) = k_\alpha^b(\lambda) k_\alpha^b(\alpha)^{-1} (R_\alpha^* g)(\alpha).$$

But, with the aid of (7.88) written for g, this implies that if (1) is in force, then

$$g(\lambda) = \xi - b(\lambda)\eta$$

for some pair of vectors $\xi, \eta \in \mathbb{C}^p$. But this implies that $\xi - b\eta \in L_2^p(\mathbb{R})$ and hence, as

$$\|\xi - b(\mu)\eta\| \geq | \, \|\xi\| - \|\eta\| \, | \quad \text{a.e. on } \mathbb{R}$$

by the triangle inequality, that

$$\int_{-R}^{R} \left(\|\xi\| - \|\eta\| \right)^2 d\mu \leq \int_{-R}^{R} \|\xi - b(\mu)\eta\|^2 d\mu.$$

Consequently, $\|\xi\| = \|\eta\|$, which in turn insures that the vectors ξ and η are nonzero, since g is nonzero, i.e., (1) $\Longrightarrow$ (2). The converse, (2) $\Longrightarrow$ (1), is obtained by observing that

$$g(\lambda) = \xi - b(\lambda)\eta \implies (R_\alpha^* g)(\lambda) = -2\pi i k_\alpha^b(\lambda)\xi$$

and then running the argument backwards.

The implication (2) $\Longrightarrow$ (3) is self-evident. On the other hand, if (3) is in force, then $\xi - b\eta \in L_2^p(\mathbb{R}) \cap \mathcal{N}_+^p$ and hence, by the Smirnov maximum principle, $\xi - b\eta \in H_2^p$. However, (3) also implies that $b^*\xi - \eta \in L_2^p(\mathbb{R})$ and hence, by similar considerations based on the Smirnov maximum principle, that $b^*\xi - \eta \in (H_2^p)^\perp$ and hence that $\xi - b\eta \in \mathcal{H}(b)$. Thus, (3) implies (2) and the proof that (3) is equivalent to (2) and (1) is complete. □

Lemma 7.23 *Let $b \in \mathcal{S}_{\text{in}}^{p\times p}$ and suppose that there exists at least one ray $\omega + re^{i\theta}$, $r \geq 0$, in $\mathbb{C}_+$ such that*

$$b(re^{i\theta} + \omega)^* b(re^{i\theta} + \omega) \leq \gamma I_p \tag{7.89}$$

for $r \geq r_0$ and some constant γ, $0 < \gamma < 1$. Then

$$\overline{\mathcal{M}_b} = \mathcal{H}(b) \, .$$

Proof If $\overline{\mathcal{M}_b} \neq \mathcal{H}(b)$, then by the preceding lemma, there exists a pair of nonzero vectors $\xi, \eta \in \mathbb{C}^p$ with $\|\xi\| = \|\eta\|$ such that

$$g(\lambda) = \xi - b(\lambda)\eta$$

belongs to H_2^p. Therefore, by standard estimates,

$$\|g(\mu + i\nu)\| \leq \|g\|_{\text{st}} \frac{1}{\sqrt{4\pi\nu}} \quad \text{for } \mu \in \mathbb{R} \text{ and } \nu > 0,$$

and hence

$$\|g(\omega + re^{i\theta})\| = \|\xi - b(\omega + re^{i\theta})\eta\| \longrightarrow 0$$

as $r \uparrow \infty$ if $0 < \theta < \pi$. However, if (7.89) is in force, then

$$\|\xi - b(\omega + re^{i\theta})\eta\| \geq \|\xi\| - \gamma^{\frac{1}{2}} \|\eta\| > 0$$

for $r > r_0$, which leads to a contradiction. Therefore $\overline{\mathcal{M}_b} = \mathcal{H}(b)$ if (7.89) holds for at least one choice of $\theta \in (0, \pi)$. The conclusions for the cases $\theta = 0$ and $\theta = \pi$ are obtained in much the same way from the fact that

$$\int_{-\infty}^{\infty} \|g(\mu + i\nu)\|^2 d\mu = \int_{-\infty}^{\infty} \|\xi - b(\mu + i\nu)\eta\|^2 d\mu < \infty. \qquad \square$$

The preceding two lemmas have immediate implications for de Branges spaces $\mathcal{B}(\mathfrak{E})$ based on a de Branges matrix $\mathfrak{E}(\lambda) = [E_-(\lambda) \quad E_+(\lambda)]$, since $\chi(\lambda) = E_+(\lambda)^{-1}E_-(\lambda)$ is an inner mvf.

Lemma 7.24 *If $\mathfrak{E} = \begin{bmatrix} E_- & E_+ \end{bmatrix}$ is an entire de Branges matrix, $\chi = E_+^{-1}E_-$ and $\mathcal{M}_{\mathfrak{E}}$ is defined by (7.83), then the following conditions are equivalent:*

(1) $\overline{\mathcal{M}_{\mathfrak{E}}} = \mathcal{B}(\mathfrak{E})$.
(2) $\overline{\mathcal{M}_{\chi}} = \mathcal{H}(\chi)$.
(3) $\xi - \chi(\lambda)\eta \notin \mathcal{H}(\chi)$ *for any pair of unit vectors* $\xi, \eta \in \mathbb{C}^p$.
(4) $\xi - \chi(\lambda)\eta \notin L_2^p(\mathbb{R})$ *for any pair of unit vectors* $\xi, \eta \in \mathbb{C}^p$.
(5) $E_+(\lambda)\xi - E_-(\lambda)\eta \notin \mathcal{B}(\mathfrak{E})$ *for any pair of unit vectors* $\xi, \eta \in \mathbb{C}^p$.

If there exists a ray $\omega + re^{i\theta}$, $r \geq 0$, in $\mathbb{C}_+$ such that

$$\chi(\omega + re^{i\theta})^*\chi(\omega + re^{i\theta}) \leq \gamma I_p \tag{7.90}$$

for $r \geq r_0$ and some constant γ, $0 < \gamma < 1$, then $\overline{\mathcal{M}_{\mathfrak{E}}} = \mathcal{B}(\mathfrak{E})$.

Proof The equivalence of statements (2), (3) and (4) follows from Lemma 7.22. The equivalences (4) $\Longleftrightarrow$ (5) and (1) $\Longleftrightarrow$ (2) follow from the formula

$$f \in \mathcal{B}(\mathfrak{E}) \iff E_+^{-1}f \in \mathcal{H}(\chi).$$

The rest then follows from Lemma 7.23. $\square$

Theorem 7.25 *Let $A \in \mathcal{E} \cap \mathcal{U}_{rsR}(J_p)$, $\{b_3, b_4\} \in ap_{II}(A)$, $\mathfrak{E}(\lambda) = \sqrt{2}\, N_2^*A(\lambda)\mathfrak{V}$ and let $\chi_1(\lambda) = b_4(\lambda)b_3(\lambda)$. Then*

(1) $\mathcal{B}(\mathfrak{E}) = \mathcal{H}_*(b_4) \oplus \mathcal{H}(b_3)$ *(as linear spaces of vvf's with equivalent norms).*
(2) *(C2) holds if and only if (C4) holds.*
(3) *The constraint (C3) is equivalent to each of the following conditions:*
  (a) $\overline{\mathcal{M}_{\chi_1}} = \mathcal{H}(\chi_1)$.
  (b) $\xi - \chi_1(\lambda)\eta \notin \mathcal{H}(\chi_1)$ *for any pair of unit vectors* $\xi, \eta \in \mathbb{C}^p$.
  (c) $\xi - \chi_1(\lambda)\eta \notin L_2^p(\mathbb{R})$ *for any pair of unit vectors* $\xi, \eta \in \mathbb{C}^p$.
(4) *If there exists a ray $\omega + re^{i\theta}$, $r \geq 0$, in $\mathbb{C}_+$ such that*

$$\chi_1(\omega + re^{i\theta})^*\chi_1(\omega + re^{i\theta}) \leq \gamma I_p \tag{7.91}$$

for $r \geq r_0$ and some constant γ, $0 < \gamma < 1$, then (C3) is in force.

Proof Under the given assumptions, the two Hilbert spaces $\mathcal{H}_*(b_4) \oplus \mathcal{H}(b_3)$ and $\mathcal{B}(\mathfrak{E})$ contain the same set of elements and have equivalent norms; see Theorem 4.78. Therefore, $\mathcal{M}_{\mathfrak{E}}$ is dense in $\mathcal{B}(\mathfrak{E})$ if and only if

$$\mathcal{M}_{b_4,b_3} = \{f \in \mathcal{H}_*(b_4) \oplus \mathcal{H}(b_3) : \lambda f(\lambda) \in \mathcal{H}_*(b_4) \oplus \mathcal{H}(b_3)\}$$

is dense in $\mathcal{H}_*(b_4) \oplus \mathcal{H}(b_3)$. Thus, as the operator M_{b_4} of multiplication by $b_4(\lambda)$ is a unitary map of $\mathcal{H}_*(b_4) \oplus \mathcal{H}(b_3)$ onto $\mathcal{H}(X_1)$, it follows that (1) is equivalent to (2). The remaining equivalences then follow from Lemma 7.24. □

Remark 7.26 *In view of the last assertion in Lemma 7.22, the constraint "for any pair of unit vectors" in (3)–(5) of Lemma 7.24 and (3)–(4) of Theorem 7.25 may be replaced by "for any pair of nonzero vectors."*

Remark 7.27 *The condition (7.91) will be met for every* $\theta \in (0, \pi)$ *if* $b_4(\lambda)b_3(\lambda)$ *has a scalar entire inner divisor* $e^{i\lambda\tau}$ *for some* $\tau > 0$.

7.9 Parametrization of the set $\Sigma_{\rm psf}^d(dM)$ for regular canonical integral systems

If $A(\lambda) = A_d(\lambda)$ is the monodromy matrix for a regular canonical integral system (7.1) and $\mathfrak{E}(\lambda) = \sqrt{2}N_2^*A(\lambda)\mathfrak{V}$, then $A \in \mathcal{E} \cap \mathcal{U}(J_p)$ and $A(0) = I_m$. Thus, as $\mathcal{F}_2 = U_2\mathcal{F}$, where $\mathcal{F}$ is a coisometry from $L_2^m(dM; [0, d])$ onto $\mathcal{H}(A)$, and U_2 is a coisometry from $\mathcal{H}(A)$ onto $\mathcal{B}(\mathfrak{E})$, it follows that

$$f \in L_2^m(dM; [0, d]) \ominus \ker \mathcal{F}_2 \Longleftrightarrow \mathcal{F}f \in \mathcal{H}(A) \ominus \ker U_2.$$

Consequently, the set $\Sigma_{\rm psf}^d(dM)$ of pseudospectral functions of the canonical system with mass function $M(t)$ coincides with the set $(\mathcal{H}(A))_{\rm psf} = (\mathfrak{E}))_{\rm sf}$. Moreover,

$$\mathcal{C}_{\rm imp}^d(dM) = \mathcal{C}(A)$$

and the lower right-hand block $M_{22}(d) = N_2^*M(d)N_2$ of the full mass $M(d)$ is also given by the formula

$$M_{22}(d) = \pi K_0^{\mathfrak{E}}(0).$$

Theorem 7.28 *If* $A(\lambda)$ *is the monodromy matrix for a regular canonical integral system (7.1) on* $[0, d]$, *and if* B, $\mathfrak{E}$, c_0 *and* β_{c_0} *are defined in terms of* $A(\lambda)$ *as in Theorem 7.21, then:*

(1) *Each of the conditions (1)–(8) in Lemma 4.67 is equivalent to the condition*

$$M_{22}(d) > 0. \tag{7.92}$$

(2) *If (7.92) is in force, then every* $\sigma \in \Sigma_{\rm psf}^d(dM)$ *meets the condition (3.43) and there exists a Hermitian matrix* $\alpha \in \mathbb{C}^{p\times p}$ *and a positive semidefinite matrix*

$\beta \in \mathbb{C}^{p\times p}$ such that the mvf $c(\lambda)$ defined by (7.85) belongs to $\mathcal{C}_{imp}(dM)$. Moreover, $\mathcal{C}_{imp}^d(dM) = \mathcal{C}(A)$, $\mathcal{C}(A) = T_{\mathfrak{B}}[\mathcal{S}^{p\times p}]$ and we can choose $\beta = \beta_{c_0}$.

Proof Under the given assumptions, Theorem 7.21 is applicable and condition (C2) is equivalent to (7.92). □

Theorem 7.29 *If $A(\lambda)$ is the monodromy matrix of a regular canonical integral system (7.1) on $[0, d]$ and $A \in \mathcal{U}_{AR}(J_p)$, then:*

(1) $\ker \mathcal{F} = \ker \mathcal{F}_2 = \{0\}$ *and hence*

$$\Sigma_{sf}^d(dM) = \Sigma_{psf}^d(dM) = (\mathcal{H}(A))_{sf} = (\mathcal{B}(\mathfrak{E}))_{sf}. \tag{7.93}$$

(2) *If also (7.90) or (7.91) hold, then the constraints (C1)–(C4) are all in force and hence*

$$\mathcal{C}_{imp}^d(dM) = \mathcal{C}(A) = T_{\mathfrak{B}}[\mathcal{S}^{p\times p}]$$

and formulas (7.84) or (7.85) define a one-to-one correspondence between the sets $\mathcal{S}^{p\times p}$, $\mathcal{C}_{imp}^d(dM)$ and $\Sigma_{sf}^d(dM)$.

Proof Without loss of generality, we can assume that $M(t) = \int_0^t H(s)ds$, where $H \in L_1^{m\times m}([0, d])$ and $H(s) \geq 0$ a.e. on $[0, d]$. Let $v \in \mathbb{C}^m$ and let

$$v_\tau(s) = \begin{cases} v & \text{for } 0 \leq s \leq \tau \\ 0 & \text{for } \tau < s < d. \end{cases}$$

Then $v_\tau \in L_2^m(dM)$ and

$$\begin{aligned} (\mathcal{F}v_\tau)(\lambda) &= \frac{1}{\sqrt{2\pi}} \int_0^\tau A(s, \lambda)H(s)v ds = \frac{1}{i\lambda\sqrt{2\pi}}\{A(\tau, \lambda) - I_m\}Jv \\ &= \sqrt{2\pi} K_0^{A_\tau}(\lambda)v \quad \text{for } 0 \leq \tau \leq d. \end{aligned}$$

Consequently, $\mathcal{F}v_\tau \in \mathcal{H}(A_\tau)$ and

$$\|\mathcal{F}v_\tau\|_{\mathcal{H}(A_\tau)}^2 = 2\pi v^* K_0^{A_\tau}(0)v = v^* M(\tau)v = \|v_\tau\|_{L^2(dM)}^2.$$

Moreover, since $A \in \mathcal{U}_{AR}(J_p)$, Theorem 4.56 guarantees that $\mathcal{H}(A_\tau)$ is isometrically included in $\mathcal{H}(A)$. Thus,

$$\mathcal{F}v_\tau \in \mathcal{H}(A) \quad \text{and} \quad \|\mathcal{F}v_\tau\|_{\mathcal{H}(A)} = \|v_\tau\|_{L_2^m(dM)}.$$

Thus, if $f \in L_2^m(dM; [0, d]) \cap \ker \mathcal{F}$, then

$$\int_0^\tau v^* H(s) f(s) ds = 0$$

for every $\tau \in [0, d]$ and $v \in \mathbb{C}^m$. Therefore, $f = 0$ in $L_2^{m\times m}(Hdt; [0, d])$, i.e., $\ker \mathcal{F} = \{0\}$ and hence $\mathcal{F}$ is a unitary map of $L_2^m(dM; [0, d])$ onto $\mathcal{H}(A)$. Consequently,

$$\Sigma_{\rm sf}^d(dM) = \Sigma_{\rm psf}^d(dM).$$

Moreover, since $A \in \mathcal{U}_{AR}(J_p)$, (C1) is in force and hence (1) of Theorem 7.21 implies that

$$\ker \mathcal{F}_2 = \{0\}, \quad \ker U_2 = \{0\} \quad \text{and} \quad (\mathcal{H}(A))_{\rm sf} = (\mathcal{B}(\mathfrak{E}))_{\rm sf}.$$

Furthermore, since $\mathcal{F}$ is a unitary map from $L_2^m(dM)$ onto $\mathcal{H}(A)$, $\Sigma_{\rm sf}^d(dM) = (\mathcal{H}(A))_{\rm sf}$ and hence (7.93) holds. This completes the proof of (1).

If also at least one of the conditions (7.90) or (7.91) is in force, then, in view of Lemma 7.24 and Theorem 7.25, the constraints (C2)–(C4) are also met. The rest follows from (8) of Theorem 7.21. □

Theorem 7.30 *Let $A(\lambda)$ be the monodromy matrix for the regular canonical integral system (7.1) on $[0, d]$ and suppose that (7.90) is in force. Then the constraints (C2) and (C3) are in force and hence formulas (7.84) and (7.85) define a correspondence between the sets $\mathcal{C}_{\rm imp}(dM)$, $\Sigma_{\rm psf}^d(dM)$ and $\mathcal{S}^{p\times p}$.*

Proof This is an immediate consequence of Lemmas 4.67 and 7.24 and Theorems 7.28 and 7.21. □

7.10 Pseudospectral and spectral functions for singular systems

If the canonical integral system (7.1) is singular with $d \le \infty$, then the generalized Fourier transforms $\mathcal{F}$ and $\mathcal{F}_2$ are defined by formulas (7.69) and (7.70) for vvf's from the linear space

$$\mathcal{L} = \{f \in L_2^m(dM; [0, d)) : f \text{ has support in a closed subinterval } [0, d_f] \text{ of } [0, d)\}. \quad (7.94)$$

Just as in the case of regular canonical systems, the vvf's $(\mathcal{F}f)(\lambda)$ and $(\mathcal{F}_2 f)(\lambda)$ are entire for every $f \in \mathcal{L}$, but $\ker \mathcal{F}$ and $\ker \mathcal{F}_2$ may not be closed subspaces in $L_2^m(dM; [0, d))$ (since here $\mathcal{F}$ and $\mathcal{F}_2$ are only defined on $\mathcal{L}$).

A nondecreasing $p \times p$ mvf $\sigma(\mu)$ on $\mathbb{R}$ is a spectral function of the system (7.1) if the Parseval equality (7.71) holds for every vvf $f \in \mathcal{L}$. If $\tau \in (0, d)$, then we can consider $L_2^m(dM; [0, \tau])$ as a subspace of $L_2^m(dM; [0, d))$ by identifying a vvf $f \in L_2^m(dM; [0, \tau])$ with its extension onto the interval $[0, d)$ with $f(t) = 0$ for $t \in (\tau, d)$. Then we can consider the restrictions $\mathcal{F}^\tau$ and $\mathcal{F}_2^\tau$ of the transforms

$\mathcal{F}$ and $\mathcal{F}_2$ onto the subspaces $L_2^m(dM;[0,\tau]) \subset \mathcal{L}$. The restriction of the system (7.1) onto the interval $[0,\tau]$ yields a regular canonical integral system with end point τ and we can consider the sets $\Sigma_{\rm psf}^{\tau}(dM)$ and $\Sigma_{\rm sf}^{\tau}(dM)$ of pseudospectral and spectral functions of this restricted system. It is clear that

$$\Sigma_{\rm sf}^{\tau_2}(dM) \subseteq \Sigma_{\rm sf}^{\tau_1}(dM) \quad \text{for every } 0<\tau_1<\tau_2\le d \quad \text{and}$$

$$\Sigma_{\rm sf}^{d}(dM) = \bigcap_{0<\tau<d} \Sigma_{\rm sf}^{\tau}(dM). \tag{7.95}$$

We remark that the equality in (7.95) is also in force for regular canonical systems. We cannot pursue the same strategy to define $\Sigma_{\rm psf}^{d}(dM)$ if (7.1) is not a regular canonical integral system because we cannot guarantee the validity of the inclusion

$$\Sigma_{\rm psf}^{\tau_2}(dM) \subseteq \Sigma_{\rm psf}^{\tau_1}(dM) \quad \text{if } 0<\tau_1<\tau_2<d. \tag{7.96}$$

Nevertheless, if

$$\mathcal{S}^{p\times p} \subset \mathcal{D}(T_{B_\tau}) \quad \text{and} \quad \Sigma_{\rm psf}^{\tau}(dM) = (\mathcal{C}(A_\tau))_{\rm sf}$$

for every $\tau \in [\tau_0, d)$, then

$$\Sigma_{\rm psf}^{\tau_2}(dM) \subseteq \Sigma_{\rm psf}^{\tau_1}(dM) \quad \text{if } \tau_0\le\tau_1<\tau_2<d. \tag{7.97}$$

If condition (7.97) is satisfied, then we will say that a nondecreasing $p\times p$-mvf $\sigma(\mu)$ on $\mathbb{R}$ is a pseudospectral function of the system (7.1) and we will write $\sigma \in \Sigma_{\rm psf}^{d}(dM)$ if $\sigma(\mu) \in \Sigma_{\rm psf}^{\tau}(dM)$ for every $\tau\in[\tau_0,d)$. Thus

$$\Sigma_{\rm psf}^{d}(dM) = \bigcap_{\tau_0\le\tau<d} \Sigma_{\rm psf}^{\tau}(dM), \tag{7.98}$$

by definition.

Now let

$$\begin{gathered} B_t(\lambda) = A_t(\lambda)\mathfrak{V}, \quad \mathfrak{E}_t(\lambda) = [E_-^t(\lambda) \quad E_+^t(\lambda)] = \sqrt{2}N_2^* B_t(\lambda), \\ c_0^t(\lambda) = T_{B_t}[0] = T_{A_t}[I_p]), \quad \beta_0^t = \lim_{\nu\uparrow\infty} \nu^{-1}\Re c_0^t(i\nu), \\ \{b_3^t, b_3^t\} \in ap_{II}(A_t), \quad b_3^t(0) = b_4^t(0) = I_p, \\ \chi_1^t(\lambda) = b_4^t(\lambda)b_3^t(\lambda), \quad \chi^t(\lambda) = E_+^t(\lambda)^{-1}E_-^t(\lambda), \\ \mathcal{M}_{\mathfrak{E}_t} = \{f \in \mathcal{B}(\mathfrak{E}_t):\ \lambda f(\lambda) \in \mathcal{B}(\mathfrak{E}_t)\} \end{gathered}$$

and consider the constraints

($C_\tau 1$) $\beta_0^\tau = 0$,
($C_\tau 2$) $-\mathfrak{E}_\tau(\omega) j_p \mathfrak{E}_\tau(\omega)^* > 0 \quad$ for $\omega \in \mathbb{C}_+$,
($C_\tau 3$) $\overline{\mathcal{M}_{\mathfrak{E}_\tau}} = \mathcal{B}(\mathfrak{E}_\tau)$.

Clearly,

$$\mathcal{H}(A_{t_1}) \subseteq \mathcal{H}(A_{t_2}) \quad \text{and } K_\omega^{A_{t_2}}(\lambda) - K_\omega^{A_{t_1}}(\lambda) \text{ is a positive kernel on } \mathbb{C} \times \mathbb{C} \text{ for } \quad 0 \le t_1 \le t_2 < d \tag{7.99}$$

and

$$\mathcal{B}(\mathfrak{E}_{t_1}) \subseteq \mathcal{B}(\mathfrak{E}_{t_2}) \quad \text{and } K_\omega^{\mathfrak{E}_{t_2}}(\lambda) - K_\omega^{\mathfrak{E}_{t_1}}(\lambda) \text{ is a positive kernel on } \mathbb{C} \times \mathbb{C} \text{ for } \quad 0 \le t_1 \le t_2 < d. \tag{7.100}$$

Consequently,

$$f \in \mathcal{B}(\mathfrak{E}_{t_1}) \Longrightarrow f \in \mathcal{B}(\mathfrak{E}_{t_2}) \text{ and } \|f\|_{\mathcal{B}(\mathfrak{E}_{t_1})} \le \|f\|_{\mathcal{B}(\mathfrak{E}_{t_2})} \text{ for } 0 < t_1 \le t_2 < d. \tag{7.101}$$

We will have interest in the case when there exists $\tau_0 \in (0, d)$ such that

$$\|f\|_{\mathcal{B}(\mathfrak{E}_{t_1})} = \|f\|_{\mathcal{B}(\mathfrak{E}_{t_2})} \quad \text{for every } f \in \mathcal{B}(\mathfrak{E}_{t_1}) \tag{7.102}$$

if $\tau_0 \le t_1 < t_2 < d$.

We remark that if the constraints $(C_\tau 1)$–$(C_\tau 3)$ are in force, then properties (7.97) and (7.102) are equivalent. Indeed, if (7.102) holds and if $\tau_0 \le \tau_1 < \tau_2 < d$ and $\sigma \in \Sigma_{\rm psf}^{\tau_2}(dM)$, then

$$\|f\|_{L_2^p(d\sigma)} = \|f\|_{\mathcal{B}(\mathfrak{E}_{\tau_2})} = \|f\|_{\mathcal{B}(\mathfrak{E}_{\tau_1})} \quad \text{for every } f \in \mathcal{B}(\mathfrak{E}_{t_1}) \tag{7.103}$$

i.e., $\sigma \in \Sigma_{\rm psf}^{\tau_1}(dM)$. Conversely, if $\Sigma_{\rm psf}^{\tau_2}(dM) \subseteq \Sigma_{\rm psf}^{\tau_1}(dM)$, then, because the spectral function $\sigma_{c_0^{\tau_2}}(\mu)$ of the mvf $c_0^{\tau_2}(\lambda)$ belongs to $\Sigma_{\rm psf}^{\tau_2}(dM)$, we obtain that $\sigma_{c_0^{\tau_2}} \in \Sigma_{\rm psf}^{\tau_1}(dM)$ and

$$\|f\|_{\mathcal{B}(\mathfrak{E}_{\tau_2})} = \|f\|_{L_2^p(d\sigma_{c_0^{\tau_2}})} = \|f\|_{\mathcal{B}(\mathfrak{E}_{\tau_1})}$$

for every $f \in \mathcal{B}(\mathfrak{E}_{\tau_1})$.

Theorem 7.31 *Let the matrizant $A_t(\lambda) = A(t, \lambda)$ of the canonical integral system (7.1) satisfy the condition*

$$A_t \in \mathcal{U}_{rsR}(J_p) \quad \textit{for every } t \in [0, d). \tag{7.104}$$

Then the property (7.102) holds for every $0 < t_1 < t_2 < d$. Moreover,

$$\Sigma_{\rm sf}^d(dM) = \Sigma_{\rm psf}^d(dM).$$

Proof Under the condition (7.104) we have:

(1) The operators U_2^t, defined by the formula

$$(U_2^t f)(\lambda) = \sqrt{2}\, N_2^* f(\lambda), \quad f \in \mathcal{H}(A_t),$$

are unitary maps from $\mathcal{H}(A_t)$ onto $\mathcal{B}(\mathfrak{E}_t)$, because the condition (C_t1) holds and $A_t(\lambda)$ is entire.

(2) The inclusions in (7.99) are isometric.

This justifies (7.102) for every $0 < t_1 < t_2 < d$.

To verify the last assertion, it suffices to check that $\Sigma_{sf}^{\tau}(dM) = \Sigma_{psf}^{\tau}(dM)$ for every $\tau \in (0, d)$. But this is covered by Theorem 7.29. □

Theorem 7.32 *If $\mathfrak{E}_t = \sqrt{2}N_2^* A_t \mathfrak{V}$ and $\{b_3^t, b_4^t\} \in ap_{II}(A_t)$ for $0 \le t < d$ in the setting of Theorem 7.31, then*

$$\Sigma_{sf}^{d}(dM) = \bigcap_{0 \le t < d} (\mathcal{B}(\mathfrak{E}_t))_{sf}$$

and hence $\sigma \in \Sigma_{sf}^{d}(dM)$ if and only if

$$\int_{-\infty}^{\infty} f(\mu)^* d\sigma(\mu) f(\mu) = \int_{-\infty}^{\infty} ((E_+^t)^{-1} f)(\mu)^* ((E_+^t)^{-1} f)(\mu) d\mu \qquad (7.105)$$

for every $f \in \mathcal{H}_*(b_4^t) \oplus \mathcal{H}(b_3^t)$ for $0 \le t < d$.

Proof This follows from Theorems 7.31 and 4.78. □

Lemma 7.33 *If the condition $(C_{\tau_0}2)$ holds for some $\tau_0 \in (0, d)$ then condition $(C_\tau 2)$ holds for every $\tau \in [\tau_0, d)$.*

Proof This is immediate from the monotonicity of $M_{22}(t)$ and (1) of Theorem 7.28. □

Lemma 7.34 *If the condition*

$$\|\chi_1^t(re^{i\theta} + \omega)\| \le \gamma < 1 \qquad (7.106)$$

holds for $t = a$ for some $a \in (0, d)$ and some fixed choice of $\theta \in [0, \pi]$, $\omega \in \mathbb{C}_+$, $\gamma \in (0, 1)$, and all $r \ge r_0 > 0$, then the same inequality holds for all $t \in [a, d)$.

Proof Let $t \ge a$. Then, for every point $\lambda \in \mathbb{C}_+$,

$$\|\chi_1^t(\lambda)\| = \|b_4^t(\lambda) b_4^a(\lambda)^{-1} \chi_1^a(a) b_3^a(\lambda)^{-1} b_3^t(\lambda)\| \le \|\chi_1^a(\lambda)\|,$$

since

$$b_4^t (b_4^a)^{-1} \in \mathcal{S}_{in}^{p\times p} \quad \text{and} \quad (b_3^a)^{-1} b_3^t \in \mathcal{S}_{in}^{p\times p}. \qquad \square$$

Theorem 7.35 *If (7.1) is a singular canonical integral system on $[0, d)$ and the conditions $(C_\tau 2)$ and $(C_\tau 3)$ are in force for all points $\tau \in [\tau_0, d)$, then*

$$\Sigma_{psf}^{d}(dM) = (C_{imp}^{d}(dM))_{sf},$$

i.e., $\sigma \in \Sigma_{\rm psf}^{d}(dM)$ *if and only if there exists a mvf* $c \in \mathcal{C}_{\rm imp}^{d}(dM)$ *with spectral function* $\sigma(\mu)$*:*

$$c(\lambda) = i\alpha - i\beta\lambda + \frac{1}{\pi i}\int_{-\infty}^{\infty}\left\{\frac{1}{\mu-\lambda} - \frac{\mu}{1+\mu^2}\right\} d\sigma(\mu) \tag{7.107}$$

$$c \in \mathcal{C}_{\rm imp}^{d}(dM). \tag{7.108}$$

Proof Under the given assumptions, Theorem 7.21 guarantees that the formulas

$$c^t(\lambda) = i\alpha_t - i\beta_t\lambda + \frac{1}{\pi i}\int_{-\infty}^{\infty}\left[\frac{1}{\mu-\lambda} - \frac{\mu}{1+\mu^2}\right] d\sigma_{c^t}(\mu),$$

$$c^t \in T_{B_t}[\mathcal{S}^{p\times p}]$$

define a correspondence between the sets $\Sigma_{\rm psf}^{t}(dM)$ and $T_{B_t}[\mathcal{S}^{p\times p}]$, for $t \in [\tau_0, d)$. If $\sigma(\mu) \in \Sigma_{\rm psf}^{\tau}(dM)$ for every $\tau \in [\tau_0, d)$, then there exists a mvf $c^\tau \in T_{B_\tau}[\mathcal{S}^{p\times p}]$ for every $\tau \in [\tau_0, d)$ with spectral function $\sigma(\mu)$. Let $s^\tau(\lambda) = T_{\mathfrak{V}}[c^\tau]$. Then $s^\tau \in T_{\mathfrak{V}B_\tau}[\mathcal{S}^{p\times p}]$ and there exists a sequence $\tau_n \uparrow d$ such that

$$s(\lambda) = \lim_{n\to\infty} s^{\tau_n}(\lambda)$$

and $s \in \mathcal{S}_{\rm scat}$. In the present setting, $\mathcal{S}_{\rm scat} \subset \mathcal{D}(T_{\mathfrak{V}})$ and $\mathcal{C}_{\rm imp}^{d}(dM) = T_{\mathfrak{V}}[\mathcal{S}_{\rm scat}]$. Therefore, for a suitably chosen sequence of points $\tau_1, \tau_2, \ldots$ that tend to d,

$$c(\lambda) = \lim_{n\to\infty} c^{\tau_n}(\lambda), \quad \lambda \in \mathbb{C}_+.$$

Thus, $c(\lambda) \in \mathcal{C}_{\rm imp}^{d}(dM)$, and $\sigma(\mu)$ is the spectral function of a mvf $c(\lambda)$. Conversely, let $\sigma(\mu)$ be the spectral function of a mvf $c(\lambda) \in \mathcal{C}_{\rm imp}^{d}(dM)$. Then, because $c \in T_{B_\tau}[\mathcal{S}^{p\times p}]$ for every $\tau_0 \le \tau < d$, by Theorem 7.21, we obtain that $\sigma \in \Sigma_{\rm psf}^{\tau}(dM)$ for every $\tau_0 \le \tau < d$. In the present setting, we also have property (7.97). Consequently, $\sigma \in \Sigma_{\rm psf}^{d}(dM)$. □

Theorem 7.36 *Let* $A_t(\lambda)$ *denote the matrizant of a canonical integral system of the form (7.1) and suppose that condition (7.104) is in force and that the condition (7.106) holds for some* $t \in (0, d)$. *Then the formulas (7.107) and (7.108) give a one-to-one correspondence between the sets* $\Sigma_{\rm sf}^{d}(dM)$ *and* $\mathcal{C}_{\rm imp}^{d}(dM)$.

Proof The conditions (7.104) and (7.106) guarantee that conditions ($C_\tau 2$) and ($C_\tau 3$) are in force for every $\tau \in [\tau_0, d)$. The conclusion of the theorem then follows from Theorem 7.21. □

Theorem 7.37 *Let* $A_t(\lambda) = A(t, \lambda),\ 0 \le t < d$, *be the matrizant of a canonical integral system of the form (7.1) with a mass function* $M(t)$ *that meets the constraints (1.21). Let* $\{b_3^t, b_4^t\} \in ap_{II}A(t)$ *for every* $t \in [0, d)$ *and assume that:*

(i) $A_t \in \mathcal{U}_{rsR}(J_p)$ *for every* $t \in [0, d)$.
(ii) *There exists a number* $a > 0$ *and a point* $t_0 \in [0, d)$ *such that* $e_a^{-1} b_4^{t_0} b_3^{t_0} \in \mathcal{S}_{\text{in}}^{p\times p}$.

Then:

(1) $(C_{\text{imp}}^d(dM))_{\text{sf}} = \Sigma_{\text{sf}}^d(dM)$.
(2) $c \in C_{\text{imp}}^d(dM) \Longrightarrow \lim_{\nu\uparrow\infty} \nu^{-1}\Re c(i\nu) = 0$.
(3) $\sigma \in \Sigma_{\text{sf}}^d(dM) \Longrightarrow$ *there is only one* $c \in C_{\text{imp}}^d(dM)$ *with spectral function equal to* $\sigma(\mu)$. *Moreover,* $\beta_c = 0$.

Proof Assertion (1) follows from condition (i) by Theorem 7.31. Next, if $c \in C_{\text{imp}}^d(dM)$, then $c \in \mathcal{C}(A_{t_0})$, and therefore, since $A_{t_0} \in \mathcal{U}_{rsR}(J_p)$ and (ii) is in force, conditions $(\text{C}_{t_0}1) - (\text{C}_{t_0}3)$ are satisfied. Thus, if $c_0^{t_0} = T_{A_{t_0}}[I_p]$, then

$$\beta_c = \beta_{c_0}^{t_0} = \lim_{\nu\uparrow\infty} \nu^{-1} c_0^{t_0}(i\nu) = 0$$

by Theorem 7.21. Finally, (3) follows from Theorem 7.36. □

Remark 7.38 *Statements (2) and (3) of the last theorem guarantee that for each* $\sigma \in \Sigma_{\text{sf}}^d(dM)$ *there is exactly one Hermitian* $p \times p$ *matrix* α *such that the mvf*

$$c^{(\alpha)}(\lambda) = i\alpha + \frac{1}{\pi i}\int_{-\infty}^{\infty} \left\{ \frac{1}{\mu - \lambda} - \frac{\mu}{1+\mu^2} \right\} d\sigma(\mu) \tag{7.109}$$

belongs to $C_{\text{imp}}^d(dM)$

7.11 Supplementary notes

Most of the results in this chapter are adapted from [ArD03a], [ArD05a] and [ArD04b]. The spectral theory of canonical systems of differential and integral equations and differential equations of higher order that can be reduced to such systems has a long history and has been studied by many authors. A number of basic results in the theory of spectral functions for canonical systems (the direct problem) are discussed in the monographs [At64] by F.V. Atkinson and [Sak99] by L.A. Sakhnovich. The notes in [At64] give a useful history of some of the earlier literature.

In the literature, the mvf $c(\lambda)$ that is called the input impedance in this book is often referred to as the Weyl function or the Weyl–Titchmarsh function. The classification of canonical differential and integral systems (and differential equations of higher order that can be reduced to such systems) by consideration of the limits of a nested family of matrix balls originated with H. Weyl who introduced the limit-point, limit-circle classification for singular second-order systems in [We10] and then extended these investigations in [We35].

The general first-order system studied in chapter 7 of [At64] may be reduced to the system (2.75) with Hamiltonian subject to (2.76) and potential subject to (2.73). The spectrum of the regular canonical integral system (2.75) with boundary conditions

$$y(0,\lambda) = v^\tau M \quad \text{and} \quad y(d,\lambda) = v^\tau N \quad \text{with } v \in \mathbb{C}^m$$

and

$$MJM^* = NJN^* \tag{7.110}$$

is discrete. Atkinson studied this spectrum in terms of the monodromy matrix $Y(\lambda) = Y_d(\lambda)$ of the system and the mvf

$$F_{M,N}(\lambda) = \frac{1}{2}(MY(\lambda) + N)(MY(\lambda) - N)^{-1}J,$$

which he called the *characteristic function* of this boundary problem. If $M = I_m$ and $N = -I_m$, then the mvf $F_{M,N}(\lambda)$ is essentially the same as the mvf $C(\lambda)$ that is defined in (3.135). There is some overlap with Theorem 4.39.

Atkinson also studied the matrix circle generated by the values of $F_{M,N}(\omega)$ for fixed $\omega \in \mathbb{C}_+$ as M and N vary, subject to (7.110). He also studied the limits of the corresponding family of nested matrix balls as the right-hand end point $d \uparrow \infty$. Atkinson used techniques that are very similar to those used in this monograph, even though he did not use the theory of RKHS's. Atkinson's results for Weyl–Titchmarsh balls and spectral functions for systems of the form (2.75) with boundary conditions were developed further in [HS81] and [HS83]. The facts that the ranks $\mathfrak{n}_\ell(\omega)$ and $\mathfrak{n}_r(\omega)$ of the left and right semiradii of the limit balls $\{s(\omega) : s \in \mathcal{S}_{scat}(dM)\}$ considered in Section 6.4 and $\{c(\omega) : c \in \mathcal{C}^d_{\rm imp}(dM)\}$ considered in Section 7.2 are independent of the choice of $\omega \in \mathbb{C}_+$ were obtained earlier by S.A. Orlov [Or76], [Or89a] and [Or89b] for differential systems on $[0,\infty)$ that are more general than the system (1.1) considered here. Orlov also obtained analogous results for the limit balls based on the family of functions $F_{M,N}(\omega)$ that were studied by Atkinson. Orlov obtained these results from his *fundamental theorem*:

Theorem 7.39 (S.A. Orlov) *If U_x, $0 \le x < \infty$, is a nondecreasing $\curvearrowright$ chain of mvf's in the Potapov class $\mathcal{P}^\circ(J)$ such that*

$$G(x,\lambda) \stackrel{\rm def}{=} J - U_x(\lambda)JU_x(\lambda)^* > 0 \quad \textit{for every } \lambda \in \mathfrak{h}^+_{U_x} \textit{ and every } x \in \mathbb{R}_+,$$

then the rank of the mvf $R(\lambda)$ that is defined as the positive semidefinite square root of

$$R(\lambda)^2 = \lim_{x\uparrow\infty} G(x,\lambda)^{-1} \quad \textit{if } \lambda \in \bigcap_{x>0} \mathfrak{h}^+_{U_x}$$

is constant. Analogous conclusions hold for the mvf $R_1(\lambda)$ that is defined in the same way as $R(\lambda)$ but with $-J$ instead of J and $U_x(\lambda)^{-1}$ instead of $U_x(\lambda)$.

L.Z. Grossman [Gr85] gave another proof of Theorem 7.39 based on his development of the theory of contractive extensions of an isometric operator. Grossman's theorem is formulated for operator valued functions.

The description (7.93) of the set of spectral functions of a regular canonical differential system with monodromy matrix $A \in \mathcal{E} \cap \mathcal{U}^\circ(J)$ was obtained earlier by A.L. Sakhnovich [SakA92] under the assumption that A is perfect and $\det a_{21}(\lambda) \not\equiv 0$. These conditions are less restrictive than the conditions that are imposed in Theorem 7.29: If $A \in \mathcal{E} \cap \mathcal{U}^\circ_{AR}(J)$ then A is automatically perfect and if either (7.90) or (7.91) holds, then $\det a_{21}(\lambda) \not\equiv 0$. However, Theorem 7.29 gives stronger conclusions than are available in [SakA92].

The mapping $f \in \mathcal{H}(A) \to \sqrt{2}N_2^* f \in \mathcal{B}(\mathfrak{E})$ is a special case of multiplication by a matrix $L \in \mathbb{C}^{m\times r}$ with $r = p = m/2$, $L^*J_pL = 0$ and $L^*L = 2I_p$. The identity

$$L^*(J_p - A(\lambda)J_pA(\omega)^*)L = E_+^L(\lambda)E_+^L(\omega)^* - E_-^L(\lambda)E_-^L(\omega)^*$$

with

$$\mathfrak{E}^L = [E_-^L \quad E_+^L] = L^*A(\lambda)\mathfrak{V}$$

for such matrices L implies that $\mathfrak{E}^L$ is a de Branges matrix and the map $f \in \mathcal{H}(A) \to L^*f \in \mathcal{B}(\mathfrak{E}^L)$ is a coisometry. This mapping and multiplications by more general matrices L and corresponding generalized Fourier transforms, *L-spectral functions* and *L-pseudospectral functions* are discussed in [ArD04b], and earlier in [Ia86].

The terminology pseudospectral function is consistent with the usage in [SakA92]; pseudospectral functions are called quasispectral functions in [Ka02] and are simply referred to as measures associated with the space $\mathcal{H}(E)$ (alias $\mathcal{B}(\mathfrak{E})$ in our terminology) in [Br68a].

Direct (and inverse) spectral problems for the special class of canonical differential systems that correspond to Dirac systems have also been investigated by A.L. Sakhnovich in [SakA02]. A number of other references for direct and inverse problems for differential systems with potential will be provided in the Supplementary Notes to Chapter 12.

A mvf $\sigma \in \Sigma_{sp}^d(Hdt)$ is called an **orthogonal spectral function** if the transform $\mathcal{F}_2$ maps $L_2^m(Hdt; [0, \ell))$ unitarily onto $L_2^p(d\sigma)$. If the system is regular, then, under some conditions, the set of orthogonal spectral functions is parametrized by restricting $\varepsilon \in \mathcal{S}^{p\times p}$ to be constant unitary matrices in Theorem 7.21; for additional discussion, see e.g., [Sak99]. Orthogonal spectral functions for S_1-strings with mass function $m(x)$, $0 \le x < \ell$, are defined analogously, but with the transform $\mathcal{G}_2$ in place of $\mathcal{F}_2$. Every S_1-string with heavy right end point, i.e., with $m(x) < m(\ell)$ for every $x < \ell$ has exactly two orthogonal spectral functions

τ with support in $\mathbb{R}_+$ if $\ell < \infty$ and $m(\ell) = m(\ell - 0)$; and exactly one such τ if $m(\ell) > m(\ell - 0)$ or if the string is singular. Moreover, every such τ satisfies the condition (2.68). Conversely, every nondecreasing function $\tau(\mu)$ on $\mathbb{R}$ with support in $\mathbb{R}_+$ that satisfies the constraint (2.68) is an orthogonal spectral function of exactly one S_1-string with heavy right end point; see the discussion in Sections 2.6 and 7.7, and [KaKr74b] for the direct problem and [DMc76] for the inverse problem.

In view of Remark 7.18, these results for strings have implications for 2×2 canonical differential systems with $J = J_1$ and normalized diagonal Hamiltonians. However, we shall not enter into that here.

8

Inverse monodromy problems

This chapter is devoted to the **inverse monodromy problem** for canonical integral and differential systems. The given data for this problem is a mvf

$$U \in \mathcal{E} \cap \mathcal{U}^{\circ}(J) \tag{8.1}$$

and the objective is either to find a regular canonical integral system with monodromy matrix U or a regular canonical differential system with monodromy matrix U.

A fundamental theorem of V.P. Potapov (Theorem 2.11) guarantees that both of these problems have at least one solution. In particular, Potapov's theorem guarantees that there exists a continuous nondecreasing $m \times m$ mvf $M(t)$ on an interval $[0, d]$ with $M(0) = 0$ such that $U(\lambda)$ is equal to the monodromy matrix $U_d(\lambda)$ of the corresponding canonical integral system (2.1). Moreover, this solution $M(t)$ can even be chosen to be absolutely continuous on $[0, d]$ with a derivative $M'(t) = H(t)$ a.e. on $[0, d]$ that satisfies the normalization condition (2.46) and is a solution of the inverse monodromy problem for regular canonical differential systems (1.1) on the interval $[0, d]$ with monodromy matrix $U_d(\lambda) = U(\lambda)$ that satisfies the condition (1.19). However, even under these extra restrictions, there may be infinitely many solutions. To insure that there is only one solution of the inverse monodromy problem it is necessary to impose additional constraints:

If $J = \pm I_m$, a necessary and sufficient condition for uniqueness (due to Brodskii and Kisilevskii) that will be discussed in more detail below is that

$$\delta_U = \tau_U,$$

i.e., $\det U(\lambda)$ and $U(\lambda)$ have the same exponential type.

If $J \neq \pm I_m$ and the monodromy matrix

$$U \in \mathcal{E} \cap \mathcal{U}_{AR}^{\circ}(J), \tag{8.2}$$

then the solutions may be parametrized by normalized continuous nondecreasing chains $\{b_1^t, b_2^t\}$, $0 \le t \le d$, of entire inner mvf's such that

$$\{b_1^d, b_2^d\} \in ap_I(U).$$

In our formulation of the inverse monodromy problem we shall include such a chain as part of the given data and shall show that there exists exactly one continuous nondecreasing $m \times m$ mvf $M(t)$ on the interval $[0, d]$ with $M(0) = 0$ such that the matrizant $U_t(\lambda)$, $0 \le t \le d$, of the corresponding integral system (2.1) meets the constraints

$$\{b_1^t, b_2^t\} \in ap_I(U_t) \quad \text{for } 0 \le t \le d \quad \text{and} \quad U_d(\lambda) = U(\lambda) \quad \text{for } \lambda \in \mathbb{C}. \tag{8.3}$$

To ease the discussion, suppose that $J = j_{pq}$ and, in the usual notation, that the monodromy matrix $W \in \mathcal{E} \cap \mathcal{U}_{AR}^{\circ}(j_{pq})$, $\{b_1^t, b_2^t\}, 0 \le t \le d$, is a normalized nondecreasing continuous chain of pairs of entire inner mvf's such that $\{b_1^d, b_2^d\} \in ap(W)$ and $s^{\circ} \in T_W[\mathcal{S}^{p\times q}]$. Then for each $t \in [0, d]$ the GSIP$(b_1^t, b_2^t; s^{\circ})$ is completely indeterminate. Therefore, there exists exactly one mvf $W_t \in \mathcal{E} \cap \mathcal{U}_{rR}^{\circ}(j_{pq})$ such that

$$\{b_1^t, b_2^t\} \in ap(W_t) \quad \text{and} \quad T_{W_t}[\mathcal{S}^{p\times q}] = \mathcal{S}(b_1^t, b_2^t; s^{\circ}). \tag{8.4}$$

This family W_t, $0 \le t \le d$, of resolvent matrices is the matrizant of a canonical integral system (6.1) with mass function

$$M(t) = -i\left(\frac{\partial}{\partial\lambda} W_t\right)(0) j_{pq}$$

and monodromy matrix $W = W_d$. This gives the existence of exactly one solution of the inverse problem with matrizant W_t that meets the conditions (8.4).

If $W \in \mathcal{E} \cap \mathcal{U}_{rsR}^{\circ}(j_{pq})$ then the GSIP$(b_1^t, b_2^t; s^{\circ})$ is strictly completely indeterminate for each $t \in [0, d]$ and the formulas that were obtained in Section 4.12 are applicable.

If $W \in \mathcal{E} \cap \mathcal{U}_{AR}^{\circ}(j_{pq})$, then the GSIP$(b_1^t, b_2^t; s^{\circ})$ may not be strictly completely indeterminate. However, in view of Theorems 4.19 and 4.13, it is completely indeterminate and the matrizant $W_t \in \mathcal{E} \cap \mathcal{U}_{rR}^{\circ}(j_{pq})$ of the corresponding canonical system is the unique resolvent matrix for this problem with $\{b_1^t, b_2^t\} \in ap(W_t)$. Moreover, if $W_t^{(\rho)}$ denotes the unique resolvent matrix of the modified problem GSIP$(b_1^t, b_2^t; \rho s^{\circ})$ with $0 < \rho < 1$ such that

$$\{b_1^t, b_2^t\} \in ap(W_t^{(\rho)}) \quad \text{and} \quad W_t^{(\rho)} \in \mathcal{E} \cap \mathcal{U}^{\circ}(j_{pq}), \tag{8.5}$$

then

$$W_t(\lambda) = \lim_{\rho\uparrow 1} W_t^{(\rho)}(\lambda)$$

and, since $W_t^{(\rho)} \in \mathcal{U}_{rsR}^{\circ}(j_{pq})$, it may be obtained by the formulas that are presented in Theorem 4.75.

The bitangential inverse monodromy problem for canonical integral systems with $J = j_{pq}$ may be solved as above by identifying the matrizant with a family of resolvent matrices W_t, $0 \le t < d$, of suitably chosen interpolation problems under less restrictive conditions: that the monodromy matrix $W \in \mathcal{E} \cap \mathcal{U}_{rR}^{\circ}(j_{pq})$, and the chain of resolvent matrices

$$W_t \in \mathcal{E} \cap \mathcal{U}_{rR}^{\circ}(j_{pq}), \quad 0 \le t \le d.$$

These conditions are automatically satisfied, if $W \in \mathcal{E} \cap \mathcal{U}_{AR}^{\circ}(j_{pq})$.

Uniqueness theorems for the inverse monodromy problem with a monodromy matrix $U \in \mathcal{E} \cap \mathcal{U}_{AR}^{\circ}(J)$ that do not specify chains $\{b_1^t, b_2^t\}$ of entire inner mvf's as part of the given data will be obtained by restricting attention to matrizants in special subclasses of $\mathcal{U}(J)$ in which only one such chain is possible.

If J is unitarily equivalent to j_{pq}, then a mvf $U \in \mathcal{E} \cap \mathcal{U}(J)$ is said to belong to:

(1) The class $\mathcal{U}^H(J)$ of **homogeneous** mvf's if

$$\{e_{a_1} I_p, e_{a_2} I_q\} \in ap_I(U) \quad \text{for some } a_1 \ge 0 \text{ and } a_2 \ge 0. \tag{8.6}$$

(2) The class *Symp* of **symplectic** mvf's if

$$q = p \quad \text{and} \quad U(\lambda)^{\tau} \mathcal{J}_p U(\lambda) = \mathcal{J}_p. \tag{8.7}$$

(3) The class *Real* of **real** mvf's if

$$(\mathcal{I} U)(\lambda) = U(\lambda), \tag{8.8}$$

where

$$(\mathcal{I} f)(\lambda) = \overline{f(-\overline{\lambda})}. \tag{8.9}$$

Moreover, if $U_x(\lambda)$, $0 \le x \le d$, is the matrizant of a canonical differential system (2.14) with $J = \mathcal{J}_p$, then

$$U_x \in \mathcal{E} \cap \mathcal{U}^{\circ}(\mathcal{J}_p) \cap Symp \Longleftrightarrow \overline{H(x)} = H(x) \quad \text{a.e. on } [0, d]. \tag{8.10}$$

Within these classes there are two principal uniqueness results (the second of which rests on the result of Brodskii and Kisilevskii that was cited earlier):

Theorem 8.1 *If $W \in \mathcal{E} \cap \mathcal{U}_{AR}^{H}(j_p) \cap Symp$ and $W(0) = I_m$, then there exists exactly one canonical differential system with normalized Hamiltonian $H(x)$ and matrizant $W_x \in \mathcal{U}^H(j_p) \cap Symp$.*

Theorem 8.2 *If $W \in \mathcal{E} \cap \mathcal{U}_{AR}^{\circ}(j_p) \cap Symp$, then there exists exactly one canonical differential system with normalized Hamiltonian $H(x)$ and matrizant $W_x \in \mathcal{U}^{\circ}(j_p) \cap Symp$ if and only if $\tau_W^+ = \delta_{w_{22}}^+$ (or, equivalently $\tau_W^- = \delta_{w_{11}}^-$).*

If $q = p = 1$, then $\mathcal{U}^H(J) = \mathcal{U}(J)$, the constraint $\tau_W^+ = \delta_{w_{22}}^+$ is automatically met and a fundamental theorem of de Branges establishes uniqueness under less restrictive conditions on the monodromy matrix:

Theorem 8.3 (L. de Branges) *If $U \in \mathcal{E} \cap \mathcal{U}^\circ(\mathcal{J}_1) \cap Symp$, then it is the monodromy matrix of exactly one canonical differential system (2.14) with $J = \mathcal{J}_1$, and a real Hamiltonian $H(x)$ that is subject to the normalization (1.19).*

The original proof of this remarkable result may be found in [Br68a]. The main difficulty in the proof is to show that the symplectic left divisors of U are ordered. An expanded version of de Branges' proof is given in [DMc76].

In this chapter we shall also extend the Brodskii–Kisilevskii criteria to monodromy matrices $U \in \mathcal{E} \cap \mathcal{U}^\circ_{AR}(J)$ and shall discuss connections of these results with Livsic triangular models for Volterra operators with finite dimensional imaginary part. Solutions with special properties will be studied: extremal type solutions, solutions with symplectic, real and homogeneous matrizants.

8.1 Some simple illustrative examples

This section is devoted to a number of simple examples that are intended to illustrate the main themes of this rather long chapter.

Let $U \in \mathcal{E} \cap \mathcal{U}^\circ(J)$ and let

$$d = -i \,\mathrm{trace}\, \{U'(0)J\}. \tag{8.11}$$

Then, by Potapov's theorem (Theorem 2.11), there exists an $m \times m$ mvf $H \in L_1^{m\times m}([0, d])$ with

$$H(t) \geq 0 \quad \text{and} \quad \mathrm{trace}\, H(t) = 1 \quad \text{a.e. on } [0, d] \tag{8.12}$$

such that the solution $U_t(\lambda)$ of the equation

$$U_t(\lambda) = I_m + i\lambda \int_0^t U_s(\lambda)H(s)dsJ \quad \text{for } 0 \leq t \leq d \tag{8.13}$$

satisfies the condition

$$U_d(\lambda) = U(\lambda). \tag{8.14}$$

Example 8.4 *Let*

$$U(\lambda) = \begin{bmatrix} e^{i\lambda a} & 0 \\ 0 & 1 \end{bmatrix}, \quad J = I_2 \quad \text{and} \quad a > 0.$$

If $U(\lambda)$ is the monodromy matrix of the canonical system (8.13) with Hamiltonian $H \in L_1^{2\times 2}([0, d])$ that is subject to (8.12) and matrizant $U_t(\lambda)$ for $0 \leq t \leq d$, then $d = a$ by formula (8.11), and, since $U_t(\lambda)$ is a divisor of $U(\lambda)$, it must be of the form

$$U_t(\lambda) = \begin{bmatrix} e^{i\lambda\varphi(t)} & 0 \\ 0 & 1 \end{bmatrix} \quad \text{for } 0 \leq t \leq a$$

for some continuous nondecreasing function $\varphi(t)$ *on* $[0, a]$ *with* $\varphi(0) = 0$ *and* $\varphi(a) = a$. *The formula*

$$\begin{bmatrix} \varphi(t) & 0 \\ 0 & 0 \end{bmatrix} = -i\frac{\partial U_t}{\partial \lambda}(0) = \int_0^t H(s)ds \quad \text{for } 0 \le t \le a \tag{8.15}$$

implies that $\varphi(t)$ *is absolutely continuous and that*

$$H(t) = \begin{bmatrix} \varphi'(t) & 0 \\ 0 & 0 \end{bmatrix} \quad \text{a.e. on } [0, a]. \tag{8.16}$$

Thus, in view of the normalization of the trace imposed in (8.12), $\varphi'(t) = 1$ *a.e. on* $[0, a]$. *Therefore,*

$$H(t) = \begin{bmatrix} 1 & 0 \\ 0 & 0 \end{bmatrix} \quad \text{a.e. on } [0, a] \tag{8.17}$$

is the one and only normalized solution to this inverse monodromy problem. Since $\tau_U = \delta_U$ *in this example, this conclusion is consistent with the general criterion of Brodskii and Kisilevskii that is presented in Theorem 8.32.*

Example 8.5 *Let*

$$W(\lambda) = \begin{bmatrix} e^{i\lambda a_1} & 0 \\ 0 & e^{-i\lambda a_2} \end{bmatrix}, \quad J = j_1, \quad a_1 \ge 0, \quad a_2 \ge 0 \quad \text{and} \quad a_1 + a_2 > 0.$$

If $W(\lambda)$ *is the monodromy matrix of the canonical system (8.13) with matrizant* $W_t(\lambda)$ *for* $0 \le t \le d$, *then* $d = a_1 + a_2$ *by formula (8.11), and, since* $W_t(\lambda)$ *is a divisor of* $W(\lambda)$, *it must be of the form*

$$W_t(\lambda) = \begin{bmatrix} e^{i\lambda\varphi_1(t)} & 0 \\ 0 & e^{-i\lambda\varphi_2(t)} \end{bmatrix} \quad \text{for } 0 \le t \le a_1 + a_2$$

for some pair of continuous nondecreasing functions $\varphi_1(t)$ *and* $\varphi_2(t)$ *on* $[0, d]$ *with* $\varphi_j(0) = 0$ *and* $\varphi_j(d) = a_j$ *for* $j = 1, 2$. *The formula*

$$\begin{bmatrix} \varphi_1(t) & 0 \\ 0 & \varphi_2(t) \end{bmatrix} = -i\frac{\partial W_t}{\partial \lambda}(0)J = \int_0^t H(s)ds \quad \text{for } 0 \le t \le a \tag{8.18}$$

implies that the functions $\varphi_j(t)$ *are absolutely continuous on* $[0, d]$ *and that*

$$H(t) = \begin{bmatrix} \varphi_1'(t) & 0 \\ 0 & \varphi_2'(t) \end{bmatrix} \quad \text{a.e. on } [0, d]. \tag{8.19}$$

with

$$\varphi_1'(t) + \varphi_2'(t) = 1 \quad \text{a.e. on } [0, d], \tag{8.20}$$

in view of the normalization of the trace imposed in (8.12). If $a_1 > 0$ and $a_2 > 0$, then there are infinitely many normalized solutions $H(t)$ for this inverse monodromy problem. Two such possibilities are

$$H_1(t) = \begin{cases} \begin{bmatrix} 1 & 0 \\ 0 & 0 \end{bmatrix} & \text{for } 0 \le t < a_1 \\ \begin{bmatrix} 0 & 0 \\ 0 & 1 \end{bmatrix} & \text{for } a_1 < t \le a_1 + a_2 \end{cases}$$

and

$$H_2(t) = \begin{cases} \begin{bmatrix} 0 & 0 \\ 0 & 1 \end{bmatrix} & \text{for } 0 \le t < a_2 \\ \begin{bmatrix} 1 & 0 \\ 0 & 0 \end{bmatrix} & \text{for } a_2 < t \le a_1 + a_2. \end{cases}$$

There are of course infinitely more choices. Since $\tau_W < \delta_W$ in this example, this conclusion is also consistent with the general criterion of Brodskii and Kisilevskii that is presented in Theorem 8.32.

Example 8.6 *Let*

$$W(\lambda) = \begin{bmatrix} e^{i\lambda a_1} & 0 \\ 0 & e^{-i\lambda a_2} \end{bmatrix}, \quad J = j_1, \quad a_1 \ge 0, \quad a_2 \ge 0 \quad \text{and} \quad a_1 + a_2 > 0.$$

If $W(\lambda)$ is the monodromy matrix of the canonical system (8.13) with matrizant $W_t(\lambda)$ for $0 \le t \le d$, then $d = a_1 + a_2$ by formula (8.11), and, since $W_t(\lambda)$ is a divisor of $W(\lambda)$, it must be of the form

$$W_t(\lambda) = \begin{bmatrix} e^{i\lambda\varphi_1(t)} & 0 \\ 0 & e^{-i\lambda\varphi_2(t)} \end{bmatrix} \quad \text{for } 0 \le t \le a_1 + a_2$$

for some pair of continuous nondecreasing functions $\varphi_1(t)$ and $\varphi_2(t)$ on $[0, d]$ with $\varphi_j(0) = 0$ and $\varphi_j(d) = a_j$ for $j = 1, 2$. Formula (8.18) implies that the functions $\varphi_j(t)$ are absolutely continuous on $[0, d]$ and that $H(t)$ will be of the form (8.19), subject to the constraint (8.20). If $a_1 > 0$ and $a_2 > 0$, then there are infinitely many normalized solutions $H(t)$ for this inverse monodromy problem.

Example 8.7 *Let*

$$W(\lambda) = \begin{bmatrix} e^{i\lambda a} & 0 \\ 0 & e^{-i\lambda a} \end{bmatrix}, \quad J = j_1 \quad \text{and} \quad a > 0.$$

This is a special case of Example 8.6 and again there will be infinitely many solutions of the inverse monodromy problem unless additional restrictions are imposed. In particular, $d = 2a$, and, if it is assumed that the matrizant $W_t(\lambda)$ is *symplectic for every* $t \in [0, d]$, *then it must be of the form*

$$W_t(\lambda) = \begin{bmatrix} e^{i\lambda\varphi(t)} & 0 \\ 0 & e^{-i\lambda\varphi(t)} \end{bmatrix} \quad \text{for } 0 \le t \le d$$

for some continuous nondecreasing function $\varphi(t)$ *on* $[0, d]$ *with* $\varphi(0) = 0$ *and* $\varphi(d) = a$. *Formula (8.18) implies that the function* $\varphi(t)$ *is absolutely continuous and that*

$$H(t) = \begin{bmatrix} \varphi'(t) & 0 \\ 0 & \varphi'(t) \end{bmatrix} = \begin{bmatrix} 1/2 & 0 \\ 0 & 1/2 \end{bmatrix} \quad \text{a.e. on } [0, d], \tag{8.21}$$

in view of the constraint (8.20) with $\varphi_1(t) = \varphi_2(t) = \varphi(t)$ *on* $[0, d]$. *Therefore, there is exactly one normalized solution* $H(t)$ *for this inverse monodromy problem with symplectic matrizant.*

Example 8.8 *Let* $J = j_{pq}$,

$$W(\lambda) = \begin{bmatrix} e^{i\lambda a_1} I_p & 0 \\ 0 & e^{-i\lambda a_2} I_q \end{bmatrix}, \quad a_1 \ge 0, \quad a_2 \ge 0 \quad \text{and} \quad a_1 + a_2 > 0.$$

Then $d = pa_1 + qa_2$ *and, much as in Example 8.6, the inverse monodromy problem for the system (8.13) will have infinitely many normalized solutions* $H(t)$. *Uniqueness may be achieved by restricting attention to matrizants of the form*

$$W_t(\lambda) = \begin{bmatrix} e^{i\lambda\varphi_1(t)} I_p & 0 \\ 0 & e^{-i\lambda\varphi_2(t)} I_q \end{bmatrix} \quad \text{for } 0 \le t \le pa_1 + qa_2$$

for some pair of absolutely continuous nondecreasing functions $\varphi_1(t)$ *and* $\varphi_2(t)$ *on* $[0, d]$ *with* $p\varphi_1'(t) + q\varphi_2'(t) = 1$ *a.e. on* $[0, d]$, $\varphi_j(0) = 0$ *and* $\varphi_j(d) = a_j$ *for* $j = 1, 2$ *by setting either* $a_2 = 0$ *or* $a_1 = 0$. *This corresponds to the case of real homogeneous matrizants that will be considered in Section 8.8.*

If $a_1 = a_2 = a$ *and* $q = p$, *then* $W \in \mathcal{E} \cap \mathcal{U}^H(j_p) \cap Symp$, $W(0) = I_{2p}$ *and* $H(t) = (1/2p)I_{2p}$ *is the only solution of the inverse monodromy problem for this monodromy matrix* W *with* trace $H(t) = 1$ *a.e. on* $[0, a]$ *and matrizant* $W_t \in \mathcal{U}^H(j_p) \cap Symp$ *for* $0 \le t \le a$. *Moreover,* $\varphi_1(t) = \varphi_2(t) = t/(2p)$ *for* $0 \le t \le a$.

8.2 Extremal solutions when $J = I_m$

In this section we present solutions of the inverse monodromy problem for canonical differential and integral systems with $J = I_m$ and $m = p$ and show that there is only one extremal solution in a sense that will be explained below. In particular, this yields a proof of Potapov's theorem (Theorem 2.11) for $J = I_m$ that is different from the original proof by Potapov [Po60] and the proof given by Brodskii [Bro72].

Theorem 8.9 *If $b \in \mathcal{E} \cap \mathcal{S}_{\text{in}}^{p\times p}$, $a = \tau_b > 0$, $b(0) = I_p$ and*

$$\mathcal{L}_t = \mathcal{H}(e_t I_p) \cap \mathcal{H}(b) \quad \text{for } 0 \leq t \leq a, \tag{8.22}$$

then there exists exactly one family of mvf's

$$b_t \in \mathcal{E} \cap \mathcal{S}_{\text{in}}^{p\times p} \quad \text{with } b_t(0) = I_p \text{ for } 0 \leq t \leq a \text{ and } b_0(\lambda) = I_p \tag{8.23}$$

such that $\mathcal{L}_t = \mathcal{H}(b_t)$ for $0 \leq t \leq a$. Morever:

1. *$b_t(\lambda)$, $0 \leq t \leq a$, is a normalized nondecreasing $\curvearrowright$ continuous chain of entire inner $p \times p$ mvf's with $b_a(\lambda) = b(\lambda)$.*
2. *$b(\lambda)$ is the monodromy matrix of the canonical integral system (2.1) with $J = I_p$, $d = a$ and matrizant $b_t(\lambda)$, $0 \leq t \leq a$, i.e., b_t is the unique continuous solution of the integral equation*

$$b_t(\lambda) = I_p + i\lambda \int_0^t b_s(\lambda) dm(s) \quad \text{for } 0 \leq t \leq a \tag{8.24}$$

with a continuous nondecreasing $p \times p$ mass function $m(t)$ on $[0, a]$ with $m(0) = 0$. Moreover, this mass function can be recovered from $b_t(\lambda)$ by the formula

$$m(t) = -i\left(\frac{\partial b_t}{\partial \lambda}\right)(0) = 2\pi k_0^{b_t}(0) \quad \text{for } 0 \leq t \leq a. \tag{8.25}$$

3. *$\tau_{b_t} = t$, $\mathcal{H}(b_t) = \{f \in \mathcal{H}(b) : \tau_f \leq t\}$ for $0 \leq t \leq a$ and b_t, $0 \leq t \leq a$, is a strictly increasing chain.*
4. *If $\widetilde{b}_t(\lambda)$, $0 \leq t \leq a$, is the matrizant of another solution $\widetilde{m}(t)$, $0 \leq t \leq a$, of the inverse monodromy problem with monodromy matrix $b(\lambda)$ such that $\text{type}(\widetilde{b}_t) \leq t$ for $0 \leq t \leq a$, then $\widetilde{b}_t^{-1} b_t \in \mathcal{E} \cap \mathcal{S}_{\text{in}}^{p\times p}$ and $\widetilde{m}(t) \leq m(t)$ for $0 \leq t \leq a$.*

Proof Since $\mathcal{L}_t$ is R_α invariant for every point $\alpha \in \mathbb{C}$, Lemma 3.6 guarantees the existence of an essentially unique mvf $b_t \in \mathcal{S}_{\text{in}}^{p\times r}$ such that $\mathcal{L}_t = \mathcal{H}(b_t)$. The inclusion $\mathcal{H}(b_t) \subseteq \mathcal{H}(e_t I_p)$ insures that $r = p$ and $b_t \in \mathcal{E} \cap \mathcal{S}_{\text{in}}^{p\times p}$, thanks to Corollary 3.8 and Lemma 3.35. Thus, $b_t(\lambda)$ may (and will) be normalized so that $b_t(0) = I_p$. The formula

$$\mathcal{H}(b_t) = \mathcal{H}(e_t I_p) \cap \mathcal{H}(b) \quad \text{for } 0 \leq t \leq a \tag{8.26}$$

guarantees that $b_t(\lambda)$, $0 \leq t \leq a$, is a normalized nondecreasing $\curvearrowright$ chain of entire inner mvf's on the interval $[0, a]$. The continuity of this chain then follows from the identities

$$\bigvee_{0<\varepsilon\leq t} \mathcal{H}(b_{t-\varepsilon}) = \mathcal{H}(b_t) \quad \text{for } 0 < t \leq a,$$

$$\bigcap_{0<\varepsilon<a-t} \mathcal{H}(b_{t+\varepsilon}) = \mathcal{H}(b_t) \quad \text{for } 0 < t \leq a$$

and Theorem 5.1. This completes the verification of (1).

Assertion (2) follows from Theorem 5.6.

Next, to verify (3), let $\tau(t) = \tau_{b_t}$ and fix a point $s \in (0, a)$. Then, since $\tau(t)$ is a continuous nondecreasing function on the interval $[0, a]$ with $\tau(0) = 0$ and $\tau(a) = a$ (by Lemma 5.16), there exists a point $t \in (0, a)$ such that $\tau(t) = s$. Thus, $\mathcal{H}(b_t) \subseteq \mathcal{H}(b_s)$. On the other hand, the opposite inclusion $\mathcal{H}(b_s) \subseteq \mathcal{H}(b_t)$ must also hold, since $s = \tau(t)$ and $\tau(t) \leq t$. Therefore, $\mathcal{H}(b_t) = \mathcal{H}(b_s)$ and hence $b_t = b_s$. But this in turn implies that $\tau(s) = \tau(t) = s$.

The second assertion in (3) is now self-evident.

To verify (4), observe that if $\{\widetilde{b}_t : 0 \leq t \leq a\}$ is a normalized nondecreasing continuous chain of divisors of $b(\lambda)$ with type $\widetilde{b}_t \leq t$, then, since $\mathcal{H}(\widetilde{b}_t) \subseteq \mathcal{H}(e_t I_p) \cap \mathcal{H}(b) = \mathcal{H}(b_t)$, it follows that $\approx{b}_t = \widetilde{b}_t^{-1} b_t$ belongs to $\mathcal{E} \cap \mathcal{S}_{in}^{p\times p}$ and hence (in a self-evident notation) that

$$\begin{aligned} m(t) &= -i\left(\frac{\partial b_t}{\partial \lambda}\right)(0) = -i\left(\frac{\partial \widetilde{b}_t}{\partial \lambda}\right)(0)\widetilde{\widetilde{b}}_t(0) - i\widetilde{b}_t(0)\left(\frac{\partial \widetilde{\widetilde{b}}_t}{\partial \lambda}\right)(0) \\ &= \widetilde{m}(t) + \widetilde{\widetilde{m}}(t) \geq \widetilde{m}(t). \end{aligned}$$

□

The chain b_t, $0 \leq t \leq a$, considered in (1) of Theorem 8.9 will be called an **extremal type-chain** (of left divisors of b).

Lemma 8.10 *Let $b \in \mathcal{E} \cap \mathcal{S}_{\text{in}}^{p\times p}$ with $b(0) = I_p$ and $\tau_b = a$, let $\{b_t : 0 \leq t \leq a\}$ denote the extremal type-chain of left inner divisors of $b(\lambda)$ that was introduced in Theorem 8.9 and let $m(t)$ denote the mass function in (8.24). Then*

$$\begin{aligned} b \in Real &\iff b_t \in Real \quad \text{for every } 0 \leq t \leq a \\ &\iff m(t) = \overline{m(t)} \quad \text{for every } 0 \leq t \leq a. \end{aligned}$$

Proof The implication

$$f(\lambda) = \int_0^\infty e^{i\lambda x} f^\vee(x)dx \Longrightarrow (\mathcal{I}f)(\lambda) = \int_0^\infty e^{i\lambda x}\overline{f^\vee(x)}dx \tag{8.27}$$

for $\lambda \in \mathbb{C}_+$ and $f^\vee \in L_2^p(\mathbb{R}_+)$ guarantees that H_2^p is closed under the action of $\mathcal{I}$. Moreover, since

$$\langle(\mathcal{I}f), (\mathcal{I}g)\rangle_{\text{st}} = \langle f, g\rangle_{\text{st}}, \tag{8.28}$$

$b(\lambda)$ and $e_t(\lambda)I_p$ are both real and $\mathcal{I}(fg) = \mathcal{I}(f)\mathcal{I}(g)$, it is readily checked that $\mathcal{I}\mathcal{H}(b_t) \subseteq \mathcal{H}(b_t)$ and hence, as $\mathcal{I}$ is an involution, that $\mathcal{I}\mathcal{H}(b_t) = \mathcal{H}(b_t)$. Thus, if

$$k_0^t(\lambda) = \frac{I_p - b_t(\lambda)}{-2\pi i\lambda}$$

denotes the RK for $\mathcal{H}(b_t)$ (that evaluates at the point zero) and if Π_t denotes the orthogonal projection of $\mathcal{H}(b_a)$ onto $\mathcal{H}(b_t)$, then

$$\mathcal{I}k_0^t\xi = \mathcal{I}\,\Pi_t k_0^a\xi = \Pi_t k_0^a\xi = k_0^t\xi$$

for every $\xi \in \mathbb{C}^p$. Therefore, $b \in Real \Longrightarrow b_t \in Real$ (since $\mathcal{I}k_0^a\xi$ is real, because $b_a = b$). The converse is self-evident.

Next, if $\overline{m(t)} = m(t)$ for $0 \leq t \leq a$, then for each $\lambda \in \mathbb{C}$, $b_t(\lambda)$ and $\mathcal{I}b_t(\lambda)$ are both continuous solutions of the same integral equation (8.24). Therefore, they coincide. The converse is an easy consequence of formula (8.25). □

Much the same sort of analysis leads to the following auxiliary result for right inner divisors of $b(\lambda)$.

Lemma 8.11 *If* $b \in \mathcal{E} \cap \mathcal{S}_{\text{in}}^{p\times p}$, $a = \tau_b$ *and*

$$\mathcal{L}_t = \mathcal{H}_*(e_t I_p) \cap \mathcal{H}_*(b) \quad \text{for } 0 \leq t \leq a,$$

then:

(1) *There exists exactly one family of mvf's*

$$b_t \in \mathcal{E} \cap \mathcal{S}_{\text{in}}^{p\times p} \quad \text{with } b_t(0) = I_p \text{ for } 0 \leq t \leq a \text{ and } b_0(\lambda) = I_p$$

such that $\mathcal{L}_t = \mathcal{H}_*(b_t)$ *for* $0 \leq t \leq a$. *Moreover, it is a normalized nondecreasing* $\curvearrowleft$ *continuous chain of entire inner mvf's and is the only continuous solution of the integral equation*

$$b_t(\lambda) = I_p + i\lambda \int_0^t dm(s) b_s(\lambda) \quad \text{for } 0 \leq t \leq a$$

where $m(t)$ *is the continuous nondecreasing* $p \times p$ *mvf on* $[0, a]$ *with* $m(0) = 0$ *that may be defined by the formula* $m(t) = -i\frac{\partial b_t}{\partial \lambda}(0)$.

(2) *The following equivalences hold between the monodromy matrix* b, *the matrizant* b_t *and the mass function* $m(t)$:

$$b \in Real \Longleftrightarrow b_t \in Real \quad \text{for every } t \in [0, a]$$
$$\Longleftrightarrow \overline{m(t)} = m(t) \quad \text{for every } t \in [0, a].$$

Proof The proof is similar to the proof of Theorem 8.9 and Lemma 8.10. (The reader might also find it helpful to identify b_t^{-1} with the matrizant U_t of (2.1) with $J = -I_p$.) □

Theorem 8.12 *Let* $b \in \mathcal{E} \cap \mathcal{S}_{\text{in}}^{p\times p}$ *with* $b(0) = I_p$ *and* $\tau_b = a$ *and let* $\varphi(t)$ *be any continuous nondecreasing function on the interval* $[0, d]$ *with* $\varphi(0) = 0$ *and* $\varphi(d) = a$. *Then there exists one and only one solution* $m_\varphi(t)$, $0 \leq t \leq d$, *of the inverse monodromy problem for the integral equation (8.24) with monodromy matrix* $b(\lambda)$ *and matrizant* $b_t(\lambda)$ *such that the following two conditions are met:*

(1) $\tau_{b_t} = \varphi(t)$ *for every* $t \in [0, d]$.
(2) *If* $\widetilde{m}(t)$, $0 \leq t \leq d$, *is any solution of this inverse monodromy problem such that the exponential type of the corresponding matrizant* $\widetilde{b}_t(\lambda)$ *satisfies the inequality* $\text{type}(\widetilde{b}_t) \leq \varphi(t)$ *for every* $t \in [0, d]$, *then* $\widetilde{m}(t) \leq m_\varphi(t)$ *for every* $t \in [0, d]$.

Moreover, if $\widetilde{m}(t)$ is any solution of this inverse monodromy problem with monodromy matrix $b(\lambda)$ and matrizant $\widetilde{b}_t(\lambda)$ and if $\widetilde{\varphi}(t) = \text{type}(\widetilde{b}_t)$, then $\widetilde{\varphi}(t)$ is a continuous nondecreasing function on $[0, d]$, with $\widetilde{\varphi}(0) = 0$ and $\widetilde{\varphi}(d) = a$ and

$$\widetilde{m}(t) \leq m_{\widetilde{\varphi}}(t) \quad \textit{for every } t \in [0, d].$$

Proof Let $\{\mathfrak{b}_t;\ 0 \leq t \leq a\}$ be the extremal type-chain of left inner divisors of $b(\lambda)$ and let $b_t = \mathfrak{b}_{\varphi(t)}$ for $0 \leq t \leq d$. Then $\{b_t;\ 0 \leq t \leq d\}$ is a continuous chain of left inner divisors of $b(\lambda)$ such that $\text{type}(b_t) = \varphi(t)$ for every $t \in [0, d]$. Now let $\widetilde{m}(t),\ 0 \leq t \leq d$, be any solution of the inverse monodromy problem for the integral equation (8.24) with monodromy matrix $b(\lambda)$ such that the inequality

$$\text{type}(\widetilde{b}_t) \leq \varphi(t) \quad \text{for every} \quad t \in [0, d]$$

holds for the matrizant $\widetilde{b}_t$ corresponding to $\widetilde{m}(t)$. Then

$$\mathcal{H}(\widetilde{b}_t) \subseteq \mathcal{H}(b) \cap \mathcal{H}(e_{\varphi(t)} I_p) = \mathcal{H}(\mathfrak{b}_{\varphi(t)})$$

and hence

$$\widetilde{b}_t^{-1} \mathfrak{b}_{\varphi(t)} \in \mathcal{E} \cap \mathcal{S}_{\text{in}}^{p \times p}.$$

Thus,

$$\widetilde{m}(t) = -i \left(\frac{\partial \widetilde{b}_t}{\partial \lambda} \right)(0) \leq -i \left(\frac{\partial \mathfrak{b}_{\varphi(t)}}{\partial \lambda} \right)(0) = m_{\varphi}(t),$$

just as in the proof of Theorem 8.9.

The final assertion is verified in much the same way after first invoking Lemma 5.16 to establish the continuity of $\widetilde{\varphi}(t)$. □

We remark that $x(t) = \text{trace } m(t)$ is continuous and strictly increasing on the interval $[0, a]$ when $m(t)$ is defined by the extremal type-chain as in Theorem 8.9. Therefore the corresponding inverse function $t(x)$ is a well-defined continuous strictly increasing function on the interval $[0, \ell]$, where $\ell = x(a)$. Moreover, since $m(t)$ is absolutely continuous with respect to $x(t)$, it follows (just as in Section 2.2) that $b_x(\lambda) = \mathfrak{b}(t(x), \lambda)$ is the solution of a canonical system of differential equations of the form (1.1) with $J = I_p$ and Hamiltonian

$$H(x) = \frac{d}{dx} m(t(x)) \quad \text{for a.e. } x \in (0, \ell)$$

with trace $H(x) = 1$ a.e. in the interval $[0, \ell]$. This particular solution of the inverse monodromy problem is maximal in the following sense:

Theorem 8.13 *Let $b \in \mathcal{E} \cap \mathcal{S}_{\text{in}}^{p \times p}$ with $b(0) = I_p$. Then there exists a unique normalized solution $H(x)$ of the inverse monodromy problem for the canonical system (1.1) with $J = I_p$ and monodromy matrix $b(\lambda)$ that satisfies the following extremal property:*

If $b_x(\lambda)$ is the matrizant for $H(x)$ and if $\widetilde{H}(x)$, $0 \le x \le \ell$, is any solution of the inverse monodromy problem for the canonical differential system (1.1) with $J = I_p$ and monodromy matrix $b(\lambda)$ with matrizant $\widetilde{b}_x(\lambda)$, then the condition

$$\text{type}(\widetilde{b}_{\widetilde{x}}) \le \text{type}(b_x) \quad \textit{for some pair of points } \widetilde{x},\ x \in [0, \ell],$$

implies that $\widetilde{b}_{\widetilde{x}}^{-1} b_x \in \mathcal{S}^{p\times p}$.

If

$$\text{type}\ (\widetilde{b}_x) \le \text{type}(b_x) \ \ \text{for every} \ \ x \in [0, \ell], \tag{8.29}$$

then

$$\int_0^x \widetilde{H}(s)ds \le \int_0^x H(s)ds \quad \text{for every} \quad x \in [0, \ell]. \tag{8.30}$$

Thus, if (8.29) holds and trace $\widetilde{H}(s) =$ trace $H(s)$ *a.e. on* $[0, \ell]$, *then* $\widetilde{H}(x) = H(x)$ *a.e. on* $[0, \ell]$. *Consequently, if $\widetilde{H}(x)$ is some other normalized solution of this inverse monodromy problem with matrizant $\widetilde{b}_x$, then*

$$\text{type}\, \widetilde{b}_x > \text{type}\, b_x \quad \textit{for some } x \in [0, \ell].$$

Proof Let $H(x)$ be the solution of the inverse monodromy problem with matrizant $b_x(\lambda)$ that was constructed in the paragraph preceding the statement of the theorem. Let $\widetilde{H}(x)$, $0 \le x \le \ell$, be a second solution of this inverse problem with matrizant $\widetilde{b}_x(\lambda)$ and suppose that

$$\text{type}(\widetilde{b}_{\widetilde{x}}) \le \text{type}(b_x) = t \quad \text{for} \quad \widetilde{x},\ x \in [0, \ell].$$

Then, since $\widetilde{b}_{\widetilde{x}}^{-1} b \in \mathcal{S}_{\text{in}}^{p\times p}$,

$$\mathcal{H}\widetilde{b}_{\widetilde{x}}) \subseteq \mathcal{H}(b) \cap \mathcal{H}(e_t I_p) = \mathcal{H}(b_x).$$

Consequently, $\widetilde{b}_{\widetilde{x}}^{-1} b_x \in \mathcal{S}_{\text{in}}^{p\times p}$.

Suppose next that (8.29) is in force. Then, in view of the preceding discussion,

$$\widetilde{b}_x^{-1} b_x \in \mathcal{E} \cap \mathcal{S}_{\text{in}}^{p\times p}$$

and hence, just as in the proof of Theorem 8.12, that

$$\int_0^x \widetilde{H}(s)ds = -i\left(\frac{\partial \widetilde{b}_x}{\partial \lambda}\right)(0) \le -i\left(\frac{\partial b_x}{\partial \lambda}\right)(0) = \int_0^x H(s)ds,$$

for every $x \in [0, \ell]$, as claimed.

Next, if also trace $\widetilde{H}(x) = 1$ a.e. on $[0, \ell]$, then

$$\int_0^x \{H(s) - \widetilde{H}(s)\}ds \ge 0 \quad \text{and} \quad \text{trace} \int_0^x \{H(s) - \widetilde{H}(s)\}ds = 0$$

imply that

$$\int_0^x \{H(s) - \widetilde{H}(s)\}ds = 0 \quad \text{for every } x \in [0, \ell]$$

and hence that $\widetilde{H}(x) = H(x)$ a.e. on $[0, \ell]$.

Finally, if $\widetilde{H}(x)$ is some other normalized solution of this inverse monodromy problem with matrizant $\widetilde{b}_x$, then the preceding discussion implies that

$$\text{type}\,\widetilde{b}_x > \text{type}\,b_x \quad \text{for some } x \in [0, \ell]. \qquad \square$$

8.3 Solutions for $U \in \mathcal{U}_{AR}(J)$ when $J \neq \pm I_m$

The next theorem guarantees the existence of at least one solution to the inverse monodromy problem for canonical integral systems when the monodromy matrix U belongs to the class $\mathcal{U}_{AR}(J)$ for $J \neq \pm I_m$ and gives a parametrization of the set of all solutions to this problem. Since J is unitarily equivalent to j_{pq}, when $J \neq \pm I_m$, it suffices to focus on mondromy matrices $W \in \mathcal{U}_{AR}^{\circ}(j_{pq})$.

Two preliminary lemmas are needed.

Lemma 8.14 *If $W \in \mathcal{E} \cap \mathcal{U}^{\circ}(j_{pq})$ and $W \notin \mathcal{U}_s(j_{pq})$, then there exists a normalized strictly increasing continuous chain of pairs $\{b_1^t, b_2^t\}$, $0 \le t \le d$, of entire inner mvf's of sizes $p \times p$ and $q \times q$, respectively, such that $\{b_1^d, b_2^d\} \in ap(W)$.*

Proof Fix $d > 0$, let $\{b_1, b_2\} \in ap(W)$ with $b_1(0) = I_p$, $b_2(0) = I_q$ and let $\varphi_1(t)$ and $\varphi_2(t)$ be a pair of continuous nondecreasing functions on the interval $[0, d]$ such that $\varphi_1(0) = \varphi_2(0) = 0$, $\varphi_1(d) = \tau_{b_1}$, $\varphi_2(d) = \tau_{b_2}$ and $\varphi_j(t)$ is strictly increasing if $\tau_{b_j} > 0$. Then for each $t \in [0, d]$ there exists a unique pair of mvf's $b_1^t \in \mathcal{E} \cap \mathcal{S}_{\text{in}}^{p\times p}$ and $b_2^t \in \mathcal{E} \cap \mathcal{S}_{in}^{q\times q}$ such that

$$\mathcal{H}(b_1^t) = \mathcal{H}(e_{\varphi_1(t)} I_p) \cap \mathcal{H}(b_1), \quad b_1^t(0) = I_p, \quad b_1^0(\lambda) = I_p,$$
$$\mathcal{H}_*(b_2^t) = \mathcal{H}_*(e_{\varphi_2(t))} I_q) \cap \mathcal{H}_*(b_2), \quad b_2^t(0) = I_q \quad \text{and} \quad b_2^0(\lambda) = I_q,$$

by Theorem 8.9 applied to b_1 and its dual version applied to b_2. This family of pairs meets the stated conditions. $\square$

Lemma 8.15 *If U_t, $0 \le t \le d$, is a normalized nondecreasing $\curvearrowright$ chain of entire J-inner mvf's, then the following statements are equivalent:*

(1) *$U_t(\lambda)$ is left (resp., right) continuous on $[0, d]$ for each point $\lambda \in \mathbb{C}$.*
(2) *$2\pi K_0^{U_t}(0)$ is left (resp., right) continuous on $[0, d]$.*
(3) *trace $2\pi K_0^{U_t}(0)$ is left (resp., right) continuous on $[0, d]$.*

Proof The first two equivalences follow readily from Theorem 5.1 and its proof; the third rests in part on the fact that if $0 \le t_1 \le t_2 \le d$, then the matrix

$$Q = 2\pi K_0^{U_{t_2}}(0) - 2\pi K_0^{U_{t_1}}(0)$$

is positive semidefinite and hence the ij entry q_{ij} of Q is subject to the bounds

$$|q_{ij}|^2 \le q_{ii} q_{jj} \le \{\text{trace}\, Q\}^2.$$

The rest is easy and is left to the reader. $\square$

Theorem 8.16 *If $W \in \mathcal{E} \cap \mathcal{U}_{rR}^{\circ}(j_{pq})$, $W \not\equiv I_m$, $\{b_1^t, b_2^t\}$, $0 \le t \le d$, is a normalized nondecreasing continuous chain of pairs of entire inner mvf's of sizes $p \times p$ and $q \times q$, respectively, such that $\{b_1^d, b_2^d\} \in ap(W)$ and if $s^{\circ} \in T_W[\mathcal{S}^{p\times q}]$, then:*

(1) *For each $t \in [0, d]$ there exists exactly one mvf $W_t \in \mathcal{U}^{\circ}(j_{pq})$ such that*

$$\mathcal{S}(b_1^t, b_2^t; s^{\circ}) = T_{W_t}[\mathcal{S}^{p\times q}] \quad \text{and} \quad \{b_1^t, b_2^t\} \in ap(W_t). \tag{8.31}$$

The family of resolvent matrices W_t, $0 \le t \le d$, that is obtained this way is independent of the choice of $s^{\circ} \in T_W[\mathcal{S}^{p\times q}]$ and enjoys the following properties:

(2a) *W_t, $0 \le t \le d$, is a normalized nondecreasing $\curvearrowright$ chain of mvf's in the class $\mathcal{E} \cap \mathcal{U}_{rR}^{\circ}(j_{pq})$ with $W_0 = I_m$ and $W_d = W$.*

(2b) *$W_t(\lambda)$ is left continuous on $[0, d]$ for each $\lambda \in \mathbb{C}$.*

(2c) *If $W_t(\lambda)$ is continuous on $[0, d]$ for each $\lambda \in \mathbb{C}$ then it is the matrizant of the canonical integral system (6.1) with mass function that meets the constraints (1.22) and is given by $M(t) = 2\pi K_0^{W_t}(0)$.*

(2d) *If $W \in \mathcal{E} \cap \mathcal{U}_{AR}(j_{pq})$, then $W_t(\lambda)$ is continuous on $[0, d]$ for each $\lambda \in \mathbb{C}$.*

If $M(t)$ is any continuous nondeceasing $m \times m$ mvf on $[0, d]$ with $M(0) = 0$ and if the matrizant $W_t(\lambda)$, $0 \le t \le d$, of the canonical integral system (6.1) with this mass function belongs to the class $\mathcal{U}_{rR}^{\circ}(j_{pq})$ for every $t \in [0, d]$, then $M(t)$ is the solution of the inverse monodromy problem with data

$$W(\lambda) = W_d(\lambda) \quad \text{and} \quad \{b_1^t, b_2^t\} \in ap(W_t) \quad \text{for } 0 \le t \le d,$$

where $b_1^t(0) = I_p$ and $b_2^t(0) = I_q$ for every $t \in [0, d)$ and $M(t)$ may be obtained from this data via (1) and (2c).

Proof The proof is broken into steps:

1. Verification of (1) and (2a): The GSIP$(b_1^t, b_2^t; s^{\circ})$ is completely indeterminate for every $t \in [0, d]$ by Theorem 4.13. Therefore, Theorem 4.12 guarantees that for each $t \in [0, d]$ there exists exactly one mvf $W_t \in \mathcal{E} \cap \mathcal{U}_{rR}^{\circ}(j_{pq})$ such that (8.31) holds. Moreover,

$$W_0(\lambda) = I_m \quad \text{and} \quad W_d(\lambda) = W(\lambda),$$

since

$$T_{I_m}[\mathcal{S}^{p\times q}] = \mathcal{S}(I_p, I_q; s^{\circ}) \quad \text{and} \quad T_W[\mathcal{S}^{p\times q}] = \mathcal{S}(b_1^d, b_2^d; s^{\circ}) = T_{W_d}[\mathcal{S}^{p\times q}],$$

by Theorem 4.13. The asserted monotonicity of the chain W_t, $0 \le t \le d$, follows from the monotonicity of the chain of pairs $\{b_1^t, b_2^t\}$ and Theorem 3.64. Theorem 4.13 insures that this chain of resolvent matrices is independent of the choice of $s^{\circ} \in T_W[\mathcal{S}^{p\times q}]$.

2. Verification of (2b): In view of Theorem 5.1, it suffices to show that

$$\mathcal{L}_t^- \stackrel{\text{def}}{=} \bigvee_{0<\varepsilon\leq t} \mathcal{H}(W_{t-\varepsilon}) = \mathcal{H}(W_t) \quad \text{for every } t \in (0, d]. \tag{8.32}$$

Since the inclusion $\mathcal{H}(W_t) \subseteq \mathcal{H}(W)$ is isometric for $0 \leq t \leq d$ by Theorem 4.56, it is readily checked that $\mathcal{L}_t^-$ is a closed R_α-invariant subspace of $\mathcal{H}(W)$ for every $\alpha \in \mathbb{C}$. Consequently, Theorem 4.46 guarantees the existence of a mvf $\widetilde{W}_t \in \mathcal{U}(j_{pq})$ such that $\mathcal{L}_t^- = \mathcal{H}(\widetilde{W}_t)$, $\widetilde{W}_t^{-1}W \in \mathcal{U}(j_{pq})$ and $\mathcal{H}(\widetilde{W}_t)$ is isometrically included in $\mathcal{H}(W)$. Let $\{\widetilde{b}_1^t, \widetilde{b}_2^t\} \in ap(\widetilde{W}_t)$ for $t \in (0, d]$. Then, in view of Theorem 4.50, the inclusions

$$\mathcal{H}(W_{t-\varepsilon}) \subseteq \mathcal{H}(\widetilde{W}_t) \subseteq \mathcal{H}(W_t) \quad \text{for } 0 < t \leq d \text{ and } \varepsilon > 0,$$

imply that $\widetilde{W}_t$ is entire and that

$$\mathcal{H}(b_1^{t-\varepsilon}) \subseteq \mathcal{H}(\widetilde{b}_1^t) \subseteq \mathcal{H}(b_1^t) \quad \text{and} \quad \mathcal{H}_*(b_2^{t-\varepsilon}) \subseteq \mathcal{H}_*(\widetilde{b}_2^t) \subseteq \mathcal{H}_*(b_2^t)$$

for $t \in (0, d)$ and $0 < \varepsilon < t$. Therefore, since the chain of pairs $\{b_1^t, b_2^t\}$ is continuous by assumption, Theorem 5.1 (with J replaced by I_p) implies that

$$\mathcal{H}(b_1^t) = \mathcal{H}(\widetilde{b}_1^t) \quad \text{and} \quad \mathcal{H}_*(b_2^t) = \mathcal{H}_*(\widetilde{b}_2^t) \quad \text{for } t \in (0, d)$$

and hence, if the entire inner mvf's $\widetilde{b}_1^t$ and $\widetilde{b}_2^t$ are normalized,

$$b_1^t = \widetilde{b}_1^t \quad \text{and} \quad b_2^t = \widetilde{b}_2^t \quad \text{for } t \in (0, d],$$

i.e.,

$$\{b_1^t, b_2^t\} \in ap(\widetilde{W}_t) \quad \text{for } t \in (0, d]. \tag{8.33}$$

Moreover, since $\widetilde{W}_t^{-1}W_t \in \mathcal{U}(j_{pq})$, another application of Theorem 4.13 supplies the inclusions

$$\mathcal{S}(b_1^t, b_2^t; s^\circ) \supseteq T_{\widetilde{W}_t}[\mathcal{S}^{p\times q}] \supseteq T_{W_t}[\mathcal{S}^{p\times q}] = \mathcal{S}(b_1^t, b_2^t; s^\circ),$$

which yields the identity

$$\mathcal{S}(b_1^t, b_2^t; s^\circ) = T_{\widetilde{W}_t}[\mathcal{S}^{p\times q}] \quad \text{for } t \in [0, d). \tag{8.34}$$

But since there is only one mvf in the class $\mathcal{U}_{rR}^\circ(j_{pq})$ that meets the constraints (8.33) and (8.34), this implies that $\widetilde{W}_t = W_t\widetilde{W}_t(0)$ for $0 \leq t < d$ and hence justifies (8.32).

3. Verification of (2c): This is immediate from Theorems 4.56 and 5.6.

4. Verification of (2d): In view of (2b), it remains only to prove that $W_t(\lambda)$ is right continuous on $[0, d]$ for each $\lambda \in \mathbb{C}$. In view of Theorem 5.1, it suffices to show that

$$\mathcal{L}_t^+ \stackrel{\text{def}}{=} \bigcap_{0<\varepsilon\leq d-t} \mathcal{H}(W_{t+\varepsilon}) = \mathcal{H}(W_t) \quad \text{for every } t \in [0, d). \tag{8.35}$$

Since the inclusion $\mathcal{H}(W_t) \subseteq \mathcal{H}(W)$ is isometric for $0 \le t \le d$ by Theorem 4.56, it is readily checked that $\mathcal{L}_t^+$ is a closed R_α-invariant subspace of $\mathcal{H}(W)$ for every $\alpha \in \mathbb{C}$. Consequently, Theorem 4.46 guarantees the existence of a mvf $\widetilde{W}_t \in \mathcal{U}(j_{pq})$ such that $\mathcal{L}_t^+ = \mathcal{H}(\widetilde{W}_t)$, $\widetilde{W}_t^{-1} W \in \mathcal{U}(j_{pq})$ and $\mathcal{H}(\widetilde{W}_t)$ is isometrically included in $\mathcal{H}(W)$. Let $\{\widetilde{b}_1^t, \widetilde{b}_2^t\} \in ap(\widetilde{W}_t)$ for $t \in [0, d)$. Then, in view of Theorem 4.50, the inclusions

$$\mathcal{H}(W_t) \subseteq \mathcal{H}(\widetilde{W}_t) \subseteq \mathcal{H}(W_{t+\varepsilon}) \quad \text{for } 0 \le t \le d - \varepsilon \text{ and } \varepsilon > 0,$$

imply that

$$\mathcal{H}(b_1^t) \subseteq \mathcal{H}(\widetilde{b}_1^t) \subseteq \mathcal{H}(b_1^{t+\varepsilon}) \quad \text{and} \quad \mathcal{H}_*(b_2^t) \subseteq \mathcal{H}_*(\widetilde{b}_2^t) \subseteq \mathcal{H}_*(b_2^{t+\varepsilon})$$

for $t \in [0, d)$ and $0 < \varepsilon \le d - t$. Therefore, since the chain of pairs $\{b_1^t, b_2^t\}$ is continuous by assumption, Theorem 5.1 (with J replaced by I_p) implies that

$$\mathcal{H}(b_1^t) = \mathcal{H}(\widetilde{b}_1^t) \quad \text{and} \quad \mathcal{H}_*(b_2^t) = \mathcal{H}_*(\widetilde{b}_2^t) \quad \text{for } t \in [0, d)$$

and hence, as all the indicated inner mvf's are entire and may be normalized,

$$b_1^t = \widetilde{b}_1^t \quad \text{and} \quad b_2^t = \widetilde{b}_2^t \quad \text{for } t \in [0, d),$$

i.e., (8.33) holds. Moreover, since $\widetilde{W}_t^{-1} W \in \mathcal{U}(j_{pq})$ and $W \in \mathcal{U}_{AR}(j_{pq})$, $\widetilde{W}_t \in \mathcal{U}_{rR}(j_{pq})$ and hence, by Theorem 4.13, (8.34) is in force. But since there is only one mvf in the class $\mathcal{U}_{rR}^\circ(j_{pq})$ that meets the constraints (8.33) and (8.34), this implies that $\widetilde{W}_t = W_t$ for $0 \le t < d$ and hence justifies (8.35).

5. Verification of the rest: By Theorem 2.5, the matrizant W_t, $0 \le t \le d$, is a normalized nondecreasing $\curvearrowright$ continuous chain of entire J-inner mvf's. Moreover, since $W_t \in \mathcal{U}_{rR}(j_{pq})$ for every $t \in [0, d]$ by assumption, Theorem 5.13 insures that the normalized chain of associated pairs $\{b_1^t, b_2^t\} \in ap(W_t)$, $0 \le t \le d$, is a continuous nondecreasing chain of pairs on $[0, d]$. Furthermore,

$$\mathcal{S}(b_1^t, b_2^t; s^\circ) = T_{W_t}[\mathcal{S}^{p \times q}] \quad \text{and} \quad \{b_1^t, b_2^t\} \in ap(W_t) \quad \text{for every } t \in [0, d].$$

Therefore, the matrizant W_t, $0 \le t \le d$, coincides with the family of resolvent matrices that is obtained in Step 1. □

Theorem 8.17 *Let $U \in \mathcal{E} \cap \mathcal{U}_{AR}^\circ(J)$, where $J = V^* j_{pq} V$ for some unitary matrix V, $U(\lambda) \not\equiv I_m$ and let $\{b_1^t, b_2^t\}$, $0 \le t \le d$, be a normalized strictly increasing continuous chain of pairs of entire inner mvf's such that $\{b_1^d, b_2^d\} \in ap(VUV^*)$. Then there exists exactly one normalized (as in (1.19)) solution $H(x)$, $0 \le x \le \ell$, of the bitangential inverse monodromy problem for canonical differential system (1.1) with monodromy matrix U for which the matrizant Ω_x satisfies the condition $\{b_1^{t(x)}, b_2^{t(x)}\} \in ap(\Omega_x)$ for every $x \in [0, \ell]$, where $\ell = -\text{itrace}\,\{U'(0)J\}$ and $t(x)$ is a function from $[0, \ell]$ into $[0, d]$ such that $t(0) = 0$ and $t(\ell) = d$. The function $t(x)$ is uniquely defined by the given data of the bitangential inverse monodromy problem and it is a strictly increasing continuous function on $[0, \ell]$.*

Proof Let $W(\lambda) = VU(\lambda)V^*$ and fix $s^\circ \in T_W[\mathcal{S}^{p\times q}]$. Then, since $W \in \mathcal{U}_{AR}(j_{pq})$, Theorem 8.16 guarantees that the family of resolvent matrices W_t, $0 \le t \le d$, that is obtained in (1) of that theorem is the matrizant of the canonical integral system (6.1) with mass function $M(t) = 2\pi K_0^{W_t}(0)$ for $0 \le t \le d$. Moreover, by Lemmas 5.14 and 8.15, $x(t) = \text{trace}\, M(t)$ is a strictly increasing continuous function on $[0, d]$ with $x(0) = 0$ and $x(d) = \ell$. Therefore, there exists exactly one strictly increasing continuous function $t(x)$ on $[0, \ell]$ such that $t(x(t)) = t$ for every $t \in [0, d]$. Moreover, $\widetilde{M}(x) = M(t(x))$ is absolutely continuous with respect to x on $[0, \ell]$. Let

$$H(x) = \widetilde{M}'(x) \quad \text{and} \quad \Omega_x(\lambda) = W_{t(x)}(\lambda).$$

Then $H(x)$ is a normalized solution of the stated inverse problem with matrizant $\Omega_x(\lambda)$, i.e.,

$$\Omega_x(\lambda) = I_m + i\lambda \int_0^x \Omega_u(\lambda)H(u)du\, j_{pq} \quad \text{for } 0 \le u \le \ell.$$

This establishes the existence of at least one solution.

Suppose next that $H^{(1)}(x)$ is a second normalized solution of the stated inverse problem with matrizant $\Omega_x^{(1)}(\lambda) = W_{\varphi(x)}(\lambda)$ for some function $\varphi(x)$ from $[0, \ell]$ into $[0, d]$ such that $\varphi(0) = 0$ and $\varphi(\ell) = d$. Then

$$\Omega_x^{(1)}(\lambda) = I_m + i\lambda \int_0^x \Omega_u^{(1)}(\lambda)H^{(1)}(u)du\, j_{pq} \quad \text{for } 0 \le u \le \ell.$$

Suppose further that $\varphi(x_0) \le t(x_0)$ at some point $x_0 \in [0, \ell]$. Then $\Omega_{x_0}^{(1)} = W_{\varphi(x_0)}$ is a left divisor of $\Omega_{x_0} = W_{t(x_0)}$ and hence,

$$\begin{aligned}\int_0^{x_0} H^{(1)}(u)du = M^{(1)}(x_0) &= -i\left(\frac{\partial}{\partial\lambda}\Omega_{x_0}^{(1)}\right)(0)\, j_{pq} \\ &\le -i\left(\frac{\partial}{\partial\lambda}\Omega_{x_0}\right)(0)\, j_{pq} = M(x_0) = \int_0^{x_0} H(u)du.\end{aligned}$$

But this implies that

$$M^{(1)}(x_0) - M(x_0) \ge 0 \quad \text{and} \quad \text{trace}\,\{M^{(1)}(x_0) - M(x_0)\} = 0.$$

Therefore, the two mass functions must be equal, i.e.,

$$\varphi(x_0) \le t(x_0) \Longrightarrow \int_0^{x_0} H^{(1)}(u)du = \int_0^{x_0} H(u)du.$$

But this serves to complete the proof since the same conclusion holds if $\varphi(x_0) \ge t(x_0)$ and x_0 is any point in $[0, \ell]$. $\square$

8.4 Connections with the Livsic model of a Volterra node

The representation of a mvf $U \in \mathcal{E} \cap \mathcal{U}^\circ(J)$ as the monodromy matrix of a canonical integral system (2.1) is intimately connected with the Livsic triangular functional model of a Volterra operator node with characteristic function $U(\lambda)$.

Let $\mathcal{L}(X, Y)$ denote the space of bounded linear operators acting from a Hilbert space X into a Hilbert space Y, and let $\mathcal{L}(X) = \mathcal{L}(X, X)$. The ordered set $\mathfrak{N} = (K, F; X, Y; J)$ of two separable Hilbert spaces X and Y, two bounded linear operators $K \in \mathcal{L}(X)$, $F \in \mathcal{L}(X, Y)$ and a signature operator $J = J^* = J^{-1} \in \mathcal{L}(Y)$, is called an **LB** (acronym for Livsic–Brodskii) J-node if

$$K - K^* = iF^*JF. \tag{8.36}$$

The function

$$U_{\mathfrak{N}}(\lambda) = I + i\lambda F(I - \lambda K)^{-1}F^*J \tag{8.37}$$

is called the **characteristic function** of the node $\mathfrak{N}$. It is defined and holomorphic on the set

$$\Lambda_K = \{\, \lambda \in \mathbb{C} : (I - \lambda K)^{-1} \in \mathcal{L}(X)\}.$$

However, $U_{\mathfrak{N}}(\lambda)$ may be extended to a holomorphic mvf on a domain $\mathfrak{h}_{U_{\mathfrak{N}}} \supseteq \Lambda_K$. Clearly, $0 \in \Lambda_K$ and $U_{\mathfrak{N}}(0) = I$. Moreover,

$$J - U_{\mathfrak{N}}(\lambda)JU_{\mathfrak{N}}(\omega)^* = -i(\lambda - \overline{\omega})F(I - \lambda K)^{-1}(I - \overline{\omega}K^*)^{-1}F^* \tag{8.38}$$

and

$$J - U_{\mathfrak{N}}(\omega)^*JU_{\mathfrak{N}}(\lambda) = -i(\lambda - \overline{\omega})JF(I - \overline{\omega}K^*)^{-1}(I - \lambda K)^{-1}F^*J \tag{8.39}$$

for points $\lambda, \omega \in \Lambda_K$ and consequently $U_{\mathfrak{N}}(\lambda)$ and $U_{\mathfrak{N}}(\lambda)^*$ are J-contractive in $\Lambda_K \cap \mathbb{C}_+$ and J-unitary on $\Lambda_K \cap \mathbb{R}$.

The node $\mathfrak{N}$ is said to be a **simple node** if

$$\bigcap_{n=0}^{\infty} \ker\{FK^n\} = \{0\}. \tag{8.40}$$

A simple LB J-node is uniquely defined up to unitary equivalence by its characteristic function; see e.g., [Liv73] and [Bro72].

Throughout the rest of this section, we shall fix $Y = \mathbb{C}^m$, endowed with the usual inner product $\langle u, v\rangle = v^*u$ for $u, v \in \mathbb{C}^m$. Then the operators in $\mathcal{L}(Y)$ will be defined by $m \times m$ matrices, J will be an $m \times m$ signature matrix and $U_{\mathfrak{N}}(\lambda)$ will be an $m \times m$ mvf, which is called the **characteristic mvf** of the LB J-node

$$\mathfrak{N} = (K, F; X, \mathbb{C}^m; J).$$

Every mvf $U \in \mathcal{U}^\circ(J)$ may be identified as the characteristic mvf of the simple LB J-node

$$\mathfrak{N}_0 = \{R_0, F_0; \mathcal{H}(U), \mathbb{C}^m; J\},$$

where $\mathcal{H}(U)$ is the RKHS with RK given by (4.73),

$$R_0 : f \in \mathcal{H}(U) \longrightarrow \frac{f(\lambda) - f(0)}{\lambda} \quad \text{and } F_0 : f \in \mathcal{H}(U) \longrightarrow \sqrt{2\pi} f(0) \in \mathbb{C}^m, \tag{8.41}$$

see, e.g., theorem 6.2 in [ArD08b]. Moreover, if $U \in \mathcal{E} \cap \mathcal{U}^\circ(J)$, then as noted in Section 4.6, R_0 is a Volterra operator and its real part is a compact operator that belongs to the von Neumann–Schatten class $\mathcal{S}_p$ for every $p > 1$. To put this conclusion in perspective, recall that if $\{E_\mu\}$, $-\infty < \mu < \infty$, is a family of spectral projectors associated with a self-ajoint operator

$$A = \int_{-\infty}^{\infty} \mu dE_\mu \quad \text{in a Hilbert space } X,$$

then A is said to have purely singular spectrum if

$$\sigma_x(\mu) = \langle E_\mu x, x \rangle_X$$

is singular with respect to Lebesgue measure (i.e., $\sigma_x'(\mu) = 0$ a.e. on $\mathbb{R}$) for every vector $x \in X$.

Theorem 8.18 *Every mvf $U \in \mathcal{U}^\circ(J)$ is the characteristic mvf of a simple LB J-node $\mathfrak{N} = (K, F; X, \mathbb{C}^m; J)$ in which $K_R = (K + K^*)/2$ has purely singular spectrum.*

Conversely, if $\mathfrak{N} = (K, F; X, \mathbb{C}^m; J)$ is a simple LB J-node and K_R has purely singular spectrum, then the characteristic mvf U of $\mathfrak{N}$ belongs to the class $\mathcal{U}^\circ(J)$. Moreover,

$$K \text{ is a Volterra operator} \Longleftrightarrow U \in \mathcal{E} \cap \mathcal{U}^\circ(J).$$

Proof The proof of the first half of the theorem rests on the observation that if $U = U_{\mathfrak{N}}$ is the characteristic mvf of the LB J-node $\mathfrak{N} = (K, F; X, \mathbb{C}^m; J)$, then the mvf $C = J(I_m - U)(I_m + U)^{-1}$ that is introduced in Lemma 3.48 admits the representation

$$C(\lambda) = \frac{-i\lambda}{2} F(I - \lambda K_R)^{-1} F^* \quad \text{for } \lambda \in \mathbb{C}_+$$

and if $\mathfrak{N}$ is a simple node, then

$$\bigcap_{n \geq 0} F K_R^n = \{0\}.$$

For additional details see pp. 18–30 in [Bro72] and lemma 6.3 and theorem 6.4 in [ArD08b]. □

The generalized Fourier transform $\mathcal{F}_{\mathfrak{N}}$ for the node $\mathfrak{N} = (K, F; X, \mathbb{C}^m; J)$ is defined by the formula

$$(\mathcal{F}_{\mathfrak{N}}x)(\lambda) = \frac{1}{\sqrt{2\pi}}F(I - \lambda K)^{-1}x \quad \text{for } x \in X; \tag{8.42}$$

it maps X into $\mathcal{H}(U_{\mathfrak{N}})$.

Theorem 8.19 *Let $\mathfrak{N} = (K, F; X, \mathbb{C}^m; J)$ be an LB J-node with characteristic mvf $U_{\mathfrak{N}} \in \mathcal{U}^\circ(J)$ and let $\mathfrak{N}_0 = \{R_0, F_0; \mathcal{H}(U), \mathbb{C}^m; J\}$ be the simple LB node defined by formula (8.41) for $U(\lambda) = U_{\mathfrak{N}}(\lambda)$. Then:*

(1) *Formula (8.42) defines a coisometry $\mathcal{F}_{\mathfrak{N}}$ acting from X onto $\mathcal{H}(U)$.*
(2) *The operator $\mathcal{F}_{\mathfrak{N}}$ intertwines K and R_0:*

$$\mathcal{F}_{\mathfrak{N}}K = R_0\mathcal{F}_{\mathfrak{N}}. \tag{8.43}$$

Moreover,

$$F = F_0\mathcal{F}_{\mathfrak{N}} \quad \text{and} \quad \ker\, \mathcal{F}_{\mathfrak{N}} = \bigcap_{n\geq 0} \ker(FK^n). \tag{8.44}$$

(3) *The operator $\mathcal{F}_{\mathfrak{N}}$ is unitary if and only if the LB-node $\mathfrak{N}$ is simple, i.e., if and only if condition (8.40) is met.*
(4) *If the LB J-node is simple, then the operator $\mathcal{F}_{\mathfrak{N}}$ establishes a unitary equivalence between the simple LB J-nodes*

$$\mathfrak{N} = (K, F; X, \mathbb{C}^m; J) \tag{8.45}$$

and $\mathfrak{N}_0$.

Proof Let

$$x = \frac{1}{\sqrt{2\pi}}\sum_{j=1}^{n}(I - \overline{\omega}_j K^*)^{-1}F^*u_j, \quad \text{with } \omega_j \in \mathfrak{h}_U,\, u_j \in \mathbb{C}^m \text{ and } n \geq 1.$$

Then formulas (8.38), (4.73) and (8.42) yield the relations

$$f = \sum_{j=1}^{n} K_{\omega_j}^U u_j = \mathcal{F}_{\mathfrak{N}}x \quad \text{and} \quad \|f\|^2_{\mathcal{H}(U)} = \|x\|^2_X.$$

Therefore, since the linear manifold of the considered vvf's $f \in \mathcal{H}(U)$ is dense in the space $\mathcal{H}(U)$, $\mathcal{F}_{\mathfrak{N}}$ is an isometry from the closure of the linear manifold of the corresponding vectors x in X onto $\mathcal{H}(U)$. The closure of the latter is the orthogonal complement of

$$\bigcap_{n\geq 0} \ker(FK^n) = \ker\, \mathcal{F}_{\mathfrak{N}}.$$

Thus, $\mathcal{F}_{\mathfrak{N}}$ is a coisometry from X onto $\mathcal{H}(U)$ as claimed. This establishes (1) and (3). Statement (2) may be verified by direct calculation. Finally, (4) follows from (1)–(3). □

Theorem 8.20 *Let $\mathfrak{N} = (K, F; X, \mathbb{C}^m; J)$ be an LB J-node with characteristic mvf $U \in \mathcal{U}^\circ(J)$ and spectrum of K equal to $\{0\}$. Then $\mathfrak{N}$ is a simple node if and only if*

$$\ker K \cap \ker F = \{0\}.$$

Proof This follows from theorem 5.2 in [ArD04b]. □

Products and projections of LB J-nodes

The **product** $\mathfrak{N} = \mathfrak{N}_1 \times \mathfrak{N}_2$ of two LB J-nodes $\mathfrak{N}_j = (K_j, F_j; X_j, \mathbb{C}^m; J)$, $j = 1, 2$, is defined as the LB J-node $\mathfrak{N} = (K, F; X, \mathbb{C}^m; J)$, where $X = X_1 \oplus X_2$,

$$K = \begin{bmatrix} K_1 & iF_1^*JF_2 \\ 0 & K_2 \end{bmatrix} \quad \text{and} \quad F = \begin{bmatrix} F_1 & F_2 \end{bmatrix}.$$

If $\mathfrak{N} = (K, F; X, \mathbb{C}^m; J)$ is an LB J-node, $\mathcal{L}$ is a proper nonzero closed subspace of X and if

$$K_{\mathcal{L}} = \Pi_{\mathcal{L}} K|_{\mathcal{L}} \quad \text{and} \quad F_{\mathcal{L}} = F|_{\mathcal{L}},$$

then

$$(K_{\mathcal{L}})^* = \Pi_{\mathcal{L}} K^*|_{\mathcal{L}}, \quad (F_{\mathcal{L}})^* = \Pi_{\mathcal{L}} F^*$$

and

$$K_{\mathcal{L}} - K_{\mathcal{L}}^* = iF_{\mathcal{L}}^* J F_{\mathcal{L}},$$

i.e.,

$$\mathfrak{N}_{\mathcal{L}} = (K_{\mathcal{L}}, F_{\mathcal{L}}; \mathcal{L}, \mathbb{C}^m; J) \quad \text{is an LB } J\text{-node.}$$

The node $\mathfrak{N}_{\mathcal{L}}$ is called the **projection of the node $\mathfrak{N}$ onto the subspace $\mathcal{L}$.** Moreover, if $K\mathcal{L} \subseteq \mathcal{L}$, then $K^*\mathcal{L}^\perp \subseteq \mathcal{L}^\perp$,

$$\begin{aligned} K &= \Pi_{\mathcal{L}} K \Pi_{\mathcal{L}} + \Pi_{\mathcal{L}^\perp} K \Pi_{\mathcal{L}} + \Pi_{\mathcal{L}} K \Pi_{\mathcal{L}^\perp} + \Pi_{\mathcal{L}^\perp} K \Pi_{\mathcal{L}^\perp} \\ &= K_{\mathcal{L}} \Pi_{\mathcal{L}} + 0 + \Pi_{\mathcal{L}}(K - K^*)\Pi_{\mathcal{L}^\perp} + K_{\mathcal{L}^\perp} \Pi_{\mathcal{L}^\perp} \end{aligned}$$

and $\mathfrak{N} = \mathfrak{N}_{\mathcal{L}} \times \mathfrak{N}_{\mathcal{L}^\perp}$ is the product of the two LB J-nodes. Therefore,

$$K = K_{\mathcal{L}} \Pi_{\mathcal{L}} + K_{\mathcal{L}^\perp} \Pi_{\mathcal{L}^\perp} + iF_{\mathcal{L}}^* J F_{\mathcal{L}^\perp} \quad \text{and} \quad F = F_{\mathcal{L}} + F_{\mathcal{L}^\perp}.$$

Theorem 8.21 *If $\mathfrak{N} = (K, F; X, \mathbb{C}^m; J)$ is an LB J-node and $\mathcal{L}$ is a closed subspace of X that is invariant under K, then*

$$\mathfrak{N} = \mathfrak{N}_{\mathcal{L}} \times \mathfrak{N}_{\mathcal{L}^\perp} \quad \text{and} \quad U_{\mathfrak{N}}(\lambda) = U_{\mathfrak{N}_{\mathcal{L}}}(\lambda) U_{\mathfrak{N}_{\mathcal{L}^\perp}}(\lambda) \quad \text{for } \lambda \in \Lambda_K.$$

Moreover, if $\mathfrak{N}$ is a simple node, then the nodes $\mathfrak{N}_{\mathcal{L}}$ and $\mathfrak{N}_{\mathcal{L}^\perp}$ are both simple.

Proof See, e.g., theorem 2.2 on p. 8 of [Bro72]. □

Theorem 8.22 *Let $\mathfrak{N} = \mathfrak{N}_1 \times \mathfrak{N}_2$ be the product of the LB J-nodes $\mathfrak{N}_1 = (K_1, F_1; X_1, \mathbb{C}^m; J)$ and $\mathfrak{N}_2 = (K_2, F_2; X_2, \mathbb{C}^m; J)$. Then*

(1) *$X_1 \oplus \{0\}$ is a closed subspace of X that is invariant under K, and $\mathfrak{N}_1$ and $\mathfrak{N}_2$ are projections of $\mathfrak{N}$ onto the subspaces $X_1 \oplus \{0\}$ and $\{0\} \oplus X_2$, respectively.*
(2) *If $\mathfrak{N}$ is a simple LB J-node then the nodes $\mathfrak{N}_1$ and $\mathfrak{N}_2$ are simple too.*
(3) *$U_{\mathfrak{N}}(\lambda) = U_{\mathfrak{N}_1}(\lambda) U_{\mathfrak{N}_2}(\lambda)$ for $\lambda \in \Lambda_K$.*

Proof (1) follows from the definitions of products and projections of a node; (2) and (3) follow from Theorem 8.21. □

A left divisor U_1 of $U \in \mathcal{U}^\circ(J)$ belongs to the Brodskii class $\mathfrak{B}_U$ of left divisors of U if it is the characteristic mvf of the projection of a simple LB J-node $\mathfrak{N} = (K, F; X, \mathbb{C}^m; J)$ with characteristic mvf $U \in \mathcal{U}^\circ(J)$ onto a closed K-invariant subspace of X.

Remark 8.23 *The converse of Assertion (2) in Theorem 8.22 is false: if $U_{\mathfrak{N}} \in \mathcal{U}^\circ(J)$ and $U_j = U_{\mathfrak{N}_j}$, $j = 1, 2$, and $\mathfrak{N}_j$ are simple LB J-nodes then their product $\mathfrak{N}$ is simple if and only if $U_1 \in \mathfrak{B}_U$; see, e.g., theorem 2.3 in [Bro72] and remark 6.9 in [ArD08b].*

Theorem 8.24 *If $U \in \mathcal{U}^\circ(J)$, $U_1 \in \mathcal{U}^\circ(J)$ and $U_1^{-1}U \in \mathcal{U}(J)$, then the inclusion $\mathcal{H}(U_1) \subseteq \mathcal{H}(U)$ is isometric if and only if $U_1 \in \mathfrak{B}_U$.*

Proof Let $\mathfrak{N}_0 = (R_0, F_0; \mathcal{H}(U), \mathbb{C}^m; J)$ be the functional model of a simple LB J-node with characteristic mvf $U(\lambda)$ that is considered in Theorem 8.19 and suppose that $U_1 \in \mathfrak{B}_U$, i.e., U_1 is the characteristic mvf of a simple node $\mathfrak{N}_{\mathcal{L}}$ for some closed subspace $\mathcal{L}$ of X that is invariant under K. Then, by Theorem 8.22, $\mathfrak{N}_{\mathcal{L}}$ is also a simple node and hence the generalized Fourier transform $\mathcal{F}_{\mathfrak{N}_{\mathcal{L}}}$ for $\mathfrak{N}_{\mathcal{L}}$ onto its functional model $(\mathfrak{N}_{\mathcal{L}})_0 = (R_0, F_0; \mathcal{H}(U_1), \mathbb{C}^m; J)$ is unitary. Therefore, the inclusion $\mathcal{H}(U_1) \subseteq \mathcal{H}(U)$ is isometric.

Conversely, if the inclusion $\mathcal{H}(U_1) \subseteq \mathcal{H}(U)$ is isometric, then $\mathcal{L} = \mathcal{H}(U_1)$ is a closed subspace of $\mathcal{H}(U)$ that is invariant under R_0 and $U_1 = U_{\mathfrak{N}_{\mathcal{L}}}$. □

Theorem 8.25 *If $U \in \mathcal{U}^\circ(J)$ and $U_1 \in \mathcal{U}_{rR}^\circ(J)$ is a left divisor of U, then $U_1 \in \mathfrak{B}_U$, i.e., $U_1 = U_{\mathfrak{N}_{\mathcal{L}}}$ for some R_0 invariant closed subspace $\mathcal{L}$ of $\mathcal{H}(U)$.*

Proof This follows from Theorems 8.24 and 4.56. □

An LB J-node $\mathfrak{N} = (K, F; X, \mathbb{C}^m; J)$ is said to be a **Volterra node** if K is a Volterra operator, i.e., if K is compact and 0 is the only point in its spectrum.

Theorem 8.26 *If $U \in \mathcal{E} \cap \mathcal{U}_{\ell R}^\circ(J)$ is the characteristic mvf of a simple Volterra node $\mathfrak{N} = (K, F; X, \mathbb{C}^m; J)$, then K is a completely non-self-adjoint operator, i.e., there are no nonzero closed subspaces $\mathcal{L}$ of X that are invariant under K such that $K|_{\mathcal{L}}$ is self-adjoint.*

Proof Let $\mathcal{L}$ be a closed subspace of X that is invariant under K such that $K_{\mathcal{L}}$ is self-adjoint. Then since K is a Volterra operator, $K_{\mathcal{L}}$ is also a Volterra, operator, and the spectral theorem implies that $K_{\mathcal{L}} = 0$. Thus $\mathfrak{N}_{\mathcal{L}} = (0, F_{\mathcal{L}}; \mathcal{L}, \mathbb{C}^m; J)$ is a simple Volterra node with characteristic mvf

$$U_{\mathfrak{N}_{\mathcal{L}}}(\lambda) = I_m + i\lambda F_{\mathcal{L}}(I - \lambda K_{\mathcal{L}})^{-1}F_{\mathcal{L}}^*J = I_m + i\lambda F_{\mathcal{L}}F_{\mathcal{L}}^*J.$$

Moreover, since

$$iF_{\mathcal{L}}^*JF_{\mathcal{L}} = K_{\mathcal{L}} - K_{\mathcal{L}}^* = 0,$$

it follows that

$$U_{\mathfrak{N}_{\mathcal{L}}}(\lambda)^{-1} = I_m - i\lambda F_{\mathcal{L}}F_{\mathcal{L}}^*J$$

and hence that $U_{\mathfrak{N}_{\mathcal{L}}} \in \mathcal{U}_S(J)$ which is not possible when $U \in \mathcal{U}_{\ell R}(J)$, unless $F_{\mathcal{L}}F_{\mathcal{L}}^* = 0$. Therefore, $F_{\mathcal{L}}^* = 0$ and thus, as $\mathcal{L}$ is a simple node, $\mathcal{L} = 0$. □

Remark 8.27 *The conclusions of Theorem 8.26 remain valid if $U \in \mathcal{E} \cap \mathcal{U}^\circ(J)$ does not admit a left divisor of the form $I_m + i\lambda VV^*J$ with $V^*JV = 0$ and $VV^* \neq 0$.*

A family $\mathfrak{C} = \{\mathcal{L}_\alpha : \alpha \in \mathcal{A}\}$ of closed invariant subspaces of an operator K in a Hilbert space X is called a **chain of invariant subspaces** if it is totally ordered by inclusion, i.e., if $\mathcal{L}_1 \in \mathfrak{C}$ and $\mathcal{L}_2 \in \mathfrak{C}$, then either $\mathcal{L}_1 \subseteq \mathcal{L}_2$ or $\mathcal{L}_2 \subseteq \mathcal{L}_1$. A chain $\mathfrak{C}$ is called **maximal** if it is not contained in any other chain of closed invariant subspaces of K. It is known that a compact operator K in a separable Hilbert space X has at least one nontrivial closed invariant subspace and that every chain $\mathfrak{C}$ of closed invariant subspaces of K is contained in a maximal chain of closed invariant subspaces of K; see, e.g., theorems 15.1 and 15.3 in [Bro72]. Correspondingly, a family $\{U_\alpha : \alpha \in \mathcal{A}\}$ of mvf's in $\mathfrak{B}_U$ is called a **chain** if either $U_\alpha^{-1}U_\beta \in \mathcal{U}(J)$ or $U_\beta^{-1}U_\alpha \in \mathcal{U}(J)$ for every choice of $\alpha, \beta \in \mathcal{A}$. Such a chain is called a **maximal chain** if it is not contained in any other chain of mvf's in $\mathfrak{B}_U$.

The stated connections between the left divisors of the characteristic mvf $U_{\mathfrak{N}}$ of a simple LB J-node $\mathfrak{N} = (K, F; X, \mathbb{C}^m; J)$ and the closed subspaces of X that are invariant under K lead to the following conclusions:

If $U \in \mathcal{E} \cap \mathcal{U}^\circ(J)$ and $U(\lambda) \not\equiv I_m$, then:

(1) U has at least one nontrivial left divisor in the Brodskii class $\mathfrak{B}_U$.
(2) Every chain in $\mathfrak{B}_U$ is contained in a maximal chain in $\mathfrak{B}_U$.
(3) Every mvf $\widetilde{U} \in \mathfrak{B}_U$ may be parametrized by the index

$$\alpha_{\widetilde{U}} = 2\pi \operatorname{trace} K_0^{\widetilde{U}}(0). \tag{8.46}$$

Then $\alpha_{\widetilde{U}} \in [0, d]$, where $d = \alpha_U$ and if $U_1 \in \mathfrak{B}_U$ and $U_2 \in \mathfrak{B}_U$ belong to the same chain, then

$$U_1^{-1}U_2 \in \mathcal{U}^\circ(J) \Longleftrightarrow \mathcal{H}(U_1) \subseteq \mathcal{H}(U_2) \Longleftrightarrow \alpha_{U_1} \leq \alpha_{U_2}$$

and

$$U_1 = U_2 \iff \mathcal{H}(U_1) = \mathcal{H}(U_2) \iff \alpha_{U_1} = \alpha_{U_2}.$$

(4) If $\mathfrak{N} = (K, F; X, \mathbb{C}^m; J)$ is a simple LB J-node with characteristic mvf $U \in \mathcal{U}^\circ(J)$, then there exists a one-to-one correspondence between maximal chains of closed subspaces of X that are invariant under K and maximal chains of left divisors of U in $\mathfrak{B}_U$.

The **Livsic triangular model** of a Volterra node is based on the $m \times m$ mvf $M(t)$ that appears in the Potapov multiplicative representation (2.40) of a mvf $U \in \mathcal{E} \cap \mathcal{U}^\circ(J)$. This is equivalent to viewing $U(\lambda)$ as the monodromy matrix of a canonical integral system (2.1) with a continuous nondecreasing $m \times m$ mvf $M(t)$ on the interval $[0, d]$ with $M(0) = 0$:

To each nondecreasing continuous $m \times m$ mvf $M(t)$ on $[0, d]$ with $M(0) = 0$, associate the set $\mathfrak{N}_M = (K_M, F_M; X_M, \mathbb{C}^m; J)$, where

$$X_M = L_2^m(dM; [0, d]), \tag{8.47}$$

$$(K_M f)(t) = iJ \int_t^d dM(s) f(s) \text{ and } F_M f = \int_0^d dM(s) f(s) \quad \text{for } f \in X_M. \tag{8.48}$$

It is easy to check that K_M and F_M are well defined by these formulas as operators in $\mathcal{L}(X_M)$ and $\mathcal{L}(X_M, \mathbb{C}^m)$, respectively. Then, since

$$F_M^* : \xi \in \mathbb{C}^m \longrightarrow f_\xi(t) \in X_M, \text{ where } f_\xi(t) = \xi \text{ on } [0, d]$$

and

$$(K_M^* f)(t) = -iJ \int_0^t dM(s) f(s) \quad \text{for } 0 \le t \le d,$$

it is readily seen that

$$K_M - K_M^* = iF_M^* J F_M$$

and hence that $\mathfrak{N}_M$ is an LB J-node.

Theorem 8.28 *Let $M(t)$ be a continuous nondecreasing $m \times m$ mvf on $[0, d]$ with $M(0) = 0$ and let $U_t(\lambda)$, $0 \le t \le d$, be the matrizant of the corresponding integral system (2.1). Let $\mathfrak{N}_M = (K_M, F_M; X_M, \mathbb{C}^m; J)$ be the LB J-node defined by formulas (8.47) and (8.48). Then:*

(1) *The generalized Fourier transform (8.42) for the LB J-node $\mathfrak{N}_M$ coincides with the generalized Fourier transform*

$$(\mathcal{F} f)(\lambda) = \frac{1}{\sqrt{2\pi}} \int_0^d U_s(\lambda) dM(s) f(s) \quad \textit{for every } f \in X_M, \tag{8.49}$$

for the canonical integral system (2.1).

(2) *The characteristic mvf* $U_{\mathfrak{N}_M}(\lambda)$ *of the node* $\mathfrak{N}_M$ *coincides with the monodromy matrix* $U_d(\lambda)$ *of this system.*

(3) *The family*

$$X_M^t = \left\{ f \in X_M : \int_t^d f(s)^* dM(s) f(s) = 0 \right\} \quad \text{for } 0 \le t \le d$$

is a maximal chain of closed subspaces of X_M *that are invariant under* K_M.

(4) *The mvf's* $U_t(\lambda)$ *are the characteristic mvf's of the projections* $\mathfrak{N}_M^t = (K_M^t, F_M^t; X_M^t, \mathbb{C}^m; J)$ *of the node* $\mathfrak{N}_M$ *onto the subspaces* X_M^t.

(5) *The node* $\mathfrak{N}_M$ *is simple if and only if the inclusions* $\mathcal{H}(U_t) \subseteq \mathcal{H}(U_d)$ *are isometric for every* $t \in [0, d]$.

(6) *If* $U_d \in \mathcal{U}_{BR}(J)$, *then the node* $\mathfrak{N}_M$ *is simple.*

Proof Our first objective is to show that the two transforms (8.42) and (8.49) coincide, i.e.,

$$F_M(I - \lambda K_M)^{-1} f = \int_0^d U(t, \lambda) dM(t) f(t) \quad \text{for every } f \in L_2^m(dM; [0, d]).$$

To this end, let $\eta \in \mathbb{C}^m$ and

$$g = (I - \overline{\lambda} K_M^*)^{-1} F_M^* \eta = (I - \overline{\lambda} K_M^*)^{-1} f_\eta,$$

and observe that

$$\eta^* F_M (I - \lambda K_M)^{-1} f = \langle f, g \rangle_{X_M} = \int_0^d g(t)^* dM(t) f(t) \quad \text{for } f \in X_M.$$

Now fix $\lambda \in \mathbb{C}$ and let $u_\eta(t) = U(t, \lambda)^* \eta$ for $t \in [0, d]$. Then, as the matrizant $U_t(\lambda) = U(t, \lambda)$, $0 \le t \le d$, is the continuous solution of the integral system (2.1), it follows that

$$u_\eta(t) = f_\eta(t) - i\overline{\lambda} J \int_0^t dM(s) u_\eta(s) = f_\eta(t) + \overline{\lambda}(K_M^* u_\eta)(t),$$

i.e.,

$$u_\eta(t) = ((I - \overline{\lambda} K_M^*)^{-1} f_\eta)(t) = g(t).$$

Thus,

$$\eta^* F_M (I - \lambda K_M)^{-1} f = \eta^* \int_0^d U(t, \lambda) dM(t) f(t),$$

which justifies (1). Moreover,

$$I_m + i\lambda F_M (I - \lambda K_M)^{-1} F_M^* J = I_m + i\lambda \int_0^d U(t, \lambda) dM(t) J = U_d(\lambda),$$

i.e., the characteristic function of the operator node $\mathfrak{N}_M$ based on $M(t)$ is equal to the monodromy matrix of the canonical integral system based on $M(t)$, which justifies (2).

The spaces X_M^t, $0 \le t \le d$, are clearly closed subspaces of X_M that are invariant under K_M. To verify the asserted maximality, let $\mathcal{L}$ be a closed subspace of X_M that is invariant under K_M such that either $\mathcal{L} \subseteq X_M^t$ or $X_M^t \subseteq \mathcal{L}$ for each choice of $t \in [0, d]$. Let $\mathcal{A}_- = \{t \in [0, d] : X_M^t \subseteq \mathcal{L}\}$ and let $\mathcal{A}_+ = \{t \in [0, d] : \mathcal{L} \subseteq X_M^t\}$. Then $0 \in \mathcal{A}_-$, $d \in \mathcal{A}_+$, $\mathcal{A}_- \cup \mathcal{A}_+ = [0, d]$ and, since

$$\bigvee_{0 \le s < t} X_M^s = X_M^t = \bigcap_{t < s \le d} X_M^s,$$

$\mathcal{A}_-$ and $\mathcal{A}_+$ are closed nonempty subsets of the connected set $[0, d]$. Therefore, there exists at least one point $t_0 \in \mathcal{A}_- \cap \mathcal{A}_+$. Clearly $\mathcal{L} = X_M^{t_0}$. (Since $M(t)$ may be constant on a subinterval of $[0, d]$, there may be many points in $\mathcal{A}_- \cap \mathcal{A}_+$.) This completes the proof of (3)

The identification of $U_t(\lambda)$ with the characteristic mvf $U_{\mathfrak{N}_M^t}$ of $\mathfrak{N}_M^t$ is similar to the proof that $U_d(\lambda) = U_{\mathfrak{N}_M}(\lambda)$. Therefore, (4) holds.

Suppose next that $\mathfrak{N}_M$ is a simple node. Then $\mathcal{F}_{\mathfrak{N}_M}$ is a unitary map from X_M onto $\mathcal{H}(U_d)$, since (8.40) holds if and only if $\ker \mathcal{F}_{\mathfrak{N}_M} = \{0\}$. Therefore, the restrictions $\mathcal{F}_{\mathfrak{N}_M^t}$ are unitary maps of X_M^t onto $\mathcal{H}(U_t)$ for $0 \le t \le d$. Thus, the spaces $\mathcal{H}(U_t)$ are contained isometrically in $\mathcal{H}(U_d)$. This proves one direction in (5).

To establish the other direction, suppose that the spaces $\mathcal{H}(U_t)$ are contained isometrically in $\mathcal{H}(U_d)$ for $0 \le t \le d$ and let $U = U_d$ and

$$f_t(s) = \begin{cases} I & \text{if } 0 \le s \le t \\ 0 & \text{if } t < s \le d. \end{cases}$$

Then

$$\left(\mathcal{F}_{\mathfrak{N}_M} \sum_{j=1}^n f_{t_j} v_j\right)(\lambda) = \sum_{j=1}^n \frac{1}{\sqrt{2\pi}} \int_0^{t_j} U_s(\lambda) dM(s) v_j = \sum_{j=1}^n \sqrt{2\pi} K_0^{t_j}(\lambda) v_j$$

and, because of the presumed isometric inclusions,

$$\begin{aligned} \left\| \mathcal{F}_{\mathfrak{N}_M} \sum_{j=1}^n f_{t_j} v_j \right\|_{\mathcal{H}(U)}^2 &= \left\langle \sum_{j=1}^n \sqrt{2\pi} K_0^{t_j} v_j, \sum_{i=1}^n \sqrt{2\pi} K_0^{t_i} v_i \right\rangle_{\mathcal{H}(U)} \\ &= 2\pi \sum_{i,j=1}^n v_i^* K_0^{t_j \wedge t_i}(0) v_j \quad (\text{where } t_j \wedge t_i = \min\{t_j, t_i\}) \\ &= \left\| \sum_{j=1}^n f_{t_j} v_j \right\|_{X_M}^2 . \end{aligned}$$

Since the linear space

$$\left\{ \sum_{j=1}^n f_{t_j} v_j : t_j \in [0, d],\ v_j \in \mathbb{C}^m \text{ and } n \ge 1 \right\}$$

is dense in $L_2^m(dM; [0, d])$ (see, e.g., Section 72 in [AkGl63]), $\mathcal{F}_M$ maps X_M isometrically onto $\mathcal{H}(U)$. Therefore, $\mathfrak{N}_M$ is a simple node. This completes the proof of (5); (6) then follows from (5) and Theorem 8.24. □

The formulas for F_M, F_M^*, $F_M^t = \Pi_{X_M^t} F_M$ and $(F_M^t)^*$ imply that

$$\text{trace}\,(F_M F_M^*) = \text{trace}\, M(d) = 2\pi \,\text{trace}\, K_0^U(0)$$

and

$$\text{trace}\,(F_M^t (F_M^t)^*) = \text{trace}\, M(t) = 2\pi \,\text{trace}\, K_0^{U_t}(0). \tag{8.50}$$

Thus, if $U \in \mathcal{E} \cap \mathcal{U}_{BR}^\circ(J)$ is the monodromy matrix of a canonical differential system (1.1) with Hamiltonian $H(t)$ that satisfies the conditions (1.19) and $U_t(\lambda)$, $0 \le t \le d$, is the matrizant of this system, then $\{U_t : t \in [0, d]\}$ is a maximal chain in $\mathfrak{B}_U$ and

$$t = 2\pi \,\text{trace}\, K_0^{U_t}(0). \tag{8.51}$$

If $M(t)$ is assumed to be absolutely continuous on the interval $[0, d]$ and $H(t) = M'(t)$ a.e. on $[0, d]$, then the space X_M and the operators K_M and F_M can be reexpressed in terms of $H(t)$:

$$X_M = L_2^m(Hdt; [0, d]) = \left\{\text{measurable } f : \int_0^d f(t)^* H(t) f(t) dt < \infty\right\}, \tag{8.52}$$

$$(K_M f)(t) = iJ \int_t^d H(s) f(s) ds \text{ and } F_M f = \int_0^d H(s) f(s) ds, \quad \text{for } f \in X_M. \tag{8.53}$$

Theorem 8.29 *If $U \in \mathcal{E} \cap \mathcal{U}_{BR}^\circ(J)$, $d = \text{trace}\, 2\pi K_0^U(0)$ and $\{U_t : t \in \mathcal{A}\}$ is a maximal chain of normalized left divisors of U that is parametrized by formula (8.51), then $\mathcal{A} = [0, d]$, $U_t(\lambda)$, $0 \le t \le d$, is the matrizant of a canonical differential system (1.1) with a Hamiltonian $H(t)$ that satisfies the conditions (1.19) and*

$$M(t) = \int_0^t H(s) ds = 2\pi K_0^{U_t}(0) \quad \textit{for } 0 \le t \le d. \tag{8.54}$$

Proof The proof is broken into steps.

1. $\mathcal{A}$ is a closed subset of $[0, d]$**:** Let $t_1 \le t_2 \le \cdots$ be a monotone nondecreasing sequence of points in $\mathcal{A}$. Then, since $\mathcal{A} \subseteq [0, d]$, this sequence tends to a limit t_0 as $n \uparrow \infty$, $\mathcal{H}(U_{t_n}) \subseteq \mathcal{H}(U_{t_{n+1}})$ and

$$\mathcal{L} = \bigvee_{n \ge 1} \mathcal{H}(U_{t_n})$$

is a closed R_0-invariant subspace of $\mathcal{H}(U)$. Therefore, by Theorem 4.46, there exists a mvf $\widetilde{U} \in \mathcal{U}(J)$ such that $\widetilde{U}^{-1}U \in \mathcal{U}(J)$ and $\mathcal{L} = \mathcal{H}(\widetilde{U})$. Clearly

$$\mathcal{H}(U_t) \subseteq \mathcal{H}(\widetilde{U}) \quad \text{for } t < t_0 \quad \text{and} \quad \mathcal{H}(\widetilde{U}) \subseteq \mathcal{H}(U_t) \quad \text{for } t_0 < t.$$

Thus, if $\mathcal{L} \neq \mathcal{H}(U_{t_0})$, then the chain of closed R_0-invariant subspaces $\{\mathcal{H}(U_t) : t \in \mathcal{A}\}$ is not maximal, which contradicts the assumption on the chain $\{U_t : t \in \mathcal{A}\}$. Therefore, $t_0 \in \mathcal{A}$. A similar argument shows that if $t_1 \geq t_2 \geq \cdots$ is a sequence of points in $\mathcal{A}$, then $t_0 = \lim_{n\uparrow\infty} t_n$ exists and $t_0 \in \mathcal{A}$.

2. $\mathcal{A} = [0, d]$**:** If $\mathcal{A} \neq [0, d]$, then there exists an interval $[\alpha, \beta] \subset [0, d]$ such that $\alpha \in \mathcal{A}$, $\beta \in \mathcal{A}$, but $(\alpha, \beta) \cap \mathcal{A} = \emptyset$. Moreover, since U_β is the characteristic mvf of the simple Volterra LB J-node $\mathfrak{N} = (R_0, F_0; \mathcal{H}(U_\beta), \mathbb{C}^m; J)$ and $\mathcal{H}(U_\alpha)$ is a closed R_0-invariant subspace of $\mathcal{H}(U_\beta)$, the restriction of the adjoint R_0^* to $\mathcal{H}(U_\beta) \ominus \mathcal{H}(U_\alpha)$ has a closed nontrivial invariant subspace $\mathcal{L}^*$, since it is also a Volterra operator. Then $\mathcal{L} = \mathcal{H}(U_\beta) \ominus \mathcal{L}^*$ is a closed R_0 invariant subspace of $\mathcal{H}(U_\beta)$ that contains $\mathcal{H}(U_\alpha)$. The characteristic mvf $U_{\mathfrak{N}_\mathcal{L}}(\lambda)$ of the projection $\mathfrak{N}_\mathcal{L}$ of the node $\mathfrak{N}$ onto the subspace $\mathcal{L}$ belongs to the Brodskii class $\mathfrak{B}_{U_\beta}$. But this leads to a contradiction of the presumed maximality of the given chain $\{U_t : t \in \mathcal{A}\}$.

3. *The normalized nondecreasing family* $\{U_t(\lambda) : t \in [0, d]\}$ *is continuous on* $[0, d]$ *for each* $\lambda \in \mathbb{C}$.

This follows from the maximality of the given chain and (6) of Theorem 5.1.

4. $\{U_t(\lambda)\}$, $0 \leq t \leq d$ *is the matrizant of the canonical differential system (1.1) with Hamiltonian* $H(t)$ *that satisfies the conditions (1.19).*

This follows from Theorem 5.8 and the imposed normalization (8.51). □

The LB J-node $\mathfrak{N}_M$ that was considered above is unitarily equivalent to the LB J-node $\mathfrak{N}_\Lambda = (K_\Lambda, F_\Lambda; X_\Lambda, \mathbb{C}^m; J)$ that is based on the factorization

$$H(t) = \Lambda(t)\Lambda(t)^* \quad \text{a.e. on } [0, d],$$

where $\Lambda(t) \in L_2^{m\times r}([0, d])$ for some $r \leq m$. The formula

$$(T_\Lambda f)(t) = \Lambda(t)^* f(t)$$

defines an isometric operator from X_M into $L_2^r([0, d])$ that yields the unitary equivalence of the nodes $\mathfrak{N}_M$ and $\mathfrak{N}_\Lambda$, in which

$$X_\Lambda = \text{range } T_\Lambda, \; K_\Lambda = T_\Lambda K_M T_\Lambda^*|_{X_\Lambda} \text{ and } F_\Lambda = F_M T_\Lambda^*|_{X_\Lambda}$$

and, for $g \in X_\Lambda$,

$$(K_\Lambda g)(t) = i\Lambda(t)^* J \int_t^d \Lambda(s)g(s)ds \tag{8.55}$$

$$(F_\Lambda g)(t) = \int_0^d \Lambda(s)g(s)ds. \tag{8.56}$$

Then

$$F_\Lambda^* \xi = \Lambda(t)^* \xi \quad \text{a.e. on } [0, d] \text{ for every } \xi \in \mathbb{C}^m. \tag{8.57}$$

It is easy to check that if K and F are defined by formulas (8.55) and (8.56) on the full space $L_2^r([0, d])$, then X_Λ is invariant under K, $\mathfrak{N} = (K, F; L_2^r([0, d]), \mathbb{C}^m; J)$ is an LB J-node and $\mathfrak{N}_\Lambda$ is the projection of this node onto the subspace X_Λ. If $H(t) > 0$ a.e. on $[0, d]$, then $r = m$ and $X_\Lambda = L_2^m([0, d])$, i.e., the node $\mathfrak{N}_M$ is unitarily equivalent to the node $\mathfrak{N} = (K_\Lambda, F_\Lambda; L_2^m([0, d]), \mathbb{C}^m; J)$, where K_Λ and F_Λ are defined on the space $L_2^m([0, d])$ by formulas (8.55) and (8.56).

Theorem 8.30 *If the Hamiltonian $H(t) = \Lambda(t)\Lambda(t)^*$ is positive definite a.e. on $[0, d]$, then the LB J-node $\mathfrak{N}_\Lambda = (K_\Lambda, F_\Lambda; L_2^m([0, d]), \mathbb{C}^m; J)$ based on $\Lambda(t)$ is simple.*

Proof See, e.g., theorem 5.15 in [ArD04b]. □

8.5 Conditions for the uniqueness of normalized Hamiltonians

A Volterra operator K in a separable Hilbert space X is called **unicellular** if the set of closed subspaces in X that are invariant under K is **totally ordered**, i.e., if X_1 and X_2 are closed subspaces of X that are invariant under K, then either $X_1 \subseteq X_2$ or $X_2 \subseteq X_1$. This is equivalent to the statement that there exists only one maximal chain of closed subspaces of X that are invariant under K. Analogously, a mvf $U \in \mathcal{E} \cap \mathcal{U}^\circ(J)$ is called a **unicellular mvf** if the set of left divisors of U in $\mathcal{U}^\circ(J)$ is totally ordered, i.e., if $U_1, U_2 \in \mathcal{U}^\circ(J)$, $U_1^{-1}U \in \mathcal{U}^\circ(J)$ and $U_2^{-1}U \in \mathcal{U}^\circ(J)$, then either $U_1^{-1}U_2 \in \mathcal{U}^\circ(J)$ or $U_2^{-1}U_1 \in \mathcal{U}^\circ(J)$. A totally ordered family of normalized left divisors of U is called **maximal** if it is not a proper subset of another totally ordered family of normalized left divisors of U. A mvf $U \in \mathcal{E} \cap \mathcal{U}^\circ(J)$ is unicellular if and only if there exists exactly one maximal normalized nondecreasing $\curvearrowright$ chain of entire J-inner mvf's that are left divisors of U.

Theorem 8.31 *Let $U \in \mathcal{E} \cap \mathcal{U}_{BR}^\circ(J)$ and let $\mathfrak{N} = (K, F; X, \mathbb{C}^m; J)$ be a simple Volterra LB J-node with characteristic mvf $U(\lambda)$. Then the following three assertions are equivalent:*

1. *The Volterra operator K is unicellular.*
2. *The mvf $U(\lambda)$ is unicellular.*
3. *$U(\lambda)$ is the monodromy matrix of exactly one canonical differential system (1.1) with a Hamiltonian $H(x)$ that satisfies the conditions (1.19).*

Proof Let $\mathfrak{N}_0 = (R_0, F_0; \mathcal{H}(U), \mathbb{C}^m; J)$ be the functional model of $\mathfrak{N}$. Then, in view of Theorem 8.24 there is a one-to-one correspondence between the closed subspaces of $\mathcal{H}(U)$ that are invariant under R_0 and the left divisors of U. Thus, (1) and (2) are equivalent. Moreover, by Theorem 8.29, any maximal chain of normalized left divisors of U may be parametrized by formula (8.46):

$$\{U_t : t \in [0, d]\} \quad \text{is a maximal chain of normalized left divisors of } U \tag{8.58}$$

where $d = 2\pi \operatorname{trace} K_0^U(0)$, and $U_t(\lambda)$, $0 \le t \le d$, is the matrizant of exactly one canonical differential system (1.1) with monodromy matrix $U = U_d$ and a Hamiltonian $H(t)$ that satisfies the conditions (1.19). Consequently, (2) and (3) are equivalent. □

We remark that if $U \in \mathcal{U}_{BR}(J)$, then (8.58) is equivalent to

$$\{\mathcal{H}(U_t) : t \in [0, d]\} \quad \text{is a maximal chain of closed } R_0\text{-invariant subspaces of } \mathcal{H}(U). \tag{8.59}$$

Theorem 8.32 (Brodskii–Kisilevskii) *If $b \in \mathcal{E} \cap \mathcal{S}_{\text{in}}^{p\times p}$ and $b(0) = I_p$, then there exists exactly one canonical differential system (1.1) with $J = I_p$, monodromy matrix $b(\lambda)$ and Hamiltonian $H(x)$ that satisfies the conditions in (1.19) (with m replaced by p) if and only if $\tau_b = \delta_b$.*

Proof Suppose first that $\tau_b = \delta_b$ and let $H(x)$, $0 \le x \le \ell$, be the normalized solution of the inverse monodromy problem that is considered in Theorem 8.13 for the monodromy matrix $b(\lambda)$. Then $\ell = \delta_b$ and, by Lemma 3.40, $\tau_{b_x} = \delta_{b_x} = x$ for $0 \le x \le \ell$ for the matrizant $b_x(\lambda)$ of this solution. Suppose further that $\widetilde{H}(x)$, $0 \le x \le \ell$, is also a normalized solution of this inverse monodromy problem and let $\widetilde{b}_x$, $0 \le x \le \ell$, denote the corresponding matrizant. Then the preceding argument implies that $\delta_{\widetilde{b}_x} = x = \tau_{\widetilde{b}_x}$ for $0 \le x \le \ell$. Therefore, $\tau_{\widetilde{b}_x} = \tau_{b_x}$ for $0 \le x \le \ell$ and hence, by the extremal properties of the first solution, $\widetilde{H}(x) = H(x)$ a.e. on $[0, \ell]$, by Theorem 8.13. Suppose next that the inverse monodromy problem for a given mvf $b \in \mathcal{E} \cap \mathcal{S}_{\text{in}}^{p\times p}$ with $b(0) = I_p$ has exactly one normalized solution $H(x)$, $0 \le x \le d$, and let $b_x(\lambda)$, $0 \le x \le d$, denote the corresponding matrizant. It remains to show that $\delta_b = \tau_b$. The proof of this implication is more complicated. We shall only outline the main steps and refer the reader to the proof of theorem 30.1 on p. 186 of [Bro72] for the details.

1. If $b^\circ \in \mathcal{E} \cap \mathcal{S}_{\text{in}}^{p\times p}$, $b^\circ(0) = I_p$, and $(b^\circ)^{-1} b \in \mathcal{S}_{\text{in}}^{p\times p}$, then $b^\circ(\lambda) = b_x(\lambda)$ for some point $x \in [0, d]$, since the uniqueness of the normalized solution is equivalent to the fact that the family $\{b_x\}$, $0 \le x \le d$, is the set of all normalized divisors of b.

2. Let $v \in \mathbb{C}^p$, $v^* v = 1$, $P = vv^*$, $Q = I_p - P$, $a = \tau_b$ and $s_Q(\lambda) = e^{ia\lambda Q}$. Then $s_Q \in \mathcal{E} \cap \mathcal{S}_{\text{in}}^{p\times p}$ and $s_Q(0) = I_p$. Therefore, since $\mathcal{H}(b) \cap \mathcal{H}(s_Q)$ is a closed subspace of

H_2^p that is invariant under R_α for $\alpha \in \mathbb{C}_+$ and $\mathcal{H}(s_Q) \subseteq \mathcal{H}(e_a I_p)$, Lemma 3.6 and Theorem 3.33 guarantee that there exists exactly one mvf $b_P \in \mathcal{E} \cap \mathcal{S}_{\text{in}}^{p\times p}$ with $b_P(0) = I_p$ such that

$$\mathcal{H}(b_P) = \mathcal{H}(b) \cap \mathcal{H}(s_Q).$$

By Theorem 3.9 and Corollary 3.10, b_P is the unique greatest common left divisor of s_Q and b that meets the normalization condition $b_P(0) = I_p$. The main fact is

$$\delta_{b_P} \geq \delta_b - \tau_b. \tag{8.60}$$

The verification of this inequality is the most difficult part of the proof; see, e.g., lemma 29.3 on p. 184 of [Bro72].

3. By Theorem 3.9, there exists an essentially unique greatest common left divisor $\widetilde{b}$ of all the mvf's b_P. By Lemma 3.35, $\widetilde{b} \in \mathcal{E} \cap \mathcal{S}_{\text{in}}^{p\times p}$ and so may be normalized $\widetilde{b}(0) = I_p$. Moreover,

$$\delta_{\widetilde{b}} = \inf\{\delta_{b_P}\} \geq \delta_b - \tau_b.$$

However,

$$k_0^{\widetilde{b}}(0) \leq k_0^{s_Q}(0) = \frac{1}{2\pi} aQ$$

for every choice of the orthogonal projection P, since $\widetilde{b}$ is a left divisor of s_Q. But this implies that

$$k_0^{\widetilde{b}}(0)v = 0 \quad \text{for every } v \in \mathbb{C}^p.$$

Therefore, $\widetilde{b}(\lambda) = I_p$ and $\delta_{\widetilde{b}} = 0$. Thus, $\delta_b - \tau_b = 0$ and the proof is complete. □

Remark 8.33 *If $\{b_t(\lambda) : 0 \leq t \leq d\}$ is a normalized nondecreasing continuous ↶ chain, then the transposed chain $\{b_t(\lambda)^\tau : 0 \leq t \leq d\}$ is a normalized nondecreasing continuous ↷ chain, and the integral equation*

$$b_t(\lambda) = I_p + i\lambda \int_0^t dm(s) b_s(\lambda), \quad 0 \leq t \leq d,$$

is equivalent to the integral equation

$$b_t(\lambda)^\tau = I_p + i\lambda \int_0^t b_s(\lambda)^\tau dm(s)^\tau, \quad 0 \leq t \leq d.$$

Consequently, Theorems 8.9–8.32 all have dual versions for nondecreasing ↶ chains.

If $J = -I_m$, then

$$U \in \mathcal{E} \cap \mathcal{U}(-I_m) \Longleftrightarrow U^\# \in \mathcal{E} \cap \mathcal{S}_{\text{in}}^{m\times m}.$$

Thus, $\{U_t\}$, $0 \le t < d$, is a normalized nondecreasing continuous $\curvearrowright$ chain of entire $-I_m$-inner mvf's if and only if $\{U_t^\#\}$, $0 \le t < d$, is a normalized nondecreasing continuous $\curvearrowleft$ chain of entire I_m-inner mvf's. Consequently, the dual theorems to Theorems 8.9–8.32 apply to the inverse monodromy problem when $J = -I_m$.

The next theorem reformulates the Brodskii–Kisilevskii condition for $J = -I_m$.

Theorem 8.34 *If $U \in \mathcal{E} \cap \mathcal{U}^\circ(-I_m)$, then the inverse monodromy problem for canonical differential systems (1.1) with $J = -I_m$ and monodromy matrix $U(\lambda)$ has exactly one solution $H(x)$, $0 \le x \le d$, that satisfies the conditions in (1.19) if and only if*

$$\tau_U = \delta_U \quad \text{or, equivalently,} \quad \tau_U^+ = \delta_U^+. \tag{8.61}$$

Proof This is an immediate corollary of the Brodskii–Kisilevskii criterion since $U(\lambda)$ is the monodromy matrix of the system (1.1) with $J = -I_m$ and normalized Hamiltonian $H(x)$ if and only if $\overline{U(\overline{\lambda})}$ is the monodromy matrix of the system (1.1) with $J = I_m$ and normalized Hamiltonian $\overline{H(x)}$. □

Theorem 8.35 *Let $b \in \mathcal{E} \cap \mathcal{S}_{\text{in}}^{p\times p}$ with $b(0) = I_p$ and $\tau_b = \delta_b = a$ and let $\{b_t : 0 \le t \le a\}$ be the extremal type-chain of left (resp., right) inner divisors of $b(\lambda)$. Then $\{\widetilde{b}_u : 0 \le u \le d\}$ is a normalized nondecreasing $\curvearrowright$ continuous chain (resp., $\curvearrowleft$ chain) of entire inner mvf's such that $\widetilde{b}_d = b$ if and only if there exists a continuous nondecreasing function $\varphi(u)$ on $[0, d]$ such that $\varphi(0) = 0$, $\varphi(d) = a$ and*

$$\widetilde{b}_u(\lambda) = b_{\varphi(u)}(\lambda) \quad \textit{for every } u \in [0, d].$$

Moreover, in this case,

$$\tau_{\widetilde{b}_u}^- = \delta_{\widetilde{b}_u}^- = \varphi(u) \quad \textit{for every } u \in [0, d].$$

Proof Under the given assumptions, consider first the case of nondecreasing $\curvearrowright$ chains. Then, since the condition $\tau_b = \delta_b$ is equivalent to the unicellularity of b by Theorems 8.31 and 8.32, it follows that $\widetilde{b}_u = b_{\varphi(u)}$ for some function $\varphi(u)$ on $[0, d]$ such that $\varphi(0) = 0$ and $\varphi(d) = a$. Moreover,

$$\tau_{\widetilde{b}_u}^- = \tau_{b_{\varphi(u)}}^- = \delta_{b_{\varphi(u)}}^- = \varphi(u) = \delta_{\widetilde{b}_u}^-.$$

In view of Lemma 5.16, the function $\varphi(u)$ is continuous and nondecreasing on the interval $[0, d]$.

The same conclusions hold for nondecreasing $\curvearrowleft$ chains by Remark 8.33. □

Theorem 8.36 *Let $b \in \mathcal{E} \cap \mathcal{S}_{\text{in}}^{p\times p}$ meet the conditions*

$$\tau_b^- = \delta_b^- = a \tag{8.62}$$

and let $\mathcal{M}$ be a closed R_0-invariant subspace of $\mathcal{H}(b)$. Then

$$\mathcal{M} = \mathcal{H}(e_t I_p) \cap \mathcal{H}(b) \tag{8.63}$$

for some $t \in [0, a]$. Conversely, if every closed R_0-invariant subspace $\mathcal{M}$ of $\mathcal{H}(b)$ is of the form (8.63) for some $t \in [0, a]$, then $b(\lambda)$ meets condition (8.62).

Proof Since $\tau_b = \tau_b^-$ and $\delta_b = \delta_b^-$, Theorem 8.32 guarantees that item (3) of Theorem 8.31 is in force when $\tau_b^- = \delta_b^-$. Therefore, by (1) of that theorem, R_0 is a unicellular operator in $\mathcal{H}(b)$ and hence the set of spaces $\mathcal{H}(b_t)$, $0 \le t \le d$, based on the extremal type-chain b_t of left divisors of b is equal to the set of all closed subspaces of $\mathcal{H}(b)$ that are invariant under R_0. The same two theorems serve to justify the converse statement. □

Theorem 8.37 *If $U \in \mathcal{E} \cap \mathcal{U}_{AR}^{\circ}(J)$, $J \neq \pm I_m$, and if $\tau_U^+ > 0$ and $\tau_U^- > 0$, then the inverse monodromy problem for canonical differential systems (1.1) with monodromy matrix $U(\lambda)$ has at least two solutions $H(x)$, $0 \le x \le d$, that satisfy the conditions (1.19).*

Proof It suffices to prove the theorem for $J = j_{pq}$ and $U(\lambda) = W(\lambda)$. Let $\{b_1, b_2\} \in ap(W)$. Then, by Theorem 3.56, $b_1 \in \mathcal{E} \cap \mathcal{S}_{\rm in}^{p\times p}$, $b_2 \in \mathcal{E} \cap \mathcal{S}_{\rm in}^{q\times q}$, $\tau_{b_2}^- = \tau_W^+ > 0$ and $\tau_{b_1}^- = \tau_W^- > 0$. Therefore, $b_1(\lambda)$ and $b_2(\lambda)$ are nonconstant mvf's that will be normalized by the conditions $b_1(0) = I_p$ and $b_2(0) = I_q$. Let $a_1 = \tau_{b_1}$ and $a_2 = \tau_{b_2}$. In view of Theorem 8.9 and Remark 8.33, there exists a normalized strictly increasing $\curvearrowright$ continuous chain of entire inner mvf's b_1^t, $0 \le t \le a_1$ such that $b_1^{a_1} = b_1$ and a normalized strictly increasing $\curvearrowleft$ continuous chain of entire inner mvf's b_2^t, $0 \le t \le a_2$ such that $b_2^{a_2} = b_2$. Let $a = a_1 + a_2$ and set

$$\widetilde{b}_1^t = \begin{cases} b_1^t & \text{for } 0 \le t \le a_1 \\ b_1 & \text{for } a_1 \le t \le a \end{cases}$$

and

$$\widetilde{b}_2^t = \begin{cases} I_q & \text{for } 0 \le t \le a_1 \\ b_2^{t-a_1} & \text{for } a_1 \le t \le a. \end{cases}$$

Then $\{\widetilde{b}_1^t, \widetilde{b}_2^t\}$, $0 \le t \le a$, is a normalized strictly increasing continuous chain of pairs of entire inner mvf's such that $\widetilde{b}_1^a = b_1$ and $\widetilde{b}_2^a = b_2$. Therefore, by Theorem 8.17, there exists a Hamiltonian $H_1(x)$, $0 \le x \le \ell$, where $\ell = -i\text{trace}\,(W'(0)j_{pq})$, that satisfies the conditions (1.19) and a strictly increasing function $t_1(x)$, $0 \le x \le \ell_W$, such that the matrizant $W_x^{(1)}$, $0 \le x \le \ell$, of the canonical differential system (1.1) with $J = j_{pq}$ and Hamiltonian $H(x) = H_1(x)$ satisfies the conditions $W_\ell^{(1)} = W$, $\{\widetilde{b}_1^{t_1(x)}, \widetilde{b}_2^{t_1(x)}\} \in ap(W_x^{(1)})$, $W_x^{(1)}$ in $\mathcal{U}_{rR}^{\circ}(j_{pq})$, $0 \le x \le \ell$. For this solution of the inverse monodromy problem $\tau_{W_x^{(1)}}^+ = 0$ for $x \in [0, x_1]$ and $W_{x_1}^{(1)}(\lambda) \not\equiv I_m$, where x_1 is such that $t_1(x_1) = a_1$.

A second solution $H_2(x)$, $0 \le x \le d$, $d = \ell$ of the inverse monodromy problem that is normalized by (1.19) may be obtained in the same way by considering the normalized strictly increasing continuous chain of pairs $\{(b_1^\circ)^t, (b_2^\circ)^t\}$ for $0 \le t \le a$, where

$$(b_1^\circ)^t = \begin{cases} I_p & \text{for} \quad 0 \le t \le a_2 \\ b_1^{t-a_2} & \text{for} \quad a_2 \le t \le a \end{cases}$$

and

$$(b_2^\circ)^t = \begin{cases} b_2^t & \text{for} \quad 0 \le t \le a_2 \\ b_2 & \text{for} \quad a_2 \le t \le a. \end{cases}$$

The matrizant $W_x^{(2)}(\lambda)$, $0 \le x \le \ell$, of the solution of the inverse monodromy problem corresponding to this normalized nondecreasing continuous chain of pairs of entire inner mvf's will be such that

$$W_x^{(2)} \in \mathcal{U}_{rR}(j_{pq}) \quad \text{and} \quad \tau^-_{W_x^{(2)}} = 0 \quad \text{for } 0 \le x \le x_2 \quad \text{and} \quad W_{x_2}^{(2)}(\lambda) \not\equiv I_m,$$

where $t_2(x_2) = a_2$. Let $x_0 = \min\{x_1, x_2\}$. Then $W_x^{(1)}(\lambda) \not\equiv W_x^{(2)}(\lambda)$ for every $x \in (0, x_0)$, since

$$\tau^+_{W_x^{(1)}} = 0 \quad \text{for } 0 \le x \le x_0,$$

whereas

$$\tau^-_{W_x^{(2)}} = 0 \quad \text{for } 0 \le x \le x_0$$

and at least one of the mvf's $W_{x_0}^{(1)}(\lambda)$, $W_{x_0}^{(2)}(\lambda)$ is not a constant matrix. Consequently, $W_{x_0}^{(1)} \not\equiv W_{x_0}^{(2)}(\lambda)$, because otherwise $W_{x_0}^{(2)}(\lambda)$ would have minimal exponential type, which, in view of Theorem 3.56, would imply that $W_{x_0}^{(2)} \in \mathcal{U}_S(J)$. But this is not possible, since $W_{x_0}^{(2)} \in \mathcal{U}_{rR}(j_{pq})$. □

The next theorem summarizes and extends a number of the main conclusions on uniqueness that have been obtained to this point.

Theorem 8.38 *Let $U \in \mathcal{E} \cap \mathcal{U}_{BR}^\circ(J)$ and let $\mathfrak{N} = (K, F; X, \mathbb{C}^m; J)$ be a simple Volterra node with characteristic mvf $U(\lambda)$. Then the following three conditions are equivalent:*

(1) *The operator K is unicellular.*
(2) *The mvf U is unicellular.*
(3) *There exists exactly one differential system (1.1) with monodromy matrix $U(\lambda)$ and Hamiltonian $H(x)$ that meets the conditions in (1.19).*

Moreover,

(4a) *if $J = \pm I_m$, then (1) —(3) hold $\Longleftrightarrow \tau_U = \delta_U$,*
(4b) *if $J \ne \pm I_m$ and $U \in \mathcal{U}_{AR}^\circ(J)$, then (1) —(3) hold $\Longleftrightarrow \tau_U = \delta_U$ and either $\tau_U^- = 0$ or $\tau_U^+ = 0$.*

Proof The equivalence of (1), (2) and (3) is covered by Theorem 8.31. The equivalence of (1) and (4a) is justified in Theorems 8.32 and 8.34. It remains only to verify the equivalence of (4b) with (1). Since $J \neq \pm I_m$, it suffices to focus on the case $W \in \mathcal{E} \cap \mathcal{U}^{\circ}_{AR}(j_{pq})$, with $\{b_1, b_2\} \in ap(W)$, $b_1 \in \mathcal{E} \cap \mathcal{S}_{\text{in}}^{p\times p}$, $b_1(0) = I_p$, $b_2 \in \mathcal{E} \cap \mathcal{S}_{in}^{q\times q}$ and $b_2(0) = I_q$. In view of Theorem 4.24,

$$\tau_W^+ = \tau_{b_2}^- \quad \text{and} \quad \tau_W^- = \tau_{b_1}^-.$$

If $\tau_{b_1} > 0$, $\tau_{b_2} > 0$ and $U \in \mathcal{U}^{\circ}_{AR}(J)$, then (3) is false by Theorem 8.37. The rest of the proof is broken into steps.

1. $W \in \mathcal{E} \cap \mathcal{U}^{\circ}_{AR}(j_{pq})$ **and** $\tau_W^+ = 0$**:** Let W_x, $0 \le x \le \ell$, $\ell = \ell_W$, be the matrizant of the canonical differential system (1.1) with $J = j_{pq}$, Hamiltonian $H(x)$ subject to (1.19) and monodromy matrix $W = W_\ell$. Then, since $b_2(\lambda) \equiv I_q$,

$$\tau_{b_1} = \tau_W = \tau_W^- \quad \text{and, as } \det W = \det b_1 \text{ by Lemma 3.49,} \quad \delta_W = \delta_{b_1}.$$

Then the chain $\{b_1^x, I_q\} \in ap(W_x)$, $0 \le x \le \ell$, is a normalized strictly increasing continuous chain of pairs of entire inner mvf's. Moreover, $W_x \in \mathcal{E} \cap \mathcal{U}^{\circ}_{rR}(j_{pq})$, since $W \in \mathcal{U}_{AR}(j_{pq})$.

Let $\tau_W^- \neq \delta_W^-$. Then $\tau_{b_1} \neq \delta_{b_1}$, and by the Brodskii–Kiselevskii criteria, applied to b_1 there exists a normalized strictly increasing absolutely continuous chain of of entire inner mvf's $\widetilde{b}_x$, $0 \le x \le \ell$, such that $\widetilde{b}_\ell = b_1$ and it cannot be obtained by replacing the variable x from the chain b_1^x, $0 \le x \le \ell$. By Theorem 1.19, there exists a unique Hamiltonian $\widetilde{H}(x)$, $0 \le x \le \ell$, that satisfies the condition (1.19) such that the canonical differential system (1.1) with $J = j_{pq}$ and this Hamiltonian has matrizant $\widetilde{W}_x$, $0 \le x \le \ell$, such that $\widetilde{W}_\ell = W$, $\{\widetilde{b}_x^{t(x)}, I_q\} \in ap(\widetilde{W}_x)$ and $\widetilde{W}_x \in \mathcal{E} \cap \mathcal{U}^{\circ}_{rR}(j_{pq})$, $0 \le x \le \ell$. Since the chain $\widetilde{b}_x^{t(x)}$ is different from the chain b_1^x, $0 \le x \le \ell$, $\widetilde{H}$ is different from H, i.e., (3) is false.

Conversely, if (3) is false, then there exist two normalized solutions H and $\widetilde{H}$ of the inverse monodromy problem for the monodromy matrix W with matrizants W_x and $\widetilde{W}_x$, respectively, with $\{b_1^x, I_q\} \in ap(W_x)$ and $\{\widetilde{b}_1^x, I_q\} \in ap(\widetilde{W}_x)$ for $0 \le x \le \ell$.

Let $\tau_{b_1} = \delta_{b_1}$. Then $\widetilde{b}_x = b_1^{\varphi(x)}$, $0 \le x \le \ell$, for some strictly increasing continuous function $\varphi(x)$ on $[0, \ell]$ with $\varphi(0) = 0$ and $\varphi(\ell) = \ell$. Then $\widetilde{W}_x = W_{\varphi(x)}$, $0 \le x \le \ell$. Since $\operatorname{trace} \widetilde{H}(x) = \operatorname{trace} H(x) = 1$ a.e. on $[0, \ell]$, it follows that $\widetilde{H} = H$.

2. $W \in \mathcal{E} \cap \mathcal{U}^{\circ}_{AR}(j_{pq})$ **and** $\tau_W^- = 0$**:** The proof is similar to the first case. □

Remark 8.39 *The family*

$$U_t(\lambda) = I_m + i\lambda t VV^*J \quad \text{with } V^*JV = 0,\ V^*V = I_k,\ 0 \le t \le 1$$

considered in Example 5.7 is a chain of left divisors of $U_1(\lambda) = U(\lambda)$ such that

$$\mathcal{H}(U_t) = \{VV^*\xi : \xi \in \mathbb{C}^m\}$$

and

$$\langle VV^*\xi, VV^*\eta\rangle_{\mathcal{H}(U_t)} = \frac{2\pi}{t}\eta^*VV^*VV^*\xi = \frac{2\pi}{t}\eta^*VV^*\xi$$

for $0 < t \leq 1$. *Thus the inclusions* $\mathcal{H}(U_t) \subseteq \mathcal{H}(U_1)$, $0 < t < 1$, *are contractive, but not isometric. Consequently, these divisors do not belong to the Brodskii class* $\mathfrak{B}_U$ *and hence* $U \notin \mathcal{U}_{BR}(J)$. *Since* $F_0 f = \sqrt{2\pi} f(0)$ *and* $R_0 f = 0$ *for* $f \in \mathcal{H}(U)$, *i.e., for* $f = VV^*\xi$ *for some* $\xi \in \mathbb{C}^m$, $U(\lambda)$ *is the characteristic mvf of the simple LB Volterra node* $\mathfrak{N}_0 = (0, M_{\sqrt{2\pi}}; \mathcal{H}(U), \mathbb{C}^m; J)$, *where* $M_{\sqrt{2\pi}}$ *denotes multiplication by* $\sqrt{2\pi}$. *Moreover, although every space* $\mathcal{H}(U_t)$, $0 \leq t \leq 1$, *is invariant under* R_0, *only* $\{0\}$ *and* $\mathcal{H}(U)$ *itself are included isometrically in* $\mathcal{H}(U)$. *This is consistent with de Branges' theorem (Theorem 4.47), since*

$$\mathcal{H}(U_t) \cap U_t\mathcal{H}(U_t^{-1}U_1) = \{VV^*\xi : \xi \in \mathbb{C}^m\} \neq \{0\} \quad \text{if } 0 < t < 1.$$

In this example $U \in \mathcal{U}_S(J)$. *Moreover,* U *is unicellular if and only if* $k = 1$, *i.e., the inverse monodromy problem for the canonical differential system (2.14) with monodromy matrix* $U(\lambda) = I_m + i\lambda VV^*J$ *has exactly one solution* $H(t)$ *that meets the conditions (1.19) if and only if* $k = 1$.

If $k > 1$, *and*

$$V = \begin{bmatrix} V_1 & V_2 \end{bmatrix} \quad \text{with } V_1 \in \mathbb{C}^{m\times k_1} \quad \text{and} \quad V_2 \in \mathbb{C}^{m\times k_2},$$

then, in view of the formulas

$$\begin{aligned}(I_m + aV_1V_1^*J + bV_2V_2^*J) &= (I_m + aV_1V_1^*J)(I_m + bV_2V_2^*J)\\ &= (I_m + bV_2V_2^*J)(I_m + aV_1V_1^*J)\end{aligned}$$

and

$$V_1V_1^* + V_2V_2^* = VV^*,$$

it is easy to exhibit many solutions of the inverse monodromy problem. Thus, for example, if

$$\Omega_t(\lambda) = (I_m + i\lambda\varphi_1(t)V_1V_1^*J)(I_m + i\lambda\varphi_2(t)V_2V_2^*J),$$

where $\varphi_j(t)$ *are absolutely continuous nondecreasing functions on the interval* $[0, k]$, *with* $\varphi_j(0) = 0$ *and* $\varphi_j(k) = 1$ *for* $j = 1, 2$, *then* Ω_t, $0 \leq t \leq k$, *is a normalized nondecreasing* $\curvearrowright$ *continuous chain of divisors of* $U(\lambda)$ *with* $\Omega_k = U$. *The supplementary constraint*

$$\varphi_1'(t)k_1 + \varphi_2'(t)k_2 = 1 \quad \text{a.e. on } [0, k],$$

which comes from the formula

$$-i\left(\frac{\partial\Omega_t}{\partial\lambda}\right)(0)J = \int_0^t H(s)ds \quad \text{for } 0 \leq t \leq k,$$

and the normalization trace $H(s) = 1$ *a.e. on* $[0, k]$, *yields the solution*

$$H(t) = \varphi_1'(t)V_1V_1^* + \varphi_2'(t)V_2V_2^* \quad \text{a.e. on } [0, k].$$

Two such distinct solutions of the inverse monodromy problem for the canonical differential system (2.14) with monodromy matrix $U(\lambda)$ that arise this way are

$$H_1(t) = \begin{cases} \frac{1}{k_1} V_1 V_1^* & \text{for } 0 \le t < k_1 \\ \frac{1}{k_2} V_2 V_2^* & \text{for } k_2 < t \le k_1 + k_2 \end{cases}$$

and

$$H_2(t) = \frac{1}{k} V V^* \text{ for } 0 \le t \le k_1 + k_2.$$

8.6 Solutions with symplectic and/or real matrizants

In this section we shall consider the inverse monodromy problem for canonical integral and differential systems with symplectic and/or real monodromy matrices and shall focus on those solutions $M(t)$ and $H(x)$ of the inverse problem for which the matrizant is also symplectic and/or real. We shall work primarily with antisymplectic signature matrices J, i.e., with signature matrices J for which

$$J^\tau \mathcal{J}_p J = -\mathcal{J}_p. \tag{8.64}$$

It is readily seen that the signature matrices $\pm j_p$, $\pm J_p$ and $\pm \mathcal{J}_p$ are all antisymplectic. However, it is convenient to work first with $J = j_p$ and then, in the second part of the section, adapt the results obtained for $J = j_p$ to other choices of J.

Lemma 8.40 *If $W \in \mathcal{E} \cap \mathcal{U}^\circ(j_p)$, $s_{12} = T_W[0]$, $\{b_1, b_2\} \in ap(W)$, $b_1(0) = b_2(0) = I_p$, then:*

(1) $W(\lambda)$ *is symplectic* $\Longleftrightarrow \overline{W(\overline{\lambda})} = J_p W(\lambda) J_p$
$\Longleftrightarrow b_1(\lambda) = b_2(\lambda)^\tau$ *and* $s_{12}(\lambda) = s_{12}(\lambda)^\tau$.

If W is also assumed to be symplectic, then

(2) $\tau_W^+ = \tau_W^-$ *and* $\delta_{w_{11}}^- = \delta_{w_{22}}^+$.
(3) $\det W(\lambda) = 1$ *for every* $\lambda \in \mathbb{C}$.

If W is symplectic, $\widetilde{W} \in \mathcal{E} \cap \mathcal{U}_{rR}^\circ(j_p)$, $\{\widetilde{b}_1, \widetilde{b}_2\} \in ap(\widetilde{W})$, $\widetilde{b}_1(0) = \widetilde{b}_2(0) = I_p$, and $\widetilde{W}^{-1}W \in \mathcal{E} \cap \mathcal{U}(j_p)$, then

(4) $\widetilde{W}(\lambda)$ *is symplectic* $\Longleftrightarrow \widetilde{b}_1(\lambda) = \widetilde{b}_2(\lambda)^\tau$.

Proof The first equivalence in (1) is verified by direct calculation with the help of the identity $W^\#(\lambda) j_p W(\lambda) = j_p$. The second equivalences in (1) and (4) are covered by lemmas 5.8 and 5.11 in [ArD00b]. Assertion (2) follows from (1) and Theorem 3.56, whereas (3) is immediate from the definition of a symplectic mvf, since det $W^\tau(\lambda) =$ det $W(\lambda)$ and $W(0) = I_m$. □

Lemma 8.41 *If $W \in \mathcal{E} \cap \mathcal{U}^\circ(j_{pq})$, $s_{12} = T_W[0]$, $\{b_1, b_2\} \in ap(W)$, $b_1(0) = I_p$, $b_2(0) = I_q$, and if $\widetilde{W} \in \mathcal{E} \cap \mathcal{U}_{rR}^\circ(j_{pq})$, $\{\widetilde{b}_1, \widetilde{b}_2\} \in ap(\widetilde{W})$, $\widetilde{b}_1(0) = I_p$, $\widetilde{b}_2(0) = I_q$, and $\widetilde{W}^{-1}W \in \mathcal{E} \cap \mathcal{U}(j_{pq})$, then:*

(1) *$W(\lambda)$ is real if and only if $b_1(\lambda)$, $b_2(\lambda)$ and s_{12} are all real.*

If W is is also assumed to be real, then

(2) *$\widetilde{W}(\lambda)$ is real if and only if $\widetilde{b}_1(\lambda)$ and $\widetilde{b}_2(\lambda)$ are both real.*

Proof The first assertion follows from lemma 6.3 on pp. 164–166 in [ArD00b], but note that p. 165 and p. 166 have been printed in reverse order.

Next, to justify (2), let $s^\circ = s_{12}$. Then, as $s^\circ \in T_W[\mathcal{S}^{p\times q}]$ and $T_W[\mathcal{S}^{p\times q}] \subset T_{\widetilde{W}}[\mathcal{S}^{p\times q}]$, Theorem 4.13 insures that the GSIP $(\widetilde{b}_1, \widetilde{b}_2; s^\circ)$ is completely indeterminate. Thus, Theorem 4.12 guarantees that $\widetilde{W}$ is the only resolvent matrix of this problem in the class $\mathcal{E} \cap \mathcal{U}_{rR}^\circ(j_{pq})$ with $\{\widetilde{b}_1, \widetilde{b}_2\} \in ap(\widetilde{W})$. The same conclusion holds if $\widetilde{W}$, $\widetilde{b}_1$, $\widetilde{b}_2$ and s° are replaced by $\mathcal{I}\widetilde{W}$, $\mathcal{I}\widetilde{b}_1$, $\mathcal{I}\widetilde{b}_2$ and $\mathcal{I}s^\circ$, respectively. Therefore, since $s^\circ = \mathcal{I}s^\circ$, it follows that $\widetilde{W} = \mathcal{I}\widetilde{W}$ if and only if $\widetilde{b}_1 = \mathcal{I}\widetilde{b}_1$ and $\widetilde{b}_2 = \mathcal{I}\widetilde{b}_2$. □

Theorem 8.42 *If $W \in \mathcal{E} \cap \mathcal{U}_{AR}^\circ(j_{pq})$ and $\{b_1^t, b_2^t\}$, $0 \le t \le d$, is a normalized nondecreasing continuous chain of pairs of entire inner mvf's such that $\{b_1^d, b_2^d\} \in ap(W)$, then there exists exactly one canonical integral system (6.1) with a continuous nondecreasing mass function $M(t)$ on $[0, d]$ with $M(0) = 0$ and matrizant W_t, $0 \le t \le d$, such that*

$$\{b_1^t, b_2^t\} \in ap(W_t) \quad \text{for every } t \in [0, d] \quad \text{and} \quad W_d = W. \tag{8.65}$$

Moreover, if it is also assumed that:

(1) *W is symplectic, then*

$$W_t \quad \text{is symplectic} \iff b_2^t = (b_1^t)^\tau \iff \overline{M(t)} = J_pM(t)J_p.$$

(2) *W is real, then*

$$W_t \quad \text{is real} \iff b_1^t \text{ and } b_2^t \text{ are both real} \iff \overline{M(t)} = M(t).$$

(3) *W is real and symplectic, then*

$$\begin{aligned} W_t \quad \text{is real and symplectic} &\iff b_2^t = (b_1^t)^\tau \text{ and } b_1^t \text{ is real} \\ &\iff \overline{M(t)} = M(t) = J_pM(t)J_p. \end{aligned}$$

Proof This follows from Theorem 8.16, Lemmas 8.40 and 8.41 and an easy calculation based on the formula

$$W_t(\lambda) = I_m + i\lambda \int_0^t W_s(\lambda)dM(s)j_{pq} \quad \text{for } 0 \le t \le d. \tag{8.66}$$

□

Remark 8.43 *The conditions on $M(t)$ in Theorem 8.42 can be reexpressed in terms of the $p \times p$ blocks $m_{ij}(t)$ in the four-block decomposition of $M(t)$, since for any positive semidefinite matrix $M \in \mathbb{C}^{m\times m}$ with $m = 2p$,*

$$\overline{M} = J_pMJ_p \Longleftrightarrow m_{22} = m_{11}^{\tau} \quad \text{and} \quad m_{12} = m_{12}^{\tau}$$

$$\overline{M} = M = J_pMJ_p \Longleftrightarrow m_{22} = m_{11}^{\tau} = m_{11} = \overline{m_{11}} \text{ and } m_{12} = \overline{m_{12}} = m_{12}^{\tau}.$$

Theorem 8.44 *If $W \in \mathcal{E} \cap \mathcal{U}_{AR}^{\circ}(j_{pq})$, $\ell = \ell_W = -i\ \operatorname{trace}\{W'(0)j_{pq}\}$, and $\{b_1^t, b_2^t\}$, $0 \le t \le d$, is a normalized strictly increasing continuous chain of pairs of entire inner mvf's such that $\{b_1^d, b_2^d\} \in ap(W)$, then there exists exactly one normalized solution $H(x)$ of the inverse monodromy problem for the canonical differential system (2.14) with $J = j_{pq}$, monodromy matrix $W(\lambda)$ and matrizant $\Omega_x(\lambda) = \Omega(x, \lambda)$ such that*

$$\{b_1^{t(x)}, b_2^{t(x)}\} \in ap(\Omega_x) \quad \text{for every } x \in [0, \ell]. \tag{8.67}$$

for some function $t = t(x)$ on $[0, \ell]$. This function $t(x)$ is unique, continuous and strictly increasing on $[0, \ell]$ with $t(0) = 0$ and $t(\ell) = d$. Moreover, if it is also assumed that:

(1) *W is symplectic, then Ω_x is symplectic for every $x \in [0, \ell]$*

$$\Longleftrightarrow b_2^t = (b_1^t)^{\tau} \text{ for every } t \in [0, d]$$
$$\Longleftrightarrow \overline{H(x)} = J_pH(x)J_p \text{ a.e. on } [0, \ell].$$

(2) *W is real, then Ω_x is real for every $x \in [0, \ell]$*

$$\Longleftrightarrow b_1^t \text{ and } b_2^t \text{ are both real for every } t \in [0, d]$$
$$\Longleftrightarrow \overline{H(x)} = H(x) \text{ a.e. on } [0, \ell].$$

(3) *W is real and symplectic, then*

$$\Omega_x \quad \text{is real and symplectic for every } x \in [0, \ell]$$
$$\Longleftrightarrow b_2^t = (b_1^t)^{\tau} \text{ and } b_1^t \text{ is real for every } t \in [0, d]$$
$$\Longleftrightarrow \overline{H(x)} = H(x) = J_pH(x)J_p \text{ a.e. on } [0, \ell].$$

Proof This follows from Theorems 8.17 and 8.42. □

Symplectic and/or real matrizants for $J \neq j_p$

It is easily checked that a mvf $U \in \mathcal{U}(J)$ is symplectic if and only if

$$\overline{U(\overline{\lambda})} = \overline{\mathfrak{J}}\mathcal{J}_pU(\lambda)\mathcal{J}_p\mathfrak{J} \tag{8.68}$$

for every point $\lambda \in \mathfrak{h}_U \cap \mathfrak{h}_{U^\#}$. Thus, if U_t is the matrizant of the canonical integral system (2.1), then, in view of formula (2.25),

$$U_t \in Symp \text{ for every } t \in [0, d] \Longleftrightarrow \overline{M(t)} = \overline{J}\,\overline{\mathcal{J}}_p M(t) J \mathcal{J}_p \quad \text{for every } t \in [0, d]. \tag{8.69}$$

Similarly, if U_x is the matrizant of the canonical differential system (1.1), then

$$U_x \in Symp \text{ for every } x \in [0, \ell] \Longleftrightarrow \overline{H}(x) = \overline{J}\,\overline{\mathcal{J}}_p H(x) J \mathcal{J}_p \quad \text{a.e. on } [0, \ell]. \tag{8.70}$$

It is also readily checked that the matrizant of the canonical integral system (2.1) is real for every $x \in [0, d]$ if and only if the mass function $M(t)$ meets the condition

$$\overline{M(t)}\,\overline{J} = M(t)J \quad \text{for every } t \in [0, d].$$

Correspondingly, the matrizant of the canonical differential system (1.1) is real for every $x \in [0, \ell]$ if and only if the Hamiltonian $H(x)$ meets the condition

$$\overline{H(x)}\,\overline{J} = H(x)J \quad \text{a.e. on } [0, \ell].$$

Thus, if $\overline{J} = -J$, then $H(x) = 0$ a.e. on $[0, \ell]$ (since $H \geq 0$ if and only if $\overline{H} \geq 0$). This is consistent with the general fact that.

$$\text{if } U \in \mathcal{U}^\circ(J) \text{ and } \overline{J} = -J, \text{ then } U \in Real \Longleftrightarrow U(\lambda) \equiv I_m. \tag{8.71}$$

Consequently,

$$U \in \mathcal{U}^\circ(\mathcal{J}_p) \cap Real = \{I_m\}.$$

On the other hand, in view of (1.7),

$$\{U \in \mathcal{U}(\mathcal{J}_p) : \mathfrak{V}_1 U \mathfrak{V}_1^* \in Real\} = \{U \in \mathcal{U}(\mathcal{J}_p) : (\mathcal{I}U)(\lambda) = j_p U(\lambda) j_p\}.$$

Moreover:

(1) If $J = J_p$, then

$$U_x \in Symp \quad \text{for every } x \in [0, \ell] \Longleftrightarrow \overline{H(x)} = j_p H(x) j_p \quad \text{a.e. on } [0, \ell];$$

and

$$U_x \in Real \quad \text{for every } x \in [0, \ell] \Longleftrightarrow \overline{H(x)} = H(x) \quad \text{a.e. on } [0, \ell];$$

(2) If $J = \mathcal{J}_p$, then

$$U_x \in Symp \quad \text{for every } x \in [0, \ell] \Longleftrightarrow \overline{H(x)} = H(x) \quad \text{a.e. on } [0, \ell].$$

and

$$\mathfrak{V}_1 U_x \mathfrak{V}_1^* \in Real \quad \text{for every } x \in [0, \ell] \Longleftrightarrow \overline{H(x)} = j_p H(x) j_p \quad \text{a.e. on } [0, \ell].$$

Remark 8.45 *The condition $j_pH(x)j_p = H(x)$ a.e. on $[0, \ell]$ holds if and only if $H(x)$ is block diagonal.*

Lemma 8.46 *Let J be an $m \times m$ signature matrix with $m = 2p$. Then:*

(1) *J is antisymplectic if and only if there exists a unitary symplectic matrix V such that $V^*JV = j_p$.*
(2) *J is real and antisymplectic if and only if there exists a real unitary symplectic matrix V such that $V^*JV = j_p$.*
(3) *If V is a unitary symplectic matrix such that $V^*JV = j_p$, then*

$$U \in \mathcal{E} \cap \mathcal{U}^\circ(J) \cap Symp \Longleftrightarrow V^*UV \in \mathcal{E} \cap \mathcal{U}^\circ(j_p) \cap Symp.$$

(4) *If V is a real unitary symplectic matrix such that $V^*JV = j_p$, then*

$$U \in \mathcal{E} \cap \mathcal{U}^\circ(J) \cap Symp \cap Real \Longleftrightarrow V^*UV \in \mathcal{E} \cap \mathcal{U}^\circ(j_p) \cap Symp \cap Real.$$

Proof See, e.g., lemma 5.5 on pp. 156–157 in [ArD00b] for (1); the rest is straightforward. □

If $J = J_p$, then we can choose $V = i\mathfrak{V}$ (since $\mathfrak{V}^\tau \mathcal{J}_p \mathfrak{V} = -\mathcal{J}_p$).
If $J = \mathcal{J}_p$, then we can choose

$$V = \frac{1-i}{2}\begin{bmatrix} iI_p & -iI_p \\ I_p & I_p \end{bmatrix}.$$

8.7 Entire homogeneous resolvent matrices

If $J \neq \pm I_m$, then a mvf $U \in \mathcal{U}(J)$ is said to belong to the class $\mathcal{U}^H(J)$ of **homogeneous** J-inner functions if there exists a pair of scalar inner functions $\beta_1(\lambda)$ and $\beta_2(\lambda)$ such that

(1) $\beta_1^{-1}U \in \mathcal{U}(J)$ and $\beta_1 U^\# \in \mathcal{N}_+^{m\times m}$;
(2) $\beta_2 U \in \mathcal{U}(J) \cap \mathcal{N}_+^{m\times m}$.

If $U \in \mathcal{U}(J)$ and J is unitarily equivalent to j_{pq}, then (1) and (2) hold if and only if $\{\beta_1 I_p, \beta_2 I_q\} \in ap_I(U)$. If also $q = p$, then this is equivalent to the the condition $\{\beta_1 I_p, \beta_2 I_p\} \in ap_{II}(U)$. If $U \in \mathcal{E} \cap \mathcal{U}^H(J)$, then $\beta_1 = e_{a_1}$ and $\beta_2 = e_{a_2}$, with $a_1 = \tau_U^-$ and $a_2 = \tau_U^+$, i.e., (8.6) holds.

Correspondingly, a mvf $U \in \mathcal{U}(I_m)$ (resp., $U \in \mathcal{U}(-I_m)$) is said to be homogeneous if $U = \beta V$ (resp., $U = \beta^{-1}V$) for some scalar inner function $\beta(\lambda)$ and some unitary $V \in \mathbb{C}^{m\times m}$.

If $J \neq \pm I_m$, then we shall set

$$\mathcal{U}_{rR}^H(J) = \mathcal{U}_{rR}(J) \cap \mathcal{U}^H(J), \qquad \mathcal{U}_{rsR}^H(J) = \mathcal{U}_{rsR}(J) \cap \mathcal{U}^H(J),$$

$$\mathcal{U}_{AR}^H(J) = \mathcal{U}_{AR}(J) \cap \mathcal{U}^H(J) \quad \text{and} \quad \mathcal{U}_{BR}^H(J) = \mathcal{U}_{BR}(J) \cap \mathcal{U}^H(J).$$

Lemma 8.47 *Let $U_1, U_2 \in \mathcal{U}(J)$, $J \neq \pm I_m$, $U = U_1U_2$ and assume that one of the factors belongs to $\mathcal{U}^H(J)$. Then $U \in \mathcal{U}^H(J)$ if and only if the other factor also belongs to $\mathcal{U}^H(J)$. Moreover, if α_1, α_2, β_1 and β_2 are scalar inner functions such that $\{\alpha_1 I_p, \alpha_2 I_q\} \in ap_I(U_1)$ and $\{\beta_1 I_p, \beta_2 I_q\} \in ap_I(U_2)$, then $\{\alpha_1\beta_1 I_p, \alpha_2\beta_2 I_q\} \in ap_I(U)$*

Proof It suffices to consider the lemma for $J = j_{pq}$. Then, upon writing W, W_1 and W_2 in place of U, U_1 and U_2, and letting S, S_1 and S_2 denote the corresponding PG transforms, the stated results follow from (3.150) and (3.151), since the middle factors on the right-hand side are outer mvf's by (4) of Theorem 3.11 and the inequalities (3.118). □

We remark that

$$b \in \mathcal{E} \cap \mathcal{U}^H(I_p) \Longleftrightarrow b(\lambda) = e_a b(0) \quad \text{for some } a \geq 0.$$

Lemma 8.48 *If $U \in \mathcal{E} \cap \mathcal{U}(J)$ and J is unitarily equivalent to j_{pq}, then:*

(1) $U \in \mathcal{E}(a_1, a_2) \cap \mathcal{U}^H(J) \Longleftrightarrow \{e_{a_1} I_p, e_{a_2} I_q\} \in ap_I(U)$.
(2) $U \in \mathcal{E}(a_1, a_2) \cap \mathcal{U}^H(J) \Longleftrightarrow a_1 = \tau_U^- = p^{-1}\delta_{U_P}^-$ *and* $a_2 = \tau_U^+ = q^{-1}\delta_{U_Q}^+$.
(3) *If* $q = p$, *then* $U \in \mathcal{E}(a_1, a_2) \cap \mathcal{U}^H(J) \Longleftrightarrow \{e_{a_1} I_p, e_{a_2} I_q\} \in ap_{II}(U)$.

Proof Assertions (1) and (3) are immediate from the definitions of the classes $\mathcal{U}^H(J)$ and $\mathcal{E}(a_1, a_2) \cap \mathcal{U}^H(J)$ (see (5.13) for the latter) and Theorems 4.24 and 3.36. Assertion (2) then follows easily from (1) and items (5) and (6) in Theorem 5.18. □

Lemma 8.49 *If $U_t \in \mathcal{E} \cap \mathcal{U}^H(J)$, $0 \leq t < d$, is a normalized nondecreasing $\curvearrowright$ chain, J is unitarily equivalent to j_{pq} and $\{e_{\tau_1(t)} I_p, e_{\tau_2(t)} I_q\} \in ap_I(U_t)$ for $0 \leq t \leq d$, then*

$$\{e_{\tau_1(t)-\tau_1(s)} I_p, e_{\tau_2(t)-\tau_2(s)} I_q\} \in ap_I U_s^{-1} U_t. \tag{8.72}$$

Proof This follows easily from Lemmas 8.47 and 8.48. □

Since

$$\mathcal{S}(e_{a_1} I_p, e_{a_2} I_q; s^\circ) = \mathcal{S}(e_a I_p, I_q; s^\circ) \quad \text{and} \quad \mathcal{C}(e_{a_1} I_p, e_{a_2} I_p; c^\circ) = \mathcal{C}(e_a I_p, I_p; c^\circ),$$

with $a = a_1 + a_2$, it is convenient to introduce the shorter notation:

SP$(e_a; s^\circ)$ for the GSIP$(e_a I_p, I_q; s^\circ)$,
CP$(e_a; c^\circ)$ for the GCIP$(e_a I_p, I_p; c^\circ)$,
$\mathcal{S}(e_a; s^\circ)$ for $\mathcal{S}(e_a I_p, I_q; s^\circ)$, and
$\mathcal{C}(e_a; c^\circ)$ for $\mathcal{C}(e_a I_p, I_p; c^\circ)$.

Thus, if $a > 0$, $s^\circ \in \mathcal{S}^{p\times q}$ and $c^\circ \in \mathcal{C}^{p\times p}$, then

$$\mathcal{S}(e_a; s^\circ) = \{s \in \mathcal{S}^{p\times q} : e_{-a}(s - s^\circ) \in H_\infty^{p\times q}\}$$

and

$$\mathcal{C}(e_a; c^\circ) = \{c \in \mathcal{C}^{p\times p} : e_{-a}(c - c^\circ) \in \mathcal{N}_+^{p\times p}\}.$$

Theorem 8.50 *If the $SP(e_a; s^\circ)$ is completely indeterminate, then there exists exactly one mvf $W_+ \in \mathcal{E}(a; 0) \cap \mathcal{U}^\circ(j_{pq})$ such that*

$$T_{W_+}[\mathcal{S}^{p\times q}] = \mathcal{S}(e_a; s^\circ). \tag{8.73}$$

Moreover:

(1) *$W_+ \in \mathcal{U}_{rR}^H(j_{pq})$ and $\{e_a I_p, I_q\} \in ap(W_+)$.*
(2) *The set of all mvf's $W \in \mathcal{U}^\circ(j_{pq})$ such that*

$$T_W[\mathcal{S}^{p\times q}] = \mathcal{S}(e_a; s^\circ). \tag{8.74}$$

is described by the formula

$$W(\lambda) = e_{-\alpha}(\lambda)W_+(\lambda), \quad \text{where } 0 \le \alpha \le a. \tag{8.75}$$

(3) *The resolvent matrix $W = e_{-\alpha}W_+$ is uniquely defined by the conditions $W \in \mathcal{E}(a-\alpha, \alpha) \cap \mathcal{U}^\circ(j_{pq})$ and (8.74), or, equivalently, by the constraints*

$$W \in \mathcal{E} \cap \mathcal{U}^\circ(j_{pq}) \quad \text{and} \quad \{e_{a-\alpha}I_p, e_\alpha I_q\} \in ap(W) \tag{8.76}$$

and (8.74).
(4) *All the resolvent matrices W described in formula (8.75) belong to the class $\mathcal{E} \cap \mathcal{U}_{rR}^H(j_{pq})$ with $W(0) = I_m$.*
(5) *If $W \in \mathcal{E}(a_1, a_2) \cap \mathcal{U}^H(j_{pq})$, $a = a_1 + a_2 > 0$ and $s^\circ \in T_W[\mathcal{S}^{p\times q}]$, then the $SP(e_a; s^\circ)$ is completely indeterminate and $T_W[\mathcal{S}^{p\times q}] \subseteq \mathcal{S}(e_a; s^\circ)$ with equality if and only if $W \in \mathcal{U}_{rR}(j_{pq})$.*

Conversely, if $W \in \mathcal{U}_{rR}^\circ(j_{pq})$ and $\{e_{a_1}I_p, e_{a_2}I_q\} \in ap(W)$ for some choice of $a_1 \ge 0$, $a_2 \ge 0$ with $a = a_1 + a_2 > 0$ and $s^\circ \in T_W[\mathcal{S}^{p\times q}]$, then $W \in \mathcal{E}(a_1, a_2) \cap \mathcal{U}^H(j_{pq})$ and W is a resolvent matrix of the $GSIP(e_a; s^\circ)$.

Proof This follows from Theorems 4.12 and 4.13 and the observation that the constraints on β_1 and β_2 in (4.39) force $\beta_2 = 0$ and $0 \le \beta_1 \le a$, and hence the supplementary type constraint $W_+ \in \mathcal{E}(a, 0)$ forces $a - \beta_1 = a$. □

Theorem 8.51 *If the $SP(e_a; s^\circ)$ is completely indeterminate and $W_+ \in \mathcal{E}(a, 0) \cap \mathcal{U}_{rR}^\circ(j_{pq})$ is the unique resolvent matrix of this problem that was considered in Theorem 8.50, then:*

(1) *$s^\circ \in Real \Longrightarrow W_+ \in Real$.*
(2) *$(s^\circ)^\tau = s^\circ \Longrightarrow e_{-a/2}W_+ \in Symp$. Moreover, $e_{-a/2}W_+$ is the only symplectic resolvent matrix of this problem in the class $\mathcal{U}^\circ(j_{pq})$.*
(3) *$(s^\circ)^\tau = s^\circ$ and $s^\circ \in Real \Longrightarrow e_{-a/2}W_+ \in Symp \cap Real$.*

Proof If $W_+^{(1)} = \mathcal{I}W_+$, then $W_+^{(1)} \in \mathcal{E} \cap \mathcal{U}^\circ(j_{pq})$, $\mathcal{I}\mathcal{H}(W_+) = \mathcal{H}(W_+^{(1)})$ and, as follows from (2) of Theorem 4.55, $W_+^{(1)} \in \mathcal{U}_{rR}(j_{pq})$. Moreover, since $\{e_a I_p, I_p\} \in$

$ap(W_+^{(1)})$ and

$$\mathcal{S}(e_a I_p; \mathcal{I}s^\circ) = \{\mathcal{I}s : s \in \mathcal{S}(e_a I_p; s^\circ)\} = T_{W_+^{(1)}}[\mathcal{S}^{p\times q}],$$

(3) of Theorem 8.50 serves to identify $W_+^{(1)}$ with W_+ when $s^\circ \in Real$, to complete the proof of (1).

The proof of (2) is similar to the proof of (1), and (3) follows from (1) and (2). □

Lemma 8.52 *If $W \in \mathcal{U}^H(j_{pq})$, $W \notin \mathcal{U}_S(j_{pq})$ and $s \in T_W[\mathcal{S}^{p\times q}]$, then*

$$s(\lambda)^* s(\lambda) < I_q \quad \text{and} \quad s(\lambda)s(\lambda)^* < I_p \tag{8.77}$$

for every point $\lambda \in \mathbb{C}_+$.

Proof See the corollary to theorem 7.3 in [ArD97]. □

Analogous results hold for the resolvent matrices of a completely indeterminate $\mathrm{CP}(e_a; c^\circ)$.

If $A \in \mathcal{E}(a_1, a_2) \cap \mathcal{U}_{rR}^\circ(J_p)$ is a resolvent matrix of a completely indeterminate $\mathrm{CP}(e_a; c^\circ)$ and $B(\lambda) = A(\lambda)\mathfrak{V}$, then, by Lemma 8.52 and Theorem 4.21, $\mathcal{S}^{p\times p} \subset \mathcal{D}(T_B)$ and hence

$$\mathcal{C}(A) = T_B[\mathcal{S}^{p\times p}]. \tag{8.78}$$

Forumla (8.78) and Theorem 4.21 yield the following analogs of the preceding two theorems.

Theorem 8.53 *If the $\mathrm{CP}(e_a; c^\circ)$ is completely indeterminate, then there exists exactly one mvf $A_+ \in \mathcal{E}(a; 0) \cap \mathcal{U}^\circ(J_p)$ such that*

$$T_{B_+}[\mathcal{S}^{p\times p}] = \mathcal{C}(e_a; c^\circ), \quad \text{where } B_+(\lambda) = A_+(\lambda)\mathfrak{V}. \tag{8.79}$$

Moreover:

(1) *$A_+ \in \mathcal{U}_{rR}^H(J_p)$ and $\{e_a I_p, I_q\} \in ap(B_+)$.*
(2) *The set of all mvf's $A \in \mathcal{U}^\circ(J_p)$ such that*

$$T_B[\mathcal{S}^{p\times p}] = \mathcal{C}(e_a; c^\circ), \quad \text{where } B(\lambda) = A(\lambda)\mathfrak{V}, \tag{8.80}$$

is described by the formula

$$A(\lambda) = e_{-\alpha}(\lambda)A_+(\lambda), \quad \text{where } 0 \le \alpha \le a. \tag{8.81}$$

(3) *The resolvent matrix $A = e_{-\alpha}A_+$ is uniquely defined by the conditions $A \in \mathcal{E}(a-\alpha, \alpha) \cap \mathcal{U}^\circ(J_p)$ and (8.80), or, equivalently, by the constraints*

$$A \in \mathcal{E} \cap \mathcal{U}^\circ(J_p) \quad \text{and} \quad \{e_{a-\alpha}I_p, e_\alpha I_p\} \in ap(B) \tag{8.82}$$

and (8.80).

(4) *All the resolvent matrices A described in formula (8.81) belong to the classes* $\mathcal{U}_{rR}^{H}(J_p)$.
(5) *If* $A \in \mathcal{E}(a_1, a_2) \cap \mathcal{U}^H(J_p)$, $B = A\mathfrak{V}$, $a = a_1 + a_2 > 0$ *and* $c^\circ \in \mathcal{C}(A)$, *then the GCIP*$(e_a; c^\circ)$ *is completely indeterminate,* $\mathcal{S}^{p\times p} \subset \mathcal{D}(T_B)$ *and*

$$T_B[\mathcal{S}^{p\times p}] \subseteq \mathcal{C}(e_a; c^\circ) \quad \text{with equality if and only if } A \in \mathcal{U}_{rR}(J_p).$$

Conversely, if $A \in \mathcal{U}_{rR}^\circ(J_p)$, $B = A\mathfrak{V}$ *and* $\{e_{a_1}I_p, e_{a_2}I_p\} \in ap(B)$ *for some choice of* $a_1 \geq 0$, $a_2 \geq 0$ *with* $a = a_1 + a_2 > 0$ *and* $c^\circ \in T_B[\mathcal{S}^{p\times p}]$, *then* $A \in \mathcal{E}(a_1, a_2) \cap \mathcal{U}^H(J_p)$ *and A is a resolvent matrix of the GCIP*$(e_a; c^\circ)$.

Theorem 8.54 *If the GCIP*$(e_a; c^\circ)$ *is completely indeterminate and* $A_+ \in \mathcal{E}(a, 0) \cap \mathcal{U}_{rR}^\circ(J_p)$ *is the unique resolvent matrix of this problem that was considered in Theorem 8.53, then:*

(1) $c^\circ \in Real \Longrightarrow A_+ \in Real$.
(2) $(c^\circ)^\tau = c^\circ \Longrightarrow e_{-a/2}A_+ \in Symp$. *Moreover,* $e_{-a/2}A_+$ *is the only symplectic resolvent matrix of this problem in the class* $\mathcal{U}^\circ(J_p)$.
(3) $(c^\circ)^\tau = c^\circ$ *and* $c^\circ \in Real \Longrightarrow e_{-a/2}A_+ \in Symp \cap Real$.

8.8 Solutions with homogeneous matrizants

In this section we present a description of the set of all solutions to the inverse monodromy problem when the given monodromy matrix $U \in \mathcal{E} \cap \mathcal{U}_{AR}^H(J)$ and the matrizants are also homogeneous. We begin with canonical integral systems, where a solution always refers to a mass function M that meets the constraints (1.22).

Theorem 8.55 *Let* $U \in \mathcal{E}(a_1, a_2) \cap \mathcal{U}_{AR}^H(J)$ *with* $a = a_1 + a_2 > 0$ *and* $U(0) = I_m$ *and suppose that J is unitarily equivalent to* j_{pq} *and that either* $a_1 = 0$ *or* $a_2 = 0$. *Then there exists exactly one solution* $M(t)$ *of the inverse monodromy problem for the canonical integral system (2.1) with monodromy matrix* $U(\lambda)$ *and* $d = a$ *such that the corresponding matrizant* $U_t(\lambda) = U(t, \lambda)$ *enjoys the following two properties:*

$$\mathrm{type}(U_t) = t \quad \text{and} \quad U_t \in \mathcal{U}^H(J) \quad \text{for every } t \in [0, a],$$

i.e.,

$$\{e_t I_p, I_q\} \in ap_I(U_t) \quad \text{if } a_2 = 0$$

and

$$\{I_p, e_t I_q\} \in ap_I(U_t) \quad \text{if } a_1 = 0.$$

Moreover, the following estimates hold for this solution $M(t)$ when $0 \le t_1 \le t_2 \le a$:

$$p(t_2 - t_1) = \mathrm{trace}\{[M(t_2) - M(t_1)]J\} \le \mathrm{trace}\{M(t_2) - M(t_1)\} \quad \text{if } a_2 = 0. \tag{8.83}$$

$$q(t_2 - t_1) = \mathrm{trace}\{[M(t_2) - M(t_1)](-J)\} \le \mathrm{trace}\{M(t_2) - M(t_1)\} \quad \text{if } a_1 = 0. \tag{8.84}$$

Proof The first part of the theorem is an easy consequence of Theorem 8.16 upon setting $b_1^t = e_t I_p$ and $b_2^t = I_q$ if $a_2 = 0$ and $b_1^t = I_p$ and $b_2^t = e_t I_q$ if $a_1 = 0$.

Next, Theorem 2.5 yields the identities

$$pt = p\tau_{U_t}^- = \delta_{U_t}^- = \mathrm{trace}\{M(t)J\} \quad \text{if } a_2 = 0$$

and

$$qt = q\tau_{U_t}^+ = \delta_{U_t}^+ = \mathrm{trace}\{-M(t)J\} \quad \text{if } a_1 = 0.$$

This justifies the equalities in (8.83) and (8.84). The rest follows from the observation that if $A \in \mathbb{C}^{m\times m}$ and $A \ge 0$, then trace $\{AJ\} \le \mathrm{trace}\, A$. If $J = \pm I_m$, this is self-evident; if $J = V^* j_{pq} V$ for some unitary $m \times m$ matrix V, then

$$\begin{aligned}\mathrm{trace}\,\{AJ\} &= \mathrm{trace}\,\{AV^* j_{pq} V\} = \mathrm{trace}\,\{VAV^* j_{pq}\} \\ &\le \mathrm{trace}\,\{VAV^*\} = \mathrm{trace}\, A.\end{aligned}$$

□

Theorem 8.56 *Let $\widetilde{M}(t)$, $0 \le t \le d$, be any solution of the inverse monodromy problem for the canonical integral system (2.1) with monodromy matrix $U \in \mathcal{E}(a_1, a_2) \cap \mathcal{U}_{AR}^H(J)$, where $U(0) = I_m$, $J \ne \pm I_m$, $a = a_1 + a_2 > 0$, and either $a_1 = 0$ or $a_2 = 0$ and the corresponding matrizant $\widetilde{U}_t(\lambda) = \widetilde{U}(t, \lambda)$ is homogeneous for every $t \in [0, d]$. Let*

$$\tau(t) = \mathrm{type}(\widetilde{U}_t),$$

and let $M(t)$, $0 \le t \le a$, denote the particular solution considered in Theorem 8.55. Then $\tau(t)$ is a continuous nondecreasing function on the interval $[0, d]$ with $\tau(0) = 0$ and $\tau(d) = a$ and

$$\widetilde{M}(t) = M(\tau(t)) \quad \text{for every} \quad t \in [0, d]. \tag{8.85}$$

Conversely, if $\widetilde{M}$ is defined in terms of some continuous nondecreasing function $\tau(t)$ on $[0, d]$ by the preceding formulas and if $\tau(0) = 0$ and $\tau(d) = a$, then $\widetilde{M}(t)$, $0 \le t \le d$, is a solution of the considered inverse monodromy problem with homogeneous matrizant $\widetilde{U}_t$ such that $\tau(t) = \mathrm{type}(\widetilde{U}_t)$.

Proof In view of formula (2.25) the verification of (8.85) reduces to checking that $\widetilde{U}_t = U_{\tau(t)}$. Since $J \ne \pm I_m$ we can assume that $J = j_{pq}$.

If $a_2 = 0$, then

$$\widetilde{U}_t \in \mathcal{E}(\tau(t), 0) \cap \mathcal{U}^H(j_{pq}) \quad \text{and} \quad \{e_{\tau(t)} I_p, I_q\} \in ap(\widetilde{U}_t) \quad \text{for } 0 \le t \le d.$$

Therefore, $\widetilde{U}_t = U_{\tau(t)}$.

If $a_1 = 0$, then

$$\widetilde{U}_t \in \mathcal{E}(0, \tau(t)) \cap \mathcal{U}^H(j_{pq}) \quad \text{and} \quad \{I_p, e_{\tau(t)} I_q\} \in ap(\widetilde{U}_t) \quad \text{for } 0 \le t \le a_1.$$

Therefore, $\widetilde{U}_t = U_{\tau(t)}$.

The converse may be verified in much the same way. The details are left to the reader. □

Theorem 8.55 guarantees that the inverse monodromy problem for the canonical integral system (2.1) with $d = a$ and a monodromy matrix $U(\lambda)$ which belongs to the class $\mathcal{E}(a, 0) \cap \mathcal{U}^H_{AR}(J)$ (resp., $\mathcal{E}(0, a) \cap \mathcal{U}^H_{AR}(J)$) has a unique continuous solution $M(t)$, $0 \le t \le a$, when the variable t is restricted to equal the type of the matrizant U_t and the normalization condition

$$pt = \text{trace}\left\{-i\left(\frac{\partial U_t}{\partial \lambda}\right)(0)\right\} \quad \left(\text{resp., } qt = \text{trace}\left\{i\left(\frac{\partial U_t}{\partial \lambda}\right)(0)\right\}\right) \tag{8.86}$$

is imposed for every $t \in [0, a]$.

Theorem 8.57 *Let $U \in \mathcal{E}(a_1, a_2) \cap \mathcal{U}^H_{AR}(J)$ with $U(0) = I_m$, let $a = a_1 + a_2$ and suppose that $a_1 > 0$ and $a_2 > 0$. Let $M_+(t)$ (resp., $M_-(t)$), $0 \le t \le a$, denote the unique solution of the inverse monodromy problem for the canonical integral system (2.1) with monodromy matrix $U_+ = e_{a_2} U$ (resp., $U_- = e_{-a_1} U$) such that the corresponding matrizant U_t^+ belongs to the class $\mathcal{E}(t, 0) \cap \mathcal{U}^H(J)$ (resp., U_t^- belongs to the class $\mathcal{E}(0, t) \cap \mathcal{U}^H(J)$) for every $t \in [0, a]$. Then the set of all solutions $M(u)$, $0 \le u \le d$, of the inverse monodromy problem for the canonical integral system (2.1) with monodromy matrix $U(\lambda)$ and homogeneous matrizant $U_u(\lambda) = U(u, \lambda)$ is described by the formulas*

$$M(u) = M_+(\tau_1(u) + \tau_2(u)) - \tau_2(u) J \quad \textit{for} \quad u \in [0, d] \tag{8.87}$$

and

$$M(u) = M_-(\tau_1(u) + \tau_2(u)) + \tau_1(u) J \quad \textit{for} \quad u \in [0, d], \tag{8.88}$$

where $\tau_j(u)$ are arbitrary continuous nondecreasing functions on $[0, d]$ with $\tau_j(0) = 0$ and $\tau_j(d) = a_j$ for $j = 1, 2$. Moreover, the matrizant

$$U_u \in \mathcal{E}(\tau_1(u), \tau_2(u)) \cap \mathcal{U}^H_{rR}(J) \quad \textit{for } u \in [0, d].$$

Proof See theorem 4.7 in [ArD00b]. □

Remark 8.58 *Formula (8.88) also follows directly from (8.87), since*

$$M_+(t) = M_-(t) + tJ \quad \textit{for } 0 \le t \le a. \tag{8.89}$$

Theorem 8.59 *Let $U \in \mathcal{E}(a_1, a_2) \cap \mathcal{U}_{AR}^{H}(J)$ with $a = a_1 + a_2 > 0$ and $U(0) = I_m$, let $\ell = \ell_U$ and suppose that either $a_1 = 0$ or $a_2 = 0$. Then the inverse monodromy problem for the canonical differential system (2.14) with monodromy matrix $U(\lambda)$ has exactly one normalized solution $H(x)$, $0 \le x \le \ell$, for which the corresponding matrizant $U_x(\lambda)$ is homogeneous. Moreover, the exponential type*

$$t(x) = \text{type}(U_x), \quad 0 \le x \le \ell, \tag{8.90}$$

of this matrizant is an absolutely continuous strictly increasing function on the interval $[0, \ell]$ with $t(0) = 0$ and $t(\ell) = a$;

$$0 \le pt'(x) \le 1 \quad \text{a.e. on the interval } [0, \ell] \text{ if } a_2 = 0 \tag{8.91}$$

and

$$0 \le qt'(x) \le 1 \quad \text{a.e. on the interval } [0, \ell] \text{ if } a_1 = 0. \tag{8.92}$$

Proof By Theorem 8.55, there exists exactly one solution $M(t)$ of the inverse monodromy problem for the canonical integral system (2.1) with $d = a$ and monodromy matrix $U(\lambda)$ such that the corresponding matrizant $U_t(\lambda) = U(t, \lambda)$ is homogeneous and of exponential type t for every $t \in [0, a]$. The type constraint guarantees that $M(t_2) \neq M(t_1)$ for $t_1 \neq t_2$ and hence that the continuous nondecreasing function $x(t) = \text{trace}\, M(t)$ is strictly increasing on the interval $[0, a]$. Therefore, it admits an inverse $t(x)$ which is continuous and strictly increasing on the interval $[0, \ell]$, $\ell = x(a)$, and, by a standard argument that is spelled out in the proof of Lemma 2.2,

$$M(t(x)) = \int_0^x H(s)ds,$$

for some measurable $m \times m$ mvf $H(s)$ with $H(s) \ge 0$ and trace $H(s) = 1$ a.e. on the interval $[0, \ell]$. Thus, $H(x)$ is the Hamiltonian of a canonical differential system (2.14) that meets the constraints (1.19) and $t(x)$ is the type of the matrizant of this system.

Now let $\widetilde{H}(x)$ be a normalized solution of the inverse monodromy problem for the canonical differential system (2.14) with monodromy matrix $U(\lambda)$ and homogeneous matrizant $\widetilde{U}_x(\lambda) = \widetilde{U}(x, \lambda)$ and let

$$\widetilde{t}(x) = \text{type}(\widetilde{U}_x) \text{ for every } x \in [0, \ell]\,.$$

Then, by Theorem 8.56,

$$\int_0^x \widetilde{H}(s)ds = M(\widetilde{t}(x))$$

and hence

$$x = \text{trace}\, M(\widetilde{t}(x)) = \text{trace}\, M(t(x)).$$

But as

$$\text{trace}\, M(t_2) > \text{trace}\, M(t_1) \geq 0$$

for $0 \leq t_1 < t_2$, this is enough to guarantee that $\widetilde{t}(x) = t(x)$ and hence that $\widetilde{H}(x) = H(x)$ a.e., as claimed.

The inequalities (8.91) and (8.92) follow from the formula

$$\text{trace}\, M(t(x)) = \text{trace}\left\{\int_0^x H(s)ds\right\} = x$$

and the inequalities (8.83) and (8.84). □

The preceding proof relates solutions with normalized Hamiltonians to solutions with matrizants of specified type.

Lemma 8.60 *Let $U(\lambda)$ and $U_x(\lambda)$ be as in Theorem 8.59 and let $\widetilde{U} \in \mathcal{U}^H(J)$ be a J-inner divisor of $U(\lambda)$ with $\widetilde{U}(0) = I_m$. Then there exists exactly one point $x \in [0, \ell]$ such that* $\text{type}(U_x) = \text{type}(\widetilde{U})$. *Moreover, $\widetilde{U}(\lambda) = U_x(\lambda)$ at this point x.*

Proof Since $t(x) = \text{type}(U_x)$ is a strictly increasing continuous function on $[0, \ell]$ with $t(0) = 0$ and $t(\ell) = a$ and $\text{type}(\widetilde{U}) \in [0, a]$, there clearly exists exactly one point $x \in [0, \ell]$ such that $t(x) = \text{type}(\widetilde{U})$. Consequently, $\widetilde{U} \in \mathcal{E}(t(x), 0) \cap \mathcal{U}^H_{rR}(J)$ if $a_2 = 0$ and $\widetilde{U} \in \mathcal{E}(0, t(x)) \cap \mathcal{U}^H_{rR}(J)$ if $a_1 = 0$.

Now let $W(\lambda)$, $\widetilde{W}(\lambda)$ and $W_x(\lambda)$ be the j_{pq}-inner mvf's corresponding to $U(\lambda)$, $\widetilde{U}(\lambda)$ and $U_x(\lambda)$, via relation (3.140), respectively, and suppose that $a_2 = 0$. Then, since $\{e_{t(x)}I_p, I_q\} \in ap(\widetilde{W})$ and $\{e_{t(x)}I_p, I_q\} \in ap(W_x)$ and both $\widetilde{W}$ and W_x are entire right regular divisors of W with $\widetilde{W}(0) = W_x(0) = I_m$, it follows from Theorem 4.13 that $\widetilde{W}(\lambda) = W_x(\lambda)$. Therefore, $\widetilde{U}(\lambda) = U_x(\lambda)$, as claimed. The proof for $a_1 = 0$ is similar. □

Theorem 8.61 *Let $U \in \mathcal{E}(a_1, a_2) \cap \mathcal{U}^H_{AR}(J)$ with $a_1 > 0$, $a_2 > 0$, $U(0) = I_m$ and J unitarily equivalent to j_{pq}, let*

$$\ell = \ell_U = 2\pi\, \text{trace}\, K_0^U(0)$$

and let $H(x)$ be a normalized solution of the inverse monodromy problem for the canonical differential system (2.14) with monodromy matrix $U(\lambda)$ such that the matrizant $U_x(\lambda)$ is homogeneous for every $x \in [0, \ell]$. Let

$$\tau_1(x) = \tau^-_{U_x} \quad \text{and} \quad \tau_2(x) = \tau^+_{U_x} \quad \text{for } x \in [0, \ell].$$

Then:

(1) *$U_x \in \mathcal{E}(\tau_1(x), \tau_2(x)) \cap \mathcal{U}^H_{rR}(J)$ for every $x \in [0, \ell]$.*
(2) *$\tau_1(x)$ and $\tau_2(x)$ are absolutely continuous nondecreasing functions of x on the interval $[0, \ell]$.*
(3) *$\tau_1(0) = \tau_2(0) = 0$, $\tau_1(\ell) = a_1$ and $\tau_2(\ell) = a_2$.*
(4) *$\tau(x) = \tau_1(x) + \tau_2(x)$ is strictly increasing on $[0, \ell]$.*

(5) *The functions* $\tau_1(x)$ *and* $\tau_2(x)$ *are subject to the inequality*

$$p\tau_1'(x) + q\tau_2'(x) \le 1 \quad a.e. \text{ on } [0, \ell]. \tag{8.93}$$

and hence

$$pa_1 + qa_2 \le \ell. \tag{8.94}$$

Moreover, if $H_1(u)$, $0 \le u \le \ell_1$ *(resp.,* $H_2(u)$, $0 \le u \le \ell_2$*), denotes the unique normalized solution of the inverse monodromy problem for the canonical differential system with monodromy matrix* $e_{a_2}U$ *(resp.,* $e_{-a_1}U$*) and homogeneous matrizant, then*

$$H(x) = \{1 + (p-q)\tau_2'(x)\}H_1(x + (p-q)\tau_2(x)) - \tau_2'(x)J \quad a.e. \text{ on } [0, \ell], \tag{8.95}$$

$$H(x) = \{1 + (q-p)\tau_1'(x)\}H_2(x + (q-p)\tau_1(x)) + \tau_1'(x)J \quad a.e. \text{ on } [0, \ell], \tag{8.96}$$

$$\ell_1 = \ell + (p-q)a_2 \quad \text{and} \quad \ell_2 = \ell + (q-p)a_1. \tag{8.97}$$

Proof See theorem 4.9 in [ArD00b]. □

Theorem 8.62 *Let* $W \in \mathcal{E}(a_1, a_2) \cap \mathcal{U}^H(j_{pq})$ *with* $a_1 \ge 0$, $a_2 \ge 0$, $a = a_1 + a_2 > 0$ *and* $W(0) = I_m$ *and let* $s_{12} = T_W[0]$. *Then:*

(1) $W \in Symp \Longleftrightarrow p = q$, $a_1 = a_2 = a/2$ *and* $s_{12}^\tau = s_{12}$.
(2) $W \in Real \Longleftrightarrow s_{12} \in Real$.
(3) $W \in Symp \cap Real \Longleftrightarrow a_1 = a_2 = a/2$, $s_{12}^\tau = s_{12}$ *and* $s_{12} \in Real$.

Moreover, if the specified mvf W *also belongs to the class* $\mathcal{U}_{AR}(j_{pq})$ *and is the monodromy matrix of the canonical differential system (2.14) with normalized Hamiltonian* $H(x)$, $0 \le x \le \ell$, *and matrizant* $W_x \in \mathcal{E}(\tau_1(x), \tau_2(x)) \cap \mathcal{U}^H(j_{pq})$, *then* $W_x \in \mathcal{U}_{rR}(j_{pq})$ *and the following additional implications are in force:*

(a) $W \in Real \Longrightarrow W_x \in Real$ *for every* $x \in [0, \ell]$.
(b) *If* $W \in Symp$, *then there is only one canonical differential system (2.14) with* $H(x)$ *subject to (1.19) such that* $W_x \in Symp$ *for every* $x \in [0, \ell]$. *Moreover, for this system*

$$H(x) = \frac{H_1(x) + H_2(x)}{2}, \tag{8.98}$$

where $H_1(x)$ *and* $H_2(x)$ *are defined in Theorem 8.61,* $W_x \in \mathcal{E}(\tau(x), \tau(x)) \cap \mathcal{U}_{rR}^H(j_p)$ *and*

$$\tau(x)I_m = \int_0^x \{H_1(s) - H_2(s)\}j_p ds \quad \text{for } 0 \le x \le \ell. \tag{8.99}$$

(c) *If* $W \in Real \cap Symp$, *then the Hamiltonian* $H(x)$ *specified by formula (8.98) is real too.*

Proof Since

$$W \in \mathcal{E}(a_1, a_2) \cap \mathcal{U}^H(j_{pq}) \iff W \in \mathcal{U}(j_{pq}) \quad \text{and} \quad \{e_{a_1} I_p, e_{a_2} I_q\} \in ap(W)$$

the equivalence in (1) follows from Lemma 8.40. If the matrizant of the normalized solution $H(x)$, $0 \le x \le \ell$, of the inverse monodromy problem with monodromy matrix $W \in \mathcal{E}(a/2, a/2) \cap \mathcal{U}^H_{AR}(j_{pq})$ and $W(0) = I_m$ meets the condition $W_x \in \mathcal{U}^H_{rR}(j_{pq}) \cap Symp$, $0 \le x \le \ell$, then $q = p$ and formulas (8.95) and (8.96) with $\tau_1(x) = \tau_2(x) = \tau(x)/2$ imply that

$$H_1(x) - H_2(x) = \tau'(x) j_p \quad \text{for } 0 \le x \le \ell, \tag{8.100}$$

which leads easily to (8.99).

The equivalence in (2) follows from Lemma 8.41. The remaining assertions for real monodromy matrices then follow from Theorem 8.61 and Lemma 8.40; (3) follows from (1) and (2). □

8.9 Extremal solutions for $J \neq \pm I_m$

The class $\mathfrak{M}_U(\tau_1, \tau_2)$ is defined for mvf's $U \in \mathcal{E}(a_1, a_2) \cap \mathcal{U}^\circ(J)$ and a pair of continuous nondecreasing functions $\tau_j(t)$ on the interval $[0, d]$ such that $\tau_j(0) = 0$ and $\tau_j(d) = a_j$ for $j = 1, 2$ as the class of solutions $M(t)$ of the inverse monodromy problem for the canonical integral system (2.1) that satisfy (1.22) with monodromy matrix $U(\lambda)$ and matrizant $U_t(\lambda) = U(t, \lambda)$ such that

$$U_t \in \mathcal{E}(\tau_1(t), \tau_2(t)) \cap \mathcal{U}(J) \quad \text{for every} \quad t \in [0, d]. \tag{8.101}$$

Theorem 8.63 *Let $U \in \mathcal{E}(a_1, a_2) \cap \mathcal{U}^\circ_{AR}(J)$ with $a_1 \ge 0$, $a_2 \ge 0$ and $a = a_1 + a_2 > 0$. Let $\tau_j(t)$, $j = 1, 2$, be a pair of continuous nondecreasing functions on the interval $[0, d]$ such that $\tau_j(0) = 0$ and $\tau_j(d) = a_j$. Then*

$$\mathfrak{M}_U(\tau_1, \tau_2) \neq \emptyset. \tag{8.102}$$

Moreover, the following additional conclusions hold:

(1) *There exists exactly one extremal solution $M_{\tau_1,\tau_2} \in \mathfrak{M}_U(\tau_1, \tau_2)$ for which*

$$M(t) \le M_{\tau_1,\tau_2}(t) \quad \textit{for every} \quad t \in [0, d]$$

and every $M \in \mathfrak{M}_U(\tau_1, \tau_2)$.

(2) *$\mathfrak{M}_U(\tau_1, \tau_2) = \{M_{\tau_1,\tau_2}\}$ for every such choice of τ_1 and τ_2 if and only if*

$$\tau_U^- = \delta_{U_P}^- \quad \textit{and} \quad \tau_U^+ = \delta_{U_Q}^+. \tag{8.103}$$

(3) *If the monodromy matrix $U(\lambda)$ is homogeneous, i.e., if*

$$\delta_{U_P}^- = p\tau_U^- \quad \textit{and} \quad \delta_{U_Q}^+ = q\tau_U^+, \tag{8.104}$$

then $M_{\tau_1,\tau_2}(t)$ is the only solution in the class $\mathfrak{M}_U(\tau_1, \tau_2)$ for which the matrizant is also homogeneous for every $t \in [0, d]$.

(4) *If* $M(t)$, $0 \le t \le d$, *is any solution of this inverse monodromy problem for* $U(\lambda)$ *such that the corresponding matrizant* $U_t(\lambda) = U(t,\lambda)$ *meets the conditions*

$$\tau_{U_t}^- \le \tau_1(t) \text{ and } \tau_{U_t}^+ \le \tau_2(t) \text{ for every } t \in [0,d],$$

then

$$M(t) \le M_{\tau_1,\tau_2} \text{ for every } t \in [0,d].$$

Proof The proof is similar to the proof of theorem 4.4 in [ArD00b], in which the more restrictive condition $U \in \mathcal{U}_{rsR}^\circ(J)$ is imposed. □

Remark 8.64 *In view of Theorems 5.19 and 8.63, there is a one-to-one correspondence between pairs of continuous nondecreasing functions* $\tau_j(t)$ *on the interval* $[0,d]$ *with* $\tau_j(0) = 0$ *and* $\tau_j(d) = a_j$, $j = 1,2$, *and the extremal solutions* $M_{\tau_1,\tau_2} \in \mathfrak{M}_U(\tau_1,\tau_2)$ *for each* $U \in \mathcal{E}(a_1,a_2) \cap \mathcal{U}_{AR}^\circ(J)$.

Remark 8.65 *If either* $a_1 = 0$ *or* $a_2 = 0$ *in Theorem 8.63, then the family of extremal solutions* $M_{\tau_1,\tau_2}(t)$ *is effectively parametrized by only one function:* $\tau_1(t)$ *if* $a_2 = 0$ *and* $\tau_2(t)$ *if* $a_1 = 0$. *In these cases more can be said about the family of extremal solutions. In particular, suppose that* $a_1 = a$ *and* $a_2 = 0$ *and let* $\mathring{M}(t)$ *denote the extremal solution in the family* $\mathfrak{M}_U(\tau_1,\tau_2)$ *for the special case that* $\tau_1(t) = t$ *and* $\tau_2(t) = 0$ *on the interval* $[0,a]$. *Then, for any continuous nondecreasing function* $\tau(t)$ *on the interval* $[0,d]$ *with* $\tau(0) = 0$ *and* $\tau(d) = a$,

$$M_{\tau,0}(t) = \mathring{M}(\tau(t)) \quad \text{for every} \quad t \in [0,d].$$

Thus, in this instance, all extremal solutions are obtained by a change in variable in the one fixed mvf $\mathring{M}(t)$. *Similar remarks apply to the case* $a_1 = 0$.

Remark 8.66 *Suppose now that* $a_j > 0$ *for* $j = 1,2$ *in Theorem 8.63, let* $\{b_1,b_2\} \in ap(W)$, *where* $W(\lambda)$ *is defined in terms of the given monodromy matrix* $U(\lambda)$ *by (3.140) and let* $\{\mathfrak{b}_1^t;\ 0 \le t \le a_1\}$ *(resp.* $\{\mathfrak{b}_2^t;\ 0 \le t \le a_2\}$*) be the extremal type-chain of left (resp. right) inner divisors of* $b_1(\lambda)$ *(resp.* $b_2(\lambda)$*). Then for each pair of points* $(s,t) \in [0,a_1] \times [0,a_2]$, *there exists a unique mvf* $\Omega(s,t;\lambda) \in \mathcal{U}^\circ(j_{pq})$ *such that:*

(1) $\Omega(s,t;\lambda)^{-1}W \in \mathcal{U}(j_{pq})$.
(2) $\{b_1^s, b_2^t\} \in ap(\Omega(s,t;\lambda))$.

Moreover, if

$$N(s,t) = -iV\left(\frac{\partial\Omega}{\partial\lambda}(s,t;\lambda)\bigg|_{\lambda=0}\right) j_{pq}V^*,$$

then

$$M_{\tau_1,\tau_2}(t) = N(\tau_1(t),\tau_2(t)).$$

Theorem 8.67 *Let $U \in \mathcal{E}(a_1, a_2) \cap \mathcal{U}_{AR}^{\circ}(J)$ with $a = a_1 + a_2 > 0$, let $\ell = \ell_U$ and suppose that either $a_1 = 0$ or $a_2 = 0$. Then there exists exactly one normalized solution $H(x)$, $0 \leq x \leq \ell$, of the inverse monodromy problem for the canonical differential system (2.14) with monodromy matrix $U(x)$ and matrizant $U_x(\lambda) = U(x, \lambda)$ for which the following extremal property holds: If $\widetilde{H}(x)$ is any solution of this inverse monodromy problem with matrizant $\widetilde{U}_x(\lambda) = \widetilde{U}(x, \lambda)$, $0 \leq x \leq \ell$, such that*

$$\text{type}(\widetilde{U}_x) \leq \text{type}(U_x) \quad \textit{for every} \quad x \in [0, \ell],$$

then

$$\int_0^x \widetilde{H}(s)ds \leq \int_0^x H(s)ds \quad \textit{for every} \quad x \in [0, \ell].$$

Proof Let $\tau(t) = t$ for $0 \leq t \leq a$ and let

$$M(t) = \begin{cases} M_{\tau,0}(t) & \text{for} \quad t \in [0, a] \quad \textit{if} \quad a_2 = 0 \\ M_{0,\tau}(t) & \text{for} \quad t \in [0, a] \quad \textit{if} \quad a_1 = 0. \end{cases}$$

Then, by the argument used to prove Lemma 5.14, $x(t) = \text{trace}\, M(t)$ is a continuous strictly increasing function on $[0, a]$. Let $t(x)$ denote the inverse of $x(t)$ on the interval $[0, \ell]$, $\ell = x(a) = \ell_U$. Then $M(t(x))$ is absolutely continuous with respect to x on $[0, \ell]$:

$$M(t(x)) = \int_0^x H(s)ds \quad \text{for every} \quad x \in [0, \ell],$$

and its derivative $H(x)$ is a normalized solution of the inverse monodromy problem for the canonical differential system (2.14) with monodromy matrix $U(\lambda)$. Moreover, for the corresponding matrizant $U_x(\lambda) = U(x, \lambda)$ we have

$$\text{type}(U_x) = t(x) \quad \text{for every} \quad x \in [0, \ell].$$

Now let $\widetilde{H}(x)$, $0 \leq x \leq \ell$, be any solution of the inverse monodromy problem for the canonical differential system (2.14) with monodromy matrix $U(\lambda)$ and matrizant $\widetilde{U}_x(\lambda)$ such that

$$\text{type}(\widetilde{U}_x) \leq t(x) \quad \text{for every} \quad x \in [0, \ell].$$

Then

$$\widetilde{M}(x) = \int_0^x \widetilde{H}(s)ds$$

is a solution of the inverse monodromy problem for the canonical integral system (2.1) with monodromy matrix $U(\lambda)$. Therefore, by Remark 8.64,

$$\widetilde{M}(x) \leq M(t(x)) \quad \text{for every} \quad x \in [0, \ell],$$

which coincides with the claimed inequality. □

8.10 The unicellular case for $J \neq \pm I_m$

Section 8.8 focused on the case where the upper bounds $\delta^-_{U_P} = p\tau^-_{U_P}$ and $\delta^+_{U_Q} = q\tau^+_{U_Q}$ in the inequalities in (4) of Theorem 5.18 are in force for the monodromy matrix $U(\lambda)$ and are imposed on the matrizant $U_x(\lambda)$. We now turn to the other extreme: the unicellular case, where the lower bounds $\tau^-_{U_P} = \delta^-_{U_P}$ and $\tau^+_{U_Q} = \delta^+_{U_Q}$ in (4) of Theorem 5.18 are in force for a monodromy matrix $U \in \mathcal{E}(a_1, a_2) \cap \mathcal{U}^\circ_{AR}(J)$ for which either $a_1 = 0$ or $a_2 = 0$. It turns out that these constraints are automatically inherited by the matrizant $U_t(\lambda)$ for any canonical integral system (2.1) with monodromy matrix $U_d(\lambda) = U(\lambda)$:

Theorem 8.68 *Let $U \in \mathcal{E}(a_1, a_2) \cap \mathcal{U}^\circ_{AR}(J)$ with $a = a_1 + a_2 > 0$ and suppose that $U(\lambda)$ is unicellular and that either $a_1 = 0$ or $a_2 = 0$. Then there exists exactly one solution $M(t)$ of the inverse monodromy problem for the canonical integral system (2.1) with $d = a$ and monodromy matrix $U(\lambda)$ such that the corresponding matrizant $U_t(\lambda) = U(t, \lambda)$ is of exponential type t for every $t \in [0, a]$.*

In this instance, the equality

$$t = \begin{cases} \operatorname{trace}\{M(t)J\} & if\ a_2 = 0 \\ \operatorname{trace}\{-M(t)J\} & if\ a_1 = 0 \end{cases} \tag{8.105}$$

is automatically fulfilled for every $t \in [0, a]$. Moreover, if $\tau(u)$ is any continuous nondecreasing function on the interval $[0, d]$ such that $\tau(0) = 0$ and $\tau(d) = a$, then

$$\widetilde{M}(u) = M(\tau(u)), \qquad 0 \le u \le d, \tag{8.106}$$

is a solution of the inverse monodromy problem for the canonical integral system (2.1) with monodromy matrix $U(\lambda)$ and matrizant $\widetilde{U}_u(\lambda) = \widetilde{U}(u, \lambda)$ such that

$$\widetilde{U}(u, \lambda) = U(\tau(u), \lambda) \quad and \quad \text{type}\, \widetilde{U}_u = \tau(u)\ for\ every\ u \in [0, d]. \tag{8.107}$$

Conversely, every solution $\widetilde{M}(u)$, $0 \le u \le d$, of the inverse monodromy problem for the canonical integral system (2.1) with monodromy matrix $U(\lambda)$ arises in this way for some such function $\tau(u)$.

Proof The proof is similar to the proof of theorem 4.12 in [ArD00b], in which the more restrictive condition $U \in \mathcal{U}^\circ_{rsR}(J)$ is imposed. □

Theorem 8.69 *Let $U \in \mathcal{E}(a_1, a_2) \cap \mathcal{U}^\circ_{AR}(J)$ with $a = a_1 + a_2 > 0$, let $\ell = \ell_U$ and suppose that either $a_1 = 0$ or $a_2 = 0$ and that $U(\lambda)$ is unicellular. Then the inverse monodromy problem for the canonical differential system (2.14) with monodromy matrix $U(\lambda)$ has exactly one normalized solution $H(x)$, $0 \le x \le \ell$. Moreover, an $m \times m$ mvf $\widetilde{H}(x)$, $0 \le x \le \ell$, is a solution (normalized or not) of the inverse monodromy problem for canonical differential systems for the given* *monodromy matrix* $U(\lambda)$ *if and only if*

$$\widetilde{H}(x) = \tau'(x)H(\tau(x)) \quad \text{for a.e. } x \in [0, \ell] \tag{8.108}$$

for some absolutely continuous nondecreasing function $\tau(x)$ *on* $[0, \ell]$ *with* $\tau(0) = 0$ *and* $\tau(\ell) = \ell$. *The exponential type* $\widetilde{\tau}(x)$ *of the matrizant* $\widetilde{U}_x(\lambda) = \widetilde{U}(x, \lambda)$ *corresponding to* $\widetilde{H}(x)$ *is given by the formula*

$$\widetilde{\tau}(x) = \begin{cases} \text{trace}\left\{\int_0^x \widetilde{H}(s)ds J\right\} & \text{if} \quad a_2 = 0 \\ \text{trace}\left\{-\int_0^x \widetilde{H}(s)ds J\right\} & \text{if} \quad a_1 = 0 \end{cases} \tag{8.109}$$

for every $x \in [0, \ell]$; $\widetilde{\tau}(x)$ *is absolutely continuous and*

$$\widetilde{\tau}'(x) = \pm\text{trace}\{\widetilde{H}(x)J\} \quad \text{for a.e. } x \in [0, \ell], \tag{8.110}$$

with $+$ *if* $a_2 = 0$ *and* $-$ *if* $a_1 = 0$.

Proof The proof is similar to the proof of theorem 4.13 in [ArD00b], in which the more restrictive condition $U \in \mathcal{U}^{\circ}_{rsR}(J)$ is imposed. □

The case $p = q = 1$ is of special interest: If the monodromy matrix $U \in \mathcal{E}(a_1, a_2) \cap \mathcal{U}^{\circ}_{AR}(J)$ and either $a_1 = 0$ or $a_2 = 0$ and $p = q = 1$, then U is automatically unicellular. Therefore, by Theorem 8.69, there exists exactly one normalized solution $H(x)$ of the inverse monodromy problem for the canonical differential system (2.14) with monodromy matrix $U(\lambda)$ when either $U \in \mathcal{E}(a, 0) \cap \mathcal{U}^{\circ}_{AR}(J)$ or $U \in \mathcal{E}(0, a) \cap \mathcal{U}^{\circ}_{AR}(J)$.

8.11 Solutions with symmetric type

Let $U \in \mathcal{E} \cap \mathcal{U}^{\circ}_{AR}(J)$ and let $M(t)$ be a solution of the inverse monodromy problem for the canonical integral system (2.1) with monodromy matrix $U(\lambda)$, let $U_t(\lambda) = U(t, \lambda)$ denote the corresponding matrizant and let

$$\tau_1(t) = \tau^-_{U_t} \quad \text{and} \quad \tau_2(t) = \tau^+_{U_t} \quad \text{for } t \in [0, d].$$

Then $\tau_1(t)$ and $\tau_2(t)$ are continuous nondecreasing functions on $[0, d]$ and $M \in \mathfrak{M}_U(\tau_1, \tau_2)$. Thus every solution $M(t)$ of the inverse monodromy problem for canonical integral systems belongs to such a class. In general, however, the set $\mathfrak{M}_U(\tau_1, \tau_2)$ contains many solutions of the inverse monodromy problem, i.e., $\tau_1(t)$ and $\tau_2(t)$ do not uniquely specify $M(t)$. One way to obtain $M(t)$, is to pass to the $J = j_{pq}$ formalism and then to fix a normalized nondecreasing continuous chain of pairs $\{b_1^t, b_2^t\}$, $0 \le t \le d\}$, such that $\{b_1^d, b_2^d\} \in ap(W)$. The point is that the functions

$$\tau_1(t) = \text{type}(b_1^t) \quad \text{and} \quad \tau_2(t) = (b_2^t)$$

only specify the type and in general that is not enough to specify the full associated inner pair.

There are some exceptions:

$b_1(\lambda)$ and $\tau_1(t)$, $0 \le t \le d$ (resp. $b_2(\lambda)$ and $\tau_2(t)$, $0 \le t \le d$) serve to specify the continuous chain of left inner divisors $\{b_1^t;\ 0 \le t \le d\}$ of $b_1(\lambda)$ (resp. right inner divisors $\{b_2^t;\ 0 \le t \le d\}$ of $b_2(\lambda)$) up to normalization if either:

(1) $\{b_1^t\}$ (resp. $\{b_2^t\}$) is an extremal type-chain, or
(2) $b_1^t(\lambda)$ (resp. $b_2^t(\lambda)$) is homogeneous for every $t \in [0, d]$, or
(3) $b_1(\lambda)$ (resp. $b_2(\lambda)$) is unicellular.

These three categories are not mutually exclusive, since a homogeneneous chain is automatically an extremal type-chain and, in the unicellular case, the extremal type-chain is the only continuous chain of divisors of the given inner mvf.

Let $U \in \mathcal{E}(a, a) \cap \mathcal{U}_{AR}^{\circ}(J)$ with $a > 0$ and let $\tau(t)$ be a continuous nondecreasing function on the interval $[0, d]$ with $\tau(0) = 0$ and $\tau(d) = a$. Then, by Theorem 8.63, the class $\mathfrak{M}_U(\tau, \tau)$ contains exactly one extremal element $M_{\tau,\tau}(t)$ (and hence is nonempty). Moreover, if $\tau_U^- = \delta_{U_P}^-$ and $\tau_U^+ = \delta_{U_Q}^+$, then $M_{\tau,\tau}(t)$ is the only element in $\mathfrak{M}_U(\tau, \tau)$, whereas, if $U(\lambda)$ is homogeneous, then $M_{\tau,\tau}(t)$ is the only element in $\mathfrak{M}_U(\tau, \tau)$ for which the corresponding matrizant is homogeneous. Now, let $U_t(\lambda) = U(t, \lambda)$, $0 \le t \le d$, denote the matrizant corresponding to $M_{\tau,\tau}(t)$, and let $\mathring{M}(t)$, $0 \le t \le a$, denote the extremal solution of the inverse monodromy problem for the canonical integral system (2.1) with $d = a$ and monodromy matrix $U(\lambda)$ such that the corresponding matrizant $\Omega_t(\lambda) = \Omega(t, \lambda)$ belongs to the class $\mathcal{E}(t, t) \cap \mathcal{U}(J)$ for every $t \in [0, a]$. Then

$$U_t(\lambda) = \Omega_{\tau(t)}(\lambda) \quad \text{for every} \quad t \in [0, d] \tag{8.111}$$

and consequently,

$$M_{\tau,\tau}(t) = \mathring{M}(\tau(t)) \quad \text{for every} \quad t \in [0, d]. \tag{8.112}$$

The next step is to invoke Theorem 8.44 for the special case of extremal type-chains $\{\mathfrak{b}_1^t, \mathfrak{b}_2^t\}$ and Lemma 8.46 in order to conclude that:

(1) If J is antisymplectic, then

$$U \in Symp \Longleftrightarrow \Omega_t \in Symp \quad \text{for every} \quad t \in [0, a]. \tag{8.113}$$

(2) If J is real, then

$$U \in Real \Longleftrightarrow \Omega_t \in Real \quad \text{for every} \quad t \in [0, a]. \tag{8.114}$$

(3) If J is real and antisymplectic, then

$$U \in Real \cap Symp \Longleftrightarrow \Omega_t \in Real \cap Symp \quad \text{for every} \quad t \in [0, a]. \tag{8.115}$$

Theorem 8.70 *Let $U \in \mathcal{E}(a_1, a_2) \cap \mathcal{U}_{AR}^{\circ}(J)$ with $a_1 \geq 0$, $a_2 \geq 0$ and $a_1 + a_2 > 0$. Let $\tau_j(t)$, $j = 1, 2$, be continuous nondecreasing functions on $[0, d]$ with $\tau_j(0) = 0$ and $\tau_j(d) = a_j$. Let $U_t(\lambda) = U(t, \lambda)$ denote the matrizant corresponding to the extremal solution $M_{\tau_1,\tau_2}(t)$ of the inverse monodromy problem for the canonical integral system (2.1) with monodromy matrix $U(\lambda)$ that is considered in Theorem 8.63. Then*

$$U_t \in \mathcal{E}(\tau_1(t), \tau_2(t)) \cap \mathcal{U}(J)$$

and the following supplementary conclusions hold:

(1) *If J is antisymplectic, then:*
 (a) $U_t \in Symp$ *for every* $t \in [0, d]$

$$\Longleftrightarrow U \in Symp \quad \text{and} \quad \tau_1(t) = \tau_2(t) \quad \text{for every } t \in [0, d].$$

 (b) $U_t \in \mathcal{U}^H(J) \cap Symp$ *for every* $t \in [0, d]$

$$\Longleftrightarrow U \in \mathcal{U}^H(J) \cap Symp \quad \text{and} \quad \tau_1(t) = \tau_2(t) \quad \text{for every } t \in [0, d].$$

(2) *If J is real, then:*
 (a) $U_t \in Real$ *for every* $t \in [0, d] \Longleftrightarrow U \in Real$.
 (b) $U_t \in \mathcal{U}^H(J) \cap Real$ *for every* $t \in [0, d]$
 $\Longleftrightarrow U \in \mathcal{U}^H(J) \cap Real$.
(3) *If J is real and antisymplectic, then:*
 (a) $U_t \in Real \cap Symp$ *for every* $t \in [0, d]$

$$\Longleftrightarrow U \in Real \cap Symp \quad \text{and} \quad \tau_1(t) = \tau_2(t) \quad \text{for every } t \in [0, d].$$

 (b) $U_t \in Real \cap Symp \cap \mathcal{U}^H(J)$ *for every* $t \in [0, d]$

$$\Longleftrightarrow U \in Real \cap Symp \cap \mathcal{U}^H(J) \quad \text{and} \quad \tau_1(t) = \tau_2(t) \qquad \text{for every } t \in [0, d].$$

Proof See theorem 7.2 in [ArD00b]; the proof still works even though the condition $U \in \mathcal{U}_{rsR}^{\circ}(J)$ has been relaxed to $U \in \mathcal{U}_{AR}^{\circ}(J)$. □

Theorem 8.71 *Let $U \in \mathcal{E}(a, a) \cap \mathcal{U}_{AR}^{\circ}(J)$ with $a > 0$ and let $\ell = \ell_U$. Then there exists exactly one normalized solution $H(x)$ of the inverse monodromy problem for the canonical differential system (2.14) with monodromy matrix $U(\lambda)$ and matrizant $U_x(\lambda) = U(x, \lambda)$ such that the following two conditions hold:*

(1) $\tau_{U_x}^- = \tau_{U_x}^+$ *for every* $x \in [0, \ell]$.
(2) *$H(x)$ is extremal in the sense that if $\widetilde{H}(x)$ is any other solution of this monodromy problem with matrizant $\widetilde{U}_x(\lambda) = \widetilde{U}(x, \lambda)$ such that*

$$\tau_{\widetilde{U}_x}^- \leq \tau_{U_x}^- \quad \text{and} \quad \tau_{\widetilde{U}_x}^+ \leq \tau_{U_x}^+ \text{ for every } x \in [0, \ell]$$

then

$$\int_0^x \widetilde{H}(s)ds \leq \int_0^x H(s)ds \;\; for\ every \;\; x \in [0, \ell].$$

Moreover:

(1) *If J is antisymplectic, then*

$$U_x \in Symp \;\; for\ every \;\; x \in [0, \ell] \Longleftrightarrow U \in Symp.$$

(2) *If $J = \overline{J}$, then*

$$U_x \in Real \;\; for\ every \;\; x \in [0, \ell] \Longleftrightarrow U \in Real.$$

(3) *If $J = \overline{J}$ and J is antisymplectic, then*

$$U_x \in Real \cap Symp \Longleftrightarrow U \in Real \cap Symp.$$

(4) $U_x \in \mathcal{U}^H(J)$ *for every* $x \in [0, \ell] \Longleftrightarrow U \in \mathcal{U}^H(J)$.
(5) $U_x(\lambda)$ *is unicellular for every* $x \in [0, \ell] \Longleftrightarrow U(\lambda)$ *is unicellular.*

Proof See theorem 7.3 in [ArD03a]; the proof still works even though the condition $U \in \mathcal{U}_{rsR}^\circ(J)$ has been relaxed to $U \in \mathcal{U}_{AR}^\circ(J)$. □

Theorem 8.72 *Let $U \in \mathcal{E}(a, a) \cap \mathcal{U}_{AR}^\circ(J)$ with $a > 0$. Then there exists exactly one normalized solution $H(x)$ of the inverse monodromy problem for the canonical differential system (2.14) with monodromy matrix $U(\lambda)$ such that for the corresponding matrizant $U_x(\lambda) = U(x, \lambda)$ we have*

$$\tau_{U_x}^- = \tau_{U_x}^+ \quad for\ every \;\; x \in [0, \ell] \tag{8.116}$$

if and only if

$$\tau_U^- = \delta_{U_P}^- \quad and \quad \tau_U^+ = \delta_{U_Q}^+. \tag{8.117}$$

Proof See theorem 7.4 in [ArD00b]; the proof still works even though the condition $U \in \mathcal{U}_{rsR}^\circ(J)$ has been relaxed to $U \in \mathcal{U}_{AR}^\circ(J)$. □

Theorem 8.73 *Let $U \in \mathcal{E}(a, a) \cap \mathcal{U}_{AR}^\circ(J) \cap Symp$ with $a > 0$. Let $\ell = \ell_U$ and suppose that J is antisymplectic and (8.117) is in force. Then there exists one and only one normalized solution $H(x)$ of the inverse monodromy problem for the canonical differential system (2.14) with monodromy matrix $U(\lambda)$ such that*

$$\overline{H(x)} = \overline{J}\mathcal{J}_p H(x)\mathcal{J}_p\overline{J} \;\; for\ a.e. \;\; x \in [0, \ell]. \tag{8.118}$$

Proof This theorem follows from Theorems 8.72 and 8.44, Lemma 8.46 and the equivalence (8.69). □

Corollary 8.74 *Let $U \in \mathcal{E} \cap \mathcal{U}_{AR}^\circ(\mathcal{J}_1) \cap Symp$ and let $\ell = \ell_U$. Then there exists one and only one normalized solution $H(x)$ of the inverse monodromy problem* *for the canonical differential system (2.14) with* $J = \mathcal{J}_1$ *and monodromy matrix* $U(\lambda)$ *such that*

$$\overline{H(x)} = H(x) \quad \textit{for a.e. } x \in [0, \ell]. \tag{8.119}$$

Proof This is immediate from Theorem 8.73, since $\mathcal{J}_p$ is antisymplectic and (8.117) holds automatically if $p = q = 1$. □

We remark that the corollary also remains valid if $\mathcal{J}_1$ is replaced by J_1 (resp. j_1) providing that (7.13) is replaced by the condition

$$\overline{H(x)} = j_1 H(x) j_1 \quad (\text{resp. } \overline{H(x)} = J_1 H(x) J_1) \quad \text{for a.e. } x \in [0, \ell]. \tag{8.120}$$

Remark 8.75 *Theorem 8.3 (due to de Branges) yields the same conclusions as Corollary 8.74 under the less restrictive conditions on* U:

$U \in \mathcal{E} \cap \mathcal{U}^\circ(\mathcal{J}_1) \cap Symp.$

8.12 The inverse monodromy problem for 2 × 2 differential systems

The first part of this section reviews a number of results on the inverse monodromy problem for 2 × 2 canonical differential systems (1.1), that are mostly adapted from [ArD01a], where the missing proofs may be found. Many of the proofs furnished there depend essentially on the de Branges uniqueness theorem, Theorem 8.3. The main point of this collection of results is that because of de Brange's theorem, the assumption that the monodromy matrix $U \in \mathcal{U}^\circ_{AR}(J)$, which is in force for most of the earlier conclusions in this chapter, can be replaced by the weaker assumption that $U \in \mathcal{U}^\circ(J)$ when $m = 2$. It is also useful to recall that

$$\text{If } A \in \mathbb{C}^{2\times 2}, \text{ then } A \text{ is symplectic if and only if } \det A = 1. \tag{8.121}$$

Theorem 8.76 *Let* $U \in \mathcal{E}(a, a) \cap \mathcal{U}^\circ(J)$ *with* $J \neq \pm I_2$. *Then there is exactly one normalized solution* $H(x)$ *of the inverse monodromy problem for the canonical differential system (1.1) with monodromy matrix* $U(\lambda)$ *such that the corresponding matrizant* $U_x(\lambda)$ *meets the condition*

$$\tau^+_{U_x} = \tau^-_{U_x} \tag{8.122}$$

for every $x \in [0, d]$, *where* $d = -i\mathrm{trace}\,\{U'(0)J\}$. *In this instance,*

$$\tau^+_{U_x} = \tau^-_{U_x} = \tau(x) \stackrel{\text{def}}{=} \int_0^x \sqrt{\det H(s)}ds \tag{8.123}$$

and

$$\tau'(x) \leq \frac{1}{2} \tag{8.124}$$

for every $x \in [0, d]$.

Remark 8.77 *In the setting of Theorem 8.76, for each $\rho > 0$ there exists exactly one solution $H^{(\rho)}(t)$ of the inverse monodromy problem for 2×2 canonical differential systems with monodromy matrix U, matrizant $U_t^{(\rho)} \in \mathcal{E}(\rho t, \rho t) \cap \mathcal{U}^\circ(J_1)$ and* $\det H(t) = \rho^2$ *a.e. on $[0, a]$ if and only if $H(x) > 0$ a.e. on $[0, d]$: $U_t^{(\rho)}(\lambda) = U_{\varphi(t)}(\rho\lambda)$ and $H^{(\rho)}(t) = \rho\varphi'(t)H(\varphi(t))$ a.e., where $\varphi(\tau(x)) = x$ and $\tau(\varphi(t)) = t$.*

Remark 8.78 *If $J = \mathcal{J}_1$, then, in view of (4) of Lemma 8.40 and (8.70),*

$$\text{(8.122) holds} \iff \overline{H(x)} = H(x) \text{ a.e. on } [0, d]$$

$$\iff U_x \in Symp \text{ for every } x \in [0, d].$$

Thus, Theorem 8.76 implies de Branges' theorem. However, it was obtained in [ArD01a] on the basis of de Branges' theorem.

Theorem 8.79 *If $A \in \mathcal{E} \cap \mathcal{U}^\circ(J_1) \cap Real \cap Symp$, then:*

(1) *$A(\lambda)$ is the monodromy matrix of exactly one canonical differential system (1.1) with $J = J_1$ and a normalized Hamiltonian*

$$H(x) = \begin{bmatrix} h_1(x) & 0 \\ 0 & h_2(x) \end{bmatrix} = \overline{H(x)} \geq 0 \quad \text{a.e. on } [0, d] \tag{8.125}$$

(i.e., with a real symplectic matrizant). Moreover, $d = \text{trace}\,\{-i(A'(0)J_1\}$.

If, in addition,

$$a'_{21}(0) \neq 0, \quad \text{i.e., if } \int_0^d h_2(x)dx > 0, \tag{8.126}$$

then:

(2) *The set of input impedances $C^d_{\text{imp}}(Hdx)$ of the system is equal to*

$$\left\{ \frac{a_{11}(\lambda)\omega(\lambda) + a_{12}(\lambda)}{a_{21}(\lambda)\omega(\lambda) + a_{22}(\lambda)} : \omega \in \mathcal{C} \cup \{\infty\} \right\} = \mathcal{C}(A). \tag{8.127}$$

(3) *$c \in \mathcal{C}(A) \cap Real \iff \omega \in \mathcal{C} \cap Real$ or $\omega = \infty$ in (8.127).*
(4) *$c \in \mathcal{C}(A) \cap Real \iff$ its spectral function σ_c is odd, i.e., $\sigma_c(-\mu) = -\sigma_c(\mu)$.*
(5) *$\sigma_c(-\mu) = -\sigma_c(\mu)$ implies that the function $\tau(\mu)$ defined by (7.72) is a spectral function of the S_1-string that corresponds to the system under consideration as in Section 2.6 and Remark 7.18. Moreover, every spectral function $\tau(\mu)$ of this string with $\tau(\mu) = 0$ for $\mu < 0$ may be obtained from an odd spectral function $\sigma \in \Sigma^d_{\text{sf}}(Hdx)$.*

If $A \in \mathcal{E} \cap \mathcal{U}^\circ_{\text{AR}}(J_1)$ and $A(\lambda) \not\equiv I_2$, then the condition (8.126) is satisfied automatically and, in addition to (1)–(5), the following properties hold:

(6) $\Sigma_{sf}^{d}(Hdx) = (\mathcal{C}(A))_{sf}$.

(7) *If* $c \in \mathcal{C}(A)$, *then* $\beta = \lim_{\nu\uparrow\infty} \nu^{-1}c(i\nu) = 0$, $\alpha = (c(i) - \overline{c(i)})/(2i)$ *is uniquely determined by* σ.

(8) *The mass distribution function* $m(x)$ *of the* S_1*-string corresponding to the given canonical system is continuous at the end points* 0 *and* d *and meets the constraints*

$$m(x) > 0 \quad \text{and } m(d) - m(x) > 0 \text{ for every } x \in (0, d).$$

Proof Assertion (1) follows from Theorem 8.3; (2) follows from Theorems 7.29 and 7.28; (3) holds because $A \in Real$, whereas (4) follows from (3.65). Assertion (5) follows from (2), (4), formula (2.62) and the description of the set of spectral functions of an S_1-string given in [KaKr74b].

Next, (6) and (7) follow from Theorems 7.29 and 7.28.

Finally, (8) follows from the assumption that $A \in \mathcal{U}_{AR}(J_1)$, which guarantees that $A(\lambda)$ has no singular nonconstant left or right divisors in the class $\mathcal{U}(J_1)$. □

The case $m = 2, U \in \mathcal{E}(a, 0) \cap \mathcal{U}^{\circ}(J), J \neq I_2$ and $a \geq 0$

Theorem 8.80 *If* $m = 2$, $U \in \mathcal{E}(a, 0) \cap \mathcal{U}^{\circ}(J)$ *and* $J \neq I_2$, *then there exists exactly one normalized solution* $H(x)$ *of the inverse monodromy problem for the canonical differential system (1.1) with monodromy matrix* $U(\lambda)$. *Moreover, if* $U_x(\lambda) = U(x, \lambda)$, $0 \leq x \leq d$, *denotes the corresponding matrizant and* $\tau(x) = \text{type}(U_x)$, *then:*

(1) $U_x \in \mathcal{E}(\tau(x), 0) \cap \mathcal{U}(J)$ *for every* $x \in [0, d]$.

(2) $\tau(x) = \int_0^x \text{trace}\{H(s)J\}ds$ *for every* $x \in [0, d]$.

(3) $\text{trace}\{H(x)J\} \geq 0$ *for a.e.* $x \in [0, d]$.

(4) $\det H(x) = 0$ *for a.e.* $x \in [0, d]$.

Theorem 8.81 *If* $m = 2$, $U \in \mathcal{E} \cap \mathcal{U}^{\circ}(J)$ *with* $J \neq \pm I_2$ *and* $H(x)$ *is a normalized solution of the inverse monodromy problem for the canonical differential system (1.1) with monodromy matrix* $U(\lambda)$ *such that*

$$\text{trace}\{H(x)J\} \geq 0 \quad \text{and} \quad \det H(x) = 0 \quad \text{for a.e.} \quad x \in [0, d],$$

then $U \in \mathcal{E}(a, 0)$, *where*

$$a = \text{type}(U) = \int_0^d \text{trace}\{H(s)J\}ds \quad \text{and} \quad d = -i\text{trace}\,\{U'(0)J\}.$$

The case $m = 2, U \in \mathcal{E}(0, a) \cap \mathcal{U}^{\circ}(J), a \geq 0$

Theorem 8.82 *If* $m = 2$ *and* $U \in \mathcal{E}(0, a) \cap \mathcal{U}^{\circ}(J)$, *then there exists exactly one normalized solution* $H(x)$ *of the inverse monodromy problem for the canonical differential system (1.1) with monodromy matrix* $U(\lambda)$. *Moreover, if* $U_x(\lambda) =$

$U(x,\lambda)$, $0 \le x \le d$, denotes the corresponding matrizant and $\tau(x) = \text{type}(U_x)$, *then:*

1. $U_x \in \mathcal{E}(0, \tau(x)) \cap \mathcal{U}(J)$ *for every* $x \in [0, d]$.
2. $\tau(x) = -\int_0^x \text{trace}\{H(s)J\}ds$ *for every* $x \in [0, d]$.
3. $\text{trace}\{H(x)J\} \le 0$ *for a.e.* $x \in [0, d]$.
4. $\det H(x) = 0$ *for a.e.* $x \in [0, d]$.

Theorem 8.83 *If $m = 2$, $U \in \mathcal{E} \cap \mathcal{U}^\circ(J)$ and $H(x)$ is a normalized solution of the inverse monodromy problem for the canonical differential system (1.1) with monodromy matrix $U(\lambda)$ such that*

$$\text{trace}\{H(x)J\} \le 0 \quad \textit{and} \quad \det H(x) = 0 \quad \textit{for a.e.} \quad x \in [0, d],$$

then $U \in \mathcal{E}(0, a)$, where

$$a = \text{type}(U) = -\int_0^a \text{trace}\{H(s)J\}ds.$$

The case $m = 2$, $U \in \mathcal{E}(a_1, a_2) \cap \mathcal{U}^\circ(J)$, $a_1 \neq a_2$, $a_1 a_2 > 0$

Lemma 8.84 *If $U \in \mathcal{E}(a_1, a_2) \cap \mathcal{U}^\circ(J)$ and $m = 2$, then*

$$\text{type}\,\{U\} = \text{type}\,\{\det U\} \tag{8.128}$$

if and only if $a_1 a_2 = 0$, i.e., if and only if either $a_1 = 0$ or $a_2 = 0$.

Theorem 8.85 *If $U \in \mathcal{E} \cap \mathcal{U}^\circ(J)$ and $m = 2$, then a normalized solution of the inverse monodromy problem for the canonical differential system (1.1) with monodromy matrix $U(\lambda)$ is unique if and only if the condition (8.128) holds.*

Remark 8.86 *If $J = I_2$ (resp., $-I_2$), then $a_2 \le 0$ (or $a_1 \le 0$) and condition (8.128) is the Brodskii–Kiselevskii condition. Notice that in contrast with Theorem 8.38 which is formulated for $m \ge 2$, the monodromy matrix is not assumed to belong to $\mathcal{U}_{AR}(J)$ in Theorem 8.85.*

Theorem 8.87 *Let $U \in \mathcal{E}(a_1, a_2) \cap \mathcal{U}^\circ(J)$, $m = 2$ and $a_1 a_2 > 0$. Let $H(x)$, $0 \le x \le d$, be a normalized solution of the inverse monodromy problem for the canonical differential system (1.1) with monodromy matrix $U(\lambda)$. Let $U_x(\lambda) = U(x, \lambda)$ denote the corresponding matrizant and let*

$$\tau_1(x) = \tau_{U_x}^-, \quad \tau_2(x) = \tau_{U_x}^+ \quad \textit{and} \quad \tau(x) = \tau_1(x) + \tau_2(x) \textit{ for } 0 \le x \le d.$$

Then:

1. *$\tau_1(x)$ and $\tau_2(x)$ are absolutely continuous nondecreasing functions of x on the interval $[0, d]$ with $\tau_j(0) = 0$ and $\tau_j(d) = a_j$ for $j = 1, 2$.*

(2) *The function* $\tau(x)$ *is uniquely determined by the monodromy matrix* U. *Moreover, it is an absolutely continuous nondecreasing function on* $[0, d]$ *with* $\tau(0) = 0$ *and* $\tau(d) = a_1 + a_2$.

(3) *For each choice of* $s \in [0, 1]$, *the mvf*

$$U(x, \lambda : s) = e^{i\lambda(s\tau_2(x)-(1-s)\tau_1(x))}U(x, \lambda)$$

is the matrizant of the canonical differential system (1.1) with normalized Hamiltonian

$$H(x) + (s\tau_2'(x) - (1 - s)\tau_1'(x))J$$

and monodromy matrix

$$U(d, \lambda; s) = e^{i\lambda(sa_2-(1-s)a_1)}U(x, \lambda).$$

Moreover,

$$U(x, \lambda; s) \in \mathcal{E}(s\tau(x), (1 - s)\tau(x)) \cap \mathcal{U}^\circ(J).$$

(4) $\tau_1(x) - \tau_2(x) = \int_0^x \mathrm{trace}\{H(s)J\}ds \quad$ *for* $\quad 0 \le x \le d$.

Theorem 8.88 *Let* $U \in \mathcal{E}(a_1, a_2) \cap \mathcal{U}^\circ(J)$ *with* $m = 2$ *and* $a_1a_2 > 0$. *Let*

$$U_+(\lambda) = e^{i\lambda a_2}U(\lambda) \quad \text{and} \quad U_-(\lambda) = e^{-i\lambda a_1}U(\lambda)\,.$$

Then $U_+ \in \mathcal{E}(a, 0) \cap \mathcal{U}^\circ(J)$, $U_- \in \mathcal{E}(0, a) \cap \mathcal{U}^\circ(J)$, *where* $a = a_1 + a_2$.

Let $H_+(x)$ *and* $H_-(x)$ *be the unique normalized solutions of the inverse monodromy problem for canonical differential systems of the form (1.1) with monodromy matrices* $U_+(\lambda)$ *and* $U_-(\lambda)$, *respectively. Let* $U_+(x, \lambda)$ *and* $U_-(x, \lambda)$ *denote the matrizants of these systems with* $d = \ell_U$. *Then:*

(1) $U_+(x, \cdot) \in \mathcal{E}(\tau(x), 0) \cap \mathcal{U}(J)$ *and* $U_-(x, \cdot) \in \mathcal{E}(0, \tau(x)) \cap \mathcal{U}(J)$ *for some absolutely continuous nondecreasing function* $\tau(x)$ *on the interval* $[0, d]$ *with* $\tau(0) = 0$ *and* $\tau(d) = a_1 + a_2$.

(2) $U_+(x, \lambda) = e^{i\lambda\tau(x)}U_-(x, \lambda) \quad$ *for* $\quad 0 \le x \le d$.

(3) $H_+(x) = H_-(x) + \tau'(x)J \quad$ *for a.e.* $\quad x \in [0, d]$.

Let $\tau_1(x)$ *be an absolutely continuous function of* x *on the interval* $[0, d]$ *such that*

$$0 \le \tau_1'(x) \le \tau'(x) \quad \text{a.e. on} \quad [0, d], \qquad \tau_1(0) = 0 \quad \text{and} \quad \tau_1(d) = a_1.$$

Let $\tau_2(x) = \tau(x) - \tau_1(x)$. *Then the formula*

$$H(x) = H_+(x) - \tau_2'(x)J = H_-(x) + \tau_1'(x)J$$

describes the set of all normalized solutions of the inverse monodromy problem for the system (1.1) with monodromy matrix $U(\lambda)$ *and matrizant* $U_x \in \mathcal{E}(\tau_1(x), \tau_2(x))$.

If $\tau_1(x) = \tau_2(x) = \tau(x)/2$, *then*

$$H_+(x) - \frac{1}{2}\tau'(x)J = H_-(x) + \frac{1}{2}\tau'(x)J$$

is the unique normalized solution of the inverse monodromy problem for the monodromy matrix

$$e^{i\lambda(a_2-a_1)/2}U(\lambda)$$

and the corresponding matrizant

$$e^{i\lambda\tau(x)/2}U_-(x,\lambda) \quad \textit{belongs to } \mathcal{E}\left(\frac{\tau(x)}{2}, \frac{\tau(x)}{2}\right).$$

8.13 Examples of 2×2 Hamiltonians with constant determinant

Let

$$A'(x,\lambda) = i\lambda A(x,\lambda)H(x)J_1 \quad \text{for } 0 \le x < \infty,$$

where

$$A(x,\lambda) = \begin{bmatrix} a_{11}(x,\lambda) & a_{12}(x,\lambda) \\ a_{21}(x,\lambda) & a_{22}(x,\lambda) \end{bmatrix} \quad \text{and} \quad H(x) = \begin{bmatrix} h_1(x) & 0 \\ 0 & h_2(x) \end{bmatrix}$$

is a real diagonal locally absolutely continuous mvf on $\mathbb{R}_+$ with $\det H(x) = \rho^2$ on $\mathbb{R}_+$. Then

$$A_x \in \mathcal{E} \cap \mathcal{U}^\circ(J_1) \cap \mathit{Real} \cap \mathit{Symp}$$

(the last conclusion follows from (8.121) and the fact that

$$\det A_x(\lambda) = \exp\{i\lambda \text{trace}\,(H(x)J_1)\} = 1)$$

and, in view of formula (8.123),

$$\text{type}\, A_x = \int_0^x \sqrt{h_1(s)h_2(s)}ds = \rho x.$$

Moreover, since the columns of $A_x(\lambda)$ are solutions of the differential equations

$$\begin{bmatrix} a_{11} \\ a_{21} \end{bmatrix}' = -i\lambda h_2 \begin{bmatrix} a_{12} \\ a_{22} \end{bmatrix} \quad \text{and} \quad \begin{bmatrix} a_{12} \\ a_{22} \end{bmatrix}' = -i\lambda h_1 \begin{bmatrix} a_{11} \\ a_{21} \end{bmatrix},$$

it is readily seen that, at least formally,

$$h_1^{-1}\left(h_2^{-1}\begin{bmatrix} a_{11} \\ a_{21} \end{bmatrix}'\right)' = -\lambda^2 \begin{bmatrix} a_{11} \\ a_{21} \end{bmatrix} \tag{8.129}$$

and

$$h_2^{-1}\left(h_1^{-1}\begin{bmatrix} a_{12} \\ a_{22} \end{bmatrix}'\right)' = -\lambda^2 \begin{bmatrix} a_{12} \\ a_{22} \end{bmatrix}', \tag{8.130}$$

or, equivalently, that

$$0 = \begin{bmatrix} a_{11} \\ a_{21} \end{bmatrix}'' - \left(\frac{h_2'}{h_2}\right) \begin{bmatrix} a_{11} \\ a_{21} \end{bmatrix}' + \lambda^2 h_1 h_2 \begin{bmatrix} a_{11} \\ a_{21} \end{bmatrix}$$

$$= \begin{bmatrix} a_{12} \\ a_{22} \end{bmatrix}'' - \left(\frac{h_1'}{h_1}\right) \begin{bmatrix} a_{12} \\ a_{22} \end{bmatrix}' + \lambda^2 h_1 h_2 . \begin{bmatrix} a_{12} \\ a_{22} \end{bmatrix}.$$

We shall focus attention on the case

$$h_1(x) = d_1(x+\delta)^r \quad \text{and} \quad h_2(x) = \rho^2 h_1(x)^{-1},$$

where $d_1 > 0$, $\rho > 0$, $\delta > 0$ and $r \in \mathbb{R}$ are constant parameters ($\delta = 0$ is discussed in Section 11.6). Then

$$\frac{h_1'(x)}{h_1(x)} = \frac{r}{x+\delta} \quad \text{and} \quad \frac{h_2'(x)}{h_2(x)} = -\frac{r}{x+\delta}$$

and hence a_{12} and a_{22} are solutions of the Bessel equation

$$(x+\delta)^2 y''(x,\lambda) + (1-2t)(x+\delta)y'(x,\lambda) + \rho^2\lambda^2(x+\delta)^2 y(x,\lambda) = 0, \tag{8.131}$$

with $t = (1+r)/2$, whereas a_{11} and a_{21} are solutions of the same equation but with $t = (1-r)/2$; [Hi62] includes a concise summary of all the facts on Bessel functions that will be needed in this monograph. Bessel functions of the first kind of order t will be denoted $J_t(x)$.

Formulas for $A(x,\lambda)$ are presented in terms of a solution $a(x,\lambda)$ to (8.131) that meets the initial conditions

$$a(0,\lambda) = 1 \quad \text{and} \quad a'(0,\lambda) = 0 \tag{8.132}$$

and $a'(x,\lambda)$. If $\delta > 0$, then

$$a(x,\lambda) = c(x+\delta)^t \{J_{t-1}(\rho\lambda\delta)J_{-t}(\rho\lambda(x+\delta)) + J_{1-t}(\rho\lambda\delta)J_t(\rho\lambda(x+\delta))\}$$

is a solution of the Bessel equation (8.131) such that $a'(0,\lambda) = 0$ and

$$a(0,\lambda) = cb^t \{J_{t-1}(\rho\lambda\delta)J_{-t}(\rho\lambda\delta) + J_{1-t}(\rho\lambda\delta)J_t(\rho\lambda\delta)\} = cb^t \frac{2}{\pi\rho\lambda\delta} \sin t\pi.$$

Therefore, **if t is not an integer** and

$$c = \frac{\pi\rho\lambda\delta^{1-t}}{2\sin t\pi},$$

then

$$a(x,\lambda) = \frac{\pi\rho\lambda\delta^{1-t}}{2\sin t\pi}(x+\delta)^t$$
$$\times \{J_{t-1}(\rho\lambda\delta)J_{-t}(\rho\lambda(x+\delta)) + J_{1-t}(\rho\lambda\delta)J_t(\rho\lambda(x+\delta))\}$$

meets the conditions (8.132) and

$$a'(x,\lambda) = \frac{\pi(\rho\lambda)^2\delta^{1-t}}{2\sin t\pi}(x+\delta)^t$$
$$\times \{-J_{t-1}(\rho\lambda\delta)J_{1-t}(\rho\lambda(x+\delta)) + J_{1-t}(\rho\lambda\delta)J_{t-1}(\rho\lambda(x+\delta))\}.$$

Therefore,

$$A(x,\lambda) = \begin{bmatrix} a_{11}(x,\lambda) & \frac{ia'_{11}(x,\lambda)}{\lambda h_2(x)} \\ \frac{ia'_{22}(x,\lambda)}{\lambda h_1(x)} & a_{22}(x,\lambda) \end{bmatrix},$$

where

$$a_{11}(x,\lambda) = a(x,\lambda) \quad \text{with } t = (1-r)/2$$

and

$$a_{22}(x,\lambda) = a(x,\lambda) \quad \text{with } t = (1+r)/2.$$

These formulas imply that $A_x \in L_\infty^{2\times 2}$ for every $x \in [0,\infty)$ and hence, by Theorem 4.18, that $A_x \in \mathcal{U}^\circ_{\ell sR}(J) \cap \mathcal{U}^\circ_{rsR}(J)$ for all such x. Moreover, with the help of the discussion in Section 8.6, it is readily checked that $A_x \in \mathcal{E} \cap \mathcal{U}^\circ(J_1) \cap Symp \cap Real$. Therefore, $\tau^+_{A_x} = \tau^-_{A_x}$. Then, since $\det H(x) = \rho^2$ for $x \geq 0$, Theorem 8.76 implies that $\tau^\pm_{A_x} = \rho x$. Thus, $\{e_{\rho x}, e_{\rho x}\} \in ap_{II}(A_x)$ and $A_x \in \mathcal{U}^\circ_{rsR}(J_1)$, and, by Theorem 8.53, A_x may be characterized as the only resolvent matrix with these properties of the strictly completely indeterminate interpolation problem CP($e_{2\rho x}$; c°_x) with $c^\circ_x \in T_{A_x}[1]$. Consequently, the matrizant A_x is uniquely defined on each finite interval $[0,d]$ by the monodromy matrix A_d. It is also readily checked that $a_{11}(x,\lambda)$ and $a_{22}(x,\lambda)$ are even entire functions of λ, $a_{12}(x,\lambda)$ and $a_{21}(x,\lambda)$ are odd entire functions of λ and that $\det A(x,\lambda) = 1$ (as it should be, since A_x is symplectic). Furthermore, it may be shown, much as in the example treated in Section 11.6, that the system is in the limit point case:

$$C^\infty_{imp}(Hdx) = \bigcap_{x\geq 0} C(e_{2\rho x}; c^\circ_x) = \{c^\circ_\infty\},$$

where $c^\circ_\infty(\lambda) = \lim_{x\uparrow\infty} c^\circ_x(\lambda)$ for $\lambda \in \mathbb{C}_+$ and the spectral function of c°_∞ is the spectral function of the considered system.

The formulas become more tractable when r is an even integer, because then the Bessel functions can be expressed in terms of trigonometric functions. Thus, for example, if $\delta > 0$ and $r = 2$, then

$$a_{11}(x,\lambda) = \frac{\delta}{x+\delta}\left\{\cos(\rho\lambda x) + \frac{\sin(\rho\lambda x)}{\rho\lambda\delta}\right\}, \qquad a_{12}(x,\lambda) = \frac{a'_{11}(x,\lambda)}{-i\lambda h_2(x)}$$

$$a_{21}(x,\lambda) = \frac{\rho\sin(\rho\lambda x)}{id_1(x+\delta)\delta}, \qquad a_{22}(x,\lambda) = \frac{x+\delta}{\delta}\left\{\cos(\rho\lambda x) - \frac{\sin(\rho\lambda x)}{\rho\lambda(x+\delta)}\right\}.$$

Consequently, if $\nu > 0$

$$\frac{a_{11}(x, i\nu)}{a_{21}(x, i\nu)} \sim d_1 \frac{\delta^2}{\rho} + d_1 \frac{\delta}{\rho^2 \nu} \stackrel{def}{=} c_\infty^{(2)}(i\nu) \quad \text{as } x \uparrow \infty,$$

i.e.,

$$c_\infty^{(2)}(\lambda) = d_1 \frac{\delta^2}{\rho} + i d_1 \frac{\delta}{\rho^2 \lambda} = \frac{1}{\pi i} \int_{-\infty}^{\infty} \left(\frac{1}{\mu - \lambda} - \frac{\mu}{1 + \mu^2} \right) d\sigma^{(2)}(\mu)$$

with spectral function $\sigma^{(2)} = \sigma_a^{(2)} + \sigma_s^{(2)}$, where

$$\sigma_s^{(2)}(\mu) = \frac{d_1 \delta^2 \pi \mu}{2\rho^2 |\mu|} \quad \text{if } \mu \neq 0, \quad \sigma_s^{(2)}(0) = 0 \quad \text{and} \quad \sigma_a^{(2)}(\mu) = \frac{d_1 \delta^2}{\rho} \mu.$$

Analogous calculations based on the mvf $J_1 A(x, \lambda) J_1$ lead to the formulas

$$\frac{a_{22}(x, i\nu)}{a_{12}(x, i\nu)} \sim \frac{\nu \rho^2}{d_1(1 + \delta \rho \nu)} \stackrel{def}{=} c_\infty^{(1)}(i\nu) = (c_\infty^{(2)}(i\nu))^{-1} \quad \text{as } x \uparrow \infty,$$

$$c_\infty^{(1)}(\lambda) = \frac{-i\lambda \rho^2}{d_1 \delta (1 - \delta \rho i \lambda)} = \frac{1}{\pi i} \int_{-\infty}^{\infty} \left(\frac{1}{\mu - \lambda} - \frac{\mu}{1 + \mu^2} \right) d\sigma^{(1)}(\mu)$$

with spectral function $\sigma^{(1)}(\mu) = \sigma_a^{(1)}(\mu)$ and

$$(\sigma_a^{(1)})'(\mu) = \frac{\rho^3 \delta \mu^2}{d_1 \delta (1 + (\rho \mu \delta)^2)}.$$

Thus, if

$$f(x) = \begin{bmatrix} f_1(x) \\ f_2(x) \end{bmatrix} \quad \text{with} \quad \int_0^{\infty} f(x)^* H(x) f(x) dx < \infty,$$

$$(\mathcal{F} f)(\mu) = \frac{1}{\sqrt{2\pi}} \int_0^{\infty} A(x, \mu) H(x) f(x) dx,$$

$$(\mathcal{F}_1 f)(\mu) = \sqrt{2}[1 \quad 0](\mathcal{F} f)\mu) \quad \text{and} \quad (\mathcal{F}_2 f)(\mu) = \sqrt{2}[0 \quad 1](\mathcal{F} f)\mu),$$

then

$$\int_{-\infty}^{\infty} (\mathcal{F}_j f)(\mu)^* d\sigma^{(j)}(\mu) (\mathcal{F}_j f)(\mu) d\mu = \int_0^{\infty} f(x)^* H(x) f(x) dx \quad \text{for } j = 1, 2$$

and

$$\int_{-\infty}^{\infty} (\mathcal{F} f)(\mu)^* \begin{bmatrix} d\sigma^{(1)}(\mu) & 0 \\ 0 & d\sigma^{(2)}(\mu) \end{bmatrix} (\mathcal{F} f)(\mu) d\mu = \int_0^{\infty} f(x)^* H(x) f(x) dx.$$

If

$$\tau^{(j)}(\mu) = \frac{2}{\pi} \sigma^{(j)}(\sqrt{\mu}) \quad \text{for } \mu \geq 0 \quad \text{and} \quad \tau^{(j)}(\mu) = 0 \quad \text{for } \mu < 0,$$

then, in the terminology of Kac and Krein [KaKr74b], $\tau^{(2)}$ is the principal spectral function of an S_1-string and $\tau^{(1)}$ is the principal spectral function of the **dual string**.

8.14 Supplementary notes

Most of this chapter is adapted from [ArD00a] and [ArD00b]. The given data for the inverse monodromy problem is a mvf $U \in \mathcal{E} \cap \mathcal{U}^\circ(J)$. If $J = \pm I_m$, this problem is solved in Section 8.2 without any extra conditions on U. The method of proof is close in spirit to the method of *greatest scalar minorants* developed by Yu.P. Ginzburg to study multiplicative representations of matrix and operator valued functions in the Smirnov class; see, e.g., [Gi67], [GiZe90] and the discussion preceding theorem 9 in [GiSh94]. The given data for the inverse monodromy problem with $J = I_p$ is a mvf $b \in \mathcal{E} \cap \mathcal{S}_{\text{in}}^{p\times p}$ with $b(0) = I_p$. If type $(b) = a$ and $a > 0$, then e_a is such a minorant. Similarly e_t is such a minorant for each divisor b_t of b with $b_t(0) = I_p$ and type $b_t = t$.

In Section 8.2 the Beurling–Lax theorem is used to establish the existence of a maximal family of mvf's b_t of left divisors of the monodromy matrix b that is characterized by the property that $\mathcal{H}(b_t) = \mathcal{H}(b) \cap \mathcal{H}(e_t I_p)$ for $0 \le t \le$ type b.

The monograph [ABT11] is a good supplementary source of information on LB-nodes and a number of other topics that are considered in this book.

An important uniqueness theorem was obtained by L. de Branges for $m = 2$ with $U \in \mathcal{E} \cap \mathcal{U}^\circ(\mathcal{J}_1) \cap Symp$ and normalized real H. The result is presented as Theorem 8.3, without proof. The original proof of this remarkable result may be found in [Br68a]. The main difficulty in the proof is to show that the symplectic left divisors of the monodromy matrix U are ordered. An expanded version of de Branges' proof is given in [DMc76]. The applications of de Branges' theorem that are presented in Section 8.12 are adapted from [ArD01a].

If $m > 2$ the situation is more complicated and knowledge of the monodromy matrix is usually not enough to insure a unique solution to the inverse problem. One way to overcome this difficulty is to restrict attention to special classes of Hamiltonians. Thus, for example, in recent publications [Sak94], [Sak98a], [Sak98b], [Sak98c] and [Sak99], L.A. Sakhnovich obtained uniqueness theorems in the class of Hamiltonians $H(x)J = V(x)MV(x)^{-1}$ that are similar to a matrix $M \in \mathbb{C}^{m\times m}$ with special structure and $V(x)$ is suitably smooth. Some other more general classifications were considered earlier in [Sak93]. Other restrictions on the Hamiltonian that lead to uniqueness are discussed on pp. 388–391 of the monograph [GK70] by I. Gohberg and M.G. Krein. In particular it is noted there that if $U(\lambda)$ is the monodromy matrix of the canonical differential system (2.14) with $J = \mathcal{J}_p$, then there is at most one Hamiltonian $H(x)$ that is absolutely continuous, real and symplectic for every $x \in [0, \ell)$. A proof may be found in M.M. Malamud

[Mal95]; see also [Mal97]. For additional information from the point of view of RKHS's of entire vvf's, see [Br68b] and [Br83].

The example in Section 8.13 is adapted from [Dy70]; a number of additional examples are considered there; a particularly interesting one that leads to discrete spectral functions is the choice $h_1(x) = \exp(-x^2)$ and $h_2(x) = h_1(x)^{-1}$; for additional information and generalizations to $m \times m$ systems with $m = 2p$, $H(x) = \mathrm{diag}\,\{h(x), h(x)^{-1}\}$ with $p \times p$ block entries, see [Sak01].

Most of the results on the inverse monodromy problem that are presented in this chapter for $J \neq \pm I_m$ for $m \geq 2$ are obtained for monodromy matrices $U \in \mathcal{E} \cap \mathcal{U}_{AR}^{\circ}(J)$. The original papers [ArD00a] and [ArD00b] focused on the class $\mathcal{E} \cap \mathcal{U}_{rsR}^{\circ}(J)$. The proofs, however, for both of these classes are much the same, the key observation being Theorem 4.56.

The extension to the class $\mathcal{U}_{BR}(J)$ that was discussed in Section 4.9 is based on [ArD12a]. Necessary and sufficient conditions for a left divisor U_1 of U to be in $\mathfrak{B}_U$ were obtained by Ju.L. Shmuljan [Shm62]; this result was generalized by E.L. Pekarev [Pe80].

An elegant proof of a restricted version of de Branges' theorem for the inverse monodromy problem under an extra constraint on the Riemann surface

$$\{(\lambda, \omega) : \det\{A(\lambda) - \omega I_2\} = 0 \quad \text{and} \quad |\omega| < 1\}$$

generated by the monodromy matrix $A(\lambda)$ is presented by P. Yuditskii in [Yu01]. Yuditskii's approach is applicable to monodromy matrices A in the class $U_S(J_1)$.

Inverse monodromy problems with partially specified monodromy matrices have also been studied in recent years; see, e.g., M.M. Malamud [Mal05] and the references cited therein.

Comparisons between Krein's theory of resolvent matrices of entire symmetric operators and characteristic functions (resp., de Branges' theory of Hilbert spaces of entire functions) are discussed in [KrS70], [Ko80] and [TsS77] (resp., appendix 3 of [GoGo97]); for related developments, see also [AlD84] and [AlD85].

9

Bitangential Krein extension problems

In this chapter we shall discuss bitangential generalizations of matrix versions of two classical extension problems that were introduced by M.G. Krein: the helical extension problem and the accelerant extension problem, and some analogs of these problems related to the Schur class $\mathcal{S}^{p\times q}$ (instead of the Carathéodory class $\mathcal{C}^{p\times p}$). These extension problems will be identified with the generalized interpolation problems GCIP$(b_3, b_4; c^\circ)$ and GSIP$(b_1, b_2; s^\circ)$ with appropriately chosen mvf's $c^\circ \in \mathcal{C}^{p\times p}$, $s^\circ \in \mathcal{S}^{p\times q}$ and entire inner mvf's $b_1, \ldots, b_4$. We shall make use of the fact that if $b \in \mathcal{E} \cap \mathcal{S}_{\text{in}}^{p\times p}$, $b(0) = I_p$ and $\tau_b = a$, then, by the Paley–Wiener theorem,

$$b(\lambda) = I_p + i\lambda \int_0^a e^{i\lambda t} h_b(t)dt, \quad \text{where } h_b \in L_2^{p\times p}([0, a]). \tag{9.1}$$

Correspondingly, the time domain versions of the bitangential interpolation problems referred to above will be formulated in terms of the mvf's h_{b_j} instead of b_j, $j = 1, \ldots, 4$. We shall also make use of the following basic fact and variations thereof:

If $v(x)$ is a $p \times p$ measurable mvf on $[-a, a]$, then the formula

$$(Xf)(x) = \frac{d}{dx}\int_0^a v(x - u)f(u)du \quad \text{for } 0 \le x \le a$$

defines a bounded linear operator from $L_2^p([0, a])$ into itself if and only if

(1) $v \in L_2^{p\times p}([-a, a])$ and
(2) $\int_0^a v(x - u)f(u)du$ is absolutely continuous and its derivative is in $L_2^p([0, a])$ for every $f \in L_2^p([0, a])$.

This circle of ideas will be developed further below; for additional information and numerous applications, see L.A. Sakhnovich [Sak80].

9.1 Helical extension problems

A mvf $g(t)$ is said to belong to the class $\mathcal{G}_\infty^{p\times p}$ of **helical** $p \times p$ mvf's on the interval $(-\infty, \infty)$ if it meets the following three conditions:

(1) $g(t)$ is a continuous $p \times p$ mvf on the interval $(-\infty, \infty)$.
(2) $g(t)^* = g(-t)$ for every t in the interval $(-\infty, \infty)$.
(3) The kernel

$$k(t, s) = g(t - s) - g(t) - g(-s) + g(0) \tag{9.2}$$

is positive on $[0, \infty) \times [0, \infty)$.

The class $\mathcal{G}_a^{p\times p}$ of helical $p \times p$ mvf's on the closed interval $[-a, a]$ is defined for $a > 0$ by the same set of three conditions except that the finite closed intervals $[-a, a]$ and $[0, a]$ are considered in place of the intervals $(-\infty, \infty)$ and $[0, \infty)$, respectively.

The classical **Krein helical extension problem** is:

HEP(g°; a): *Given a mvf* $g^\circ \in \mathcal{G}_a^{p\times p}$, $0 < a < \infty$, *describe the set*

$$\mathcal{G}(g^\circ; a) = \{g \in \mathcal{G}_\infty^{p\times p} : g(t) = g^\circ(t) \quad \textit{for every} \quad t \in [-a, a]\}. \tag{9.3}$$

Theorem 9.1 (M.G. Krein) *A $p \times p$ mvf $g(t)$ belongs to the class $\mathcal{G}_a^{p\times p}$ if and only if it admits a representation of the form*

$$g(t) = -\beta + it\alpha + \frac{1}{\pi}\int_{-\infty}^{\infty}\left\{e^{-i\mu t} - 1 + \frac{i\mu t}{1+\mu^2}\right\}\frac{d\sigma(\mu)}{\mu^2} \tag{9.4}$$

on the interval $[-a, a]$ if $0 < a < \infty$ and on $(-\infty, \infty)$ if $a = \infty$, where $\alpha = \alpha^*$ and $\beta = \beta^*$ are constant $p \times p$ matrices and $\sigma(\mu)$ is a nondecreasing $p \times p$ mvf on $\mathbb{R}$ such that

$$\int_{-\infty}^{\infty}(1 + \mu^2)^{-1} d(\text{trace}\,\sigma(\mu)) < \infty. \tag{9.5}$$

Proof A proof for the case $p = 1$, based on Krein's unpublished lecture notes, is given in chapter 3 of [GoGo97] (see theorem 3.12 for the statement and [Kr44c]). This proof can be extended to cover the matrix case. □

Theorem 9.1 guarantees that the HEP(g°; a) is always solvable, i.e., if g° admits a representation of the form (9.4) for $t \in [-a, a]$, then the same formula considered on $(-\infty, \infty)$ yields an extension; all solutions may be obtained this way. Moreover, Theorem 9.1 leads easily to a growth estimate for mvf's $g \in \mathcal{G}_\infty^{p\times p}$.

Corollary 9.2 *If $g \in \mathcal{G}_\infty^{p\times p}$, then*

$$\|g(t)\| = O(t^2) \quad as \quad t \to \infty. \tag{9.6}$$

Proof See corollary 8.2 in [ArD08b]. □

The next theorem exhibits an important connection between the classes $\mathcal{C}^{p\times p}$ and

$$\mathcal{G}_a^{p\times p}(0) = \{g \in \mathcal{G}_a^{p\times p} : g(0) \le 0\} \tag{9.7}$$

when $a = \infty$.

Theorem 9.3 *There is a one-to-one correspondence between mvf's $c(\lambda)$ in the class $\mathcal{C}^{p\times p}$ and mvf's $g(t)$ in the subclass $\mathcal{G}_\infty^{p\times p}(0)$ that is defined by the formula*

$$c_g(\lambda) = \lambda^2 \int_0^\infty e^{i\lambda t} g(t)dt \quad \text{for} \quad \lambda \in \mathbb{C}_+. \tag{9.8}$$

Moreover, if $0 < a < \infty$, $g^\circ \in \mathcal{G}_a^{p\times p}(0)$ and $g_1 \in \mathcal{G}(g^\circ; a)$, then

$$\mathcal{C}(e_a I_p, I_p;\, c_{g_1}) = \{c_g : g \in \mathcal{G}(g^\circ; a)\}. \tag{9.9}$$

Proof See theorem 8.3 in [ArD08b]. □

In the future, the abbreviated **notation**

$$\mathcal{C}(e_a; c^\circ) = \mathcal{C}(e_a I_p, I_p;\, c^\circ) \quad \text{and} \quad \mathcal{C}(g^\circ; a) = \{c_g : g \in \mathcal{G}(g^\circ; a)\} \tag{9.10}$$

will be used. Thus, if $c^\circ \in \mathcal{C}^{p\times p}$, then

$$\mathcal{C}(e_a; c^\circ) = \{c \in \mathcal{C}^{p\times p} : e_{-a}(c - c^\circ) \in \mathcal{N}_+^{p\times p}\}. \tag{9.11}$$

Given $g^\circ \in \mathcal{G}_a^{p\times p}(0)$, $0 < a < \infty$, the HEP($g^\circ; a$) is said to be

(1) **determinate** if the problem has only one solution;
(2) **indeterminate** if the problem has more than one solution;
(3) **completely indeterminate** if for every nonzero vector $\xi \in \mathbb{C}^p$, there exist at least two mvf's $g_1, g_2 \in \mathcal{G}(g^\circ; a)$ such that $\{g_1(t) - g_2(t)\}\xi \not\equiv 0$ on $\mathbb{R}$;
(4) **strictly completely indeterminate** if there exists at least one $g \in \mathcal{G}(g^\circ; a)$ such that $c_g \in \mathring{\mathcal{C}}^{p\times p}$.

We remark that the identification (9.9) exhibits the fact that a strictly completely indeterminate HEP is automatically completely indeterminate.

Theorems 4.25 and 9.3 provide a link between completely indeterminate HEP's and the class of mvf's $A \in \mathcal{E} \cap \mathcal{U}_{rR}(J_p)$ such that $\{e_{a_3} I_p, e_{a_4} I_p\} \in ap_{II}(A)$ for some choice of $a_3 \ge 0$ and $a_4 \ge 0$.

Theorem 9.4 *If $0 < a < \infty$, $g \in \mathcal{G}_a^{p\times p}(0)$ and the HEP($g; a$) is strictly completely indeterminate, then:*

(1) *$g(t)$ is absolutely continuous in the interval $[0, a]$, $g(0) = 0$ and $g' \in L_2^p([0, a])$, i.e.,*

$$g(t) = -\int_0^t v(s)ds \quad \text{for } t \in [0, a] \text{ and some } v \in L_2^{p\times p}([0, a]).$$

(2) *The $p \times 1$ mvf $\int_0^t v(t-s)\varphi(s)ds$ is absolutely continuous on the interval $[0, a]$ for each $\varphi \in L_2^p([0, a])$ and the operator $X^\vee$ that is defined by the formula*

$$(X^\vee \varphi)(t) = \frac{d}{dt} \int_0^t v(t-s)\varphi(s)ds \tag{9.12}$$

is a bounded linear operator from $L_2^p([0, a])$ into itself.

Moreover, if v is extended to the interval $[-a, 0]$ by the formula

$$v(t) = -v(-t)^* \quad \text{for almost all points } t \in [-a, 0],$$

then the following additional conclusions hold:

(3) *The $p \times 1$ mvf $\int_t^a v(t-s)\varphi(s)ds$ is absolutely continuous on the interval $[0, a]$ and the adjoint $(X^\vee)^*$ of $X^\vee$ in $L_2^p([0, a])$ is given by the formula*

$$((X^\vee)^* \varphi)(t) = \frac{d}{dt} \int_t^a v(t-s)\varphi(s)ds \tag{9.13}$$

for every $\varphi \in L_2^p([0, a])$.

(4) *There exists a $\delta > 0$ such that the operator $Y^\vee = X^\vee + (X^\vee)^*$, which is given by the formula*

$$(Y^\vee \varphi)(t) = \frac{d}{dt} \int_0^a v(t-s)\varphi(s)ds \tag{9.14}$$

on $L_2^p([0, a])$, is bounded from below by δI, i.e.,

$$\int_0^a \varphi(t)^* \left\{ \frac{d}{dt} \int_0^a v(t-s)\varphi(s)ds \right\} dt \geq \delta \int_0^a \varphi(t)^* \varphi(t)dt \tag{9.15}$$

for every $\varphi \in L_2^p([0, a])$.

Proof This is theorem 8.39 in [ArD08b]. □

Theorem 9.5 *Let $v \in L_2^{p\times p}([-a, a])$, $0 < a < \infty$, be such that $v(-t) = -v(t)^*$ for almost all points $t \in [-a, a]$ and properties (2)–(4) in Theorem 9.4 are in force (property (3) follows from (2)). Then the mvf*

$$g(t) = -\int_0^t v(s)ds, \quad -a \leq t \leq a,$$

belongs to $\mathcal{G}_a^{p\times p}(0)$ and the HEP$(g; a)$ is strictly completely indeterminate.

Proof See theorem 8.41 in [ArD08b]. □

Theorem 9.6 *If $0 < a < \infty$, $g \in \mathcal{G}_a^{p\times p}(0)$ and the HEP$(g; a)$ is strictly completely indeterminate, then there is exactly one mvf $A \in \mathcal{E} \cap \mathcal{U}_{rsR}^\circ(J_p)$ such that*

$$\mathcal{C}(e_a; c_g) = \mathcal{C}(A) \quad \text{and} \quad \{e_a I_p, I_p\} \in ap_{II}(A). \tag{9.16}$$

Moreover,

$$\mathcal{C}(A) = T_{A\mathfrak{V}}[\mathcal{S}^{p\times p}]$$

(i.e., $\mathcal{S}^{p\times p} \subseteq \mathcal{D}(T_{A\mathfrak{V}})$) and the mvf $A(\lambda)$ is given by the formula

$$A(\lambda) = I_m + i\lambda \int_0^a e^{i\lambda t} \begin{bmatrix} \varphi_{11}(t) & \varphi_{12}(t) \\ \varphi_{21}(t) & \varphi_{22}(t) \end{bmatrix} dt J_p, \tag{9.17}$$

where φ_{21} and φ_{22} are obtained in terms of

$$v(t) = \begin{cases} -g'(t) & a.e.\ in\ [0,a] \\ -v(-t)^* & a.e.\ in\ [-a,0], \end{cases}$$

as the unique solutions in $L_2^{p\times p}([0,a])$ of the equations

$$\frac{d}{dt}\int_0^a v(t-s)\varphi_{22}(s)ds = I_p, \quad 0 \le t \le a, \tag{9.18}$$

$$\frac{d}{dt}\int_0^a v(t-s)\varphi_{21}(s)ds = -v(t), \quad 0 \le t \le a, \tag{9.19}$$

and the mvf's

$$\varphi_{1j}(t) = -\frac{d}{dt}\int_t^a v(t-s)\varphi_{2j}(s)ds, \quad j = 1,2, \quad 0 \le t \le a, \tag{9.20}$$

also belong to $L_2^{p\times p}([0,a])$.

Proof See theorems 8.39 and 8.42 in[ArD08b]. □

Corollary 9.7 *If $A(\lambda)$ is given by formula (9.17) in the setting of Theorem 9.6 and $B(\lambda) = A(\lambda)\mathfrak{V}$, then*

$$B(\lambda) = \mathfrak{V} + \frac{i\lambda}{\sqrt{2}}\int_0^a e^{i\lambda t} \begin{bmatrix} \psi_1(t) & \psi_2(t) \\ \varphi_1(t) & \varphi_2(t) \end{bmatrix} dt, \tag{9.21}$$

where φ_1 and φ_2 are solutions of the equations

$$\frac{d}{dt}\int_0^a v(t-s)\varphi_1(s)ds = v(t) + I_p \tag{9.22}$$

and

$$\frac{d}{dt}\int_0^a v(t-s)\varphi_2(s)ds = v(t) - I_p, \tag{9.23}$$

respectively, in the space $L_2^{p\times p}([0,a])$, and

$$\psi_j(t) = -\frac{d}{dt}\int_t^a v(t-s)\varphi_j(s)ds = -((X^\vee)^*\varphi_j)(t) \quad for\ j = 1,2. \tag{9.24}$$

If $g \in \mathcal{G}_a^{p\times p}(0)$ for some $a > 0$, $B \in \mathcal{U}(j_p, J_p)$ and

$$\mathcal{C}(g; a) = T_B[\mathcal{S}^{p\times p} \cap \mathcal{D}(T_B)], \tag{9.25}$$

then B is called a **B-resolvent matrix of the HEP**$(g; a)$. In view of Theorems 9.3 and 4.25, there exists a mvf $B \in \mathcal{U}(j_p, J_p)$ such that (9.25) holds if and only if the HEP$(g; a)$ is completely indeterminate. Moreover, if the HEP$(g; a)$ is completely indeterminate, then there exists exactly one B-resolvent matrix in the class $\mathcal{U}(j_p, J_p)$ such that

(i) $B \in \mathcal{U}(j_p, J_p)$,
(ii) $\{e_a I_p, I_p\} \in ap(B)$ and
(iii) $B(0) = \mathfrak{V}$.

Condition (iii) in the preceding list is meaningful because in view of (ii) and Theorem 4.25, B is an entire mvf.

Example 9.8 *The function $v(t) = \sin t$ for $0 \le t \le a$ serves to illustrate some parts of the previous theorem. In particular*

$$g(t) = -\int_0^t \sin s ds = \cos t - 1$$

is a helical function, i.e., g is continuous $g(-t) = \overline{g(t)}$ and the kernel

$$\begin{aligned} g(t-s) - g(t) - g(-s) + g(0) &= \cos(t-s) - \cos t - \cos s + 1 \\ &= \Re\{(e^{it} - 1)(e^{-is} - 1)\} \end{aligned}$$

is positive on $[0, a] \times [0, a]$. Moreover, g admits the representation (9.4) with $\alpha = \beta = 0$ and $\sigma(\mu)$ constant except for a jump of $\pi/2$ at the points $\mu = \pm 1$. Correspondingly, by formula (9.8) (or, eqivalently by the Riesz–Herglotz representation (3.42) with the same choice of α, β and σ as in (9.4)),

$$c_g(\lambda) = \frac{i\lambda}{\lambda^2 - 1} \quad \textit{for } \lambda \in \mathbb{C}_+.$$

Since $v(s) = \sin s$, equation (9.18) reduces to

$$\cos t \int_0^a \cos s\, \varphi_{22}(s) ds + \sin t \int_0^a \sin s\, \varphi_{22}(s) ds = 1 \quad \textit{for } 0 \le t \le a,$$

which does not admit a solution in $L_2([0, a])$ for any $a > 0$. Thus, in view of Theorem 9.4, the HEP$(g^\circ; a)$ based on the restriction g° of g to $[0, a]$ is not strictly completely indeterminate; in fact it is determinate by Theorem 4.35.

Example 9.9 *Let $g(t) = \frac{1}{2}t^2 - |t|$. Then*

$$g(t) = \sum_{n=0}^{\infty} a_n \cos(n\pi t/2) \quad \textit{for } t \in [-2, 2]$$

with coefficients

$$a_n = \int_0^2 \left(\frac{t^2}{2} - t\right) \cos(n\pi t/2)dt = \frac{1 + \cos n\pi}{(n\pi/2)^2} \quad \text{for } n \geq 1$$

and

$$a_0 = \frac{1}{2}\int_0^2 \left(\frac{t^2}{2} - t\right) dt = -\frac{1}{3}.$$

Moreover, since

$$\sum_{n=1}^{\infty} a_n = \frac{1}{3}$$

the Fourier series representation for $g(t)$ can be reexpressed as

$$g(t) = \sum_{n=1}^{\infty} a_n(\cos(n\pi t/2) - 1) = 2\sum_{k=1}^{\infty} \frac{\cos(k\pi t) - 1}{k^2\pi^2} \quad \text{for } t \in [-2, 2],$$

which exhibits g as a sum of the helical functions of the kind considered in Example 9.8 with positive coefficients. This sum can be expressed in terms of Krein's formula (9.4) with $\alpha = \beta = 0$ and a spectral function σ having a jump of height π at the points $k\pi$ for every nonzero integer k. Thus

$$c_g(\lambda) = 2i\sum_{k=1}^{\infty} \frac{\lambda}{\lambda^2 - (k\pi)^2} \quad \text{for } \lambda \in \mathbb{C}_+.$$

However, in contrast to the preceding example, the HEP$(g; a)$ is strictly completely indeterminate for this choice of g for $0 < a < 2$.

Invoking formulas (9.22)–(9.24) with

$$v(t) = \begin{cases} 1 - t & \text{for } t > 0 \\ -1 - t & \text{for } t < 0, \end{cases}$$

it is readily checked that

$$\varphi_2(t) - \varphi_1(t) = \frac{2}{a-2}, \quad \varphi_2(t) + \varphi_1(t) = \frac{2+a}{2} - t,$$

$$\psi_2(t) - \psi_1(t) = \frac{2(a - t - 1)}{a - 2} \quad \text{and} \quad \psi_2(t) + \psi_1(t) = \frac{a - 2 - at + t^2}{2}.$$

A lengthy but straightforward calculation for the mvf

$$A(a, \lambda) = e^{-i\lambda a/2} B(\lambda)\mathfrak{V}$$

based on formula (9.21) for $B(\lambda)$ yields the evaluation

$$A(a,\lambda) =$$

$$\begin{bmatrix} \frac{2}{2-a}\left\{\cos(\lambda a/2) - \frac{\sin(\lambda a/2)}{\lambda}\right\} & a_{12}(a,\lambda) \\ \frac{2i\sin(\lambda a/2)}{a-2} & \frac{(2-a)}{2}\cos(\lambda a/2) + \frac{\sin(\lambda a/2}{\lambda} \end{bmatrix},$$

where

$$a_{12}(a,\lambda) = \frac{i(a-2)\sin(\lambda a/2)}{2} - \frac{a\cos(\lambda a/2)}{2i\lambda} + \frac{\sin(\lambda a/2)}{i\lambda^2}.$$

Moreover,

$$\lim_{\lambda\to 0}\left\{-i\frac{A(a,\lambda)-I_2}{\lambda}\right\}J_1 = \begin{bmatrix} \frac{(a-2)^3+8}{24} & 0 \\ 0 & \frac{a}{2-a} \end{bmatrix}$$

$$= \int_0^a \begin{bmatrix} \frac{(2-s)^2}{8} & 0 \\ 0 & \frac{2}{(2-s)^2} \end{bmatrix} ds.$$

This identifies $A(a,\lambda)$ (resp., $e^{i\lambda a/2}A(a,\lambda)$) as the matrizant of a canonical differential system with Hamiltonian

$$H(a) = \begin{bmatrix} \frac{(2-a)^2}{8} & 0 \\ 0 & \frac{2}{(2-a)^2} \end{bmatrix} \quad (\text{resp., } H(a) + \frac{1}{2}J_1) \tag{9.26}$$

for points $a \in \mathbb{R}_+ \setminus \{2\}$. Thus, $H \in L_1^{2\times 2}([0,d])$ if $0 < d < 2$ and $H \notin L_1^{2\times 2}([0,2])$ if $2 \le d$, i.e., the system is regular, if $0 < d < 2$ and singular if $d = 2$. Note that $A(a,\lambda)$ is of exponential type $a/2$ in $\mathbb{C}_\pm$, which is consistent with the formula type $(A_a) = \int_0^a \sqrt{\det H(s)}ds$, *which is valid for diagonal Hamiltonians; the formula does not work for the Hamiltonian $H(a) + (1/2)J_1$.*

9.2 Bitangential helical extension problems

In this section the identification (9.9) will be generalized to tangential and bitangential Krein helical extension problems by replacing the special choices $b_3(\lambda) = e_a(\lambda)I_p$ and $b_4(\lambda) = I_p$, with an arbitrary normalized pair $b_3(\lambda), b_4(\lambda)$ of entire inner $p \times p$ mvf's. Correspondingly, the set $\mathcal{G}(g^\circ; a)$ on the right-hand side of (9.9) is replaced by the set $\mathcal{G}(g^\circ; \mathcal{F}_\ell, \mathcal{F}_r)$ of solutions of the following bitangential Krein helical extension problem that will be called the GHEP, an acronym for **generalized helical extension problem**:

GHEP($g^\circ; \mathcal{F}_\ell, \mathcal{F}_r$): *Given a mvf $g^\circ \in \mathcal{G}_\infty^{p\times p}(0)$ and a pair of sets $\mathcal{F}_\ell \subseteq L_2^p([0,\alpha_\ell])$ and $\mathcal{F}_r \subseteq L_2^p([0,\alpha_r])$, at least one of which contains a nonzero element, describe* *the set* $\mathcal{G}(g^\circ; \mathcal{F}_\ell, \mathcal{F}_r\}$ *of mvf's* $g \in \mathcal{G}_\infty^{p\times p}(0)$ *that meet the following three conditions for every choice of* $h^\ell \in \mathcal{F}_\ell$ *and* $h^r \in \mathcal{F}_r$*:*

$$\int_t^{\alpha_\ell} h^\ell(u)^*\{g(u-t) - g^\circ(u-t)\}du = 0 \quad \text{for } 0 \le t \le \alpha_\ell. \tag{9.27}$$

$$\int_t^{\alpha_r} \{g(v-t) - g^\circ(v-t)\}h^r(v)dv = 0 \quad \text{for } 0 \le t \le \alpha_r. \tag{9.28}$$

$$\int_t^{\alpha_r} \left[\int_0^{\alpha_\ell} h^\ell(u)^*\{g(u+v-t) - g^\circ(u+v-t)\}du\right] h^r(v)dv = 0 \quad \text{for } 0 \le t \le \alpha_r. \tag{9.29}$$

The three sets of conditions (9.27), (9.28) and (9.29) are equivalent to the three sets (9.27), (9.28) and

$$\int_t^{\alpha_\ell} h^\ell(u)^* \left[\int_0^{\alpha_r} \{g(u+v-t) - g^\circ(u+v-t)\}h^r(v)dv\right] du = 0 \quad \text{for } 0 \le t \le \alpha_\ell \tag{9.30}$$

(see lemma 8.12 in [ArD08b]).

If $\mathcal{F}_r = \{0\}$ or $\mathcal{F}_r = \emptyset$, then only constraint (9.27) is in effect and the GHEP is a left tangential extension problem.

If $\mathcal{F}_\ell = \{0\}$ or $\mathcal{F}_\ell = \emptyset$, then only constraint (9.28) is in effect and the GHEP is a right tangential extension problem.

Theorem 9.10 *If* $g^\circ \in \mathcal{G}_\infty^{p\times p}(0)$, $a_3 \ge 0$, $a_4 \ge 0$, $a_3 + a_4 > 0$,

$$b_j(\lambda) = I_p + i\lambda \int_0^{a_j} e^{i\lambda t} h_{b_j}(t)dt \quad \textit{belongs to } \mathcal{S}_{in}^{p\times p} \textit{ for } j = 3, 4,$$

$$\mathcal{F}_\ell \overset{\text{def}}{=} \{h_{b_3}(t)\xi : \xi \in \mathbb{C}^p\} \subseteq L_2^p([0, a_3]),$$

and (9.31)

$$\mathcal{F}_r \overset{\text{def}}{=} \{h_{b_4}(t)^*\eta : \eta \in \mathbb{C}^p\} \subseteq L_2^p([0, a_4]),$$

then

$$\mathcal{C}(b_3, b_4; c^\circ) = \{c_g : g \in \mathcal{G}(g^\circ; \mathcal{F}_\ell, \mathcal{F}_r)\} \quad \textit{with} \quad c^\circ = c_{g^\circ}. \tag{9.32}$$

Proof See theorem 8.7 in [ArD08b]. □

We remark that conditions (9.27)–(9.29) depend at most on the values of $g(t)$ and $g^\circ(t)$ on the interval $-(a_3 + a_4) \le t \le a_3 + a_4$.

If

$$b_3(\lambda) = e^{i\lambda\alpha_\ell} I_p \quad \text{and} \quad b_4(\lambda) = e^{i\lambda\alpha_r} I_p,$$

then

$$h_{b_3}(t) = I_p \quad \text{for } 0 \le t \le \alpha_\ell \quad \text{and} \quad h_{b_4}(t) = I_p \quad \text{for } 0 \le t \le \alpha_r.$$

Correspondingly, if $\alpha_\ell > 0$ and $\alpha_r > 0$, then the conditions (9.27)–(9.28) with

$$h^\ell(u) = \xi \quad \text{for } 0 \le t \le \alpha_\ell \quad \text{and} \quad h^r(t) = \eta \quad \text{for } 0 \le t \le \alpha_r$$

imply that

$$\xi^*\{g(t) - g^\circ(t)\}\eta = 0 \quad \text{for } 0 \le t \le \alpha_\ell + \alpha_r$$

and hence that

$$\{g(t) - g^\circ(t)\} = 0 \quad \text{for } 0 \le t \le \alpha_\ell + \alpha_r.$$

Similar conclusions hold if either $\alpha_\ell = 0$ or $\alpha_r = 0$.

We shall identify $L_2^p([0, \alpha])$ with the subspace of vvf's in $L_2^p([0, \infty))$ with support in the interval $[0, \alpha]$. With this identification, $L_2^p([0, \alpha])$ is invariant under the action of the semigroup T_τ, $\tau \ge 0$, of backward shift operators in the time domain that are defined in $L_2^p(\mathbb{R}_+)$ by the formula

$$(T_\tau f)(t) = f(t + \tau) \quad \text{for} \quad f \in L_2^p([0, \infty)) \quad \text{and} \quad \tau \ge 0, \tag{9.33}$$

i.e.,

$$T_\tau f = (\Pi_+ e_{-\tau} \widehat{f})^\vee.$$

Let

$$\widehat{\mathcal{H}} = \{\widehat{h} : h \in \mathcal{H}\}$$

denote the Fourier transform of a subspace $\mathcal{H}$ of $L_2^p(\mathbb{R})$.

Lemma 9.11 *If $\mathcal{F}_\ell \subseteq L_2^p([0, \alpha_\ell])$, $\mathcal{F}_r \subseteq L_2^p([0, \alpha_r])$ and $g^\circ \in \mathcal{G}_\infty^{p\times p}(0)$, then:*

(1) $\mathcal{G}(g^\circ; \mathcal{F}_\ell, \mathcal{F}_r) = \mathcal{G}(g^\circ; \bigvee_{\tau\ge 0}\{T_\tau \mathcal{F}_\ell\}, \bigvee_{\tau \ge 0}\{T_\tau \mathcal{F}_r\})$.

(2) *There exists a unique normalized pair $b_\ell, b_r \in \mathcal{E} \cap \mathcal{S}_{in}^{p\times p}$ such that*

$$\left(\bigvee_{\tau\ge 0}\{T_\tau \mathcal{F}_\ell\}\right)^\wedge = \mathcal{H}(b_\ell) \quad \text{and} \quad \left(\bigvee_{\tau\ge 0}\{T_\tau \mathcal{F}_r\}\right)^\wedge = \mathcal{H}(b_r).$$

Moreover, if $\mathcal{F}'_\ell$ and $\mathcal{F}'_r$ denote the set of columns of the mvf's $h_{b_\ell}(t)$ and $h_{b_r}(t)^*$, respectively, in the representation formulas (9.1) for $b_\ell(\lambda)$ and $b_r(\lambda)$, then

(3) $\mathcal{G}(g^\circ; \mathcal{F}_\ell, \mathcal{F}_r) = \mathcal{G}(g^\circ; \mathcal{F}'_\ell, \mathcal{F}'_r)$.

Proof See lemma 8.17 in [ArD08b]. □

Theorem 9.12 *Let $g^\circ \in \mathcal{G}_\infty^{p\times p}(0)$, $\mathcal{F}_\ell \subseteq L_2^p([0, \alpha_\ell])$ and $\mathcal{F}_r \subseteq L_2^p([0, \alpha_r])$, where $\alpha_\ell + \alpha_r > 0$. Then there exists a pair $\{b_3, b_4\}$ of normalized entire inner $p \times p$*

mvf's such that (9.32) holds. Moreover, $b_3(\lambda)$ and $b_4(\lambda)$ may be chosen so that

$$\tau(b_3) \le \alpha_\ell \quad \text{and} \quad \tau(b_4) \le \alpha_r, \tag{9.34}$$

with $b_4(\lambda) \equiv I_p$ for left tangential problems and $b_3(\lambda) \equiv I_p$ for right tangential problems.

Conversely, if a mvf $c^\circ \in \mathcal{C}^{p\times p}$ and a pair $\{b_3, b_4\}$ of normalized entire inner $p \times p$ mvf's is given, then there exists infinitely many pairs of sets $\mathcal{F}_\ell \subseteq L_2^p([0, \alpha_\ell])$ and $\mathcal{F}_r \subseteq L_2^p([0, \alpha_r])$ such that the identification (9.32) is in force and equality holds in both of the inequalities in (9.34).

Proof This follows from Theorem 9.10 and Lemma 9.11. □

A GHEP is said to be **determinate** if it has only one solution and **indeterminate** otherwise. It is said to be **completely indeterminate** or **strictly completely indeterminate** if the corresponding GCIP is completely indeterminate or strictly completely indeterminate, respectively.

9.3 The Krein accelerant extension problem

Let $\gamma \in \mathbb{C}^{p\times p}$ and let $\mathcal{A}_a^{p\times p}(\gamma)$ (for $0 < a \le \infty$) denote the class of mvf's $h \in L_1^{p\times p}((-a, a))$ for which

$$\int_0^a \varphi(t)^* \left\{ (\Re\gamma)\varphi(t) + \int_0^a h(t-s)\varphi(s)ds \right\} dt \ \ge 0 \tag{9.35}$$

for every $\varphi \in L_2^p((0, a))$. This condition forces the equality $h(t) = h(-t)^*$ for almost all points $t \in (-a, a)$. Following Krein, a function $h \in L_1^{p\times p}((-a, a))$ that meets the positivity condition (9.35) will be called an **accelerant** and the following extension problem will be called the **accelerant extension problem** (AEP):

AEP$(\gamma, h^\circ; a)$: *Given $\gamma \in \mathbb{C}^{p\times p}$, $h^\circ \in \mathcal{A}_a^{p\times p}(\gamma)$, $0 < a < \infty$, find $h \in \mathcal{A}_\infty^{p\times p}(\gamma)$ such that $h(t) = h^\circ(t)$ a.e. on the interval $[-a, a]$.*

The symbol $\mathcal{A}(\gamma, h^\circ; a)$ will be used to denote the set of solutions to this problem.

The condition (9.35) for $a = \infty$ can be reexpressed in terms of the Fourier transform $\widehat{h}$ of h.

Lemma 9.13 *Let $h \in L_1^{p\times p}(\mathbb{R})$. Then the following three conditions are equivalent:*

(1) $\displaystyle\int_0^\infty \varphi(t)^* \left\{ (\Re\gamma)\varphi(t) + \int_0^\infty h(t-s)\varphi(s)ds \right\} dt \ge 0$ *for every* $\varphi \in L_2^p(\mathbb{R}_+)$.

(2) $\displaystyle\int_{-\infty}^\infty \widehat{\varphi}(\mu)^* \{\Re\gamma + \widehat{h}(\mu)\}\widehat{\varphi}(\mu)d\mu \ge 0$ *for every* $\widehat{\varphi} \in L_2^p(\mathbb{R})$.

(3) $\Re\gamma + \widehat{h}(\mu) \ge 0$ *for every* $\mu \in \mathbb{R}$.

Moreover, the equivalence between these three statements continues to be valid if the inequality ≥ 0 *is replaced by* $\geq \delta\|\varphi\|_{\rm st}^2$ *in (1), by* $\geq \delta\|\widehat{\varphi}\|_{\rm st}^2$ *in (2) and* $\geq \delta I_p$ *in (3) for some* $\delta > 0$.

Proof See lemma 8.36 in [ArD08b]. □

Condition (3) of Lemma 9.13 implies that if $h \in \mathcal{A}_\infty^{p\times p}(\gamma)$, then

$$c(\lambda) = \gamma + 2\int_0^\infty e^{i\lambda t} h(t)dt \quad \text{for } \lambda \in \overline{\mathbb{C}_+} \tag{9.36}$$

belongs to $\mathcal{C}^{p\times p} \cap \mathcal{W}_+^{p\times p}(\gamma)$ (as is easily checked with the help of the Poisson formula (3.7) applied to $c(\lambda)$). Conversely, every mvf $\mathcal{C}^{p\times p} \cap \mathcal{W}_+^{p\times p}(\gamma)$ has a unique representation (9.36) with $h \in \mathcal{A}_\infty^{p\times p}(\gamma)$.

Assertion (3) of Lemma 9.13 implies that $\Re\gamma \geq 0$ if $\mathcal{A}_a^{p\times p}(\gamma) \neq \emptyset$. If $\Re\gamma > 0$, then it is possible to renormalize the data so that $\gamma = I_p$. However, even in this case, there exist mvf's $h^\circ \in \mathcal{A}_a^{p\times p}(I_p)$ for which the AEP$(I_p; h^\circ; a)$ is not solvable; see section 4 of [KrMA86] for an example. Because of this difficulty, we shall focus on mvf's h in the subclass $\mathring{\mathcal{A}}_a^{p\times p}$ of $\mathcal{A}_a^{p\times p}(I_p)$:

$h \in \mathring{\mathcal{A}}_a^{p\times p}$ if $h \in \mathcal{A}_a^{p\times p}(I_p)$ and there exists a $\delta > 0$ such that

$$\int_0^a \varphi(t)^* \left\{\varphi(t) + \int_0^a h(t-s)\varphi(s)ds\right\} dt \geq \delta\|\varphi\|_{\rm st}^2 \tag{9.37}$$

for every $\varphi \in L_2^p((0,a))$, $0 < a \leq \infty$.

$\mathring{\rm A}$EP$(h^\circ; a)$: *Given* $h^\circ \in \mathring{\mathcal{A}}_a^{p\times p}$ *with* $0 < a < \infty$, *find* $h \in \mathring{\mathcal{A}}_\infty^{p\times p}$ *such that* $h(t) = h^\circ(t)$ *for almost all points* $t \in [-a, a]$.

The symbol $\mathring{\mathcal{A}}(h^\circ; a)$ will be used to designate the set of solutions to this problem.

If $h \in \mathring{\mathcal{A}}_\infty^{p\times p}$, then the mvf $c(\lambda)$ which is defined by formula (9.36) with $\gamma = I_p$ satisfies the following properties:

(1) c is continuous on $\mathbb{R}$.
(2) $(\Re c)(\mu) = I_p + \widehat{h}(\mu) > 0$ for every point $\mu \in \mathbb{R}$.
(3) $(\Re c)(\mu) \to I_p$ as $|\mu| \to \infty$.
(4) $c \in H_\infty^{p\times p}$.

Thus,

$$c \in \mathring{\mathcal{C}}^{p\times p} \cap \mathcal{W}_+^{p\times p}(I_p). \tag{9.38}$$

Conversely, if (9.38) holds and c is expressed in the form (9.36) with $\gamma = I_p$, then $h \in \mathring{\mathcal{A}}_\infty^{p\times p}$. These facts will play an important role in the sequel.

Theorem 9.14 *Let* $g \in \mathcal{G}_a^{p\times p}$ *enjoy the following properties:*

(1) *g is locally absolutely continuous on* $(-a, a)$ *and* $g(0) = 0$.

(2) *g' is absolutely continuous on $(-a, 0) \cup (0, a)$, $g'(0+) = -\gamma$ and $g'(0-) = \gamma^*$.*

(3) $g'' \in L_1^{p\times p}((-a, a))$.

Then $h = -g''/2$ belongs to $\mathcal{A}_a^{p\times p}(\gamma)$ and

$$g(t) = \begin{cases} -t\gamma - 2\int_0^t (t-s)h(s)ds & \text{for} \quad 0 \le t < a \\ t\gamma^* + 2\int_t^0 (t-s)h(-s)^*ds & \text{for} \quad -a < t \le 0. \end{cases} \tag{9.39}$$

Conversely, if $h \in \mathcal{A}_a^{p\times p}(\gamma)$, then the function g defined by (9.39) belongs to $\mathcal{G}_a^{p\times p}$ and enjoys the properties (1)–(3) and $h = -g''/2$.

Proof See [KrMA86]. □

If $0 < a < \infty$ and $g \in \mathcal{G}_a^{p\times p}$ admits a representation of the form (9.39) on the interval $[-a, a]$, then h is called the **accelerant of** g **on** $[-a, a]$. Similarly, if $g \in \mathcal{G}_\infty^{p\times p}$ admits a representation of the form (9.39) on $\mathbb{R}$, then h is called the **accelerant of** g **on** $\mathbb{R}$. Moreover, if $g \in \mathcal{G}_\infty^{p\times p}$, $c = c_g$ and $h \in \mathcal{A}_a^{p\times p}(\gamma)$ is the accelerant of the restriction of g to $[-a, a]$, then h is also called the **accelerant of** c **on** $[-a, a]$.

Corollary 9.15 *Let $\gamma \in \mathbb{C}^{p\times p}$, $0 < a < \infty$, and suppose that $g^\circ \in \mathcal{G}_a^{p\times p}$ has an accelerant $h^\circ \in \mathcal{A}_a^{p\times p}(\gamma)$. Then*

$$\mathcal{A}(\gamma, h^\circ; a) = \{-g''/2 : \; g \in \mathcal{G}(g^\circ; a) \quad \textit{and has an accelerant on } \mathbb{R}\}.$$

Formulas (9.8), (9.36) and (9.39) imply that if $g \in \mathcal{G}_\infty^{p\times p}$ admits an accelerant $h \in \mathcal{A}_\infty^{p\times p}(\gamma)$, then $c_g \in \mathcal{C}_a^{p\times p}$ and

$$\Re c_g(\mu) = \Re\gamma + \widehat{h}(\mu) \quad \text{for every } \mu \in \mathbb{R}. \tag{9.40}$$

Conversely, if $c_g \in \mathcal{C}_a^{p\times p}$ and (9.40) holds for some $\gamma \in \mathbb{C}^{p\times p}$ and $h \in L_1^{p\times p}(\mathbb{R})$, then g has accelerant $h \in \mathcal{A}_\infty^{p\times p}(\gamma)$.

If $0 < a < \infty$ and $g^\circ \in \mathcal{G}_a^{p\times p}(0)$ has an accelerant $h^\circ \in \mathcal{A}_a^{p\times p}(\gamma)$, then

$$g^\circ(t) = -\int_0^t v^\circ(s)ds \quad \text{for } -a \le t \le a, \tag{9.41}$$

where

$$v^\circ(t) = \begin{cases} \gamma + 2\int_0^t h^\circ(u)du & \text{for } 0 < t \le a \\ -\gamma^* - 2\int_t^0 h^\circ(u)du & \text{for } -a \le t < 0, \text{ and} \\ h^\circ(-u) = h^\circ(u)^* & \text{a.e. on } [-a, a]. \end{cases} \tag{9.42}$$

In this case Theorem 9.6 is applicable and formulas (9.18)–(9.20) may be rewritten in terms of the accelerant h°, since

$$\begin{aligned}\frac{d}{dt}\int_0^t v^\circ(t-s)\varphi(s)ds &= \frac{d}{dt}\int_0^t \left\{\gamma + 2\int_0^{t-s} h^\circ(u)du\right\}\varphi(s)ds \\ &= \gamma\varphi(t) + 2\int_0^t h^\circ(t-s)\varphi(s)ds\end{aligned} \tag{9.43}$$

and

$$\begin{aligned}\frac{d}{dt}\int_t^a v^\circ(t-s)\varphi(s)ds &= -\frac{d}{dt}\int_t^a \left\{\gamma^* + 2\int_0^{s-t} h^\circ(u)^*du\right\}\varphi(s)ds \\ &= \gamma^*\varphi(t) + 2\int_t^a h^\circ(s-t)^*\varphi(s)ds \\ &= \gamma^*\varphi(t) + 2\int_t^a h^\circ(t-s)\varphi(s)ds.\end{aligned} \tag{9.44}$$

Lemma 9.16 *If $h^\circ \in \mathcal{A}_a^{p\times p}(I_p)$, $0 < a < \infty$, and*

$$K : \varphi \in L_2^p([0,a]) \longrightarrow \int_0^a h^\circ(t-s)\varphi(s)ds, \tag{9.45}$$

then

$$\langle (I+K)\varphi, \varphi\rangle_{\rm st} > 0 \quad \text{for every nonzero } \varphi \in L_2^p([0,a]) \tag{9.46}$$

if and only if there exists a $\delta > 0$ such that

$$\langle (I+K)\varphi, \varphi\rangle_{\rm st} \geq \delta\langle \varphi, \varphi\rangle_{\rm st} \quad \text{for every } \varphi \in L_2^p([0,a]). \tag{9.47}$$

Proof Under assumption (9.46), the operator $I+K$ is a bounded positive operator that is a perturbation of the identity in $L_2^p([0,a])$ by the compact operator K, and 0 is not an eigenvalue of $I+K$. Therefore 0 does not belong to the spectrum of $I+K$. Thus, $I+K$ is a bounded invertible operator in $L_2^p([0,a])$ and (9.47) holds. The converse is self-evident. □

Theorem 9.17 *If $0 < a < \infty$ and $g^\circ \in \mathcal{G}_a^{p\times p}(0)$ has an accelerant $h^\circ \in \mathcal{A}_a^{p\times p}(I_p)$, then the following three conditions are equivalent:*

(1) *The HEP$(g^\circ; a)$ is completely indeterminate.*
(2) *The HEP$(g^\circ; a)$ is strictly completely indeterminate.*
(3) $h^\circ \in \mathring{\mathcal{A}}_a^{p\times p}$.

Proof If (1) is in force and $g \in \mathcal{G}(g^\circ; a)$, then the GSIP$(e_aI_p, I_p; s)$ based on $s = T_{\mathfrak{V}}[c_g]$ is completely indeterminate. Therefore, the NP(Γ) with $\Gamma = \Pi_- M_f|_{H_2^q}$ and $f = e_{-a}s$ is completely indeterminate and hence $I - \Gamma^*\Gamma > 0$ by Remark 4.4. But this implies that (9.46) holds and hence that (1) $\Longrightarrow$ (2) by Lemma 9.16 and

Theorem 9.5. The converse implication follows from Theorem 9.3. Finally the equivalence of (2) and (3) follows from Lemma 9.16 and Theorems 9.14 and 9.5. □

Theorem 9.18 *If $0 < a < \infty$, $g \in \mathcal{G}_a^{p\times p}(0)$ has an accelerant $h \in \mathcal{A}_a^{p\times p}(I_p)$ and K denotes the operator defined in (9.45), then:*

(1) *-1 is an eigenvalue of $K \Longleftrightarrow$ the HEP$(g; a)$ is not completely indeterminate.*
(2) *If $p = 1$, then -1 is an eigenvalue of $K \Longleftrightarrow$ the HEP$(g; a)$ is determinate.*

Proof If -1 is an eigenvalue of K, then $h \notin \mathring{\mathcal{A}}^{p\times p}$. Therefore, the implication $\Longrightarrow$ in (1) follows from Theorem 9.17. The other direction follows from the same theorem, since K is compact and $I + K \geq 0$.

Assertion (2) is immediate from (1), because if $p = 1$, then the HEP$(g; a)$ is not completely indeterminate if and only if it is determinate. □

Example 9.19 *The helical functions g considered in Examples 9.8 and 9.9 both have locally summable accelerants $h = -g''/2$ that can be classified with the help of Theorem 9.14:*

(1) *If $g(t) = \cos t - 1$, then $h \in \mathcal{A}_a^{1\times 1}(0)$ for every $a > 0$.*
(2) *If $g(t) = (1/2)t^2 - |t|$, then $h \in \mathcal{A}_a^{1\times 1}(1)$ for every $a \in (0, 2]$.*

Moreover, as noted earlier in Example 9.8, the HEP$(g; a)$ is determinate for every $a > 0$ when $g(t) = \cos t - 1$. On the other hand, Theorem 9.18 implies that if $g(t) = (1/2)t^2 - |t|$, then the HEP$(g; a)$ is strictly completely indeterminate if $0 < a < 2$ and determinate if $a = 2$.

Theorem 9.20 *If $0 < a < \infty$ and $g^\circ \in \mathcal{G}_a^{p\times p}(0)$ has an accelerant $h^\circ \in \mathring{\mathcal{A}}_a^{p\times p}$, then:*

(1) *The HEP$(g^\circ; a)$ is strictly completely indeterminate.*
(2) *The B-resolvent matrix $B(\lambda)$ of the HEP$(g^\circ; a)$ that was defined in Corollary 9.7 by formula (9.21) can be expressed in terms of h°: The mvf's $\varphi_j(t)$ and $\psi_j(t)$, $j = 1, 2$, in (9.21) are absolutely continuous on $[0, a]$, φ_1 and φ_2 are solutions of the equations*

$$\varphi_1(t) + \int_0^a h^\circ(t-s)\varphi_1(s)ds = I_p + \int_0^t h^\circ(s)ds \tag{9.48}$$

and

$$\varphi_2(t) + \int_0^a h^\circ(t-s)\varphi_2(s)ds = \int_0^t h^\circ(s)ds \tag{9.49}$$

in $L_2^{p\times p}([0,a])$, *respectively,* $h(-s) = h(s)^*$, *and*

$$\psi_j(t) = -\varphi_j(t) - 2\int_t^a h^\circ(t-s)\varphi_j(s)ds \quad for\ j = 1, 2. \tag{9.50}$$

Proof (1) and (2) follow from Theorem 9.17 and Corollary 9.7, respectively, with the help of formulas (9.42)–(9.44). (The assumption $h^\circ \in \mathring{\mathcal{A}}_a^{p\times p}$ fixes $\gamma = I_p$ in (9.42) by definition.) □

Since the operator K that is defined in (9.45) is compact in $L_2^p([0,a])$ when $h^\circ \in L_1^{p\times p}([0,a])$, lemma 7.1 in [AAK71b] guarantees that it has the same nonzero spectrum in $L_2^p([0,a])$ as in the Banach space $\mathcal{X}$ of absolutely continuous $p \times 1$ vvf's on $[0,a]$ with norm

$$\|\varphi\|_{\mathcal{X}} = \max_{t\in[0,a]} \|\varphi(t)\| + \int_0^a \|\varphi'(t)\|dt.$$

Therefore, since $Y^\vee \geq \delta I$ in $L_2^p([0,a])$, it also has a bounded inverse in $\mathcal{X}$. Thus, as the right-hand sides of equations (9.48) and (9.49) are absolutely continuous on $[0,a]$, $\varphi_1(t)$ and $\varphi_2(t)$ are also absolutely continuous on $[0,a]$, as is the mvf

$$\mathcal{B}(t) = \begin{bmatrix} \psi_1(t) & \psi_2(t) \\ \varphi_1(t) & \varphi_2(t) \end{bmatrix} \tag{9.51}$$

due to the preceding discussion and formulas (9.50). Consequently, by (9.21),

$$B(\lambda) = \mathfrak{V} + \frac{1}{\sqrt{2}}(e^{i\lambda a}\mathcal{B}(a) - \mathcal{B}(0)) - \frac{1}{\sqrt{2}}\int_0^a e^{i\lambda t}\mathcal{B}'(t)dt, \tag{9.52}$$

where $\mathcal{B}' \in L_1^{m\times m}([0,a])$ and, in view of formulas (9.48)–(9.50),

$$\mathcal{B}(0) = \begin{bmatrix} \varphi_1(0) - 2I_p & \varphi_2(0) \\ \varphi_1(0) & \varphi_2(0) \end{bmatrix} \quad \text{and} \quad \mathcal{B}(a) = \begin{bmatrix} -\varphi_1(a) & -\varphi_2(a) \\ \varphi_1(a) & \varphi_2(a) \end{bmatrix}. \tag{9.53}$$

A straightforward calculation yields the formula

$$\begin{bmatrix} e^{-i\lambda a}I_p & 0 \\ 0 & I_p \end{bmatrix} \mathfrak{V}\left(\mathfrak{V} + \frac{1}{\sqrt{2}}(e^{i\lambda a}\mathcal{B}(a) - \mathcal{B}(0))\right) = \begin{bmatrix} \varphi_1(a) & \varphi_2(a) \\ I_p - \varphi_1(0) & I_p - \varphi_2(0) \end{bmatrix}.$$

Thus, by the Riemann–Lebesgue lemma, the matrix

$$V = \begin{bmatrix} \varphi_1(a) & \varphi_2(a) \\ I_p - \varphi_1(0) & I_p - \varphi_2(0) \end{bmatrix} = \lim_{\mu\uparrow\infty} \begin{bmatrix} e^{-i\mu a}I_p & 0 \\ 0 & I_p \end{bmatrix} \mathfrak{V}B(\mu) \tag{9.54}$$

as μ tends to infinity through real values. Therefore,

$$V^* j_p V = \lim_{\mu\uparrow\infty} B(\mu)^*\mathfrak{V} \begin{bmatrix} e^{i\mu a}I_p & 0 \\ 0 & I_p \end{bmatrix} j_p \begin{bmatrix} e^{-i\mu a}I_p & 0 \\ 0 & I_p \end{bmatrix} \mathfrak{V}B(\mu) = j_p$$

and hence the mvf

$$B_1(\lambda) = B(\lambda)V^{-1} = \mathfrak{V}\begin{bmatrix} e^{i\lambda a}I_p & 0 \\ 0 & I_p \end{bmatrix} - \frac{1}{\sqrt{2}}\int_0^a e^{i\lambda t}\mathcal{B}'(t)V^{-1}dt \tag{9.55}$$

is also a resolvent matrix for the HEP(g°; a) under consideration. Moreover, $B_1(\lambda)$ is uniquely defined by the conditions

(i) $B_1 \in \mathcal{E} \cap \mathcal{U}(j_p, J_p)$,
(ii) $\{e_a I_p, I_p\} \in ap(B_1)$

and a normalization condition at infinity:

(iii) The mvf

$$\mathfrak{A}_1(\mu) = \begin{bmatrix} e^{-ia\mu}I_p & 0 \\ 0 & I_p \end{bmatrix}\mathfrak{V}B_1(\mu) \tag{9.56}$$

tends to I_m as $\mu \in \mathbb{R}$ tends to ∞ (instead of the condition $B(0) = \mathfrak{V}$).

Theorem 9.21 *If $0 < a < \infty$ and $g^\circ \in \mathcal{G}_a^{p\times p}(0)$ has an accelerant $h^\circ \in \mathring{\mathcal{A}}_a^{p\times p}$, then there exists a unique B-resolvent matrix $B_1(\lambda)$ of the HEP(g°; a) that has the properties (i), (ii) and (iii) listed just above:*

$$B_1(\lambda) = \mathfrak{V}\begin{bmatrix} e^{i\lambda a}I_p & 0 \\ 0 & I_p \end{bmatrix} + \frac{1}{\sqrt{2}}\int_0^a e^{i\lambda t}\mathcal{B}_1(t)dt, \tag{9.57}$$

where

$$\mathcal{B}_1(t) = \begin{bmatrix} (X^\vee e)(t) & -((X^\vee)^* f)(t) \\ e(t) & f(t) \end{bmatrix} \quad \textit{belongs to } L_1^{m\times m}([0,a]) \tag{9.58}$$

and $e(t)$ and $f(t)$ are solutions of the equations

$$e(t) + \int_0^a h^\circ(t-s)e(s)ds = -h^\circ(t-a), \tag{9.59}$$

$$f(t) + \int_0^a h^\circ(t-s)f(s)ds = -h^\circ(t), \tag{9.60}$$

in $L_1^{p\times p}([0,a])$, respectively.

Proof See theorem 8.48 in [ArD08b]. □

Example 9.22 *Let $g(t) = \frac{1}{2}t^2 - |t|$, just as in Example 9.9. Then*

$$h(t) = -\frac{g''(t)}{2} = -\frac{1}{2};$$

and the Cauchy–Schwarz inequality guarantees that

$$\int_0^a \overline{\varphi(t)}\{\varphi(t) + \int_0^a h(t-s)\varphi(s)ds\}dt \geq (1 - a/2)\int_0^a |\varphi(t)|^2 dt,$$

and hence that $h \in \mathring{\mathcal{A}}_a^{1\times 1}$ *if* $0 < a < 2$. *Next, upon invoking the formulas (9.59), (9.41), (9.12), (9.60) and (9.13) it is readily seen that for* $0 \le t \le a < 2$,

$$e(t) = \frac{1}{2-a}, \quad (X^\vee e)(t) = \frac{d}{dt}\int_0^t (1-t+s)e(s)ds = \frac{1-t}{2-a},$$

$$f(t) = \frac{1}{2-a} \quad \text{and} \quad (X^\vee f)^*(t) = \frac{d}{dt}\int_t^a (1-t+s)f(s)ds = \frac{1+t-a}{2-a}.$$

Therefore, the mvf

$$A_1(a,\lambda) \stackrel{\text{def}}{=} B_1(\lambda)\mathfrak{V}$$

can be calculated explicitly from the formulas in Theorem 9.21; an easier calculation yields the formula

$$A_1(a,0) = \begin{bmatrix} \frac{2-a}{2} & 0 \\ 0 & \frac{2}{2-a} \end{bmatrix}$$

and hence, in view of (2.79) (or, looking ahead a bit, (12.6)), the Hamiltonian of the canonical differential system corresponding to the DK-system with matrizant $A_1(a,\lambda)$ *for* $0 < a < 2$ *is*

$$H(a) = A_1(a,0)\mathfrak{V}\begin{bmatrix} 1 & 0 \\ 0 & 0 \end{bmatrix}\mathfrak{V}A_1(a,0)^* = \frac{1}{2}A_1(a,0)\begin{bmatrix} 1 & -1 \\ -1 & 1 \end{bmatrix}A_1(a,0)^*$$

$$= \frac{1}{2}\begin{bmatrix} \left(\frac{2-a}{2}\right)^2 & -1 \\ -1 & \left(\frac{2}{2-a}\right)^2 \end{bmatrix} = \begin{bmatrix} \frac{(2-a)^2}{8} & 0 \\ 0 & \frac{2}{(2-a)^2} \end{bmatrix} + \frac{1}{2}J_1,$$

which coincides with the second formula in (9.26).

Theorem 9.23 *If* $0 < a < \infty$ *and* $g^\circ \in \mathcal{G}_a^{p\times p}(0)$ *has an accelerant* $h^\circ \in \mathring{\mathcal{A}}_a^{p\times p}$, *and if* $B(\lambda)$ *and* $B_1(\lambda)$ *are the B-resolvent matrices for the HEP*$(g^\circ; a)$ *considered in Theorems 9.20 and 9.21, respectively, then the formulas*

$$\mathcal{G}(g^\circ; a) = \{g \in \mathcal{G}_\infty^{p\times p}(0) : c_g \in T_{B_1}[\mathcal{S}^{p\times p}]\} \tag{9.61}$$

and

$$T_{B_1}[\mathring{\mathcal{S}}^{p\times p} \cap \mathcal{W}_+^{p\times p}(0)] = T_{B_1}[\mathring{\mathcal{S}}^{p\times p}] \cap \mathcal{W}_+^{p\times p}(I_p)$$

define a one-to-one correspondence between the set $\mathring{\mathcal{A}}(h^\circ; a)$ *of solutions* h *of the* $\mathring{\text{A}}$EP$(h^\circ; a)$ *and the set of mvf's* $\varepsilon \in \mathring{\mathcal{S}}^{p\times p} \cap \mathcal{W}_+^{p\times p}(0)$, *i.e.,*

$$\mathring{\mathcal{A}}(h^\circ; a) = \{h \in \mathcal{A}_\infty^{p\times p}(0) : I_p + 2\int_0^\infty e^{i\lambda t}h(t)dt \in T_{B_1}[\mathring{\mathcal{S}}^{p\times p} \cap \mathcal{W}_+^{p\times p}(0)]\}. \tag{9.62}$$

Moreover, formula (9.62) may be rewritten as

$$\mathring{\mathcal{A}}(h^\circ; a) = \left\{ h \in \mathcal{A}_\infty^{p\times p}(0) : I_p + 2\int_0^\infty e^{i\lambda t} h(t)dt \in T_B[\mathring{\mathcal{S}}^{p\times p} \cap \mathcal{W}_+^{p\times p}(\gamma_{h^\circ})] \right\}, \tag{9.63}$$

where $\gamma_{h^\circ} \in \mathring{\mathcal{S}}_{\text{const}}^{p\times p}$ *and is given by the formula*

$$\gamma_h = -\{I_p - \varphi_1(0)^*\}\{I_p - \varphi_2(0)^*\}^{-1}, \tag{9.64}$$

in which φ_1 *and* φ_2 *are solutions of (9.48) and (9.49), respectively.*

Proof See theorem 8.49 in [ArD08b]. □

Theorem 9.24 *Let* $B_1(\lambda)$ *be any* $m \times m$ *mvf that satisfies properties (i) and (ii) in the list just below (9.55) and is such that*

$$\mathfrak{A}_1(\mu) = \begin{bmatrix} e^{-ia\mu} I_p & 0 \\ 0 & I_p \end{bmatrix} \mathfrak{V} B_1(\mu) \quad \text{belongs to } \mathcal{W}^{m\times m}(I_m).$$

Then

$$T_{B_1}[0] = I_p + 2\int_0^\infty e^{i\lambda t}\widetilde{h}(t)dt, \quad \text{where } \widetilde{h} \in \mathring{\mathcal{A}}_\infty^{p\times p}$$

and B_1 *is the B-resolvent matrix of the HEP(g; a) based on the mvf* $g \in \mathcal{G}_a^{p\times p}(0)$ *with accelerant h equal to the restriction of* $\widetilde{h}$ *to the interval* $[0, a]$.

Proof See theorem 8.50 in [ArD08b]. □

9.4 Continuous analogs of the Schur extension problem

This section is devoted to an analog of the helical extension problem in which the Carathéodory class $\mathcal{C}^{p\times p}$ is replaced by the Schur class $\mathcal{S}^{p\times q}$. The formulation of the problem rests on the fact that the formula

$$s(\lambda) = -i\lambda \int_0^\infty e^{i\lambda t} v(t)dt \quad \text{for } \lambda \in \mathbb{C}_+, \tag{9.65}$$

defines a one-to-one correspondence between the class of mvf's $s \in \mathcal{S}^{p\times q}$ and the mvf's v that belong to the class $\mathfrak{V}_\infty^{p\times q}$ of $p \times q$ mvf's in $L_{2,loc}^{p\times q}(\mathbb{R}_+)$ that meet the following conditions:

(1) The $p \times 1$ mvf $\int_0^t v(t-s)\varphi(s)ds$ is absolutely continuous on the interval $[0, a]$ and its derivative

$$(X^\vee \varphi)(t) = \frac{d}{dt}\int_0^t v(t-s)\varphi(s)ds \tag{9.66}$$

belongs to $L_2^p([0, a])$ for every $a > 0$ and every choice of $\varphi \in L_2^q([0, \infty))$.

(2) The operator $X^\vee$ from $L_2^q([0, \infty))$ into $L_2^p([0, \infty))$ that is defined by formula (9.66) is contractive.

These properties will be justified in Theorem 9.27.

The given data for the problem of interest in this section is a mvf $v^\circ \in L_2^{p\times q}([0,a])$ and the objective is to describe the set

$$\mathfrak{V}(v^\circ; a) = \{v \in \mathfrak{V}_\infty^{p\times q} : v(t) = v^\circ(t) \quad \text{a.e. on } [0,a]\}. \tag{9.67}$$

Since the restrictions of mvf's $v \in \mathfrak{V}_\infty^{p\times q}$ to the interval $[0,a]$ belong to the class $\mathfrak{V}_a^{p\times q}$ of $v \in L_2^{p\times q}([0,a])$ such that

(1) $\int_0^t v(t-s)\varphi(s)ds$ is absolutely continuous on $[0,a]$ and
(2) the formula

$$(X^\vee\varphi)(t) = \frac{d}{dt}\int_0^t v(t-s)\varphi(s)ds \quad \text{for } \varphi \in L_2^q([0,a]) \tag{9.68}$$

defines a contractive operator from $L_2^q([0,a])$ into $L_2^p([0,a])$,

it is clear that if $\mathfrak{V}(v^\circ; a) \neq \emptyset$, then $v^\circ \in \mathfrak{V}_a^{p\times q}$. The converse statement will be established below in Theorem 9.28.

For future use we note that the adjoint $(X^\vee)^*$ of the operator $X^\vee$ in (9.68) is given by the formula

$$((X^\vee)^*\psi)(t) = -\frac{d}{dt}\int_t^a v(u-t)^*\psi(u)du \quad \text{for } \psi \in L_2^p([0,a]). \tag{9.69}$$

The next lemma presents some useful connections between the classes $\mathcal{S}^{p\times q}$ and $\mathcal{C}^{m\times m}$ for $m = p+q$ that will be exploited to obtain conclusions for the problem of extending mvf's in $\mathfrak{V}_a^{p\times q}$ from the results obtained earlier for the HEP.

Lemma 9.25 *If*

$$\widetilde{c}(\lambda) = \frac{1}{2}\begin{bmatrix} I_p & 2s(\lambda) \\ 0 & I_q \end{bmatrix}, \tag{9.70}$$

then:

(1) $s \in \mathcal{S}^{p\times q} \Longleftrightarrow \widetilde{c} \in \mathcal{C}^{m\times m}$.
(2) $s \in \mathring{\mathcal{S}}^{p\times q} \Longleftrightarrow \widetilde{c} \in \mathring{\mathcal{C}}^{m\times m}$.
(3) $s \in \mathcal{W}_+ \cap \mathcal{S}^{p\times q} \Longleftrightarrow \widetilde{c} \in \mathcal{W}_+ \cap \mathcal{C}^{m\times m}$.
(4) $s \in \mathcal{W}_+ \cap \mathring{\mathcal{S}}^{p\times q} \Longleftrightarrow \widetilde{c} \in \mathcal{W}_+ \cap \mathring{\mathcal{C}}^{m\times m}$.

Proof Clearly $s(\lambda)$ is holomorphic in $\mathbb{C}_+$ if and only if $\widetilde{c}(\lambda)$ is holomorphic in $\mathbb{C}_+$. The rest is an easy consequence of the identities

$$\widetilde{c}(\lambda) + \widetilde{c}(\lambda)^* = \begin{bmatrix} I_p & s(\lambda) \\ s(\lambda)^* & I_q \end{bmatrix}$$

$$= \begin{bmatrix} I_p & 0 \\ s(\lambda)^* & I_q \end{bmatrix}\begin{bmatrix} I_p & 0 \\ 0 & I_q - s(\lambda)^*s(\lambda) \end{bmatrix}\begin{bmatrix} I_p & s(\lambda) \\ 0 & I_q \end{bmatrix}. \tag{9.71}$$

□

Next, we will establish an integral representation formula for $s \in \mathcal{S}^{p\times q}$ analogous to the integral representation formula (9.8) for $c \in \mathcal{C}^{p\times p}$. To this end, it is convenient to imbed each continuous $p \times q$ mvf $g(t)$ on $[0, \infty)$ into the $m \times m$ mvf (with $m = p + q$)

$$\widetilde{g}(t) = \frac{1}{2}\begin{bmatrix} -tI_p & 2g(t) \\ 0 & -tI_q \end{bmatrix} \quad \text{for } t \geq 0, \tag{9.72}$$

and to set

$$\widetilde{g}(t) = \widetilde{g}(-t)^* \quad \text{for } t < 0 \tag{9.73}$$

and let

$$\mathfrak{S}_{\infty}^{p\times q} = \{\text{continuous } p \times q \text{ mvf's } g(t) \text{ on } [0, \infty) \text{ for which the kernel}$$
$$\widetilde{k}(t, s) = \widetilde{g}(t - s) - \widetilde{g}(t) - \widetilde{g}(-s) \tag{9.74}$$
$$\text{is positive on}[0, \infty) \times [0, \infty)\}.$$

We remark that if $g \in \mathfrak{G}_{\infty}^{p\times q}$, then automatically $g(0) = 0$. Thus,

$$g \in \mathfrak{S}_{\infty}^{p\times q} \Longleftrightarrow \widetilde{g} \in \mathcal{G}_{\infty}^{m\times m} \quad \text{and} \quad \widetilde{g}(0) = 0.$$

Let $\mathfrak{G}_a^{p\times q}$ denote the class of continuous $p \times q$ mvf's $g(t)$ on $[0, a]$ such that the mvf $\widetilde{g}$ defined by (9.72) and (9.73) on $[-a, a]$ belongs to $\mathcal{G}_a^{m\times m}(0)$ with $m = p + q$ and hence that the kernel defined by (9.74) is positive on $[0, a] \times [0, a]$.

Theorem 9.26 *The formula*

$$s(\lambda) = \lambda^2 \int_0^{\infty} e^{i\lambda t} g(t)dt \quad \textit{for } \lambda \in \mathbb{C}_+ \tag{9.75}$$

defines a one-to-one mapping from the class of $g \in \mathfrak{S}_{\infty}^{p\times q}$ *onto* $\mathcal{S}^{p\times q}$.

Proof If $g \in \mathfrak{S}_{\infty}^{p\times q}$, then the $m \times m$ mvf $\widetilde{g}$ defined by (9.72) and (9.73) belongs to the class $\mathcal{G}_{\infty}^{m\times m}(0)$ and therefore, by Theorem 9.3 the mvf

$$\widetilde{c}(\lambda) = \lambda^2 \int_0^{\infty} e^{i\lambda t}\widetilde{g}(t)dt \quad \text{for } \lambda \in \mathbb{C}_+$$

belongs to the Carathéodory class $\mathcal{C}^{m\times m}$. Moreover, since

$$\lambda^2 \int_0^{\infty} e^{i\lambda t} t dt = -1 \quad \text{for } \lambda \in \mathbb{C}_+, \tag{9.76}$$

it follows from the special form of $\widetilde{g}$ that $\widetilde{c}$ is of the form (9.70) and hence that (9.75) defines a one-to-one mapping of $\mathfrak{S}^{p\times q}$ into $\mathcal{S}^{p\times q}$.

However, it is easily seen that the argument can be reversed to show that every $s \in \mathcal{S}^{p\times q}$ can be obtained this way and hence that the mapping defined by (9.75) is one-to-one onto, as claimed. □

Theorem 9.27 *If $s \in H_\infty^{p\times q}$, then it admits exactly one representation of the form (9.75) with $\|s\|_\infty^{-1} g \in \mathfrak{S}_\infty^{p\times q}$ if $\|s\|_\infty > 0$ and $g(t) \equiv 0$ if $\|s\|_\infty = 0$. Moreover, in this representation,*

(1) *$g(t)$ is locally absolutely continuous on $\mathbb{R}_+$ and meets the even stronger condition*

$$g(t) = -\int_0^t v(s)ds \quad \text{for } t \geq 0 \text{ and some } v \in L_{2,\mathrm{loc}}^{p\times q}(\mathbb{R}_+). \tag{9.77}$$

(2) *The mvf $s(\lambda)$ can be reexpressed in terms of v as*

$$s(\lambda) = -i\lambda \int_0^\infty e^{i\lambda t} v(t)dt \quad \text{for } \lambda \in \mathbb{C}_+. \tag{9.78}$$

(3) *The $p \times 1$ mvf $\int_0^t v(t-s)\varphi(s)ds$ is absolutely continuous on the interval $[0, a]$ and its derivative*

$$(X^\vee \varphi)(t) = \frac{d}{dt}\int_0^t v(t-s)\varphi(s)ds \tag{9.79}$$

belongs to $L_2^p([0, a])$ for every $a > 0$ and every choice of $\varphi \in L_2^q([0, \infty))$.

(4) *The operator $X^\vee$ from $L_2^q([0, \infty))$ into $L_2^p([0, \infty))$ that is defined by formula (9.79) is bounded.*

(5) *The operator $X^\vee$ is unitarily equivalent to the operator*

$$X = M_s|_{H_2^q} \tag{9.80}$$

of multiplication by $s(\lambda)$ acting between H_2^q and H_2^p in the sense that

$$(M_s\widehat{\varphi})(\lambda) = \int_0^\infty e^{i\lambda t}(X^\vee\varphi)(t)dt \quad \text{for } \varphi \in L_2^q([0, \infty)) \tag{9.81}$$

(i.e., strictly speaking $XT_1 = T_2X^\vee$, where T_1 (resp., T_2) denotes the one-sided Fourier transform from $L_2^q(\mathbb{R}_+)$ onto H_2^q (resp., from $L_2^p(\mathbb{R}_+)$ onto H_2^p) and hence

$$\|X^\vee\| = \|M_s\| = \|s\|_\infty.$$

(6) *The adjoints of the operators X and $X^\vee$ are given by the formulas*

$$X^*\widehat{\psi} = \Pi_+ M_{s^*}\widehat{\psi} = \int_0^\infty e^{i\lambda t}((X^\vee)^*\psi)(t)dt \tag{9.82}$$

and

$$((X^\vee)^*\psi)(t) = -\frac{d}{dt}\int_t^\infty v(s-t)^*\psi(s)ds, \tag{9.83}$$

where the latter is valid for the dense set of $\psi \in L_2^p([0, \infty))$ with compact support.

Conversely, if $v(t)$ is a mvf that enjoys the properties (3)–(4) and s is defined by (9.78), then $s \in H_\infty^{p\times q}$ and admits a representation of the form (9.75). Moreover,

$$s \in \mathcal{S}^{p\times q} \Longleftrightarrow g \in \mathfrak{G}_\infty^{p\times q} \Longleftrightarrow v \in \mathfrak{V}_\infty^{p\times q} \Longleftrightarrow \|X^\vee\| \le 1$$
$$s \in \mathring{\mathcal{S}}^{p\times p} \Longleftrightarrow \|X\| < 1$$
$$s \in H_\infty^{p\times q} \cap \mathcal{C}^{p\times p} \Longleftrightarrow X^\vee + (X^\vee)^* \ge 0$$
$$s \in \mathring{\mathcal{C}}^{p\times p} \Longleftrightarrow X^\vee + (X^\vee)^* \ge \delta I \quad \text{for some } \delta > 0.$$

Proof See theorems 5.2 and 5.3 in [ArD02a]. □

Clearly the restrictions of mvf's $v \in \mathfrak{V}_\infty^{p\times q}$ to the interval $[0, a]$ belong to the class $\mathfrak{V}_a^{p\times q}$. The converse is also true:

Theorem 9.28 *If $0 < a < \infty$ and $v^\circ \in L_2^{p\times q}([0, a])$, then*

$$\mathfrak{V}(v^\circ; a) \ne \emptyset \Longleftrightarrow v^\circ \in \mathfrak{V}_a^{p\times q}. \tag{9.84}$$

Proof If $v^\circ \in \mathfrak{V}_a^{p\times q}$ the linear operator X defined by the formula

$$(X\widehat{\varphi})(\lambda) = \begin{cases} \int_0^a e^{i\lambda t}\left(\frac{d}{dt}\int_0^t v^\circ(t-u)\varphi(u)du\right)dt & \text{for } \varphi \in L_2^q([0, a]) \\ 0 & \text{for } \varphi \in L_2^q([a, \infty)) \end{cases} \tag{9.85}$$

is a contractive operator from H_2^q into $\mathcal{H}(e_a I_p)$ that meets the condition (4.33). Therefore, by Remark 4.11, there exists a mvf $s \in \mathcal{S}^{p\times q}$ such that

$$X\widehat{\varphi} = \Pi_{\mathcal{H}(e_a I_p)} s\widehat{\varphi} \quad \text{for every } \varphi \in L_2^q([0, a]).$$

However, by Theorem 9.27 the mvf s admits a representation of the form (9.78) in terms of some mvf $v \in \mathfrak{V}_\infty^{p\times q}$. But this in turn implies that

$$(X\widehat{\varphi})(\lambda) = \int_0^a e^{i\lambda t}\left(\frac{d}{dt}\int_0^t v(t-u)\varphi(u)du\right)dt \quad \text{for every } \varphi \in L_2^q([0, a]).$$

Thus,

$$\frac{d}{dt}\int_0^t v(t-u)\varphi(u)du = \frac{d}{dt}\int_0^t v^\circ(t-u)\varphi(u)du \quad \text{for a.a. } t \in [0, a].$$

The choice $\varphi(u) = \eta \in \mathbb{C}^q$ on $[0, a]$ yields $v(t) = v^\circ(t)$ a.e. on $[0, a]$. □

The continuous analog of the **Schur extension problem** is formulated as follows:

SEP$(v^\circ; a)$: *Given $v^\circ \in L_2^{p\times q}([0, a])$, $a > 0$, describe the set*

$$\mathcal{S}(v^\circ; a) = \left\{-i\lambda\int_0^\infty e^{i\lambda t} v(t)dt \in \mathcal{S}^{p\times q} : v \in \mathfrak{V}(v^\circ; a)\right\}.$$

In view of Theorem 9.28,

$$\mathcal{S}(v^\circ; a) \neq \emptyset \Longleftrightarrow \mathfrak{V}(v^\circ; a) \neq \emptyset \Longleftrightarrow v^\circ \in \mathfrak{V}_a^{p\times q}.$$

Moreover,

$$\mathcal{S}(v^\circ; a) = \mathcal{S}(e_a I_p, I_q; X), \tag{9.86}$$

where X is defined in terms of v° by formula (9.85). In view of the identification (9.86), all results that were obtained earlier for the GSIP$((b_1, b_2; X)$ are applicable to the SEP$(v^\circ; a)$. Thus, Theorems 4.10 and 4.12 yield:

Theorem 9.29 *If the SEP$(v^\circ; a)$ is completely indeterminate, then:*

(1) *There exists exactly one mvf $\mathring{W} \in \mathcal{U}^\circ(j_{pq})$ such that*

$$\mathcal{S}(v^\circ; a) = T_{\mathring{W}}[\mathcal{S}^{p\times q}] \quad \textit{and} \quad \{e_a I_p, I_q\} \in ap(\mathring{W}). \tag{9.87}$$

Moreover,

$$\mathring{W} \in \mathcal{E} \cap \mathcal{U}_{rR}^\circ(j_{pq}). \tag{9.88}$$

(2) *If*

$$\mathcal{S}(v^\circ; a) = T_W[\mathcal{S}^{p\times q}] \quad \textit{for some mvf } W \in \mathcal{P}^\circ(j_{pq}), \tag{9.89}$$

then

$$\begin{aligned} &W \in \mathcal{E} \cap \mathcal{U}_{rR}(j_{pq}) \quad \textit{and} \quad \{e_\alpha I_p, e_\beta I_q\} \in ap(W) \\ &\textit{for some choice of } \alpha \geq 0 \textit{ and } \beta \geq 0 \textit{ such that } \alpha + \beta = a. \end{aligned} \tag{9.90}$$

Moreover, in this case, if $W(0) = I_m$, then

$$W = e_{-\beta}\mathring{W}.$$

Conversely, if (9.90) is in force for some mvf $W \in \mathcal{U}(j_{pq})$, then there exists exactly one mvf $v^\circ \in \mathfrak{V}_a^{p\times q}$ such that (9.89) holds.

Moreover, the formulas for the resolvent matrix of a strictly completely indeterminate GSIP$(b_1, b_2; X)$ that were established in Theorem 4.75 simplify, since $b_1 = e_a I_p$ and $b_2 = I_q$, and hence

$$L_X = \begin{bmatrix} I \\ X^* \end{bmatrix} : \mathcal{H}(e_a I_p) \longrightarrow \mathcal{H}(W),$$

$$\Delta_X = I - XX^* : \mathcal{H}(e_a I_p) \longrightarrow \mathcal{H}(e_a I_p)$$

and

$$F_0^X(\lambda) = [k_0^{e_a I_p}(\lambda) \quad X k_0^{e_a I_q}(\lambda)],$$

since we may choose $\widehat{b}_1 = e_a I_q$. Thus,

$$\mathring{W}(\lambda) = I_m + i\lambda \int_0^a e^{i\lambda t} \begin{bmatrix} e_1(t) & f_1(t) \\ e_2(t) & f_2(t) \end{bmatrix} dt, \tag{9.91}$$

where $e_1 \in L_2^{p\times p}([0,a])$, $e_2 \in L_2^{q\times p}([0,a])$, $f_1 \in L_2^{p\times q}([0,a])$ and $f_2 \in L_2^{q\times q}([0,a])$ are solutions of the following two systems of equations:

$$\begin{aligned} e_1(t) - (X^\vee e_2)(t) &= I_p \\ -((X^\vee)^* e_1)(t) + e_2(t) &= 0 \quad \text{for } 0 \le t \le a \end{aligned} \tag{9.92}$$

and

$$\begin{aligned} f_1(t) - (X^\vee f_2)(t) &= v^\circ(t) \\ -((X^\vee)^* f_1)(t) + f_2(t) &= 0 \quad \text{for } 0 \le t \le a \end{aligned} \tag{9.93}$$

and the operators $X^\vee$ and $(X^\vee)^*$ that are defined by formulas (9.68) and (9.69) with $v(t) = v^\circ(t)$ a.e. on $[0,a]$ act on mvf's column by column.

9.5 A bitangential generalization of the Schur extension problem

In this section we exploit the integral representation (9.75) with $g \in \mathfrak{S}_\infty^{p\times q}$ for $s \in \mathcal{S}^{p\times q}$ to formulate extension problems analogous to the helical extension problems considered in Section 9.2. The given data for the problem will be a mvf $g^\circ \in \mathfrak{S}_\infty^{p\times q}$, a set $\mathcal{F}_\ell \subseteq L_2^p([0,\alpha_\ell])$ and a set $\mathcal{F}_r \subseteq L_2^q([0,\alpha_r])$. The problem is to describe the set

$$\mathfrak{S}(g^\circ; \mathcal{F}_\ell, \mathcal{F}_r) = \{g \in \mathfrak{S}_\infty^{p\times q} : \ (9.27) - (9.29) \quad \text{hold} \\ \text{for every } h^\ell \in \mathcal{F}_\ell \text{ and } h^r \in \mathcal{F}_r\}.$$

There is a natural analog of Lemma 9.11 in this setting that is formulated in terms of the semigroups T_τ^ℓ (resp., T_τ^r), $\tau \ge 0$, defined in (9.33), but now acting on $L_2^p([0,\alpha_\ell])$ (resp., $L_2^q([0,\alpha_r])$). Since the domains of these two semigroups will always be clear from the context, the superscripts ℓ and r will be dropped in order to simplify the notation.

Lemma 9.30 *Let $\mathcal{F}_\ell \subseteq L_2^p([0,\alpha_\ell])$, $\mathcal{F}_r \subseteq L_2^q([0,\alpha_r])$ and $g^\circ \in \mathfrak{S}_\infty^{p\times q}(0)$. Then:*

(1) $\mathfrak{S}(g^\circ; \mathcal{F}_\ell, \mathcal{F}_r) = \mathfrak{S}(g^\circ; \bigvee_{\tau\ge 0}\{T_\tau \mathcal{F}_\ell\}, \bigvee_{\tau\ge 0}\{T_\tau \mathcal{F}_r\})$.

(2) *There exists a unique pair of normalized entire inner mvf's $b_1 \in \mathcal{E} \cap \mathcal{S}_{\text{in}}^{p\times p}$ and $b_2 \in \mathcal{S}_{\text{in}}^{q\times q}$ such that*

$$\left(\bigvee_{\tau\ge 0}\{T_\tau \mathcal{F}_\ell\}\right)^\wedge = \mathcal{H}(b_1) \quad \text{and} \quad \left(\bigvee_{\tau\ge 0}\{T_\tau \mathcal{F}_r\}\right)^\wedge = \mathcal{H}(b_2). \tag{9.94}$$

Moreover, if $\mathcal{F}'_\ell$ and $\mathcal{F}'_r$ denote the set of columns of the mvf's $h_{b_1}(t)$ and $h_{b_2}(t)^*$, respectively, in the representation formulas (9.1) for $b_1(\lambda)$ and $b_2(\lambda)$, then

(3) $\mathfrak{S}(g^\circ; \mathcal{F}_\ell, \mathcal{F}_r) = \mathfrak{S}(g^\circ; \mathcal{F}'_\ell, \mathcal{F}'_r)$.

Proof Assertion (2) is verified in Lemma 9.11. Let

$$\widetilde{g}^\circ(t) = \frac{1}{2}\begin{bmatrix} -tI_p & 2g^\circ(t) \\ 0 & -tI_q \end{bmatrix} \quad \text{for } t \geq 0,$$

$$\widetilde{b}_3 = \begin{bmatrix} b_1 & 0 \\ 0 & I_q \end{bmatrix}, \quad \widetilde{b}_4 = \begin{bmatrix} I_p & 0 \\ 0 & b_2 \end{bmatrix},$$

$$\widetilde{\mathcal{F}}_\ell = \left\{\begin{bmatrix} h_\ell \\ 0 \end{bmatrix} : h_\ell \in \mathcal{F}_\ell\right\} \quad \text{and} \quad \widetilde{\mathcal{F}}_r = \left\{\begin{bmatrix} 0 \\ h_r \end{bmatrix} : h_r \in \mathcal{F}_r\right\}.$$

Then, in view of Lemma 9.11,

$$\mathcal{G}(\widetilde{g}^\circ; \widetilde{\mathcal{F}}_\ell; \widetilde{\mathcal{F}}_r) = \mathcal{G}(\widetilde{g}^\circ; \widetilde{\mathcal{F}}'_\ell; \widetilde{\mathcal{F}}'_r)).$$

This completes the proof, since a mvf $\widetilde{g} \in \mathcal{G}(\widetilde{g}^\circ; \widetilde{\mathcal{F}}_\ell, \widetilde{\mathcal{F}}_r)$ is of the form (9.72) for $t \geq 0$ if and only if the mvf g in this representation belongs to the set $\mathfrak{S}(g^\circ; \mathcal{F}_\ell, \mathcal{F}_r)$. □

The **notation** s_g for the mvf $s \in \mathcal{S}^{p\times q}$ that is expressed in terms of $g \in \mathfrak{S}_\infty^{p\times q}$ by formula (9.75) will be used below.

Theorem 9.31 *If $g^\circ \in \mathfrak{S}_\infty^{p\times q}$, $\mathcal{F}_\ell \subseteq L_2^{p\times p}([0, \alpha_\ell])$ and $\mathcal{F}_r \subseteq L_2^{q\times q}(([0, \alpha_r])$, $\alpha_\ell + \alpha_r > 0$, are specified, then there exists exactly one mvf $s^\circ = s_{g^\circ}$ and exactly one pair b_1, b_2 of entire inner mvf's with $b_1(0) = I_p$ and $b_2(0) = I_q$ such that (1) and (2) of Lemma 9.30 hold. Moreover, under this correspondence,*

$$\mathcal{S}(b_1, b_2; s^\circ) = \{s_g : g \in \mathfrak{S}(g^\circ; \mathcal{F}_\ell, \mathcal{F}_r)\}. \tag{9.95}$$

Conversely, if $s^\circ \in \mathcal{S}^{p\times p}$, $b_1 \in \mathcal{E} \cap \mathcal{S}_{\text{in}}^{p\times p}$ with $b_1(0) = I_p$ and $b_2 \in \mathcal{E} \cap \mathcal{S}^{q\times q}$ with $b_2(0) = I_q$ are specified, then there exists exactly one mvf $g^\circ \in \mathfrak{S}_\infty^{p\times q}$ such that $s^\circ = s_{g^\circ}$ and (in view of Lemma 9.30) infinitely many pairs of sets $\mathcal{F}_\ell \subseteq L_2^p([0, \alpha_\ell])$ and $\mathcal{F}_r \subseteq L_2^q([0, \alpha_r])$ such that (9.94) holds. Moreover, under this correspondence (9.95) also holds.

Proof This follows from Lemmas 9.11 and 9.30. □

The operators X_{ij} defined in (4.24) will now be reexpressed in terms of the mvf's h_{b_1}, h_{b_2} and v that are taken from the representation (9.1) for a normalized entire inner mvf b and the representation (9.78) for the mvf s°.

Lemma 9.32 *Let $b(\lambda)$ be an entire inner $r \times r$ mvf of exponential type a with integral representation (9.1). Then*

$$(\Pi_{\mathcal{H}(b)}\widehat{f})(\omega) = b(\omega) \int_{-a}^{0} e^{i\omega t} \left\{ \frac{d}{dt} \int_{0}^{a+t} h_b(s-t)^* f(s) ds \right\} dt \tag{9.96}$$

$$(\Pi_{\mathcal{H}_*(b)}\widehat{f})(\omega) = -b^{\#}(\omega) \int_{0}^{a} e^{i\omega t} \left\{ \frac{d}{dt} \int_{t-a}^{0} h_b(t-s) f(s) ds \right\} dt \tag{9.97}$$

for every choice of $\widehat{f} \in L_2^r(\mathbb{R})$ and $\omega \in \mathbb{C}$.

Proof Since $b(\overline{\omega})b(\omega)^* = I_r$, formula (9.1) implies that

$$\begin{aligned} k_\omega^b(\lambda)b(\overline{\omega}) &= \frac{b(\overline{\omega}) - b(\lambda)}{-2\pi i(\lambda - \overline{\omega})} = \int_0^a \frac{i\lambda e^{i\lambda s} - i\overline{\omega}e^{i\overline{\omega}s}}{2\pi i(\lambda - \overline{\omega})} h_b(s) ds \\ &= \frac{1}{2\pi} \int_0^a e^{i\lambda s} \left\{ h_b(s) + i\overline{\omega}e^{-i\overline{\omega}s} \int_s^a e^{i\overline{\omega}u} h_b(u) du \right\} ds. \end{aligned}$$

Consequently, upon using the reproducing kernel formula

$$\xi^*(\Pi_{\mathcal{H}(b)}\widehat{f})(\omega) = \langle \widehat{f}, k_\omega^b \xi \rangle_{\rm st}$$

to calculate the projection of $\widehat{f} \in H_2^r$ onto $\mathcal{H}(b)$ and invoking Parseval's formula, we find that

$$\begin{aligned} (\Pi_{\mathcal{H}(b)}\widehat{f})(\omega) &= b(\omega) \int_0^a \left\{ h_b(s)^* - i\omega e^{i\omega s} \int_s^a e^{-i\omega u} h_b(u)^* du \right\} f(s) ds \\ &= b(\omega) \int_0^a h_b(u)^* \left\{ f(u) - i\omega e^{-i\omega u} \int_0^u e^{i\omega s} f(s) ds \right\} du \\ &= b(\omega) \left\{ \text{①} + \text{②} \right\}, \end{aligned}$$

where

$$\begin{aligned} \text{②} &= -i\omega \int_0^a h_b(u)^* \left\{ e^{-i\omega u} \int_0^u e^{i\omega s} f(s) ds \right\} du \\ &= -i\omega \int_0^a h_b(u)^* \left\{ \int_0^u e^{-i\omega t} f(u-t) dt \right\} du \\ &= -i\omega \int_{-a}^0 e^{i\omega t} \left\{ \int_0^{a+t} h_b(s-t)^* f(s) ds \right\} dt. \end{aligned}$$

Next, upon identifying $i\omega e^{i\omega t}$ as the derivative of the exponential and integrating by parts, we obtain

$$\text{②} = -\int_0^a h_b(s)^* f(s) ds + \int_{-a}^0 e^{i\omega t} \left\{ \frac{d}{dt} \int_0^{a+t} h_b(s-t)^* f(s) ds \right\} dt .$$

Formula (9.96) now drops out easily upon combining terms.

Similar considerations based on (9.1) lead to the formulas

$$\begin{aligned}\ell_\omega^b(\lambda) &= \left\{\frac{b(\bar{\lambda})^* - b(\omega)^*}{-2\pi i(\lambda - \overline{\omega})}\right\} b(\omega)^{-*} \\ &= \frac{1}{2\pi}\int_{-a}^{0} e^{i\lambda u}\left\{h_b(-u)^* - i\overline{\omega}e^{-i\overline{\omega}u}\int_{-a}^{u} e^{i\overline{\omega}s}h_b(-s)^*ds\right\}du\, b(\omega)^{-*}.\end{aligned}$$

Formula (9.97) is then obtained by using the reproducing kernel formula

$$\eta^*(\Pi_{\mathcal{H}_*(b)}\widehat{f})(\omega) = \langle \widehat{f}, \ell_\omega^b\eta\rangle_{st}$$

to calculate the projection and then invoking Parseval's formula to finish, much as before. □

We remark that formulas (9.96) and (9.97) are of the form

$$F_1(\omega) = b(\omega)\int_{-a}^{0} e^{i\omega t}\psi_1(t)dt, \quad \psi_1 \in L_2^r([-a,0]), \tag{9.98}$$

$$F_2(\omega) = -b^\#(\omega)\int_{0}^{a} e^{i\omega t}\psi_2(t)dt\,, \quad \psi_2 \in L_2^r([0,a]), \tag{9.99}$$

respectively. However, since $F_1 \in H_2^r$ and $F_2 \in (H_2^r)^\perp$ both of these expressions can be reduced with the help of formula (9.1) to

$$F_1(\omega) = -\int_{0}^{a} e^{i\omega u}\left\{\frac{d}{du}\int_{u-a}^{0} h_b(u-t)\psi_1(t)dt\right\}du \tag{9.100}$$

$$F_2(\omega) = -\int_{-a}^{0} e^{i\omega u}\left\{\frac{d}{du}\int_{0}^{u+a} h_b(t-u)^*\psi_2(t)dt\right\}du. \tag{9.101}$$

Putting all these results together, taking a_j equal to the exponential type of $b_j(\lambda)$, $j = 1, 2$, and recalling the formulas for $X^\vee$ and its adjoint from Theorem 9.27, we obtain (with just a little extra effort) the following time domain versions of the operators in (4.24) and their adjoints:

$$(X_{11}\widehat{f})^\vee(t) = -\frac{d}{dt}\int_{t-a_1}^{0} h_{b_1}(t-u)\left\{\frac{d}{du}\int_{0}^{u+a_1} h_{b_1}(s-u)^*(X^\vee f)(s)ds\right\}du \tag{9.102}$$

for $\widehat{f} \in H_2^q$ and a.e. $t \in [0, a_1]$;

$$\begin{aligned}(X_{12}\widehat{f})^\vee(t) = -\frac{d}{dt}\int_{t-a_1}^{0} h_{b_1}(t-u)&\left\{\frac{d}{du}\int_{0}^{u+a_1} h_{b_1}(s-u)^*\right. \\ &\times\left.\left[\frac{d}{ds}\int_{-a_2}^{0} v(s-x)f(x)dx\right]ds\right\}du\end{aligned} \tag{9.103}$$

for $\widehat{f} \in \mathcal{H}_*(b_2)$ and $t \in [0, a_1]$;

$$(X_{22}^*\widehat{f})^\vee(t) = \frac{d}{dt}\int_0^{t+a_2} h_{b_2}(u-t)^* \left\{\frac{d}{du}\int_{u-a_2}^0 h_{b_2}(u-s)\right.$$
$$\left.\times\left[\frac{d}{ds}\int_s^0 v(x-s)^* f(x)dx\right]ds\right\}du \tag{9.104}$$

for $\widehat{f} \in (H_2^p)^\perp$ and a.a. $t \in [-a_2, 0]$;

$$(X_{12}^*\widehat{f})^\vee(t) = -\frac{d}{dt}\int_0^{t+a_2} h_{b_2}(u-t)^* \left\{\frac{d}{du}\int_{u-a_2}^0 h_{b_2}(u-s)((X^\vee)^* f)(s)ds\right\}du \tag{9.105}$$

for $\widehat{f} \in \mathcal{H}(b_1)$ and a.e. $t \in [0, a_2]$;

$$(X_{22}\widehat{f})^\vee(t) = \frac{d}{dt}\int_{-a_2}^t v(t-u)f(u)du \tag{9.106}$$

for $\widehat{f} \in \mathcal{H}_*(b_2)$ and a.e. $t \in [-a_2, 0]$;

$$(X_{11}^*\widehat{f})^\vee(t) = -\frac{d}{dt}\int_t^{a_1} v(u-t)^* f(u)du, \tag{9.107}$$

where v is taken from the representation (9.78) for $s = s^\circ$.

9.6 The Nehari extension problem for mvf's in Wiener class

Let $\mathfrak{g} \in L_1^{p\times q}(\mathbb{R}_+)$,

$$\mathfrak{g}^\vee(\mu) = \frac{1}{2\pi}\int_0^\infty e^{-i\mu t}\mathfrak{g}(t)dt \quad \text{and} \quad \Gamma = \Pi_- M_{2\pi\mathfrak{g}^\vee}|_{H_2^q}.$$

Then, since

$$\mathcal{N}(\Gamma) = \{f \in L_\infty^{p\times q} : \|f\|_\infty \le 1 \text{ and } f - 2\pi\mathfrak{g}^\vee \in H_\infty^{p\times q}\}$$

depends only upon $\mathfrak{g}$, we shall also refer to this set as $\mathcal{N}(\mathfrak{g})$ and shall let

NP($\mathfrak{g}$) denote the problem of describing $\mathcal{N}(\mathfrak{g})$; and
WNP($\mathfrak{g}$) denote the problem of describing the set

$$\mathcal{WN}(\mathfrak{g}) = \{f \in \mathcal{W}^{p\times q} : \|f\|_\infty \le 1 \text{ and } f - 2\pi\mathfrak{g}^\vee \in \mathcal{W}_+^{p\times q}\}.$$

Correspondingly, let $\Gamma^\vee$ denote the linear operator from $L_2^q(\mathbb{R}_+)$ into $L_2^p(\mathbb{R}_+)$ that is defined by the formula

$$(\Gamma^\vee f)(t) = (\Gamma\widehat{f})^\vee(t) = \int_0^\infty \mathfrak{g}(t+s)f(s)ds. \tag{9.108}$$

Then, since

$$\frac{\|\Gamma^\vee f)\|_{st}^2}{\|f\|_{st}^2} = \frac{\|(\Gamma\widehat{f})^\vee\|_{st}^2}{\|f\|_{st}^2} = \frac{\|\Gamma\widehat{f}\|_{st}^2}{2\pi\|f\|_{st}^2} = \frac{\|\Gamma\widehat{f}\|_{st}^2}{\|\widehat{f}\|_{st}^2}$$

for nonzero $f \in L_2^q(\mathbb{R}_+)$, it follows that

$$\|\Gamma\| = \|\Gamma^\vee\|$$

and hence, in view of Theorem 4.1, that

$$\|\Gamma^\vee\| \le 1 \Longleftrightarrow \mathcal{N}(\mathfrak{g}) \neq \emptyset$$

$$\|\Gamma^\vee\| < 1 \Longleftrightarrow \text{NP}(\mathfrak{g}) \text{ is strictly completely indeterminate.}$$

We remark that if

$$\|\Gamma^\vee\| = 1, \quad \text{then } \mathcal{WN}(\mathfrak{g}) \text{ is not empty,}$$

see, e.g., [DG83] and, for a description of the sets $\mathcal{N}(\mathfrak{g})$ and $\mathcal{WN}(\mathfrak{g})$, [KrMA84]. Moreover, since the operator $\Gamma^\vee$ defined by (9.108), acting from $L_2^q(\mathbb{R}_+)$ into $L_2^p(\mathbb{R}_+)$ is compact (see, e.g., corollary 8.5 in [Pe03]) and hence

$$(\Gamma^\vee)^*\Gamma^\vee < I \Longleftrightarrow \|\Gamma^\vee\| < 1$$
$$\Longleftrightarrow \mathcal{WN}(\mathfrak{g}) \cap \mathring{\mathbb{B}}^{p\times q} \neq \emptyset \Longrightarrow \mathcal{WN}(\mathfrak{g}) \neq \emptyset.$$

A description of $\mathcal{N}(\mathfrak{g})$ and $\mathcal{WN}(\mathfrak{g})$ when $\|\Gamma\| < 1$ will be furnished in terms of linear fractional transformations $T_{\mathfrak{A}}$ based on mvf's

$$\mathfrak{A}(\mu) = \begin{bmatrix} \mathfrak{a}_-(\mu) & \mathfrak{b}_-(\mu) \\ \mathfrak{b}_+(\mu) & \mathfrak{a}_+(\mu) \end{bmatrix} \tag{9.109}$$

that belong to the class $\mathcal{W}_r(j_{pq})$, which is defined by the following two sets of constraints:

(1) $\mathfrak{a}_-^{\pm 1} \in \mathcal{W}_-^{p\times p}(I_p)$, $\mathfrak{b}_- \in \mathcal{W}_-^{p\times q}(0)$, $\mathfrak{b}_+ \in \mathcal{W}_+^{q\times p}(0)$ and $\mathfrak{a}_+^{\pm 1} \in \mathcal{W}_+^{q\times q}(I_q)$.
(2) $\mathfrak{A}(\mu)$ is j_{pq}-unitary on $\mathbb{R}$, i.e.,

$$\mathfrak{A}(\mu)^* j_{pq}\mathfrak{A}(\mu) = j_{pq} \quad \text{for every point } \mu \in \mathbb{R}. \tag{9.110}$$

Thus, $\mathfrak{A} \in \mathcal{W}^{m\times m}(I_m)$, $\mathfrak{A}(\mu)j_{pq}\mathfrak{A}(\mu)^* = j_{pq}$ on $\mathbb{R}$ and the blocks of $\mathfrak{A}$ satisfy the following equations:

$$\mathfrak{a}_-(\mu)^*\mathfrak{a}_-(\mu) - \mathfrak{b}_+(\mu)^*\mathfrak{b}_+(\mu) = I_p = \mathfrak{a}_-(\mu)\mathfrak{a}_-(\mu)^* - \mathfrak{b}_-(\mu)\mathfrak{b}_-(\mu)^* \quad (9.111)$$

$$\mathfrak{a}_-(\mu)^*\mathfrak{b}_-(\mu) - \mathfrak{b}_+(\mu)^*\mathfrak{a}_+(\mu) = 0 = \mathfrak{b}_+(\mu)\mathfrak{a}_-(\mu)^* - \mathfrak{a}_+(\mu)\mathfrak{b}_-(\mu)^* \quad (9.112)$$

$$\mathfrak{a}_+(\mu)^*\mathfrak{a}_+(\mu) - \mathfrak{b}_-(\mu)^*\mathfrak{b}_-(\mu) = I_q = \mathfrak{a}_+(\mu)\mathfrak{a}_+(\mu)^* - \mathfrak{b}_+(\mu)\mathfrak{b}_+(\mu)^*. \quad (9.113)$$

Lemma 9.33 *If $\mathfrak{A} \in \mathcal{W}_r(j_{pq})$ and $\Sigma = PG(\mathfrak{A})$, then the blocks σ_{ij} of Σ are given by the formulas*

$$\sigma_{11} = (\mathfrak{a}_-^*)^{-1}, \quad \sigma_{12} = \mathfrak{b}_-\mathfrak{a}_+^{-1}, \quad \sigma_{21} = -\mathfrak{a}_+^{-1}\mathfrak{b}_+, \quad \sigma_{22} = \mathfrak{a}_+^{-1} \quad (9.114)$$

on $\mathbb{R}$ and:

(1) $\sigma_{11}^{\pm 1} \in \mathcal{W}_+^{p\times p}(I_p)$ *and* $\sigma_{11} \in \mathcal{S}_{\text{out}}^{p\times p}$.
(2) $\sigma_{22}^{\pm 1} \in \mathcal{W}_+^{q\times q}(I_q)$ *and* $\sigma_{22} \in \mathcal{S}_{\text{out}}^{q\times q}$.
(3) $\sigma_{21} \in \mathcal{W}_+^{q\times p}(0)$ *and* $\sigma_{21} \in \mathring{\mathcal{S}}^{q\times p}$.
(4) $\sigma_{12} \in \mathcal{W}^{p\times q}(0)$ *and* $\|\sigma_{12}\|_\infty < 1$.

Proof (1), (2) and the fact that $\sigma_{12} \in \mathcal{W}^{p\times q}(0)$ and $\sigma_{21} \in \mathcal{W}_+^{q\times p}(0)$ are immediate from the identifications in (9.114) and the properties of the blocks of $\mathfrak{A}$. Moreover, since $\sigma_{11}^{-1} \in H_\infty^{p\times p}$ and $\sigma_{22}^{-1} \in H_\infty^{q\times q}$, it follows that

$$\sigma_{11}(\lambda)^*\sigma_{11}(\lambda) \geq \delta I_p \quad \text{and} \quad \sigma_{22}(\lambda)^*\sigma_{22}(\lambda) \geq \delta I_q \quad \text{for } \lambda \in \mathbb{C}_+.$$

Therefore, since Σ is unitary on $\mathbb{R}$, it follows that

$$\sigma_{21}^*(\mu)\sigma_{21}(\mu) = I_p - \sigma_{11}^*(\mu)\sigma_{11}(\mu) \quad \text{and} \quad \sigma_{12}^*(\mu)\sigma_{12}(\mu) = I_q - \sigma_{22}^*(\mu)\sigma_{22}(\mu)$$

on $\mathbb{R}$ and hence that $\|\sigma_{21}\|_\infty < 1$ and $\|\sigma_{12}\|_\infty < 1$, as needed to complete the proof. □

Lemma 9.34 $\mathcal{W}_r(j_{pq}) = \mathcal{W}^{m\times m}(I_m) \cap \mathfrak{M}_r(j_{pq}) = \mathcal{W}^{m\times m}(I_m) \cap \mathfrak{M}_{rsR}(j_{pq})$.

Proof The first inclusion in

$$\mathcal{W}_r(j_{pq}) \subseteq \mathcal{W}^{m\times m}(I_m) \cap \mathfrak{M}_r(j_{pq}) \subseteq \mathcal{W}^{m\times m}(I_m) \cap \mathfrak{M}_{rsR}(j_{pq})$$

follows from Lemma 9.33; the second from the fact that $L_\infty^{m\times m} \cap \mathfrak{M}_r(j_{pq}) \subseteq \mathfrak{M}_{rsR}(j_{pq})$. Thus, as the inclusion

$$\mathcal{W}^{m\times m}(I_m) \cap \mathfrak{M}_{rsR}(j_{pq}) \subseteq \mathcal{W}^{m\times m}(I_m) \cap \mathfrak{M}_r(j_{pq})$$

is self-evident, these two sets must be equal.

It remains to show that $\mathcal{W}^{m\times m}(I_m) \cap \mathfrak{M}_r(j_{pq}) \subseteq \mathcal{W}_r(j_{pq})$. To verify this, let $\mathfrak{A} \in \mathcal{W}^{m\times m}(I_m) \cap \mathfrak{M}_r(j_{pq})$. Then $\mathfrak{a}_+^{-1} \in \mathcal{S}_{\text{out}}^{p\times q}$, $\mathfrak{a}_+ \in \mathcal{W}_+^{q\times q}(I_q)$ and $\mathfrak{a}_+(\mu)\mathfrak{a}_+(\mu)^* \geq I_q$

on $\mathbb{R}$. Then, by Theorem 3.31 (due to N. Wiener), $\mathfrak{a}_+^{-1} \in \mathcal{W}_+^{q\times q}(I_q)$. A similar argument serves to prove that $\mathfrak{a}_-^{\pm 1} \in \mathcal{W}_-^{p\times p}(I_p)$. Then,

$$\mathfrak{b}_+ = -\mathfrak{a}_+(-\mathfrak{a}_+^{-1}\mathfrak{b}_+) = -\mathfrak{a}_+\sigma_{21} \quad \text{belongs to } \mathcal{W}_+^{p\times q}(0),$$

since $\sigma_{21} \in \mathcal{S}^{q\times p}$. A similar argument based on the identities

$$\mathfrak{b}_-^\# = -\{-\mathfrak{b}_-^\#(\mathfrak{a}_-^\#)^{-1}\}\mathfrak{a}_-^\# = -\sigma_{21}\mathfrak{a}_-^\#$$

serves to prove that $\mathfrak{b}_- \in \mathcal{W}_-^{p\times q}(0)$. Thus, $\mathfrak{A} \in \mathcal{W}_r(j_{pq})$. □

Theorem 9.35 *If $\mathfrak{A} \in \mathcal{W}_r(j_{pq})$ and $\sigma_{12} = T_{\mathfrak{A}}[0]$, then there exists exactly one mvf $\mathfrak{g} \in L_1^{p\times q}(\mathbb{R}_+)$ such that*

$$\sigma_{12}(\mu) - \int_0^\infty e^{-i\mu t}\mathfrak{g}(t)dt \quad \textit{belongs to } \mathcal{W}_+^{p\times q}(0). \tag{9.115}$$

Moreover, for this mvf $\mathfrak{g}$:

(1) *The problems $NP(\mathfrak{g})$ and $WNP(\mathfrak{g})$ are both strictly completely indeterminate.*
(2) *$\mathfrak{A}$ is a resolvent matrix for $NP(\mathfrak{g})$, i.e.,*

$$\mathcal{N}(\mathfrak{g}) = T_{\mathfrak{A}}[\mathcal{S}^{p\times q}]. \tag{9.116}$$

(3) $T_{\mathfrak{A}}[\mathring{\mathcal{S}}^{p\times q}] = \mathcal{N}(\mathfrak{g}) \cap \mathring{\mathbb{B}}^{p\times q}$.
(4) *If $p \geq q$, then*

$$T_{\mathfrak{A}}[\mathcal{S}_{\text{in}}^{p\times q}] = \{f \in \mathcal{N}(\mathfrak{g}) : f(\mu)^* f(\mu) = I_q \quad \textit{a.e. on } \mathbb{R}\},$$

whereas, if $p \leq q$, then

$$T_{\mathfrak{A}}[\mathcal{S}_{*in}^{p\times q}] = \{f \in \mathcal{N}(\mathfrak{g}) : f(\mu)f(\mu)^* = I_p \quad \textit{a.e. on } \mathbb{R}\}.$$

(5) *$\mathfrak{A}$ is the only resolvent matrix of $NP(\mathfrak{g})$ in the class $\mathfrak{M}_r(j_{pq}) \cap \mathcal{W}^{m\times m}(I_m)$.*
(6) *The identity*

$$\mathcal{W}^{p\times q}(\gamma) \cap \mathcal{N}(\mathfrak{g}) = T_{\mathfrak{A}}[\mathcal{W}_+^{p\times q}(\gamma) \cap \mathcal{S}^{p\times q}] \tag{9.117}$$

holds for every choice of $\gamma \in \mathbb{C}^{p\times q}$; though both sides are empty if $\|\gamma\| > 1$.

Proof Let $f_0 = \sigma_{12}$ and $\Gamma = \Pi_- M_{f_0}|_{H_2^q}$. Then Lemma 9.33 guarantees that $f_0 \in \mathcal{W}^{p\times q}(0)$ and $\|f_0\| < 1$. Therefore, $\mathfrak{A} \in \mathfrak{M}_{rsR}(j_{pq})$ and (9.115) holds with $\mathfrak{g}(t) = \sigma_{12}^\vee(-t)$ for a.a. $t \geq 0$.

Assertions (1) and (2) follow from Theorem 4.6, since $\mathcal{N}(\Gamma) = \mathcal{N}(\mathfrak{g})$.

Let $\varepsilon \in \mathcal{S}^{p\times q}$ and set

$$f_\varepsilon(\mu) = (T_{\mathfrak{A}}[\varepsilon])(\mu) \quad \text{a.e. on } \mathbb{R}.$$

Then the identity

$$I_q - f_\varepsilon^* f_\varepsilon = \sigma_{22}^*(I_q - \sigma_{21}\varepsilon)^{-*}(I_q - \varepsilon^*\varepsilon)(I_q - \sigma_{21}\varepsilon)^{-1}\sigma_{22} \quad \text{a.e. on } \mathbb{R}, \tag{9.118}$$

which is expressed in terms of the blocks σ_{ij} of $PG(\mathfrak{A})$, yields assertion (3) and the first equality in (4).

Analogously, the second assertion in (4) follows from the identity

$$I_p - f_\varepsilon(\mu)f_\varepsilon(\mu)^* = \sigma_{11}(I_p - \varepsilon\sigma_{21})^{-1}(I_p - \varepsilon\varepsilon^*)^{-1}(I_p - \varepsilon\sigma_{21})^{-*}\sigma_{11}^* \quad \text{a.e. on } \mathbb{R},$$

which may be obtained from the identity

$$\begin{bmatrix} I_p & \varepsilon(\mu) \end{bmatrix} \mathfrak{A}(\mu)^* j_{pq} \mathfrak{A}(\mu) \begin{bmatrix} I_p \\ \varepsilon(\mu) \end{bmatrix} = 0 \quad \text{a.e. on } \mathbb{R}$$

to reexpress f_ε as

$$f_\varepsilon(\mu) = \{\varepsilon(\mu)\mathfrak{b}_-(\mu)^* + \mathfrak{a}_-(\mu)^*\}^{-1}\{\varepsilon(\mu)\mathfrak{a}_+(\mu)^* + \mathfrak{b}_+(\mu)^*\}. \tag{9.119}$$

Next, (5) follows from the description of right gamma generating matrices $\mathfrak{A}$ for a completely indeterminate NP(Γ) furnished in (1) of Theorem 4.6 and the constraint $\mathfrak{A} \in \mathcal{W}^{m\times m}(I_m)$.

To verify (6), let $\varepsilon \in \mathcal{W}_+^{p\times q}(\gamma_0) \cap \mathcal{S}^{p\times q}$ and rewrite $T_{\mathfrak{A}}[\varepsilon]$ in terms of the Redheffer transform (as in formula (3.28)) as

$$f_\varepsilon(\mu) = \sigma_{12}(\mu) + \sigma_{11}(\mu)\varepsilon(\mu)(I_q - \sigma_{21}(\mu)\varepsilon(\mu))^{-1}\sigma_{22}(\mu) \quad \text{a.e. on } \mathbb{R}.$$

Then, since $\sigma_{21} \in \mathcal{W}_+^{q\times p}(0)$ and $\|\sigma_{21}\|_\infty = \delta < 1$, $\|(I_q - \sigma_{21}\varepsilon)\| \geq 1 - \delta$ and hence, by Theorem 3.31, $(I_q - \sigma_{21}\varepsilon)^{-1} \in \mathcal{W}_+^{q\times q}(I_q)$. Thus, as $\sigma_{12} \in \mathcal{W}_+^{p\times q}(0)$, $\sigma_{11} \in \mathcal{W}_+^{p\times p}(I_p)$ and $\sigma_{22} \in \mathcal{W}_+^{q\times q}(I_q)$, it is readily seen that $f_\varepsilon \in \mathcal{W}^{p\times q}(\gamma_0) \cap \mathcal{N}(\mathfrak{g})$.

Conversely, if $f_\varepsilon \in \mathcal{W}^{p\times q}(\gamma_0) \cap \mathcal{N}(\mathfrak{g})$, then

$$f_\varepsilon - \sigma_{12} \in \mathcal{W}_+^{p\times q}(\gamma_0)$$

and, by Lemma 9.33, $\sigma_{11}^{\pm 1} \in \mathcal{W}_+^{p\times p}(I_p)$ and $\sigma_{22}^{\pm 1} \in \mathcal{W}_+^{q\times q}(I_q)$ and hence

$$\varepsilon(I_q - \sigma_{21}\varepsilon)^{-1} \in \mathcal{W}_+^{p\times q}(\gamma_0).$$

Thus,

$$\sigma_{21}\varepsilon(I_q - \sigma_{21}\varepsilon)^{-1} \in \mathcal{W}_+^{q\times q}(0)$$

and

$$(I_q - \sigma_{21}\varepsilon)^{-1} = I_q + \sigma_{21}\varepsilon(I_q - \sigma_{21}\varepsilon)^{-1} \in \mathcal{W}_+^{q\times q}(I_q).$$

However, since $\sigma_{21} \in \mathring{\mathcal{S}}^{q\times p}$, it follows that $(I_q - \sigma_{21}\varepsilon)^{-1} \in \mathring{\mathcal{C}}^{q\times q}$ and hence that

$$(I_q - \sigma_{21}\varepsilon)^{\pm 1} \in H_\infty^{q\times q}.$$

Therefore, by another application of Theorem 3.31,

$$(I_q - \sigma_{21}\varepsilon) \in \mathcal{W}_+^{q\times q}(I_q).$$

Thus, the formula

$$\varepsilon = \{\varepsilon(I_q - \sigma_{21}\varepsilon)^{-1}\}\{(I_q - \sigma_{21}\varepsilon)\}$$

displays ϵ as the product of a factor in $\mathcal{W}_+^{p\times q}(\gamma_0)$ with a factor in $\mathcal{W}_+^{q\times q}(I_q)$, which in turn implies that $\varepsilon \in \mathcal{W}_+^{p\times q}(\gamma_0)$. This completes the proof of (9.117). □

If the operator $\Gamma^\vee$, defined by formula (9.108) is considered as an operator from the Hilbert space $L_2^q(\mathbb{R}_+)$ into the Hilbert space $L_2^p(\mathbb{R}_+)$, then its adjoint coincides with the operator $\Gamma_*^\vee$ that is defined by formula

$$(\Gamma_*^\vee f)(t) = \int_0^\infty \mathfrak{g}(t+s)^* f(s) ds \tag{9.120}$$

acting from $L_2^p(\mathbb{R}_+)$ into $L_2^q(\mathbb{R}_+)$. If $\mathfrak{g} \in L_1^{p\times q}(\mathbb{R}_+)$, then the linear operators $\Gamma^\vee$ and $\Gamma_*^\vee$, defined by formulas (9.108) and (9.120), are compact from $\mathcal{X}$ into $\mathcal{Y}$ and from $\mathcal{Y}$ into $\mathcal{X}$, respectively, for each of the Banach spaces $\mathcal{X} = L_r^q(\mathbb{R}_+)$ and $\mathcal{Y} = L_r^p(\mathbb{R}_+)$ for each choice of $1 \le r < \infty$ and for the Banach spaces $\mathcal{X} = C_\ell^q(\mathbb{R}_+)$ and $\mathcal{Y} = C_\ell^p(\mathbb{R}_+)$ of vvf's φ that have absolutely continuous derivatives of order $\ell - 1$ for $\ell \ge 1$ with norm

$$\|\varphi\| = \sup\{\|\varphi(t)\| : t \in \mathbb{R}_+\} + \sum_{k=1}^{\ell} \int_0^\infty \|\varphi(t)^{(k)}\| dt < \infty,$$

and $C_0^q(\mathbb{R}_+)$ and $C_0^p(\mathbb{R}_+)$ are Banach spaces of continuous bounded vvf's φ on $\mathbb{R}_+$ with values from $\mathbb{C}^q$ and $\mathbb{C}^p$, respectively, with norm

$$\|\varphi\| = \|\varphi\|_\infty.$$

Let $\mathfrak{g} \in L_1^{p\times q}(\mathbb{R}_+)$ and let

$$\Delta = \begin{bmatrix} I_{\mathcal{Y}} & -\Gamma^\vee \\ -\Gamma_*^\vee & I_{\mathcal{X}} \end{bmatrix}. \tag{9.121}$$

Lemma 9.36 *If $\mathfrak{g} \in L_1^{p\times q}(\mathbb{R}_+)$, then the operators $\Gamma^\vee$ and $\Gamma_*^\vee$ defined by formulas (9.108) and (9.120), respectively, acting between the Banach spaces $\mathcal{X}$ and $\mathcal{Y}$ discussed above and the operator $\Delta - I$ on $\mathcal{Y} \times \mathcal{X}$ are compact. Moreover, if $\mathcal{X}$ and $\mathcal{Y}$ are the Hilbert spaces $L_2^q(\mathbb{R}_+)$, and $L_2^p(\mathbb{R}_+)$, respectively, then $\Gamma_*^\vee$ is the adjoint to $\Gamma^\vee$,*

$$\|\Gamma^\vee\| \le 1 \Longleftrightarrow \Delta \ge 0,$$

and the following conditions are equivalent:

(1) $\|\Gamma^\vee \varphi\| < \|\varphi\|$ *for every* $\varphi \in L_2^q(\mathbb{R}_+)$ *with* $\varphi \neq 0$.

(2) $\|\Gamma^\vee\| < 1$.
(3) $\Delta > 0$.
(4) $\Delta > \delta I$ *for some* $\delta > 0$.

Furthermore, if any one (and hence every one) of these last four conditions hold, then the operator Δ *is boundedly invertible on the Banach space* $\mathcal{Y} \times \mathcal{X}$ *for each choice of the Banach spaces* $\mathcal{X}$ *and* $\mathcal{Y}$ *listed above.*

Proof It is well known that the operators $\Gamma^\vee$ and $\Gamma_*^\vee$ are compact and that in the Hilbert space setting $\Gamma_*^\vee = (\Gamma^\vee)^*$. Thus, $(\Gamma^\vee)^*\Gamma^\vee$ is a compact self-adjoint operator on $L_2^q(\mathbb{R}_+)$ and hence there exists a vvf $\varphi \in L_2^q(\mathbb{R}_+)$, $\varphi \neq 0$, such that $(\Gamma^\vee)^*\Gamma^\vee\varphi = \|\Gamma^\vee\|^2\varphi$. Consequently, (1) $\Longrightarrow$ (2). The opposite implication is self-evident. The equivalences (1) $\Longleftrightarrow$ (3) and (2) $\Longleftrightarrow$ (4) follow from the Schur complement formula

$$\Delta = \begin{bmatrix} I & 0 \\ -(\Gamma^\vee)^* & I \end{bmatrix} \begin{bmatrix} I & 0 \\ 0 & I - (\Gamma^\vee)^*\Gamma^\vee \end{bmatrix} \begin{bmatrix} I & -\Gamma^\vee \\ 0 & I \end{bmatrix}.$$

Finally, (4) implies that Δ has a bounded inverse on the Hilbert space $L_2^p(\mathbb{R}_+) \oplus L_2^q(\mathbb{R}_+)$. Since the operator $\Delta - I$ is compact on the Banach space $\mathcal{Y} \times \mathcal{X}$ for each choice of the Banach spaces $\mathcal{X}$ and $\mathcal{Y}$ discussed above, it has the same nonzero spectrum as on the Hilbert space $L_2^p(\mathbb{R}_+) \oplus L_2^q(\mathbb{R}_+)$; see, e.g., lemma 7.1 in [AAK71b]. Therefore, 0 does not belong to the spectrum of Δ considered on each of these spaces. Thus, Δ is a bounded linear operator in $\mathcal{Y} \times \mathcal{X}$ with a bounded inverse for each of the considered Banach spaces $\mathcal{X}$ and $\mathcal{Y}$. □

Theorem 9.37 *If* $\mathfrak{g} \in L_1^{p\times q}(\mathbb{R}_+)$ *and the operator* $\Gamma^\vee$ *from* $L_2^q(\mathbb{R}_+)$ *into* $L_2^p(\mathbb{R}_+)$ *that is defined by formula (9.108) is strictly contractive (i.e.,* $\|\Gamma^\vee\| < 1$*), then:*

(1) *There exists exactly one mvf* $\mathfrak{A} \in \mathcal{W}_r$ *such that (9.116) is in force. Moreover, (9.117) is then also in force.*
(2) *The mvf* $\mathfrak{A}$ *referred to in (1) may be defined by the formula*

$$\mathfrak{A}(\mu) = I_m + \begin{bmatrix} \widehat{u}_{11}(-\mu) & \widehat{u}_{12}(-\mu) \\ \widehat{u}_{21}(\mu) & \widehat{u}_{22}(\mu) \end{bmatrix}, \tag{9.122}$$

where the $\widehat{u}_{ij}$ *are the Fourier transforms of the mvf's* u_{ij}, $i, j = 1, 2$, *that are the only summable solutions of the system of equations*

$$\begin{bmatrix} I & -\Gamma^\vee \\ -\Gamma_*^\vee & I \end{bmatrix} \begin{bmatrix} u_{11} & u_{12} \\ u_{21} & u_{22} \end{bmatrix} = \begin{bmatrix} 0 & \mathfrak{g} \\ \mathfrak{g}^* & 0 \end{bmatrix} \tag{9.123}$$

and the operators act on the indicated mvf's column by column.

(3) *If the columns of* $\mathfrak{g}$ *belong to the Banach space* $L_r^p(\mathbb{R}_+)$ *or to the Banach space* $C_\ell^p(\mathbb{R}_+)$*, then the columns of the mvf*

$$U(t) = \begin{bmatrix} u_{11}(t) & u_{12}(t) \\ u_{21}(t) & u_{22}(t) \end{bmatrix} \tag{9.124}$$

belong $L_r^m(\mathbb{R}_+)$ *or* $C_\ell^m(\mathbb{R}_+)$*, respectively.*

Proof The proof is long and is broken into steps.

1. *If formula (9.116) holds for some mvf* $\mathfrak{A} \in \mathcal{W}_r$*, then* $\mathfrak{A}$ *is uniquely determined by* $\mathfrak{g}$ *and may be obtained from formulas (9.122) and (9.123).*

Clearly

$$\mathfrak{A}(\mu) = \begin{bmatrix} I_p + \widehat{u}_{11}(-\mu) & \widehat{u}_{12}(-\mu) \\ \widehat{u}_{21}(\mu) & I_q + \widehat{u}_{22}(\mu) \end{bmatrix},$$

where

$$\widehat{u}_{11} \in \mathcal{W}_+^{p\times p}(0), \quad \widehat{u}_{12} \in \mathcal{W}_+^{p\times q}(0), \quad \widehat{u}_{21} \in \mathcal{W}_+^{q\times p}(0) \quad \text{and} \quad \widehat{u}_{22} \in \mathcal{W}_+^{q\times q}(0).$$

Since $T_{\mathfrak{A}}[0] \in \mathcal{W}^{p\times q}(0) \cap \mathcal{N}(\mathfrak{g})$,

$$\mathfrak{b}_-(\mu)\mathfrak{a}_+(\mu)^{-1} = \widehat{\mathfrak{g}}(-\mu) + \widehat{h_+}(\mu) \quad \text{for some } h_+ \in L_1^{p\times q}(\mathbb{R}_+). \tag{9.125}$$

Thus, in view of (9.122),

$$\widehat{u}_{12}(-\mu) = (\widehat{\mathfrak{g}}(-\mu) + \widehat{h_+}(\mu))(I_q + \widehat{u}_{22}(\mu)),$$

which upon setting $\widehat{g_-}(\mu) = \widehat{\mathfrak{g}}(-\mu)$ and noting that

$$\Pi_- \widehat{g_-}\widehat{u}_{22} = \int_0^\infty e^{-i\mu t}(\Gamma^\vee u_{22})(t)dt,$$

yields the formula

$$u_{12}(t) = \mathfrak{g}(t) + (\Gamma^\vee u_{22})(t) \quad \text{a.e. on } \mathbb{R}_+. \tag{9.126}$$

Similarly, the identity

$$\mathfrak{b}_-(\mu)\mathfrak{a}_+(\mu)^{-1} = \mathfrak{a}_-(\mu)^{-*}\mathfrak{b}_+(\mu)^* \quad \text{on } \mathbb{R}$$

implies that

$$\{\widehat{\mathfrak{g}}(\mu)^* + \widehat{h_+}(-\mu)^*\}\{I_p + \widehat{u}_{11}(\mu)\} = \widehat{u}_{21}(-\mu) \quad \text{on } \mathbb{R}$$

and hence that

$$u_{21}(t) = \mathfrak{g}(t)^* + (\Gamma_*^\vee u_{11})(t) \quad \text{a.e. on } \mathbb{R}_+. \tag{9.127}$$

Next, since $f_\varepsilon \in \mathcal{W}^{p\times q}(\varepsilon) \cap \mathcal{N}(\mathfrak{g})$ for every $\varepsilon \in \mathcal{S}_{\text{const}}^{p\times q}$, it follows that

$$f_\varepsilon - f_0 \in \mathcal{W}_+^{p\times q}(\varepsilon) \quad \text{for every } \varepsilon \in \mathcal{S}_{\text{const}}^{p\times q}.$$

Therefore,

$$(\mathfrak{a}_-\varepsilon + \mathfrak{b}_-)(\mathfrak{b}_+\varepsilon + \mathfrak{a}_+)^{-1} - \mathfrak{b}_-\mathfrak{a}_+^{-1} = (\mathfrak{a}_- - \mathfrak{b}_-\mathfrak{a}_+^{-1}\mathfrak{b}_+)\varepsilon(\mathfrak{b}_+\varepsilon + \mathfrak{a}_+)^{-1}$$

and

$$(\mathfrak{a}_- - \mathfrak{b}_-\mathfrak{a}_+^{-1}\mathfrak{b}_+)\varepsilon$$

also belong to $\mathcal{W}_+^{p\times q}(\varepsilon)$ for every $\varepsilon \in \mathcal{S}_{\text{const}}^{p\times q}$ and

$$(\mathfrak{a}_- - I_p - \mathfrak{b}_-\mathfrak{a}_+^{-1}\mathfrak{b}_+)\varepsilon$$

belongs to $\mathcal{W}_+^{p\times q}(0)$ for every $\varepsilon \in \mathcal{S}_{\text{const}}^{p\times q}$. Thus,

$$(\mathfrak{a}_- - I_p - \mathfrak{b}_-\mathfrak{a}_+^{-1}\mathfrak{b}_+) \in \mathcal{W}_+^{p\times p}(0).$$

But, in view of (9.125), this in turn implies that

$$\widehat{u_{11}}(-\mu) - \widehat{\mathfrak{g}}(-\mu)\widehat{u_{21}}(\mu) = \widehat{g_+}(\mu) \quad \text{for some mvf } g_+ \in L_1^{p\times p}(\mathbb{R}_+)$$

and hence that

$$u_{11}(t) - (\Gamma^\vee u_{21})(t) = 0 \quad \text{a.e. on } \mathbb{R}_+. \tag{9.128}$$

The supplementary formula

$$-(\Gamma_*^\vee u_{12})(t) + u_{22}(t) = 0 \quad \text{a.e. on } \mathbb{R}_+ \tag{9.129}$$

is established in much the same way by exploiting the left representation (9.119) for f_ε.

2. *If the block entries $\mathfrak{a}_\pm$ and $\mathfrak{b}_\pm$ of $\mathfrak{A}$ are defined by (9.122), where the u_{ij} are the solutions of the system of matrix equations (9.123), then $\mathfrak{A}$ is j_{pq}-unitary on $\mathbb{R}$.*

It is readily checked that:

(1) (9.128) $\Longrightarrow$ $\mathfrak{a}_-(\mu) - \widehat{\mathfrak{g}}(-\mu)\mathfrak{b}_+(\mu)$ belongs to $\mathcal{W}_+^{p\times p}(I_p)$.
(2) (9.127) $\Longrightarrow$ $\mathfrak{b}_+(\mu)^* - \mathfrak{a}_-(\mu)^*\widehat{\mathfrak{g}}(-\mu)$ belongs to $\mathcal{W}_+^{p\times q}(0)$.
(3) (9.126) $\Longrightarrow$ $\mathfrak{b}_-(\mu) - \widehat{\mathfrak{g}}(-\mu)\mathfrak{a}_+(\mu)$ belongs to $\mathcal{W}_+^{p\times q}(0)$.
(4) (9.129) $\Longrightarrow$ $\mathfrak{a}_+(\mu)^* - \mathfrak{b}_-(\mu)^*\widehat{\mathfrak{g}}(-\mu)$ belongs to $\mathcal{W}_+^{q\times q}(I_q)$.

Thus, if

$$\Theta(\mu) = \begin{bmatrix} \theta_{11}(\mu) & \theta_{12}(\mu) \\ \theta_{21}(\mu) & \theta_{22}(\mu) \end{bmatrix} = \mathfrak{A}(\mu)^* j_{pq}\mathfrak{A}(\mu),$$

it is readily seen that

$$\theta_{11}(\mu) = \mathfrak{a}_-(\mu)^*\mathfrak{a}_-(\mu) - \mathfrak{b}_+(\mu)^*\mathfrak{b}_+(\mu)$$

can be written as

$$\mathfrak{a}_-(\mu)^*\{\mathfrak{a}_-(\mu) - \widehat{\mathfrak{g}}(-\mu)\mathfrak{b}_+(\mu)\} + \left\{\mathfrak{a}_-(\mu)^*\widehat{\mathfrak{g}}(-\mu) - \mathfrak{b}_+(\mu)^*\right\}\mathfrak{b}_+(\mu).$$

But, in view of (1) and (2), this implies that $\theta_{11} \in \mathcal{W}_+^{p\times p}(I_p)$. However, since $\theta_{11}(\mu)^* = \theta_{11}(\mu)$, it follows that $\theta_{11} \in \mathcal{W}_+^{p\times p}(I_p) \cap \mathcal{W}_-^{p\times p}(I_p)$, and hence that $\theta_{11}(\mu) = I_p$. Similar considerations serve to complete the proof that $\Theta(\mu) = j_{pq}$.

3. $\mathfrak{a}_-^{-1} \in \mathcal{W}_-^{p\times p}(I_p)$ *and* $\mathfrak{a}_+^{-1} \in \mathcal{W}_+^{q\times q}(I_q)$.

The formulas $\theta_{11}(\mu) = I_p$ on $\mathbb{R}$ and $\theta_{22}(\mu) = -I_q$ on $\mathbb{R}$ imply that $\mathfrak{a}_-(\mu)$ and $\mathfrak{a}_+(\mu)$ are both invertible on $\mathbb{R}$, and, since $\theta_{12}(\mu) = 0$ on $\mathbb{R}$,

$$\mathfrak{a}_-(\mu)^{-*}\mathfrak{b}_+(\mu)^* = \mathfrak{b}_-(\mu)\mathfrak{a}_+(\mu)^{-1} \quad \text{belongs to } \mathcal{N}(\mathfrak{g}).$$

Therefore,

$$\begin{aligned}\mathfrak{a}_-(\mu)^{-*} &= \mathfrak{a}_-(\mu) - \mathfrak{a}_-(\mu)^{-*}\mathfrak{b}_+(\mu)^*\mathfrak{b}_+(\mu)\\ &= \mathfrak{a}_-(\mu) - \{\widehat{\mathfrak{g}}(-\mu) + \widehat{h_+}(\mu)\}\mathfrak{b}_+(\mu),\end{aligned}$$

thanks to (9.125). But, in view of (1), the last term on the right belongs to $\mathcal{W}_+^{p\times p}(I_p)$. Therefore, $\mathfrak{a}_-(\mu)^{-1} \in \mathcal{W}_-^{p\times p}(I_p)$ as claimed. The proof that $\mathfrak{a}_+^{-1} \in \mathcal{W}_+^{q\times q}(I_q)$ is obtained from formula (9.125) and the identity

$$\mathfrak{a}_+(\mu)^{-1} = \mathfrak{a}_+(\mu)^* - \mathfrak{b}_-(\mu)^*\mathfrak{b}_-(\mu)\mathfrak{a}_+(\mu)^{-1}$$

in much the same way.

4. $\mathfrak{b}_-\mathfrak{a}_+^{-1} \in \mathcal{N}(\mathfrak{g})$.

This follows from (9.125).

In view of Theorem 9.35, the proof is complete. □

9.7 Continuous analogs of the Schur extension problem for mvf's in the Wiener class

In this section we shall consider analogs of the accelerant extension problem in which the class $\mathcal{C}^{p\times p} \cap \mathcal{W}_+^{p\times p}(\gamma)$ is replaced by the class $\mathcal{S}^{p\times q} \cap \mathcal{W}_+^{p\times q}(\gamma)$. If $s \in \mathcal{S}^{p\times q} \cap \mathcal{W}_+^{p\times q}(\gamma)$, then the Riemann–Lebesgue lemma implies that $\|\gamma\| \le 1$. We shall in fact restrict attention to the case $\|\gamma\| < 1$. Then there is no loss of generality in taking $\gamma = 0$, since the linear fractional transformation $T_{W_1}[s]$ based on the matrix $W_1 \in \mathcal{U}_{\rm const}(j_{pq})$ defined by (3.127) with $k = -\gamma$ maps $s \in \mathcal{S}^{p\times q} \cap \mathcal{W}_+^{p\times q}(\gamma)$ with $\|\gamma\| < 1$ into $\mathcal{S}^{p\times q} \cap \mathcal{W}_+^{p\times q}(0)$. There are two main problems of interest:

SP$(h^\circ; a)$**:** *Given a mvf* $h^\circ \in L_1^{p\times q}([0, a])$, *find a mvf* $s \in \mathcal{S}^{p\times q}$ *such that*

$$e_{-a}\left(s - \int_0^a e^{i\lambda t}h^\circ(t)dt\right) \in H_\infty^{p\times q}.$$

The set of solutions of this problem will be denoted $\mathcal{S}(h^\circ; a)$.

WSP$(h^\circ; a)$**:** *Given a mvf* $h^\circ \in L_1^{p\times q}([0, a])$, *find a mvf* $s \in \mathcal{W}_+^{p\times q}(0) \cap \mathcal{S}^{p\times q}$ *such that*

$$e_{-a}\left(s - \int_0^a e^{i\lambda t}h^\circ(t)dt\right) \in \mathcal{W}_+^{p\times q}(0).$$

The set of solutions of this problem will be denoted $\mathcal{WS}(h^\circ; a)$.

Let $h^\circ \in L_1^{p\times q}([0,a])$, $0 < a < \infty$, and

$$\mathfrak{g}(t) = \begin{cases} h^\circ(a-t) & \text{for } 0 \le t \le a \\ 0 & \text{for } t > a. \end{cases} \tag{9.130}$$

Then $\mathcal{S}(h^\circ; a) = e_a\mathcal{N}(\mathfrak{g})$, i.e.,

$$s \in \mathcal{S}(h^\circ; a) \Longleftrightarrow e_{-a}s \in \mathcal{N}(\mathfrak{g}). \tag{9.131}$$

The correspondence (9.131) reduces the $SP(h^\circ; a)$ to a corresponding $NP(\mathfrak{g})$ with $\mathfrak{g}(t)$ defined by formula (9.130).

If $0 < a < \infty$ and the mvf $\mathfrak{g}(t)$ is defined in terms of the mvf $h \in L_1^{p\times p}([0,a])$ by formula (9.130), then the operators $\Gamma^\vee$ and $\Gamma_*^\vee$ act between the subspaces $\mathcal{X}_a$ and $\mathcal{Y}_a$ of vvf's $\varphi(t)$ from the Banach spaces $\mathcal{X}$ and $\mathcal{Y}$ with support on the interval $[0,a]$. Let the operators $X^\vee$ and $X_*^\vee$ be defined between $\mathcal{X}_a$ and $\mathcal{Y}_a$ by the formulas

$$(X^\vee x)(t) = \int_0^t h^\circ(t-s)x(s)ds \text{ and } (X_*^\vee y)(t) = \int_t^a h^\circ(s-t)^* y(s)ds, \tag{9.132}$$

for $0 \le t \le a$, respectively, and let $(R\varphi)(t) = \varphi(a-t)$. Then it is readily checked that

$$RX^\vee = \Gamma^\vee \quad \text{and} \quad X_*^\vee R = \Gamma_*^\vee \tag{9.133}$$

and hence that formula (9.123) can be reexpressed in terms of $X^\vee$, $X_*^\vee$, R, h°, $v_j = Ru_{1j}$ and $y_j = u_{2j}$ as

$$\begin{bmatrix} I & -X^\vee \\ -X_*^\vee & I \end{bmatrix} \begin{bmatrix} v_1 & v_2 \\ y_1 & y_2 \end{bmatrix} = \begin{bmatrix} 0 & h^\circ \\ (Rh^\circ)^* & 0 \end{bmatrix}. \tag{9.134}$$

Remark 9.38 *If $h \in L_1^{p\times q}([0,a])$ and the operators $X^\vee$ from $L_2^q([0,a])$ into $L_2^p([0,a])$ and $X_*^\vee$ from $L_2^p([0,a])$ into $L_2^q([0,a])$ are as defined in (9.132), then the following assertions are equivalent:*

(1) $\|X^\vee x\| < \|x\|$ *for every nonzero vector* $x \in L_2^q([0,a])$.

(2) $\|X^\vee\| < 1$.

(3) $\begin{bmatrix} I & -X^\vee \\ -(X^\vee)^* & I \end{bmatrix} > 0.$

(4) $\begin{bmatrix} I & -X^\vee \\ -(X^\vee)^* & I \end{bmatrix} \ge \delta I$ *for some* $\delta > 0$.

Moreover, if any one of these properties is in force, then the operator

$$\begin{bmatrix} I & -X^\vee \\ -X_*^\vee & I \end{bmatrix}$$

has a bounded inverse in each of the Banach spaces $\mathcal{Y}_a \times \mathcal{X}_a$, where $\mathcal{Y}_a$ is $L_r^p([0,a])$ or $C_\ell^p([0,a])$ and $\mathcal{X}_a$ is $L_r^q([0,a])$ or $C_\ell^q([0,a])$, respectively, $1 \le r < \infty$

and $\ell = 0, 1, \ldots$. The verification of these assertions is similar to the proof of Lemma 9.36 and may obtained on the basis of this lemma and the relations (9.133).

Theorem 9.39 *Let $0 < a < \infty$ and let $h^\circ \in L_1^{p\times q}([0, a])$. Let the operators $X^\vee$ and $X_*^\vee$ be defined by formula (9.132), and let $\|X^\vee\|$ be the norm of the operator $X^\vee$ from the space $L_2^q([0, a])$ into $L_2^p([0, a])$. Then:*

(1) *$\mathcal{S}(h^\circ; a) \neq \emptyset$ if and only if $\|X^\vee\| \leq 1$.*
(2) *If $\|X^\vee\| \leq 1$, then $\|X^\vee\| = \min\{\|s\|_\infty : s \in \mathcal{S}(h^\circ; a)\}$.*
(3) *If $\|X^\vee\| < 1$, then there exists exactly one mvf W such that*

$$W \in \mathcal{E} \cap \mathcal{U}(j_{pq}), \quad W = \begin{bmatrix} e_a I_p & 0 \\ 0 & I_q \end{bmatrix} \mathfrak{A} \quad \text{with } \mathfrak{A} \in \mathcal{W}_r(j_{pq}) \tag{9.135}$$

and

$$\mathcal{S}(h^\circ; a) = T_W[\mathcal{S}^{p\times q}]. \tag{9.136}$$

This resolvent mvf W may be obtained from the formulas

$$W(\lambda) = \begin{bmatrix} e_a I_p & 0 \\ 0 & I_q \end{bmatrix} + \begin{bmatrix} \widehat{v}_1(\lambda) & \widehat{v}_2(\lambda) \\ \widehat{y}_1(\lambda) & \widehat{y}_2(\lambda) \end{bmatrix}, \tag{9.137}$$

where v_1, v_2, y_1 and y_2 are solutions of the system (9.134).
(4) *The mvf W considered in (3) is also a resolvent matrix for $WSP(h^\circ; a)$, i.e.,*

$$\mathcal{WS}(h^\circ; a) = T_W[\mathcal{W}_+^{p\times q}(0) \cap \mathcal{S}^{p\times q}]. \tag{9.138}$$

(5) *If a mvf W meets the constraints in (9.135), then there exists a unique mvf $h^\circ \in L_1^{p\times p}([0, a])$ such that $\mathcal{S}(h^\circ; a) = T_W[\mathcal{S}^{p\times p}]$: $h^\circ(t) = h(t)$ on $[0, a]$, where $\widehat{h} = T_w[0]$. Moreover, the corresponding Toeplitz operator $X^\vee$ is strictly contractive and $SP(h^\circ; a)$ is strictly completely indeterminate.*
(6) *If the columns of the mvf h° belong to one of the Banach spaces $L_r^p([0, a])$ for $1 \leq r < \infty$ or $C_\ell^p([0, a])$ for $\ell = 0, 1, \ldots$ then the columns of the $m \times m$ mvf's that are solutions of the equation (9.134) belong to the Banach spaces $L_r^m([0, a])$ and $C_\ell^m([0, a])$, respectively.*

Proof The main effort is to verify that the mvf W defined in (3) belongs to the class $\mathcal{U}(j_{pq})$, since just about everything else follows from Theorem 9.37 and the relations $RX^\vee = \Gamma^\vee$ and $X_*^\vee R = \Gamma_*^\vee$. It suffices to show that the Potapov–Ginzburg transform

$$S = \begin{bmatrix} s_{11} & s_{12} \\ s_{21} & s_{22} \end{bmatrix} = \begin{bmatrix} w_{11} & w_{12} \\ 0 & I_q \end{bmatrix} \begin{bmatrix} I_p & 0 \\ w_{21} & w_{22} \end{bmatrix}^{-1}$$

belongs to the class $\mathcal{S}^{m\times m}$. However, since $s_{11} = e_a(a_-^\#)^{-1} \in \mathcal{S}^{p\times p}$, $s_{12} = T_W[0] \in \mathcal{S}(h; a) \subseteq \mathcal{S}^{p\times q}$, $s_{21} = (a_+)^{-1}b_+ \in \mathcal{S}^{q\times p}$, $s_{22} = (a_+)^{-1} \in \mathcal{S}^{q\times q}$ and $S(\mu)$ is unitary on $\mathbb{R}$ this follows from the maximum principle. □

Remark 9.40 *Let $\mathring{W}$ and W denote the resolvent matrices of $\mathcal{S}(v^\circ; a)$ and $\mathcal{S}(h^\circ; a)$ that are subject to the constraints (9.87) and (9.135), repectively. Then*

$$W,\ \mathring{W} \in \mathcal{E} \cap \mathcal{U}(j_{pq}), \quad \{e_a I_p, I_q\} \in ap(W), \quad \text{and} \quad \{e_a I_p, I_q\} \in ap(\mathring{W}).$$

Moreover, since $\mathcal{S}(v^\circ; a) = \mathcal{S}(h^\circ; a)$, these two mvf's agree up to a right constant j_{pq}-unitary multiplier. Therefore, since $\mathring{W}(0) = I_m$, it follows that

$$W(\lambda) = \mathring{W}(\lambda) W(0).$$

The connection between W and $\mathring{W}$ is analogous to the connection between the two B-resolvent matrices for the accelerant extension problem that was considered after Theorem 9.20.

9.8 Bitangential Schur extension problems in the Wiener class

The main conclusion of this section is that if $s^\circ \in \mathcal{W}_+^{p\times q}(\gamma) \cap \mathcal{S}^{p\times q}$ and $b_1(\lambda)$ and $b_2(\lambda)$ are both entire mvf's, then a sufficient condition for (4.35) is that $\|\gamma\| < 1$. If either

$$\lim_{\nu\uparrow\infty} \|b_1(i\nu)\| = 0 \text{ or } \lim_{\nu\uparrow\infty} \|b_2(i\nu)\| = 0, \tag{9.139}$$

then this condition is necessary. We begin with the proof of the necessity of the condition $\|\gamma\| < 1$, because it is simple and applicable to a wider set of circumstances than indicated above.

If $s \in \mathcal{W}(\gamma) \cap \mathcal{S}^{p\times q}$, then the formulas for the operator $X^\vee$ and (its adjoint) $(X^\vee)^*$ that are defined in formulas (9.79) and (9.83), respectively, simplify, thanks to the following lemma:

Lemma 9.41 *Let*

$$s(\lambda) = \lambda^2 \int_0^\infty e^{i\lambda t} g(t)dt \quad \text{for } \lambda \in \mathbb{C}_+,$$

where $g \in \mathfrak{S}_\infty^{p\times q}$. Then the following statements are equivalent:

(1) $s \in \mathcal{W}_+^{p\times q}(\gamma) \cap \mathcal{S}^{p\times q}$.

(2) *$g(t)$ is absolutely continuous on $[0, a]$ for every $a > 0$, $g'(t)$ is absolutely continuous on $\mathbb{R}_+$ and $g'(0+) = -\gamma$, i.e.,*

$$g'(t) = -\gamma - \int_0^t h(y)dy \quad \text{for } t \geq 0 \text{ and some } h \in L_1^{p\times q}(\mathbb{R}_+).$$

(3) *$g(t)$ admits a representation of the form*

$$g(t) = -\gamma t - \int_0^t (t-y)h(y)dy \quad \text{for } t \geq 0 \text{ and some } h \in L_1^{p\times q}(\mathbb{R}_+). \tag{9.140}$$

Moreover, if any one (and hence every one) of these three conditions is in force and if $g(t)$ is expressed in the form (9.140), then

$$h = -g'' \quad \text{a.e. on } \mathbb{R}_+ \quad \text{and} \tag{9.141}$$

$$s(\lambda) = \gamma + \int_0^\infty e^{i\lambda t} h(t)dt. \tag{9.142}$$

Conversely, if $\gamma \in \mathbb{C}^{p\times q}$, $h \in L_1^{p\times q}(\mathbb{R}_+)$, g is defined by (9.140) and the mvf s defined by (9.142) belongs to $\mathcal{S}^{p\times q}$, then $g \in \mathfrak{G}_\infty^{p\times q}$ and $s \in \mathcal{W}_+^{p\times q}(\gamma) \cap \mathcal{S}^{p\times q}$.

Proof This follows by applying Theorem 9.14 to the mvf's $\widetilde{c}(\lambda)$ and $\widetilde{g}(t)$ that are defined in terms of $s(\lambda)$ and $g(t)$ by formulas (9.70) and (9.72), respectively. Correspondingly, in view of (9.70) and (9.142),

$$\widetilde{\gamma} = \frac{1}{2}\begin{bmatrix} I_p & 2\gamma \\ 0 & I_q \end{bmatrix} \quad \text{and} \quad \widetilde{h}(t) = \begin{bmatrix} 0_{p\times p} & h(t) \\ 0_{q\times p} & 0_{q\times q} \end{bmatrix}.$$

□

Lemma 9.42 *If $s \in \mathcal{W}^{p\times q}(\gamma)$ admits the representation (9.142), then the operators $X^\vee$ and $X_*^\vee$ defined by (9.79) and (9.83), respectively, can be expressed in terms of h:*

$$(X^\vee \varphi)(t) = \gamma\varphi(t) + \int_0^t h(t-y)\varphi(y)dy \tag{9.143}$$

and

$$((X^\vee)^*\psi)(t) = \gamma^*\psi(t) + \int_t^\infty h(y-t)^*\psi(y)dy. \tag{9.144}$$

Proof The proof rests on the observation that under the given assumptions the mvf v in formulas (9.79) and (9.83) can be expressed as

$$v(t) = \gamma + \int_0^t h(u)du.$$

This relation is obtained by comparing formula (9.78) with (9.142) written as

$$\begin{aligned} s(\lambda) &= \int_0^\infty e^{i\lambda t}\left(\frac{d}{dt}\int_0^t h(u)du\right)dt \\ &= \gamma + \left(e^{i\lambda t}\int_0^t h(u)du\right)\Big|_0^\infty - i\lambda\int_0^\infty e^{i\lambda t}\left(\int_0^t h(u)du\right)dt \\ &= -i\lambda\int_0^\infty e^{i\lambda t}\left\{\gamma + \int_0^t h(u)du\right\}dt \quad \text{for } \lambda \in \mathbb{C}_+. \end{aligned}$$

□

Lemma 9.43 *Let $s^\circ \in \mathcal{S}^{p\times q}$ and suppose that*

$$\gamma = \lim_{\nu\uparrow\infty} s^\circ(i\nu) \quad \text{exists and} \quad \|\gamma\| = 1\,.$$

Let $b_1 \in \mathcal{S}_{\text{in}}^{p\times p}$ and $b_2 \in \mathcal{S}_{\text{in}}^{q\times q}$ be such that (9.139) is in force. Then the $GSIP(b_1, b_2; s^\circ)$ is not strictly completely indeterminate.

Proof Let $s \in \mathcal{S}(b_1, b_2; s^\circ)$. Then

$$s(i\nu) = s^\circ(i\nu) + b_1(i\nu)f(i\nu)b_2(i\nu)$$

for some mvf $f \in H_\infty^{p\times q}$. Therefore, under the given assumptions,

$$\lim_{\nu\uparrow\infty} s(i\nu) = \lim_{\nu\uparrow\infty} s^\circ(i\nu) = \gamma\,, \quad \text{i.e.,} \quad \|s\|_\infty = \sup_{\lambda\in\mathbb{C}_+} \|s(\lambda)\| = 1. \qquad \square$$

Lemma 9.44 *Let $s^\circ \in \mathcal{W}_+^{p\times q}(\gamma) \cap \mathcal{S}^{p\times q}$ and let $b_1 \in \mathcal{S}_{\rm in}^{p\times p}$ and $b_2 \in \mathcal{S}_{\rm in}^{q\times q}$ be such that condition (9.139) is in force. Then:*

(1) $\gamma = \lim_{\nu\uparrow\infty} s^\circ(i\nu)$ *and* $\|\gamma\| \le 1$.

(2) *If $\|\gamma\| = 1$, then the GSIP $(b_1, b_2; s^\circ)$ is not strictly completely indeterminate.*

(3) *If $s^\circ \in \mathcal{W}_+^{p\times q}(\gamma) \cap \mathring{\mathcal{S}}^{p\times q}$, then $\|\gamma\| < 1$.*

Proof The first item is an easy consequence of the representation formula (9.142) and the Riemann–Lebesgue lemma; the second then follows from the last lemma. Item (3) is immediate from (1) and (2). $\square$

Lemma 9.45 *Let $s^\circ \in \mathcal{W}_+^{p\times q}(\gamma) \cap \mathcal{S}^{p\times q}$ and assume that $\|\gamma\| < 1$. Then for every $a > 0$ the operator*

$$T_a = \Pi_{\mathcal{H}(e_a I_p)} M_{s^\circ}|_{\mathcal{H}(e_a I_q)} \tag{9.145}$$

is strictly contractive: $\|T_a\| < 1$ and hence the GSIP$(e_a I_p, I_q; s^\circ)$ is strictly completely indeterminate.

Proof By assumption

$$s^\circ(\lambda) = \gamma + \int_0^\infty e^{i\lambda t} h(t)dt$$

for some $h \in L_1^{p\times q}([0, \infty))$. Thus, for $g \in \mathcal{H}(e_a I_q)$,

$$T_a g = \gamma g + K_a g,$$

where the operator K_a from $\mathcal{H}(e_a I_q)$ into $\mathcal{H}(e_a I_p)$ defined by the formula

$$(K_a g)(\lambda) = \int_0^a e^{i\lambda t} \left\{ \int_0^t h(t-s) g^\vee(s) ds \right\} dt$$

is compact. Therefore,

$$T_a^* T_a = A + B,$$

where $A = M_{\gamma^*\gamma}|_{\mathcal{H}(e_a I_q)}$ and B are bounded self-adjoint operators from $\mathcal{H}(e_a I_q)$ into itself such that: (1) $A + B \ge 0$, (2) $\|A + B\| \le 1$ and (3) B is compact. Consequently,

$$\|A + B\| = \max\{\lambda : \lambda \in \sigma(A + B)\},$$

the essential spectrum $\sigma_e(A+B) = \sigma_e(A)$, and

$$\sigma_d(A+B): \ = \sigma(A+B)\backslash\sigma_e(A+B)$$

consists only of eigenvalues of $A+B$ of finite multiplicity; see, e.g., chapter 9 of [BS87]. Moreover, since $\mu \in \sigma_e(A)$ if and only if μ is an eigenvalue of the matrix $\gamma^*\gamma$, it follows that

$$\max\{\lambda : \lambda \in \sigma_e(A+B)\} = \|\gamma\|^2 < 1\,.$$

Thus, in order to complete the proof, it suffices to show that 1 is not an eigenvalue of $T_a^*T_a$. Suppose to the contrary that 1 is an eigenvalue of $T_a^*T_a$ and let $g \in \mathcal{H}(e_aI_q)$ be an eigenfunction corresponding to this eigenvalue. Then

$$\|g\| = \|\Pi_{\mathcal{H}(e_aI_p)}s^\circ g\| \leq \|s^\circ g\| \leq \|g\|$$

and hence

$$g(\mu)^*\{I_q - s^\circ(\mu)^*s^\circ(\mu)\}g(\mu) = 0$$

for a.e. $\mu \in \mathbb{R}$. However, since

$$I_q - s^\circ(\mu)^*s^\circ(\mu) \geq \delta I_q > 0$$

for $|\mu| > R$ by the Riemann–Lebesgue lemma, this forces

$$g(\mu) = 0 \ \text{ for } \ |\mu| > R,$$

which in turn implies that

$$g(\lambda) \equiv 0,$$

because $g(\lambda)$ is an entire vvf. Thus 1 is not an eigenvalue of $T_a^*T_a$ and hence $\|T_a\| < 1$, as claimed. □

Before stating the next lemma it is useful to recall that if $b \in \mathcal{E} \cap \mathcal{S}_{\text{in}}^{p\times p}$, then $b(\lambda)$ is of exponential type and the type is given by the formula

$$\text{type}(b) \;=\; \limsup_{\nu\uparrow\infty} \frac{\ln \|b(i\nu)^{-1}\|}{\nu}. \tag{9.146}$$

Lemma 9.46 *Let $s^\circ \in \mathcal{S}^{p\times q}$, $b_1 \in \mathcal{E} \cap \mathcal{S}_{\text{in}}^{p\times p}$, $b_2 \in \mathcal{E} \cap \mathcal{S}_{\text{in}}^{q\times q}$ and $f^\circ = b_1^{-1}s^\circ b_2^{-1}$, let $a_j = \text{type}(b_j)$, $j = 1, 2$, and let $a = a_1 + a_2 > 0$. Then*

$$\|\Gamma\| \leq \|T_a\|, \tag{9.147}$$

where the operators are defined by formulas (4.27) and (9.145), respectively. If $b_1(\lambda)$ and $b_2(\lambda)$ are both homogeneous (i.e., if $b_j(\lambda) = e_{a_j}(\lambda)b_j(0)$, $j = 1, 2$), then equality prevails in (9.147).

Proof It is readily checked that

$$M_{b_1}\Pi_- M_{b_1^{-1}} = \Pi_{b_1(H_2^p)^\perp}$$

and hence that

$$\|\Gamma\| = \|\Pi_{b_1(H_2^p)^\perp} M_{s^\circ}|_{b_2^\# H_2^q}\|.$$

Thus, as

$$b_2^\# H_2^q \subseteq e_{-a_2} H_2^q \text{ and } b_1 (H_2^p)^\perp \subseteq e_{a_1} (H_2^p)^\perp, \tag{9.148}$$

it follows that

$$\begin{aligned}\|\Gamma\| &\leq \|\Pi_{e_{a_1}(H_2^p)^\perp} M_{s^\circ}|_{e_{-a_2} H_2^q}\| = \|\Pi_{(H_2^p)^\perp} e_{-a_1} M_{s^\circ} e_{-a_2}|_{H_2^q}\| \\ &= \|\Pi_{(H_2^p)^\perp} e_{-a} M_{s^\circ}|_{H_2^q}\| = \|\Pi_{e_a (H_2^p)^\perp} M_{s^\circ}|_{H_2^q}\| \\ &= \|\Pi_{\mathcal{H}(e_a I_p)} M_{s^\circ}|_{H_2^q}\| = \|\Pi_{\mathcal{H}(e_a I_p)} M_{s^\circ}|_{\mathcal{H}(e_a I_q)}\|.\end{aligned}$$

The inequality in this chain stems from the inclusions (9.148). If $b_1(\lambda)$ and $b_2(\lambda)$ are both homogeneous, then the inclusions in (9.148) are in fact equalities. Therefore, equality prevails in (9.147) also. □

Theorem 9.47 *Let* $s^\circ \in \mathcal{W}^{p\times q}(\gamma) \cap \mathcal{S}^{p\times q}$ *where* $\|\gamma\| < 1$ *and let* $b_1 \in \mathcal{E} \cap \mathcal{S}_{\rm in}^{p\times p}$ *and* $b_2 \in \mathcal{E} \cap \mathcal{S}_{\rm in}^{q\times q}$. *Then the GSIP*$(b_1, b_2; s^\circ)$ *is strictly completely indeterminate.*

Proof In view of assertion (2) in Theorems 4.1 and 4.10, it suffices to show that the Hankel operator Γ considered in Lemma 9.46 is strictly contractive. If the number $a = \text{type}(b_1) + \text{type}(b_2)$ is equal to zero, then $\Gamma = 0$, which is certainly strictly contractive. If $a > 0$, then the desired conclusion is immediate from Lemmas 9.45 and 9.46. □

Remark 9.48 *Notice that in Theorem 9.47 it was not assumed that* $I_q - s^\circ(\mu)^* s^\circ(\mu) > 0$ *on* $\mathbb{R}$. *The conclusion of the theorem remains correct even if* $\det\{I_q - s^\circ(\mu)^* s^\circ(\mu)\} = 0$ *for some points* $\mu \in \mathbb{R}$.

9.9 Supplementary notes

Most of the material in this section is adapted from [ArD98] and chapter 8 of [ArD08b]. A good source of information for the connection between extension problems and 2×2 canonical systems is [KrL85].

Assertion (2) of Theorem 9.18 was obtained earlier in [KrL85]; for additional investigations on this theme, see [MiPo81], [KoPo82], [Kat85a] and [Kat85b]; for additional perspective on extensions of the restriction of the function $g(t) = (1/2)t^2 - |t|$ that is considered in Examples 9.9 and 9.22, see [LLS04].

Section 9.4 is adapted from the treatment of continuous analogs of Schur extension problems in [ArD12b].

Section 9.7 is adapted from [KrMA86].

10

Bitangential inverse input scattering problems

In our formulation of the bitangential inverse input scattering problem, the given data are:

(1) A mvf $s^\circ \in \mathcal{S}^{p\times q}$.

(2) A normalized nondecreasing continuous chain of pairs of entire inner mvf's $\{b_1^t, b_2^t\}$ (with $b_1^t \in \mathcal{S}_{\text{in}}^{p\times p}$ and $b_2^t \in \mathcal{S}_{\text{in}}^{q\times q}$) for $0 \le t < d$.

The problem is to find a continuous nondecreasing $m \times m$ mvf $M(t)$ on the interval $[0, d)$ with $M(0) = 0$ such that the matrizant $W_t(\lambda)$, $0 \le t < d$, of the corresponding canonical integral system (6.1) meets the conditions

(S1) $s^\circ \in \bigcap_{0\le t<d} T_{W_t}[\mathcal{S}^{p\times q}]$ (i.e., $s^\circ \in \mathcal{S}_{\text{scat}}^d(dM)$).

(S2) $\{b_1^t, b_2^t\} \in ap(W_t)$ for every $t \in [0, d)$.

If this problem is solvable, then the GSIP$(b_1^t, b_2^t; s^\circ)$ is completely indeterminate for every $t \in [0, d)$. Therefore, there exists exactly one normalized nondecreasing chain of mvf's $W_t \in \mathcal{E} \cap \mathcal{U}^\circ(j_{pq})$ such that (8.31) holds. If $W_t \in \mathcal{E} \cap \mathcal{U}_{AR}^\circ(j_{pq})$ for every $t \in [0, d)$, then W_t can be identified as the matrizant of the canonical integral system (6.1) with mass function

$$M(t) = -i\left(\frac{\partial W_t}{\partial \lambda}\right)(0)\, j_{pq}$$

just as in the proof of Theorem 8.16. A sufficient condition for this is that

$$\mathcal{S}(b_1^t, b_2^t; s^\circ) \cap \mathring{\mathcal{S}}^{p\times q} \neq \emptyset \quad \text{for } 0 \le t < d. \tag{10.1}$$

It is helpful to keep in mind that the GSIP$(b_1, b_2; s^\circ)$ is

$$\text{completely indeterminate} \iff \mathcal{S}(b_1, b_2; s^\circ) \cap \mathcal{S}_{sz}^{p\times q} \neq \emptyset \tag{10.2}$$

$$\text{strictly completely indeterminate} \iff \mathcal{S}(b_1, b_2; s^\circ) \cap \mathring{\mathcal{S}}^{p\times q} \neq \emptyset; \tag{10.3}$$

(10.2) is by Theorem 4.16; (10.3) is by definition.

Clearly (10.1) holds if the given mvf $s^\circ \in \mathring{\mathcal{S}}^{p\times q}$; but this is not a necessary condition for (10.1) to hold. Thus, for example, if $s^\circ \in \mathcal{W}_+(\gamma) \cap \mathcal{S}^{p\times q}$ with $\gamma^*\gamma < I_q$, condition (10.1) holds by Theorem 9.47. This case will be discussed in Section 10.3.

To obtain another example, let $\{b_1^t, b_2^t\}$, $0 \le t \le d$, be a normalized nondecreasing continuous chain of pairs of entire inner mvf's such that the GSIP$(b_1^t, b_2^t; s^\circ)$ is strictly completely indeterminate. Then

$$\mathcal{S}(b_1^d, b_2^d; s^\circ) = T_{W_d}[\mathcal{S}^{p\times q}]$$

for exactly one mvf $W_d \in \mathcal{E} \cap \mathcal{U}_{rsR}^\circ(j_{pq})$ such that $\{b_1^d, b_2^d\} \in ap(W_d)$. Moreover,

$$\mathcal{S}(b_1^t, b_2^t; s^\circ) = \mathcal{S}(b_1^t, b_2^t; T_{W_d}[\varepsilon]) \quad \text{for every } \varepsilon \in \mathcal{S}^{p\times q}.$$

Therefore, since $T_{W_d}[\mathcal{S}_{\text{in}}^{p\times q}] \subseteq \mathcal{S}_{\text{in}}^{p\times q}$ if $p \ge q$, there exist mvf's $\widetilde{s} \in \mathcal{S}_{\text{in}}^{p\times q}$ such that

$$\mathcal{S}(b_1^t, b_2^t; \widetilde{s}) = \mathcal{S}(b_1^t, b_2^t; s^\circ) \quad \text{and} \quad \mathcal{S}(b_1^t, b_2^t; \widetilde{s}) \cap \mathring{\mathcal{S}}^{p\times q} \ne \emptyset \quad \text{for } 0 \le t \le d.$$

Thus, the GSIP$(b_1^t, b_2^t; \widetilde{s})$ is strictly completely indeterminate for every $t \in [0, d]$ even though $\|\widetilde{s}\|_\infty = 1$.

10.1 Existence and uniqueness of solutions

The next two theorems are analogs of Theorems 8.16 and 8.17. Before tackling them, it is probably helpful to review Lemmas 8.14 and 8.15.

Theorem 10.1 *If $s^\circ \in \mathcal{S}^{p\times q}$, $\{b_1^t, b_2^t\}$, $0 \le t < d$, is a normalized nondecreasing continuous chain of pairs of entire inner mvf's of sizes $p \times p$ and $q \times q$, respectively, and the GSIP$(b_1^t, b_2^t; s^\circ)$ is completely indeterminate for every $t \in [0, d)$, then:*

(1) *For each $t \in [0, d)$ there exists exactly one mvf $W_t \in \mathcal{U}^\circ(j_{pq})$ such that*

$$\mathcal{S}(b_1^t, b_2^t; s^\circ) = T_{W_t}[\mathcal{S}^{p\times q}] \quad \text{and} \quad \{b_1^t, b_2^t\} \in ap(W_t). \tag{10.4}$$

The family of resolvent matrices W_t, $0 \le t < d$, that is obtained this way enjoys the following properties:

(2a) *W_t, $0 \le t < d$, is a normalized nondecreasing $\curvearrowright$ chain of mvf's in the class $\mathcal{E} \cap \mathcal{U}_{rR}^\circ(j_{pq})$ with $W_0 = I_m$ and $s^\circ \in \bigcap_{0\le t<d} T_{W_t}[\mathcal{S}^{p\times q}]$.*

(2b) *$W_t(\lambda)$ is left continuous on $[0, d)$ for each $\lambda \in \mathbb{C}$.*

(2c) *If $W_t(\lambda)$ is continuous on $[0, d)$ for each $\lambda \in \mathbb{C}$ then it is the matrizant of the canonical integral system (6.1) with mass function that meets the constraints (1.21) and is given by $M(t) = 2\pi K_0^{W_t}(0)$.*

(2d) *If $W_t \in \mathcal{U}_{AR}(j_{pq})$ for every $t \in [0, d)$, then $W_t(\lambda)$ is continuous on $[0, d)$ for each $\lambda \in \mathbb{C}$.*

(3) *If $\widetilde{M}(t)$ is any continuous nondeceasing $m \times m$ mvf on $[0, d)$ with $\widetilde{M}(0) = 0$ and if the matrizant $\widetilde{W}_t(\lambda)$, $0 \le t < d$, of the canonical integral system (6.1)*

with this mass function belongs to the class $\mathcal{U}^{\circ}_{rR}(j_{pq})$ *for every* $t \in [0, d)$, *then* $\widetilde{M}(t)$ *is the solution of the inverse input scattering problem with data*

$$s^{\circ} \in \bigcap_{0 \le t < d} T_{\widetilde{W}_t}[\mathcal{S}^{p\times q}] \quad \text{and} \quad \{b_1^t, b_2^t\} \in ap(\widetilde{W}_t) \quad \text{for } 0 \le t < d,$$

where $b_1^t(0) = I_p$ *and* $b_2^t(0) = I_q$ *for every* $t \in [0, d)$ *and* $\widetilde{M}(t)$ *may be obtained from this data via (1) and (2c).*

Proof The proof of (1)–(2d) is essentially the same as the proof of the corresponding statements in Theorem 8.16.

The assumption on s° in the setting of (3) insures that $s^{\circ} \in T_{\widetilde{W}_t}[\mathcal{S}^{p\times q}]$ for every $t \in [0, d)$ and hence, as $\{b_1^t, b_2^t\} \in ap(\widetilde{W}_t)$, that the GSIP$(b_1^t, b_2^t; s^{\circ})$ is completely indeterminate for every $t \in [0, d)$ and that $\widetilde{W}_t$ is the one and only resolvent matrix in the class $\mathcal{U}^{\circ}(j_{pq})$ for this problem. □

Theorem 10.2 *In the setting of Theorem 10.1, assume that the chain* $\{b_1^t, b_2^t\}$, $0 \le t < d$, *is strictly increasing and the family of resolvent matrices* $W_t(\lambda)$, $0 \le t < d$, *that is specified in (1) of that theorem is continuous in* t *on* $[0, d)$ *for each* $\lambda \in \mathbb{C}$. *Then there exists exactly one normalized solution* $H(x)$, $0 \le x < \ell$, *of the bitangential input scattering problem for the canonical differential system (2.14) with* $J = j_{pq}$, *input scattering matrix* s° *and matrizant* Ω_x, $0 \le x < \ell$, *such that*

$$\Omega_x \in \mathcal{U}_{rR}(j_{pq}) \quad \text{and} \quad \{b_1^{t(x)}, b_2^{t(x)}\} \in ap(\Omega_x) \text{ for every } x \in [0, \ell) \tag{10.5}$$

for some function $t(x)$ *that maps* $[0, \ell)$ *into* $[0, d)$ *with* $t(0) = 0$ *and* $\lim_{x\uparrow\ell} t(x) = d$. *Moreover, there is only one such function* $t(x)$: *it is the inverse of the function* $x(t) = \operatorname{trace} M(t)$ *and is a strictly increasing continuous function on* $[0, \ell)$. *The mvf's* $H(x)$ *and* $\Omega_x(\lambda)$ *may be obtained from* $M(t)$ *and* W_t *by the formulas*

$$H(x) = \frac{d}{dx} M(t(x)) \quad \text{a.e. on } [0, \ell) \quad \text{and} \quad \Omega_x = W_{t(x)} \text{ for every } x \in [0, \ell).$$

Proof This follows from Theorem 8.17 by first taking $0 < d_1 < d$, $W = W_{d_1}$ and W_t is the resolvent matrix in $\mathcal{U}^{\circ}(j_{pq})$ with $\{b_1^t, b_2^t\} \in ap(W_t)$ of the GSIP$(b_1^t, b_2^t; s^{\circ})$. □

10.2 Formulas for the solution of the inverse input scattering problem

In this section we shall obtain formulas for the solution of the bitangential inverse input scattering problem when the given data is a mvf $s^{\circ} \in \mathcal{S}^{p\times q}$ and a normalized nondecreasing continuous chain of pairs $\{b_1^t, b_2^t\}$, $0 \le t < d$, of entire inner mvf's of sizes $p \times p$ and $q \times q$, respectively, such that the GSIP$(b_1^t, b_2^t; s^{\circ})$ is strictly completely indeterminate for every $t \in [0, d)$. Theorem 4.12 guarantees that for each $t \in [0, d)$ there exists exactly one mvf $W_t \in \mathcal{E} \cap \mathcal{U}^\circ(j_{pq})$ such that

$$T_{W_t}[\mathcal{S}^{p\times q}] = \mathcal{S}(b_1^t, b_2^t; s^\circ) \quad \text{and} \quad \{b_1^t, b_2^t\} \in ap(W_t). \tag{10.6}$$

Moreover, this mvf $W_t(\lambda)$ is automatically right strongly regular.

Theorem 4.75 gives a formula for $W_t(\lambda)$ in terms of operators that are defined in terms of $b_1^t(\lambda)$, $b_2^t(\lambda)$ and $s^\circ(\lambda)$, i.e., in terms of the given data. Our next objective is to reexpress this formula as a system of equations in terms of the mvf's $v(t)$, $h_{b_1^t}$ and $h_{b_2^t}$ that are taken from the integral representations of $s(\lambda) = s^\circ(\lambda)$, $b_1^t(\lambda)$ and $b_2^t(\lambda)$. To this end it is convenient to write L_t, Δ_t, X_{11}^t, X_{22}^t and X_{12}^t for the operators based on the normalized associated pair $\{b_1^t, b_2^t\}$ of W_t; and G_t in place of $F_\omega^{X_t}$ with $\omega = 0$. Thus,

$$X_{11}^t = \Pi_{\mathcal{H}(b_1^t)} M_{s^\circ}\big|_{H_2^q}, \quad X_{22}^t = \Pi_- M_{s^\circ}\big|_{\mathcal{H}_*(b_2^t)} \quad \text{and}$$
$$X_{12}^t = \Pi_{\mathcal{H}(b_1^t)} M_{s^\circ}\big|_{\mathcal{H}_*(b_2^t)} \quad \text{for } t \in [0, d). \tag{10.7}$$

In the interest of simplicity, we shall abuse this notation a little and allow each of the five operators listed above (and their adjoints) to act on mvf's in the obvious way, a column at a time (i.e., the k'th column of the "input" gets mapped into the k'th column of the "output").

Thus,

$$L_t = \begin{bmatrix} I & X_{22}^t \\ (X_{11}^t)^* & I \end{bmatrix} : \mathcal{H}(b_1^t, b_2^t) \to \mathcal{H}(W_t), \tag{10.8}$$

$$\Delta_t = L_t^* L_t = \begin{bmatrix} I - X_{11}^t (X_{11}^t)^* & -X_{12}^t \\ -(X_{12}^t)^* & I - (X_{22}^t)^* X_{22}^t \end{bmatrix} : \mathcal{H}(b_1^t, b_2^t) \to \mathcal{H}(b_1^t, b_2^t), \tag{10.9}$$

$$G_t = \begin{bmatrix} k_0^{b_1^t}(\lambda) & (X_{11}^t k_0^{\widehat{b}_1^t})(\lambda) \\ ((X_{22}^t)^* \ell_0^{\widehat{b}_2^t})(\lambda) & \ell_0^{b_2^t}(\lambda) \end{bmatrix} : \mathbb{C}^m \to \mathcal{H}(b_1^t, b_2^t), \tag{10.10}$$

where $\widehat{b}_1^t = e_{\tau_1(t)} I_p$, $\widehat{b}_2^t = e_{\tau_2(t)} I_q$ and $\tau_j(t)$ denotes the exponential type of the entire inner mvf's b_j^t for $j = 1, 2$. Then, in view of (4.153) and (4.154), the RK K_ω^t of the RKHS $\mathcal{H}(W_t)$ at the point $\omega = 0$ is given by the formula

$$K_0^t(\lambda) u = L_t \Delta_t^{-1} G_t u \quad \text{for } u \in \mathbb{C}^m \text{ and } t \in [0, d). \tag{10.11}$$

Thus,

$$\begin{aligned} u^* M(t) u &= 2\pi \langle K_0^t u, K_0^t u \rangle_{\mathcal{H}(A_t)} = 2\pi \langle L_t \Delta_t^{-1} G_t u, L_t \Delta_t^{-1} G_t u \rangle_{\mathcal{H}(A_t)} \\ &= 2\pi \langle \Delta_t^{-1} G_t u, G_t u \rangle_{\text{st}}. \end{aligned} \tag{10.12}$$

Therefore,

$$M(t) = 2\pi G_t^* \Delta_t^{-1} G_t \quad \text{for } 0 \le t < d. \tag{10.13}$$

Theorem 10.3 *Let $s^\circ \in \mathcal{S}^{p\times q}$ and let $\{b_1^t, b_2^t\}$, $0 \le t < d$, be a normalized nondecreasing continuous chain of pairs of entire inner mvf's of sizes $p \times p$ and*

$q \times q$, respectively, such that $GSIP(b_1^t, b_2^t; s^\circ)$ is strictly completely indeterminate for every $t \in [0, d)$. Then there exists exactly one mvf $W_t \in \mathcal{E} \cap \mathcal{U}^\circ(j_{pq})$ that meets the conditions (10.6) for every t in the interval $0 \le t < d$. The RK $K_\omega^t(\lambda)$ of the RKHS $\mathcal{H}(W_t)$ evaluated at $\omega = 0$ is given by the formula

$$K_0^t(\lambda) = \frac{1}{2\pi}\begin{bmatrix} \widehat{x}_1^t(\lambda) + \widehat{v}_1^t(\lambda) & \widehat{x}_2^t(\lambda) + \widehat{v}_2^t(\lambda) \\ \widehat{u}_1^t(\lambda) + \widehat{y}_1^t(\lambda) & \widehat{u}_2^t(\lambda) + \widehat{y}_2^t(\lambda) \end{bmatrix}, \tag{10.14}$$

where the mvf's that appear on the right-hand side of (10.14) are solutions of the systems of equations

$$\begin{bmatrix} I & -X_{11}^t & -X_{12}^t & 0 \\ -(X_{11}^t)^* & I & 0 & 0 \\ -(X_{12}^t)^* & 0 & I & -(X_{22}^t)^* \\ 0 & 0 & -X_{22}^t & I \end{bmatrix}\begin{bmatrix} \widehat{x}_j^t \\ \widehat{y}_j^t \\ \widehat{u}_j^t \\ \widehat{v}_j^t \end{bmatrix} = \begin{bmatrix} \widehat{\varphi}_j^t(\lambda) \\ 0 \\ \widehat{\psi}_j^t(\lambda) \\ 0 \end{bmatrix} \tag{10.15}$$

for $j = 1, 2$, in which

$$\widehat{\varphi}_1^t(\lambda) = \frac{I_p - b_1^t(\lambda)b_1^t(0)^*}{-i\lambda} = \frac{b_1^t(\lambda) - I_p}{i\lambda}, \tag{10.16}$$

$$\widehat{\varphi}_2^t(\lambda) = \left(X_{11}^t\left(\int_0^{\tau_1(t)} e_s I_q ds\right)\right)(\lambda), \tag{10.17}$$

$$\widehat{\psi}_1^t(\lambda) = \left((X_{22}^t)^*\left(\int_{-\tau_2(t)}^{0} e_s I_p ds\right)\right)(\lambda), \tag{10.18}$$

$$\widehat{\psi}_2^t(\lambda) = \frac{b_2^t(\lambda)^{-1}b_2^t(0)^{-*} - I_q}{-i\lambda} = \frac{b_2^t(\lambda)^{-1} - I_q}{-i\lambda}, \tag{10.19}$$

$$\tau_1(t) = \text{type}\{b_1^t\}, \quad \tau_2(t) = \text{type}\{b_2^t\} \tag{10.20}$$

and

the columns of $\widehat{x}_j^t(\lambda)$ belong to $\mathcal{H}(b_1^t)$,
the columns of $\widehat{y}_j^t(\lambda)$ belong to $(X_{11}^t)^*\mathcal{H}(b_1^t) \subseteq \mathcal{H}(e_{\tau_1(t)}I_q)$,
the columns of $\widehat{u}_j^t(\lambda)$ belong to $\mathcal{H}_*(b_2^t)$,
the columns of $\widehat{v}_j^t(\lambda)$ belong to $X_{22}^t\mathcal{H}_*(b_2^t) \subseteq \mathcal{H}_*(e_{\tau_2(t)}I_p)$.

Proof By formulas (10.10) and (10.11), the RK $K_\omega^t(\lambda)$ for the RKHS $\mathcal{H}(W_t)$ for the point $\omega = 0$ is given by the formula

$$K_0^t\begin{bmatrix} \xi \\ \eta \end{bmatrix} = \frac{1}{2\pi}L_t\Delta_t^{-1}\begin{bmatrix} \widehat{\varphi}_1^t\xi + \widehat{\varphi}_2^t\eta \\ \widehat{\psi}_1^t\xi + \widehat{\psi}_2^t\eta \end{bmatrix}.$$

Let

$$\begin{bmatrix} \widehat{x}_j^t \\ \widehat{u}_j^t \end{bmatrix} = \Delta_t^{-1} \begin{bmatrix} \widehat{\varphi}_j^t \\ \widehat{\psi}_j^t \end{bmatrix}, \tag{10.21}$$

$$\widehat{y}_j^t = (X_{11}^t)^* \widehat{x}_j^t \quad \text{and} \quad \widehat{v}_j^t = X_{22}^t \widehat{u}_j^t, \tag{10.22}$$

for $j = 1, 2$. Then it is readily seen that

$$K_0^t \begin{bmatrix} \xi \\ 0 \end{bmatrix} = \frac{1}{2\pi} L_t \begin{bmatrix} \widehat{x}_1^t \xi \\ \widehat{u}_1^t \xi \end{bmatrix} = \frac{1}{2\pi} \begin{bmatrix} I & X_{22}^t \\ (X_{11}^t)^* & I \end{bmatrix} \begin{bmatrix} \widehat{x}_1^t \xi \\ \widehat{u}_1^t \xi \end{bmatrix} = \frac{1}{2\pi} \begin{bmatrix} \widehat{x}_1^t \xi + \widehat{v}_1^t \xi \\ \widehat{y}_1^t \xi + \widehat{u}_1^t \xi \end{bmatrix}.$$

This yields the first block column of formula (10.14). Similar considerations lead to the second. Moreover, by formula (10.21),

$$\begin{bmatrix} \widehat{\varphi}_j^t \\ \widehat{\psi}_j^t \end{bmatrix} = \Delta_t \begin{bmatrix} \widehat{x}_j^t \\ \widehat{u}_j^t \end{bmatrix} = \begin{bmatrix} I - X_{11}^t (X_{11}^t)^* & -X_{12}^t \\ -(X_{12}^t)^* & I - (X_{22}^t)^* X_{22}^t \end{bmatrix} \begin{bmatrix} \widehat{x}_j^t \\ \widehat{u}_j^t \end{bmatrix},$$

which, with the aid of (10.22), is readily seen to be equivalent to the system of equations (10.15). □

Theorem 10.4 *In the setting of Theorem 10.3, the mvf $M(t) = 2\pi K_0^t(0)$ is given by the formulas*

$$\begin{aligned} M(t) &= \begin{bmatrix} \widehat{x}_1^t(0) + \widehat{v}_1^t(0) & \widehat{x}_2^t(0) + \widehat{v}_2^t(0) \\ \widehat{u}_1^t(0) + \widehat{y}_1^t(0) & \widehat{u}_2^t(0) + \widehat{y}_2^t(0) \end{bmatrix} \\ &= \int_0^{\tau_1(t)} \begin{bmatrix} x_1^t(a) & x_2^t(a) \\ y_1^t(a) & y_2^t(a) \end{bmatrix} da + \int_{-\tau_2(t)}^0 \begin{bmatrix} v_1^t(a) & v_2^t(a) \\ u_1^t(a) & u_2^t(a) \end{bmatrix} da \end{aligned} \tag{10.23}$$

and

$$M(t) = \int_0^{\tau_1(t)} \begin{bmatrix} I_p \\ v(a)^* \end{bmatrix} [x_1^t(a) \quad x_2^t(a)] da + \int_{-\tau_2(t)}^0 \begin{bmatrix} v(-b) \\ I_q \end{bmatrix} [u_1^t(b) \quad u_2^t(b)] db, \tag{10.24}$$

where $\widehat{x}_j^t$, $\widehat{y}_j^t$, $\widehat{u}_j^t$ and $\widehat{v}_j^t$, $j = 1, 2$, are solutions of the system of equations (10.15) with right-hand sides specified by (10.16)–(10.19) and $s^\circ \in \mathcal{S}^{p\times q}$ is given by formula (9.78) with $s = s^\circ$.

The mvf $M(t)$ specified by formulas (10.23) and (10.24) is the unique solution of the bitangential inverse input scattering problem for the canonical system of integral equations (6.1), given data $s^\circ \in \mathcal{S}^{p\times q}$ and $\{b_1^t, b_2^t\}$, $0 \le t < d$, as in Theorem 10.3.

Proof Formula (10.23) is an immediate corollary of Theorem 10.3, whereas formula (10.24) is then an easy consequence of formulas (9.106) and (9.107).

Thus, for example, formula (9.107 serves to exhibit

$$\int_0^{\tau_1(t)} y_j^t(a)da = \int_0^{\tau_1(t)} \left\{ -\frac{d}{da} \int_a^{\tau_1(t)} v(u-a)^* x_j^t(u)du \right\} da$$

as the integral of a derivative, which is easily evaluated to yield the first asserted formulas. The rest is similar. □

10.3 Input scattering matrices in the Wiener class

In this section the bitangential input scattering problem for input scattering matrices $s^\circ \in \mathcal{S}^{p\times q}$ in the Wiener class $\mathcal{W}_+^{p\times q}(\gamma)$ will be studied. In view of the Riemann–Lebesgue lemma,

$$s^\circ \in \mathcal{W}_+^{p\times q}(\gamma) \cap \mathcal{S}^{p\times q} \Longrightarrow \|\gamma\| \leq 1.$$

Moreover, if the GSIP$(b_1, b_2; s^\circ)$ based on $s^\circ \in \mathcal{W}_+^{p\times q}(\gamma) \cap \mathcal{S}^{p\times q}$ and a pair of inner mvf's b_1 and b_2 that have the property (9.139) is strictly completely indeterminate, then, by Lemma 9.43, $\|\gamma\| < 1$. On the other hand, if $\|\gamma\| < 1$, then, by Theorem 9.47, the GSIP$(b_1, b_2; s^\circ)$ with entire inner mvf's b_1 and b_2 is strictly completely indeterminate, even if (9.139) is not assumed. These facts, taken together with Theorem 10.1, lead to the next result.

Theorem 10.5 *Let* $\{b_1^t, b_2^t\}$, $0 \leq t < d$, *be a normalized nondecreasing continuous chain of pairs of entire inner mvf's of sizes* $p \times p$ *and* $q \times q$, *respectively, and let* $s^\circ \in \mathcal{W}_+^{p\times q}(\gamma) \cap \mathcal{S}^{p\times q}$ *with* $\|\gamma\| < 1$. *Then there exists exactly one continuous nondecreasing* $m \times m$ *mvf* $M(t)$ *on the interval* $[0, d)$ *with* $M(0) = 0$ *such that the matrizant* $W_t(\lambda) = W(t, \lambda)$, $0 \leq t < d$, *of the corresponding canonical integral system (6.1) satisfies the following three conditions*

(1) $s^\circ \in \bigcap_{0\leq t<d} T_{W_t}[\mathcal{S}^{p\times q}]$.
(2) $\{b_1^t, b_2^t\} \in ap(W_t)$ *for every* $t \in [0, d)$.
(3) $W_t \in \mathcal{U}_{rsR}^\circ(j_{pq})$ *for every* $t \in [0, d)$.

Moreover, if

$$\lim_{\nu\uparrow\infty} b_1^{t_0}(i\nu) = 0 \quad or \quad \lim_{\nu\uparrow\infty} b_2^{t_0}(i\nu) = 0$$

for some t_0, $0 < t_0 < d$, *then the condition* $\|\gamma\| < 1$ *is necessary for the existence of a canonical system (6.1) with a matrizant* $W_t(\lambda)$, $0 \leq t < d$, *that meets the stated three conditions*

Proof In view of Theorem 9.47, the GSIP$(b_1^t, b_2^t; s^\circ)$ is strictly completely indeterminate for every $t \in [0, d)$ and all the asserted conclusions except for the last assertion follow from Theorem 10.1; the last assertion follows from Lemma 9.43. □

Corollary 10.6 *Let $s^\circ \in \mathcal{W}_+^{p\times q}(\gamma)\cap \mathcal{S}^{p\times q}$ and let $\tau_1(t)$ and $\tau_2(t)$ be a pair of nondecreasing functions on the interval $[0,d)$ with $\tau_1(0)=\tau_2(0)=0$ and $\tau_1(t)+\tau_2(t)>0$ for some point $t\in(0,d)$. Then there exists a continuous nondecreasing $m\times m$ mvf $M(t)$ on $[0,d)$ with $M(0)=0$ such that the matrizant $W_t(\lambda)$ of the corresponding integral system (6.1) satisfies the three conditions*

1. $s^\circ \in \bigcap_{0\le t<d} T_{W_t}[\mathcal{S}^{p\times q}]$;
2. $\{e_{\tau_1(t)}I_p, e_{\tau_2(t)}I_q\}\in ap(W_t)$ *for every* $t\in[0,d)$;
3. $W_t\in \mathcal{U}_{rsR}(j_{pq})$ *for every* $t\in[0,d)$;

   if and only if

$$\gamma^*\gamma < I_q. \tag{10.25}$$

Moreover, if $\gamma^*\gamma<I_q$, then there exists exactly one mass function $M(t)$ that meets the conditions stated above.

Proof Since the condition (9.139) is in force for $b_1^t=e_{\tau_1(t)}I_p$ and $b_2^t=e_{\tau_2(t)}I_q$, this follows immediately from Theorem 10.5. □

In view of Theorem 8.57, applied to compact subintervals of $(0,d)$, the solution $M(t)$ of the inverse input scattering problem considered in the corollary can be expressed in terms of the solution $M_+(t)$ (resp., $M_-(t)$) of the same problem with the same input scattering matrix s°, but with $b_1^t=e_tI_p$ and $b_2^t=I_q$ (resp., $b_1^t=I_p$ and $b_2^t=e_tI_q$ for $0\le t\le \ell$ and $\ell=\tau_1(d)+\tau_2(d)$.

10.4 Examples with diagonal mvf's b_1^t and b_2^t

In this section we shall consider a number of examples of the bitangential inverse input scattering problem using the formulas developed above. In all of these examples it will be assumed that the given data: $s^\circ\in\mathcal{S}^{p\times q}$, and a set of pairs of normalized mvf's $b_1^t\in\mathcal{E}\cap\mathcal{S}_{\rm in}^{p\times p}$ and $b_2^t\in\mathcal{E}\cap\mathcal{S}_{\rm in}^{q\times q}$, is such that the GSIP$(b_1^t,b_2^t;s^\circ)$ is strictly completely indeterminate and hence that the general formula for the mass function $M(t)$ that was presented in Theorem 10.4 is applicable.

We begin with the simplest cases when $b_1^t(\lambda)$ and $b_2^t(\lambda)$ are homogeneous. The following lemma allows us to evaluate $\widehat{\varphi}_2^t(\lambda)$ and $\widehat{\psi}_1^t(\lambda)$ in this setting.

Lemma 10.7 *Let $b_1(\lambda)=e_{\alpha_1}(\lambda)I_p$, $b_2(\lambda)=e_{\alpha_2}(\lambda)I_q$, let $(R_0f)(\lambda)=\{f(\lambda)-f(0)\}/\lambda$ and let $s^\circ\in\mathcal{S}^{p\times q}$ be given by formula (9.78), where $v(t)$ is a $p\times q$ mvf that meets the constraints set forth in conditions (2)–(4) of Theorem 9.27. Then*

$$\frac{1}{i}\Pi_{\mathcal{H}(b_1)}\{M_{s^\circ}R_0e_{\alpha_1}I_q\}=\int_0^{\alpha_1}e^{i\lambda s}v(s)ds$$

and

$$-\frac{1}{i}\Pi_{\mathcal{H}_*(b_2)}\{(M_{s^\circ})^*R_0e_{-\alpha_2}I_p\}=\int_{-\alpha_2}^{0}e^{i\lambda s}v(-s)^*ds.$$

Proof By formula (9.78),

$$\frac{1}{i}(M_{s^\circ}R_0e_{\alpha_1}I_q)(\lambda) = \left\{-i\lambda\int_0^\infty e^{i\lambda s}v(s)ds\right\}\frac{e_{\alpha_1}(\lambda)-1}{i\lambda}I_q$$
$$= \int_0^\infty e^{i\lambda s}v(s)ds - e^{i\alpha_1\lambda}\int_0^\infty e^{i\lambda s}v(s)ds$$
$$= \int_0^{\alpha_1} e^{i\lambda s}v(s)ds + \int_{\alpha_1}^\infty e^{i\lambda s}\{v(s)-v(s-\alpha_1)\}ds.$$

The first projection formula now drops out easily since the columns of the first term on the right in the last line belong to $\mathcal{H}(b_1)$ and the columns of the second term are orthogonal to $\mathcal{H}(b_1)$.

The second projection formula is verified in much the same way. □

Remark 10.8 *The last formula shows that even though $v \in L_{2,\mathrm{loc}}^{p\times q}$,*

$$\int_0^\infty \|v(u+a)-v(u)\|^2du < \infty \quad \textit{for every } a>0.$$

Example 10.9 *Let $b_1^t(\lambda) = e_{\tau_1(t)}I_p$ and $b_2^t(\lambda) = I_q$ for $0 \le t < d$, where $\tau_1(t)$ is a continuous nondecreasing function of t on the given interval $[0,d)$ with $\tau_1(0) = 0$. Then*

$$M(t) = \int_0^{\tau(t)}\begin{bmatrix} x_1^t(a) & x_2^t(a) \\ v(a)^*x_1^t(a) & v(a)^*x_2^t(a)\end{bmatrix}da \tag{10.26}$$

for $0 \le t < d$, where

$$\tau(t) = \tau_1(t). \tag{10.27}$$

$x_j^t(a)$ and $y_j^t(a)$ are solutions of the following systems of equations

$$\begin{aligned} x_j^t(a) - \frac{d}{da}\int_0^a v(a-s)y_j^t(s)ds &= \varphi_j^t(a) \\ \frac{d}{da}\int_a^{\tau_1(t)} v(s-a)^*x_j^t(s)ds + y_j^t(a) &= 0 \end{aligned} \tag{10.28}$$

for $0 \le a \le \tau_1(t)$ and $j = 1,2$, where

$$\varphi_1^t(a) = I_p \quad \text{and} \quad \varphi_2^t(a) = v(a) \tag{10.29}$$

for $0 \le a \le \tau_1(t)$

Discussion If $b_2^t(\lambda) = I_q$ for $0 \le t < d$ in the setting of Theorem 10.3, then formulas (10.24) and (10.23) reduce to (10.26) and

$$M(t) = \begin{bmatrix} \widehat{x}_1^t(0) & \widehat{x}_2^t(0) \\ \widehat{y}_1^t(0) & \widehat{y}_2^t(0)\end{bmatrix}, \tag{10.30}$$

respectively, and (10.15) simplifies to

$$\begin{bmatrix} I & -X_t \\ -X_t^* & I \end{bmatrix} \begin{bmatrix} \widehat{x}^t_j \\ \widehat{y}^t_j \end{bmatrix} = \begin{bmatrix} \widehat{\varphi}^t_j(\lambda) \\ 0 \end{bmatrix} \quad \text{for } j = 1, 2 \tag{10.31}$$

with $X_t = X^t_{11}$ and $\widehat{\varphi}^t_1(\lambda)$ (resp., $\widehat{\varphi}^t_2(\lambda)$) given by formula (10.16) (resp., (10.17)).

In the case at hand,

$$\widehat{\varphi}^t_1(\lambda) = \left(\int_0^{\tau_1(t)} e^{i\lambda s} ds \right) I_p \tag{10.32}$$

and, by Lemma 10.7,

$$\widehat{\varphi}^t_2(\lambda) = \int_0^{\tau_1(t)} e^{i\lambda s} v(s) ds. \tag{10.33}$$

This justifies (10.29). Next, by Theorem 9.27,

$$M_{s^\circ} \widehat{y}^t_j = \int_0^\infty e^{i\lambda a} \left\{ \frac{d}{da} \int_0^a v(a-s) y^t_j(s) ds \right\} da \tag{10.34}$$

and

$$\Pi_+ (M_{s^\circ})^* \widehat{x}^t_j = - \int_0^\infty e^{i\lambda a} \left\{ \frac{d}{da} \int_a^\infty v(s-a)^* x^t_j(s) ds \right\} da.$$

For the given choice of $b^t_1(\lambda)$,

$$X_t \widehat{y}^t_j = \int_0^{\tau_1(t)} e^{i\lambda a} \left\{ \frac{d}{da} \int_0^a v(a-s) y^t_j(s) ds \right\} da \tag{10.35}$$

and, as

$$x^t_j(s) = 0 \ \text{ for } \ s > \tau_1(t),$$

$$X^*_t \widehat{x}^t_j = - \int_0^{\tau_1(t)} e^{i\lambda a} \left\{ \frac{d}{da} \int_a^{\tau_1(t)} v(s-a)^* x^t_j(s) ds \right\} da. \tag{10.36}$$

The equations for $x^t_j(a)$ and $y^t_j(a)$ now emerge easily upon substituting (10.32), (10.33), (10.35) and (10.36) into formula (10.31) and then taking inverse Fourier transforms.

The final formula for $M(t)$ exploits the observation that

$$\begin{aligned} \widehat{y}^t_j(0) &= (X^*_t \widehat{x}^t_j)(0) \\ &= - \int_0^{\tau_1(t)} \left\{ \frac{d}{da} \int_a^{\tau_1(t)} v(s-a)^* x^t_j(s) ds \right\} da \\ &= \int_0^{\tau_1(t)} v(s)^* x^t_j(s) ds. \end{aligned}$$

Example 10.10 The two-sided homogeneous case
Let $b_1^t(\lambda) = e_{\tau_1(t)}(\lambda)I_p$ and $b_2^t(\lambda) = e_{\tau_2(t)}(\lambda)I_q$ for $0 \le t < d$, where $\tau_1(t)$ and $\tau_2(t)$ are continuous nondecreasing functions of t on the interval $[0, d)$ with $\tau_1(0) = \tau_2(0) = 0$. Then

$$M(t) = M_1(t) - \tau_2(t) j_{pq}, \tag{10.37}$$

where $M_1(t)$ is given by formula (10.26) with

$$\tau(t) = \tau_1(t) + \tau_2(t), \tag{10.38}$$

for $0 \le t < d$.

Discussion This follows from Theorem 8.57.

Example 10.11 A one-sided diagonal case
Let $b_1^t(\lambda) = \mathrm{diag}\{e_{\alpha_1(t)}(\lambda), \ldots, e_{\alpha_p(t)}(\lambda)\}$, where $\alpha_j(t)$ are continuous nondecreasing functions on the interval $[0, d)$ with $\alpha_j(0) = 0$, for $j = 1, \ldots, p$ and let $b_2^t(\lambda) = I_q$. Then $M(t)$ is given by formula (10.26), where now

$$\tau(t) = \max\{\alpha_1(t), \ldots, \alpha_p(t)\} \tag{10.39}$$

for $0 \le t < d$ and $x_j^t(a)$ and $y_j^t(a)$ are solutions of the following systems of equations

$$\begin{aligned} x_j^t(a) - \chi_t(a)\frac{d}{da}\int_0^a v(a-s)y_j^t(s)ds &= \varphi_j^t(a) \\ \frac{d}{da}\int_a^{\tau(t)} v(s-a)^* x_j^t(s)ds + y_j^t(a) &= 0 \end{aligned} \tag{10.40}$$

for $0 \le a \le \tau(t)$ and $j = 1, 2$, where

$$\varphi_1^t(a) = \chi_t(a), \quad \varphi_2^t(a) = \chi_t(a)v(a), \quad \chi_t(a) = \mathrm{diag}\{\chi_t^{(1)}(a), \ldots, \chi_t^{(p)}(a)\}$$

and

$$\chi_t^{(j)}(a) = \begin{cases} 1 & \text{for } 0 \le a \le \alpha_j(t) \\ 0 & \text{otherwise.} \end{cases}$$

Discussion The requisite calculations are easily adapted from Example 10.9. The main difference is that now the orthogonal projection of $\widehat{f} \in H_2^p$ onto $\mathcal{H}(b_1^t)$ is given by the formula

$$(\Pi_{\mathcal{H}(b_1^t)}\widehat{f})(\lambda) = \int_0^{\tau(t)} e^{i\lambda a}\chi_t(a)f(a)da. \tag{10.41}$$

Applying this recipe for the projection to formula (10.34) we obtain

$$(X_t\widehat{y}_j^t)(\lambda) = \int_0^{\tau(t)} e^{i\lambda a}\chi_t(a)\left\{\frac{d}{da}\int_0^a v(a-s)y_j^t(s)ds\right\}da \tag{10.42}$$

and, by a similar calculation,

$$\widehat{\varphi}_2(\lambda) = \Pi_{\mathcal{H}(b_1^t)}\{M_{s^\circ}R_0(e_{\tau(t)}I_q)\} = (\Pi_{\mathcal{H}(b_1^t)}\widehat{v})(\lambda)$$
$$= \int_0^{\tau(t)} e^{i\lambda a}\chi_t(a)v(a)da.$$

Correspondingly,

$$(X_t^*\widehat{x}_j^t)(\lambda) = -\int_0^{\tau(t)} e^{i\lambda a}\left\{\frac{d}{da}\int_a^{\tau(t)} v(s-a)^* x_j^t(s)ds\right\}da. \tag{10.43}$$

The remaining calculations go through much as before.

Example 10.12 Two-sided diagonal case
Let

$$\begin{aligned} b_1^t(\lambda) &= \mathrm{diag}\{e_{\alpha_1(t)}(\lambda), \ldots, e_{\alpha_p(t)}(\lambda)\} \quad \textit{and} \\ b_2^t(\lambda) &= \mathrm{diag}\{e_{\beta_1(t)}(\lambda), \ldots, e_{\beta_q(t)}(\lambda)\}, \end{aligned} \tag{10.44}$$

where $\alpha_j(t)$, $j = 1, \ldots, p$ and $\beta_k(t)$, $k = 1, \ldots, q$, are continuous nondecreasing functions on the interval $[0, d)$ with $\alpha_j(0) = \beta_k(0) = 0$. Let

$$\tau_1(t) = \max\{\alpha_1(t), \ldots, \alpha_p(t)\} \quad \text{and} \quad \tau_2(t) = \max\{\beta_1(t), \ldots, \beta_q(t)\}. \tag{10.45}$$

Discussion In order to obtain explicit formulas for the equations we need to evaluate the six operators that appear in formula (10.15) in more concrete form. Although we have dealt with two of them before, we repeat the conclusions in order to obtain a complete statement in the lemma. The following supplementary notation will prove convenient:

$$\chi_t^+(a) = \mathrm{diag}\{\chi_t^{(1)}(a), \ldots, \chi_t^{(p)}(a)\},$$
$$\chi_t^-(a) = \mathrm{diag}\{\chi_t^{(p+1)}(-a), \ldots, \chi_t^{(p+q)}(-a)\}$$
$$\chi_t^{(j)}(a) = \begin{cases} 1 & \text{for } 0 \le a \le \alpha_j(t) \\ 0 & \text{elsewhere,} \end{cases}$$

for $j = 1, \ldots, m$ and $\alpha_{p+j}(t) = \beta_j(t)$ for $j = 1, \ldots, q$.

Lemma 10.13 *If $b_1^t(\lambda)$ and $b_2^t(\lambda)$ are given by formula (10.44) and $s^\circ(\lambda) \in \mathcal{S}^{p\times q}$ is given by formula (9.78), then:*

(1) $\widehat{y} \in \mathcal{H}(e_{\tau_1(t)}I_q) \Longrightarrow$

$$(X_{11}^t\widehat{y})(\lambda) = \int_0^{\tau_1(t)} e^{i\lambda a}\chi_t^+(a)\left\{\frac{d}{da}\int_0^a v(a-s)y(s)ds\right\}da.$$

(2) $\widehat{u} \in \mathcal{H}_*(b_2^t) \Longrightarrow$

$$(X_{12}^t\widehat{u})(\lambda) = \int_0^{\tau_1(t)} e^{i\lambda a}\chi_t^+(a)\left\{\frac{d}{da}\int_{-\tau_2(t)}^0 v(a-s)u(s)ds\right\}da.$$

(3) $\widehat{u} \in \mathcal{H}_*(b_2^t) \Longrightarrow$

$$(X_{22}^t \widehat{u})(\lambda) = \int_{-\tau_2(t)}^{0} e^{i\lambda a} \left\{ \frac{d}{da} \int_{-\tau_2(t)}^{a} v(a-s)u(s)ds \right\} da.$$

(4) $\widehat{x} \in \mathcal{H}(b_1^t) \Longrightarrow$

$$((X_{11}^t)^* \widehat{x})(\lambda) = -\int_{0}^{\tau_1(t)} e^{i\lambda a} \left\{ \frac{d}{da} \int_{a}^{\tau_1(t)} v(s-a)^* x(s)ds \right\} da.$$

(5) $\widehat{x} \in \mathcal{H}(b_1^t) \Longrightarrow$

$$((X_{12}^t)^* \widehat{x})(\lambda) = -\int_{-\tau_2(t)}^{0} e^{i\lambda a} \chi_t^-(a) \left\{ \frac{d}{da} \int_{a}^{\tau_1(t)} v(s-a)^* x(s)ds \right\} da.$$

(6) $\widehat{w} \in \mathcal{H}_*(e_{\tau_2} I_p) \Longrightarrow$

$$((X_{22}^t)^* \widehat{w})(\lambda) = -\int_{-\tau_2(t)}^{0} e^{i\lambda a} \chi_t^-(a) \left\{ \frac{d}{da} \int_{a}^{0} v(s-a)^* w(s)ds \right\} da.$$

Proof Formulas (1) and (4) have been established in the discussion of Example 10.11; see also Lemma 10.7. The remaining formulas can be established in much the same way, or by simple changes of variable. Thus, for example, to obtain (2), observe that

$$\mathcal{H}_*(b_2^t) \subseteq \mathcal{H}_*(e_{\tau_2(t)} I_q)$$

and hence that if $\widehat{u} \in \mathcal{H}_*(b_2^t)$, then $\widehat{u_1} = e_{\tau_2(t)} \widehat{u}$ belongs to $\mathcal{H}(e_{\tau_2(t)} I_q)$. Therefore, by (9.79) and (9.81),

$$\begin{aligned}
(s^\circ \widehat{u})(\lambda) &= e_{-\tau_2(t)}(\lambda) \int_0^\infty e^{i\lambda a} \left\{ \frac{d}{da} \int_0^{\tau_2(t)} v(a-s)u_1(s)ds \right\} da \\
&= e_{-\tau_2(t)}(\lambda) \int_0^\infty e^{i\lambda a} \left\{ \frac{d}{da} \int_0^{\tau_2(t)} v(a-s)u(s-\tau_2(t))ds \right\} da \\
&= e_{-\tau_2(t)}(\lambda) \int_0^\infty e^{i\lambda a} \left\{ \frac{d}{da} \int_{-\tau_2(t)}^{0} v(a-\tau_2(t)-s)u(s)ds \right\} da \\
&= \int_{-\tau_2(t)}^{\infty} e^{i\lambda a} \left\{ \frac{d}{da} \int_{-\tau_2(t)}^{0} v(a-s)u(s)ds \right\} da.
\end{aligned}$$

Formula (2) now drops out easily by projecting the last right-hand side onto $\mathcal{H}(b_1^t)$. This can be done in two stages: first by projecting onto H_2^p and then invoking formula (10.41).

Next, in a similar vein, the fact that $\widehat{w} \in \mathcal{H}_*(e_{\tau_2(t)}I_q)$ is exploited in order to break the calculation of formula (6) into two convenient steps. Thus,

$$\begin{aligned}(X_{22}^t)^*\widehat{w} &= \Pi_{\mathcal{H}_*(b_2^t)}(s^\circ)^*\widehat{w} = \Pi_{\mathcal{H}_*(b_2^t)}\Pi_{\mathcal{H}_*(e_{\tau_2(t)}I_q)}(s^\circ)^*\widehat{w} \\ &= \Pi_{\mathcal{H}_*(b_2^t)}e_{-\tau_2(t)}\Pi_{\mathcal{H}(e_{\tau_2(t)}I_q)}(s^\circ)^*\widehat{w}_1,\end{aligned}$$

where the columns of $\widehat{w}_1 = e_{\tau_2(t)}\widehat{w}$ belong to $\mathcal{H}(e_{\tau_2(t)}I_p)$. Now, to save space in the next calculation, let $\Pi_{\tau_2} = \Pi_{\mathcal{H}(e_{\tau_2(t)}I_q)}$. Then, by formulas (9.82) and (9.83),

$$\begin{aligned}\Pi_{\tau_2}(s^\circ)^*\widehat{w}_1 &= -\Pi_{\tau_2}\int_0^\infty e^{i\lambda a}\left\{\frac{d}{da}\int_a^{\tau_2(t)} v(s-a)^*w_1(s)ds\right\}da \\ &= -\int_0^{\tau_2(t)} e^{i\lambda a}\left\{\frac{d}{da}\int_a^{\tau_2(t)} v(s-a)^*w(s-\tau_2(t))ds\right\}da \\ &= -\int_0^{\tau_2(t)} e^{i\lambda a}\left\{\frac{d}{da}\int_{a-\tau_2(t)}^0 v(s+\tau_2(t)-a)^*w(s)ds\right\}da \\ &= -e_{\tau_2(t)}(\lambda)\int_{-\tau_2(t)}^0 e^{i\lambda a}\left\{\frac{d}{da}\int_a^0 v(s-a)^*w(s)ds\right\}da.\end{aligned}$$

Thus,

$$\begin{aligned}((X_{22}^t)^*\widehat{w})(\lambda) &= -\Pi_{\mathcal{H}_*(b_2^t)}\int_{-\tau_2(t)}^0 e^{i\lambda a}\left\{\frac{d}{da}\int_a^0 v(s-a)^*w(s)ds\right\}da \\ &= -\int_{-\tau_2(t)}^0 e^{i\lambda a}\chi_t^-(a)\left\{\frac{d}{da}\int_a^0 v(s-a)^*w(s)ds\right\}da,\end{aligned}$$

as claimed.

The justification of the other two formulas is left to the reader. □

Lemma 10.14 *In the setting of Example 10.12,*

$$\widehat{\varphi}_1^t(\lambda) = \int_0^{\tau_1(t)} e^{i\lambda a}\chi_t^+(a)da, \quad \widehat{\varphi}_2^t(\lambda) = \int_0^{\tau_1(t)} e^{i\lambda a}\chi_t^+(a)v(a)da,$$

$$\widehat{\psi}_1^t(\lambda) = \int_{-\tau_2(t)}^0 e^{i\lambda a}\chi_t^-(a)v(-a)^*da \quad \textit{and} \quad \widehat{\psi}_2^t(\lambda) = \int_{-\tau_2(t)}^0 e^{i\lambda a}\chi_t^-(a)da.$$

Proof The stated conclusions are immediate from formulas (10.16)–(10.19) and the evaluations

$$\widehat{\varphi}_2^t(\lambda) = \Pi_{\mathcal{H}(b_1^t)}\{\widehat{v} - e_{\tau_1(t)}\widehat{v}\} = \Pi_{\mathcal{H}(b_1^t)}\widehat{v} = \int_0^{\tau_1(t)} e^{i\lambda a}\chi_t^+(a)v(a)da$$

and

$$\widehat{\psi}_1^t(\lambda) = \Pi_{\mathcal{H}_*(b_2^t)}\left\{(s^\circ)^{\#}\frac{1-e_{-\tau_2(t)}}{i\lambda}\right\} = \Pi_{\mathcal{H}_*(b_2^t)}\int_{-\infty}^{0} e^{i\lambda s}v(-s)^*ds$$
$$= \int_{-\tau_2(t)}^{0} e^{i\lambda a}\chi_t^-(a)v(-a)^*da.$$

□

In view of Lemmas 10.13 and 10.14, the system of equations (10.15) can be rewritten as

$$x_j^t(a) - \chi_t^+(a)\frac{d}{da}\int_0^a v(a-s)y_j^t(s)ds \qquad (10.46)$$
$$- \chi_t^+(a)\frac{d}{da}\int_{-\tau_2(t)}^{0} v(a-s)u_j^t(s)ds = \varphi_j^t(a),$$

$$\frac{d}{da}\int_a^{\tau_1(t)} v(s-a)^*x_j^t(s)ds + y_j^t(a) = 0, \qquad (10.47)$$

$$\chi_t^-(b)\frac{d}{db}\int_b^{\tau_1(t)} v(s-b)^*x_j^t(s)ds + u_j^t(b) \qquad (10.48)$$
$$+ \chi_t^-(b)\frac{d}{db}\int_b^{0} v(s-b)^*v_j^t(s)ds = \psi_j^t(b),$$

$$-\frac{d}{db}\int_{-\tau_2(t)}^{b} v(b-s)u_j^t(s)ds + v_j^t(b) = 0, \qquad (10.49)$$

for $j = 1, 2$, where,

$$\varphi_1^t(a) = \chi_t^+(a), \quad \varphi_2^t(a) = \chi_t^+(a)v(a), \qquad (10.50)$$
$$\psi_1^t(b) = \chi_t^-(b)v(-b)^*, \quad \psi_2^t(b) = \chi_t^-(b), \qquad (10.51)$$

and the variables a and b are restricted to the intervals $0 \le a \le \tau_1(t)$ and $-\tau_2(t) \le b \le 0$, respectively.

Remark 10.15 *If the given input scattering matrix s° belongs to the Wiener class $\mathcal{W}_+^{p\times q}(0)$, i.e., if*

$$s^\circ(\lambda) = \int_0^\infty e^{i\lambda t}h(t)dt \quad \textit{belongs to } \mathcal{S}^{p\times q} \textit{ and } h \in L_1^{p\times q}(\mathbb{R}_+),$$

then the mvf v in the integral representation for s° in Lemma 10.7 can be written in terms of h as

$$v(t) = \int_0^t h(s)ds \quad \textit{for } t \ge 0.$$

Consequently,

$$\frac{d}{da}\int_0^a v(a-s)y(s)ds = \int_0^a h(a-s)y(s)ds \quad \textit{for } a > 0 \textit{ and } y \in L_2^q(\mathbb{R}_+)$$

and

$$-\frac{d}{da}\int_a^\infty v(s-a)^*x(s)ds = \int_a^\infty h(s-a)^*x(s)ds \quad \text{for } a > 0 \text{ and } x \in L_2^p(\mathbb{R}_+).$$

Thus, the system (10.28) can be rewritten in terms of h as

$$\begin{aligned} x_j^t(a) - \int_0^a h(a-s)y_j^t(s)ds &= \varphi_j^t(a) \\ -\int_a^{\tau_1(t)} h(s-a)^*x_j^t(s)ds + y_j^t(a) &= 0 \end{aligned} \tag{10.52}$$

for $0 \le a \le \tau_1(t)$ *and* $j = 1, 2$, *where*

$$\varphi_1^t(a) = I_p \quad \text{and} \quad \varphi_2^t(a) = \int_0^a h(s)ds \tag{10.53}$$

for $0 \le a \le \tau_1(t)$. *Moreover, the system (10.40) may be rewritten in terms of the mvf* $\chi_t(a)$ *that is defined in Example 10.11 as*

$$\begin{aligned} x_j^t(a) - \chi_t(a)\int_0^a h(a-s)y_j^t(s)ds &= \varphi_j^t(a) \\ -\int_a^{\tau_1(t)} h(s-a)^*x_j^t(s)ds + y_j^t(a) &= 0 \end{aligned}$$

for $0 \le a \le \tau_1(t)$ *and* $j = 1, 2$, *where*

$$\varphi_1^t(a) = \chi_t(a) \quad \text{and} \quad \varphi_2^t(a) = \chi_t(a)\int_0^a h(s)ds.$$

Remark 10.16 *If* $s^\circ \in \mathcal{W}_+^{p\times q}(\gamma) \cap \mathcal{S}^{p\times q}$ *for some nonzero matrix* $\gamma \in \mathbb{C}^{p\times q}$ *with* $\gamma^*\gamma < I_q$ *and* W° *is the constant* j_{pq}*-unitary matrix defined by (3.127) with* $k = -\gamma$, *and if* W_t *is the matrizant of the canonical integral system (6.1) with input scattering matrix* s°, *then*

$$\widetilde{W}_t = W^\circ W_t (W^\circ)^{-1}$$

is the matrizant of (6.1) with mass function $\widetilde{M}(t) = W^\circ M(t)(W^\circ)^{-1}$ *and*

$$\{e_{\tau_1(t)}I_p, e_{\tau_2(t)}I_q\} \in ap(W_t) \iff \{e_{\tau_1(t)}I_p, e_{\tau_2(t)}I_q\} \in ap(\widetilde{W}_t) \quad \text{for } 0 \le t \le d$$

and

$$\mathcal{S}_{\text{scat}}^d(\widetilde{M}) \cap \mathcal{W}_+^{p\times q}(0) = T_{W^\circ}[\mathcal{S}_{\text{scat}}^d(M) \cap \mathcal{W}_+^{p\times q}(\gamma)].$$

Let

$$s^\circ \in \mathcal{S}_{\text{scat}}^d(M) \cap \mathcal{W}_+^{p\times q}(\gamma) \quad \text{with} \quad \gamma^*\gamma < I_q$$

and let $\widetilde{s^\circ} = T_{W^\circ}[s^\circ]$. *Then*

$$\widetilde{s^\circ} \in \mathcal{S}_{\text{scat}}^d(\widetilde{M}) \cap \mathcal{W}_+^{p\times q}(0)$$

and $M(t) = (W^\circ)^{-1}\widetilde{M}(t)W^\circ$, $0 \le t < d$, *is the solution of the original inverse scattering problem with* $\gamma \neq 0$.

10.5 Supplementary notes

Most of the material in this chapter is adopted from [ArD02a] and [ArD02b]. The latter contains explicit formulas for the solution of the inverse input scattering problem for rational input scattering matrices; there is some overlap with the formulas that were obtained from a different circle of ideas by D. Alpay and I. Gohberg in the earlier papers [AG95] and [AG01]; see also [AKvdM00].

The strategy of identifying the matrizant $W_t(\lambda)$ of a canonical integral system with $\{b_1^t, b_2^t\} \in ap(W_t)$ as a normalized resolvent matrix of GSIP$(b_1^t, b_2^t; s^\circ)$ for $0 \le t < d$, is a bitangential generalization of the method that M.G. Krein used to solve the inverse input scattering problem for the systems that we refer to as DK-systems with locally summable potential. In such systems $b_1^t = e_{\alpha t}I_p$ and $b_2^t = e_{\beta t}I_q$ with $\alpha \ge 0$, $\beta \ge 0$ and $\kappa = \alpha + \beta > 0$. In particular, Krein reduced the system with $\alpha = \beta = \kappa/2$ to the system with $\alpha = \kappa$ and $\beta = 0$ and identified the matrizant W_t, $0 \le t < \infty$, of this new system with a family of resolvent matrices of the SEP$(h; \kappa t)$, $0 \le t < \infty$, that have exponential type zero in $\mathbb{C}_+$.

Additional details on Krein's method for solving the inverse input scattering problem for DK-systems will be provided in the Supplementary Notes to Chapter 12.

11

Bitangential inverse input impedance and spectral problems

The data for our formulation of the **bitangential inverse input impedance problem** is a mvf $c^\circ \in \mathcal{C}^{p\times p}$ and a normalized nondecreasing continuous chain of pairs $\{b_3^t, b_4^t\}$, $0 \le t < d$, of entire inner $p \times p$ mvf's. The problem is to find a continuous nondecreasing $m \times m$ mvf $M(t)$ on the interval $[0, d)$ with $m = 2p$ and $M(0) = 0$ such that the matrizant $A_t(\lambda)$, $0 \le t < d$, of the canonical integral system (7.1) with this mass function $M(t)$ meets the following two conditions:

(C1) $c^\circ \in \bigcap_{0\le t<d} \mathcal{C}(A_t) \stackrel{\text{def}}{=} \mathcal{C}^d_{\text{imp}}(dM)$.
(C2) $\{b_3^t, b_4^t\} \in ap_{II}(A_t)$ for every $t \in [0, d)$.

Analogously, the data for the **bitangential inverse spectral problem** is a nondecreasing $p \times p$ mvf $\sigma(\mu)$ that meets the constraint (3.43) and a normalized nondecreasing continuous chain of pairs $\{b_3^t, b_4^t\}$, $0 \le t < d$, of entire inner $p \times p$ mvf's. The problem is the same as for the inverse impedance problem except that now (C1) is replaced by

(C1a) $\sigma \in \Sigma^d_{\text{sp}}(dM)$.

Under appropriate conditions on the data, this second problem may be reduced to the first in which σ is the spectral function of a mvf $c^\circ \in \mathcal{C}^{p\times p}$.

The conditions

(C3) $A_t \in \mathcal{U}_{rsR}(J_p)$ for every $t \in [0, d)$,

which guarantees that $A_t \in \mathcal{U}_{AR}(J_p)$ for every $t \in [0, d)$, and

(C4) $A_t \in \mathcal{U}_{rR}(J_p)$ for every $t \in [0, d)$

which is weaker than (C3), will play a significant role.

It is also useful to keep in mind that the GCIP$(b_3, b_4; c^\circ)$ is

$$\text{completely indeterminate} \iff \mathcal{C}(b_3, b_4; c^\circ) \cap \mathcal{C}_{sz}^{p\times p} \neq \emptyset; \quad (11.1)$$

$$\text{strictly completely indeterminate} \iff \mathcal{C}(b_3, b_4; c^\circ) \cap \mathring{\mathcal{C}}^{p\times p} \neq \emptyset; \quad (11.2)$$

(11.1) is by Theorem 4.22); (11.2) is by definition.

Conditions for the existence and uniqueness of solutions to the inverse input impedance problem are furnished in Section 11.1, which contains analogs of Theorems 8.16 and 8.17. In particular if $c^\circ \in \mathcal{C}_{sz}^{p\times p}$, then there will exist exactly one normalized nondecreasing chain of mvf's $A_t \in \mathcal{E} \cap \mathcal{U}_{rR}^\circ(J_p)$ of resolvent matrices for the GCIP$(b_3^t, b_4^t; c^\circ)$ for which (C2) holds. If that chain is continuous, then it is the matrizant of exactly one solution to the inverse impedance problem such that (C4) holds. A sufficient, but not necessary, condition for this is (C3). Formulas for the solution under condition (C3) are developed in Section 11.2. In Section 11.3 it is shown that $c^\circ \in \mathcal{W}_+^{p\times p}(\gamma) \cap \mathcal{C}^{p\times p}$, then condition (C3) will be in force if $\gamma + \gamma^* > 0$, which is an easy condition to check. Section 11.5 discusses the inverse spectral problem. Sections 11.4 and 11.6 are devoted to examples.

11.1 Existence and uniqueness of solutions

The next two theorems are analogs of Theorems 10.1 and 10.2. Before tackling them, it is probably helpful to review Lemmas 8.14 and 8.15. The first of these two theorems supplies conditions for the existence of a solution to the inverse input impedance problem for canonical integral systems and a parametrization of all the solutions that exist in terms of normalized nondecreasing continuous chains of pairs of entire inner $p \times p$ mvf's. The second deals with canonical differential systems.

Theorem 11.1 *If $c^\circ \in \mathcal{C}^{p\times p}$ and $\{b_3^t, b_4^t\}$, $0 \le t < d$, is a normalized nondecreasing continuous chain of pairs of entire inner $p \times p$ mvf's such that the GCIP$(b_3^t, b_4^t; c^\circ)$ is completely indeterminate for every $t \in [0, d)$, then:*

(1) *For each $t \in [0, d)$ there exists exactly one mvf $A_t \in \mathcal{U}^\circ(J_p)$ such that*

$$\mathcal{C}(b_3^t, b_4^t; c^\circ) = \mathcal{C}(A_t) \quad \text{and} \quad \{b_3^t, b_4^t\} \in ap_{II}(A_t). \quad (11.3)$$

The family of resolvent matrices A_t, $0 \le t < d$, that is obtained this way enjoys the following properties:

(2a) *A_t, $0 \le t < d$, is a normalized nondecreasing $\curvearrowright$ chain of mvf's in the class $\mathcal{E} \cap \mathcal{U}_{rR}^\circ(J_p)$.*

(2b) *$A_t(\lambda)$ is left continuous on $[0, d)$ for each $\lambda \in \mathbb{C}$.*

(2c) *If $A_t(\lambda)$ is continuous on $[0, d)$ for each $\lambda \in \mathbb{C}$. then it is the matrizant of the canonical integral system (7.1) with mass function given by $M(t) = 2\pi K_0^{A_t}(0)$ that meets the constraints (1.21).*

(2d) *If* $A_t \in \mathcal{E} \cap \mathcal{U}_{AR}^{\circ}(J_p)$ *for every* $t \in [0, d)$, *then* $A_t(\lambda)$ *is continuous on* $[0, d)$ *for each* $\lambda \in \mathbb{C}$.

(3) *If* $M(t)$ *is any continuous nondecreasing* $m \times m$ *mvf on* $[0, d)$ *with* $M(0) = 0$ *and if the matrizant* $A_t(\lambda)$, $0 \le t < d$, *of the canonical integral system (7.1) with this mass function belongs to the class* $\mathcal{U}_{rR}^{\circ}(J_p)$ *for every* $t \in [0, d)$ *and* $\mathcal{C}_{\rm imp}^{d}(dM) \neq \emptyset$, *then* $M(t)$ *is the solution of the inverse input impedance problem with data*

$$c^\circ \in \bigcap_{0 \le t < d} \mathcal{C}(A_t) \quad and \quad \{b_3^t, b_4^t\} \in ap_{II}(A_t) \quad for\ 0 \le t < d,$$

where $b_3^t(0) = b_4^t(0) = I_p$ *for* $\le t < d$ *and* $M(t)$ *may be obtained from this data via (1) and (2c).*

Proof The proof of (1)–(2d) is essentially the same as the proof of the corresponding statements in Theorem 8.16 except that Theorem 4.13 is replaced by Theorem 4.25. The fact that the chain of normalized pairs $\{b_3^t, b_4^t\}$, $0 \le t < d$, is nondecreasing and continuous follows from Theorem 5.15 applied to closed subintervals $[0, d_1]$ of $[0, d)$.

The assumption on c° in the setting of (3) insures that $c^\circ \in \mathcal{C}(A_t)$ for every $t \in [0, d)$ and hence, as $\{b_3^t, b_4^t\} \in ap_{II}(A_t)$ and $A_t \in \mathcal{U}_{rR}(J_p)$, that the GCIP$(b_3^t, b_4^t; c^\circ)$ is completely indeterminate for every $t \in [0, d)$. Thus, A_t is the one and only resolvent matrix in the class $\mathcal{U}^\circ(J_p)$ for this problem with $\{b_3^t, b_4^t\} \in ap_{II}(A_t)$ for $t \in [0, d)$. □

Theorem 11.2 *In the setting of Theorem 11.1, assume that the chain* $\{b_3^t, b_4^t\}$, $0 \le t < d$, *is strictly increasing and the family of resolvent matrices* $A_t(\lambda)$, $0 \le t < d$, *that is specified in (1) of that theorem is continuous in* t *on* $[0, d)$ *for each* $\lambda \in \mathbb{C}$. *Then there exists exactly one normalized solution* $H(x)$, $0 \le x < \ell$, *of the bitangential input impedance problem for the canonical differential system (2.14) with* $J = J_p$, *input impedance matrix* c° *and matrizant* Ω_x, $0 \le x < \ell$, *such that*

$$\Omega_x \in \mathcal{U}_{rR}(J) \quad and \quad \{b_3^{t(x)}, b_4^{t(x)}\} \in ap_{II}(\Omega_x)\ for\ every\ x \in [0, \ell) \qquad (11.4)$$

for some function $t(x)$ *that maps* $[0, \ell)$ *onto* $[0, d)$ *with* $t(0) = 0$ *and* $\lim_{x \uparrow \ell} t(x) = d$. *Moreover, there is only one such function* $t(x)$: *it is the inverse of the function* $x(t) = \operatorname{trace} 2\pi K_0^{A_t}(0)$ *and is a strictly increasing continuous function on* $[0, \ell)$. *The mvf's* $H(x)$ *and* $\Omega_x(\lambda)$ *may be obtained from* $M(t) = 2\pi K_0^{A_t}(0)$ *and* A_t *by the formulas*

$$H(x) = \frac{d}{dx} M(t(x)) \quad a.e.\ on\ [0, \ell) \quad and \quad \Omega_x = A_{t(x)}\ for\ every\ x \in [0, \ell).$$

Proof This follows from Theorem 8.17 with $J = J_p$ by first taking $0 < d_1 < d$, $A = A_{d_1}$ and A_t is the resolvent matrix in $\mathcal{U}^\circ(J_p)$ with $\{b_3^t, b_4^t; c^\circ\} \in ap_{II}(A_t)$ of the GCIP$(b_3^t, b_4^t; c^\circ)$. □

11.2 Formulas for the solutions

The calculations in this section are based on the observation that the mass function in the canonical system (7.1) is given by the formula

$$M(t) = -i\left(\frac{\partial A_t}{\partial \lambda}\right)(0)J_p = -i \lim_{\lambda \to 0} \frac{A_t(\lambda) - A_t(0)}{\lambda} J_p = 2\pi K_0^{A_t}(0) \qquad (11.5)$$

for $0 \le t < d$. Our strategy is to express the reproducing kernel on the right in formula (11.5) in terms of the data $\{c(\lambda); b_3^t(\lambda), b_4^t(\lambda),\ 0 \le t < d\}$. (To avoid confusion with the notation c^t which will be used below, we now write c instead of c°.)

We shall assume that

$$\mathcal{C}(b_3^t, b_4^t; c) \cap \mathring{\mathcal{C}}^{p\times p} \neq \emptyset \ \text{ for every } \ t \in [0, d). \qquad (11.6)$$

In the setting of Theorem 11.1, this condition guarantees that there exists exactly one mvf $A_t \in \mathcal{E} \cap \mathcal{U}_{rsR}^\circ(J_p)$ such that (11.3) holds. Therefore, there exists a mvf

$$c^t \in \mathcal{C}(A_t) \cap H_\infty^{p\times p} \qquad (11.7)$$

for each of these uniquely specified mvf's A_t, $t \in [0, d)$, and hence, upon writing Φ_{ij}^t for the operators Φ_{ij} defined in formula (4.157) with $b_3^t(\lambda)$ in place of $b_3(\lambda)$, $b_4^t(\lambda)$ in place of $b_4(\lambda)$ and $c^t(\lambda)$ in place of $c(\lambda)$, we obtain

$$\Phi_{11}^t = \Pi_{\mathcal{H}(b_3^t)} M_{c^t}\Big|_{H_2^p}, \quad \Phi_{22}^t = \Pi_- M_{c^t}\Big|_{\mathcal{H}_*(b_4^t)} \quad \text{and} \quad \Phi_{12}^t = \Pi_{\mathcal{H}(b_3^t)} M_{c^t}\Big|_{\mathcal{H}_*(b_4^t)}. \qquad (11.8)$$

Correspondingly, let

$$Y_1^t = \Pi_{\mathcal{H}(b_3^t)}\left\{M_{c^t} + (M_{c^t})^*\right\}\Big|_{\mathcal{H}(b_3^t)} = 2\Re\left(\Phi_{11}^t\Big|_{\mathcal{H}(b_3^t)}\right), \qquad (11.9)$$

$$Y_2^t = \Pi_{\mathcal{H}_*(b_4^t)}\left\{M_{c^t} + (M_{c^t})^*\right\}\Big|_{\mathcal{H}_*(b_4^t)} = 2\Re\left(\Pi_{\mathcal{H}_*(b_4^t)}\Phi_{22}^t\right), \qquad (11.10)$$

$$\tau_j(t) = \limsup_{\nu\uparrow\infty} \frac{\ln \|b_j^t(-i\nu)\|}{\nu} \quad \text{for } j = 3, 4. \qquad (11.11)$$

Lemma 11.3 *Let $A \in \mathcal{E} \cap \mathcal{U}(J_p)$, let $\{b_3, b_4\} \in ap_{II}(A)$ and $\{\mathring{b}_3, \mathring{b}_4\} \in ap_{II}(J_pAJ_p)$. Then the exponential type $\mathring{\tau}_3$ of $\mathring{b}_3(\lambda)$ (respectively, $\mathring{\tau}_4$ of $\mathring{b}_4(\lambda)$) is equal to the exponential type τ_3 of $b_3(\lambda)$ (respectively, τ_4 of $b_4(\lambda)$).*

Proof This is an immediate consequence of Theorem 4.24 and the equalities $\tau_A^{\pm 1} = \tau_{\mathring{A}}^{\pm 1}$, where $\mathring{A} = J_pAJ_p$. □

Next, in view of Lemma 4.79 and the fact that

$$\mathcal{H}(b) \subseteq \mathcal{H}(e_\tau I_p)$$

for an entire inner $p \times p$ mvf $b(\lambda)$ of exponential type τ, we can conclude that

$$(\Phi_{11}^t)^*\mathcal{H}(b_3^t) \subseteq \mathcal{H}(e_{\tau_3(t)}I_p) \quad \text{and} \quad \Phi_{22}^t\mathcal{H}_*(b_4^t) \subseteq \mathcal{H}_*(e_{\tau_4(t)}I_p),$$

and hence that we can choose

$$\widetilde{b}_3(\lambda) = \widetilde{b}_3^t(\lambda) = e^{i\lambda\tau_3(t)}I_p \text{ and } \widetilde{b}_4(\lambda) = \widetilde{b}_4^t(\lambda) = e^{i\lambda\tau_4(t)}I_p$$

in Theorem 4.83.

The formula for $M(t)$ will be obtained by applying Theorem 4.83 to $A(\lambda) = A_t(\lambda)$. Then, upon writing L_t for L_{Φ_t}, Δ_t for Δ_{Φ_t} and $K_\omega^t(\lambda)$ for $K_\omega^{A_t}(\lambda)$, we arrive at a recipe for calculating $K_0^t(\lambda)$. In particular, in view of Theorem 4.77 and formula (4.170), the RK $K_0^t(\lambda)u$ can be expressed in the form

$$K_0^t(\lambda)u = L_t\Delta_t^{-1}G_tu \quad \text{for every } u \in \mathbb{C}^m, \tag{11.12}$$

where

$$G_t = \begin{bmatrix} -\Phi_{11}^t k_0^{\widetilde{b}_3^t} & k_0^{b_3^t} \\ (\Phi_{22}^t)^*\ell_0^{\widetilde{b}_4^t} & \ell_0^{b_4^t} \end{bmatrix} : \mathbb{C}^m \rightarrow \begin{array}{c} \mathcal{H}(b_3^t) \\ \oplus \\ \mathcal{H}_*(b_4^t) \end{array}, \tag{11.13}$$

$$\Delta_t = 2\Re \begin{bmatrix} \Phi_{11}^t|_{\mathcal{H}(b_3^t)} & \Phi_{12}^t \\ 0 & \Pi_{\mathcal{H}_*(b_4^t)}\Phi_{22}^t \end{bmatrix} : \begin{array}{c} \mathcal{H}(b_3^t) \\ \oplus \\ \mathcal{H}_*(b_4^t) \end{array} \longrightarrow \begin{array}{c} \mathcal{H}(b_3^t) \\ \oplus \\ \mathcal{H}_*(b_4^t) \end{array} \tag{11.14}$$

and hence (for $A_t \in \mathcal{U}_{rsR}(J_p)$)

$$\begin{aligned} u^*M(t)u = 2\pi\langle K_0^t u, K_0^t u\rangle_{\mathcal{H}(A_t)} &= 2\pi\langle L_t\Delta_t^{-1}G_tu, L_t\Delta_t^{-1}G_tu\rangle_{\mathcal{H}(A_t)} \\ &= 2\pi\langle \Delta_t^{-1}G_tu, G_tu\rangle_{\text{st}}. \end{aligned} \tag{11.15}$$

Formula (11.15) may be rewritten as

$$M(t) = 2\pi G_t^*\Delta_t^{-1}G_t \quad \text{for } 0 \le t < d. \tag{11.16}$$

In order to keep the notation relatively simple, an operator T that acts in the space of $p \times 1$ vvf's will be applied to $p \times p$ mvf's with columns $f_1, \ldots, f_p$ column by column: $T[f_1 \cdots f_p] = [Tf_1 \cdots Tf_p]$.

Theorem 11.4 *Let $c \in \mathcal{C}^{p\times p}$, let $\{b_3^t(\lambda), b_4^t(\lambda)\}$, $0 \le t < d$, be a normalized nondecreasing continuous chain of pairs of entire inner $p \times p$ mvf's and let assumption (11.6) be in force. Then there exists exactly one canonical integral system (7.1) with matrizant $A_t(\lambda) = A(t, \lambda)$, $0 \le t < d$, that satisfies the conditions (C1), (C2) and (C4). This matrizant automatically satisfies the condition (C3) also.*

Let $\tau_3(t)$ *and* $\tau_4(t)$ *be defined by formula (11.11), and recalling that* $(R_0 f)(\lambda) = \{f(\lambda) - f(0)\}/\lambda$, *let*

$$\widehat{y}^t_{11}(\lambda) = i\Big(\Phi^t_{11}(R_0 e_{\tau_3(t)} I_p)\Big)(\lambda), \quad \widehat{y}^t_{12}(\lambda) = -i(R_0 b^t_3)(\lambda) \tag{11.17}$$

$$\begin{aligned} \widehat{y}^t_{21}(\lambda) &= i\Big((\Phi^t_{22})^*(R_0 e_{-\tau_4(t)} I_p)\Big)(\lambda) \quad \text{and} \\ \widehat{y}^t_{22}(\lambda) &= i(R_0 (b^t_4)^{-1})(\lambda). \end{aligned} \tag{11.18}$$

Then the RK $K^t_\omega(\lambda)$ *of the RKHS* $\mathcal{H}(A_t)$ *evaluated at* $\omega = 0$ *is given by the formula*

$$K^t_0(\lambda) = \frac{1}{2\pi}\begin{bmatrix} \widehat{x}^t_{11}(\lambda) + \widehat{x}^t_{21}(\lambda) & \widehat{x}^t_{12}(\lambda) + \widehat{x}^t_{22}(\lambda) \\ \widehat{u}^t_{11}(\lambda) + \widehat{u}^t_{21}(\lambda) & \widehat{u}^t_{12}(\lambda) + \widehat{u}^t_{22}(\lambda) \end{bmatrix}, \tag{11.19}$$

where:

(1) *The* $\widehat{u}^t_{ij}(\lambda)$ *are* $p \times p$ *mvf's such that the columns of* $\widehat{u}^t_{1j}(\lambda)$ *belong to* $\mathcal{H}(b^t_3)$ *and the columns of* $\widehat{u}^t_{2j}(\lambda)$ *belong to* $\mathcal{H}_*(b^t_4)$. *The* $\widehat{u}^t_{ij}(\lambda)$ *may be defined as the solutions of the systems of equations:*

$$\begin{aligned} Y^t_1 \widehat{u}^t_{1j} + \Phi^t_{12}\widehat{u}^t_{2j} &= \widehat{y}^t_{1j}(\lambda) \\ (\Phi^t_{12})^*\widehat{u}^t_{1j} + Y^t_2 \widehat{u}^t_{2j} &= \widehat{y}^t_{2j}(\lambda), \quad j = 1, 2. \end{aligned} \tag{11.20}$$

(2) *The mvf's* $\widehat{x}^t_{ij}(\lambda)$ *are defined by the formulas*

$$\begin{aligned} \widehat{x}^t_{1j}(\lambda) &= -(\Phi^t_{11})^*\widehat{u}^t_{1j}, \\ \widehat{x}^t_{2j}(\lambda) &= \Phi^t_{22}\widehat{u}^t_{2j}, \quad j = 1, 2. \end{aligned} \tag{11.21}$$

Proof This theorem is an immediate consequence of Theorem 4.83. □

Remark 11.5 *In the one-sided cases, when either* $b^t_4(\lambda) = I_p$ *or* $b^t_3(\lambda) = I_p$, *the formulas for recovering* $M(t)$ *are simpler. For example, if* $b^t_4(\lambda) = I_p$, *then* $\tau_4(t) = 0$ *and* $\mathcal{H}_*(b^t_4) = \{0\}$ *and hence equations (11.20) and (11.21) simplify to*

$$Y^t_1 \widehat{u}^t_{1j} = \widehat{y}^t_{1j}(\lambda), \quad j = 1, 2, \tag{11.22}$$

and

$$\widehat{x}^t_{1j} = -(\Phi^t_{11})^*\widehat{u}^t_{1j} \quad j = 1, 2. \tag{11.23}$$

Theorem 11.6 *The unique solution* $M(t)$ *of the inverse input impedance problem considered in the setting of Theorem 11.4, is given by the formula*

$$\begin{aligned} M(t) &= 2\pi K^t_0(0) \\ &= \int_0^{\tau_3(t)} \begin{bmatrix} x^t_{11}(a) & x^t_{12}(a) \\ u^t_{11}(a) & u^t_{12}(a) \end{bmatrix} da + \int_{-\tau_4(t)}^0 \begin{bmatrix} x^t_{21}(a) & x^t_{22}(a) \\ u^t_{21}(a) & u^t_{22}(a) \end{bmatrix} da \end{aligned} \tag{11.24}$$

and the corresponding matrizant may be defined by the formula

$$A_t(\lambda) = I_m + 2\pi i\lambda K_0^t(\lambda)J_p, \tag{11.25}$$

where $K_0^t(\lambda)$ *is specified by formula (11.19) and* $x_{ij}^t(a)$ *and* $u_{ij}^t(a)$ *designate the inverse Fourier transforms of* $\widehat{x}_{ij}^t(\lambda)$ *and* $\widehat{u}_{ij}^t(\lambda)$, *respectively.*

Proof Formula (11.24) follows from formulas (11.5) and (11.19). Formula (11.25) follows from the definition of the RK $K_\omega^t(\lambda) = K_\omega^{A_t}(\lambda)$ at $\omega = 0$ and the fact that $A_t(0) = I_m$. □

The next result uses a time domain version of the operator $X = M_c|_{H_2^p}$ that is considered in Theorem 9.27.

Lemma 11.7 *If* $c \in H_\infty^{p\times p}$ *is expressed in the form (9.78) with* c *in place of* s *and* $g \in \mathcal{H}(e_tI_p)$ *for some* $t > 0$, *then*

$$(\Pi_{\mathcal{H}(e_tI_p)}M_c g)(\lambda) = \int_0^t e^{i\lambda u}\left\{\frac{d}{du}\int_0^u v(u-s)g^\vee(s)ds\right\}du \tag{11.26}$$

and, upon setting

$$v(-t) = -v(t)^* \quad for\ a.e.\ t < 0, \tag{11.27}$$

$$(\Pi_{\mathcal{H}(e_tI_p)}M_{c^*} g)(\lambda) = \int_0^t e^{i\lambda u}\left\{\frac{d}{du}\int_u^t v(u-s)g^\vee(s)ds\right\}du \tag{11.28}$$

Proof The proof follows from Theorem 9.27. □

Lemma 11.8 *If* c *is as in Lemma 11.7 and* $h \in \mathcal{H}_*(e_tI_p)$ *for some* $t > 0$, *then*

$$(\Pi_{\mathcal{H}_*(e_tI_p)}M_c h)(\lambda) = \int_{-t}^0 e^{i\lambda u}\left\{\frac{d}{du}\int_{-t}^u v(u-s)h^\vee(s)ds\right\}du \tag{11.29}$$

$$(\Pi_{\mathcal{H}_*(e_tI_p)}M_{c^*} h)(\lambda) = \int_{-t}^0 e^{i\lambda u}\left\{\frac{d}{du}\int_u^0 v(u-s)h^\vee(s)ds\right\}du. \tag{11.30}$$

Proof The stated formulas follow from (11.26) and (11.28) by translation, since

$$\Pi_{\mathcal{H}_*(e_tI_p)}f = e_{-t}\Pi_{\mathcal{H}(e_tI_p)}e_tf$$

for $f \in L_2^p$. □

11.3 Input impedance matrices in the Wiener class

In this section we shall focus on the case when the input impedance $c \in \mathcal{C}^{p\times p}$ belongs to the Wiener class $\mathcal{W}_+^{p\times p}(\gamma)$, i.e., when

$$c(\lambda) = \gamma + 2\int_0^\infty e^{i\lambda t}h(t)dt, \tag{11.31}$$

where $\gamma \in \mathbb{C}^{p\times p}$ and $h \in L_1^{p\times p}(\mathbb{R}_+)$. In this instance, $c(\lambda)$ can also be written as

$$c(\lambda) = -i\lambda \int_0^\infty e^{i\lambda t} v(t)dt \quad \text{for } \lambda \in \mathbb{C}_+ \tag{11.32}$$

with

$$v(t) = \gamma + 2\int_0^t h(s)ds, \tag{11.33}$$

and the time domain versions $X^\vee$ and $X^{\vee *}$ of the operators $X = M_c|_{H_2^p}$ and $X^* = \Pi_+ M_{c^*}|_{H_2^p}$ from H_2^p into H_2^p take the form

$$\begin{aligned}(X^\vee\varphi)(t) &= \gamma\varphi(t) + 2\int_0^t h(t-s)\varphi(s)ds \\ (X^{\vee *}\varphi)(t) &= \gamma^*\varphi(t) + 2\int_t^\infty h(s-t)^*\varphi(s)ds\end{aligned}$$

for $\varphi \in L_2^p(\mathbb{R}_+)$. Therefore, upon setting

$$h(-t) = h(t)^* \quad \text{for a.a. } t \in \mathbb{R}_+,$$

we see that

$$((X^\vee + X^{\vee *})\varphi)(t) = (\gamma + \gamma^*)\varphi(t) + 2\int_0^\infty h(t-s)\varphi(s)ds \tag{11.34}$$

for every $\varphi \in L_2^p(\mathbb{R}_+)$.

The operator $X^\vee + X^{\vee *}$ is positive semidefinite:

$$2\pi\langle (X^\vee + X^{\vee *})\varphi, \varphi\rangle_{\rm st} = \langle (X + X^*)\widehat{\varphi}, \widehat{\varphi}\rangle_{\rm st} = \langle (c + c^*)\widehat{\varphi}, \widehat{\varphi}\rangle_{\rm st} \geq 0. \tag{11.35}$$

If $c(\mu) + c(\mu)^* \geq \delta I_p$ for some $\delta > 0$ a.e. on $\mathbb{R}$, then in fact

$$2\pi\langle (X^\vee + X^{\vee *})\varphi, \varphi\rangle_{\rm st} \geq \delta\langle \widehat{\varphi}, \widehat{\varphi}\rangle_{\rm st} = 2\pi\delta\langle \varphi, \varphi\rangle_{\rm st},$$

i.e.,

$$X^\vee + X^{\vee *} \geq \delta I.$$

Moreover, in view of the representation formula (11.31), it follows readily from the Riemann–Lebesgue lemma that

$$\gamma + \gamma^* \geq \delta I_p. \tag{11.36}$$

The last condition is significant in the investigation of the bitangential inverse input impedance problem for input impedances that belong to $\mathcal{W}_+^{p\times p}(\gamma)$.

Two basic theorems

The condition (11.36) is central to the next two theorems, which can be established in much the same way as Theorem 10.5. We begin, however, with a lemma, that is the analog of Lemma 9.43 and serves to illustrate how to pass from one setting to the other.

Lemma 11.9 *Let $c^\circ \in \mathcal{C}^{p\times p} \cap H_\infty^{p\times p}$, $b_3 \in \mathcal{S}_{\text{in}}^{p\times p}$, $b_4 \in \mathcal{S}_{\text{in}}^{p\times p}$ and suppose that the following two conditions are also met:*

(1) $\gamma = \lim_{\nu\uparrow\infty} c^\circ(i\nu)$ *exists and* $\det(\gamma + \gamma^*) = 0$.
(2) *Either* $\lim_{\nu\uparrow\infty} b_3(i\nu) = 0$ *or* $\lim_{\nu\uparrow\infty} b_4(i\nu) = 0$.

Then the GCIP $(b_3, b_4; c^\circ)$ is not strictly completely indeterminate.

Proof Suppose that the conditions of the lemma are in force and that there exists a mvf $c \in \mathcal{C}(b_3, b_4; c^\circ) \cap \mathring{\mathcal{C}}^{p\times p}$. Then, since c and c° both belong to $H_\infty^{p\times p}$, the mvf $h = b_3^{-1}(c - c^\circ)b_4^{-1}$ belongs to $\mathcal{N}_+^{p\times p} \cap L_\infty^{p\times p}$. Therefore, by the Smirnov maximum principle, $h \in H_\infty^{p\times p}$. Consequently,

$$\lim_{\nu\uparrow\infty} c(i\nu) = \lim_{\nu\uparrow\infty}\{c^\circ(i\nu) + b_3(i\nu)h(i\nu)b_4(i\nu)\} = \gamma.$$

But if $c \in \mathring{\mathcal{C}}^{p\times p}$, there exists a $\delta > 0$ such that $c(\lambda) + c(\lambda)^* \geq \delta I_p$ for $\lambda \in \mathbb{C}_+$. Thus we must have $\gamma + \gamma^* \geq \delta I_p$, which contradicts the second assumption imposed in (1). □

Theorem 11.10 *Let $c \in \mathcal{W}_+(\gamma) \cap \mathcal{C}^{p\times p}$, $b_3 \in \mathcal{E} \cap \mathcal{S}_{\text{in}}^{p\times p}$, $b_4 \in \mathcal{E} \cap \mathcal{S}_{\text{in}}^{p\times p}$ and let $\gamma + \gamma^* > 0$. Then the GCIP$(b_3, b_4; c)$ is strictly completely indeterminate, i.e.,*

$$\gamma + \gamma^* > 0 \Longrightarrow \mathcal{C}(b_3, b_4; c) \cap \mathring{\mathcal{C}}^{p\times p} \neq \emptyset. \tag{11.37}$$

Proof Let $a = \text{type}(b_3) + \text{type}(b_4)$. If $a = 0$, then $b_3(\lambda)$ and $b_4(\lambda)$ are constant unitary matrices and hence $\mathcal{C}(b_3, b_4; c) = \mathcal{C}^{p\times p}$. Thus, (11.37) holds.

If $a > 0$, then $\mathcal{C}(b_3, b_4; c) \supseteq \mathcal{C}(e_a I_p, I_p; c)$. Therefore, it is enough to prove that

$$\mathcal{C}(e_a I_p, I_p; c) \cap \mathring{\mathcal{C}}^{p\times p} \neq \emptyset. \tag{11.38}$$

But (11.38) will hold if and only if the real part $Y_a = (X_a + X_a^*)/2$ of the bounded linear operator

$$X_a = \Pi_{\mathcal{H}(e_a I_p)} M_c|_{\mathcal{H}(e_a I_p)}$$

is strictly positive, i.e., if and only if $Y_a \geq \delta I$ for some $\delta > 0$; see Theorem 9.17. However, in view of (11.34),

$$Y_a^\vee = (\Re\gamma)I + K_a^\vee, \quad \text{where } (K_a^\vee\varphi)(t) = \int_0^a h(t-s)\varphi(s)ds$$

and hence K_a is a compact self-adjoint operator on $\mathcal{H}(e_a I_p)$. Consequently, Y_a is a positive semidefinite self-adjoint Fredholm operator. Thus, in order to complete the proof, it suffices to show that 0 is not an eigenvalue of Y_a. Suppose to the contrary that $Y_a f = 0$ for some $f \in \mathcal{H}(e_a I_p)$. Then

$$\langle Y_a f, f\rangle = \frac{1}{2}\int_{-\infty}^{\infty} f(\mu)^*(c(\mu)+c(\mu)^*)f(\mu)d\mu = 0.$$

Therefore, since the integrand is continuous and nonnegative on $\mathbb{R}$, we must have

$$f(\mu)^*(c(\mu)+c(\mu)^*)f(\mu) = 0$$

for every point $\mu \in \mathbb{R}$. By the Riemann–Lebesgue lemma,

$$c(\mu)+c(\mu)^* = \gamma+\gamma^*+2\int_{-\infty}^{\infty} e^{i\mu t}h(t)dt \longrightarrow \gamma+\gamma^*$$

as $|\mu| \longrightarrow \infty$ along $\mathbb{R}$. Thus,

$$c(\mu)+c(\mu)^* \geq (\gamma+\gamma^*)/2 \geq \delta I_p, \qquad \delta > 0,$$

for $\mu \geq R$. But this in turn implies that $f(\mu) = 0$ for $\mu > R$ and hence, since $f(\lambda)$ is an entire function (of exponential type), that $f(\lambda) \equiv 0$. Consequently, 0 is not an eigenvalue of Y_a, i.e., Y_a is a positive definite operator on $\mathcal{H}(e_a I_p)$ as claimed.

□

Remark 11.11 *It is important to note that in the formulation of the preceding theorem, we did not assume that* $c(\mu)+c(\mu)^* > 0$ *for a.e. point* $\mu \in \mathbb{R}$. *The point is that if* $c \in \mathcal{W}_+(\gamma) \cap \mathcal{C}^{p\times p}$ *and condition (11.36) is in force for some* $\delta > 0$, *then the operator* Y_a *will be strictly positive even though* $c(\lambda)$ *itself need not belong to* $\mathring{\mathcal{C}}^{p\times p}$.

Theorem 11.12 *If* $c \in \mathcal{W}_+(\gamma) \cap \mathcal{C}^{p\times p}$, $\gamma+\gamma^* > 0$, *and* $\{b_3^t(\lambda), b_4^t(\lambda)\}$, $0 \leq t < d$, *is a normalized nondecreasing continuous chain of pairs of entire inner* $p \times p$ *mvf's, then there exists one and only one continuous nondecreasing* $m \times m$ *mvf* $M(t)$ *on the interval* $[0, d)$ *with* $M(0) = 0$ *such that (C1), (C2) and (C4) hold for the matrizant* $A_t(\lambda)$, $0 \leq t < d$, *of the corresponding integral system (7.1). Moreover, in this setting, the matrizant automatically satisfies the condition (C3).*

If either

$$\lim_{\nu\uparrow\infty} b_3^{t_0}(i\nu) = 0 \quad or \quad \lim_{\nu\uparrow\infty} b_4^{t_0}(i\nu) = 0$$

for some point $t_0 \in [0, d)$, *then the condition* $\gamma+\gamma^* > 0$ *is necessary for the existence of a canonical system (7.1) with a matrizant* $A_t(\lambda)$, $0 \leq t < d$, *that meets the conditions (C1)–(C3).*

Proof This follows from Theorems 11.10, 11.1 and Lemma 11.9. □

Reduction to the class of impedances in $\mathcal{W}_+^{p\times p}(I_p)$

Theorem 11.13 *Let* $c \in \mathcal{W}_+(\gamma) \cap \mathcal{C}^{p\times p}$ *and* $\delta = \Re\gamma > 0$. *Let* $c_I(\lambda) = \delta^{-1/2}(c(\lambda) - i\Im\gamma)\delta^{-1/2}$. *Then* $c_I \in \mathcal{C}^{p\times p} \cap \mathcal{W}_+(I_p)$. *Let* $\{b_3^t, b_4^t\}$, $0 \leq t < d$, *be* *a normalized nondecreasing continuous chain of pairs of entire inner $p \times p$ mvf's that commute with $\delta^{1/2}$:*

$$b_3^t \delta^{1/2} = \delta^{1/2} b_3^t, \quad b_4^t \delta^{1/2} = \delta^{1/2} b_4^t \quad \textit{for every } t \in [0, d). \tag{11.39}$$

Let $M(t)$, $0 \le t < d$, and $M_I(t)$, $0 \le t < d$, be solutions of the bitangential inverse input impedance problems with given data

$$\{c; b_3^t, b_4^t, \ 0 \le t < d\} \text{ and } \{c_I; b_3^t, b_4^t, \ 0 \le t < d\}$$

respectively. Then

$$M(t) = L_\gamma M_I(t) L_\gamma^*, \text{ where } L_\gamma = \begin{bmatrix} I_p & i\mathfrak{I}\gamma \\ 0 & I_p \end{bmatrix} \begin{bmatrix} \delta^{1/2} & 0 \\ 0 & \delta^{-1/2} \end{bmatrix}, \quad 0 \le t < d. \tag{11.40}$$

Proof Let $M_\gamma(t) = L_\gamma M_I(t) L_\gamma^*$, $0 \le t < d$, where L_γ is defined in (11.40). Then $M_\gamma(t)$, $0 \le t < d$, is a nondecreasing continuous $m \times m$ mvf on $[0, d)$ and $M_\gamma(0) = 0$. Let $A_t^{(\gamma)}(\lambda)$, $0 \le t < d$, and $A_t^{(I)}(\lambda)$, $0 \le t < d$, be the matrizants of systems of the form (7.1) with mass functions $M_\gamma(t)$, $0 \le t < d$, and $M_I(t)$, $0 \le t < d$, respectively. Then, since the matrix L_γ is J_p-unitary (as are both of its factors), it is readily seen that

$$A_t^{(\gamma)}(\lambda) = L_\gamma A_t^{(I)}(\lambda) L_\gamma^{-1}, \quad 0 \le t < d.$$

The next step is to check that

$$\{b_3^t, b_4^t\} \in ap_{II}(A_t^{(\gamma)}) \text{ for every } t \in [0, d). \tag{11.41}$$

To this end, let

$$B_t^{(\gamma)}(\lambda) = A_t^{(\gamma)}(\lambda)\mathfrak{V} \text{ and } B_t^{(I)}(\lambda) = A_t^{(I)}(\lambda)\mathfrak{V}, \quad 0 \le t < d.$$

Then, by assumption,

$$\{b_3^t, b_4^t\} \in ap(B_t^{(I)}) \text{ for every } t \in [0, d).$$

Moreover,

$$B_t^{(\gamma)}(\lambda) = L_\gamma B_t^{(I)}(\lambda) U_\gamma,$$

where

$$U_\gamma = \mathfrak{V} L_\gamma^{-1} \mathfrak{V}$$

is a constant j_p-unitary matrix. Therefore, the mvf's

$$\widetilde{B}_t(\lambda) = B_t^{(I)}(\lambda) U_\gamma \quad \text{and} \quad B_t^{(I)}(\lambda)$$

have the same associated pairs $\{b_3^t(\lambda), b_4^t(\lambda)\}$ for every $t \in [0, d)$. Thus, as

$$N_2^* B_t^{(\gamma)}(\lambda) = \delta^{-1/2} N_2^* \widetilde{B}_t(\lambda),$$

(11.41) follows from the characterization

$$\{b_3, b_4\} \in ap(B) \iff b_{21}^{\#} b_3 \in \mathcal{N}_{\text{out}}^{p\times p} \text{ and } b_4 b_{22} \in \mathcal{N}_{\text{out}}^{p\times p}$$

furnished in Section 4.3 and assumption (11.39) of the theorem.

Finally, it remains to check that $c(\lambda)$ is an input impedance matrix of a canonical integral system of the form (7.1) with mass function $M_\gamma(t)$, $0 \le t < d$, i.e., that

$$c \in \mathcal{C}(A_t^{(\gamma)}) \text{ for every } t \in [0, d). \tag{11.42}$$

The proof rests on the fact that $c_I(\lambda)$ is an input impedance matrix of a canonical integral system of the form (7.1) with mass function $M_I(t)$, $0 \le t < d$, and hence that

$$c_I(\lambda) = T_{B_t^{(I)}}[\varepsilon_t], \quad \varepsilon_t \in \mathcal{S}^{p\times p} \cap \mathcal{D}(B_t^{(I)}),$$

for every $t \in [0, d)$. Therefore, since U_γ is a constant j_p-unitary matrix

$$\varepsilon_t^{(\gamma)} = T_{U_\gamma^{-1}}[\varepsilon_t]$$

belongs to $\mathcal{S}^{p\times p} \cap \mathcal{D}(B_t^{(I)} U_\gamma)$, and

$$T_{B_t^{(\gamma)}}[\varepsilon_t^{(\gamma)}] = T_{L_\gamma} T_{B_t^{(I)}}[\varepsilon_t] = T_{L_\gamma}[c_I] = \delta^{1/2} c_I \delta^{1/2} + i\Im\gamma = c.$$

Thus, by the uniqueness of the solution to this inverse impedance problem, (11.42) is in force. □

11.4 Examples with diagonal mvf's b_3^t and $b_4^t = I_p$

In this section the formulas developed in Section 11.2 will be used to obtain the solution $M(t)$ of the (left tangential) inverse impedance problem with data $\{c(\lambda); b_3^t(\lambda), b_4^t(\lambda), 0 \le t < \infty\}$ when:

(1) The input impedance matrix $c \in \mathcal{C}^{p\times p} \cap \mathcal{W}_+^{p\times p}(I_p)$ is of the form

$$c(\lambda) = I_p + 2\int_0^\infty e^{i\lambda a} h(a)da \quad \text{with } h \in L_1^{p\times p}(\mathbb{R}_+), \tag{11.43}$$

and h is extended to $\mathbb{R}$ by the formula

$$h(a) = h(-a)^* \quad \text{for a.e. point } a \in (-\infty, 0). \tag{11.44}$$

(2)
$$b_3^t(\lambda) = e^{i\lambda t D} \quad \text{and} \quad b_4^t(\lambda) = I_p \quad \text{for} \quad t \ge 0, \tag{11.45}$$

where

$$D = \operatorname{diag}\{\alpha_1, \ldots, \alpha_p\}, \quad \alpha_j > 0, \tag{11.46}$$

is a positive definite diagonal matrix and

$$\alpha = \max\{\alpha_1, \ldots, \alpha_p\}. \tag{11.47}$$

It is readily seen that the family $\{e^{i\lambda tD}\}$, $t \geq 0$, is a normalized continuous nondecreasing $\curvearrowright$ chain of entire inner $p \times p$ mvf's and $b^t(\lambda) = e^{i\lambda tD}$ is of exponential type αt. Let $\mathfrak{e}_k$, $k = 1, \ldots, p$, denote the standard basis for $\mathbb{C}^p$ and, for any $p \times 1$ measurable vvf f on $\mathbb{R}$ or $\mathbb{R}_+$, let $f_D(a)$ be the $p \times 1$ vvf that is defined by the formula

$$\mathfrak{e}_k^* f_D(a) = \mathfrak{e}_k^* f(\alpha_k a) \quad \text{for } k = 1, \ldots, p. \tag{11.48}$$

The next theorem summarizes a number of facts and formulas that will be useful in the sequel.

Theorem 11.14 *If $f \in L_2^p$, $t \geq 0$ and $b^t(\lambda) = e^{i\lambda tD}$, then:*

$$f_D(a) = \lim_{R \uparrow \infty} \frac{1}{2\pi} \int_{-R}^{R} e^{-ia\mu D} \widehat{f}(\mu) d\mu \ \ in\ L_2^p \tag{11.49}$$

and

(1) $\|\widehat{f}\|_{\mathrm{st}}^2 = 2\pi \|D^{1/2} f_D\|_{\mathrm{st}}^2 = 2\pi \|f\|_{\mathrm{st}}^2$.
(2) $\widehat{f} \in \mathcal{H}(b^t) \iff f_D(a) = 0$ *for* $a < 0$ *and* $a > t$.
(3) $\widehat{f} \in \mathcal{H}_*(b^t) \iff f_D(a) = 0$ *for* $a < -t$ *and* $a > 0$.
(4) *The reproducing kernel $k_\omega^{b^t}(\lambda)$ of the RKHS $\mathcal{H}(b^t)$ is given by the formula*

$$k_\omega^{b^t}(\lambda) = \frac{1}{2\pi} \int_0^t e^{i\lambda aD} D e^{-i\overline{\omega} aD} da.$$

(5) *The reproducing kernel $\ell_\omega^{b^t}(\lambda)$ of the RKHS $\mathcal{H}_*(b^t)$ is given by the formula*

$$\ell_\omega^{b^t}(\lambda) = \frac{1}{2\pi} \int_{-t}^0 e^{i\lambda aD} D e^{-i\overline{\omega} aD} da.$$

(6) $(\Pi_{\mathcal{H}(b^t)} \widehat{f})(\lambda) = \int_0^t e^{i\lambda aD} D f_D(a) da.$
(7) $(\Pi_{\mathcal{H}_*(b^t)} \widehat{f})(\lambda) = \int_{-t}^0 e^{i\lambda aD} D f_D(a) da.$

Proof The verification of (11.49) is straightforward. Consequently,

$$\begin{aligned}
\|D^{1/2} f_D\|_{\mathrm{st}}^2 &= \textstyle\sum_{k=1}^p \|\mathfrak{e}_k^* D^{1/2} f_D\|_{\mathrm{st}}^2 \\
&= \textstyle\sum_{k=1}^p \int_{-\infty}^{\infty} (\mathfrak{e}_k^* \alpha_k^{1/2} f(\alpha_k a))^* \mathfrak{e}_k^* \alpha_k^{1/2} f(\alpha_k a) da \\
&= \textstyle\sum_{k=1}^p \int_{-\infty}^{\infty} f(a)^* \mathfrak{e}_k \mathfrak{e}_k^* f(a) da = \|f\|_{\mathrm{st}}^2.
\end{aligned}$$

Thus, as $\|\widehat{f}\|_{\mathrm{st}}^2 = 2\pi \|f\|_{\mathrm{st}}^2$, by the Plancherel theorem, the proof of (1) is complete.
Next, $\widehat{f} \in \mathcal{H}(b^t)$ if and only if

$$(\mathfrak{e}_k^* \widehat{f})(\lambda) = \int_0^{\alpha_k t} e^{i\lambda a} (\mathfrak{e}_k^* f)(a) da$$

for $k = 1, \ldots, p$. But this is the same as to say that

$$\begin{aligned}\widehat{f}(\lambda) &= \textstyle\sum_{k=1}^{p} \mathfrak{e}_k(\mathfrak{e}_k^*\widehat{f})(\lambda) = \sum_{k=1}^{p} \mathfrak{e}_k \int_0^t e^{i\lambda a\alpha_k}\alpha_k \mathfrak{e}_k^* f(\alpha_k a)da \\ &= \textstyle\sum_{k=1}^{p} \mathfrak{e}_k\mathfrak{e}_k^* \int_0^t e^{i\lambda aD} D f_D(a)da,\end{aligned}$$

which serves to verify (2). The proofs of (3), (6) and (7) are similar, while (4) and (5) are special cases of (2.20). □

Remark 11.15 *In future calculations it will be convenient to apply the transformation (11.48) to mvf's. It is readily checked that $f \in L_2^{p\times q}(\mathbb{R}_+)$, then (11.49) and (1) of Theorem 11.14 are still valid.*

In view of Theorem 11.12, the inverse input impedance problem based on the given data $\{c(\lambda); b_3^t(\lambda), b_4^t(\lambda),\ \ 0 \le t < \infty\}$ has a unique solution $M(t)$ in the class of canonical integral systems (7.1) with matrizants $A_t(\lambda)$ that fulfill the condition $A_t \in \mathcal{U}_{rR}(J_p)$ for $0 \le t < \infty$. Moreover, $A_t \in \mathcal{U}_{rsR}(J_p)$ for $0 \le t < \infty$, and hence the formulas of Theorems 11.4 and 11.6 are applicable. The calculation is now carried out in a number of separate lemmas in which formulas are obtained for the mvf's $\widehat{y}_{1j}^t(\lambda)$, $\widehat{u}_{1j}^t(\lambda)$ and $\widehat{x}_{1j}^t(\lambda)$, $j = 1, 2$, that figure in Theorem 11.4.

Lemma 11.16 *Let (11.43)–(11.45) be in force and let*

$$v_1(a) = -I_p - 2\int_0^a h(b)db \ \text{ and } \ v_2(a) = I_p \ \text{ for } \ a \ge 0. \tag{11.50}$$

Then the mvf's $\widehat{y}_{1j}^t(\lambda)$ defined by formula (11.17) are:

$$\widehat{y}_{1j}^t(\lambda) = \int_0^t e^{i\lambda aD} D(v_j)_D(a)da \ \text{ for } \ j = 1, 2. \tag{11.51}$$

Proof Let

$$\widehat{f}(\lambda) = c(\lambda)\int_0^{\alpha t} e^{i\lambda a} I_p da.$$

Then

$$(\Pi_{\mathcal{H}(e_{\alpha t}I_p)}\widehat{f})(\lambda) = \int_0^{\alpha t} e^{i\lambda a}\left\{I_p + 2\int_0^a h(a-b)db\right\}da = -\int_0^{\alpha t} e^{i\lambda a} v_1(a)da.$$

The formula for $\widehat{y}_{11}^t(\lambda)$ now follows easily by (7) of Theorem 11.14. The formula for $\widehat{y}_{12}^t(\lambda)$ is obtained in much the same way, though the calculations are easier. □

Lemma 11.17 *Let (11.43)–(11.45) be in force, let $h_D(a, b)$ denote the $p \times p$ mvf with entries*

$$\mathfrak{e}_k^* h_D(a, b)\mathfrak{e}_\ell = \mathfrak{e}_k^* h(\alpha_k a - \alpha_\ell b)\mathfrak{e}_\ell \ \text{ for } \ k, \ell = 1, \ldots, p \tag{11.52}$$

and let $\widehat{u}^t_{1j}(\lambda)$ *be defined by the equations in (11.20). Then*

$$h_D(b,a) = h_D(a,b)^* \tag{11.53}$$

and

$$\begin{aligned}&\left(\Pi_{\mathcal{H}(b_3^t)}(c+c^*)\widehat{u}^t_{1j}\right)(\lambda)\\ &\quad = 2\int_0^t e^{i\lambda a D} D\left\{(u^t_{1j})_D(a) + \int_0^t h_D(a,b)D(u^t_{1j})_D(b)db\right\} da\end{aligned}$$

for $j = 1, 2$.

Proof Formula (11.53) is an easy consequence of the definition (11.49) and (11.44).

Next, let

$$\widehat{f}(\lambda) = (\Pi_{\mathcal{H}(e_{\alpha t}I_p)}\ (c+c^*)\widehat{u}^t_{1j})(\lambda).$$

Then

$$\widehat{f}(\lambda) = 2\widehat{u}^t_{1j}(\lambda) + 2\widehat{f}_1(\lambda),$$

where

$$\widehat{f}_1(\lambda) = \int_0^{\alpha t} e^{i\lambda a}\left\{\int_0^{\alpha t} h(a-b)u^t_{1j}(b)db\right\} da.$$

Therefore,

$$\begin{aligned}\mathfrak{e}_k^*\left(\Pi_{\mathcal{H}(b_3^t)}\widehat{f}_1\right)(\lambda) &= \int_0^{\alpha_k t} e^{i\lambda a}\left\{\int_0^{\alpha t} \mathfrak{e}_k^* h(a-b)u^t_{1j}(b)db\right\} da\\ &= \int_0^t e^{i\lambda a\alpha_k}\alpha_k\left\{\int_0^{\alpha t} \mathfrak{e}_k^* h(\alpha_k a-b)u^t_{1j}(b)db\right\} da.\end{aligned}$$

Moreover, since the columns of $\widehat{u}^t_{1j}(\lambda)$ belong to $\mathcal{H}(b_3^t)$, the inner integral can be reexpressed as

$$\begin{aligned}\{\cdots\} &= \sum_{\ell=1}^p \int_0^{\alpha_\ell t} \mathfrak{e}_k^* h(\alpha_k a-b)\mathfrak{e}_\ell\mathfrak{e}_\ell^* u^t_{1j}(b)db\\ &= \sum_{\ell=1}^p \int_0^t \mathfrak{e}_k^* h(\alpha_k a-\alpha_\ell b)\alpha_\ell\mathfrak{e}_\ell\mathfrak{e}_\ell^* u^t_{1j}(\alpha_\ell b)db\\ &= \sum_{\ell=1}^p \int_0^t \mathfrak{e}_k^* h_D(a,b)\mathfrak{e}_\ell\mathfrak{e}_\ell^* D(u^t_{1j})_D(b)db\\ &= \mathfrak{e}_k^*\int_0^t h_D(a,b)D(u^t_{1j})_D(b)db.\end{aligned}$$

Thus,

$$\mathfrak{e}_k^* \left(\Pi_{\mathcal{H}(b_3^t)}\widehat{f}_1\right)(\lambda) = \mathfrak{e}_k^* \int_0^t e^{i\lambda a D} D \left\{ \int_0^t h_D(a,b) D (u_{1j}^t)_D(b) db \right\} da$$

and the stated formula now drops out easily upon combining terms. □

Lemma 11.18 *Let (11.43)–(11.45) be in force, let v_1 and v_2 be defined by formula (11.50) and let $\gamma_D^t(a,b)$ denote the resolvent kernel for the Fredholm integral operator with kernel $D^{1/2}h_D(a,b)D^{1/2}$ on the square $[0,t] \times [0,t]$, where $h_D(a,b)$ is defined by formula (11.52):*

$$D^{1/2}h_D(a,b)D^{1/2} + \gamma_D^t(a,b) + \int_0^t D^{1/2}h_D(a,c)D^{1/2}\gamma_D^t(c,b)dc = 0$$

and

$$D^{1/2}h_D(a,b)D^{1/2} + \gamma_D^t(a,b) + \int_0^t \gamma_D^t(a,c)D^{1/2}h_D(c,b)D^{1/2}dc = 0.$$

Then

$$\gamma_D^t(b,a) = \gamma_D^t(a,b)^* \tag{11.54}$$

and

$$2D^{1/2}(u_{1j}^t)_D(a) = D^{1/2}(v_j)_D(a) + \int_0^t \gamma_D^t(a,c)D^{1/2}(v_j)_D(c)dc \tag{11.55}$$

for a.e. point $a \in [0,t]$ and $j = 1, 2$.

Proof Since $D = D^*$, the identity (11.54) follows from (11.53) and the resolvent equations (that are recorded just above (11.54)).

Next, in view of the preceding calculations, the equations for $\widehat{u}_{1j}^t(\lambda)$ in (11.22) can be reexpressed as

$$2\int_0^t e^{i\lambda a D} D \left\{ (u_{1j}^t)_D(a) + \int_0^t h_D(a,b) D (u_{1j}^t)_D(b) db \right\} da$$
$$= \int_0^t e^{i\lambda a D} D (v_j)_D(a) da.$$

Therefore,

$$2D^{1/2}\left\{ (u_{1j}^t)_D(a) + \int_0^t h_D(a,b) D (u_{1j}^t)_D(b) db \right\} = D^{1/2}(v_j)_D(a)$$

for a.e. point $a \in [0,t]$ and $j = 1, 2$, and hence $2D^{1/2}(u_{1j}^t)_D$ is seen to be a solution of a Fredholm equation of the first kind with kernel $D^{1/2}h_D(a,b)D^{1/2}$. Thus the solution is given by formula (11.55). □

Lemma 11.19 *Let (11.43)–(11.45) be in force and let*

$$\varphi_1(a,\lambda) = -I_p - 2\int_0^a e^{i\lambda b}h(b)db \ \text{ and } \ \varphi_2(a,\lambda) = I_p \ \text{ for } a \geq 0.$$

Then

$$\widehat{x}^t_{1j}(\lambda) = \int_0^t (\varphi_1)_D(a,\overline{\lambda})^* e^{i\lambda a D} D (u^t_{1j})_D(a)da$$

and

$$\widehat{u}^t_{1j}(\lambda) = \int_0^t (\varphi_2)_D(a,\overline{\lambda})^* e^{i\lambda a D} D (u^t_{1j})_D(a)da$$

for $j = 1, 2$.

Proof Since the columns of $\widehat{x}^t_{1j}(\lambda)$ belong to $\mathcal{H}(b^t_3)$ and $\mathcal{H}(b^t_3) \subseteq \mathcal{H}(e_{\alpha t}I_p)$, formula (11.23) implies that

$$\widehat{x}^t_{1j}(\lambda) = -\widehat{u}^t_{1j}(\lambda) - 2\int_0^{\alpha t} e^{i\lambda a}\left\{\int_a^{\alpha t} h(a-b)u^t_{1j}(b)db\right\}da.$$

Moreover, upon changing orders of integration, the double integral can be rewritten as

$$\int_0^{\alpha t}\left\{\int_0^b e^{i\lambda a}h(a-b)da\right\}u^t_{1j}(b)db = \int_0^{\alpha t} e^{i\lambda b}\left\{\int_0^b e^{-i\lambda a}h(-a)da\right\}u^t_{1j}(b)db.$$

Therefore, since the columns of $\widehat{u}^t_{1j}(\lambda)$ also belong to $\mathcal{H}(b^t_3)$,

$$\begin{aligned}
\widehat{x}^t_{1j}(\lambda) &= \Pi_{\mathcal{H}(b^t_3)} \textstyle\int_0^{\alpha t} e^{i\lambda a}\varphi_1^\#(a,\lambda)u^t_{1j}(a)da \\
&= \sum_{k=1}^p \mathfrak{e}_k\mathfrak{e}_k^* \int_0^{\alpha_k t} e^{i\lambda a}\varphi_1^\#(a,\lambda)u^t_{1j}(a)da \\
&= \sum_{k=1}^p \mathfrak{e}_k\mathfrak{e}_k^* \int_0^t \varphi_1^\#(\alpha_k a,\lambda)e^{i\lambda a\alpha_k}\alpha_k u^t_{1j}(\alpha_k a)da \\
&= \sum_{k=1}^p \mathfrak{e}_k\mathfrak{e}_k^* \int_0^t (\varphi_1)_D(a,\overline{\lambda})^* e^{i\lambda a D} D (u^t_{1j})_D(a)da,
\end{aligned}$$

which coincides with the first stated formula, since $\sum_{k=1}^p \mathfrak{e}_k\mathfrak{e}_k^* = I_p$.

The second formula is verified in much the same way, but with far less work.

□

Theorem 11.20 *Let (11.43)–(11.45) be in force and let*

$$\varphi_D(a,\lambda) = [(\varphi_1)_D(a,\lambda) \quad (\varphi_2)_D(a,\lambda)]$$

and

$$Q_D(t,a) = D^{1/2}\varphi_D(a,0) + \int_0^t \gamma_D^t(a,b)D^{1/2}\varphi_D(b,0)db.$$

Then

$$\varphi_D(a,0) = v_D(a) = [(v_1)_D(a) \quad (v_2)_D(a)]$$

and the following formulas hold:

$$K_0^t(\lambda) = \frac{1}{4\pi}\int_0^t \varphi_D(a,\overline{\lambda})^* e^{i\lambda aD} D^{1/2} Q_D(t,a)da \tag{11.56}$$

$$M(t) = \frac{1}{2}\int_0^t \varphi_D(a,0)^* D^{1/2} Q_D(t,a)da. \tag{11.57}$$

Proof The preceding evaluations applied to formula (11.19) for the reproducing kernel for $\mathcal{H}(A_t)$ imply that

$$K_0^t(\lambda) = \frac{1}{2\pi}\int_0^t \varphi_D(a,\overline{\lambda})^* e^{i\lambda aD} D[(u_{11}^t)_D(a) \quad (u_{12}^t)_D(a)]da,$$

which leads easily to (11.56), in view of Lemma 11.18. The last formula follows from the fact that $M(t) = 2\pi K_0^t(0)$. □

Theorem 11.21 *In the setting of Theorem 11.20, assume that $h(t)$ is continuous on $\mathbb{R}$. Then:*

(1) *For each $t > 0$ the equation*

$$\varphi(a) + \int_0^t h_D(a,b)\varphi(b)db = f(a) \quad \text{for } 0 \le a \le t \tag{11.58}$$

has exactly one continuous $p \times 1$ vector-valued solution φ on $[0,t]$ for each continuous $p \times 1$ vector-valued right-hand side f on $[0,t]$.

(2) *The resolvent kernel satisfies the identity*

$$\frac{\partial}{\partial t}\gamma_D^t(a,b) = \gamma_D^t(a,t)\gamma_D^t(t,b). \tag{11.59}$$

(3) *$M(t)$ is locally absolutely continuous and*

$$M'(t) = Y_1(t)Y_1(t)^*, \tag{11.60}$$

where

$$Y_1(t)^* = \frac{Q_D(t,t)}{\sqrt{2}} = \frac{1}{\sqrt{2}}\left(D^{1/2}v_D(t) + \int_0^t \gamma_D^t(t,b)D^{1/2}v_D(b)db\right). \tag{11.61}$$

Proof The integral operator with kernel $h_D(a,b)$ in (11.58) is a compact self-adjoint operator in $L_2^p([0,t])$. Moreover, since -1 is not an eigenvalue of this operator, the equation (11.58) is uniquely solvable in $L_2^p([0,t])$ for each $t > 0$.

Lemma 7.1 in [AAK71b] guarantees that it is also uniquely solvable in the space of continuous $p \times 1$ vvf's on $[0, t]$. This completes the proof of (1).

Formula (11.59) is verified in [GK85] on the basis of (1). The rest is a straightforward calculation. □

Formula (11.59) is called the **Krein–Sobolev identity** in [GK85]; it is also sometimes referred to as the Bellman–Krein–Sobolev–Siegert identity, because Bellman and Siegert also obtained this formula independently, just a little later. There is another form that is useful:

$$\frac{\partial}{\partial t}\gamma_D^t(t-a, t-b) = \gamma_D^t(t-a, 0)\gamma_D^t(0, t-b). \tag{11.62}$$

Remark 11.22 *In the setting of Theorem 11.21, let*

$$\sqrt{2}Y_2(t)^* = D^{1/2}[I_p \quad I_p] + \int_0^t \gamma_D^t(0, b)D^{1/2}v_D(b)db.$$

Then, with the help of formulas (11.59) and (11.62), it is readily checked that

$$Y_2'(t)^* = \gamma_D^t(0, t)Y_1(t)^* \tag{11.63}$$

and

$$\sqrt{2}Y_1'(t)^* = \gamma_D^t(t, 0)\left\{D^{1/2}v_D(0) + \int_0^t \gamma_D^t(0, b)D^{1/2}v_D(b)db\right\}$$
$$+ D^{1/2}v_D'(t) + \int_0^t \gamma_D^t(t, b)D^{1/2}v_D'(b)db.$$

Since

$$v_D'(t) = [-2Dh_D(t) \quad 0],$$

the mvf

$$Y(t) = \begin{bmatrix} Y_1(t) & Y_2(t) \end{bmatrix} \tag{11.64}$$

is a solution of the Cauchy problem

$$Y'(t) = Y(t)\begin{bmatrix} 0 & \gamma_D^t(t, 0) \\ \gamma_D^t(0, t) & 0 \end{bmatrix} + \begin{bmatrix} \Theta_D(t) & 0 \\ 0 & 0 \end{bmatrix}, \quad 0 \le t < d, \tag{11.65}$$

with initial condition

$$Y(0) = \mathfrak{V}\mathcal{D}^{1/2} = \mathcal{D}^{1/2}\mathfrak{V},$$

where

$$\Theta_D(t)^* = -\sqrt{2}\left\{\gamma_D^t(t, 0)D^{1/2} + D^{3/2}h_D(t) + \int_0^t \gamma_D^t(t, b)D^{3/2}h_D(b)db\right\} \tag{11.66}$$

and

$$\mathcal{D} = \begin{bmatrix} D & 0 \\ 0 & D \end{bmatrix}$$

At this moment, we do not have a convenient way of evaluating $\Theta_D(t)$. However, if

$$Dh_D = h_D D, \tag{11.67}$$

then

$$\begin{aligned} \Theta_D(t)^* = -\sqrt{2}\,\{\gamma_D^t(t,0) + D^{1/2}h_D(t)D^{1/2} \\ + \int_0^t \gamma_D^t(t,b)D^{1/2}h_D(b)D^{1/2}db\Big\}\, D^{1/2} = 0 \end{aligned} \tag{11.68}$$

by the resolvent identity and

$$Y(t)j_pY(t)^* = \mathcal{D}^{1/2}\, J_p\, \mathcal{D}^{1/2} \text{ for every } t \in [0,d). \tag{11.69}$$

To justify (11.69) it suffices to verify that

$$Y(t)j_pY(t)^* = Y(0)j_pY(0)^*,$$

which is obtained by verifying that the derivative of $Y(t)j_pY(t)^*$ is equal to zero.

Lemma 11.23 *If (11.67) is in force, then:*

(1) $D\gamma_D^t(a,b) = \gamma_D^t(a,b)D$ *for* $0 \le a, b \le t < d$.
(2) $D^{1/2}h_D(a,b) = h_D(a,b)D^{1/2}$ *and* $D^{1/2}\gamma_D^t(a,b) = \gamma_D^t(a,b)D^{1/2}$ *for* $0 \le a, b \le t < d$.
(3) $Y_1(t)^*\mathcal{D}^{1/2} = D^{1/2}Y_1(t)^*$ *for* $0 \le t < d$.
(4) $\mathcal{D}^{-1/2}Y(t)\begin{bmatrix} D & 0 \\ 0 & 0 \end{bmatrix}Y(t)^*\mathcal{D}^{-1/2} = Y(t)\begin{bmatrix} I_p & 0 \\ 0 & 0 \end{bmatrix}Y(t)^*$ *for* $0 \le t < d$.

Proof To verify (1), fix $t > 0$ and let H_D (resp., Γ_D) denote the integral operator on $L_2^p([0,t])$ with kernel $D^{1/2}h_D(a,b)D^{1/2}$ (resp., $\gamma_D(a,b)$) for $0 \le a, b \le t$. Then, since $DH_D = H_D D$, the resolvent identity in terms of the operators H_D and Γ_D implies that

$$D = D(I + H_D)(I + \Gamma_D) = (I + H_D)D(I + \Gamma_D)$$

and hence that

$$(I + \Gamma_D)D = (I + \Gamma_D)(I + H_D)D(I + \Gamma_D) = D(I + \Gamma_D).$$

To verify (2), let $\alpha = \max\{\alpha_{11}, \dots, \alpha_{pp}\}$, $\beta = \min\{\alpha_{11}, \dots, \alpha_{pp}\}$, $\gamma = (\alpha + \beta)/2$ and $D = \gamma I_p - \Delta$. Then, since $||\Delta/\gamma|| < 1$,

$$D^{1/2} = \gamma^{1/2}\left(I_p - \frac{\Delta}{\gamma}\right)^{1/2} = \gamma^{1/2}\sum_{k=0}^{\infty} c_k\left(\frac{\Delta}{\gamma}\right)^k = \gamma^{1/2}\sum_{k=0}^{\infty} c_k\left(\frac{\gamma I_p - D}{\gamma}\right)^k,$$

where the c_k are the coefficients in the power series

$$(1-x)^{1/2} = \sum_{k=0}^{\infty} c_k x^k \quad \text{for } |x| < 1.$$

This representation for $D^{1/2}$ in terms of D^k, $k = 0, 1, \ldots$, serves to verify (2).

Finally, (4) follows from (3) and (3) is immediate from (1), (2) and formula (11.61). □

A generalized Krein system

If (11.58) is in force, then in view of formulas (11.65) and (11.68), Lemma 11.23 and the formula

$$A_t'(\lambda) = i\lambda A_t(\lambda)Y(t)\begin{bmatrix} I_p & 0 \\ 0 & 0 \end{bmatrix} Y(t)^* J_p,$$

it is readily checked that the mvf $U_t(\lambda) = \mathfrak{V}A_t(\lambda)\mathcal{D}^{-1/2}Y(t)$ is the matrizant of the generalized Krein system

$$u_t'(\lambda) = i\lambda u_t(\lambda)\begin{bmatrix} D & 0 \\ 0 & 0 \end{bmatrix} j_p + u_t(\lambda)\mathcal{V}(t) \quad \text{for } 0 \le t < d, \tag{11.70}$$

with potential

$$\mathcal{V}(t) = \begin{bmatrix} 0 & \gamma_D^t(t,0) \\ \gamma_D^t(0,t) & 0 \end{bmatrix} \quad \text{for } 0 \le t < d. \tag{11.71}$$

Moreover, since $\mathfrak{V}U_t = B_t\mathfrak{V}\mathcal{D}^{-1/2}Y(t)$ and $\mathfrak{V}\mathcal{D}^{-1/2}Y(t) \in \mathcal{U}_{const}(j_p)$, it follows that

$$T_{\mathfrak{V}U_t}[S^{p\times p}] = T_{B_t}[S^{p\times p}]$$

and hence that the mvf $c(\lambda)$ is also an input impedance for this system:

$$c \in \bigcap_{0\le t<d} T_{\mathfrak{V}U_t}[\mathcal{S}^{p\times p}].$$

Thus, Theorem 11.21 yields a solution of the inverse input impedance problem for a differential system of the form (11.70), with potential of the form (11.71) that is continuous on $[0, d)$. The verification of uniqueness of the solution of the inverse impedance problem for such a system is similar to that in Chapter 12.

Similar considerations show that if $W_t(\lambda) = \mathfrak{V}A_t(\lambda)\mathfrak{V}$, then $W_0(\lambda) = I_m$ and

$$W_t'(\lambda) = i\lambda W_t(\lambda)\mathfrak{V}\mathcal{D}^{-1/2}Y(t)\begin{bmatrix} D & 0 \\ 0 & 0 \end{bmatrix} Y(t)^*\mathcal{D}^{-1/2}\mathfrak{V}j_p \quad \text{for } 0 \le t < d. \tag{11.72}$$

Moreover, $\{e^{i\lambda tD}, I_p\} \in ap(W_t)$ for $0 \le t < d$.

11.5 The bitangential inverse spectral problem

In this section the bitangential inverse spectral problem for canonical integral systems of the form (7.1) with data $\{\sigma; b_3^t, b_4^t, 0 \le t < d\}$, will be reduced to the bitangential inverse impedance problem with data $\{c^{(0)}; b_3^t, b_4^t, 0 \le t < d\}$, where

$$c^{(\alpha)}(\lambda) = i\alpha + \frac{1}{\pi i}\int_{-\infty}^{\infty}\left\{\frac{1}{\mu-\lambda} - \frac{\mu}{1+\mu^2}\right\} d\sigma(\mu) \tag{11.73}$$

for every $p \times p$ Hermitian matrix α. Formula (11.73) defines a one-to-one correspondence between the set of Hermitian matrices $\alpha \in \mathbb{C}^{p\times p}$ and the set of mvf's $c \in \mathcal{C}^{p\times p}$ with spectral function $\sigma_c = \sigma$ and $\beta_c = \lim_{\nu\uparrow\infty} \nu^{-1}\Re c(i\nu) = 0$.

It is also not difficult to show that if $c \in \mathcal{C}^{p\times p}$ is of the form (11.31) with $\gamma \in \mathbb{C}^{p\times p}$, $h \in L_1^{p\times p}(\mathbb{R}_+)$ and $h(-t) = h(t)^*$ a.e. on $\mathbb{R}$, then

$$c(\lambda) = c^{(\alpha)}(\lambda) \quad \text{with } \alpha = \frac{\gamma - \gamma^*}{2i} - i\int_0^{\infty} e^{-t}\{h(t) - h(t)^*\}dt$$

and

$$d\sigma(\mu) = \left\{\frac{\gamma+\gamma^*}{2} + \widehat{h}(\mu)\right\} d\mu.$$

Theorem 11.24 *Let the data $\{\sigma; b_3^t, b_4^t,\ 0 \le t < d\}$ for the bitangential inverse spectral problem satisfy the following conditions:*

(1) *$\sigma(\mu)$ is a nondecreasing $p \times p$ mvf on $\mathbb{R}$ such that (3.43) holds.*
(2) *$\{b_3^t(\lambda), b_4^t(\lambda)\}$, $0 \le t < d$, is a normalized nondecreasing continuous chain of entire inner $p \times p$ mvf's such that the mvf $\chi_1^t(\lambda) = b_4^t(\lambda)b_3^t(\lambda)$ satisfies the condition (7.106).*
(3) *The set*

$$\mathcal{C}(b_3^t, b_4^t; c^{(0)}) \cap \mathring{\mathcal{C}}^{p\times p} \neq \emptyset \quad \textit{for every } t \in [0, d),$$

Then for each Hermitian matrix $\alpha \in \mathbb{C}^{p\times p}$, there exists exactly one canonical integral system (7.1) with continuous nondecreasing $m \times m$ mvf

$$M^{(\alpha)}(t) = \begin{bmatrix} I_p & i\alpha \\ 0 & I_p \end{bmatrix} M^{(0)}(t) \begin{bmatrix} I_p & 0 \\ -i\alpha & I_p \end{bmatrix}, \quad 0 \le t < d, \tag{11.74}$$

with $M^{(\alpha)}(0) = 0$ and matrizant

$$A_t^{(\alpha)}(\lambda) = \begin{bmatrix} I_p & i\alpha \\ 0 & I_p \end{bmatrix} A_t^{(0)}(\lambda) \begin{bmatrix} I_p & -i\alpha \\ 0 & I_p \end{bmatrix}, \quad 0 \le t < d, \tag{11.75}$$

such that (C1)–(C3) hold for $c = c^{(\alpha)}$ and $A_t = A_t^{(\alpha)}$ for $0 \le t < d$. Moreover, $\sigma \in \Sigma_{sf}^d(dM^{(\alpha)})$ and formula (11.74) defines a one-to-one correspondence between *the set of all solutions $M(t)$ of the bitangential inverse spectral problem with data $\{\sigma; b_3^t, b_4^t, 0 \le t < d\}$ that have matrizants $A_t \in \mathcal{U}_{rsR}(J_p)$ for every $t \in [0, d)$ and the set of all Hermitian $p \times p$ matrices α.*

Proof The first conclusion is immediate from Theorem 11.1, since the given assumption (3) is in force if and only if

$$\mathcal{C}(b_3^t, b_4^t; c^{(\alpha)}) \cap \mathring{\mathcal{C}}^{p\times p} \neq \emptyset \text{ for every } t \in [0, d).$$

Thus, Theorems 7.30 (for the case of a bounded mass function), 7.35 (for the case of unbounded mass functions) and Theorem 7.31 imply that

$$\Sigma_{\rm sf}^d(dM^{(\alpha)}) = (\mathcal{C}_{\rm imp}^d(dM^{(\alpha)}))_{\rm sf}$$

and hence that $\sigma \in \Sigma_{\rm sf}^d(dM^{(\alpha)})$. Moreover, it is readily checked that the right-hand side of formula (11.75) is also the matrizant of a system (7.1) with mass function $M(t)$ given by formula (11.74) and that this mass function is also a solution of the bitangential inverse input impedance problem with the same data. Furthermore, this matrizant satisfies the condition (C3) as does the matrizant $A_t^{(\alpha)}(\lambda)$. Therefore, since there is only one such matrizant, by Theorem 11.1, the mass function $M(t)$ defined by formula (11.74) coincides with $M^{(\alpha)}(t)$. Moreover, since $M_{22}(t) > 0$ for every $t \in (0, d)$ under the assumptions of the theorem, it is readily checked (by looking at the 21 block of both sides) that

$$M^{(\alpha_1)}(t) = M^{(\alpha_2)}(t) \Longleftrightarrow \alpha_1 = \alpha_2.$$

It remains to show that every solution $M(t)$ of the bitangential inverse spectral problem with data $\{\sigma; b_3^t, b_4^t, 0 \le t < d\}$ and matrizant $A_t \in \mathcal{U}_{rsR}(J_p)$ can be expressed in the form (11.74) for some Hermitian $p \times p$ matrix α. Theorems 7.30, 7.31 and 7.35 guarantee that for this system, $\Sigma_{\rm sf}(dM) = (\mathcal{C}_{\rm imp}(dM))_{\rm sf}$ and hence that there exists a mvf $c \in \mathcal{C}_{\rm imp}(dM)$ with spectral function $\sigma(\mu)$. Moreover, by Theorems 7.35, 7.36 and 7.21, $c(\lambda) = c^{(\alpha)}(\lambda)$ for some Hermitian $p \times p$ matrix α. Consequently, by Theorem 11.1, $M(t) = M^{(\alpha)}(t)$, where $M^{(\alpha)}(t)$ is defined by formula (11.74). □

Theorem 11.25 *If, in the setting of Theorem 11.24, there exists a solution $H^{(\alpha)}$ of the bitangential inverse input impedance problem with data $\{c^{(\alpha)}; b_3^t, b_4^t, 0 \le t < d\}$ for some Hermitian matrix $\alpha \in \mathbb{C}^{p\times p}$ for the canonical differential system (2.8) with $J = J_p$, then:*

(1) *A solution $H^{(\alpha)}$ of this problem exists for every Hermitian matrix $\alpha \in \mathbb{C}^{p\times p}$.*

(2) $H^{(\alpha)}(t) = \begin{bmatrix} I_p & i\alpha \\ 0 & I_p \end{bmatrix} H^{(0)}(t) \begin{bmatrix} I_p & 0 \\ -i\alpha & I_p \end{bmatrix}.$

(3) $\sigma \in \Sigma_{\rm sf}^d(H^{(\alpha)}dt)$ *for every Hermitian matrix* $\alpha \in \mathbb{C}^{p\times p}$.

(4) *For each fixed Hermitian matrix $\alpha \in \mathbb{C}^{p\times p}$, there exists only one mvf $c \in \mathcal{C}_{\rm imp}^d(H^{(\alpha)}dt)$ with $\sigma_c = \sigma : c(\lambda) = c^{(\alpha)}(\lambda)$.*

(5) *If $H(t)$ is any solution of the bitangential inverse spectral problem such that (C4) holds, then $H = H^{(\alpha)}$ for some Hermitian matrix $\alpha \in \mathbb{C}^{p\times p}$.*

Proof This follows from Theorem 11.24. □

Example 11.26 *Let $\sigma(\mu) = \mu I_p$ and let $\{b_3^t, b_4^t\}$, $0 \le t < d$, be a normalized nondecreasing continuous chain of pairs of entire inner $p \times p$ mvf's that satisfies the condition (7.106).*

Discussion The mvf $\sigma(\mu) = \mu I_p$ is the spectral function of the mvf $c^{(0)}(\lambda) = I_p$, which belongs to $\mathring{\mathcal{C}}^{p\times p}$. Thus, Theorem 11.24 is applicable and guarantees the existence of a solution $M^{(0)}(t)$ to the bitangential inverse spectral problem for the given data: $M^{(0)}(t)$ is the solution of the bitangential inverse impedance problem with data $\{I_p; b_3^t, b_4^t, 0 \le t < d\}$ and may be obtained by invoking the formulas in Section 11.2. In particular, it follows readily from Theorem 11.4 that the RK $K_0^t(\lambda)$ of the RKHS $\mathcal{H}(A_t)$ based on the matrizant $A_t(\lambda)$ of the system (7.1) with mass function $M^{(0)}(t)$ is given by the formula

$$K_0^t(\lambda) = \sqrt{2}\mathfrak{V}\frac{1}{2\pi}\begin{bmatrix} \widehat{u}_{11}^t(\lambda) & \widehat{u}_{12}^t(\lambda) \\ \widehat{u}_{21}^t(\lambda) & \widehat{u}_{22}^t(\lambda) \end{bmatrix},$$

where

$$\widehat{u}_{11}^t(\lambda) = -\frac{b_3^t(\lambda) - I_p}{2i\lambda} = -\widehat{u}_{12}^t(\lambda), \quad \widehat{u}_{21}^t(\lambda) = -\frac{b_4^t(\lambda)^{-1} - I_p}{2i\lambda} = \widehat{u}_{22}^t(\lambda)$$

(and the mvf's $\widehat{x}_{ij}^t(\lambda)$ that appear in formula (11.19) are: $\widehat{x}_{1j}^t(\lambda) = -\widehat{u}_{1j}^t(\lambda)$ and $\widehat{x}_{2j}^t(\lambda) = \widehat{u}_{2j}^t(\lambda)$, $j = 1, 2$). Therefore,

$$M^{(0)}(t) = 2\pi K_0^t(0) = \sqrt{2}\mathfrak{V}\begin{bmatrix} \widehat{u}_{11}^t(0) & \widehat{u}_{12}^t(0) \\ \widehat{u}_{21}^t(0) & \widehat{u}_{22}^t(0) \end{bmatrix},$$

which can be reexpressed in terms of the mvf

$$m_j(t) = -i\frac{\partial b_j^t}{\partial \lambda}(0) = 2\pi k_0^{b_j^t}(0), \quad j = 3, 4,$$

as

$$M^{(0)}(t) = \mathfrak{V}\begin{bmatrix} m_3(t) & 0 \\ 0 & m_4(t) \end{bmatrix}\mathfrak{V}.$$

Moreover, since

$$B_t(\lambda) = A_t(\lambda)\mathfrak{V} = (I_m + 2\pi i\lambda K_0^t(\lambda)J_p)\mathfrak{V} = \frac{1}{\sqrt{2}}\begin{bmatrix} -b_3^t(\lambda) & b_4^t(\lambda)^{-1} \\ b_3^t(\lambda) & b_4^t(\lambda)^{-1} \end{bmatrix},$$

$$\mathfrak{E}_t(\lambda) = \sqrt{2}N_2^*B_t(\lambda) = [b_3^t(\lambda) \quad b_4^t(\lambda)^{-1}].$$

Thus, in this case $\|f\|^2_{\mathcal{B}(\mathfrak{E}_t)} = \|f\|^2_{\rm st}$ for $f \in \mathcal{B}(\mathfrak{E}_t)$ and

$$\mathcal{B}(\mathfrak{E}_t) = \mathcal{H}(b_3^t) \oplus \mathcal{H}_*(b_4^t)$$

as Hilbert spaces. Moreover

$$\chi^t(\lambda) = b_4^t(\lambda)b_3^t(\lambda) = \chi_1^t(\lambda)$$

and statement (5) of Theorem 7.19 implies that $\beta_{c^{(0)}} = 0$ and that

$$\Sigma_{sf}(\mathfrak{E}_t) = (\mathcal{C}(A_t))_{\rm sf} \quad \text{and} \quad \mathcal{C}(A_t) = \{(I_p - b_3^t\varepsilon b_4^t)(I_p + b_3^t\varepsilon b_4^t)^{-1} : \varepsilon \in \mathcal{S}^{p\times p}\}.$$

Remark 11.27 *There is a one-to-one correspondence between normalized nondecreasing continuous chains of pairs $\{b_3^t(\lambda), b_4^t(\lambda)\}, 0 \le t < d$, of entire inner $p \times p$ mvf's and the pairs $\{m_3(t), m_4(t)\}$ of continuous nondecreasing $p \times p$ mvf's on $[0, d)$ with $m_3(0) = m_4(0) = 0$: $\{b_3^t(\lambda), b_4^t(\lambda)\}, 0 \le t < d$, are the unique continuous solutions of the integral equations*

$$b_3^t(\lambda) = I_p + i\lambda \int_0^t b_3^s(\lambda)dm_3(s), \quad b_4^t(\lambda) = I_p + i\lambda \int_0^t dm_4(s)b_4^s(\lambda),\ 0 \le t < d,$$

and

$$m_3(t) = -i\frac{\partial b_3^t}{\partial \lambda}(0) \quad \text{and} \quad m_4(t) = -i\frac{\partial b_4^t}{\partial \lambda}(0).$$

This follows from Theorems 5.8 and 2.5.

11.6 An example

In this section we extend the discussion of one of the examples considered in Section 8.13: the canonical differential system

$$u'(x, \lambda) = i\lambda u(x, \lambda)\begin{bmatrix} x^r & 0 \\ 0 & x^{-r}\end{bmatrix} J_1 \quad \text{for } 0 \le x < \infty \tag{11.76}$$

and $-1 < r < 1$ with matrizant

$$A_x(\lambda) = A(x, \lambda) = \begin{bmatrix} a_{11}(x,\lambda) & a_{12}(x,\lambda) \\ a_{21}(x,\lambda) & a_{22}(x,\lambda)\end{bmatrix} \quad \text{for } 0 \le x < \infty, \tag{11.77}$$

and shall establish the following facts:

(1) *The matrizant can be expressed in terms of the Gamma function and Bessel functions* $J_p(x)$ *of the first kind as*

$$A(x,\lambda) = \begin{bmatrix} \left(\frac{x\lambda}{2}\right)^{(1-r)/2}\Gamma(\frac{1+r}{2}) & 0 \\ 0 & \left(\frac{x\lambda}{2}\right)^{(1+r)/2}\Gamma(\frac{1-r}{2}) \end{bmatrix} \times \begin{bmatrix} J_{\frac{r-1}{2}}(x\lambda) & -ix^r J_{\frac{1+r}{2}}(x\lambda) \\ -ix^{-r}J_{\frac{1-r}{2}}(x\lambda) & J_{-\frac{1+r}{2}}(x\lambda) \end{bmatrix}, \tag{11.78}$$

or, equivalently, as

$$A(x,\lambda) = \begin{bmatrix} \Gamma(\frac{1+r}{2}) & 0 \\ 0 & \Gamma(\frac{1-r}{2}) \end{bmatrix} \times \begin{bmatrix} \boldsymbol{F}_{\frac{r-1}{2}}(x\lambda) & -i\frac{x\lambda}{2}x^r\boldsymbol{F}_{\frac{1+r}{2}}(x\lambda) \\ -i\frac{x\lambda}{2}x^{-r}\boldsymbol{F}_{\frac{1-r}{2}}(x\lambda) & \boldsymbol{F}_{-\frac{1+r}{2}}(x\lambda) \end{bmatrix}, \tag{11.79}$$

where

$$\boldsymbol{F}_p(\lambda) = (\lambda/2)^{-p}J_p(\lambda) = \sum_{k=0}^{\infty} \frac{(-1)^k(\lambda/2)^{2k}}{\Gamma(k+1)\Gamma(k+1+p)}. \tag{11.80}$$

Moreover, $A_x(\lambda)$ *is both symplectic and real;* $a_{11}(x,\lambda)$ *and* $a_{22}(x,\lambda)$ *are even functions of* λ*, whereas* $a_{12}(x,\lambda)$ *and* $a_{21}(x,\lambda)$ *are odd functions of* λ *and*

$$|a_{jk}(x,\lambda)| \le a_{jk}(x, i|\lambda|) \quad \text{for } \lambda \in \mathbb{C},\ x > 0 \text{ and } j,k = 1,2. \tag{11.81}$$

(2) *The de Branges matrices* $\mathfrak{E}(x,\lambda) = \begin{bmatrix} E_-(x,\lambda) & E_+(x,\lambda) \end{bmatrix}$ *with components*

$$\begin{aligned} E_\pm(x,\lambda) &= a_{22}(x,\lambda) \pm a_{21}(x,\lambda) \\ &= \left(\frac{x\lambda}{2}\right)^{(1+r)/2} \Gamma\left(\frac{1-r}{2}\right) \left\{ J_{-\frac{1+r}{2}}(x\lambda) \mp ix^{-r}J_{\frac{1-r}{2}}(x\lambda) \right\} \end{aligned} \tag{11.82}$$

are entire functions of exponential type x,

$$|E_+(x,\mu)| = |E_+(x,-\mu)| \quad \text{for every point } \mu \in \mathbb{R}, \tag{11.83}$$

$$E_-(x,\lambda) = E_+^{\#}(x,\lambda) \quad \text{for every point } \lambda \in \mathbb{C} \tag{11.84}$$

and $E_\pm(x,\lambda) \neq 0$ for every point $\lambda \in \overline{\mathbb{C}_\pm}$.

(3) $\{e_x, e_x\} \in ap_{II}(A_x)$ *for* $0 \le x < \infty$.

(4) *The function*

$$c_x(\lambda) = T_{A_x}[1] = \frac{a_{11}(x,\lambda) + a_{12}(x,\lambda)}{a_{21}(x,\lambda) + a_{22}(x,\lambda)} \tag{11.85}$$

belongs to the class $\mathcal{C}_a$, i.e., it meets the condition

$$\lim_{\nu\uparrow\infty} \nu^{-1} c_x(i\nu) = 0 \tag{11.86}$$

and its spectral function $\sigma_x(\mu)$ is locally absolutely continuous; in fact

$$c_x(\lambda) = i\alpha + \frac{1}{\pi i}\int_{-\infty}^{\infty} \left\{ \frac{1}{\mu - \lambda} - \frac{\mu}{1+\mu^2} \right\} |E_+(x,\mu)|^{-2} d\mu \tag{11.87}$$

for $\lambda \in \mathbb{C}_+$ and appropriate $\alpha \in \mathbb{R}$.

(5) *The function*

$$\Delta_x(\mu) = |E_+(x,\mu)|^{-2} = \sigma_x'(\mu)$$

satisfies the Muckenhoupt (A_2) condition

$$\sup \left\{ \frac{1}{b-a}\int_a^b \Delta_x(\mu) d\mu \, \frac{1}{b-a}\int_a^b \Delta_x(\mu)^{-1} d\mu \, : \, a < b \right\} < \infty.$$

(6) $A_x \in \mathcal{E} \cap \mathcal{U}^\circ_{\ell sR}(J_1)$ *for* $0 \le x < \infty$.

(7) $\mathcal{C}(A_x) = \mathcal{C}(e_{2x}; c_x)$ *for* $0 \le x < \infty$.

(8) *If $g_x \in \mathcal{G}_\infty(0)$ is defined by the integral representation*

$$c_x(\lambda) = \lambda^2 \int_0^\infty e^{i\lambda t} g_x(t) dt \quad \text{for } \lambda \in \mathbb{C}_+ \text{ and } 0 \le x < \infty,$$

and if $0 < x_1 \le x_2$, then

$$g_{x_2}(t) = g_{x_1}(t) \quad \text{for } t \in [-2x_1, 2x_1]$$

and there exists exactly one function $g_\infty \in \mathcal{G}_\infty(0)$ such that

$$g_\infty(t) = g_x(t) \quad \text{for } t \in [-2x, 2x] \text{ for every } x > 0.$$

(9) *The function*

$$c_\infty(\lambda) = \lambda^2 \int_0^\infty e^{i\lambda t} g_\infty(t) dt \quad \text{for } \lambda \in \mathbb{C}_+,$$

can be evaluated explicitly as

$$c_\infty(\lambda) = \kappa_r \left(\frac{i}{\lambda}\right)^r \quad \text{for } \lambda \in \mathbb{C}_+, \text{ where } \quad \kappa_r = 2^r \frac{\Gamma\left(\frac{1+r}{2}\right)}{\Gamma\left(\frac{1-r}{2}\right)}. \tag{11.88}$$

Moreover,

$$\mathcal{C}^\infty_{\rm imp}(Hdx) = \{c_\infty\},$$

$$g_\infty(t) = -\frac{\kappa_r}{\Gamma(r+2)} |t|^{r+1} \quad \text{for } t \in \mathbb{R}, \tag{11.89}$$

$$g'_\infty \notin L_2([0,a]) \quad \text{for } -1 < r < -\frac{1}{2} \text{ and every } a > 0$$

and

$$\mathcal{C}(A_x) = \mathcal{C}(e_{2x}; c_\infty) = \{c_g : g \in \mathcal{G}(g_\infty; 2x)\}.$$

(10) *The function $c_\infty(\lambda)$ admits the representations*

$$c_\infty(\lambda) = -i\lambda \int_0^\infty \frac{1}{\mu - \lambda^2} d\tau(\mu) \quad \textit{for } \lambda \in \mathbb{C}_+ \tag{11.90}$$

where τ is *locally absolutely continuous on* $\mathbb{R}_+$,

$$\tau'(\mu) = \frac{2^r}{\Gamma\left(\frac{1-r}{2}\right)^2} \frac{1}{\mu^{(r+1)/2}} \quad \textit{for } \mu > 0 \tag{11.91}$$

and

$$c_\infty(\lambda) = \frac{1}{\pi i} \int_{-\infty}^\infty \left\{ \frac{1}{\mu - \lambda} - \frac{\mu}{1 + \mu^2} \right\} d\sigma(\mu) \quad \textit{for } \lambda \in \mathbb{C}_+ \tag{11.92}$$

where σ is *locally absolutely continuous on* $\mathbb{R}$ *and*

$$\sigma'(\mu) = \frac{\pi\, 2^r}{\Gamma(\frac{1-r}{2})^2} \frac{1}{|\mu|^r} = \kappa_r \cos(\pi r/2) \frac{1}{|\mu|^r} \quad \textit{for } \mu \in \mathbb{R} \setminus \{0\}. \tag{11.93}$$

(11) *The de Branges space $\mathcal{B}(\mathfrak{E}_x)$ is the space of entire functions f of exponential type less than or equal to x for which*

$$\int_{-\infty}^\infty \frac{|f(\mu)|^2}{|\mu|^r} d\mu < \infty. \tag{11.94}$$

(12) *If $r = 0$, then*

$$A_x(\lambda) = \begin{bmatrix} \cos(\lambda x) & -i\sin(\lambda x) \\ -i\sin(\lambda x) & \cos(\lambda x) \end{bmatrix}, \quad \mathfrak{E}_x(\lambda) = [e^{i\lambda x} \quad e^{-i\lambda x}],$$

the de Branges space $\mathcal{B}(\mathfrak{E}_x)$ is equal to the Paley–Wiener space

$$\mathcal{H}(e_x) \oplus \mathcal{H}_*(e_x) = \left\{ \int_{-x}^x e^{i\lambda t} f(t) dt : f \in L_2([-x, x]) \right\} \tag{11.95}$$

and $A_x \in \mathcal{U}_{rsR}(J_1)$.

(13) *If $-1 < r < 0$, then $\mathcal{B}(\mathfrak{E}_x)$ is a proper subspace of $\mathcal{H}(e_x) \oplus \mathcal{H}_*(e_x)$. If $0 < r < 1$, then $\mathcal{H}(e_x) \oplus \mathcal{H}_*(e_x)$ is a proper subspace of $\mathcal{B}(\mathfrak{E}_x)$.*

(14) $A_x \notin \mathcal{U}_{rsR}(J_1)$ if $x > 0$ and $r \in (-1, 0) \cup (0, -1)$

Proof The proof is divided into steps.

1. *Verification of (1).*

The top row of the matrizant is a solution of the system

$$\left[a_{11}'(x, \lambda) \quad a_{12}'(x, \lambda)\right] = -i\lambda \left[x^{-r} a_{12}(x, \lambda) \quad x^r a_{11}(x, \lambda)\right]$$

with initial condition

$$\left[a_{11}(0, \lambda) \quad a_{12}(0, \lambda)\right] = \left[1 \quad 0\right].$$

Thus, $a_{11}(x, \lambda)$ is a solution of the Bessel equation

$$a_{11}''(x, \lambda) + \frac{r}{x} a_{11}'(x, \lambda) + \lambda^2 a_{11}(x, \lambda) = 0, \quad 0 \le x < \infty,$$

with initial conditions

$$a_{11}(0, \lambda) = 1 \quad \text{and} \quad a_{11}'(0, \lambda) = 0,$$

i.e.,

$$a_{11}(x, \lambda) = \left(\frac{x\lambda}{2}\right)^{(1-r)/2} \Gamma\left(\frac{1+r}{2}\right) J_{(r-1)/2}(x\lambda) \quad \text{for } 0 \le x < \infty. \qquad (11.96)$$

This justifies the formula for a_{11}. The formula for a_{12} follows from the fact that

$$\frac{d}{dx} x^p J_p(x\lambda) = \lambda x^p J_{p-1}(x\lambda) \quad \frac{d}{dx} x^{-p} J_p(x\lambda) = -\lambda x^{-p} J_{p+1}(x\lambda)$$

The remaining entries in (11.78) can be verified in much the same way, since

$$\left[a_{21}'(x, \lambda) \quad a_{22}'(x, \lambda)\right] = -i\lambda \left[x^{-r} a_{22}(x, \lambda) \quad x^r a_{21}(x, \lambda)\right]$$

with initial condition

$$\left[a_{21}(0, \lambda) \quad a_{22}(0, \lambda)\right] = \left[0 \quad 1\right].$$

Furthermore, $A_x(\lambda)$ is real and symplectic, since

$$\overline{A_x(-\overline{\lambda})} = A_x(\lambda) \quad \text{and} \quad \det A_x(\lambda) = 1.$$

Finally, the inequalities in (11.81) are immediate from (11.79) and (11.80).

2. *(11.83) and (11.84) hold and* $|E_\pm(x, \lambda)| > 0$ *for every point* $\lambda \in \overline{\mathbb{C}_\pm}$.

The two cited formulas are obvious from the definitions; the rest follows from (1) of Theorem 4.70 and (4.123).

3. *If* $x > 0$ *and* $\mu > 0$, *then there exist a pair of positive constants* γ_1 *and* γ_2 *that depend on* x *and* r *such that*

$$\gamma_1(1 + |\mu|^r) \le |E_+(x, \mu)|^2 \le \gamma_2(1 + |\mu|^r) \quad \textit{for } \mu \in \mathbb{R}. \qquad (11.97)$$

Since

$$J_p(x) \sim \sqrt{\frac{2}{\pi x}} \cos\left(x - \frac{\pi}{4} - p\frac{\pi}{2}\right) \quad \text{as } x \uparrow \infty,$$

it is readily checked that

$$|E_+(x, \mu)|^2 \sim \frac{1}{\pi}\left(\frac{x\mu}{2}\right)^r \Gamma\left(\frac{1-r}{2}\right)^2$$
$$\times \left\{\cos^2(x\mu + r(\pi/4)) + x^{-2r} \sin^2(x\mu + r(\pi/4))\right\} \quad \text{as } \mu \uparrow \infty. \qquad (11.98)$$

Thus, if

$$a_x = \frac{1}{\pi}\Gamma\left(\frac{1-r}{2}\right)^2 \min\{1, x^{-2r}\} \quad \text{and} \quad b_x = \frac{1}{\pi}\Gamma\left(\frac{1-r}{2}\right)^2 \max\{1, x^{-2r}\},$$

then

$$\begin{aligned}\left(\frac{x\mu}{2}\right)^r a_x(1+O(1/\mu)) &\leq |E_+(x,\mu)|^2 \\ &\leq \left(\frac{x\mu}{2}\right)^r b_x(1+O(1/\mu))\end{aligned}$$

as $\mu \uparrow \infty$. This serves to establish (11.97), since $E_+(x,\lambda)$ is an entire function of λ with no real zeros.

4. $E_+(x, i\nu) \sim \left(\frac{x\nu}{2}\right)^{(1+r)/2} \Gamma\left(\frac{1-r}{2}\right)(1+x^{-r})\frac{e^{x\nu}}{\sqrt{2\pi x\nu}}$ *as* $\nu \uparrow \infty$.

This follows from formula (11.82), and the relations

$$J_p(ix\nu) = i^p I_p(x\nu) \quad \text{and} \quad I_p(x\nu) \sim \frac{e^{x\nu}}{\sqrt{2\pi\nu}} \quad \text{as } \nu \uparrow \infty. \tag{11.99}$$

5. *Verification of (2) and (3).*

Assertion (2) follows from (1) and Step 2. Next, since $A_x(\lambda)$ is a symplectic 2×2 mvf, $b_3^x = b_4^x$. Assertion (3) then follows from Step 4, which implies that $b_4^x = e_x$ for $x > 0$.

6. *Verification of (4).*

In view of (11.78) and (11.99),

$$a_{11}(x, i\nu) + a_{12}(x, i\nu) = \left(\frac{x\nu}{2}\right)^{(1-r)/2} \Gamma\left(\frac{1+r}{2}\right)\left\{I_{\frac{r-1}{2}}(x\nu) + x^r I_{\frac{1+r}{2}}(x\nu)\right\}.$$

Therefore, by (11.85),

$$\begin{aligned} c_x(i\nu) &= \left(\frac{x\nu}{2}\right)^{-r} \frac{\Gamma\left(\frac{1+r}{2}\right)\{I_{\frac{r-1}{2}}(x\nu) + x^r I_{\frac{1+r}{2}}(x\nu)\}}{\Gamma\left(\frac{1-r}{2}\right)\{x^{-r} I_{\frac{1-r}{2}}(x\nu) + I_{-\frac{1+r}{2}}(x\nu)\}} \\ &\sim \left(\frac{x\nu}{2}\right)^{-r} \frac{\Gamma\left(\frac{1+r}{2}\right)(1+x^r)}{\Gamma\left(\frac{1-r}{2}\right)(1+x^{-r})} \\ &= \left(\frac{\nu}{2}\right)^{-r} \frac{\Gamma\left(\frac{1+r}{2}\right)}{\Gamma\left(\frac{1-r}{2}\right)} \quad \text{as } x \uparrow \infty. \end{aligned}$$

Thus, (11.86) holds, since $r > -1$. The rest follows from the fact that $c_x(\lambda)$ is holomorphic on $\mathbb{R}$ and $(\Re c_x)(\mu) = |E_+(x,\mu)|^{-2}$ for $\mu \in \mathbb{R}$.

7. *Verification of (5)–(7).*

(5) follows from the bounds in (11.97); (6) follows from (4), (5) and theorem 10.9 in [ArD08b]. (6) insures that $A_x \in \mathcal{U}_{rR}(J_1)$, since $\mathcal{U}_{rR}(J_1) \subset \mathcal{U}_{\ell sR}(J_1)$ by Theorem 4.19. Thus, as (3) is also in force, Theorem 4.25 is applicable to justify (7).

8. *Verification of (8) and (9)*

Theorem 9.3 implies that

$$\mathcal{C}(e_{2x}; c_x) = \{c_g : g \in \mathcal{G}(g_x; 2x)\} \quad \text{for } 0 < x < \infty.$$

Thus, (8) follows from the fact that $\mathcal{C}(A_{x_2}) \subseteq \mathcal{C}(A_{x_1})$ if $0 \le x_1 \le x_2 < \infty$; and (9) follows from (8). Formula (11.88) is first obtained for $\lambda = i\nu$ with $\nu > 0$ by invoking asymptotic formulas much as in Step 4, except that now ν is fixed and $x \uparrow \infty$. Then, since c_∞ is holomorphic on $\mathbb{C}_+$ its values on the positive imaginary axis determine its values in $\mathbb{C}_+$.

9. *Verification of (10)*

Since c_∞ belongs to the Carathеódory class and meets the reality condition $c_\infty(\lambda) = \overline{c_\infty(-\overline{\lambda})}$, the function

$$\frac{c_\infty(\lambda)}{-i\lambda} = \frac{c_\infty(\lambda) + \overline{c_\infty(-\overline{\lambda})}}{-2i\lambda} = \kappa_r \left(\frac{i}{\lambda}\right)^{r+1}$$

admits a representation of the form

$$\kappa_r \left(\frac{i}{\lambda}\right)^{r+1} = \beta + \int_0^\infty \frac{1}{\mu - \lambda^2} d\tau(\mu) = G(\lambda^2)$$

for some nondecreasing function $\tau(\mu)$ on $\mathbb{R}_+$ that is subject to the constraint (2.68). Therefore, since

$$\beta = \lim_{\nu \uparrow \infty} G(i\nu) = 0,$$

$$\kappa_r \frac{i^{r+1}}{\lambda^{(r+1)/2}} = \int_0^\infty \frac{1}{\mu - \lambda} d\tau(\mu) = G(\lambda).$$

Thus, upon writing $\lambda = \mu e^{i\theta}$ with $\mu > 0$, it follows from the Stieltjes inversion formula, that τ is locally absolutely continuous on $\mathbb{R}_+$ and

$$\begin{aligned} \tau'(\mu) &= \lim_{\theta \downarrow 0} \frac{G(\mu e^{i\theta}) - \overline{G(\mu e^{i\theta})}}{2\pi i} \\ &= \lim_{\theta \downarrow 0} \frac{\kappa_r}{\pi \mu^{(r+1)/2}} \sin\{(\pi - \theta)(r+1)/2\} = \frac{\sin\{(\pi)(r+1)/2\}}{\mu^{(r+1)/2}} \end{aligned}$$

when $\mu > 0$. The final formula then follows from the identity

$$\Gamma((1-r)/2)\Gamma((1+r)/2) = \frac{\pi}{\sin \pi(r+1)/2}.$$

10. *Verification of (11)*

If $f \in \mathcal{B}(\mathfrak{E}_x)$, then it is entire of exponential type at most x by Theorems 4.72 and 4.43; and

$$\|f\|^2_{\mathcal{B}(\mathfrak{E}_x)} = \int_{-\infty}^{\infty} \frac{|f(\mu)|^2}{|E_+(x,\mu)|^2} d\mu < \infty. \tag{11.100}$$

Thus, in view of the bounds in (11.97), (11.94) holds.

Conversely, if f is an entire function of exponential type at most x and (11.94) holds, then (11.94) holds, i.e., $E_+(x,\cdot)^{-1}f \in L_2$ and f satisfies the Cartwright condition (4.89). Thus, by a theorem of Krein, $f \in \Pi \cap \mathcal{E}$ and consequently it admits a representation of the form

$$f(\lambda) = e^{i\lambda\delta_f} g(\lambda) \quad \text{in } \mathbb{C}_+ \quad \text{with } |\delta_f| \le x \text{ and } g \in \mathcal{N}_+.$$

Therefore, since $b_4^x = e_x$, it follows that

$$E_+(x,\lambda)^{-1} = e^{i\lambda x}\varphi_x(\lambda) \quad \text{where } \varphi_x \in \mathcal{N}_{\text{out}}$$

and hence that

$$E_+(x,\lambda)^{-1} f(\lambda) = e^{i\lambda x}\varphi_x(\lambda) e^{i\lambda\delta_f} g(\lambda)$$

belongs to the Smirnov class $\mathcal{N}_+$. Therefore, by the Smirnov maximum principle, $E_+(x,\cdot)^{-1}f \in H_2$. Similar considerations imply that $E_-(x,\cdot)^{-1}f \in H_2^\perp$ Therefore, $f \in \mathcal{B}(\mathfrak{E}_x)$.

11. *Verification of (12)–(14).*

Formula (11.95) follows from the characterizations of $\mathcal{B}(\mathfrak{E}_x)$ given in (11) and the Paley–Wiener theorem; the rest of (12) follows from the exhibited formulas for $A_x(\lambda)$ and $\mathfrak{E}_x(\lambda)$ and the inclusion $L_\infty^{2\times 2} \cap \mathcal{U}(J_1) \subseteq \mathcal{U}_{rsR}(J_1)$.

To verify (13), suppose first that $0 < r < 1$. Then

$$\int_{-\infty}^{\infty} |f(\mu)|^2 d\mu < \infty \Longrightarrow \int_{-\infty}^{\infty} \frac{|f(\mu)|^2}{|\mu|^r} d\mu < \infty.$$

Thus, in view of (11),

$$0 < r < 1 \Longrightarrow \mathcal{B}([e_x \quad e_{-x}]) \subseteq \mathcal{B}(\mathfrak{E}_x).$$

This inclusion is proper because the function $f(\lambda) = \sin\sqrt{\lambda}/\sqrt{\lambda}$ is an entire function of minimal exponential type that meets the condition (11.94). Therefore, $f \in \mathcal{B}(\mathfrak{E}_x)$ for every $x > 0$. However, $f \notin L_2(\mathbb{R})$.

On the other hand, if $-1 < r < 0$, then

$$\int_{-\infty}^{\infty} \frac{|f(\mu)|^2}{|\mu|^r} d\mu < \infty \Longrightarrow \int_{-\infty}^{\infty} |f(\mu)|^2 d\mu < \infty,$$

i.e.,

$$-1 < r < 0 \Longrightarrow \mathcal{B}(\mathfrak{E}_x) \subseteq \mathcal{B}([e_x \quad e_{-x}]). \tag{11.101}$$

Next, the formula

$$f(\lambda) = \int_{-\pi/2}^{\pi/2} (\cos t)^{a-2} e^{i\lambda t} dt = \frac{\pi \Gamma(a-1)}{2^{a-2}\Gamma(\frac{1}{2}a + \frac{1}{2}\lambda)\Gamma(\frac{1}{2}a - \frac{1}{2}\lambda)} \quad (a > 1) \tag{11.102}$$

(which is taken from p. 186 of Titchmarsh [Ti60], who credits S. Ramanujan) exhibits f as an entire function of exponential type $\pi/2$ in L_2 when

$$\int_{\pi/2}^{\pi/2} (\cos t)^{2(a-2)} dt < \infty,$$

i.e., when $a > 3/2$. However, since

$$\left\{\Gamma\left(\frac{1}{2}a + \frac{1}{2}\mu\right)\Gamma\left(\frac{1}{2}a - \frac{1}{2}\mu\right)\right\}^{-1} = O(|\mu|^{1-a}), \tag{11.103}$$

(11.94) fails for $-1 < r < 0$ when $a \leq (3-r)/2$. Therefore, the inclusion (11.101) is proper for such choices of r when $x = \pi/2$. Similarly, if

$$f_x(\lambda) = f\left(\frac{2x}{\pi}\lambda\right) \quad \text{for } x > 0,$$

then

$$f_x \in \mathcal{B}([e_x \quad e_{-x}]) \quad \text{but} \quad f_x \notin \mathcal{B}(\mathfrak{E}_x) \text{ for every } x > 0.$$

This completes the proof of (13).

Finally, (14) follows from (13) and Theorem 4.78. □

Remark 11.28 *The preceding example has a number of important implications:*

1. *If $r \in (-1, 0) \cup (0, 1)$ and $x > 0$, then the mvf A_x (resp., $A_x^{\sim}$) belongs to the class $\mathcal{E} \cap \mathcal{U}_{\ell sR}^{\circ}(J_1)$ (resp., $A_{rsR}(J_1)$) but does not belong to $L_{\infty}^{2\times 2}$.*

   This follows from the inclusion $L_{\infty}^{m\times m}(\mathbb{R}) \cap \mathcal{U}(J) \subseteq \mathcal{U}_{rsR}(J) \cap \mathcal{U}_{\ell sR}(J)$ and items (6) and (14) in the example.
2. *If $r \in (-1, 0) \cup (0, 1)$, then the function $g(t) = -|t|^{r+1}$ belongs to $\mathcal{G}_{\infty}^{1\times 1}(0)$ but does not admit an accelerant, nor does its restriction to any finite interval $[-a, a]$ admit an accelerant on this interval.*

   The function g is a helical function because it is a positive constant times the helical function g_{∞}. However, $g''(t) = -(r+1)r|t|^{r-1}$ does not belong to $L_1(\mathbb{R}_+)$. In fact if $-1 < r \leq 0$, then $g'' \notin L_1([0, a])$ for any $a > 0$.

3. *If* $-1 < r \leq -1/2$, *then* $\mathcal{C}(A_a) \cap H_\infty = \emptyset$ *for every* $a > 0$.

If $c \in \mathcal{C}(A_a) \cap H_\infty$, then $c = c_g$ for some $g \in \mathcal{G}_\infty^{1\times 1}(0)$ that coincides with g_∞ on $[0, 2a]$ and, in view of Theorem 9.27,

$$\int_0^{2a} |g'(t)|^2 dt < \infty.$$

However, this contradicts the fact that

$$\int_0^{2a} |g'_\infty(t)|^2 dt = \infty \quad \text{for every } a > 0 \text{ when } -1 < r \leq -1/2.$$

4. *If* $0 < r < 1$, $a > 0$ *and* $h^\circ(t) = -|t|^{r-1}$ *for* $|t| \leq a$, *then* $h^\circ \in \mathcal{A}_a^{1\times 1}(1)$ *but* $h^\circ \notin \mathring{\mathcal{A}}_a^{1\times 1}(1)$

Let $g^\circ(t) = -2|t|^{r+1}/(r^2 + r)$ for $|t| \leq a$. Then, as $g \in \mathcal{G}_a(0)$ and $h^\circ = -(g^\circ)''/2$, $h^\circ \in \mathcal{A}_a(1)$. However, since the HEP$(g^\circ; a)$ is only completely indeterminate and not strictly completely indeterminate, $h^\circ \notin \mathring{\mathcal{A}}_a^{1\times 1}$.

Connections with strings

The mass function $M(x)$ of the given differential system (11.76) is

$$M(x) = \int_0^x \begin{bmatrix} u^r & 0 \\ 0 & u^{-r} \end{bmatrix} du = \begin{bmatrix} \dfrac{x^{1+r}}{1+r} & 0 \\ 0 & \dfrac{x^{1-r}}{1-r} \end{bmatrix} \quad \text{for } 0 \leq x < \infty. \tag{11.104}$$

Correspondingly,

$$\psi(x) = \operatorname{trace} M(x) = \frac{x^{1+r}}{1+r} + \frac{x^{1-r}}{1-r}$$

is a continuous strictly increasing function on $[0, \infty)$ that maps $[0, \infty)$ onto $[0, \infty)$ and is differentiable on $(0, \infty)$. Thus, there exists a continuous strictly increasing function φ on $[0, \infty)$ that maps $[0, \infty)$ onto $[0, \infty)$ and is differentiable on $(0, \infty)$ such that if

$$y = \psi(x) \quad \text{then} \quad x = \varphi(y).$$

Now, let

$$\widetilde{A}(y, \lambda) = A(\varphi(y), \lambda) \quad \text{for } 0 < y < \infty.$$

Then

$$\left(\frac{\partial \widetilde{A}}{\partial y}\right)(y, \lambda) = \varphi'(y)\left(\frac{\partial A}{\partial x}\right)(\varphi(y), \lambda)$$

$$= i\lambda \widetilde{A}(y, \lambda) \begin{bmatrix} h_1(y) & 0 \\ 0 & h_2(y) \end{bmatrix} J_1,$$

where

$$h_1(y) = \varphi(y)^r \varphi'(y) \quad \text{and} \quad h_2(y) = \varphi(y)^{-r} \varphi'(y)$$

and, as follows from the identity $y = \psi(\varphi(y))$,

$$\varphi'(y) = \frac{1}{f'(\varphi(y))} = \frac{1}{\varphi(y)^r + \varphi(y)^{-r}}.$$

Thus,

$$x(a) = \int_0^a h_1(y)dy = \frac{\varphi(a)^{1+r}}{1+r},$$

$$m(x) = \int_0^a h_2(y)dy = \frac{\varphi(a)^{1-r}}{1-r}$$

and hence

$$m(x) = \frac{\{(1+r)x\}^{(1-r)/(1+r)}}{1-r} \quad \text{for } 0 \le x < \infty$$

is the distribution of mass of the string that corresponds to the canonical differential system (11.76) and $c_\infty(\lambda)$ is (in Krein's terminology) the coefficient of dynamical influence of this singular string. Notice that

$$m(x) = k_t x^t \quad \text{for } 0 \le x < \infty,$$

where

$$t = \frac{1-r}{1+r}, \quad -1 < r < 1 \quad \text{and} \quad k_t = \frac{(1+r)^t}{1-r}$$

and hence t is any positive number. Thus, the spectral function $\tau(\mu)$ for the generalized Fourier transform $\mathcal{F}_{st}$ for the string with mass function $m(x) = \kappa_t x^t$, $t > 0$, is locally absolutely continuous with density $\tau'(\mu)$ given by (11.91) with $r = (1-t)/(1+t)$.

11.7 Supplementary notes

Most of the first four sections are adapted from [ArD03a] and [ArD05a].

The cleanest derivation of the Krein–Sobolev (alias Bellman–Krein–Siegert–Sobolev) identity is probably [GK85], which references the papers of Bellman and Siegert in its bibliography. Another useful reference is [KrL85]. Appendix A of [ArD02b] gives a complete proof of this identity (and variations thereof) for rational input impedances.

An inverse spectral problem for a more general class of systems than those discussed in Section 11.4 has been treated by M. Lesch and M.M. Malamud in [LM00] from a different point of view. Their procedure is based on a Gelfand–Levitan type of equation. Related developments are in [Mal97] and [Mal99]. Other variations of the Gelfand–Levitan procedure for solving inverse spectral problems are discussed in [DK78a], [DK78b], [Wi95] and [Bo10]

Section 11.5 is adapted from [ArD04b]; an abbreviated version of the example in Section 11.6 was presented in [ArD12a].

Another approach for solving the inverse spectral problem is presented in section 5.5 of [ArD04b].

The results on direct and inverse input impedance and spectral problems that are presented in Chapter 7 and this chapter are connected with a number of results of L.A. Sakhnovich; see the monograph [Sak99] and the references cited therein. In Sakhnovich's formulation of the inverse spectral problem, the given data are a spectral function σ; a Volterra operator K acting from a Hilbert space X into itself; a maximal chain X_t, $0 \leq t \leq d$, of subspaces of X that are invariant under K; and a bounded linear operator F_2 from X into $\mathbb{C}^m$. Sakhnovich also assumes that the formula

$$S = \int_{-\infty}^{\infty} (I - \mu K^*)^{-1} F_2^* d\sigma(\mu) F_2 (I - \mu K)^{-1}, \tag{11.105}$$

defines a bounded linear strictly positive operator from X into X. Moreover, if

$$F_1^{(\alpha)} = i\alpha F_2 - \frac{1}{\pi i}\int_{-\infty}^{\infty} d\sigma(\mu) F_2 \left\{K(I - \mu K)^{-1} + \frac{\mu}{1+\mu^2} I\right\}, \tag{11.106}$$

$$\alpha \in \mathbb{C}^{p\times p}, \quad \alpha = \alpha^* \quad \text{and} \quad F^{(\alpha)} = \begin{bmatrix} F_1^{(\alpha)} \\ F_2 \end{bmatrix}, \tag{11.107}$$

then

$$KS - SK^* = i(F^{(\alpha)})^* J_p F^{(\alpha)}. \tag{11.108}$$

Sakhnovich also showed that if P_t denotes the orthogonal projection from X onto X_t,

$$\mathcal{F}_t^{(\alpha)} = F^{(\alpha)}|_{X_t} \quad \text{and} \quad S_t = P_t S|_{X_t} \quad \text{for } 0 \leq t \leq d,$$

then, under appropriate assumptions,

$$M^{(\alpha)}(t) = \mathcal{F}_t^{(\alpha)} S_t^{-1} (\mathcal{F}_t^{(\alpha)})^* \quad \text{for } 0 \leq t \leq d \tag{11.109}$$

defines a continuous nondecreasing mvf on $[0, d]$ such that $\sigma \in \Sigma_{\rm sf}(dM^{(\alpha)})$. The mvf's $M^{(\alpha)}$ are connected by formula (11.74). Moreover, for each fixed choice of α, the set $\Sigma_{sf}(dM^{(\alpha)})$ coincides with the set of spectral functions $\widetilde{\sigma}$ for which

$$S = \int_{-\infty}^{\infty} (I - \mu K^*)^{-1} F_2^* d\widetilde{\sigma}(\mu) F_2 (I - \mu K)^{-1}, \tag{11.110}$$

i.e., for which, in Sakhnovich's terminology, $\widetilde{\sigma}$ is a solution of the generalized moment problem defined by K, F_2, S and the chain X_t, $0 \leq t \leq d$.

The identity (11.108) guarantees that the colligation $(K, F^{(\alpha)}; X, \mathbb{C}^m; J_p, S)$ is an S-node in the sense of L.A. Sakhnovich. Since $S^{\pm 1} \in \mathcal{B}(X)$, by assumption, and $S > 0$, the colligation $\mathfrak{N} = (S^{-1/2}KS^{1/2}, F^{(\alpha)}S^{-1/2}; X, \mathbb{C}^m; J)$ is an LB J-node with characteristic mvf $A_{\mathfrak{N}}(\lambda) = I + i\lambda F^{(\alpha)}S^{-1}(I - \lambda K)^{-1}(F^{(\alpha)})^*$. Formula (11.14) coincides with (11.109) with $S_t = \Delta_t, X_t = \mathcal{H}(b_3^t, b_4^t)$ for $0 \le t \le d, X = \mathcal{H}(b_3^d, b_4^d)$ and $K = R_0$. For additional discussion and comparisons, see Section 5.5 of [ArD04b].

Inverse input impedance problems for regular strings and regular canonical differential systems with dissipative boundary conditions at the right end point were investigated by the Darlington method in [Ar71], [Ar73], [Ar75] and [Ar01]; therein a characterization of these impedances and a procedure for recovering the monodromy matrix of the system and the boundary conditions was presented.

In the limit point case, input impedance matrices correspond to Weyl–Titchmarsh functions. Many authors relate the solution of the inverse spectral problem to an inverse problem for a related Weyl–Titchmarsh function: see, e.g., [AG95], [AG01], [Br68a], [DI82], [DMc72], [CG01], [GKM02], [GKS98c], [HS81], [LM00], [LS75], [LS91], [Mal97], [Sak99], [Wi95] and the references cited therein. The notion of associated pairs does not appear explicitly in any of these references. Nevertheless, the restrictions imposed on the structure of the systems under consideration force the associated pairs to be of a certain form, even if they are not mentioned. We shall elaborate on this theme in Chapter 12.

Explicit formulas for the solution $M(t)$ of the inverse impedance problem when the input impedance matrix is rational are derived in [ArD02b] by state space methods. The use of such methods to study inverse problems with rational scattering matrices or input impedance matrices seems to have been initiated by D. Alpay and I. Gohberg; see [AG95], [AG96], [AG97], [AG98] and [AG01] and then investigated further by I. Gohberg, M. A. Kaashoek and A.L. Sakhnovich in [GKS98a], [GKS98b], [GKS98c] and [GKS02]; see also [AGKLS10] for later developments. There is also some overlap with the formulas in [AKvdM00].

In view of the discussion in Section 7.11, an odd function σ with $\sigma(\mu) = a\mu^{1-r}$ for $\mu \ge 0$, $a > 0$ and $-1 < r < 1$ is the spectral function of at most one 2×2 differential system with normalized diagonal Hamiltonian corresponding to an S_1-string with a heavy right end point. Since the example at the end of Section 11.6 is of this form, Remark 8.77 guarantees that the solution of the inverse spectral problem for each such σ is a constant multiple of the Hamiltonian in formula (11.75) up to an appropriate change of the independent variable. Remark 8.77 may also be used to solve the inverse spectral problem for 2×2 systems with spectral functions that are constant multiples of the spectral functions considered in the example at the end of Section 8.13.

12

Direct and inverse problems for Dirac–Krein systems

In this chapter we will study direct and inverse problems for DK-systems, i.e., for differential systems of the form

$$u'(t,\lambda) = i\lambda u(t,\lambda)N_{\alpha,\beta}J + u(t,\lambda)\mathcal{V}(t), \quad 0 \le x < d, \tag{12.1}$$

in which J is an $m \times m$ signature matrix that is unitarily equivalent to j_{pq},

$$N_{\alpha,\beta} = \alpha P_+ + \beta P_-, \quad P_\pm = \frac{1}{2}(I_m \pm J), \tag{12.2}$$

$$\alpha \ge 0, \quad \beta \ge 0, \quad \kappa \overset{\text{def}}{=} \alpha + \beta > 0,$$

$$\mathcal{V} \in L_{1,\text{loc}}^{m\times m}([0,d)) \quad \text{and} \quad \mathcal{V}(t)^*J + J\mathcal{V}(t) = 0 \quad \text{a.e. on } [0,d). \tag{12.3}$$

Moreover, in view of Lemma 2.23, we can also assume without loss of generality that

$$\mathcal{V}(t)^* = \mathcal{V}(t) \text{ a.e. on } [0,d). \tag{12.4}$$

We shall exploit the connection between the DK-system (12.1) and the cannonical differential system

$$y'(t,\lambda) = i\lambda y(t,\lambda)H(t)J, \quad 0 \le t < d, \tag{12.5}$$

with Hamiltonian

$$H(t) = U(t)N_{\alpha,\beta}U(t)^* \quad \text{a.e. on the interval } [0,d), \tag{12.6}$$

where $U(t)$ is the unique locally absolutely continuous solution of the Cauchy problem

$$\begin{aligned} U'(t) &= U(t)\mathcal{V}(t) \quad \text{for } 0 \le x < d \\ U(0) &= I_m. \end{aligned} \tag{12.7}$$

The matrizants $U_t(\lambda)$ and $Y_t(\lambda)$ of the systems (12.1) and (12.5) are connected by the relation

$$U_t(\lambda) = Y_t(\lambda)U(t) \quad \text{for } 0 \le t < d. \tag{12.8}$$

The identification $U(t) = U_t(0)$ guarantees that $U(t) \in \mathcal{U}_{\text{const}}(J)$ for each point $t \in [0, d)$.

The system (12.4) is said to be a **regular DK-system** if

$$d < \infty \quad \text{and} \quad \mathcal{V} \in L_1^{m\times m}([0, d]). \tag{12.9}$$

In this case, (12.8) is in force for $t = d$ too. Thus, the monodromy matrices $U_d(\lambda)$ and $Y_d(\lambda)$ of the two systems are connected by the formula

$$U_d(\lambda) = Y_d(\lambda)U_d(0). \tag{12.10}$$

Unique solutions for a number of inverse problems for DK-systems will be obtained in roughly three steps:

(1) Establishing a one-to-one correspondence between systems of the form (12.1) and a class of systems of the form (12.5) with $H(t)$ of the form (12.6).
(2) Showing that the matrizants of systems of the form (12.5) with $H(t)$ of the form (12.6) meet the constraints

$$Y_t \in \mathcal{E} \cap \mathcal{U}_{rsR}^{\circ}(J) \quad \text{and} \quad \{e_{\alpha t}I_p, e_{\beta t}I_q\} \in ap_I(Y_t) \quad \text{for } 0 \le t < d \tag{12.11}$$

(and $\{e_{\alpha t}I_p, e_{\beta t}I_p\} \in ap_{II}(Y_t)$ for $0 \le t < d$ if $q = p$).
(3) Invoking the theorems in previous chapters that guarantee a unique solution of the inverse problem for canonical systems of differential equations with matrizants in the class $\mathcal{U}_{rsR}^{\circ}(J)$ and a given normalized nondecreasing continuous chain of associated pairs (of the first or second kind).

The correspondence referred to in (1) depends heavily upon a factorization principle that will be established in Section 12.1. The properties of the matrizant of the corresponding canonical differential system (12.5) are explored in Section 12.2. Subsequent sections will be devoted to direct and inverse input scattering, input impedance, spectral and monodromy problems and asymptotic scattering problems for DK-systems. It is convenient to let $J = j_{pq}$ for scattering problems and to let $J = J_p$ for impedance and spectral problems. In keeping with previous usage, we shall let

$W_t(\lambda)$ (resp., $A_t(\lambda)$), $0 \le t < d$, denote the matrizant of the DK-system (12.1) when $J = j_{pq}$ (resp., $J = J_p$)

and

$\widetilde{W}_t(\lambda)$ (resp., $\widetilde{A}_t(\lambda)$) denote the matrizant of
the canonical system (12.5) when $J = j_{pq}$ (resp., $J = J_p$).

Correspondingly, we set

$$B_t(\lambda) = A_t(\lambda)\mathfrak{V} \quad \text{and} \quad \widetilde{B}_t(\lambda) = \widetilde{A}_t(\lambda)\mathfrak{V}.$$

If $J = j_{pq}$, then

$$W_t^0 = \begin{bmatrix} e_{\alpha t} I_p & 0 \\ 0 & e_{-\beta t} I_q \end{bmatrix} \tag{12.12}$$

is the matrizant of the system (12.1) with $\mathcal{V}(t) = 0$, and, as will be shown in Section 12.2, the mvf

$$\mathfrak{A}_r^t(\lambda) = W_t^0(\lambda)^{-1} W_t(\lambda) \tag{12.13}$$

belongs to the class $\mathcal{W}_r(j_{pq})$ (that was defined in Section 9.6) for every $t \in [0, d)$. Analogous results will also be established for the mvf

$$\mathfrak{A}_\ell^t(\lambda) = W_t(\lambda) W_t^0(\lambda)^{-1}. \tag{12.14}$$

Most of the last few sections will be devoted to an asymptotic scattering problem for the system (12.1) with $d = \infty$ and $\mathcal{V} \in L_1^{m\times m}(\mathbb{R}_+)$.

The asymptotic scattering matrix $s_\varepsilon^\infty(\mu)$ will be defined in Section 12.10 for the boundary problem for the DK-system (12.1) with $J = j_{pq}$ and boundary condition

$$u_1(0,\lambda) = u_2(0,\lambda)\varepsilon(\lambda) \quad \text{with } \varepsilon \in \mathcal{S}^{q\times p} \tag{12.15}$$

for solutions

$$u(t,\lambda) = \begin{bmatrix} u_1(t,\lambda) & u_2(t,\lambda) \end{bmatrix}$$

of the system with blocks $u_1(t,\lambda)$ and $u_2(t,\lambda)$ of sizes $q \times p$ and $q \times q$, respectively. The inverse problem in this setting is to recover $\mathcal{V}$ and ε from s_ε^∞.

12.1 Factoring Hamiltonians corresponding to DK-systems

Let $\mathfrak{U}(J; d)$ denote the class of mvf's $U(t)$ that are locally absolutely continuous solutions of the Cauchy problem (12.7) for some mvf $\mathcal{V}(t)$ that meets the constraints in (12.3). Then $\mathfrak{U}(J; d)$ may be characterized as the class of locally absolutely continuous J-unitary valued mvf's $U(t)$ on $[0, d)$ with $U(0) = I_m$. If $U \in \mathfrak{U}(J; d)$ and $U^{-1}U' = \mathcal{V}$, then $\mathcal{V}(t)$ meets the constraint (12.4) if and only if

$$U'(t)U(t)^* = U(t)U'(t)^* \quad \text{a.e. on } [0, d). \tag{12.16}$$

Our next objective is to study Hamiltonians $H(t)$ that admit factorizations of the form (12.5) with $U \in \mathfrak{U}(J; d)$.

Theorem 12.1 *Let $J \neq \pm I_m$. An $m \times m$ mvf $H(t)$ on $[0, d)$ admits the factorization*

$$H(t) = U(t)U(t)^* \quad \text{for some mvf } U \in \mathfrak{U}(J; d) \tag{12.17}$$

if and only if

$$H(t) > 0 \quad \text{for every } t \in [0, d) \text{ and } H \in \mathfrak{U}(J; d). \tag{12.18}$$

If the conditions in (12.18) hold, then $U(t) = H(t)^{1/2}$ is a solution of the factorization problem (12.17) and in the set of the solutions of this problem there is one and only one solution $U(t) \in \mathfrak{U}(J; d)$ that also satisfies the extra condition (12.16).

Proof The implication (12.17) $\Longrightarrow$ (12.18) is obvious. To prove the converse, it suffices to consider the case $J = j_{pq}$. The first objective is to show that if a j_{pq}-unitary matrix H is positive definite, then $H^{1/2}$, the positive definite square root of H, is also a j_{pq}-unitary matrix. The verification of this assertion and the rest of the proof is broken into steps.

1. *Motivation for the next step.*

If the positive definite matrix $H^{1/2}$ is j_{pq}-unitary, then in view of Lemma 3.46,

$$H^{1/2} = \begin{bmatrix} (I_p - KK^*)^{-1/2} & K(I_q - K^*K)^{-1/2} \\ K^*(I_p - KK^*)^{-1/2} & (I_q - K^*K)^{-1/2} \end{bmatrix} \tag{12.19}$$

for some strictly contractive matrix $K \in \mathbb{C}^{p \times q}$. Thus, if $H = H^{1/2}H^{1/2}$ is written in block form as

$$H = \begin{bmatrix} H_{11} & H_{12} \\ H_{21} & H_{22} \end{bmatrix}$$

with blocks H_{11} and H_{22} of sizes $p \times p$ and $q \times q$, respectively, it is readily checked that

$$(I_p - KK^*) = 2(I_p + H_{11})^{-1}, \quad (I_q - K^*K) = 2(I_q + H_{22})^{-1}$$

and

$$K = (H_{11} + I_p)^{-1}H_{12}. \tag{12.20}$$

2. *If H is a positive definite j_{pq}-unitary matrix, then the matrix K that is defined in terms of the blocks of H by formula (12.20) is strictly contractive and the following identities hold:*

$$I_p - KK^* = 2(I_p + H_{11})^{-1}, \tag{12.21}$$

$$H_{21}(I_p + H_{11})^{-1} = (I_q + H_{22})^{-1} \quad \text{and} \quad I_q - K^*K = 2(I_q + H_{22})^{-1}. \tag{12.22}$$

All three of the stated identities follow readily from the fact that $H > 0$ and $Hj_{pq}H = j_{pq}$, especially if carried out in the indicated order. Since $H_{11} > 0$, the identity (12.21) guarantees that K is strictly contractive.

3. *If H is a positive definite j_{pq}-unitary matrix, U is set equal to the right-hand side of (12.19) and K is defined in terms of the blocks of H by (12.20), then*

$$UU^* = U^2 = H.$$

This is a straightforward calculation based on the identities (12.21) and (12.22).

4. *The mvf $U(t) = H(t)^{1/2}$ is absolutely continuous on every closed subinterval $[0, a]$ of the interval $[0, d)$.*

Let $\lambda_{\min}(t)$ and $\lambda_{\max}(t)$ denote the minimal and maximal eigenvalues of $H(t)$ on the interval $[0, a]$, i.e.,

$$\lambda_{\max}(t) = \|H(t)\| \quad \text{and} \quad \lambda_{\min}(t) = \|H(t)^{-1}\|^{-1} \quad \text{for } t \in [0, a].$$

Since these two functions are continuous on $[0, a]$, there exists a closed interval $[\gamma, \delta]$ such that

$$0 < \gamma \leq \lambda_{\min}(t) \leq \lambda_{\max}(t) \leq \delta \quad \text{for } t \in [0, a].$$

Then, by the Riesz integral representation formula,

$$H(t)^{1/2} = \frac{1}{2\pi i} \oint_{\Gamma} \sqrt{\lambda}(\lambda I - H(t))^{-1} d\lambda,$$

where Γ is a circle in the open right half plane that contains the interval $[\gamma, \delta]$ and $\sqrt{\lambda} > 0$ for $\lambda > 0$. Thus, $U(t) = H(t)^{1/2}$ is an absolutely continuous mvf on the interval $[0, a]$.

5. *Verification of the last statement of the theorem.*

The proof of the existence of a solution that meets the condition (12.16) is established by setting

$$\widetilde{U}(t) = U(t)F(t),$$

where $U(t) \in \mathfrak{U}(J; d)$ and $F(t)$ is the solution of the Cauchy problem

$$F'(t) = \frac{1}{2}\{U'(t)^* JU(t)J - JU^*(t)JU'(t)\}F(t), \qquad 0 \leq t < d,$$

$$F(0) = I_m.$$

Then, as

$$(F^*KF)'(t) = 0 \quad \text{for every } K \in \mathbb{C}^{m\times m},$$

it is readily checked that: (1) $F(t) \in \mathfrak{U}(J; d)$, (2) $F(t)$ is unitary for every $t \in [0, d)$, (3) $\widetilde{U}(t) \in \mathfrak{U}(J; d)$ and (4) $\widetilde{U}(t)$ meets the extra condition (12.16). Thus, it is left only to show that $H(t)$ admits only one factorization of the form (12.17) when $U(t) \in \mathfrak{U}(J; d)$ meets the extra condition (12.16). But in this instance,

$$H'(t) = 2U'(t)U(t)^* = 2U'(t)JU(t)^{-1}J.$$

Consequently, the components $U(t)P_+$ and $U(t)P_-$ of $U(t) = U(t)P_+ + U(t)P_-$ are uniquely determined as the solutions of the systems of equations

$$U'(t)P_+ = \frac{1}{2}H'(t)J(U(t)P_+) \quad \text{and} \quad U'(t)P_- = -\frac{1}{2}H'(t)J(U(t)P_-),$$
$$U(0)P_+ = P_+ \quad \text{and} \quad U(0)P_- = P_-,$$

respectively. □

Corollary 12.2 *Let $H(t)$ be an $m \times m$ mvf that meets the condition (12.18) and let $U \in \mathfrak{U}(J; d)$ be the unique solution of the factorization of $H(t)$ in (12.17) that satisfies the extra condition (12.16) and suppose further that $J = \pm\overline{J}$. Then*

$$H(t) = \overline{H(t)} \quad \text{for every} \quad t \in [0, d) \Longleftrightarrow U(t) = \overline{U(t)} \quad \text{for every} \quad t \in [0, d).$$

Proof This is an immediate consequence of Theorem 12.1 and the observation that if $J = \pm\overline{J}$, then $U \in \mathfrak{U}(J; d)$ if and only if $\overline{U} \in \mathfrak{U}(J; d)$. □

The next result serves to characterize the Hamiltonians that correspond to $N_{\alpha,\beta}$.

Theorem 12.3 *Let $J \neq \pm I_m$, let $\alpha \geq 0$, $\beta \geq 0$ and $\kappa = \alpha + \beta > 0$. An $m \times m$ mvf $H(t)$ on $[0, d)$ admits the factorization*

$$H(t) = U(t)N_{\alpha,\beta}U(t)^* \quad \text{for some mvf } U(t) \in \mathfrak{U}(J; d) \tag{12.23}$$

if and only if

$$\frac{2H(t) - (\alpha - \beta)J}{\alpha + \beta} \in \mathfrak{U}(J; d) \text{ and } 2H(t) - (\alpha - \beta)J > 0 \text{ for every } t \in [0, d), \tag{12.24}$$

or, equivalently, if and only if the following four conditions are met:

(1) *$H(t)$ is absolutely continuous on every closed subinterval of $[0, d)$.*
(2) *$H(t)JH(t) = (\alpha - \beta)H(t) + \alpha\beta J$ for every $t \in [0, d)$.*
(3) *$2H(t) - (\alpha - \beta)J \geq 0$ for every $t \in [0, d)$.*
(4) *$H(0) = N_{\alpha,\beta}$.*

If the conditions (1)–(4) are in force, then there is one and only one solution $U(t) \in \mathfrak{U}(J; d)$ of the factorization problem (12.23) that also satisfies the extra condition (12.16).

Proof The proof that (12.23) implies (1)–(4) is self-evident. The verification of the form of H in the converse assertion rests on (3) and the observation that the identity in (2) implies that

$$\left\{H(t) - \frac{(\alpha-\beta)}{2}J\right\} J \left\{H(t) - \frac{(\alpha-\beta)}{2}J\right\} = \frac{(\alpha+\beta)^2}{4}J, \tag{12.25}$$

and hence, Theorem 12.1 implies that the mvf

$$\frac{2H(t) - (\alpha-\beta)J}{\alpha+\beta} = U(t)U(t)^*$$

for some mvf $U \in \mathfrak{U}(J; d)$. But this in turn implies that

$$2H(t) = (\alpha+\beta)U(t)U(t)^* + (\alpha-\beta)J = (\alpha+\beta)U(t)U(t)^* + (\alpha-\beta)U(t)JU(t)^*,$$

which leads easily to the factorization (12.23). □

Corollary 12.4 *An $m \times m$ mvf $H(t)$ on $[0, d)$ admits the factorization*

$$H(t) = \alpha U(t)P_+U(t)^* \quad with\ \alpha > 0 \quad and \quad U \in \mathfrak{U}(J; d) \tag{12.26}$$

if and only if:

(1) *$H(t)$ is absolutely continuous on every closed subinterval of* $[0, d)$.
(2) $H(t)JH(t) = \alpha H(t)$ *for every* $t \in [0, d)$.
(3) *$H(t)$ is positive semidefinite for every* $t \in [0, d)$.
(4) $\operatorname{rank} H(t) = \operatorname{rank} P_+$ *for every* $t \in [0, d)$.
(5) $H(0) = \alpha P_+$.

If the conditions (1)–(5) are in force, then there is one and only one solution $U(t) \in \mathfrak{U}(J; d)$ of the factorization problem (12.26) that also satisfies the extra condition (12.16).

Proof It is readily checked that formula (12.26) implies the validity of (1)–(5). Conversely, if (1)–(5) are in force, then the $\mathcal{X}(t) = \operatorname{range} H(t)$ is a positive subspace of $\mathbb{C}^m$ with respect to the indefinite inner product induced by J: If $x = H(t)u$, then

$$\langle Jx, x\rangle = \langle JH(t)u, H(t)u\rangle = \alpha\langle H(t)u, u\rangle \geq 0,$$

with equality if and only if $x = H(t)u = 0$. Now let $\mathcal{Y}(t)$ be any q dimensional subspace of $\mathbb{C}^m$ that is negative with respect to the inner product induced by J. Then, since $\dim \mathcal{X} = p$, $\mathbb{C}^m = \mathcal{X}(t)\dot{+}\mathcal{Y}(t)$. Moreover, since $(H - \alpha J)JH = 0$,

$$\begin{aligned} \langle (H-\alpha J)J(x+y), J(x+y)\rangle &= \langle (H-\alpha J)Jy, Jy\rangle \\ &= \langle HJy, Jy\rangle - \alpha\langle Jy, y\rangle \\ &\geq 0. \end{aligned}$$

Thus, (1)–(5) of the corollary imply (1)–(4) of Theorem 12.3 in the case $\beta = 0$, which in turn yields the desired result. □

Corollary 12.5 *An $m \times m$ mvf $H(t)$ on $[0, d)$ admits the factorization*

$$H(t) = \beta U(t)P_{-}U(t)^* \text{ with } \beta > 0 \text{ and } U \in \mathfrak{U}(J; d) \tag{12.27}$$

if and only if

(1) *$H(t)$ is absolutely continuous on every closed subinterval of $[0, d)$.*
(2) *$H(t)JH(t) = -\beta H(t)$ for every $t \in [0, d)$.*
(3) *$H(t)$ is positive semidefinite for every $t \in [0, d)$.*
(4) rank $H(t) =$ rank P_- *for every $t \in [0, d)$.*
(5) *$H(0) = \beta P_-$.*

If the conditions (1)–(5) are in force, then there is one and only one solution $U \in \mathfrak{U}(J; d)$ of the factorization problem (12.27) that also satisfies the extra condition (12.16).

Proof The proof is analogous to the proof of Corollary 12.4. □

12.2 Matrizants of canonical differential systems corresponding to DK-systems

In this section we shall show that if U_t is the matrizant of a DK-system (12.1) with J unitarily equivalent to j_{pq} and a potential $\mathcal{V}$ that is subject to (12.3)–(12.4), then

$$U_t \in \mathcal{E} \cap \mathcal{U}_{rsR}^{\circ}(J) \quad \text{and} \quad \{e_{\alpha t}I_p, e_{\beta t}I_q\} \in ap_I(U_t) \quad \text{for } 0 \le t < d. \tag{12.28}$$

In view of (12.8), this means that the matrizant Y_t of the corresponding canonical differential system (12.5) meets the conditions in (12.11).

Theorem 12.6 *If $U_t(\lambda) = U(t, \lambda)$, $0 \le t < d$, is the matrizant of a DK-system (12.1) with a potential $\mathcal{V}(t)$ that meets the conditions (12.3)–(12.4) and with $\alpha \ge 0$, $\beta \ge 0$ and $\kappa = \alpha + \beta > 0$, then:*

(1) *The matrizant admits the integral representations*

$$U(t, \lambda) = e^{i\lambda t N_{\alpha,\beta} J} \left\{ I_m + \int_0^t e^{-i\lambda s \kappa J} \omega_{[r]}(t, s) ds \right\} \tag{12.29}$$

and

$$U(t, \lambda) = \left\{ I_m + \int_0^t \omega_{[\ell]}(t, s) e^{-i\lambda s \kappa J} ds \right\} e^{i\lambda t N_{\alpha,\beta} J}, \tag{12.30}$$

where $\omega_{[r]}(t, s)$ and $\omega_{[\ell]}(t, s)$ are $m \times m$ mvf's that are defined in the triangular domain

$$\{(t, s) : 0 \le s \le t, \ 0 \le t < d\}$$

such that $\omega_{[r]}(t, \cdot)$, $\omega_{[\ell]}(t, \cdot) \in L_1^{m \times m}([0, t])$ for every $t \in [0, d)$.

(2) $\{e_{\alpha t}I_p, e_{\beta t}I_q\} \in ap_I(U_t)$ *for every* $t \in [0, d)$ *and, if* $q = p$, *then* $\{e_{\alpha t}I_p, e_{\beta t}I_p\} \in ap_{II}(U_t)$ *for every* $t \in [0, d)$.

(3) $U_t \in \mathcal{E} \cap \mathcal{U}_{rsR}(J)$ *for every* $t \in [0, d)$.

(4) *If (12.1) is a regular DK-system, then assertions (1)–(3) are also in force for* $t = d$.

Proof The proof is divided into steps.

1. *Verify the identity*

$$\mathcal{V}(t)N_{\alpha,\beta}J + N_{\beta,\alpha}J\mathcal{V}(t) = 0 \quad \text{a.e. in} \quad [0, d). \tag{12.31}$$

In view of (12.3) and (12.4)

$$\mathcal{V}(t)J \pm \mathcal{V}(t) = -J\mathcal{V}(t) \pm \mathcal{V}(t) \quad \text{a.e. in} \quad [0, d),$$

and hence,

$$\mathcal{V}(t)P_+ = P_-\mathcal{V}(t) \quad \text{and} \quad \mathcal{V}(t)P_- = P_+\mathcal{V}(t) \quad \text{a.e. in} \quad [0, d).$$

Thus,

$$\begin{aligned}\mathcal{V}(t)N_{\alpha,\beta}J &= \mathcal{V}(t)(\alpha P_+ - \beta P_-)\\ &= (\alpha P_- - \beta P_+)\mathcal{V}(t) = -N_{\beta,\alpha}J\mathcal{V}(t) = 0 \quad \text{a.e. in} \quad [0, d),\end{aligned}$$

as claimed.

2. *Verify formula (12.30).*

With the help of Step 1, it is readily checked that if

$$\Omega(t, \lambda) = U(t, \lambda)e^{-i\lambda t N_{\alpha,\beta}J}, \tag{12.32}$$

then

$$\begin{aligned}\Omega'(t, \lambda) &= \Omega(t, \lambda)e^{i\lambda t N_{\alpha,\beta}J}\mathcal{V}(t)e^{-i\lambda t N_{\alpha,\beta}J}\\ &= \Omega(t, \lambda)\mathcal{V}(t)e^{-i\lambda t(N_{\beta,\alpha}+N_{\alpha,\beta})J}\end{aligned}$$

and hence that the mvf $\Omega(t, \lambda)$ is a solution of the equation

$$\Omega'(t, \lambda) = \Omega(t, \lambda)\mathcal{V}(t)e^{-i\lambda t\kappa J} \quad \text{a.e. in} \quad [0, d).$$

Thus, as $\Omega(0, \lambda) = I_m$,

$$\Omega(t, \lambda) = I_m + \int_0^t \Omega(s, \lambda)\mathcal{V}(s)e^{-i\lambda s\kappa J}ds \quad \text{for} \quad 0 \le t < d. \tag{12.33}$$

Proceeding recursively, let

$$\begin{aligned}\Omega_0(t, \lambda) &= I_m \quad \text{and}\\ \Omega_{j+1}(t, \lambda) &= \int_0^t \Omega_j(s, \lambda)\mathcal{V}(s)e^{-i\lambda s\kappa J}ds \quad \text{for} \quad j = 0, 1, \ldots.\end{aligned}$$

The standard estimate

$$\|\Omega_j(t,\lambda)\| \leq \frac{1}{j!}\left[\int_0^t \|\mathcal{V}(s)\| e^{|\Im \lambda| s\kappa} ds\right]^j, \; j \geq 0, \tag{12.34}$$

guarantees that the sum

$$\Omega(t,\lambda) = \sum_{j=0}^{\infty} \Omega_j(t,\lambda) \tag{12.35}$$

converges absolutely to a continuous solution of the integral equation (12.33) for each fixed point $\lambda \in \mathbb{C}$.

Suppose next that there exists an integer $j \geq 1$, such that

$$\Omega_j(t,\lambda) = \int_0^t \omega_j(t,s) e^{-i\lambda s\kappa J} ds \quad \text{for} \quad 0 \leq t < d,$$

for some $m \times m$ mvf $\omega_j(t,\cdot) \in L_1^{m\times m}([0,t])$ that is subject to the bound

$$\int_0^t \|\omega_j(t,s)\| ds \leq \frac{1}{j!}\left(\int_0^t \|\mathcal{V}(s)\| ds\right)^j \quad \text{for} \quad 0 \leq t < d. \tag{12.36}$$

Then

$$\begin{aligned}
\Omega_{j+1}(t,\lambda) &= \int_0^t \Omega_j(s,\lambda)\mathcal{V}(s) e^{-i\lambda s\kappa J} ds \\
&= \int_0^t \left\{\int_0^s \omega_j(s,u) e^{-i\lambda u\kappa J} du\right\} \mathcal{V}(s) e^{-i\lambda s\kappa J} ds \\
&= \int_0^t \left\{\int_0^s \omega_j(s,u)\mathcal{V}(s) e^{-i\lambda(s-u)\kappa J} du\right\} ds \\
&= \int_0^t \left\{\int_0^s \omega_j(s,s-v)\mathcal{V}(s) e^{-i\lambda v\kappa J} dv\right\} ds \\
&= \int_0^t \left\{\int_v^t \omega_j(s,s-v)\mathcal{V}(s) ds\right\} e^{-i\lambda v\kappa J} dv \\
&= \int_0^t \omega_{j+1}(t,v) e^{-i\lambda v\kappa J} dv,
\end{aligned}$$

where

$$\omega_{j+1}(t,s) = \int_s^t \omega_j(u,u-s)\mathcal{V}(u) du, \; 0 \leq s \leq t.$$

Thus, the bound (12.36) propagates and, as

$$\omega_1(t,s) = \mathcal{V}(s) \quad \text{for} \quad s \in [0,t]$$

meets the imposed conditions for each fixed $t \in [0, d)$, the sum

$$\omega_{[\ell]}(t, s) = \sum_{j=1}^{\infty} \omega_j(t, s) \tag{12.37}$$

converges a.e. in the interval $[0, t]$ to a mvf $\omega_{[\ell]}(t, \cdot) \in L_1^{m\times m}([0, t])$. Moreover,

$$\begin{aligned}\Omega(t, \lambda) &= I_m + \sum_{j=1}^{\infty} \Omega_j(t, \lambda) = I_m + \sum_{j=1}^{\infty} \int_0^t \omega_j(t, s)e^{-i\lambda s\kappa J}ds \\ &= I_m + \int_0^t \sum_{j=1}^{\infty} \omega_j(t, s)e^{-i\lambda s\kappa J}ds \\ &= I_m + \int_0^t \omega_{[\ell]}(t, s)e^{-i\lambda s\kappa J}ds .\end{aligned}$$

The order of integration and summation in the third equality can be interchanged, since

$$\int_0^t \|\omega_j(t, s)e^{-i\lambda s\kappa J}\|ds \leq \int_0^t \|\omega_j(t, s)\|ds e^{|\Im\lambda|t\kappa} \leq \frac{1}{j!}\left\{\int_0^t \|\mathcal{V}(t)\|ds\right\}^j e^{|\Im\lambda|t\kappa} .$$

Thus, the proof of Step 2 is complete.

3. *Obtain the representation (12.29).*

The asserted representation follows from (12.30) and the observation that

$$\{\omega_{2j}(t, s)e^{-i\lambda s\kappa J}\}e^{i\lambda tN_{\alpha,\beta}J} = e^{i\lambda tN_{\alpha,\beta}J}\{e^{-i\lambda s\kappa J}\omega_{2j}(t, s)\}$$

and

$$\begin{aligned}\{\omega_{2j-1}(t, s)e^{-i\lambda s\kappa J}\}e^{i\lambda tN_{\alpha,\beta}J} &= e^{-i\lambda tN_{\beta,\alpha}J}\{e^{i\lambda s\kappa J}\omega_{2j-1}(t, s)\} \\ &= e^{i\lambda tN_{\alpha,\beta}J}\{e^{-i\lambda(t-s)\kappa J}\omega_{2j-1}(t, s)\}\end{aligned}$$

a.e. in $[0, t]$. Thus,

$$\omega_{[r]}(t, s) = \sum_{j=1}^{\infty} \omega_{2j}(t, s) + \sum_{j=1}^{\infty} \omega_{2j-1}(t, (t-s)) . \tag{12.38}$$

4. *Verify the first assertion in (2).*

Without loss of generality we may assume that $J = j_{pq}$ and then, upon denoting the matrizant by $W_t(\lambda) = W(t, \lambda)$, formula (12.29) may be written in block form as

$$\begin{aligned}W(t, \lambda) &= \begin{bmatrix} w_{11}(t, \lambda) & w_{12}(t, \lambda) \\ w_{21}(t, \lambda) & w_{22}(t, \lambda) \end{bmatrix} \\ &= \begin{bmatrix} e^{i\lambda t\alpha}I_p & 0 \\ 0 & e^{-i\lambda t\beta}I_q \end{bmatrix}\begin{bmatrix} \mathfrak{a}_-(t, \lambda) & \mathfrak{b}_-(t, \lambda) \\ \mathfrak{b}_+(t, \lambda) & \mathfrak{a}_+(t, \lambda) \end{bmatrix},\end{aligned}$$

where $\mathfrak{a}_-(t,\cdot) \in \mathcal{W}_-^{p\times p}(I_p)$, $\mathfrak{b}_-(t,\cdot) \in \mathcal{W}_-^{p\times q}(0)$, $\mathfrak{b}_+(t,\cdot) \in \mathcal{W}_+^{q\times p}(0)$ and $\mathfrak{a}_+(t,\cdot) \in \mathcal{W}_+^{q\times q}(I_q)$. Moreover, since $W_t \in \mathcal{U}(j_{pq})$, it is readily checked that

$$w_{11}(t,\lambda)w_{11}(t,\lambda)^* \geq I_p \quad \text{for} \quad \Im\lambda \leq 0$$

and

$$w_{22}(t,\lambda)w_{22}(t,\lambda)^* \geq I_q \quad \text{for} \quad \Im\lambda \geq 0\,.$$

Therefore,

$$\mathfrak{a}_-(t,\cdot)^{-1} \in \mathcal{W}_-^{p\times p}(I_p) \quad \text{and} \quad \mathfrak{a}_+(t,\cdot)^{-1} \in \mathcal{W}_+^{q\times q}(I_q)\,.$$

Thus, the first part of assertion (2) follows from the representation formulas

$$w_{11}(t,\lambda) = e^{i\lambda t\alpha}\mathfrak{a}_-(t,\lambda) \quad \text{and} \quad w_{22}(t,\lambda) = e^{-i\lambda t\beta}\mathfrak{a}_+(t,\lambda).$$

5. *Prove that if $q = p$, then $\{e_{\alpha t}I_p, e_{\beta t}I_p\} \in ap_{II}(U_t)$ for every $t \in [0,d)$.*

This follows from Step 4 and Corollary 3.73.

6. *Verify assertion (3) and complete the proof of the theorem.*

The representation (12.29) guarantees that $U_t \in L_\infty^{m\times m}(\mathbb{R})$ for $0 \leq t < d$. Therefore (3) holds by Theorem 4.18. □

Remark 12.7 *The mvf's $\omega_{[\ell]}(t,s)$ and $\omega_{[r]}(t,s)$ in the representation formulas (12.30) and (12.29) depend only upon the potential $\mathcal{V}(t)$, and not upon α or β (or λ). In fact the former is a solution of the integral equation*

$$\omega_{[\ell]}(t,s) = \mathcal{V}(s) + \int_s^t \omega_{[\ell]}(u, u-s)\mathcal{V}(u)du \tag{12.39}$$

and the latter can be expressed in terms of the components $\omega_j(t,s)$ of $\omega_{[\ell]}(t,s)$ in the sum (12.37) by formula (12.38).

Remark 12.8 *In general the sets $ap_I(U_t)$ and $ap_{II}(U_t)$ depend upon the unitary matrices V_1 and V_2 that intervene in the formulas*

$$J = V_1^* j_{pq} V_1 \quad \text{and} \quad J = V_2^* J_p V_2 \ \text{if}\ q = p. \tag{12.40}$$

However, (3.142) guarantees that $ap_I(U_t)$ is independent of V_1, since the inner mvf's that come into play are of the form $b_1 = \beta_1 I_p$ and $b_2 = \beta_2 I_q$, where β_1 and β_2 are scalar inner functions. Corollary 3.73 then guarantees that $ap_{II}(U_t)$ is independent of V_2. It is also of interest to note that

$$V_1 P_+ V_1^* = \begin{bmatrix} I_p & 0 \\ 0 & 0 \end{bmatrix} \quad \text{and} \quad V_1 P_- V_1^* = \begin{bmatrix} 0 & 0 \\ 0 & I_q \end{bmatrix}.$$

Corollary 12.9 *If condition (12.9) is in force in the setting of Theorem 12.6, then the following conclusions hold for the monodromy matrix $U_d(\lambda)$:*

(1) $U_d(\lambda)$ *admits the integral representations*

$$U_d(\lambda) = e^{i\lambda d N_{\alpha,\beta} J}\left\{I_m + \int_0^d e^{-i\lambda s\kappa J}\omega_{[r]}(s)ds\right\} \tag{12.41}$$

and

$$U_d(\lambda) = \left\{I_m + \int_0^d \omega_{[\ell]}(s)e^{-i\lambda s\kappa J}ds\right\} e^{i\lambda t N_{\alpha,\beta} J}, \tag{12.42}$$

where $\omega_{[r]}(s)$ *and* $\omega_{[\ell]}(s)$ *are* $m \times m$ *mvf's that are summable on* $[0, d]$ *and depend only on* $\mathcal{V}(t)$ *and not on* $N_{\alpha,\beta}$.

(2) $\{e_{\alpha d}I_p, e_{\beta d}I_q\} \in ap_I(U_d)$ *and, if* $q = p$, *then* $\{e_{\alpha d}I_p, e_{\beta d}I_p\} \in ap_{II}(U_d)$.

(3) $U_d \in \mathcal{U}_{rsR}(J)$.

Theorem 12.10 *If the conditions (12.3) and (12.4) are in force, then, for every* $t \in [0, d)$, *the matrizant* $U_t(\lambda) = U(t, \lambda)$ *of the system (12.1) and the matrizant* $Y_t(\lambda) = Y(t, \lambda)$ *of the corresponding system (12.5) are subject to the bounds*

$$\|U_t(\mu)^{\pm 1}\| \leq \exp\left\{\int_0^t \|\mathcal{V}(s)\|ds\right\} \quad \text{and} \quad \|Y_t(\mu)^{\pm 1}\| \leq \exp\left\{2\int_0^t \|\mathcal{V}(s)\|ds\right\} \tag{12.43}$$

for every $\mu \in \mathbb{R}$.

Proof The bound for $U_t(\lambda)$ in (12.43) follows from formulas (12.32) and (12.35) and the estimate (12.34), since $\exp\{-i\mu t N_{\alpha,\beta} J\}$ is unitary for $\mu \in \mathbb{R}$. The rest follows from the identities $U_t(\mu)^{-1} = JU_t(\mu)^* J$ and (12.8). □

Theorem 12.11 *Let* $Y_t(\lambda)$, $0 \leq t < d$, *denote the matrizant of a canonical differential system (12.5) with a Hamiltonian* $H(t)$ *that meets the conditions (12.24). Then:*

(1) $\{e_{\alpha t}I_p, e_{\beta t}I_q\} \in ap_I(Y_t)$ *for every* $t \in [0, d)$ *and, if* $q = p$, *then* $\{e_{\alpha t}I_p, e_{\beta t}I_p\} \in ap_{II}(Y_t)$ *also.*

(2) $Y_t \in \mathcal{U}_{rsR}(J)$ *for every* $t \in [0, d)$.

Proof This follows from assertions (2) and (3) of Theorem 12.6, since the matrizant Y_t is connected to the matrizant U_t of a system of the form (12.5) by (12.8), in which the factor $U(t) \in \mathcal{U}_{\text{const}}(J)$ for each $t \in [0, d)$. □

Let $\mathcal{W}_\ell(j_{pq})$ denote the subclass of $\mathcal{W}^{m\times m}(I_m)$ of mvf's $\mathfrak{A}_\ell$ of the form

$$\mathfrak{A}_\ell(\mu) = \begin{bmatrix} \mathfrak{d}_-(\mu) & \mathfrak{c}_+(\mu) \\ \mathfrak{c}_-(\mu) & \mathfrak{d}_+(\mu) \end{bmatrix} \quad \text{for } \mu \in \mathbb{R} \tag{12.44}$$

such that

(a) $\mathfrak{A}_\ell(\mu)^* j_{pq} \mathfrak{A}_\ell(\mu) = j_{pq}$ for every $\mu \in \mathbb{R}$.

(b) $(\mathfrak{d}_+)^{\pm 1} \in \mathcal{W}_+^{q\times q}(I_q)$, $(\mathfrak{d}_-)^{\pm 1} \in \mathcal{W}_-^{p\times p}(I_p)$, $\mathfrak{c}_+ \in \mathcal{W}_+^{p\times q}(0)$ and $\mathfrak{c}_- \in \mathcal{W}_-^{q\times p}(0)$.

It is readily checked that

$$\mathfrak{A}_\ell \in \mathcal{W}_\ell(j_{pq}) \Longleftrightarrow \mathfrak{A}_\ell^\tau \in \mathcal{W}_r(j_{pq}) \tag{12.45}$$

$$\mathfrak{A}_\ell \in \mathcal{W}_\ell(j_{pq}) \Longleftrightarrow \mathcal{T}^*\mathfrak{A}_\ell^*\mathcal{T} \in \mathcal{W}_r(j_{qp}), \tag{12.46}$$

where $\mathcal{W}_r(j_{pq})$ denotes the class of mvf's introduced in and below (9.109) and

$$\mathcal{T} = \begin{bmatrix} 0_{p\times q} & I_p \\ I_q & 0_{q\times p} \end{bmatrix}.$$

Moreover, the right and left linear fractional transformations introduced in (3.21) and (3.25), respectively, are related by the formula

$$T^\ell_{\mathfrak{A}_\ell}[\varepsilon] = T_{\mathcal{T}^*\mathfrak{A}_\ell^*\mathcal{T}}[\varepsilon] \quad \text{a.e. on } \mathbb{R} \text{ for every } \varepsilon \in \mathbb{B}^{q\times p}. \tag{12.47}$$

Lemma 12.12 *If the potential $\mathcal{V}(t)$ of the DK-system (12.1) meets the conditions (12.3) and (12.4), and if $J = j_{pq}$ and W_t^0 denotes the matrizant of (12.1) with zero potential, i.e.,*

$$W_t^0(\lambda) = \begin{bmatrix} e_{\alpha t}I_p & 0 \\ 0 & e_{-\beta t}I_q \end{bmatrix} \quad \textit{for } 0 \le t < d, \tag{12.48}$$

then the mvf's

$$\mathfrak{A}_r^t(\lambda) = W_t^0(\lambda)^{-1}W_t(\lambda) \quad \textit{and} \quad \mathfrak{A}_\ell^t(\lambda) = W_t(\lambda)W_t^0(\lambda)^{-1} \tag{12.49}$$

belong to the classes $\mathcal{W}_r(j_{pq})$ and $\mathcal{W}_\ell(j_{pq})$, respectively.

Proof The right factor $\mathfrak{A}_r^t$ in (12.29) and the left factor $\mathfrak{A}_\ell^t$ in (12.30) are of the form

$$\mathfrak{A}_r^t = \begin{bmatrix} \mathfrak{a}_-^t & \mathfrak{b}_-^t \\ \mathfrak{b}_+^t & \mathfrak{a}_+^t \end{bmatrix} \quad \text{and} \quad \mathfrak{A}_\ell^t = \begin{bmatrix} \mathfrak{d}_-^t & \mathfrak{c}_+^t \\ \mathfrak{c}_-^t & \mathfrak{d}_+^t \end{bmatrix}, \tag{12.50}$$

where $\mathfrak{a}_-^t \in \mathcal{W}_-^{p\times p}(I_p)$, $\mathfrak{b}_-^t \in \mathcal{W}_-^{p\times q}(0)$, $\mathfrak{b}_+^t \in \mathcal{W}_+^{q\times p}(0)$, $\mathfrak{a}_+^t \in \mathcal{W}_+^{q\times q}(I_q)$, $\mathfrak{d}_-^t \in \mathcal{W}_-^{p\times p}(I_p)$, $\mathfrak{c}_+^t \in \mathcal{W}_+^{p\times q}(0)$, $\mathfrak{c}_-^t \in \mathcal{W}_-^{q\times p}(0)$ and $\mathfrak{d}_+^t \in \mathcal{W}_+^{q\times q}(0)$. Moreover, since $W_t \in \mathcal{E} \cap \mathcal{U}(j_{pq})$,

$$(w_{11}^t)^{-\#} \in \mathcal{S}^{p\times p} \quad \text{and} \quad (w_{22}^t)^{-1} \in \mathcal{S}^{q\times q}.$$

Therefore,

$$e_{\alpha t}(\mathfrak{a}_-^t)^{-\#} \in \mathcal{S}^{p\times p}, \quad e_{\alpha t}(\mathfrak{d}_-^t)^{-\#} \in \mathcal{S}^{p\times p},$$

$$e_{\beta t}(\mathfrak{a}_+^t)^{-1} \in \mathcal{S}^{q\times q} \quad \text{and} \quad e_{\beta t}(\mathfrak{d}_+^t)^{-1} \in \mathcal{S}^{q\times q}.$$

Thus, as

$$\lim_{|\lambda|\uparrow\infty} \det \mathfrak{a}_\pm^t(\lambda) = \lim_{|\lambda|\uparrow\infty} \det \mathfrak{d}_\pm^t(\lambda) = 1$$

(when $|\lambda| \uparrow \infty$ in the appropriate half plane), it follows that $\det \mathfrak{a}_{\pm}^t(\lambda) \neq 0$ in $\overline{\mathbb{C}_{\pm}} \cup \{\infty\}$ and $\det \mathfrak{d}_{\pm}^t(\lambda) \neq 0$ in $\overline{\mathbb{C}_{\pm}} \cup \{\infty\}$; and hence, by Theorem 3.31,

$$(\mathfrak{a}_-^t)^{-1} \in \mathcal{W}_-^{p\times p}(I_p), \quad (\mathfrak{d}_-^t)^{-1} \in \mathcal{W}_-^{p\times p}(I_p)$$
$$(\mathfrak{a}_+^t)^{-1} \in \mathcal{W}_+^{q\times q}(I_q) \quad \text{and} \quad (\mathfrak{d}_+^t)^{-1} \in \mathcal{W}_+^{q\times q}(I_q).$$ □

Lemma 12.13 *In the setting of Lemma 12.12, the blocks of the mvf's in (12.50) have the following properties:*

(1) $\mathfrak{a}_{\pm}^t(\mu) = \mathfrak{d}_{\pm}^t(\mu)$ *and* $\mathfrak{b}_{\pm}^t(\mu) = e^{\pm i\mu\kappa t}\mathfrak{c}_{\mp}^t(\mu)$ *for every* $\mu \in \mathbb{R}$ *and* $t \geq 0$.
(2) $\mathfrak{c}_+^t(\mathfrak{d}_+^t)^{-1} \in \mathring{\mathcal{S}}^{p\times q}$ *and* $e^{i\kappa\lambda t}(\mathfrak{d}_+^t)^{-1}\mathfrak{c}_-^t \in \mathcal{S}^{q\times p}$.

Proof Assertion (1) follows from formulas (12.49); (2) then follows from the fact that the mvf $S_t = PG(W_t)$ belongs to the class $\mathcal{S}^{m\times m}$. □

12.3 Direct and inverse monodromy problems for DK-systems

In this section it will be shown that the inverse monodromy problem for a DK-system with a potential $\mathcal{V}$ that satisfies the constraints (12.3) and (12.4) has at most one solution.

Theorem 12.14 *Let* $U \in \mathcal{E} \cap \mathcal{U}(J)$, $J \neq \pm I_m$ *and let* $0 < d < \infty$. *Then there exists at most one* $m \times m$ *mvf* $\mathcal{V}(t)$ *that meets the constraints (12.3)–(12.4) such that* $U(\lambda)$ *is the monodromy matrix of the system (12.1) on the interval* $[0, d]$ *with this potential* $\mathcal{V}(t)$.

Proof Let $U_t(\lambda)$ be the matrizant of a DK-system (12.1) that meets the stated conditions. Then

$$Y_t(\lambda) = U_t(\lambda)U_t(0)^{-1}, \quad 0 \leq t < d,$$

is the matrizant of a canonical system (12.5) with Hamiltonian $H(t) = U_t(0)N_{\alpha,\beta}U_t(0)^*$ and, by Theorem 12.11,

$$Y_d \in \mathcal{U}_{rsR}^{\circ}(J) \quad \text{and} \quad \{e_{\alpha t}I_p, e_{\beta t}I_q\} \in ap_I(Y_t) \quad \text{for every } t \in [0, d].$$

Moreover, as noted in Remark 12.8, this chain does not depend upon the choice of the unitary matrix V_1 in formula (12.40). Therefore, by Theorem 8.16, the Hamiltonian $H(t)$ of the system (12.5) is uniquely defined by the monodromy matrix $Y_d(\lambda)$ and consequently by $U_d(\lambda) = U(\lambda)$, since $Y_d(\lambda) = U(\lambda)U(0)^{-1}$.

Next, by Theorem 12.3, the mvf $U(t) = U_t(0)$, is uniquely defined by $H(t)$ and consequently by $U(\lambda)$. Finally, formula (12.7) implies that $\mathcal{V}(t)$ is also uniquely defined by $U(\lambda)$. □

12.4 Direct and inverse input scattering problems for DK-systems

In this section it will be assumed that $J = j_{pq}$ and that the potential $\mathcal{V}$ of the DK-system (12.1) satisfies the conditions (12.3) and (12.4), i.e.,

$$\mathcal{V}(t) = \begin{bmatrix} 0 & \mathfrak{v}(t) \\ \mathfrak{v}(t)^* & 0 \end{bmatrix} \quad \text{a.e. on } [0, d) \quad \text{and} \quad \mathfrak{v} \in L_{1,\mathrm{loc}}^{p\times q}([0, d)) \qquad (12.51)$$

and hence the system (12.1) can be written more explicitly as

$$u'(t,\lambda) = i\lambda u(t,\lambda)\begin{bmatrix} \alpha I_p & 0 \\ 0 & -\beta I_q \end{bmatrix} + u(t,\lambda)\begin{bmatrix} 0 & \mathfrak{v}(t) \\ \mathfrak{v}(t)^* & 0 \end{bmatrix} \qquad (12.52)$$

$$\text{for } 0 \le t < d,\ \alpha \ge 0,\ \beta \ge 0 \text{ and } \kappa = \alpha + \beta > 0.$$

The matrizant W_t of this system depends upon both α and β. However, formulas (12.48) and (12.49) imply that

$$T_{W_t}[\mathcal{S}^{p\times q}] = e_{\kappa t} T_{\mathfrak{A}_r^t}[\mathcal{S}^{p\times q}] \quad \text{for } 0 \le t < d, \qquad (12.53)$$

and hence that the set $T_{W_t}[\mathcal{S}^{p\times q}]$ depends only upon $\kappa = \alpha + \beta$ and the potential $\mathcal{V}$ on the interval $[0, t]$. Thus, the sets of **input scattering matrices** for the DK-system considered on $[0, t]$ with $t < d$ and on $[0, d)$ are

$$\mathcal{S}_{\mathrm{scat}}^{t}(N_{\alpha,\beta}; \mathcal{V}) = T_{W_t}[\mathcal{S}^{p\times q}] = \mathcal{S}_{\mathrm{scat}}^{t}(N_{\kappa,0}; \mathcal{V}) \quad \text{for } 0 \le t < d \qquad (12.54)$$

and

$$\mathcal{S}_{\mathrm{scat}}^{d}(N_{\alpha,\beta}; \mathcal{V}) = \bigcap_{0 \le t < d} T_{W_t}[\mathcal{S}^{p\times q}] = \mathcal{S}_{\mathrm{scat}}^{d}(N_{\kappa,0}; \mathcal{V}), \qquad (12.55)$$

respectively.

The input scattering matrices will be described in terms of the classes

$$\mathcal{B}_{\infty}^{p\times q} = \Big\{ h \in L_1^{p\times q}(\mathbb{R}_+) : \text{the operator } (X^{\vee}\varphi)(t) = \int_0^t h(t-u)\varphi(u)du \text{ is contractive from } L_2^q(\mathbb{R}_+) \text{ into } L_2^p(\mathbb{R}_+) \Big\}$$

$$= \Big\{ h \in L_1^{p\times q}(\mathbb{R}_+) : \int_0^{\infty} e^{i\lambda t} h(t)dt \in \mathcal{S}^{p\times q} \Big\}$$

$$\mathcal{B}_{a}^{p\times q} = \Big\{ h \in L_1^{p\times q}([0, a]) : (X^{\vee}\varphi)(t) = \int_0^t h(t-u)\varphi(u)du \text{ is contractive from } L_2^q([0, a]) \text{ into } L_2^p([0, a]) \Big\},$$

$$\mathring{\mathcal{B}}_{a}^{p\times q} = \{ h \in \mathcal{B}_{a}^{p\times q} : \|X^{\vee}\| < 1 \} \quad \text{for } 0 < a \le \infty,$$

and

$$\mathring{\mathcal{B}}^{p\times q}_{a,\mathrm{loc}} = \{h \in L^{p\times q}_{1,\mathrm{loc}}([0,a)) : \text{ the restriction of } h \text{ to the interval } [0,c] \text{ belongs to } \mathring{\mathcal{B}}^{p\times q}_{c} \text{ for every } c \in [0,a)\} \quad \text{for } 0 < a \leq \infty.$$

Theorem 12.15 *If the potential $\mathcal{V}$ of the DK-system (12.52) meets the condition (12.51), then:*

(1) *There exists exactly one mvf $h \in \mathring{\mathcal{B}}^{p\times q}_{\kappa d,\mathrm{loc}}$ such that*

$$\mathcal{S}^{t}_{\mathrm{scat}}(N_{\alpha,\beta};\mathcal{V}) = \left\{s \in \mathcal{S}^{p\times q} : e_{-\kappa t}\left\{s - \int_0^{\kappa t} e^{i\lambda u}h(u)du\right\} \in H^{p\times q}_{\infty}\right\} \quad \textit{for } 0 < t < d. \tag{12.56}$$

(2) *If condition (12.9) is in force, then (12.56) is also valid for $t = d$ and $h \in \mathring{\mathcal{B}}^{p\times q}_{\kappa d}$.*

(3) *If $d = \infty$, then $\mathcal{S}^{d}_{\mathrm{scat}}(N_{\alpha,\beta};\mathcal{V}) = \{s^\circ\}$ and s° is the unique mvf in the class $\mathcal{S}^{p\times q}$ such that*

$$e_{-t}(\lambda)\left\{s^\circ(\lambda) - \int_0^t e^{i\lambda u}h(u)du\right\} \quad \textit{belongs to } H^{p\times q}_{\infty} \textit{ for every } t \in \mathbb{R}_+. \tag{12.57}$$

Moreover,

$$s^\circ(\lambda) = \lim_{a\uparrow\infty}\int_0^a (e^{i\lambda t} - e^{i\lambda a})h(t)dt \quad \textit{for } \lambda \in \mathbb{C}_+ \textit{ and every } a > 0 \tag{12.58}$$

and hence

$$s^\circ(\lambda) = \int_0^\infty e^{i\lambda t}h(t)dt \quad \textit{if } h \in L^{p\times q}_1(\mathbb{R}_+) \textit{ and } \lambda \in \overline{\mathbb{C}_+}. \tag{12.59}$$

(4) *If $\mathcal{V} \in L^{m\times m}_1(\mathbb{R}_+)$, then*

$$h \in \mathring{\mathcal{B}}^{p\times q}_{\infty}, \quad s^\circ \in \mathcal{W}^{p\times q}_{+}(0) \cap \mathring{\mathcal{S}}^{p\times q} \quad \textit{and} \quad \textit{(12.59) is in force.}$$

Proof Assertion (5) of Theorem 9.39 and the first formula in (12.49) guarantee that for each choice of $t \in [0,d)$ there exists exactly one mvf $h^t \in \mathring{\mathcal{B}}^{p\times q}_{\kappa t}$ such that (12.56) holds with $h = h^t$. Since

$$\mathcal{S}(h^{t_2};\kappa t_2) = T_{W_{t_2}}[\mathcal{S}^{p\times q}] \subseteq T_{W_{t_1}}[\mathcal{S}^{p\times q}] = \mathcal{S}(h^{t_1};\kappa t_1) \quad \text{for } 0 \leq t_1 \leq t_2 < d,$$

it follows that $h^{t_1}(s) = h^{t_2}(s)$ a.e. on $[0, t_1\kappa]$. Thus, the mvf

$$h(s) = h^t(s) \quad \text{for } s \in [0,\kappa t]$$

is well defined on $[0,\kappa d)$ and belongs to the class $\mathring{\mathcal{B}}^{p\times q}_{\kappa d,\mathrm{loc}}$.

If (12.1) is a regular DK-system on $[0,d]$, then $h \in \mathring{\mathcal{B}}^{p\times q}_{\kappa d}$ and

$$\mathcal{S}^{d}_{\mathrm{scat}}(N_{\alpha,\beta};\mathcal{V}) = T_{W_d}[\mathcal{S}^{p\times q}] = \mathcal{S}(h;\kappa d).$$

If $d = \infty$, then $h \in L_{1,\mathrm{loc}}^{p\times q}(\mathbb{R}_+)$ and a mvf $s \in \mathcal{S}_{\mathrm{scat}}^{\infty}(N_{\alpha,\beta}; \mathcal{V})$ admits a representation of the form (9.78), where

$$v(t) = \int_0^t h(s)ds \quad \text{for } 0 \le t < \infty. \tag{12.60}$$

Thus, the set $\mathcal{S}_{\mathrm{scat}}^{\infty}(N_{\alpha,\beta}; \mathcal{V})$ contains exactly one mvf. Formula (12.58) follows from (9.78):

$$s^\circ(\lambda) = -i\lambda \lim_{a\uparrow\infty} \int_0^a e^{i\lambda t} v(t)dt = -i\lambda \lim_{a\uparrow\infty} \int_0^a e^{i\lambda t} \left\{ \int_0^t h(u)du \right\} dt$$

$$= -i\lambda \lim_{a\uparrow\infty} \int_0^a h(u) \left\{ \int_u^a e^{i\lambda t} dt \right\} du \quad \text{for } \lambda \in \mathbb{C}_+,$$

which is easily seen to coincide with the right-hand side of (12.58). Formula (12.59) follows directly from (12.58). The proof of assertion (4) is postponed to Theorem 12.42. □

The unique mvf $h \in \mathring{\mathcal{B}}_{\kappa d,\mathrm{loc}}^{p\times q}$ that is characterized by (12.56) will be denoted $h_{\mathcal{V},\kappa}^{d}$. It depends upon $\kappa = \alpha + \beta$, but not upon the individual choice of $\alpha \ge 0$ or $\beta \ge 0$.

Remark 12.16 *Theorem 12.15 guarantees that the matrizant W_t, $0 < t < d$, of the DK-system considered therein is a resolvent matrix for the strictly completely indeterminate extension problem $SP(h; \kappa t)$ in which the given data is the unique mvf $h \in \mathring{\mathcal{B}}_{\kappa d,\mathrm{loc}}^{p\times q}$ specified in (1) of the theorem.*

Remark 12.17 *If $d < \infty$ and $h \in \mathring{\mathcal{B}}_{\kappa d}^{p\times q}$, then $SP(h; \kappa d)$ is strictly completely indeterminate. Therefore, $\mathring{\mathcal{S}}^{p\times q} \cap \mathcal{S}(h; \kappa d) \neq \emptyset$. Thus, as*

$$\mathring{\mathcal{S}}^{p\times q} \cap \mathcal{S}(h; \kappa d) \subseteq \mathcal{S}_{\mathrm{scat}}^{d}(N_{\alpha,\beta}; \mathcal{V}) = \mathcal{S}_{\mathrm{scat}}^{d}(Hdt).$$

Theorem 6.12 is applicable and implies that (12.5) is a regular DK-system.

Theorem 12.18 *Let $W_t(\lambda)$, $0 \le t < d$, denote the matrizant of the DK-system (12.52) that meets the condition (12.51) and let $s^\circ \in \mathcal{S}_{\mathrm{scat}}^{d}(N_{\alpha,\beta}; \mathcal{V})$. Then:*

(1) $\mathcal{S}_{\mathrm{scat}}^{t}(N_{\alpha,\beta}; \mathcal{V}) = T_{W_t}[\mathcal{S}^{p\times q}] = \{s \in \mathcal{S}^{p\times q} : e_{-\kappa t}(s - s^\circ) \in H_\infty^{p\times q}\}$ *for every* $t \in [0, d)$.
(2) $\mathcal{S}_{\mathrm{scat}}^{t}(N_{\alpha,\beta}; \mathcal{V}) \cap \mathring{\mathcal{S}}^{p\times q} = T_{W_t}[\mathring{\mathcal{S}}^{p\times q}]$ *for every* $t \in [0, d)$.
(3) $\mathcal{S}_{\mathrm{scat}}^{t}(N_{\alpha,\beta}; \mathcal{V}) \cap \mathcal{W}_+^{p\times q}(0) = T_{W_t}[\mathcal{S}^{p\times q} \cap \mathcal{W}_+^{p\times q}(0)]$ *for every* $t \in [0, d)$.
(4) *If (12.9) is in force, then conclusions (1)–(3) are also valid for $t = d$.*

Proof This follows from Theorems 12.15 and 9.39. □

Theorem 10.2, combined with the analysis in the preceding sections, serves to justify the following uniqueness theorems for the inverse input scattering problem for DK-systems (12.1) and canonical systems (12.5) that are related in an appropriate way to DK-systems.

Theorem 12.19 *Let $J = j_{pq}$, $s \in \mathcal{S}^{p\times q}$, $0 < d \le \infty$ and $\alpha \ge 0$, $\beta \ge 0$ with $\kappa = \alpha + \beta > 0$ be given. Then there exists at most one canonical differential system of the form (12.5) with a Hamiltonian $H(t)$ that meets the conditions (12.24) in Theorem 12.3 such that $s \in \mathcal{S}^d_{\rm scat}(Hdt)$.*

Proof This follows from Theorems 12.11 and 10.2. □

Theorem 12.20 *Let $\kappa > 0$ and $0 < d \le \infty$ be specified. Then for every $s \in \mathcal{S}^{p\times q}$ there exists at most one potential $\mathcal{V}(t)$ that meets the condition (12.51) such that $s \in \mathcal{S}^d_{\rm scat}(N_{\alpha,\beta}; \mathcal{V})$ for the system (12.52) with $\alpha \ge 0$, $\beta \ge 0$ and $\alpha + \beta = \kappa$.*

Proof In view of Theorems 12.15 and 9.39, there exists at most one matrizant W_t of the DK-system (12.1) of the form $W_t(\lambda) = W_t^\circ(\lambda)\mathfrak{A}_r^t(\lambda)$ with $\mathfrak{A}_r^t \in \mathcal{W}_r(j_{pq})$ such that (1) of Theorem 12.18 holds. Correspondingly, the potential

$$\mathcal{V}(t) = \mathcal{W}_t(0)^{-1}\frac{d}{dt}\{W_t(0)\} = \mathfrak{A}_r^t(0)^{-1}\frac{d}{dt}\{W_t(0)\}. \tag{12.61}$$

□

Remark 12.21 *The system of equations in Theorem 9.39 for the resolvent matrix for $SP(h; \kappa t)$, i.e., for $e_{\beta t}W_t$, yields an algorithm for computing $\mathcal{V}(t)$ via formula (12.61).*

12.5 Direct and inverse input impedance problems for DK-systems

To study input impedance and spectral problems for DK-systems it is convenient to let $J = J_p$. Then the system (12.1) that satisfies the conditions (12.3) and (12.4) may be written as

$$u'(t,\lambda) = i\lambda u(t,\lambda)N_{\alpha,\beta}J_p + u(t,\lambda)\mathcal{V}(t) \quad \text{for } 0 \le t < d, \tag{12.62}$$

where

$$N_{\alpha,\beta} = \mathfrak{V}\begin{bmatrix}\alpha I_p & 0 \\ 0 & \beta I_p\end{bmatrix}\mathfrak{V} \quad \text{with } \alpha \ge 0,\ \beta \ge 0,\ \kappa = \alpha + \beta > 0 \tag{12.63}$$

and

$$\mathcal{V}(t) = \mathfrak{V}\begin{bmatrix}0 & \mathfrak{v}(t) \\ \mathfrak{v}(t)^* & 0\end{bmatrix}\mathfrak{V} \quad \text{for } 0 \le t < d \quad \text{with } \mathfrak{v} \in L^{p\times p}_{1,\rm loc}([0,d)). \tag{12.64}$$

Let $A_t(\lambda)$, $0 \le t < d$, denote the matrizant of this system. Then

$$\widetilde{A}_t(\lambda) = A_t(\lambda)A_t(0)^{-1}, \quad 0 \le t < d,$$

is the matrizant of the corresponding canonical differential system (12.5),

$$\mathcal{C}^d_{\rm imp}(N_{\alpha,\beta}; \mathcal{V}) = \bigcap_{0\le t<d} \mathcal{C}(A_t) \tag{12.65}$$

is the set of input impedances for the DK-system (12.62) and

$$\mathcal{C}^d_{\mathrm{imp}}(Hdt) = \bigcap_{0\le t<d} \mathcal{C}(\widetilde{A}_t) \tag{12.66}$$

is the set of input impedances for the canonical differential system (12.5) with Hamiltonian $H(t) = A_t(0)N_{\alpha,\beta}A_t(0)^*$.

Lemma 12.22 *Let $A_t(\lambda)$ be the matrizant of a DK-system (12.62) with a potential $\mathcal{V}$ that meets the conditions in (12.64); let $B_t(\lambda) = A_t(\lambda)\mathfrak{V}$ and $W_t(\lambda) = \mathfrak{V}B_t(\lambda)$ for $0 \le t < d$. Let $\widetilde{A}_t(\lambda) = A_t(\lambda)A_t(0)^{-1}$ be the matrizant of the corresponding canonical differential system (12.5) with $J = J_p$, $\widetilde{B}_t(\lambda) = \widetilde{A}_t(\lambda)\mathfrak{V}$ and $\widetilde{W}_t(\lambda) = \mathfrak{V}\widetilde{B}_t(\lambda)$ for $0 \le t < d$. Then:*

(1) $\mathcal{C}(A_t) = T_{B_t}[\mathcal{S}^{p\times p}] = T_{\widetilde{B}_t}[\mathcal{S}^{p\times p}] = \mathcal{C}(\widetilde{A}_t)$ *for* $0 \le t < d$.
(2) $\mathcal{C}^d_{\mathrm{imp}}(N_{\alpha,\beta};\mathcal{V}) = \bigcap_{0\le t<d} T_{B_t}[\mathcal{S}^{p\times p}] = \mathcal{C}^d_{\mathrm{imp}}(Hdt)$.
(3) $\mathcal{C}^d_{\mathrm{imp}}(N_{\alpha,\beta};\mathcal{V}) = T_{\mathfrak{V}}[\cap_{0\le t<d} T_{W_t}[\mathcal{S}^{p\times p}]] \ne \emptyset$.
(4) $\mathcal{C}^d_{\mathrm{imp}}(Hdt) = T_{\mathfrak{V}}[\cap_{0\le t<d} T_{\widetilde{W}_t}[\mathcal{S}^{p\times p}]] \ne \emptyset$.

Proof Since $\widetilde{B}_t(\lambda) = B_t(\lambda)W_t(0)^{-1}$,

$$\mathcal{D}(T_{\widetilde{B}_t}) = \mathcal{D}(T_{B_t}) \quad \text{for } 0 \le t < d$$

and

$$\mathcal{C}(A_t) = T_{B_t}[\mathcal{S}^{p\times p} \cap \mathcal{D}(T_{B_t})] = T_{\widetilde{B}_t}[\mathcal{S}^{p\times p} \cap \mathcal{D}(T_{\widetilde{B}_t})] = \mathcal{C}(\widetilde{A}_t).$$

Moreover, since $\widetilde{A}_t \in \mathcal{U}_{rsR}(J_p)$ and $\{e_{\alpha t}I_p, e_{\beta t}I_p\} \in ap_{II}(\widetilde{A}_t)$ for $0 \le t < d$, Lemma 7.1 implies that $\mathcal{S}^{p\times p} \subseteq \mathcal{D}(T_{\widetilde{B}_t})$ and that (4) holds. Thus, (1) holds; (2) follows from (1) and (3) follows from (2) and (4). □

Theorem 12.23 *Let $c \in \mathcal{C}^{p\times p}$, $0 < d \le \infty$, and $\alpha \ge 0$, $\beta \ge 0$ with $\kappa = \alpha + \beta > 0$ be given. Then there exists at most one canonical differential system (12.5) with $J = J_p$ and a Hamiltonian $H(t)$ that meets the conditions (12.24) such that $c \in \mathcal{C}^d_{\mathrm{imp}}(Hdt)$.*

Proof This follows from assertion (4) of Lemma 12.22 and Theorem 12.19. □

Theorem 12.24 *Let $J = J_p$, let $N = N_{\alpha,\beta}$ be defined as in (12.2) and let $0 < d \le \infty$. Then for every $c \in \mathcal{C}^{p\times p}$ there exists at most one potential $\mathcal{V}(t)$, $0 \le t < d$, that satisfies the properties (12.62)–(12.64) such that $c \in \mathcal{C}^d_{\mathrm{imp}}(N_{\alpha,\beta};\mathcal{V})$.*

Proof This follows from Lemma 12.22 and Theorem 12.20. □

Theorem 12.25 *Let $A_t(\lambda)$ be the matrizant of a DK-system (12.62) with a potential $\mathcal{V}$ that meets the conditions in (12.64); let $B_t = A_t\mathfrak{V}$ for $0 \le t < d$ and let $c^\circ \in \mathcal{C}^d_{\mathrm{imp}}(N_{\alpha,\beta};\mathcal{V})$. Then:*

(1) $\mathcal{C}^t_{\text{imp}}(N_{\alpha,\beta};\mathcal{V}) = T_{B_t}[\mathcal{S}^{p\times p}] = \{c \in \mathcal{C}^{p\times p} : e_{-\kappa t}(c - c^\circ) \in \mathcal{N}_+^{p\times p}\}$ *for every* $t \in [0, d)$.
(2) $\mathcal{C}^t_{\text{imp}}(N_{\alpha,\beta};\mathcal{V}) \cap \mathring{\mathcal{C}}^{p\times p} = T_{B_t}[\mathring{\mathcal{S}}^{p\times p}]$ *for every* $t \in [0, d)$.
(3) $\mathcal{C}^t_{\text{imp}}(N_{\alpha,\beta};\mathcal{V}) \cap \mathcal{W}_+^{p\times p}(I_p) = T_{B_t}[\mathcal{S}^{p\times p} \cap \mathcal{W}_+^{p\times p}(0)]$ *for every* $t \in [0, d)$.
(4) *If (12.9) is in force, then conclusions (1)–(3) are also valid for* $t = d$.

Proof The proof, which exploits Lemma 7.1, is analogous to the proof of Theorem 12.18. □

The classes $\mathring{\mathcal{A}}_a^{p\times p}$ defined in (9.37) and

$$\mathring{\mathcal{A}}_{a,\text{loc}}^{p\times p} = \{h \in L_{1,\text{loc}}^{p\times p}((-a,a)) : \text{the restriction of } h \text{ to } [-c,c] \text{ belongs to } \mathring{\mathcal{A}}_c^{p\times p} \text{ for every } c \in (0,a)\} \tag{12.67}$$

will enter into the description of the sets of impedance matrices (12.65) and (12.66).

Theorem 12.26 *If the potential* $\mathcal{V}$ *of the DK-system (12.62) meets the conditions in (12.64), then:*

(1) *There exists exactly one mvf* $h \in \mathring{\mathcal{A}}_{\kappa d,\text{loc}}^{p\times p}$ *such that*

$$\mathcal{C}^t_{\text{imp}}(N_{\alpha,\beta};\mathcal{V}) = \left\{c \in \mathcal{C}^{p\times p} : e_{-\kappa t}\left\{c - I_p - 2\int_0^{\kappa t} e^{i\lambda u}h(u)du\right\} \in \mathcal{N}_+^{p\times p}\right\} \quad \text{for } 0 < t < d. \tag{12.68}$$

(2) *If the condition (12.9) is in force, then* $h \in \mathring{\mathcal{A}}_{\kappa d}^{p\times p}$. *Moreover, in this case, (12.68) is also valid for* $t = d$.
(3) *If* $d = \infty$, *then* $\mathcal{C}^\infty_{\text{imp}}(N_{\alpha,\beta};\mathcal{V}) = \{c^\circ\}$ *and* c° *may be characterized as the unique mvf in the class* $\mathcal{C}^{p\times p}$ *such that*

$$e_{-t}(\lambda)\left\{c^\circ(\lambda) - I_p - 2\int_0^t e^{is\lambda}h(s)ds\right\} \in \mathcal{N}_+^{p\times p} \text{ for every } t \in \mathbb{R}_+. \tag{12.69}$$

Moreover,

$$c^\circ(\lambda) = I_p + 2\lim_{a\uparrow\infty}\int_0^a (e^{i\lambda t} - e^{iat})h(t)dt \quad \text{for } \lambda \in \mathbb{C}_+ \tag{12.70}$$

and

$$c^\circ(\lambda) = I_p + 2\int_0^\infty e^{i\lambda t}h(t)dt \quad \text{if } h \in L_1^{p\times p}(\mathbb{R}_+) \text{ and } \lambda \in \mathbb{C}_+. \tag{12.71}$$

(4) *If* $\mathcal{V} \in L_1^{m\times m}(\mathbb{R}_+)$, *then*

$$h \in \mathring{\mathcal{A}}_\infty^{p\times p}, \quad c^\circ \in \mathcal{W}_+^{p\times p}(I_p) \cap \mathring{\mathcal{C}}^{p\times p}$$

and (12.71) is in force, i.e., h is the accelerant of c°.

Proof The proof of (1)–(3) is analogous to the proof of (1)–(3) of Theorem 12.15, but uses Theorems 9.23 and 9.24 instead of 9.39. Formula (12.70) depends upon Corollary 9.2, which guarantees that

$$g(t) = O(t^2) \quad \text{as } t \uparrow \infty,$$

and hence that

$$c^\circ(\lambda) = \lambda^2 \int_0^a e^{i\lambda t} g(t)dt + o(a) \quad \text{as } a \uparrow \infty.$$

Thus, upon invoking the fact that

$$g(t) = -tI_p - 2\int_0^t (t-s)h(s)ds \quad \text{for some } h \in L_{1,\text{loc}}^{p\times p}(\mathbb{R}_+),$$

it follows that if $\lambda \in \mathbb{C}_+$, then

$$\begin{aligned}
\lambda^2 \int_0^a e^{i\lambda t} g(t)dt &= \lambda^2 \frac{1}{i\lambda} e^{i\lambda t} g(t)|_0^a - i\lambda \int_0^a e^{i\lambda t} \left\{ I_p + 2\int_0^t h(s)ds \right\} dt \\
&= I_p - 2i\lambda \int_0^a e^{i\lambda t} \left\{ \int_0^t h(s)ds \right\} dt \quad + o(a) \quad \text{as } a \uparrow \infty \\
&= I_p - 2i\lambda \int_0^a \left\{ \int_s^a e^{i\lambda t} dt \right\} h(s)ds \quad + o(a) \quad \text{as } a \uparrow \infty.
\end{aligned}$$

But this yields (12.70), much as in the verification of formula (12.58); (12.71) then follows from (12.70) and the Riemann–Lebesgue lemma.

The proof of (4) follows from the relation $c^\circ = T_{\mathfrak{V}}[s^\circ]$, (4) of Theorem 12.15 and Theorem 3.31. □

The unique mvf $h \in \mathring{\mathcal{A}}_{\kappa d,\text{loc}}^{p\times p}$ that is characterized by (12.68) will be denoted $\mathfrak{a}_{\mathcal{V},\kappa}^{d}$. It depends upon $\kappa = \alpha + \beta$, but not upon the choice of $\alpha \geq 0$ or $\beta \geq 0$.

Remark 12.27 *Theorem 12.26 guarantees that the modified matrizant $A_t(\lambda)\mathfrak{V}$, $0 < t < d$, of the DK-system considered therein is a B-resolvent matrix (see Theorem 9.21) for the following strictly completely indeterminate extension problem, in which the given data is the unique mvf $h \in \mathring{\mathcal{A}}_{\kappa d,\text{loc}}$ specified in (1) of the theorem: Describe the set of all $g \in \mathcal{G}_\infty^{p\times p}(0)$ such that h is the accelerant of g on the interval $[-\kappa t, \kappa t]$.*

Remark 12.28 *Theorems 12.6, 9.24 and 9.21 yield a constructive algorithm to obtain the mvf $B_t(\lambda) = A_t(\lambda)\mathfrak{V}$ for $0 \leq t < d$ when $A_t(\lambda)$ is the matrizant of the DK-system (12.1) with $J = J_p$, $\beta = 0$, $\kappa = \alpha$ from the corresponding accelerant $h \in \mathring{\mathcal{A}}_{\kappa d,\text{loc}}^{p\times p}$:*

$$B_t(\lambda) = \mathfrak{V}\begin{bmatrix} e^{i\lambda\kappa t} I_p & 0 \\ 0 & I_p \end{bmatrix} + \frac{1}{\sqrt{2}} \int_0^{\kappa t} e^{i\lambda s} \mathcal{B}^t(s)ds, \tag{12.72}$$

where

$$\mathcal{B}^t(s) = \begin{bmatrix} (X_t^\vee e^t)(s) & -((X_t^\vee)^* f^t)(s) \\ e^t(s) & f^t(s) \end{bmatrix} \quad \text{for } 0 \le s \le \kappa t, \tag{12.73}$$

and $e^t(s)$ *and* $f^t(s)$ *are solutions of the equations*

$$e^t(s) + \int_0^{\kappa t} h(s-u) e^t(u) du = -h(s - \kappa t), \tag{12.74}$$

$$f^t(s) + \int_0^{\kappa t} h(s-u) f^t(u) du = -h(s), \tag{12.75}$$

in $L_1^{p\times p}([0, \kappa t])$, *respectively,*

$$(X_t^\vee \varphi)(s) = \varphi(s) + 2\int_0^s h(s-u)\varphi(u)du, \ 0 \le s \le \kappa t,$$

and

$$((X_t^\vee)^* \varphi)(s) = \varphi(s) + 2\int_s^{\kappa t} h(s-u)\varphi(u)du, \ 0 \le s \le \kappa t.$$

Then the potential

$$\mathcal{V}(t) = A_t(0)^{-1}(A_t(0))'.$$

12.6 Direct and inverse spectral problems for DK-systems

In this section, the general results on direct and inverse spectral problems that were obtained for canonical differential systems in Chapters 7 and 11 will be applied first to the canonical differential system (12.5) with $J = J_p$, Hamiltonian $H(t)$ of the form (12.6) and matrizant $\widetilde{A}_t$, $0 \le t < d$, and then to the corresponding DK-system (12.62).

Recall that a nondecreasing $p \times p$ mvf $\sigma(\mu)$ on $\mathbb{R}$ is a **spectral function** for such a canonical system if the generalized Fourier transform

$$(\mathcal{F}_2 f)(\lambda) = \sqrt{2}[0\ I_p]\frac{1}{\sqrt{2\pi}}\int_0^d \widetilde{A}_t(\lambda)H(t)f(t)dt \tag{12.76}$$

based on the matrizant $\widetilde{A}_t(\lambda)$ of the system maps $L_2^m(Hdt; [0, d))$ isometrically into $L_2^p(d\sigma; \mathbb{R})$, i.e.,

$$\int_{-\infty}^{\infty} (\mathcal{F}_2 f)(\mu)^* d\sigma(\mu)(\mathcal{F}_2 f)(\mu) = \int_0^d f(t)^* H(t) f(t) dt \tag{12.77}$$

for every $f \in L_2^m(Hdt; [0, d))$; for more precise definitions see Section 7.7. The set of spectral functions of this system is denoted by $\Sigma_{\text{sf}}^d(Hdt)$.

Similarly, a non-decreasing $p \times p$ mvf $\sigma(\mu)$ on $\mathbb{R}$ is said to be a **spectral function of the DK-system** (12.62) if the generalized Fourier transform based on the matrizant $A_t(\lambda)$ of the system

$$(\mathcal{G}_2 g)(\lambda) = \sqrt{2}[0 \;\; I_p]\frac{1}{\sqrt{2\pi}}\int_0^d A_t(\lambda)N_{\alpha,\beta}\, g(t)dt \tag{12.78}$$

maps $L_2^m(N_{\alpha,\beta}dt; [0,d))$ isometrically into $L_2^p(d\sigma; \mathbb{R})$:

$$\int_{-\infty}^{\infty} (\mathcal{G}_2 g)(\mu)^* d\sigma(\mu)(\mathcal{G}_2 g)(\mu) = \int_0^d g(t)^* N_{\alpha,\beta}\, g(t)dt \tag{12.79}$$

for every $g \in L_2^m(N_{\alpha,\beta}dt; [0,d))$. This set of spectral functions is denoted by the symbol $\Sigma_{\rm sf}^d(N_{\alpha,\beta}; \mathcal{V})$.

The spaces $L_2^m(N_{\alpha,\beta}dt; [0,d))$ and $L_2^m(Hdt; [0,d))$ will be abbreviated by $L_2^m(N_{\alpha,\beta})$ and $L_2^m(H)$, respectively, below.

If the Hamiltonian $H(t)$ of the system (12.5) is related to the potential of the system (12.1) by formula (12.6), then the operator

$$K: \; f \in L_2^m(H) \longrightarrow A_t(0)^* f(t) \in L_2^m(N_{\alpha,\beta})$$

is unitary from $L_2^m(H)$ onto $L_2^m(N_{\alpha,\beta})$. Thus,

$$\|Kf\|^2_{L_2^m(N_{\alpha,\beta})} = \int_0^d f(t)^* A_t(0) N_{\alpha,\beta} A_t(0)^* f(t)dt = \|f\|^2_{L_2^m(H)}.$$

Moreover, since $\widetilde{A}_t(\lambda) = A_t(\lambda)A_t(0)^{-1}$,

$$(\mathcal{G}_2 Kf)(\lambda) = [0 \quad I_p]\frac{1}{\sqrt{\pi}}\int_0^d A_t(\lambda)N_{\alpha,\beta}A_t(0)^* f(t)dt = (\mathcal{F}_2 f)(\lambda)\,.$$

Therefore,

$$\Sigma_{\rm sf}^d(N_{\alpha,\beta}; \mathcal{V}) = \Sigma_{\rm sf}^d(H).$$

Theorem 12.29 *If the canonical differential system (12.5) corresponds to a DK-system (12.1) with $J = J_p$ and a potential $\mathcal{V}$ that meets the conditions (12.3) and (12.4), then:*

(1) *The matrizant $\widetilde{A}_t(\lambda)$, $0 \le t < d$, of the corresponding canonical differential system satisfies the conditions*

$$\widetilde{A}_t \in \mathcal{U}_{rsR}^{\circ}(J_p) \quad \textit{and} \quad \{e_{\alpha t}I_p, e_{\beta t}I_p\} \in ap_{II}(\widetilde{A}_t) \quad \textit{for } 0 \le t < d. \tag{12.80}$$

(2) $\Sigma_{\rm sf}^d(Hdt) = \Sigma_{\rm sf}^d(N_{\alpha,\beta}; \mathcal{V}) = (\mathcal{C}_{\rm imp}^d(Hdt))_{\rm sf} = (\mathcal{C}_{\rm imp}^d(N_{\alpha,\beta}; \mathcal{V}))_{\rm sf}$.

(3) $c \in \mathcal{C}_{\rm imp}^d(Hdt) \Longrightarrow \lim_{\nu\uparrow\infty} \nu^{-1}\Re c(i\nu) = 0$.

(4) $\sigma \in \Sigma_{\rm sf}^d(Hdt) \Longrightarrow$ *there is exactly one Hermitian matrix δ such that*

$$i\delta + \frac{1}{\pi i}\int_{-\infty}^{\infty}\left\{\frac{1}{\mu-\lambda} - \frac{\mu}{1+\mu^2}\right\}d\sigma(\mu) \quad \textit{belongs to } \mathcal{C}_{\rm imp}^d(Hdt). \tag{12.81}$$

Moreover, this is the only mvf in $\mathcal{C}_{\rm imp}^{d}(Hdt)$ with spectral function equal to σ and formula (12.81) defines a one-to-one correspondence between the sets $\Sigma_{\rm sf}^{d}(H)$ and $\mathcal{C}_{\rm imp}^{d}(Hdt)$.

(5) *If $d = \infty$, then there is exactly one spectral function σ for both systems, i.e.,*

$$\Sigma_{\rm sf}^{\infty}(N_{\alpha,\beta}; \mathcal{V}) = \Sigma_{\rm sf}^{\infty}(H) = \{\sigma\}.$$

If also $\mathcal{V} \in L_1^{m\times m}(\mathbb{R}_+)$, then this spectral function is locally absolutely continuous on $\mathbb{R}$ and its density $f(\mu) = \sigma'(\mu)$ belongs to $\mathcal{W}^{p\times p}(I_p)$. Moreover, $f(\mu) > 0$ on $\mathbb{R}$ and $f^{-1} \in \mathcal{W}^{p\times p}(I_p)$.

Proof Assertion (1) follows from Theorem 12.6 and the formula $\widetilde{A}_t(\lambda) = A_t(\lambda)A_t(0)^{-1}$, which guarantees that $\mathcal{H}(\widetilde{A}_t) = \mathcal{H}(A_t)$ for every $t \in [0, d)$ since the RK's coincide.

The next three assertions follow from Theorem 7.31 and the preceding discussion. The first assertion in (5) follows from (2) and from (3) of Theorem 12.26. The rest follows from (4) of Theorem 12.26 and the Wiener theorem. □

Theorem 12.30 *Let $\sigma \in (\mathcal{C}^{p\times p})_{\rm sf}$, $J = J_p$, $0 < d \le \infty$, $\alpha \ge 0$, $\beta \ge 0$ and $\kappa = \alpha + \beta > 0$. Then there exists at most one canonical differential system (12.5) with Hamiltonian $H(t)$ that meets the constraints (12.24) such that $\sigma \in \Sigma_{\rm sf}^{d}(Hdt)$.*

Proof If $\sigma \in \Sigma_{\rm sf}^{d}(H_j dt)$ for two Hamiltonians $H_j(t)$, $j = 1, 2$, that satisfy the condition (12.24), then, by Theorem 11.25, $H_j(t) = H^{(\delta_j)}(t)$ for a pair of Hermitian matrices $\delta_1, \delta_2 \in \mathbb{C}^{p\times p}$. Thus, the condition $H^{(\delta_j)}(0) = N_{\alpha,\beta}$ leads to the identity

$$\begin{bmatrix} I_p & i(\delta_2 - \delta_1) \\ 0 & I_p \end{bmatrix} N_{\alpha,\beta} \begin{bmatrix} I_p & 0 \\ -i(\delta_2 - \delta_1) & I_p \end{bmatrix} = N_{\alpha,\beta}.$$

But this identity holds if and only if $(\alpha + \beta)(\delta_2 - \delta_1) = 0$. Therefore, $\delta_2 = \delta_1$, since $\alpha + \beta > 0$, by assumption. □

Theorem 12.31 *Let $\sigma \in (\mathcal{C}^{p\times p})_{\rm sf}$, let $J = J_p$, let $0 < d \le \infty$ and let $N_{\alpha,\beta}$ be defined by (12.2). Then there exists at most one potential $\mathcal{V}(t)$ that meets the conditions (12.3)–(12.4) such that $\sigma \in \Sigma_{\rm sf}^{d}(N_{\alpha,\beta}; \mathcal{V})$.*

Proof This follows from assertions (2)–(4) of Theorem 12.29 and Theorems 12.24 and 12.30. □

12.7 The Krein algorithms for the inverse input scattering and impedance problems

Theorem 12.32 *Let $0 < d \le \infty$, $N = N_{\alpha,\beta}$ with $\alpha \ge 0$, $\beta \ge 0$ and $\kappa = \alpha + \beta > 0$, let V_2 be a unitary matrix such that $V_2 J V_2^* = J_p$ and let $\mathfrak{a} \in \mathring{\mathcal{A}}_{\kappa d}^{p\times p}$ be continuous on the interval $(0, \kappa d)$ with finite limit $a(0+)$. Then there exists one and only one* *potential $\mathcal{V}(t)$ on the interval $[0, d)$ that meets the properties (12.3)–(12.4) such that*

$$\mathfrak{a}^d_{\mathcal{V},\kappa} = \mathfrak{a}(t) \quad \text{a.e. in the interval} \quad [0, \kappa d). \tag{12.82}$$

Moreover, the potential $\mathcal{V}(t)$ is continuous on the interval $[0, d)$ and may be obtained from the formula

$$\mathcal{V}(t) = V_2^* \mathfrak{V} \begin{bmatrix} 0 & \mathfrak{v}(t) \\ \mathfrak{v}(t)^* & 0 \end{bmatrix} \mathfrak{V} V_2, \tag{12.83}$$

where $\mathfrak{v}(t) = \gamma^t(t, 0)$ for $0 < t < d$, $\mathfrak{v}(0) = \mathfrak{v}(0+)$ and $\gamma^t(a, b)$ denotes the solution of the integral equation

$$\gamma^t(a, b) + \mathfrak{a}(a - b) + \int_0^t \gamma^t(a, c)\mathfrak{a}(c - b)dc = 0 \quad \text{for} \quad 0 \le a, b \le t$$

and $0 < t < \kappa d$.

Proof This is theorem 1.2 of [AGKLS10], but the story begins much earlier; some historical notes are provided in Section 12.13. □

Remark 12.33 *The mvf*

$$\begin{aligned} u(t, \lambda) &= [u_1(t, \lambda) \quad u_2(t, \lambda)] \\ &= \left[e^{i\lambda t} \left(I_p + \int_0^t e^{-i\lambda s} \gamma^t(s, 0) ds \right) \quad I_p + e^{i\lambda t} \int_0^t e^{-i\lambda s} \gamma^t(s, t) ds \right] \end{aligned} \tag{12.84}$$

is the unique solution of the Krein system with $J = j_p$, $\alpha = 1$ and $\beta = 0$ such that

$$[u_1(0, \lambda) \quad u_2(0, \lambda)] = [I_p \quad I_p] \tag{12.85}$$

and formulas (12.82) and (12.83) with $V_2 = \mathfrak{V}$ and $\mathfrak{v}(t) = \gamma^t(0, t)$ hold. The mvf's $u_1(t, \lambda)$ and $u_2(t, \lambda)$ were viewed as continuous analogs of orthogonal polynomials that are generated by the first and last columns of the inverse to a positive definite Toeplitz (or block Toeplitz) matrix by M.G Krein [Kr55].

Theorem 12.34 *Let $0 < d < \infty$, $\kappa > 0$ and $J \neq \pm I_m$ be given, let V_1 be a unitary matrix such that $V_1 J V_1^* = j_{pq}$ and let $h \in \mathring{\mathcal{B}}^{p \times q}_{\kappa d}$ be continuous on the interval $[0, \kappa d]$. Then there exists one and only one potential $\mathcal{V}(t)$ on the interval $[0, d]$ that meets the properties (12.3)–(12.4) such that*

$$h^d_{\mathcal{V},\kappa} = h(t) \quad \text{a.e. in the interval} \quad [0, \kappa d]. \tag{12.86}$$

Moreover, the potential $\mathcal{V}(t)$ is continuous on the interval $[0, d]$ and may be obtained from the formula

$$\mathcal{V}(t) = V_1^* \begin{bmatrix} 0 & \mathfrak{v}(t) \\ \mathfrak{v}(t)^* & 0 \end{bmatrix} V_1 \quad \text{for } t > 0$$

where

$$\mathfrak{v}(t) = \begin{cases} \gamma^t_{12}(t,0) & \textit{for } t > 0 \\ \mathfrak{v}(0+) & \textit{for } t = 0, \end{cases}$$

and $\gamma^t_{12}(a,b)$ *is the* 12 *block of the solution* $\gamma^t(a,b)$ *of the integral equation*

$$\gamma^t(a,b) + \mathcal{H}(a-b) + \int_0^t \mathcal{H}(a-c)\gamma^t(c,b)dc = 0 \quad \textit{for } 0 < a,\ b < t,$$

where

$$\mathcal{H}(c) = \begin{cases} \begin{bmatrix} 0 & h(c) \\ 0 & 0 \end{bmatrix} & \textit{for } 0 < c < \kappa d \\ \begin{bmatrix} 0 & 0 \\ h(-c) & 0 \end{bmatrix} & \textit{for } -\kappa d < c < 0. \end{cases}$$

Moreover, the mvf

$$U(t) = I_m + \int_0^t \begin{bmatrix} \gamma^t_{11}(t,b) & \gamma^t_{12}(b,0) \\ \gamma^t_{21}(t,b) & \gamma^t_{22}(b,0) \end{bmatrix} db$$

is the unique solution of the differential equation

$$U'(t) = U(t) \begin{bmatrix} 0 & \mathfrak{v}(t) \\ \mathfrak{v}(t)^* & 0 \end{bmatrix} \quad \textit{for } 0 \le t \le d$$

with $U(0) = I_m$.

Proof See, e.g., theorem 3.5 of [ArD02b]. □

12.8 The left transform $T^{\ell}_{\mathfrak{A}_\ell}$ for $\mathfrak{A}_\ell \in \mathcal{W}_\ell(j_{pq})$

If

$$\Omega(\mu) = \begin{bmatrix} \Omega_{11}(\mu) & \Omega_{12}(\mu) \\ \Omega_{21}(\mu) & \Omega_{22}(\mu) \end{bmatrix}$$

is a measurable mvf on $\mathbb{R}$ that is j_{pq}-unitary a.e. on $\mathbb{R}$, then the left linear fractional transformations

$$T^{\ell}_{\Omega}[\varepsilon] = (\varepsilon\Omega_{12} + \Omega_{22})^{-1}(\varepsilon\Omega_{11} + \Omega_{21}) \tag{12.87}$$

maps $\varepsilon \in \mathbb{B}^{q\times p}$ onto $\mathbb{B}^{q\times p}$ and the right linear fractional transformation

$$T^{r}_{\Omega}[\varepsilon] = (\Omega_{11}\varepsilon + \Omega_{12})(\Omega_{21}\varepsilon + \Omega_{22})^{-1} \tag{12.88}$$

map $\varepsilon \in \mathbb{B}^{p\times q}$ onto $\mathbb{B}^{p\times q}$. Moreover,

$$\begin{aligned} &\text{if } \varepsilon \in \mathbb{B}^{q\times p}, \text{ then } \varepsilon(\mu)\varepsilon(\mu)^* = I_q \text{ a.e. on } \mathbb{R} \\ &\iff (T^{\ell}_{\Omega}[\varepsilon])(\mu)(T^{\ell}_{\Omega}[\varepsilon])(\mu)^* = I_q \text{ a.e. on } \mathbb{R} \end{aligned} \tag{12.89}$$

whereas,

$$\begin{aligned} &\text{if } \varepsilon \in \mathbb{B}^{p\times q}, \text{ then } \varepsilon(\mu)^*\varepsilon(\mu) = I_q \text{ a.e. on } \mathbb{R} \\ &\iff (T^{r}_{\Omega}[\varepsilon])(\mu)^*(T^{r}_{\Omega}[\varepsilon])(\mu) = I_q \text{ a.e. on } \mathbb{R}. \end{aligned} \tag{12.90}$$

If $E \subseteq \mathbb{B}^{q\times p}$, then

$$T^{\ell}_{\Omega}[E] = \{T^{\ell}_{\Omega}[\varepsilon] : \varepsilon \in E\}$$

and if $E \subseteq \mathbb{B}^{p\times q}$, then

$$T^{r}_{\Omega}[E] = \{T^{r}_{\Omega}[\varepsilon] : \varepsilon \in E\}.$$

Let

$$\psi_\varepsilon = \varepsilon\mathfrak{c}_+ + \mathfrak{d}_+, \quad \varphi_\varepsilon = \mathfrak{c}_-^{\#}\varepsilon + \mathfrak{d}_-^{\#}, \tag{12.91}$$

$$s_{12} = T^{r}_{\mathfrak{A}_\ell}[0] = \mathfrak{c}_+\mathfrak{d}_+^{-1} \quad \text{and} \quad f^{\ell}_{\varepsilon} = T^{\ell}_{\mathfrak{A}_\ell}[\varepsilon] = (\varepsilon\mathfrak{c}_+ + \mathfrak{d}_+)^{-1}(\varepsilon\mathfrak{d}_- + \mathfrak{c}_-) \tag{12.92}$$

be defined in terms of the block entries in the mvf $\mathfrak{A}_\ell \in \mathcal{W}_\ell(j_{pq})$ that is introduced in (12.44).

Lemma 12.35 *Let $\mathfrak{A}_\ell \in \mathcal{W}_\ell(j_{pq})$. Then*

(1) $s_{12} \in \mathring{\mathcal{S}}^{p\times q} \cap \mathcal{W}_+^{p\times q}(0)$ *and* $s_{12} = \{\mathfrak{d}_-^{\#}\}^{-1}\mathfrak{c}_-^{\#}$.
(2) $f^{\ell}_0 \in \mathcal{W}^{q\times p}(0)$, $\|f^{\ell}_0\|_\infty < 1$ *and* $f^{\ell}_0 = \mathfrak{d}_+^{-1}\mathfrak{c}_- = \mathfrak{c}_+^{\#}(\mathfrak{d}_-^{\#})^{-1}$.
(3) $\mathfrak{d}_+^{-1} \in \mathcal{S}^{q\times q}_{\text{out}}$ *and* $(\mathfrak{d}_-^{\#})^{-1} \in \mathcal{S}^{p\times p}_{\text{out}}$.
(4) *If* $\varepsilon \in \mathcal{S}^{q\times p}$, *then* $\psi_\varepsilon^{-1} \in H_\infty^{q\times q}$, $\|\psi_\varepsilon^{-1}\|_\infty \le (1 - \|s_{12}\|_\infty)^{-1}$, $\varphi_\varepsilon^{-1} \in H_\infty^{p\times p}$ *and* $\|\varphi_\varepsilon^{-1}\|_\infty \le (1 - \|s_{12}\|_\infty)^{-1}$.
(5) *If* $\varepsilon \in \mathcal{W}_+^{q\times p}(\gamma) \cap \mathcal{S}^{q\times p}$, *then* $\psi_\varepsilon^{-1} \in \mathcal{W}_+^{q\times q}(I_q)$ *and* $\varphi_\varepsilon^{-1} \in \mathcal{W}_+^{p\times p}(I_p)$.

Proof Assertions (1)–(3) follow from the equivalences (12.46) and Lemma 9.33 for mvf's in the class $\mathcal{W}_r(j_{pq})$. Next, (4) follows from (1) and (3), whereas (5) follows from (1) and Theorem 3.31, since

$$\psi_\varepsilon^{-1} = \mathfrak{d}_+^{-1}(I_q + \varepsilon s_{12})^{-1} \quad \text{and} \quad \varphi_\varepsilon^{-1} = (I_p + s_{12}\varepsilon)^{-1}\mathfrak{d}_-^{-\#}. \qquad \square$$

Lemma 12.36 *Let $\mathfrak{A}_\ell \in \mathcal{W}_\ell(j_{pq})$ be given by (12.44) and let*

$$\mathfrak{A}_r(\mu) = \begin{bmatrix} \mathfrak{d}_+(\mu)^* & \mathfrak{c}_+(\mu)^* \\ \mathfrak{c}_-(\mu)^* & \mathfrak{d}_-(\mu)^* \end{bmatrix}.$$

Then:

(1) $\mathfrak{A}_r \in \mathcal{W}_r(j_{qp})$ *and*

$$T^{\ell}_{\mathfrak{A}_\ell}[\varepsilon] = T^{r}_{\mathfrak{A}_r}[\varepsilon] \quad \text{for every mvf } \varepsilon \in \mathbb{B}^{q\times p}. \tag{12.93}$$

(2) *The transformations* $T^{\ell}_{\mathfrak{A}_\ell}$ *and* $T^{r}_{\mathfrak{A}_r}$ *are one-to-one maps of* $\mathbb{B}^{q\times p}$ *onto itself.*

(3) *If* $\varepsilon \in \mathbb{B}^{q\times p}$, $\varepsilon_\circ \in \mathbb{B}^{q\times p}$ *and* ψ_ε *and* φ_ε *are defined in formula (12.91), then*

$$T^{\ell}_{\mathfrak{A}_\ell}[\varepsilon] - T^{\ell}_{\mathfrak{A}_\ell}[\varepsilon_\circ] = \psi_\varepsilon^{-1}(\varepsilon - \varepsilon_\circ)\varphi_{\varepsilon_\circ}^{-1}. \tag{12.94}$$

Proof Since $\mathfrak{A}_\ell(\mu) j_{pq} \mathfrak{A}_\ell(\mu)^* = j_{pq}$ for $\mu \in \mathbb{R}$, a short calculation yields

$$\begin{aligned} T^{\ell}_{\mathfrak{A}_\ell}[\varepsilon] - T^{r}_{\mathfrak{A}_r}[\varepsilon_\circ] = \psi_\varepsilon^{-1}[&(\varepsilon\mathfrak{d}_- + \mathfrak{c}_-)(c_-^*\varepsilon_\circ + d_-^*) \\ &-(\varepsilon\mathfrak{c}_+ + \mathfrak{d}_+)(d_+^*\varepsilon_\circ + c_+^*)]\varphi_{\varepsilon_\circ}^{-1} \end{aligned}$$

which reduces to (12.94). Formula (12.93) is obtained by setting $\varepsilon = \varepsilon_\circ$ in (12.94).

Finally, if $f \in \mathbb{B}^{q\times p}$ and $\varepsilon = T^{\ell}_{\mathfrak{A}_\ell^{-1}}[f]$, then $\varepsilon \in \mathbb{B}^{q\times p}$, since $\mathfrak{A}_\ell^{-1}$ is j_{pq}-unitary on $\mathbb{R}$. □

Theorem 12.37 *Let* $\mathfrak{A}_\ell \in \mathcal{W}_\ell(j_{pq})$.

(1) *If* f° *is a fixed mvf in the set* $T^{\ell}_{\mathfrak{A}_\ell}[\mathcal{S}^{q\times p}]$, *then*

$$T^{\ell}_{\mathfrak{A}_\ell}[\mathcal{S}^{q\times p}] = \{f \in \mathbb{B}^{q\times p} : f - f^\circ \in H_\infty^{q\times p}\} \tag{12.95}$$

and

$$T^{\ell}_{\mathfrak{A}_\ell}[\mathcal{W}_+^{q\times p}(\gamma) \cap \mathcal{S}^{q\times p}] = \mathcal{W}^{q\times p}(\gamma) \cap T^{\ell}_{\mathfrak{A}_\ell}[\mathcal{S}^{q\times p}]. \tag{12.96}$$

(2) *There exists a unique mvf* $\mathfrak{g} \in L_1^{q\times p}(\mathbb{R}_+)$ *such that*

$$f - \int_0^\infty e^{-i\mu t}\mathfrak{g}(t)dt \in H_\infty^{q\times p} \tag{12.97}$$

for every $f \in T^{\ell}_{\mathfrak{A}_\ell}[\mathcal{S}^{q\times p}]$.

(3) *The Hankel operator* $\Gamma^\vee$ *that is defined by the formula*

$$(\Gamma^\vee\varphi)(t) = \int_0^\infty \mathfrak{g}(t+s)\varphi(s)ds, \ \varphi \in L_2^p(\mathbb{R}_+) \tag{12.98}$$

is a strictly contractive operator from $L_2^p(\mathbb{R}_+)$ *into* $L_2^q(\mathbb{R}_+)$, *i.e.,* $\|\Gamma^\vee\| < 1$.

Proof This follows from the fact that

$$\begin{bmatrix} \mathfrak{d}_- & \mathfrak{c}_+ \\ \mathfrak{c}_- & \mathfrak{d}_+ \end{bmatrix} \in \mathcal{W}_\ell(j_{pq}) \Longleftrightarrow \begin{bmatrix} \mathfrak{d}_+^* & \mathfrak{c}_+^* \\ \mathfrak{c}_-^* & \mathfrak{d}_-^* \end{bmatrix} \in \mathcal{W}_r(j_{qp})$$

and Theorem 9.35. □

12.9 Asymptotic equivalence matrices

Theorem 12.38 *If $\mathfrak{v} \in L_1^{p\times q}(\mathbb{R}_+)$ in the DK-system (12.52) and $\mathfrak{A}_\ell^t$ is defined by (12.49) for $0 \le t < \infty$, then there exists a mvf $\mathfrak{A}_\ell^\infty \in \mathcal{W}_\ell(j_{pq})$ such that*

$$\lim_{t\uparrow\infty} \|\mathfrak{A}_\ell^t - \mathfrak{A}_\ell^\infty\|_{\mathcal{W}} = 0, \tag{12.99}$$

i.e., if

$$\mathfrak{A}_\ell^\infty(\mu) = \begin{bmatrix} \mathfrak{d}_-^\infty(\mu) & \mathfrak{c}_+^\infty(\mu) \\ \mathfrak{c}_-^\infty(\mu) & \mathfrak{d}_+^\infty(\mu) \end{bmatrix} = I_m + \int_0^\infty \omega_\ell^\infty(s) e^{-i\mu s\kappa j_{pq}} ds, \tag{12.100}$$

and

$$\mathfrak{A}_\ell^t(\mu) = I_m + \int_0^t \omega_\ell^t(s) e^{-i\mu s\kappa j_{pq}} ds,$$

where $\omega_\ell^\infty \in L_1^{m\times m}(\mathbb{R}_+)$,

$$\omega_\ell^t(s) = \begin{cases} \omega_{[\ell]}(t,s) & \text{for } 0 \le s \le t \\ 0 & \text{for } s > t \end{cases}$$

and $\omega_{[\ell]}(t,s)$ is defined in Theorem 12.6, then

$$\lim_{t\uparrow\infty} \int_0^\infty \|\omega_\ell^t(s) - \omega_\ell^\infty(s)\| ds = 0. \tag{12.101}$$

Proof The assertions follow from standard estimates; see, e.g., Lemma 2.1 and the verification of (2.21) in [DI84]. □

The mvf $\mathfrak{A}_\ell^\infty$ is of special interest because of the relation

$$\lim_{t\uparrow\infty} \|W_t - \mathfrak{A}_\ell^\infty W_t^0\|_\infty = 0$$

between the matrizant W_t of the DK-system (12.52) with $\mathfrak{v} \in L_1^{p\times q}(\mathbb{R}_+)$ and the matrizant W_t^0 of the same system, but with $\mathfrak{v} = 0$ that is given explicitly by formula (12.12).

The mvf $\mathfrak{A}_\ell^\infty(\mu)^{-1}$ is called the asymptotic equivalence matrix (or A-matrix) of the equation (12.52); see [DaKr74], pp. 142–143.

12.10 Asymptotic scattering matrices (S-matrices)

In this section we shall focus on solutions $u(t,\lambda) = [u_1(t,\lambda) \quad u_2(t,\lambda)]$ of the system (12.52) with blocks $u_1(t,\lambda)$ and $u_2(t,\lambda)$ of sizes $q \times p$ and $q \times q$, respectively, that meet the asymptotic condition

$$\lim_{t\uparrow\infty} \|u_2(t,\cdot) - e_{-\beta t} I_q\|_\infty = 0 \tag{12.102}$$

and the boundary condition (12.15) at $t = 0$.

Theorem 12.39 *If $\mathfrak{A}_\ell^\infty$ is the mvf defined on $\mathbb{R}$ by formulas (12.99) and (12.100) for a DK-system (12.52) with $\mathfrak{v} \in L_1^{p\times q}(\mathbb{R}_+)$, then for each $\varepsilon \in \mathcal{S}^{q\times p}$ there exists exactly one solution*

$$u(t\,\lambda) = [u_1(t,\lambda) \quad u_2(t,\lambda)]$$

of the Dirac–Krein system (12.52) that satisfies the boundary condition (12.15) and the asymptotic condition (12.102). Moreover, if $u(t,\lambda)$ is this unique solution, then:

(1) *There exists exactly one mvf $s_\varepsilon^\infty \in \mathbb{B}^{q\times p}$ such that*

$$\lim_{t\uparrow\infty} \|u_1(t,\cdot) - e_{\alpha t} s_\varepsilon^\infty\|_\infty = 0. \tag{12.103}$$

(2) *The formulas*

$$u_2(0,\lambda) = (\psi_\varepsilon^\infty(\lambda))^{-1} = (\varepsilon(\lambda)\mathfrak{c}_+^\infty(\lambda) + \mathfrak{d}_+^\infty(\lambda))^{-1} \tag{12.104}$$

and

$$\begin{aligned} s_\varepsilon^\infty(\mu) = T_{\mathfrak{A}_\ell^\infty}^\ell[\varepsilon] &= (\varepsilon(\mu)\mathfrak{c}_+^\infty(\mu) + \mathfrak{d}_+^\infty(\mu))^{-1}(\varepsilon(\mu)\mathfrak{d}_-^\infty(\mu) + \mathfrak{c}_-^\infty(\mu)) \\ &= u(0,\mu)\begin{bmatrix} \mathfrak{d}_-^\infty(\mu) \\ \mathfrak{c}_-^\infty(\mu) \end{bmatrix} \end{aligned} \tag{12.105}$$

hold for $\lambda \in \mathbb{C}_+$ and a.e. on $\mathbb{R}$, respectively.

(3) *If $\varepsilon(\lambda) = \gamma$ is a constant contractive $q \times p$ matrix, then $s_\gamma^\infty \in \mathcal{W}^{q\times p}(\gamma)$ and it admits a factorization of the form*

$$s_\gamma^\infty(\mu) = \psi_\gamma^+(\mu)^{-1}\psi_\gamma^-(\mu) \text{ on } \mathbb{R}, \tag{12.106}$$

where

$$(\psi_\gamma^+)^{\pm 1} \in \mathcal{W}_+^{q\times q}(I_q), \quad \psi_\gamma^- \in \mathcal{W}_-^{q\times p}(\gamma), \tag{12.107}$$

and the factors $\psi_\gamma^\pm$ may be chosen by the formulas

$$\psi_\gamma^+(\mu) = \gamma\mathfrak{c}_+^\infty(\mu) + \mathfrak{d}_+^\infty(\mu) \text{ and } \psi_\gamma^-(\mu) = \gamma\mathfrak{d}_-^\infty(\mu) + \mathfrak{c}_-^\infty(\mu) \tag{12.108}$$

for $\mu \in \mathbb{R}$.

(4) *If $q = p$ and $\varepsilon(\lambda) = \gamma$ is a constant unitary matrix, then the mvf $s_\gamma^\infty(\mu)$ is unitary on $\mathbb{R}$ and admits a unique factorization of the form (12.106) with factors*

$$(\psi_\gamma^+)^{\pm 1} \in \mathcal{W}_+^{p\times p}(I_p) \quad \text{and} \quad (\psi_\gamma^-)^{\pm 1} \in \mathcal{W}_-^{p\times p}(\gamma^{\pm 1}). \tag{12.109}$$

The factors are given by formula (12.108).

Proof The proof is divided into steps.

1. Proof of (2): Every solution $u(t,\lambda) = [u_1(t,\lambda) \quad u_2(t,\lambda)]$ of the system (12.52) that meets the initial condition (12.15) is given in terms of the matrizant $U_t(\lambda) =$

$\mathfrak{A}_\ell^t(\lambda)e^{i\lambda N_{\alpha,\beta}j_{pq}}$ by the formula

$$
\begin{aligned}
u(t,\lambda) &= u(0,\lambda)\mathfrak{A}_\ell^t(\lambda)e^{i\lambda tN_{\alpha,\beta}j_{pq}} \\
&= u_2(0,\lambda)[\varepsilon(\lambda) \quad I_q]\begin{bmatrix} e^{i\lambda t\alpha}\mathfrak{d}_-^t(\lambda) & e^{-i\lambda t\beta}\mathfrak{c}_+^t(\lambda) \\ e^{i\lambda t\alpha}\mathfrak{c}_-^t(\lambda) & e^{-i\lambda t\beta}\mathfrak{d}_+^t(\lambda) \end{bmatrix}.
\end{aligned} \tag{12.110}
$$

Thus, the asymptotic condition (12.102) may be rewritten as the limit relation

$$
\lim_{t\to\infty} \|u_2(0,\mu)(\varepsilon(\mu)\mathfrak{c}_+^t(\mu)+\mathfrak{d}_+^t(\mu)) - I_q\|_\infty = 0, \tag{12.111}
$$

which, in view of Theorem 12.38, implies that

$$
u_2(0,\mu)[\varepsilon(\mu)\mathfrak{c}_+^\infty(\mu)+\mathfrak{d}_+^\infty(\mu)] = I_q \quad \text{a.e. on } \mathbb{R}. \tag{12.112}
$$

Therefore, (12.102) implies that (12.105) is valid a.e. on $\mathbb{R}$, which in turn implies that formula (12.104) holds on $\mathbb{C}_+$ and guarantees the unique dependence of $u(t,\lambda)$ on ε that is asserted in the first paragraph of the theorem.

2. Proof of (1) and (3): In view of (12.110), (12.104) and (12.105),

$$
\begin{aligned}
\|u_1(t,\cdot) - e_{\alpha t}s_\varepsilon^\infty\|_\infty &= \|(\psi_\varepsilon^\infty)^{-1}\{(\varepsilon\mathfrak{d}_-^t + \mathfrak{c}_-^t) - (\varepsilon\mathfrak{d}_-^\infty + \mathfrak{c}_-^\infty)\}\|_\infty \\
&\le \|(\psi_\varepsilon^\infty)^{-1}\|_\infty\{\|\mathfrak{d}_-^t - \mathfrak{d}_-^\infty\|_\infty + \|\mathfrak{c}_-^t - \mathfrak{c}_-^\infty\|_\infty\}.
\end{aligned}
$$

Therefore, (12.103) holds, and the mvf s_ε^∞ that satisfies (12.103) is clearly unique. Assertion (3) follows from (2) and the fact that the mvf $\mathfrak{A}_\ell^\infty$ belongs to the class $\mathcal{W}_\ell(j_{pq})$.

3. Proof of (4): If γ is unitary, then s_γ^∞ is unitary by (12.89). Moreover, $\psi_\gamma^- = \gamma\varphi_\gamma^\#$ and $(\psi_\gamma^-)^{-1} \in \mathcal{W}_-^{p\times p}(\gamma^{-1})$ by (5) of Lemma 12.35. The uniqueness of the factors in (12.106) that meet the constraints (12.109) may be easily checked. □

Remark 12.40 *If s is a unitary mvf on $\mathbb{R}$ that admits a factorization of the form $s = (\psi^+)^{-1}\psi^-$, where $(\psi^+)^{\pm 1} \in \mathcal{W}_+^{p\times p}(I_p)$ and $(\psi^-)^{\pm 1} \in \mathcal{W}_-^{p\times p}(\gamma^{\pm 1})$ for some invertible matrix γ, then γ is unitary and there exists exactly one mvf $\mathfrak{A}_\ell \in \mathcal{W}_\ell(j_p)$ such that $s = T_{\mathfrak{A}_\ell}^\ell[\gamma]$ and the factors $\psi^\pm$ are uniquely specified as $\psi^\pm = \psi_\gamma^\pm$, where $\psi_\gamma^\pm$ are defined in (12.108).*

Formula (12.105) yields a description of the set

$$
\mathcal{S}_{\text{ascat}}^\infty(N_{\alpha,\beta};\mathcal{V}) = \{s_\varepsilon^\infty : \varepsilon \in \mathcal{S}^{q\times p}\} \tag{12.113}
$$

of **asymptotical scattering matrices** of the DK-system (12.52) with $\mathfrak{v} \in L_1^{p\times q}(\mathbb{R}_+)$ as the mvf ε in condition (12.15) varies over $\mathcal{S}^{q\times p}$.

Theorem 12.41 *If $d = \infty$ and $\mathfrak{v} \in L_1^{p\times q}(\mathbb{R}_+)$ in the DK-system (12.52), $\varepsilon \in \mathcal{S}^{q\times p}$,*

$$
s_\varepsilon^t = T_{\mathfrak{A}_\ell^t}^\ell[\varepsilon] \quad \textit{and} \quad s_{12}^t = T_{\mathfrak{A}_\ell^t}^r[0] = \mathfrak{c}_+^t(\mathfrak{d}_+^t)^{-1} \quad \textit{for } 0 < t \le \infty,
$$

then $s_{12}^t = T_{w_t}[0] \in \mathcal{S}_{\text{scat}}^t(N_{\alpha,\beta}; \mathcal{V})$,

$$s_{12}^t \in \mathcal{W}_+^{p\times q}(0) \cap \mathring{\mathcal{S}}^{p\times q} \quad \textit{for } 0 < t \le \infty,$$

$$\lim_{t\uparrow\infty} \|s_\varepsilon^t - s_\varepsilon^\infty\|_\infty = 0 \quad \textit{and} \quad \lim_{t\uparrow\infty} \|s_{12}^t - s_{12}^\infty\|_{\mathcal{W}} = 0. \tag{12.114}$$

Moreover, if $\varepsilon \in \mathcal{W}^{q\times p} \cap \mathcal{S}^{q\times p}$, *then* $s_\varepsilon^t \in \mathcal{W}^{q\times p}$ *for every* t, $0 < t \le \infty$, *and*

$$\lim_{t\uparrow\infty} \|s_\varepsilon^t - s_\varepsilon^\infty\|_{\mathcal{W}} = 0. \tag{12.115}$$

Proof Using the right linear fractional representation

$$s_\varepsilon^t = T^r_{T^*(\mathfrak{A}_\ell^t)^*T}[\varepsilon] = \{(\mathfrak{d}_+^t)^*\varepsilon + (\mathfrak{c}_+^t)^*\}\{(\mathfrak{c}_-^t)^*\varepsilon + (\mathfrak{d}_-^t)^*\}^{-1}$$

for s_ε^t and the notation $\psi_\varepsilon^\infty = \varepsilon\mathfrak{c}_+^\infty + \mathfrak{d}_+^\infty$ and $\varphi_\varepsilon^t = (\mathfrak{c}_-^t)^*\varepsilon + (\mathfrak{d}_-^t)^*$, we obtain:

$$s_\varepsilon^t - s_\varepsilon^\infty = (\psi_\varepsilon^\infty)^{-1}\left\{\varepsilon(\mathfrak{c}_+^\infty\mathfrak{d}_+^{t*} - \mathfrak{d}_-^\infty\mathfrak{c}_-^{t*})\varepsilon + \varepsilon(\mathfrak{c}_+^\infty\mathfrak{c}_+^{t*} - \mathfrak{d}_-^\infty\mathfrak{d}_-^{t*})\right.$$
$$\left. + (\mathfrak{d}_+^\infty\mathfrak{d}_+^{t*} - \mathfrak{c}_-^\infty\mathfrak{c}_-^{t*})\varepsilon + (\mathfrak{d}_+^\infty\mathfrak{c}_+^{t*} - \mathfrak{c}_-^\infty\mathfrak{d}_-^{t*})\right\}(\varphi_\varepsilon^t)^{-1}.$$

Let $s_{12}^t = T^r_{\mathfrak{A}_\ell^t}[0] = \mathfrak{c}_+^t(\mathfrak{d}_+^t)^{-1}$ for $0 < t < \infty$. Then $\|s_{12}^t\|_\infty < 1$, $s_{12}^t \in \mathcal{W}_+^{p\times p}(0)$ for $0 < t \le \infty$ and

$$\|s_{12}^t - s_{12}^\infty\|_{\mathcal{W}} \le \|(\mathfrak{d}_-^{t*})^{-1}\|_{\mathcal{W}}\,\|\mathfrak{c}_-^{t*}\mathfrak{d}_+^\infty - \mathfrak{d}_-^{t*}\mathfrak{c}_+^\infty\|_{\mathcal{W}}\,\|(\mathfrak{d}_+^\infty)^{-1}\|_{\mathcal{W}}.$$

Moreover, since

$$\lim_{t\uparrow\infty} \|\mathfrak{A}_\ell^t - \mathfrak{A}_\ell^\infty\|_{\mathcal{W}} = 0 \quad \text{and} \quad \mathfrak{c}_-^{\infty*}\mathfrak{d}_+^\infty - \mathfrak{d}_-^{\infty*}\mathfrak{c}_+^\infty = 0,$$

it follows that

$$\mathfrak{c}_-^{t*}\mathfrak{d}_+^\infty - \mathfrak{d}_-^{t*}\mathfrak{c}_+^\infty = (\mathfrak{c}_-^{t*} - \mathfrak{c}_-^{\infty*})\mathfrak{d}_+^\infty + (\mathfrak{d}_-^{\infty*} - \mathfrak{d}_-^{t*})\mathfrak{c}_+^\infty$$

and hence that

$$\lim_{t\uparrow\infty} \|s_{12}^t - s_{12}^\infty\|_{\mathcal{W}} = 0 \quad \text{and} \quad \lim_{t\uparrow\infty} \|s_{12}^t\|_\infty = \|s_{12}^\infty\|_\infty < 1.$$

Therefore, there exists a $t_0 > 0$ such that $\|s_{12}^t\|_\infty \le q < 1$ for $t > t_0$. Thus, as $\varphi_\varepsilon^t(\mu) = \mathfrak{d}_-^t(\mu)^*(I_q + s_{12}^t(\mu)\varepsilon)$, a.e. on $\mathbb{R}$, it is readily seen that

$$\|(\varphi_\varepsilon^t)^{-1}\|_\infty \le (1 - \|s_{12}^t\|)^{-1} \le (1-q)^{-1} \text{ for every } t > t_0. \tag{12.116}$$

The bounds $\|\varepsilon\|_\infty \le 1$, $\|(\psi_\varepsilon^\infty)^{-1}\|_\infty < \infty$, (12.116) and the above equality for $s_\varepsilon^t - s_\varepsilon^\infty$ imply the relations (12.114) and (12.115), since $\mathfrak{A}_\ell^\infty(\mu)j_{pq}\mathfrak{A}_\ell^t(\mu)^*$ tends to j_{pq} in the $\mathcal{W}$-norm, and so too in the H_∞-norm, as $t \to \infty$. □

Theorem 12.42 *If* $d = \infty$ *and* $\mathfrak{v} \in L_1^{p\times q}(\mathbb{R}_+)$ *in the DK-system (12.52), then*

$$\mathcal{S}_{\text{scat}}^\infty(N_{\alpha,\beta}; \mathcal{V}) = \{T^r_{\mathfrak{A}_\ell^\infty}[0]\} = \{\mathfrak{c}_+^\infty(\mathfrak{d}_+^\infty)^{-1}\}. \tag{12.117}$$

Moreover,

$$\mathfrak{c}_+^\infty(\mathfrak{d}_+^\infty)^{-1} \in \mathcal{W}_+^{p\times q}(0) \cap \mathring{\mathcal{S}}^{p\times q}, \tag{12.118}$$

i.e.,

$$\mathfrak{c}_+^\infty(\lambda)\mathfrak{d}_+^\infty(\lambda)^{-1} = \int_0^\infty e^{i\lambda t} h(t)dt, \tag{12.119}$$

where $h \in \mathring{\mathcal{B}}_\infty^{p\times q}$ is the mvf that is considered in Theorem 12.15.

Proof Fix $t_0 \in (0, d)$ and let $s_{12}^t = T_{W_t}[0]$ for $t \geq t_0$. Then,

$$s_{12}^t = T_{W_{t_0}}[\varepsilon_t], \quad \text{for some } \varepsilon_t \in \mathcal{S}^{p\times q},$$

since $T_{W_t}[0] \subseteq T_{W_{t_0}}[\mathcal{S}^{p\times q}]$. Thus, there exists a sequence of points t_n that tend to ∞ as $n \uparrow \infty$ and a mvf $\varepsilon \in \mathcal{S}^{q\times p}$ such that $\varepsilon_{t_n}(\lambda) \to \varepsilon(\lambda)$ at each point $\lambda \in \mathbb{C}_+$ as $n \uparrow \infty$. Consequently,

$$\mathfrak{c}_+^{t_n}(\lambda)(\mathfrak{d}_+^{t_n}(\lambda))^{-1} = s_{12}^{t_n}(\lambda) \to T_{W_{t_0}(\lambda)}[\varepsilon(\lambda)] \quad \text{as } n \uparrow \infty$$

at each point $\lambda \in \mathbb{C}_+$. Therefore,

$$\mathfrak{c}_+^\infty(\mathfrak{d}_+^\infty)^{-1} \in T_{W_{t_0}}[\mathcal{S}^{p\times q}] \quad \text{for every } t_0 \in (0, d),$$

i.e.,

$$\mathfrak{c}_+^\infty(\mathfrak{d}_+^\infty)^{-1} \in \mathcal{S}_{\text{scat}}^\infty(N_{\alpha,\beta}; \mathcal{V}).$$

Moreover, since $(\mathfrak{d}_+^\infty)^{-1} \in \mathcal{W}_+^{q\times q}(I_q)$, $\mathfrak{c}_+^\infty \in \mathcal{W}_+^{p\times q}(0)$ and

$$I_q - s_{12}^\infty(\mu)^* s_{12}^\infty(\mu) = \mathfrak{d}_+^\infty(\mu)^{-*}\mathfrak{d}_+^\infty(\mu)^{-1} > \delta I_q \quad \text{for some } \delta > 0 \text{ on } \mathbb{R}.$$

This serves to justify (12.118); (12.119) then follows from Theorems 12.15 and 12.41. □

Let

$$(\mathcal{W}_-^{q\times p}(0) + H_\infty^{q\times p})^\circ = \{s \in \mathcal{W}_-^{q\times p}(0) + H_\infty^{q\times p} : \|s\|_\infty \leq 1 \text{ and } \|\Pi_- M_s|_{H_2^p}\| < 1\},$$

where M_s denotes the operator of multiplication by s from L_2^p into L_2^q and Π_- is the orthogonal projection from L_2^q onto $(H_2^q)^\perp$.

Theorem 12.43 *If $\mathfrak{A}_\ell^\infty$ is the mvf defined on $\mathbb{R}$ by formulas (12.99) and (12.100) for a DK-system (12.52) with $d = \infty$ and $\mathfrak{v} \in L_1^{q\times p}(\mathbb{R}_+)$, then:*

1. $\mathcal{S}_{\text{ascat}}^\infty(N_{\alpha,\beta}; \mathcal{V}) = T_{\mathfrak{A}_\ell^\infty}^\ell[\mathcal{S}^{q\times p}]$.
2. $\mathcal{S}_{\text{ascat}}^\infty(N_{\alpha,\beta}; \mathcal{V}) \subseteq (\mathcal{W}_-^{q\times p}(0) + H_\infty^{q\times p})^\circ$.
3. *There exists exactly one mvf $\mathfrak{g} \in L_1^{q\times p}(\mathbb{R}_+)$ such that $NP(\mathfrak{g})$ is strictly completely indeterminate and*

$$\mathcal{S}_{\text{ascat}}^\infty(N_{\alpha,\beta}; \mathcal{V}) = \mathcal{N}(\mathfrak{g}). \tag{12.120}$$

(4) *The Hankel operator* $\Gamma^\vee$ *acting from* $L_2^p(\mathbb{R}_+)$ *into* $L_2^q(\mathbb{R}_+)$ *that is defined in terms of* $\mathfrak{g}$ *by formula (12.98) is strictly contractive, i.e.,* $\|\Gamma^\vee\| < 1$.
(5) $\|s_\varepsilon^\infty\|_\infty < 1 \Longleftrightarrow \varepsilon \in \mathring{\mathcal{S}}^{q\times p}$.
(6) $s_\varepsilon^\infty(\mu)^* s_\varepsilon^\infty(\mu) = I_p$ *a.e. on* $\mathbb{R} \Longleftrightarrow \varepsilon \in \mathcal{S}_{\text{in}}^{q\times p}$.
(7) $s_\varepsilon^\infty(\mu) s_\varepsilon^\infty(\mu)^* = I_q$ *a.e. on* $\mathbb{R} \Longleftrightarrow \varepsilon \in \mathcal{S}_{*\text{in}}^{q\times p}$.
(8) $s_\varepsilon^\infty \in \mathcal{W}^{q\times p}(\gamma) \Longleftrightarrow \varepsilon \in \mathcal{W}_+^{q\times p}(\gamma) \cap \mathcal{S}^{q\times p}$.

Proof (1) and (5)–(8) follow from formula (12.105) and the fact that $\mathfrak{A}_\ell^\infty \in \mathcal{W}_\ell(j_{pq})$ (which is established in Theorem 12.38); (2)–(4) follow from Theorem 12.37. □

12.11 The inverse asymptotic scattering problem

The data for the inverse asymptotic scattering problem is a pair of numbers $\alpha \geq 0$, $\beta \geq 0$ with $\kappa = \alpha + \beta > 0$ and a mvf $s \in (\mathcal{W}_-^{q\times p}(0) + H_\infty^{q\times p})^\circ$, and the objective is to find a mvf $\mathfrak{v} \in L_1^{q\times p}(\mathbb{R}_+)$ and a mvf $\varepsilon \in \mathcal{S}^{q\times p}$ such that $s = s_\varepsilon^\infty$.

Theorem 12.44 *Let* $s \in (\mathcal{W}_-^{q\times p}(0) + H_\infty^{q\times p})^\circ$ *and* $\kappa > 0$ *be given. Then there exists at most one* $\mathfrak{v} \in L_1^{q\times p}(\mathbb{R}_+)$ *and at most one* $\varepsilon \in \mathcal{S}^{q\times p}$ *such that the asymptotic scattering matrix* s_ε^∞ *for the DK-system (12.52) with this choice of* $\mathfrak{v}$ *and this mvf* ε *in the condition (12.15) coincides with* s*. Moreover, if the inverse problem for the given* s *and* κ *is solvable, then* $\mathfrak{v}$ *and* ε *may be obtained by the following algorithm:*

(a) *Find* $\mathfrak{g} \in L_1^{q\times p}(\mathbb{R}_+)$ *such that*

$$s - \widehat{\mathfrak{g}}(-\cdot) \in H_\infty^{q\times p}. \tag{12.121}$$

(b) *Use formulas (9.122) and (9.123) to find*

$$\mathfrak{A}_r = \begin{bmatrix} \mathfrak{a}_- & \mathfrak{b}_- \\ \mathfrak{b}_+ & \mathfrak{a}_+ \end{bmatrix} \in \mathcal{W}_r(j_{qp}), \tag{12.122}$$

where $\Gamma^\vee$ *is the Hankel operator that is defined in terms of the* $q \times p$ *mvf* $\mathfrak{g}(t)$ *by formula (12.98).*

(c) *Use the formulas* $\mathfrak{A}_\ell^\infty(\mu)^{-1} = j_{pq} T^* \mathfrak{A}_r(\mu) T j_{pq}$ *in terms of the notation (9.109) and*

$$\varepsilon(\mu) = T^\ell_{(\mathfrak{A}_\ell^\infty)^{-1}}[s] = \{-s(\mu)\mathfrak{b}_+(\mu) + \mathfrak{a}_-(\mu)\}^{-1}\{s(\mu)\mathfrak{a}_+(\mu) - \mathfrak{b}_-(\mu)\} \tag{12.123}$$

to find the mvf $\varepsilon \in \mathcal{S}^{q\times p}$ *that defines the boundary condition (12.15).*

(d) *Compute the input scattering matrix* $s^\circ(\lambda)$ *from the formula*

$$s^\circ(\lambda) = \mathfrak{a}_+(\lambda)^{-1}\mathfrak{b}_+(\lambda) \tag{12.124}$$

and then find the mvf $h \in \mathring{\mathcal{B}}_{\infty}^{p\times q}$ such that

$$s^{\circ}(\lambda) = \int_0^{\infty} e^{i\lambda t} h(t)dt \quad \textit{for } \lambda \in \overline{\mathbb{C}_+}. \tag{12.125}$$

(e) *Find the resolvent matrix $W_a(\lambda)$ of $SP(h;\kappa a)$ via Theorem 9.39 and then use formula (12.61) to calculate $\mathcal{V}(t)$.*

Proof Under the given assumptions, s admits exactly one representation (12.121) and the Hankel operator $\Gamma^{\vee}$ defined by formula (12.98) is strictly contractive. Then, by Theorem 9.37, there exists exactly one mvf $\mathfrak{A}_r \in \mathcal{W}_r(j_{qp})$ such that $\mathcal{N}(\mathfrak{g}) = T^r_{\mathfrak{A}_r}[\mathcal{S}^{q\times p}]$; $\mathfrak{A}_r$ can be found from $\mathfrak{g}$ by formulas (9.122) and (9.123). Since $s \in \mathcal{N}(\mathfrak{g})$, there exists a mvf $\varepsilon \in \mathcal{S}^{q\times p}$ such that $s = T^r_{\mathfrak{A}_r}[\varepsilon]$. Now let

$$\mathfrak{A}_\ell = \begin{bmatrix} \mathfrak{d}_- & \mathfrak{c}_+ \\ \mathfrak{c}_- & \mathfrak{d}_+ \end{bmatrix} = \begin{bmatrix} \mathfrak{a}_+^* & \mathfrak{b}_-^* \\ \mathfrak{b}_+^* & \mathfrak{a}_-^* \end{bmatrix} \quad \text{on } \mathbb{R}.$$

Then $\mathfrak{A}_\ell \in \mathcal{W}_\ell(j_{pq})$ and, by formulas (12.87) and (12.88),

$$s = T^r_{\mathfrak{A}_r}[\varepsilon] = T^\ell_{\mathfrak{A}_\ell}[\varepsilon].$$

Thus, as

$$\mathfrak{A}^\ell(\mu)^{-1} = j_{qp}\mathfrak{A}^\ell(\mu)^* j_{qp} = \begin{bmatrix} \mathfrak{a}_+ & -\mathfrak{b}_+ \\ -\mathfrak{b}_- & \mathfrak{a}_- \end{bmatrix} \quad \text{on } \mathbb{R},$$

formula (12.123) follows from the fact that $\varepsilon = T^\ell_{\mathfrak{A}_\ell^{-1}}[s]$.

Next, formula (12.124) follows from the recipe

$$s^{\circ} = T^r_{\mathfrak{A}_\ell}[0] = \mathfrak{c}_+\mathfrak{d}_+^{-1} = \mathfrak{b}_-^*(\mathfrak{a}_-^*)^{-1} = \mathfrak{a}_+^{-1}\mathfrak{b}_+.$$

Finally, in view of Lemma 12.12 and Theorem 12.15, the matrizant $W_t, 0 < t < \infty$, of the DK-system (12.52) with $\alpha = \kappa$ and $\beta = 0$ may be identified with the resolvent matrix of SP$(h;\kappa t)$ that is considered in Theorem 9.39. The potential may then be found by formula (12.61). □

Remark 12.45 *If $q = p$, then conclusions analogous to (d) and (e) may be based on the accelerant $h_c \in L_1^{p\times p}(\mathbb{R}_+)$ of the input impedance matrix*

$$\begin{aligned} c &= (I_p - s^{\circ})(I_p + s^{\circ})^{-1} = (\mathfrak{d}_+ - \mathfrak{c}_+)(\mathfrak{d}_+ + \mathfrak{c}_+)^{-1} \\ &= I_p + 2\int_0^{\infty} e^{i\lambda t} h_c(t)dt, \end{aligned}$$

in which

$$s^{\circ} = \mathfrak{a}_+^{-1}\mathfrak{b}_+ = \mathfrak{c}_+(\mathfrak{d}_+)^{-1}.$$

Moreover, since $s^\circ \in \mathcal{W}_+^{p\times p}(0) \cap \mathring{\mathcal{S}}^{p\times p}$, $c \in \mathcal{W}_+^{p\times p}(I_p) \cap \mathring{\mathcal{C}}^{p\times p}$ *and the accelerant* $h_c \in \mathring{\mathcal{A}}_\infty^{p\times p}$ *of the DK-system (12.62) with* $N_{\alpha,\beta}$ *and* $\mathcal{V}$ *given by (12.63) and (12.64), respectively.*

The resolvent matrix $B_t(\lambda)$ *of the* $CP(h_c;\kappa t)$ *that is constructed in Theorem 9.21 with* $a = \kappa t$ *and* h *equal to the restriction of* h_c *to* $[0,\kappa t]$, *for* $0 < t < \infty$, *may be obtained from the formula* $B_t(\lambda) = A_t(\lambda)\mathfrak{V}$, *where* A_t *is the matrizant of the DK-system (12.62) with* $N_{\alpha,\beta}$ *and* $\mathcal{V}$ *given by (12.63) and (12.64), respectively, with* $\beta = 0$ *and* $\alpha = \kappa > 0$.

Let $\mathcal{C}^{p\times p}(\mathbb{R}_+)$ denote the class of continuous $p \times p$ mvf's on $\mathbb{R}_+$.

Theorem 12.46 *If* $s \in (\mathcal{W}_-^{p\times p}(0) + H_\infty^{p\times p})^\circ$ *and* $\mathfrak{g}$ *is defined by* s *as in Theorem 12.44, then:*

1. *There exists at most one* $\mathfrak{v} \in L_1^{p\times p}(\mathbb{R}_+)$ *and one* $\varepsilon \in \mathcal{S}^{p\times p}$ *such that the given mvf* s *is the asymptotic scattering matrix* s_ε^∞ *for the DK-system (12.52) with this choice of* $\mathfrak{v}$ *and this mvf* ε *in the condition (12.15) coincides with* s.
2. *If* $\mathfrak{g} \in \mathcal{C}^{p\times p}(\mathbb{R}_+)$, *then there exists exactly one mvf* $\varepsilon \in \mathcal{S}^{q\times p}$ *and exactly one mvf* $\mathfrak{v} \in L_1^{p\times p}(\mathbb{R}_+) \cap \mathcal{C}^{p\times p}(\mathbb{R}_+)$ *such that (1) holds.*
3. *If* s *is the asymptotic scattering matrix for the DK-system (12.51) with* $\mathfrak{v} \in L_1^{p\times p}(\mathbb{R}) \cap \mathcal{C}^{p\times p}(\mathbb{R}_+)$ *and some* $\varepsilon \in \mathcal{S}^{p\times p}$, *then* $\mathfrak{g} \in \mathcal{C}^{p\times p}(\mathbb{R}_+)$ *and* $s \in (\mathcal{W}_-^{p\times p}(0) + H_\infty^{p\times p})^\circ$.

Proof The first assertion is immediate from Theorem 12.44. Moreover, the condition $s \in (\mathcal{W}_-^{p\times p}(0) + H_\infty^{p\times p})^\circ$ guarantees that the mvf $\mathfrak{A}_r$ that was considered in Theorem 12.44 is defined, and the mvf's

$$s_0^\infty = (\mathfrak{a}_-^*)^{-1}\mathfrak{b}_+^* \text{ and } s_{in} = -\mathfrak{a}_+^{-1}\mathfrak{b}_+ \tag{12.126}$$

may be defined in the terms of the blocks of $\mathfrak{A}_r$. Moreover, $s_0^\infty \in \mathcal{N}(\mathfrak{g})$. Therefore, since $(\mathfrak{a}_-^\#)^{-1}$ and $\mathfrak{a}_+^{-1}$ belong to $\mathcal{W}_+^{p\times p}(I_p)$,

$$\mathfrak{g} \in \mathcal{C}^{p\times p}(\mathbb{R}_+) \Longleftrightarrow \mathfrak{b}_+^\vee \in \mathcal{C}^{p\times p}(\mathbb{R}_+) \Longleftrightarrow h \in \mathcal{C}^{p\times p}(\mathbb{R}_+) \Longleftrightarrow \mathcal{V} \in \mathcal{C}^{m\times m}(\mathbb{R}_+).$$

The rest of the theorem follows from the corresponding results on the inverse input scattering problem; see [ArD02b], and, for additional discussion of the role of continuity, [KrL85], [Dy90], [GK85] and [AGKLS10]. □

12.12 More on spectral functions of DK-systems

Theorem 12.47 *Let*

$$\mathfrak{E}_t(\lambda) = \begin{bmatrix} E_-^t(\lambda) & E_+^t(\lambda) \end{bmatrix} = \sqrt{2}N_2^* A_t(\lambda)\mathfrak{V}$$

denote the de Branges matrix based on the matrizant $A_t(\lambda)$ of the DK-system (12.62) (taking note of (12.63) and (12.64)) and let

$$\widetilde{\mathfrak{E}}_t(\lambda) = \begin{bmatrix} \widetilde{E}_-^t(\lambda) & \widetilde{E}_+^t(\lambda) \end{bmatrix} = \sqrt{2}N_2^* \widetilde{A}_t(\lambda)\mathfrak{V}$$

denote the de Branges matrix based on the matrizant $\widetilde{A}_t(\lambda) = A_t(\lambda)A_t(0)^{-1}$ of the corresponding canonical differential system (12.5). Then:

(1) $\mathcal{B}(\mathfrak{E}_t) = \mathcal{B}(\widetilde{\mathfrak{E}}_t) = \mathcal{H}_*(e_{\beta t}I_p) \oplus \mathcal{H}(e_{\alpha t}I_p)$

$$= \left\{ \int_{-\beta t}^{\alpha t} e^{i\lambda s} f^\vee(s)ds : f^\vee \in L_2^p([-\beta t, \alpha t]) \right\}$$

as linear spaces (with equivalent norms).

(2) $\sigma \in \Sigma_{\rm sf}^d(N_{\alpha,\beta}; \mathcal{V})$ *if and only if* $\sigma(\mu)$ *is a nondecreasing* $p \times p$ *mvf on* $\mathbb{R}$ *such that*

$$\int_{-\infty}^{\infty} f(\mu)^* d\sigma(\mu) f(\mu) = \int_{-\infty}^{\infty} (E_+^t(\mu)^{-1} f(\mu))^* E_+^t(\mu)^{-1} f(\mu)) d\mu \tag{12.127}$$

for every $f \in \mathcal{B}(\mathfrak{E}_t)$ *and every* $t \in [0, d)$.

(3) *If (12.62) is a regular DK-system on* $[0, d]$, *then*

$$\Sigma_{\rm sf}^d(N_{\alpha,\beta}; \mathcal{V}) = (\mathcal{B}(\mathfrak{E}_d))_{\rm sf}$$

and (2) is also in force for $t = d$.

(4) *If* $d = \infty$, $\mathfrak{v} \in L_1^{p\times p}(\mathbb{R}_+)$ *and* $\mathfrak{A}_\ell^\infty$ *is defined as in Theorem 12.38, then:*

(a) $e_{\beta t}E_+^t$ *tends to a limit* E_+^∞ *in the Wiener algebra* $\mathcal{W}^{p\times p}$, *where*

$$E_+^\infty(\lambda) = \mathfrak{c}_+^\infty(\lambda) + \mathfrak{d}_+^\infty(\lambda) \quad \text{for } \lambda \in \overline{\mathbb{C}_+}.$$

(b) $(E_+^\infty)^{\pm 1} \in \mathcal{W}_+^{p\times p}(I_p)$.

(c) *The spectral function* σ *of the DK-system (12.62) is locally absolutely continuous on* $\mathbb{R}$ *with density*

$$\sigma'(\mu) = E_+^\infty(\mu)^{-*} E_+^\infty(\mu)^{-1} = I_p + \int_{-\infty}^{\infty} e^{i\mu t} h_c(t)dt, \tag{12.128}$$

where h_c *is the accelerant of the input impedance* c *of the DK-system (12.62) considered in Remark 12.45.*

Proof The first assertion follows from Theorems 4.78, 12.6 and 12.11; whereas (2) and (3) follow from Theorem 7.32.

Finally, to verify the assertions in (4), it is convenient to first reexpress E_+^t in terms of the block entries in $\mathfrak{A}_\ell^t(\lambda)$:

$$\begin{aligned} E_+^t(\lambda) &= \sqrt{2}\begin{bmatrix}0 & I_p\end{bmatrix} A_t(\lambda)\mathfrak{V}\begin{bmatrix}0\\ I_p\end{bmatrix} = \sqrt{2}\begin{bmatrix}0 & I_p\end{bmatrix}\mathfrak{V}W_t(\lambda)\begin{bmatrix}0\\ I_p\end{bmatrix} \\ &= \begin{bmatrix}I_p & I_p\end{bmatrix}\begin{bmatrix}\mathfrak{d}_-^t(\lambda) & \mathfrak{c}_+^t(\lambda)\\ \mathfrak{c}_-^t(\lambda) & \mathfrak{d}_+^t(\lambda)\end{bmatrix}\begin{bmatrix}e_{\alpha t}I_p & 0\\ 0 & e_{-\beta t}I_p\end{bmatrix}\begin{bmatrix}0\\ I_p\end{bmatrix} \\ &= e_{-\beta t}\{\mathfrak{c}_+^t(\lambda) + \mathfrak{d}_+^t(\lambda)\}. \end{aligned}$$

Assertions (4a) and (4b) now follow from Theorem 12.38 and the fact that $\mathfrak{A}_\ell^\infty \in \mathcal{W}_\ell(j_p)$ in the present setting. The rest follows from Theorem 12.29 and Remark 12.45. □

12.13 Supplementary notes

This chapter depends heavily on [ArD05c] and [ArD07]. The presented results (see especially Theorems 12.44, 12.47, 12.42, 12.39, 12.15 and Remark 12.45) serve to guarantee that for every Dirac system with J unitarily equivalent to j_p and a summable potential $\mathcal{V}(t)$ on $\mathbb{R}_+$ that is subject to the constraints (12.3) and (12.4), there exists exactly one

(1) asymptotic scattering matrix $s(\mu) = s_\gamma^\infty(\mu)$ for each unitary $\gamma \in \mathbb{C}^{p\times p}$;
(2) input impedance matrix c;
(3) input scattering matrix s;
(4) spectral function $\sigma(\mu)$.

Moreover, each one of these mvf's defines the potential $\mathcal{V}$ uniquely and in addition has special properties:

(a) The asymptotic scattering matrix $s(\mu)$ is unitary on $\mathbb{R}$ and admits a unique factorization

$$s(\mu) = \psi_+(\mu)^{-1}\psi_-(\mu) \quad \text{on } \mathbb{R} \text{ with } \psi_+^{\pm 1} \in \mathcal{W}_+^{p\times p}(I_m) \text{ and } \psi_-^{\pm 1} \in \mathcal{W}_-^{p\times p}(\gamma^{\pm 1}).$$

(b) The input impedance matrix $c \in \mathring{\mathcal{C}}^{p\times p} \cap \mathcal{W}_+^{p\times p}(I_m)$.
(c) The input scattering matrix $s \in \mathring{\mathcal{S}}^{p\times p} \cap \mathcal{W}_+^{p\times p}(0)$.
(d) The spectral function $\sigma(\mu)$ is locally absolutely continuous on $\mathbb{R}$ and

$$\sigma'(\mu) = \Re c(\mu) = \psi_+(\mu)^{-*}\psi_+(\mu)^{-1} \quad \text{on } \mathbb{R} \text{ with } c \text{ and } \psi_+ \text{ as in (a) and (b).}$$

To the best of our knowledge, it is not known if the mappings from the class of considered potentials $\mathcal{V}$ to any one (and hence to everyone) of the three indicated classes of mvf's is onto. Thus, for example, it is not known if every mvf $c \in \mathring{\mathcal{C}}^{p\times p} \cap \mathcal{W}_+^{p\times p}(I_m)$ is the input impedance matrix of a Dirac system with summable potential $\mathcal{V}(t)$ that is subject to the constraints (12.3) and (12.4). To overcome this difficulty, F.E. Melik-Adamyan [MA89] considered the more general class of differential systems of the form

$$\left\{\frac{d}{dt}\left(y(t,\lambda)W(t)^*\right)\right\}JW(t) = i\lambda y(t,\lambda) \quad \text{for } 0 \le t < \infty, \tag{12.129}$$

where $J = V^* j_p V$, V is a unitary matrix, $W(t)$ is a continuous J-unitary mvf on $\mathbb{R}$ with $W(0) = I_m$ and there exists a solution $Y(t,\lambda)$ of (12.129) such that

$$Y(t,\lambda) = e^{i\lambda tJ} + o(1) \quad \text{for } \lambda \in \mathbb{C}_+ \text{ as } t \uparrow \infty$$

and $VY(0,\lambda)V^* \in \mathfrak{M}_\ell(j_p)$. If $W(t)$ is locally absolutely continuous on $\mathbb{R}_+$, then the system (12.129) may be rewritten as a Dirac system with $\alpha = \beta = 1$ and $\mathcal{V}(t) = JW(t)^{-1}W'(t)J$.

A number of results for 2×2 DK-systems (12.1) with potential $\mathcal{V}$ of the form (12.51) are developed by S.A. Denisov in [Den06]. In particular, he obtained the following results, in which h (resp., s) denotes the unique h considered in Theorem 12.26 (resp., input scattering matrix) for the system with $d = \infty$:

(1) (Theorem 6.3) $\mathfrak{v} \in L_{2,\mathrm{loc}}(\mathbb{R}_+) \Longleftrightarrow h \in L_{2,\mathrm{loc}}(\mathbb{R})$.
(2) (Theorem 5.13) $\mathfrak{v}$ continuous on $\mathbb{R}_+ \Longleftrightarrow h$ continuous on $\mathbb{R}_+$.
(3) (Theorem 12.11) $\mathfrak{v} \in L_2(\mathbb{R}_+) \Longleftrightarrow \ln(1-|s|) \in L_1(\mathbb{R}_+)$.

In this paper Denisov also reviews Krein's method for solving inverse problems for DK-systems with $p = q = 1$ and compares it with the methods of Gelfand–Levitan (for solving the inverse spectral problem for the Schrödinger equation) and Agranovitch–Marchenko (for solving the inverse asymptotic scattering problem). Comparisons of these three approaches may also be found in [DI82].

Denisov's review includes a discussion of the role of the continuous analogs of orthogonal polynomials introduced by Krein (and of the Wall polynomials) in the solution of inverse problems for DK-systems and some related approximation problems. An explanation of input scattering matrices in terms of wave operators is also presented. The paper includes an extensive bibliography.

The proof of the first assertion of Theorem 12.6 is adapted from section 2 of [DI84], where the case $N = I_m$ is treated for $J = \mathcal{J}_p$.

Theorem 12.32 has a long history. The connection between the accelerant extension problem and the inverse problem for differential systems goes back to M.G. Krein [Kr55], but, in the earliest formulations, he was not careful enough about the conditions under which the (in current terminology) Krein–Sobolev equation would hold. This connection between extension problems and inverse problems was exploited in [DI84], where Krein's error was repeated. The results were reproved correctly in [KrMA86] and [Dy90], this time imposing continuity of $\mathfrak{a}(t)$ on the interval $(-\kappa d, \kappa d)$. Since $\mathfrak{a}(-t) = \mathfrak{a}(t)^*$, this is equivalent to requiring $\mathfrak{a}(t)$ to be continuous on $[0, \kappa d)$ and requiring further that $\mathfrak{a}(0) = \mathfrak{a}(0)^*$. On the basis of the detailed analysis of the Krein–Sobolev equation in [GK85], it was shown in [AGKLS10] that the condition $\mathfrak{a}(0) = \mathfrak{a}(0)^*$ is superfluous and in fact, that the potential $\mathcal{V}(t)$ can be continuous, even if $\mathfrak{a}(t)$ has a jump at 0.

Corollary 12.9 supplies necessary conditions for an $m \times m$ mvf $U(\lambda)$ to be the monodromy matrix of a differential system of the form (12.1) with a potential $\mathcal{V}(t)$ that meets the conditions (12.3), (12.4) and (12.9). At the moment we do not know if these conditions are sufficient. The corollary also gives necessary conditions for an $m \times m$ mvf $Y(\lambda)$ to be the monodromy matrix of a cannonical differential system of the form (12.5) with a Hamiltonian $H(t)$ that is related to the differential system (12.1) by formula (12.6).

The solution of the inverse impedance problem for the differential system (11.70) with $D = I_p$ was first obtained by M.G. Krein in [Kr55] and [Kr56]; for additional developments and discussion, see, e.g., [Den02], [Sak99], [Sak00c] and [Sak06]. In Krein's terminology, the mvf $h(t)$ in the representation formula (11.43) is called the accelerant. Formula (11.56) is also derived in [Dy90] and [Dy94] for the special case $c \in \mathcal{W}_+^{p\times p}(I_p)$ and $D = I_p$.

Theorem 12.32 was first announced without proof for Dirac systems in [Kr55]. Analogous results were also announced without proof for the case in which $\mathfrak{a}(t)$ and $\mathcal{V}(t)$ are summable on the intervals $[0, \kappa d)$ and $[0, d)$, respectively, in [KrMA86]. However, to the best of our knowledge, a proof was never published.

DK-systems with rational potentials are discussed in [AG97], [AGKS00], [ArD02b] and [AGKLS10].

Weyl–Titchmarsh theory for Krein systems was developed by A.M. Rybalko [Ry66] and for Dirac systems by A.L. Sakhnovich [SakA02]. Uniqueness theorems for Dirac (and other) systems have been studied by other methods by F. Gesztesy, A. Kiselev and K.A. Makarov in [GKM02]. They exploit properties of a matrix-valued Weyl–Tichmarsh function, which corresponds to the inverse impedance problem studied in this book. This paper contains a good historical introduction and an extensive bibliography.

Direct and inverse spectral problems for a class of differential systems of the form (2.72) with assorted choices of N and V have been studied by a number of different authors. The case $N = \mathrm{diag}\,\{B_1, B_2\}$ with positive definite blocks B_1 and B_2 (among others), $J = j_{pq}$ and potentials with special structure was studied by M.M. Malamud in [Mal99] and (with M. Lesch) in [LM00] via a matrix version of the Gelfand–Levitan equation. Inverse monodromy and inverse spectral problems for Krein systems and for $N = \mathrm{diag}\{D, 0\}$ in which $D \in \mathbb{C}^{p\times p}$ is a positive definite diagonal matrix and $J = J_p$ have been studied by L.A. Sakhnovich in a number of publications; see, e.g., [Sak99], [Sak00a], [Sak00b], [Sak06] and the references cited therein. Another extensive study of Dirac systems is [Ga68].

REFERENCES

[Ad68] V.M. Adamjan, Canonical differential operators in Hilbert space. Dokl. Akad. Nauk SSSR **178** (1968), 9–12; English translation in Soviet Math. Dokl. **9** (1968), 1–5. (Cited on 54)

[Ad73] V.M. Adamjan, Nondegenerate unitary couplings of semiunitary operators. (Russian) Funktsional. Anal. i Prilozhen **7** (4) (1973), 1–16. (Cited on 155)

[AdAr66] V.M. Adamjan and D.Z. Arov, Unitary couplings of semi-unitary operators. (Russian) Mat. Issled. **1** (2) (1966), 3–64. (Cited on 155, 200)

[AAK68] V.M. Adamjan, D.Z. Arov and M.G. Krein, Infinite Hankel matrices and generalized problems of Carathodory–Fejr and I. Schur problems. (Russian) Funktsional. Anal. i Prilozhen **2** (4) (1968), 1–17. (Cited on 155)

[AAK71a] V.M. Adamjan, D.Z. Arov and M.G. Krein, Infinite Hankel block matrices and related problems of extension. (Russian) Izv. Akad. Nauk Armjan. SSR Ser. Mat. **6** (2–3) (1971), 87–112. (Cited on 155)

[AAK71b] V.M. Adamjan, D.Z. Arov and M.G. Krein, Analytic properties of the Schmidt pairs of a Hankel operator and the generalized Schur–Takagi problem. (Russian) Mat. Sb. (N.S.) **86** (128) (1971), 34–75; English translation in Math USSR Sbornik **15** (1971), 31–74. (Cited on 155, 325, 344, 390)

[AM63] Z.S. Agranovich and V.A. Marchenko, *The Inverse Problem of Scattering Theory*. Translated from the Russian by B.D. Seckler. Gordon and Breach, New York, 1963. (Cited on 54)

[AkGl63] N.I. Akhiezer and I.M. Glazman, *Theory of Linear Operators in Hilbert Space, Vol II.* Translated by M. Nestell. Frederick Ungar, New York, 1963. (Cited on 267)

[AKvdM00] T. Aktosun, M. Klaus and C. van der Mee, Direct and inverse scattering for selfadjoint Hamiltonian systems on the line. Integ. Equat. Oper. Theory **38** (2) (2000), 129–171. (Cited on 371, 408)

[AlD84] D. Alpay and H. Dym, Hilbert spaces of analytic functions, inverse scattering and operator models, I. Integ. Equat. Oper. Theory **7** (1984), 589–741. (Cited on 309)

[AlD85] D. Alpay and H. Dym, Hilbert spaces of analytic functions, inverse scattering and operator models, II. Integ. Equat. Oper. Theory **8** (1985), 145–180. (Cited on 309)

[AlD93] D. Alpay and H. Dym, On a new class of structured reproducing kernel spaces. J. Funct. Anal. **111** (1993), 1–28. (Cited on 156)

[ADD89] D. Alpay, P. Dewilde and H. Dym, On the existence and construction of solutions to the partial lossless inverse scattering problem with applications to estimation theory. IEEE Trans. Inform. Theory **35** (6) (1989), 1184–1205. (Cited on 200)

[AG95] D. Alpay and I. Gohberg, Inverse scattering problem for differential operators with rational scattering matrix functions, in: *Singular Integral Operators and Related Topics* (Tel Aviv, 1995), Oper. Theory Adv. Appl. 90. Birkhäuser, Basel, 1996, pp. 1–18. (Cited on 371, 408)

[AG96] D. Alpay and I. Gohberg, State space method for inverse spectral problems, in: *Systems and Control in the Twenty-first Century* (St. Louis, MO, 1996), Progr. Systems Control Theory 22. Birkhäuser, Boston, MA, 1997, pp. 1–16. (Cited on 408)

[AG97] D. Alpay and I. Gohberg, Potentials associated to rational weights, in: *New Results in Operator Theory and its Applications*, Oper. Theory Adv. Appl. 98. Birkhäuser, Basel, 1997, pp. 23–40, (Cited on 408, 449)

[AG98] D. Alpay and I. Gohberg, Inverse problem for Sturm–Liouville operators with rational reflection coefficient. Dedicated to the memory of Mark Grigorievich Krein (1907–1989). Integ. Equat. Oper. Theory **30** (3) (1998), 317–325. (Cited on 408)

[AG01] D. Alpay and I. Gohberg, Inverse problems associated to a canonical differential system, in: *Recent Advances in Operator Theory and Related Topics* (Szeged, 1999), Oper. Theory Adv. Appl. 127. Birkhäuser, Basel, 2001, pp. 1–27. (Cited on 371, 408)

[AGKS00] D. Alpay, I. Gohberg, M.A. Kaashoek and A.L. Sakhnovich, Direct and inverse scattering problem for canonical systems with a strictly pseudo-exponential potential. Math. Nachr. **215** (2000), 5–31. (Cited on 449)

[AGKLS10] D. Alpay, I. Gohberg, M.A. Kaashoek, L. Lerer and A.L. Sakhnovich, Krein systems and canonical systems on a finite interval: accelerants with a jump discontinuity at the origin and continuous potentials. Integ. Equat. Oper. Theory **68** (1) (2010), 115–150. (Cited on 408, 434, 445, 449)

[ABT11] Y. Arlinskii, S. Belyi and E. Tsekanovskii, *Conservative Realizations of Herglotz–Nevanlinna Functions*. Birkhäuser, Basel, 2011. (Cited on 308)

[Arn50] N. Aronszajn, Theory of reproducing kernels. Trans. Amer. Math. Soc. **68** (1950), 337–404. (Cited on 3, 127)

[Ar71] D.Z. Arov, Darlington's method in the study of dissipative systems. (Russian) Dokl. Akad. Nauk SSSR **201** (3) (1971), 559–562; English translation in Soviet Physics Dokl. **16** (1971), 954–956. (Cited on 200, 408)

[Ar73] D.Z. Arov, Realization of matrix-valued functions according to Darlington. (Russian) Izv. Akad. Nauk SSSR Ser. Mat. **37** (1973), 1299–1331. (Cited on 408)

[Ar75] D.Z. Arov, Realization of a canonical system with a dissipative boundary condition at one end of the segment in terms of the coefficient of dynamical compliance. (Russian) Sibirsk. Mat. **16** (3) (1975), 440–463, 643. (Cited on 408)

[Ar79] D.Z. Arov, Passive linear steady-state dynamical systems. (Russian) Sibirsk. Mat. Zh. **20** (2) (1979), 211–228, 457. (Cited on 200)

[Ar84] D.Z. Arov, *Three Problems about J-inner Matrix Functions*. Springer Lecture Notes in Mathematics, 1043 (1984), pp. 164–168. (Cited on 105, 155)

[Ar88] D.Z. Arov, Regular and singular J-inner matrix functions and corresponding extrapolation problems. (Russian) Funktsional. Anal. i Prilozhen **22** (1988), no. 1, 57–59; English translation in Funct. Anal. Appl. **22** (1) (1988), 46–48. (Cited on 105, 155)

[Ar89] D.Z. Arov, γ-generating matrices, j-inner matrix-functions and related extrapolation problems. Teor. Funktsii Funktsional. Anal. i Prilozhen, I **51** (1989), 61–67; II **52** (1989), 103–109; English translation in J. Soviet Math. I, **52** (1990), 3487–3491; III, **52** (1990), 3421–3425. (Cited on 105, 155)

[Ar90] D.Z. Arov, Regular J-inner matrix-functions and related continuation problems, in: *Linear Operators in Function Spaces* (Timişoara, 1988) (H. Helson, B. Sz.-Nagy, F.-H. Vasilescu and Gr. Arsene, eds.), Oper. Theory Adv. Appl. 43. Birkhäuser, Basel, 1990, pp. 63–87. (Cited on 105)

[Ar93] D.Z. Arov, The generalized bitangent Carathodory–Nevanlinna–Pick problem and (j, J_0)-inner matrix functions. (Russian) Izv. Ross. Akad. Nauk Ser. Mat. **57** (1) (1993), 3–32; English translation in Russian Acad. Sci. Izv. Math. **42** (1) (1994), 1–26. (Cited on 105)

[Ar95a] D.Z. Arov, γ-generating matrices, j-inner matrix-functions and related extrapolation problems, IV. Math. Phys. Anal. Geom. **2** (1995), 3–14. (Cited on 155)

[Ar95b] D.Z. Arov, A survey on passive networks and scattering systems which are lossless or have minimal losses. AEU Internat. J. Electron. and Commun. **49** (1995), 252–265. (Cited on 200)

[Ar01] D.Z. Arov, The scattering matrix and impedance of a canonical differential system with a dissipative boundary condition in which the coefficient is a rational matrix function of the spectral parameter. (Russian) Algebra i Analiz **13** (4) (2001), 26–53; English translation in St. Petersburg Math. J. **13** (4) (2002), 527–547. (Cited on 408)

[ArD97] D.Z. Arov and H. Dym, J-inner matrix functions, interpolation and inverse problems for canonical systems, I: Foundations. Integ. Equat. Oper. Theory **29** (1997), 373–454. (Cited on 105, 155, 177, 284)

[ArD98] D.Z. Arov and H. Dym, On the three Krein extension problems and some generalizations. Integ. Equat. Oper. Theory **31** (1998), 1–91. (Cited on 156, 354)

[ArD00a] D.Z. Arov and H. Dym, J-inner matrix functions, interpolation and inverse problems for canonical systems, II: The inverse monodromy problem. Integ. Equat. Oper. Theory **36** (2000), 11–70. (Cited on 177, 308, 309)

[ArD00b] D.Z. Arov and H. Dym, J-inner matrix functions, interpolation and inverse problems for canonical systems, III: More on the inverse monodromy problem. Integ. Equat. Oper. Theory **36** (2000), 127–181. (Cited on 177, 277, 278, 281, 287, 290, 292, 294, 295, 297, 298, 308, 309)

[ArD01a] D.Z. Arov and H. Dym, Some remarks on the inverse monodromy problem for 2×2 canonical differential systems, in *Operator Theory and Analysis* (Amsterdam, 1997) (H. Bart, I. Gohberg and A.C.M. Ran, eds.), Oper. Theory Adv. Appl. 122. Birkhäuser, Basel, 2001, pp. 53–87. (Cited on 299, 300, 308)

[ArD01b] D.Z. Arov and H. Dym, Matricial Nehari problems J-inner matrix functions and the Muckenhoupt condition. J. Funct. Anal. **181** (2001), 227–299. (Cited on 156)

[ArD02a] D.Z. Arov and H. Dym, J-inner matrix functions, interpolation and inverse problems for canonical systems, IV: Direct and inverse bitangential input scattering problem. Integ. Equat. Oper. Theory **43** (2002), 1–67. (Cited on 155, 193, 200, 201, 332, 371)

[ArD02b] D.Z. Arov and H. Dym, J-inner matrix functions, interpolation and inverse problems for canonical systems, V: The inverse input scattering problem for Wiener class and rational $p \times q$ input scattering matrices. Integ. Equat. Oper. Theory **43** (2002), 68–129. (Cited on 371, 406, 408, 435, 445, 449)

[ArD03a] D.Z. Arov and H. Dym, The bitangential inverse input impendance problem for canonical systems, I: Weyl–Titchmash classification, and existence and uniqueness theorems. Integ. Equat. Oper. Theory **47** (2003), 3–49. (Cited on 123, 124, 237, 298, 406)

[ArD03b] D.Z. Arov and H.Dym, Criteria for the strong regularity of J-inner functions and γ-generating functions. J. Math. Anal. Appl. **280** (2003), 387–399. (Cited on 156)

[ArD04a] D.Z. Arov and H. Dym, Strongly regular J-inner matrix functions and related problems, in: *Current Trends in Operator Theory and its Applications* (J.A. Ball, J.W. Helton, M. Klaus and L. Rodman, eds.), Oper. Theory Adv. Appl. 149. Birkhäuser, Basel, 2004, pp. 79–106. (Cited on 26)

[ArD04b] D.Z. Arov and H. Dym, The bitangential inverse spectral problem for canonical systems. J. Funct. Anal. **214** (2004), 312–385. (Cited on 53, 227, 237, 239, 261, 269, 407, 408)

[ArD05a] D.Z. Arov and H. Dym, The bitangential inverse input impedance problem for canonical systems, II: Formulas and examples. Integ. Equat. Oper. Theory **51** (2005), 155–213. (Cited on 155, 237, 406)

[ArD05b] D.Z. Arov and H. Dym, Strongly regular J-inner matrix-vaued functions and inverse problems for canonical systems, in: *Recent Advances in Operator Theory and its Applications* (M.A. Kaashoek, S. Seatzu and C. van der Mee, eds.), Oper. Theory Adv. Appl. 160, Birkhäuser, Basel, 2005, pp. 101–160. (Cited on 26)

[ArD05c] D.Z. Arov and H. Dym, Direct and inverse problems for differential systems connected with Dirac systems and related factorization problems. Indiana J. **54** (2005), 1769–1815. (Cited on 53, 447)

[ArD07] D.Z. Arov and H. Dym, Direct and inverse asymptotic scattering problems for Dirac–Krein systems. (Russian) Funktsional. Anal. i Prilozhen **41** (3), (2007), 17–33. (Cited on 447)

[ArD08a] D.Z. Arov and H. Dym, Bitangential direct and inverse problems for systems of differential equations, in: *Probability, Geometry and Integrable Systems* (M. Pinsky and B. Birnir, eds.), MSRI Publications 55. Cambridge University Press, Cambridge, 2008, pp. 1–28. (Cited on 26)

[ArD08b] D.Z. Arov and H. Dym, *J-Contractive Matrix Valued Functions and Related Topics*, Encyclopedia of Mathematics and its Applications, 116. Cambridge University Press, Cambridge, 2008. (Cited on 56, 58, 61, 62, 64, 66, 67, 74, 77, 78, 85, 91, 93, 94, 95, 96, 97, 100, 101, 103, 104, 105, 108, 109, 110, 111, 112, 116, 118, 119, 120, 121, 123, 127, 128, 134, 135, 136, 138, 143, 146, 148, 151, 152, 156, 157, 188, 208, 259, 262, 311, 312, 313, 314, 318, 319, 321, 326, 328, 354, 402)

[ArD12a] D.Z. Arov and H. Dym, B-regular J-inner matrix valued functions. Oper. Theor. Adv. Appl. **218**, 51–73. (Cited on 105, 156, 177, 309, 407)

[ArD12b] D. Z. Arov and H. Dym, Continuous analogs of Schur extension problems and bitangential generalizations, Math. Nachr., in press. (Cited on 354)

[At64] F.V. Atkinson, *Discrete and Continuous Boundary Value Problems.* Academic Press, New York, 1964. (Cited on 237, 238)

[Bel68] V. Belevich, *Classical Network Theory.* Holden Day, San Francisco, 1968. (Cited on 200)

[BS87] M.Sh. Birman and M.Z. Solomyak, *Spectral Theory of Selfadjoint Operators in Hilbert Space.* Translated from the 1980 Russian original, D. Reidel, Dordrecht, 1987. (Cited on 353)

[BoD98] V. Bolotnikov and H. Dym, On degenerate interpolation, entropy and extremal problems for matrix Schur functions. Integ. Equat. Oper. Theory **32** (4) (1998), 367–435. (Cited on 155)

[BoD06] V. Bolotnikov and H. Dym, On boundary interpolation for matrix valued Schur functions. Mem. Amer. Math. Soc. **181** (856) (2006). (Cited on 155)

[Bo10] A. Boumenir, The Gelfand–Levitan theory for strings, in: *Topics in Operator Theory. Volume 2, Systems and Mathematical Physics*, Oper. Theory Adv. Appl. 203. Birkhäuser Verlag, Basel, 2010, pp. 115–136. (Cited on 407)

[Br63] L. de Branges, Some Hilbert spaces of analytic functions I., Trans. Amer. Math. Soc. **106** (1963), 445–668. (Cited on 156)

[Br65] L. de Branges, Some Hilbert spaces of analytic functions II. J. Math. Anal. Appl. **11** (1965), 44–72. (Cited on 156)

[Br68a] L. de Branges, *Hilbert Spaces of Entire Functions.* Prentice-Hall, Englewood Cliffs, 1968. (Cited on 5, 54, 239, 244, 308, 408)

[Br68b] L. de Branges, The expansion theorem for Hilbert spaces of entire functions, in: *Entire Functions and Related Parts of Analysis.* American Mathematical Society, Providence, RI, 1968. (Cited on 309)

[Br83] L. de Branges, The comparison theorem for Hilbert spaces of entire functions. Integ. Equat. Oper. Theory **6** (1983), 603–646. (Cited on 309)

[Bro72] M.S. Brodskii, *Triangular and Jordan Representations of Linear Operators.* Transl. Math Monographs 32. American Mathematical Society, Providence, RI, 1972. (Cited on 5, 53, 247, 258, 259, 262, 263, 270, 271)

[CG01] S. Clark and F. Gesztesy, Weyl–Titchmarsh M-function asymptotics for matrix-valued Schrödinger operators. Proc. London Math. Soc. **82** (3) (2001), 701–724. (Cited on 408)

[DaKr74] Ju.L. Daleckii and M.G. Krein, *Stability of Solutions of Differential Equations in Banach Space*. Transl. Math. Monographs 43. American Mathematical Society, Providence, RI, 1974. (Cited on 438)

[Den02] S.A. Denisov, On the spectral theory of Krein systems. Integ. Equat. Oper. Theory **42** (2) (2002), 166–173. (Cited on 449)

[Den06] S.A. Denisov, Continuous analogs of polynomials orthogonal on the unit circle and Krein systems. IMRS Int. Math. Res. Surv. 2006, Art. ID 54517. (Cited on 200, 448)

[DeD84] P. Dewilde and H. Dym, Lossless inverse scattering, digital filters, and estimation theory. IEEE Trans. Inform. Theory **30** (4) (1984), 644–662. (Cited on 200)

[DF79] J.D. Dollard and C.N. Friedman, *Product Integration with Applications to Differential Equations*. With a foreword by Felix E. Browder and an appendix by P.R. Masani. Encyclopedia of Mathematics and its Applications, 10. Addison-Wesley, Reading, MA, 1979. (Cited on 38, 39, 53)

[DSS70] D. H. Shapiro and A. Shields, Cyclic vectors and invariant subspaces for the backward shift operator. Ann. Inst. Fourier (Grenoble) **20** (1) (1970), 37–76. (Cited on 78)

[Du70] P. Duren, *Theory of H^p Spaces*. Academic Press, New York, 1970. (Cited on 60)

[Dy70] H. Dym, An introduction to de Branges spaces of entire functions with applications to differential equations of the Sturm–Liouville type. Adv. Math. **5** (1970), 395–471. (Cited on 309)

[Dy89a] H. Dym, On reproducing kernel spaces, J unitary matrix functions, interpolation and displacement rank, in: *The Gohberg Anniversary Collection* (H. Dym, S. Goldberg, M.A. Kaashoek and P. Lancaster, eds.), Vol. II (Calgary, 1988). Oper. Theory Adv. Appl. 41. Birkhäuser, Basel, 1989, pp. 173–239. (Cited on 155)

[Dy89b] H. Dym, *J-contractive Matrix Functions, Reproducing Kernel Hilbert Spaces and Interpolation*. CBMS Regional Conference series, 71. American Mathematical Society, Providence, RI, 1989. (Cited on 155, 156)

[Dy89c] H. Dym, On Hermitian block Hankel matrices, matrix polynomials, the Hamburger moment problem, interpolation and maximum entropy. Integ. Equat. Oper. Theory **12** (6) (1989), 757–812. (Cited on 155)

[Dy90] H. Dym, On reproducing kernels and the continuous covariance extension problem, in: *Analysis and Partial Differential Equations: A Collection of Papers Dedicated to Mischa Cotlar* (C. Sadosky, ed.), Marcel Dekker, New York, 1990, pp. 427–482. (Cited on 155, 156, 445, 448, 449)

[Dy94] H. Dym, On the zeros of some continuous analogues of matrix orthogonal polynomials and a related extension problem with negative squares. Comm. Pure Appl. Math. **47** (1994), 207–256. (Cited on 449)

[Dy97] H. Dym, M.G. Krein's contributions to prediction theory, in: *Operator Theory and Related Topics*, Vol. II (Odessa, 1997), Oper. Theory Adv. Appl. **118**. Birkhäuser, Basel, 2000, pp. 1–15. (Cited on 54)

[Dy98] H. Dym, A basic interpolation problem, in: *Holomorphic Spaces* (S. Axler, J.E. McCarthy and D. Sarason, eds.). Cambridge University Press, Cambridge 1998, pp. 381–423. (Cited on 155)

[Dy01a] H. Dym, On Riccati equations and reproducing kernel spaces, in: *Recent Advances in Operator Theory* (Groningen, 1998) (A. Dijksma, M.A. Kaashoek and A.C.M. Ran, eds.). Oper. Theory Adv. Appl. 124, Birkhäuser, Basel, 2001, pp. 189–215. (Cited on 156)

[Dy01b] H. Dym, Reproducing kernels and Riccati equations. in: *Mathematical Theory of Networks and Systems* (Perpignan, 2000). Int. J. Appl. Math. Comput. Sci. **11** (1) (2001), 35–53. (Cited on 156)

[Dy03a] H. Dym, Riccati equations and bitangential interpolation problems with singular Pick matrices, in: *Fast Algorithms for Structured Matrices: Theory and Applications* (South Hadley, MA, 2001), Contemp. Math. **323**. American Mathematical Society, Providence, RI, 2003, pp. 361–391. (Cited on 155, 156)

[Dy03b] H. Dym, Linear fractional transformations, Riccati equations and bitangential interpolation, revisited, in: *Reproducing Kernel Spaces and Applications* (D. Alpay, ed.), Oper. Theory Adv. Appl. 143. Birkhäuser, Basel, 2003, pp. 171–212. (Cited on 116, 155)

[DG83] H. Dym and I. Gohberg, Hankel integral operators and isometric interpolants on the line. J. Funct. Anal. **54** (1983), 290–307. (Cited on 339)

[DI82] H. Dym and A. Iacob, Applications of factorization and Toeplitz operators to inverse problems, in: *Toeplitz Centennial*, Oper. Theory Adv. Appl. 4. Birkhäuser, Basel, 1982, pp. 233–260. (Cited on 408, 448)

[DI84] H. Dym and A. Iacob, Positive definite extensions, canonical equations and inverse problems, in: *Topics in Operator Theory, Systems and Networks* (H. Dym and I. Gohberg, eds.), Oper. Theory Adv. Appl. 12. Birkhäuser, Basel, 1984, pp. 141–240. (Cited on 201, 438, 448)

[DK78a] H. Dym and N. Kravitsky, On recovering the mass distribution of a string from its spectral function, in: *Topics in Functional Analysis (Essays Dedicated to M.G. Krein on the Occasion of his 70th Birthday).* Academic Press, New York, 1978, pp. 45–90. (Cited on 407)

[DK78b] H. Dym and N. Kravitsky, On the inverse spectral problem for the string equation. Integ. Equat. Oper. Theory **1/2** (1978), 270–277. (Cited on 407)

[DMc72] H. Dym and H.P. McKean, *Fourier Series and Integrals.* Academic Press, New York, 1972. (Cited on 80, 134, 408)

[DMc76] H. Dym and H.P. McKean, *Gaussian Processes, Function Theory, and the Inverse Spectral Problem.* Academic Press, New York, 1976; reprinted by Dover, New York, 2008. (Cited on 26, 54, 240, 244, 308)

[Fa59] L.D. Faddeev, The inverse problem in the quantum theory of scattering, (Russian) Uspehi Mat. Nauk **14** (4) (1959), (88), 57–119; English translation in J. Math. Phys. **4** (1963), 72–104. (Cited on 200)

[Fe55] W. Feller, On second order differential operators. Ann. Math. **61** (2) (1955), 90–105. (Cited on 54)

[Fe57] W. Feller, Generalized second order differential operators and their lateral conditions. Illinois J. Math. **1** (1957), 459–504. (Cited on 54)

[Fe59a] W. Feller, The birth and death processes as diffusion processes. J. Math. Pure Appl. **38** (9) (1959), 301–345. (Cited on 54)

[Fe59b] W. Feller, On the equation of the vibrating string. J. Math. Mech. **8** (1959), 339–348. (Cited on 54)

[Ga68] M.G. Gasymov, The inverse scattering problem for a system of Dirac equations of order $2n$. Trans. Moscow Math. Soc. (1968), 41–120. (Cited on 449)

[GRS64] I. Gelfand, D. Raikov and G. Shilov, *Commutative Normed Rings*. Translated from the Russian, with a supplementary chapter. Chelsea, New York, 1964. (Cited on 80)

[GS00] F. Gesztesy and B. Simon, A new approach to inverse spectral theory, II. General real potentials and the connection to the spectral measure. Ann. Math. **152** (2) (2000), 593–643. (Cited on 55)

[GKM02] F. Gesztesy, A. Kiselev and K.A. Makarov, Uniqueness results for matrix-valued Schrödinger, Jacobi and Dirac-type operators. Math. Nachr. **239/240** (2002), 103–145. (Cited on 408, 449)

[GJ90] R.D. Gill and S. Johansen, A survey of product-integration with a view toward application in survival analysis. Ann. Statist. **18** (4) (1990), 1501–1555. (Cited on 53)

[Gi57] Yu.P. Ginzburg, On J-contractive operator functions. (Russian) Dokl. Akad. Nauk SSSR (N.S.) **117** (1957), 171–173. (Cited on 105)

[Gi67] Yu.P. Ginzburg, Multiplicative representations and minorants of bounded analytic operator functions. (Russian) Funktsional. Anal. i Prilozhen **1** (3) (1967), 9–23. (Cited on 308)

[GiZe90] Yu.P. Ginzburg and L.M. Zemskov, Multiplicative representations of operator-functions of bounded type. (Russian) Teor. Funktsii Funktsional. Anal. i Prilozhen **53** (1990), 108–119; English translation in J. Soviet Math. **58** (6) (1992), 569–576. (Cited on 308)

[GiSh94] Yu.P. Ginzburg and L.V. Shevchuk, On the Potapov theory of multiplicative representations, in: *Matrix and Operator Valued Functions* (I. Gohberg and L.A. Sakhnovich, eds.), Oper. Theory Adv. Appl. 72. Birkhäuser, Basel, 1994, pp. 28–47. (Cited on 308)

[GK69] I.C. Gohberg and M.G. Krein, *Introduction to the Theory of Linear Nonselfadjoint Operators*, Transl. Math. Monographs, 18. American Mathematical Society, Providence, RI, 1969. (Cited on 134, 162)

[GK70] I.C. Gohberg and M.G. Krein, *Theory and Applications of Volterra Operators in Hilbert Space*. Transl. Math. Monographs, 24. American Mathematical Society, Providence, RI, 1970. (Cited on 308)

[GK85] I. Gohberg and I. Koltracht, Numerical solution of integral equations, fast algorithms and the Krein–Sobolev equation. Numer. Math. **47** (2) (1985), 237–288. (Cited on 390, 406, 445, 448)

[GKS98a] I. Gohberg, M.A. Kaashoek and A.L. Sakhnovich, Canonical systems with rational spectral densities: explicit formulas and applications. Math. Nachr. **194** (1998), 93–125. (Cited on 408)

[GKS98b] I. Gohberg, M.A. Kaashoek and A.L. Sakhnovich, Pseudo-canonical systems with rational Weyl functions: explicit formulas and applications. J. Differ. Equat. **146** (2) (1998), 375–398. (Cited on 408)

[GKS98c] I. Gohberg, M.A. Kaashoek and A.L. Sakhnovich, Sturm–Liouville systems with rational Weyl functions: explicit formulas and applications. Dedicated to the memory of Mark Grigorievich Krein (1907–1989). Integ. Equat. Oper. Theory **30** (3) (1998), 338–377. (Cited on 408)

[GKS02] I. Gohberg, M.A. Kaashoek and A.L. Sakhnovich, Scattering problems for a canonical system with a pseudo-exponential potential. Asymptot. Anal. **29** (1) (2002), 1–38. (Cited on 408)

[GoM97] L. Golinskii and I Mikhailova, Hilbert spaces of entire functions as a J theory subject [Preprint No. 28-80, Inst. Low Temp. Phys. Engrg., Kharkov, 1980]. Edited by V.P. Potapov. (Russian). English translation in Oper. Theory Adv. Appl. 95, *Topics in Interpolation Theory* (Leipzig, 1994) (H. Dym, B. Frizsche, V. Katsnelson and B. Kirstein, eds.). Birkhäuser, Basel, 1997, pp. 205–251. (Cited on 156)

[GoGo97] M.L. Gorbachuk and V.I. Gorbachuk, *M.G. Krein's Lectures on Entire Operators*. Oper. Theory Adv. Appl. 97. Birkhäuser Verlag, Basel, 1997. (Cited on 53, 54, 309, 311)

[Gr85] L. Z. Grossman, Functions of class B generated by Weyl–Krein limit operator balls, Deposited Uk-85, No 1565. (Cited on 239)

[Hel73] J.W. Helton, The characteristic functions of operator theory and electrical network realization. Indiana Univ. Math. J. **22** (1972/73), 403–414. (Cited on 200)

[Hel74] J.W. Helton, Discrete time systems, operator models, and scattering theory. J. Funct. Anal. **16** (1974), 15–38. (Cited on 200)

[Hi62] F.B. Hildebrand, *Advanced Calculus for Applications*. Prentice-Hall, Englewood Cliffs, NJ, 1962. (Cited on 305)

[HS81] D.B. Hinton and J.K. Shaw, On Titchmarsh–Weyl $M(\lambda)$-functions for linear Hamiltonian systems. J. Diff. Equat. **40** (3) (1981), 316–342. (Cited on 408)

[HS83] D.B. Hinton and J.K. Shaw, Parameterization of the $M(\lambda)$ function for a Hamiltonian system of limit circle type. Proc. Roy. Soc. Edinburgh Sect. A **93** (3–4) (1982/83), 349–360. (Cited on 238)

[Ho62] K. Hoffman, *Banach Spaces of Analytic Functions*. Prentice-Hall, Englewood Cliffs, NJ, 1962. (Cited on 156)

[Ia86] A. Iacob, *On the Spectral Theory of a Class of Canonical Systems of Differential Equations*. Ph.D. Thesis, The Weizmann Institute of Science, Rehovot 1986, Israel. (Cited on 239)

[Ka02] I.S. Kac (Kats), Linear relations generated by a canonical differential equation of dimension 2, and eigenfunction expansions. (Russian) Algebra i Analiz **14** (3) (2002), 86–120; English translation in St. Petersburg Math. J. **14** (3) (2003), 429–452. (Cited on 239)

[KaKr74a] I.S. Kac and M.G. Krein, *R*-functions-analytic functions mapping the upper halfplane into itself. Amer. Math. Soc. Transl. **103** (2) (1974), 1–18. (Cited on 70, 73)

[KaKr74b] I.S. Kac and M.G. Krein, On the spectral functions of the string. Amer. Math. Soc. Transl. **103** (2) (1974), 19–102. (Cited on 47, 54, 240, 301, 308)

[Kat85a] V.E. Katsnelson, Integral representation of Hermitian positive kernels of mixed type and the generalized Nehari problem, I. (Russian) Teor. Funktsii Funktsional. Anal. i Prilozhen **43** (1985), 54–70; English translation in J. Soviet Math. **48** (2) (1990), 162–176. (Cited on 354)

[Kat85b] V.E. Katsnelson, *Methods of J-theory in Continuous Interpolation Problems of Analysis*, Part I. Translated from the Russian and with a foreword by T. Ando. Hokkaido University, Sapporo, 1985. (Cited on 354)

[Kh90] A. Kheifets, Generalized bitangential Schur–Nevanlinna–Pick problem and the related Parseval equality. (Russian) Teor. Funktsii Funktsional Anal. i Prilozhen **54** (1990), 89–96; English translation in J. Sov. Math. **58** (1992), 358–364. (Cited on 157)

[Kh95] A. Kheifets, On regularization of C-generating pairs. J. Funct. Anal. **130** (1995), 310–333. (Cited on 157)

[Ko80] A.N. Kochubei, On extensions and characteristic functions of symmetric operators. (Russian) Izv. Akad. Nauk Armenian SSSR **15** (3) (1980), 219–232. (Cited on 309)

[KoPo82] I.V. Kovalishiva and V.P. Potapov, *Integral Representation of Hermitian Positive Functions*. Translated from the Russian by T. Ando, Hokkaido University, Sapporo, 1982. (Cited on 354)

[Kr40] M.G. Krein, Sur le problème du prolongement des fonctions hermitiennes positives et continues. (French) C. R. (Dokl.) Acad. Sci. URSS (N.S.) **26** (1940), 17–22. (Cited on 25)

[Kr44a] M.G. Krein, On the logarithm of an infinitely decomposible Hermite-positive function. C. R. (Dokl.) Acad. Sci. URSS (N.S.) **45** (1944), 91–94. (Cited on 25)

[Kr44b] M.G. Krein, On a remarkable class of Hermitian operators. C. R. (Doklady) Acad. Sci. URSS (N.S.) **44** (1944), 175–179. (Cited on 25)

[Kr44c] M.G. Krein, On the problem of continuation of helical arcs in Hilbert space. C. R. (Dokl.) Acad. Sci. URSS (N.S.) **45** (1944), 139–142. (Cited on 311)

[Kr47] M.G. Krein, On the theory of entire functions of exponential type. Izvestiya Akad. Nauk SSSR, Ser. Matem. **11** (1947), 309–326. (Cited on 76)

[Kr49] M.G. Krein, The fundamental propositions of the theory of representations of Hermitian operators with deficiency index (m, m). (Russian) Ukrain. Mat. Zh. **1** (2) (1949), 3–66. (Cited on 25)

[Kr51a] M.G. Krein, Solution of the inverse Sturm–Liouville problem. (Russian) Doklady Akad. Nauk SSSR (N.S.) **76** (1951), 21–24. (Cited on 25, 55)

[Kr51b] M.G. Krein, On the theory of entire matrix functions of exponential type. (Russian) Ukrain. Mat. Zh. **3** (1951), 164–173. (Cited on 76)

[Kr52a] M.G. Krein, On inverse problems for a nonhomogeneous string. (Russian) Doklady Akad. Nauk SSSR (N.S.) **82** (1952), 669–672. (Cited on 25, 46)

[Kr52b] M.G. Krein, On the indeterminate case of the Sturm–Liouville boundary problem in the interval $(0, \infty)$. (Russian) Izvestiya Akad. Nauk SSSR. Ser. Mat. **16** (1952), 293–324. (Cited on 25)

[Kr53] M.G. Krein, On some cases of effective determination of the density of an inhomogeneous string from its spectral function. (Russian) Doklady Akad. Nauk SSSR (N.S.) **93** (1953), 617–620. (Cited on 54)

[Kr54] M.G. Krein, On a method of effective solution of an inverse boundary problem. (Russian) Dokl. Akad. Nauk SSSR (N.S.) **94** (1954), 987–990. (Cited on 25)

[Kr55] M.G. Krein, Continuous analogues of propositions on polynomials orthogonal on the unit circle. (Russian) Dokl. Akad. Nauk SSSR (N.S.) **105** (1955), 637–640. (Cited on 25, 55, 434, 448, 449)

[Kr56] M.G. Krein, On the theory of accelerants and S-matrices of canonical differential systems. (Russian) Dokl. Akad. Nauk SSSR (N.S.) **111** (1956), 1167–1170. (Cited on 25, 54, 449)

[KrL85] M.G. Krein and H. Langer, On some continuation problems which are closely related to the theory of operators in spaces Π_κ, IV. Continuous analogues of orthogonal polynomials on the unit circle with respect to an indefinite weight and related continuation problems for some classes of functions. J. Oper. Theory **13** (2) (1985), 299–417. (Cited on 26, 354, 406, 445)

[KrMA68] M.G Krein and F.E. Melik-Adamyan, On the theory of S-matrices of canonical differential systems with summable potential. Dokl. Akad. Nauk Armyan SSR. **46** (1) (1968), 150–155. (Cited on 26)

[KrMA70] M.G Krein and F.E. Melik-Adamyan, Some applications of theorems on the factorization of a unitary matrix. Funktsional. Anal. i Prilozhen **4** (4) (1970), 73–75. (Cited on 26)

[KrMA84] M.G Krein and F.E. Melik-Adamyan, Integral Hankel operators and related continuation problems. Izv. Akad. Nauk Armyan SSR, Ser. Mat. **19** (1984), 311–332; 339–360. (Cited on 26, 339)

[KrMA86] M.G. Krein and F.E. Melik-Adamyan, Matrix-continuous analogues of the Schur and the Carathéodory–Toeplitz problem. (Russian) Izv. Akad. Nauk Armyan. SSR Ser. Mat. **21** (2) (1986), 107–141, 207. (Cited on 26, 121, 155, 156, 321, 322, 354, 448, 449)

[KrS66] M.G. Krein and Ju. L. Shmuljan, On the plus operators in a space with indefinite metric. (Russian) Mat. Issled. **1** (l) (1966), 131–161. (Cited on 106)

[KrS67] M.G. Krein and Ju. L. Shmuljan, Fractional linear transformations with operator coefficients. (Russian) Mat. Issled. **2** (3) (1967), 64–96. (Cited on 106)

[KrS70] M.G. Krein and S.N. Saakjan, The resolvent matrix of a Hermitian operator and the characteristic functions connected with it. (Russian) Funktsional. Anal. i Prilozhen **4** (3) (1970), 103–104. (Cited on 309)

[LLS04] H. Langer, M. Langer and Z. Sasvári, Continuations of Hermite indefinite functions and corresponding canonical systems: an example. Meth. Funct. Anal. Topol. **10** (1) (2004), 39–53. (Cited on 354)

[LP67] P.D. Lax and R.S. Phillips, *Scattering Theory.* Academic Press, New York, 1967. (Cited on 200)

[LM00] M. Lesch and M.M. Malamud, The inverse spectral problem for first order systems on the half line, in: *Differential Operators and Related Topics*, Vol. I (Odessa, 1997), Oper. Theory Adv. Appl. 117. Birkhäuser, Basel, 2000, pp. 199–238. (Cited on 406, 408, 449)

[Le87] B.M. Levitan, *Inverse Sturm–Liouville Problems.* Translated from the Russian by O. Efimov. VSP, Zeist, 1987. (Cited on 54)

[LS75] B.M. Levitan and S.I. Sargsjan, *Introduction to Spectral Theory: Selfadjoint Ordinary Differential Operators.* Translated from the Russian by Amiel Feinstein, Trans. Math. Monographs 39. American Mathematical Society, Providence, RI, 1975. (Cited on 54, 408)

[LS91] B.M. Levitan and S.I. Sargsjan, *Sturm–Liouville and Dirac operators.* Translated from the Russian. Mathematics and its Applications (Soviet Series) 59. Kluwer, Dordrecht, 1991. (Cited on 54, 408)

[Liv73] M.S. Livsic, *Operators, Oscillations, Waves. Open Systems.* Transl. Math. Monographs, 34. American Mathematical Society, Providence, RI, 1973. (Cited on 258)

[Mal95] M.M. Malamud, The relation between the matrix potential of a Dirac system and its Wronskian. (Russian) Dokl. Akad. Nauk **344** (5) (1995), 601–604. (Cited on 309)

[Mal97] M.M. Malamud, Spectral analysis of Volterra operators and inverse problems for systems of differential equations. SSB 288, Technical Report No. 269, Berlin. (Cited on 309, 406, 408)

[Mal99] M.M. Malamud, Uniqueness questions in inverse problems for systems of differential equations on a finite interval. (Russian) Tr. Mosk. Mat. Obs. **60** (1999), 199–258; English translation in Trans. Moscow Math. Soc. (1999), 173–224. (Cited on 406, 449)

[Mal05] M.M. Malamud, Uniqueness of the matrix Sturm–Liouville equation given a part of the monodromy matrix, and Borg type results, in: *Sturm–Liouville Theory; Past and Present.* Birkhäuser, Basel, 2005, pp. 237–270. (Cited on 309)

[Ma11] V.M. Marchenko, *Sturm–Liouville Operators and Applications.* Revised edition. AMS Chelsea Publishing, Providence, RI, 2011. (Cited on 54)

[MA89] F.E. Melik-Adamyan, A class of canonical differential operators. (Russian) Izv. Akad. Nauk Armyan. SSR Ser. Mat. **24** (6) (1989), 570–592, 620; English translation in Soviet J. Contemp. Math. Anal. **24** (6) (1989), 48–69. (Cited on 448)

[MiPo81] I.V. Mikhailova and V.P. Potapov, A criterion for Hermitian positivity. (Russian) Teor. Funktsii Funktsional. Anal. i Prilozhen (**36**) (1981), 65–89, 127; English translation in *Topics in Interpolation Theory*, Oper. Theory Adv. Appl. 95. Birkhäuser, Basel, 1997, pp. 419–451. (Cited on 354)

[Ne57] Z. Nehari, On bounded bilinear forms. Ann. Math. **65** (2) (1957), 153–162. (Cited on 155)

[Or76] S.A. Orlov, Nested matrix discs that depend analytically on a parameter, and theorems on the invariance of the ranks of the radii of the limit matrix discs. (Russian) Izv. Akad. Nauk SSSR Ser. Mat. **40** (3) (1976), 593–644, 710. (Cited on 11, 188, 193, 207, 219, 238)

[Or89a] S. A. Orlov, Description of Green functions of canonical differential systems, I. (Russian) Teor. Funktsii Funktsional. Anal. i Prilozhen **51** (1989), 78–88; English translation in J. Soviet Math. **52** (6) (1990), 3500–3508. (Cited on 238)

[Or89b] S.A. Orlov, Description of the Green functions of canonical differential systems, II. (Russian) Teor. Funktsii Funktsional. Anal. i Prilozhen **52** (1989), 33–39; English translation in J. Soviet Math. **52** (5) (1990), 3372–3377. (Cited on 238)

[PaW34] R.E.A.C. Paley and N. Wiener, *Fourier Transforms in the Complex Domain.* American Mathematical Society Colloquium Publications 19. American Mathematical Society, Providence, RI, 1934. (Cited on 80)

[Pa70] L.A. Page, Bounded and compact vectorial Hankel operators. Trans. Amer. Math. Soc. **150** (1970), 529–539. (Cited on 155)

[Pe80] E.L. Pekarev, Regular and completely nonregular systems of colligations and their characteristic functions. (Russian) Izv. Vyssh. Uchebn. Zaved. Mat. (1980) no. 2, 73–75. (Cited on 309)

[Pe03] V.V. Peller, *Hankel Operators and their Applications.* Springer-Verlag, New York, 2003. (Cited on 155, 339)

[Po60] V.P. Potapov, The multiplicative structure of J-contractive matrix functions. Amer. Math. Soc. Transl. **15** (2) (1960), 131–243. (Cited on 5, 40, 41, 53, 84, 105, 247)

[Re60] R. Redheffer, On a certain linear fractional transformation. J. Math. Phys. **39** (1960), 269–286. (Cited on 106)

[Re62] R. Redheffer, On the relation of transmission-line theory to scattering and transfer. J. Math. Phys. **41** (1962), 1–41. (Cited on 105)

[Re02] C. Remling, Schrödinger operators and de Branges spaces. J. Funct. Anal. **196** (2002), 323–394. (Cited on 55)

[Re03] C. Remling, Inverse spectral theory for one-dimensional Schrödinger operators: the A-function. Math. Z. **245** (3) (2003), 597–617. (Cited on 55)

[RR85] M. Rosenblum and J. Rovnyak, *Hardy Classes and Operator Theory.* Oxford University Press, New York, 1985; reprinted by Dover, New York, 1997. (Cited on 60, 76)

[RR94] M. Rosenblum and J. Rovnyak, *Topics in Hardy Classes and Univalent Functions.* Birkhäuser, Basel, 1994. (Cited on 71)

[Rov68] J. Rovnyak, Characterizations of spaces $\boldsymbol{K}(M)$. Unpublished manuscript, 1968; http://people.virginia.edu/ jlr5m/ (Cited on 156)

[Ry66] A.M. Rybalko, On the theory of continual analogues of orthogonal polynomials. (Russian) Teor. Funkcii Funkcional. Anal. i Prilo **3** (1966), 42–60. (Cited on 449)

[SakA92] A.L. Sakhnovich, Spectral functions of a canonical system of order $2n$. Math. USSR Sbornik. **71** (2) (1992), 355–369. (Cited on 239)

[SakA02] A.L. Sakhnovich, Dirac type and canonical systems: spectral and Weyl–Titchmarsh matrix functions, direct and inverse problems. Inverse Problems **18** (2) (2002), 331–348. (Cited on 239, 449)

[Sak80] L.A. Sakhnovich, Equations with a difference kernel on a finite interval. (Russian) Uspekhi Mat. Nauk **35** (1980), 6–129, 248. (Cited on 310)

[Sak93] L.A. Sakhnovich, The method of operator identities and problems in analysis. (Russian) Algebra i Analiz **5** (1) (1993), 3–80; English translation in St. Petersburg Math. J. **5** (1) (1994), 1–69. (Cited on 308)

[Sak94] L.A. Sakhnovich, On a conjecture concerning the Hamiltonians of canonical systems. (Russian) Ukraïn. Mat. Zh. **46** (10) (1994), 1428–1431; English translation in Ukrainian Math. J. **46** (10) (1994), 1578–1583. (Cited on 308)

[Sak96] L.A. Sakhnovich, Spectral problems on half-axis. Meth. Funct. Anal. Topol. **2** (3–4) (1996), 128–140.

[Sak98a] L.A. Sakhnovich, Classic spectral problems, in: *Differential and Integral Operators*, Oper. Theory Adv. Appl. 102. Birkhäuser, Basel, 1998, pp. 243–254. (Cited on 308)

[Sak98b] L.A. Sakhnovich, On a class of canonical systems on half-axis. Integ. Equat. Oper. Theory **31** (1) (1998), 92–112. (Cited on 308)

[Sak98c] L.A. Sakhnovich, Spectral analysis of a class of canonical differential systems. (Russian) Algebra i Analiz **10** (1) (1998), 187–201; English translation in St. Petersburg Math. J. **10** (1) (1999), 147–158. (Cited on 308)

[Sak99] L.A. Sakhnovich, *Spectral Theory of Canonical Differential Systems. Method of Operator Identities*. Translated from the Russian manuscript by E. Melnichenko. Oper. Theory Adv. Appl. 107. Birkhäuser Verlag, Basel, 1999. (Cited on 237, 239, 308, 407, 408, 449)

[Sak00a] L.A. Sakhnovich, On the spectral theory of the generalized differential system of M.G. Krein. (Russian) Ukraïn. Mat. Zh. **52** (5) (2000), 717–721; English translation in Ukrainian Math. J. **52** (5) (2000), 821–826. (Cited on 449)

[Sak00b] L.A. Sakhnovich, On the spectral theory of a class of canonical differential systems. (Russian) Funktsional. Anal. i Prilozhen **34** (2) (2000), 50–62, 96; English translation in Funct. Anal. Appl. **34** (2) (2000), 119–128. (Cited on 449)

[Sak00c] L.A. Sakhnovich, Works by M.G. Krein on inverse problems, in: *Differential Operators and Related Topics*, Vol. I (Odessa, 1997), Oper. Theory Adv. Appl. 117. Birkhäuser, Basel, 2000, pp. 59–69. (Cited on 449)

[Sak01] L.A. Sakhnovich, On reducing the canonical system to two dual differential systems. J. Math. Anal. Appl. **255** (2) (2001), 499–509. (Cited on 309)

[Sak06] L.A. Sakhnovich, On Krein's differential system and its generalization. Integ. Equat. Oper. Theory **55** (4) (2006), 561–572. (Cited on 449)

[Sar67] D. Sarason, Generalized interpolation in H^∞. Trans. Amer. Math. Soc. **127** (1967), 179–203. (Cited on 157)

[Shm62] Yu. L. Shmuljan, Some problems in the theory of operators with finite rank of nonhermitians. Mat. Sbornik **57** (1) (1962), 105–136. (Cited on 309)

[Shm68] Yu. L. Shmuljan, Operator balls. Teor. Funktsii. Funktsional. Anal. i Prilozhen **6** (1968), 68–81; English translation in Integ. Equat. Oper. Theory **13** (6) (1990), 864–882. (Cited on 188, 189, 207)

[Si74] L.A. Simakova, Plus-matrix-valued functions of bounded characteristic. (Russian) Mat. Issled. **9** (1974), 149–171, 252–253. (Cited on 105)

[Si75] L.A. Simakova, On meromorphic plus-matrix functions. Mat. Issled. **10** (1) (1975), 287–292. (Cited on 105)

[Si03] L.A. Simakova, On real and "symplectic" meromorphic plus-matrix functions and corresponding linear-fractional transformations. (Russian) Mat. Fiz. Anal. Geom. **10** (4) (2003), 557–568. (Cited on 105)

[Si99] B. Simon, A new approach to inverse spectral theory, I. Fundamental formalism. Ann. Math. (2) **150** (1999), 1029–1057. (Cited on 55)

[Ti46] E.C. Titchmarsh, *Eigenfunction Expansions Associated with Second-order Differential Equations, I.* Oxford University Press, New York, 1946. (Cited on 200)

[Ti60] E.C. Titchmarsh, *The Theory of Functions*. Second edition. Oxford University Press, London, 1960. (Cited on 404)

[TrV97] S. Treil and A. Volberg, Wavelets and the angle between past and future. J. Funct. Anal. **143** (1997) 269–308. (Cited on 156)

[TsS77] E.R. Tsekanovskii and Yu. L. Shmulyan, The theory of bi-extensions of operators on rigged Hilbert spaces. Unbounded operator colligations and characteristic functions. Russian Math. Surveys, **32** (5) (1977), 73–131. (Cited on 309)

[Vl02] V.S. Vladimorov, *Methods of the Theory of Generalized Functions (Analytical Methods and Special Functions).* CRC Press, Boca Raton, 2002. (Cited on 200)

[We10] H. Weyl, Über gewöhnliche Differentialgleichungen mit Singularitäten und die zugehörigen Entwicklungen willkürlicher Funktionen. Math. Ann. **68** (1910), 220–269. (Cited on 200, 237)

[We35] H. Weyl, Über das Pick–Nevanlinna'sche Interpolation-problem und sein infinitesimales Analogen. Ann. Math. **36** (2) (1935), 230–254. (Cited on 237)

[Wi95] H. Winkler, The inverse spectral problem for canonical systems, Integ. Equat. Oper. Theory **22** (2) (1995), 360–376. (Cited on 407, 408)

[Wo69] M.R. Wohlers, *Lumped and Distributed Passive Networks, A General and Advanced View*. Academic Press, New York, 1969. (Cited on 200)

[Yu01] P. Yuditskii, A special case of de Branges' theorem on the inverse monodromy problem. Integ. Equat. Oper. Theory **39** (2001), 229–252. (Cited on 309)

[YCC59] D.C. Youla, L.J. Castriota and H.J. Carlin, Bounded real scattering matrices and the foundations of linear passive network theory. IRE Trans. Circuit Theory, **CT-6** (1959), 102–124. (Cited on 200)

[Zol03] V.A. Zolotarev, *Analytical Methods of Spectral Representations of Nonselfadjoint and Nonunitary Operators*. (Russian) Kharkov National University, Kharkov, 2003. (Cited on 53)

SYMBOL INDEX

INDEX